SCHAUM'S OUTLINE OF

THEORY AND PROBLEMS

of

PHYSICAL CHEMISTRY

SECOND EDITION

•

by

CLYDE R. METZ, Ph.D.

Professor of Chemistry
College of Charleston

•

SCHAUM'S OUTLINE SERIES
McGRAW-HILL

New York San Francisco Washington, D.C. Auckland Bogotá Caracas Lisbon
London Madrid Mexico City Milan Montreal New Dehli
San Juan Singapore Sydney Tokyo Toronto

Clyde Metz, currently Professor of Chemistry at The College of Charleston, received his Ph.D. from Indiana University. He is a coauthor of a general chemistry textbook and related laboratory materials, solutions manuals, and study guides and has published several student research papers. Dr. Metz is active in the Division of Chemical Education of the American Chemical Society; is a member of the Electrochemical Society, Alpha Chi Sigma, Sigma Xi, Tau Beta Pi, Phi Lambda Upsilon, and the South Carolina Academy of Science; and is a Fellow of the Indiana Academy of Science.

Schaum's Outline of Theory and Problems of
PHYSICAL CHEMISTRY

7 8 9 10 11 12 13 14 15 16 17 18 19 20 PRS PRS 9 8

ISBN 0-07-041715-6

Sponsoring Editor, John Aliano
Production Editor, Leroy Young
Editing Supervisor, Marthe Grice
Project Supervision, The Total Book

Library of Congress Cataloging-in-Publication Data

Metz, Clyde R.
   Schaum's outline of theory and problems of physical chemistry.
   (Schaum's outline series)
   Includes index.
   1. Chemistry, Physical and theoretical—Problems, exercises, etc.   I. Title.   II. Title: Theory and problems of physical chemistry.
QD456.M4 1988   541.3'076   87-29839
ISBN 0-07-041715-6

McGraw-Hill

A Division of The McGraw-Hill Companies

Preface

This supplementary book has been written for students at all levels in physical chemistry and contains the solved problems and additional exercises that most texts omit because of space limitations. Each chapter is divided into three sections: The important concepts for each subject are summarized in word and equation form, usually followed by an example problem; a series of completely solved problems is presented to illustrate the concepts both singly and in combination; and a set of supplementary problems (with answers) is provided for additional drill.

Since the publication of the first edition of this book, the changeover to the SI system of units from the modified cgs system that had been used in physical chemistry for many years has been nearly completed. In fact, even a new thermodynamic standard state pressure (1 bar) has been chosen. Nearly all the data in this book are given in the newer units—the older systems are discussed briefly and conversion factors given both in the text and in the appendix.

Data for the problems have been taken from the *Handbook of Chemistry and Physics* with permission of the Chemical Rubber Company, from various original papers, and from publications of the National Bureau of Standards. Appreciation is extended to Chemical Education Resources, Inc. for permission to reprint several figures and tables found in Chapters 18 and 22; to W. H. Freeman and Co., Inc. for permission to reprint several figures in Chapter 22; to Academic Press, Inc. for permission to use tables in the Appendix; to Mr. Thomas J. Dembofsky and Mr. David Beckwith and their staff for their cooperation in making many improvements in the manuscript; to Miss Linda S. Hill for reworking all the problems in the first edition; to an anonymous reviewer who made a number of valuable suggestions, to two outstanding teachers of physical chemistry—Dr. Oran M. Knudsen and Dr. Ralph L. Seifert; and to my family—Jennie, Curtis, and Michele.

CLYDE R. METZ

Contents

CONTENTS

CONTENTS

CONTENTS

CONTENTS

Chapter 1

Gases and the Kinetic-Molecular Theory

Temperature and Pressure

1.1 TEMPERATURE

A *thermal equilibrium* exists between two systems provided no change in any observable property occurs when the systems are in thermal contact. The "zeroth law of thermodynamics" states that "two systems that are separately in thermal equilibrium with a third system are in thermal equilibrium with each other." This law implies that there must be a property of a system that signifies the existence of a condition of thermal equilibrium that is independent of the composition and size of the system. This property is called *temperature.*

The *thermodynamic temperature* (T) is defined by assigning the exact value 273.16 K to the triple point of water, and the unit of thermodynamic temperature, the *kelvin* (K), is the fraction 1/273.16 of the thermodynamic temperature of the triple point of water. The *Celsius temperature* (t) is defined as

$$t/(^\circ\text{C}) = T/(\text{K}) - 273.15 \tag{1.1}$$

EXAMPLE 1.1. The freezing point of a saturated NaCl-water solution is $-17.78\,^\circ\text{C} = 0.00\,^\circ\text{F}$, and the freezing point of pure water is $0.0\,^\circ\text{C} = 32.00\,^\circ\text{F}$. Derive an equation relating the Celsius and Fahrenheit temperature scales.

The relative sizes of the respective degrees are found by comparing the numbers of degrees for the same span, giving

$$\frac{32.00\,^\circ\text{F} - 0.00\,^\circ\text{F}}{0.00\,^\circ\text{C} - (-17.78\,^\circ\text{C})} = 1.800\,^\circ\text{F}\,^\circ\text{C}^{-1}$$

The general form of the desired relation between the scales is

$$t/(^\circ\text{F}) = 1.800t/(^\circ\text{C}) + k$$

where k corrects for the difference between the zero points of the scales. This constant can be evaluated by substituting the freezing point data for the pure water, giving

$$32.00 = (1.800)(0) + k$$

$$k = 32.00$$

The desired equation is

$$t/(^\circ\text{F}) = 1.800t/(^\circ\text{C}) + 32.00 \tag{1.2}$$

1.2 PRESSURE

Pressure (P) is defined as a force distributed over an area. The SI unit for expressing pressure is the *pascal* (Pa), which is equal to $1\,\text{N}\,\text{m}^{-2}$ or $1\,\text{kg}\,\text{m}^{-1}\,\text{s}^{-2}$. Because a pressure of 1 Pa is very small, the more convenient *bar* $(1\,\text{bar} = 10^5\,\text{Pa})$ is commonly used for expressing pressures near normal atmospheric pressure, and 1 bar has been chosen as the standard pressure for reporting various thermodynamic data. Other pressure units with appropriate conversion factors are given in Table 1-1. The *absolute pressure* of a system is defined as the gauge pressure plus the ambient atmospheric pressure.

Table 1-1

Pressure Unit	Symbol	Conversion Factor
Bar	bar	1 bar $= 10^5$ Pa*
Standard atmosphere	atm	1 atm $= 1.013\,25 \times 10^5$ Pa*
Conventional millimeter of mercury	mmHg	1 mmHg $= (13.595\,1)(9.806\,65)$ Pa*
Torr	torr	1 torr $= (1.013\,25 \times 10^5/760)$ Pa*
Pound per square inch	psi	1 psi $= 6\,894.757\,2$ Pa

* Exact conversion factor.

EXAMPLE 1.2. Calculate the height of a column of water equivalent to 1.00 bar.

The height of a mercury column equal to 1.00 bar is

$$(1.00\,\text{bar})\left(\frac{10^5\,\text{Pa}}{1\,\text{bar}}\right)\left(\frac{1\,\text{mmHg}}{(13.595\,1)(9.806\,65)\,\text{Pa}}\right)\left(\frac{1\,\text{m}}{10^3\,\text{mm}}\right) = 0.750\,\text{m}$$

The heights of various liquids that are equivalent to the same pressure are inversely proportional to the densities (ρ) of the liquids, so

$$h(\text{H}_2\text{O}) = h(\text{Hg})\,\frac{\rho(\text{Hg})}{\rho(\text{H}_2\text{O})} = (0.750\,\text{m})\,\frac{13.6 \times 10^3\,\text{kg m}^{-3}}{1.00 \times 10^3\,\text{kg m}^{-3}}$$

$$= 10.2\,\text{m} = 33.5\,\text{ft}$$

Laws for Ideal Gases

1.3 BOYLE'S OR MARIOTTE'S LAW

For a fixed amount of gas under isothermal (constant-temperature) conditions,

$$PV = \text{constant} \qquad\qquad (1.3)$$

where P is the absolute pressure and V is the volume occupied by the gas.

EXAMPLE 1.3. Calculate the pressure necessary to compress isothermally a 105-m^3 sample of air at 1.05 bar to 35 m^3.

For the fixed amount of gas, (1.3) gives $P_1 V_1 = P_2 V_2$, or

$$P_2 = P_1\,\frac{V_1}{V_2} = (1.05\,\text{bar})\,\frac{105\,\text{m}^3}{35\,\text{m}^3} = 3.2\,\text{bar}$$

EXAMPLE 1.4. Assuming the constant in (1.3) to be 22.7 dm^3 bar for 1.00 mol of gas at 273 K and 227 dm^3 bar at 2 730 K, prepare isothermal plots of V against P for values of pressure between 0 and 100 bar. *Note*: 1 cubic decimeter equals 1 liter (1 dm^3 = 10^{-3} m^3 = 1 L).

Substituting $P = 0.01, 0.05, 0.1, 1, 5, 10, 50,$ and 100 bar into $V = (22.7\,\text{dm}^3\,\text{bar})/P$ gives $V = 2\,270, 454, 227,$ 22.7, 4.5, 2.3, 0.5, and 0.2 dm^3, respectively, at 273 K. These results and the results of calculations for 2 730 K are plotted in Fig. 1-1.

1.4 CHARLES'S OR GAY-LUSSAC'S LAW

For a fixed amount of gas under isobaric (constant-pressure) conditions,

$$\frac{V}{T} = \text{constant} \qquad\qquad (1.4)$$

where V is the volume and T is the absolute temperature.

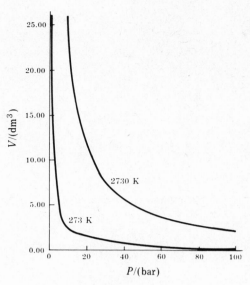

Fig. 1-1

EXAMPLE 1.5. Calculate the resulting temperature change as a 52-dm^3 sample of gas at 25 °C is isobarically expanded to 275 dm^3.

For a fixed amount of gas, (*1.4*) gives $V_1/T_1 = V_2/T_2$, or

$$T_2 = T_1 \frac{V_2}{V_1} = (25 + 273) \text{ K} \frac{275 \text{ dm}^3}{52 \text{ dm}^3} = 1\,600 \text{ K}$$

The temperature change is

$$\Delta T = T_2 - T_1 = 1\,600 \text{ K} - 298 \text{ K} = 1\,300 \text{ K}$$

EXAMPLE 1.6. Equation (*1.4*) can be stated as $V = V_0[1 + \alpha t/(°\text{C})]$, where V_0 is the volume of the gas at 0 °C, α is a constant, and t is the temperature in °C. Find the value of α.

Equations (*1.4*) and (*1.1*) give

$$V = V_0 \frac{T}{T_0} = V_0 \frac{(t + 273)\text{K}}{(0 + 273)\text{K}} = V_0 \left(1 + \frac{t}{273} \right)$$

Upon comparison of terms, $\alpha = 1/273 = 3.66 \times 10^{-3}$.

1.5 IDEAL GAS LAW

Combining (*1.3*) with (*1.4*) gives

$$\frac{PV}{T} = \text{constant} \qquad\qquad (1.5)$$

for a fixed amount of gas. Under Avogadro's hypothesis that equal volumes of all ideal gases under identical pressure and temperature conditions contain equal numbers of molecules, it can be shown that (*1.5*) becomes

$$\frac{PV}{nT} = R \qquad\qquad (1.6)$$

where n is the amount of substance and R is the gas constant ($R = 8.314\,41 \text{ J K}^{-1} \text{ mol}^{-1}$).

EXAMPLE 1.7. Consider a 42.5-dm^3 sample of gas at 25 °C and 748 torr. If the volume is expanded to 52.5 dm^3 and the pressure changed to 760.0 torr, what is the final temperature of the gas?

For a fixed amount of gas, (1.5) gives $P_1 V_1 / T_1 = P_2 V_2 / T_2$ (the *combined gas law*) or

$$T_2 = T_1 \frac{P_2}{P_1} \frac{V_2}{V_1} = (298 \text{ K}) \frac{760.0 \text{ torr}}{748 \text{ torr}} \frac{52.5 \text{ dm}^3}{42.5 \text{ dm}^3} = 374 \text{ K}$$

EXAMPLE 1.8. The value of the gas constant is usually given as $R = 8.314\,41 \text{ J K}^{-1} \text{ mol}^{-1}$, where $1 \text{ J} = 1 \text{ kg m}^2 \text{ s}^{-2}$. Although this value is convenient for calculating energy, it is rather inconvenient to use in (1.6) because of the need to repeatedly make the same unit conversions for volume and pressure. For this reason, evaluate R in units of $m^3 \text{ Pa K}^{-1} \text{ mol}^{-1}$, $dm^3 \text{ Pa K}^{-1} \text{ mol}^{-1}$, and $dm^3 \text{ bar K}^{-1} \text{ mol}^{-1}$ for use in (1.6).

The values of R in the desired units are

$$R = (8.314\,41 \text{ J K}^{-1} \text{ mol}^{-1})\left(\frac{1 \text{ kg m}^2 \text{ s}^{-2}}{1 \text{ J}}\right)\left(\frac{1 \text{ Pa}}{1 \text{ kg m}^{-1} \text{ s}^{-2}}\right) = 8.314\,41 \text{ m}^3 \text{ Pa K}^{-1} \text{ mol}^{-1}$$

$$R = (8.314\,41 \text{ m}^3 \text{ Pa K}^{-1} \text{ mol}^{-1})\left(\frac{10^3 \text{ dm}^3}{1 \text{ m}^3}\right) = 8\,314.41 \text{ dm}^3 \text{ Pa K}^{-1} \text{ mol}^{-1}$$

$$R = (8\,314.41 \text{ dm}^3 \text{ Pa K}^{-1} \text{ mol}^{-1})\left(\frac{1 \text{ bar}}{10^5 \text{ Pa}}\right) = 0.083\,144\,1 \text{ dm}^3 \text{ bar K}^{-1} \text{ mol}^{-1}$$

1.6 MOLAR MASS OF AN IDEAL GAS

Substituting $n = m/M$, where M is molar mass, into (1.6) gives

$$M = \frac{mRT}{PV} \tag{1.7}$$

or, in terms of density,

$$M = \frac{\rho RT}{P} \tag{1.8}$$

Mixtures of Ideal Gases

1.7 DALTON'S LAW OF PARTIAL PRESSURES

The *total pressure* (P_t) of a mixture of r gases is given by

$$P_t = \sum_{i=1}^{r} P_i \tag{1.9}$$

where P_i is the *partial pressure* of component i in the mixture. If P_t and the molar composition of the mixture are known, P_i can be calculated from

$$P_i = x_i P_t \tag{1.10}$$

where x_i, the *mole fraction* for component i, is defined as

$$x_i = \frac{n_i}{\displaystyle\sum_{i=1}^{r} n_i} = \frac{n_i}{n_{\text{total}}} \tag{1.11}$$

EXAMPLE 1.9. The vapor pressure (P_v) of a material can be measured by passing an inert gas over a sample of the material and analyzing the composition of the gaseous mixture. For these conditions, (1.10) and (1.11) become

$$\frac{P_v}{P_t} = \frac{n}{n + n_{\text{inert}}}$$

Calculate P_v for Hg at 23 °C if a 50.40-g sample of nitrogen-mercury vapor at 745 torr contains 0.702 mg of Hg.

The amount of each gas in the mixture is

$$n(\text{Hg}) = \frac{7.02 \times 10^{-4} \text{ g}}{200.59 \text{ g mol}^{-1}} = 3.50 \times 10^{-6} \text{ mol}$$

$$n(\text{N}_2) = \frac{[50.40 - (7.02 \times 10^{-4})]\text{g}}{28.013\,4 \text{ g mol}^{-1}} = 1.799 \text{ mol}$$

giving

$$P_v = (745 \text{ torr}) \frac{3.50 \times 10^{-6}}{(3.50 \times 10^{-6}) + 1.799} = 1.45 \times 10^{-3} \text{ torr} = 0.193 \text{ Pa}$$

1.8 AMAGAT'S LAW OF PARTIAL VOLUMES

The *total volume* (V_t) of a mixture of r gases is given by

$$V_t = \sum_{i=1}^{r} V_i \qquad (1.12)$$

where V_i is the *partial volume* of component i in the mixture. If V_t and the molar composition of the mixture are known, V_i can be calculated from

$$V_i = x_i V_t \qquad (1.13)$$

[For a mixture of real gases, (1.12) is usually more accurate than (1.9).]

1.9 MOLAR MASS OF A GASEOUS MIXTURE

The *average molar mass* ($\bar{M}$), of a mixture of r gases is given by

$$\bar{M} = \sum_{i=1}^{r} x_i M_i \qquad (1.14)$$

EXAMPLE 1.10.   The density of dry air at 0.986 6 bar and 27.0 °C is 1.146 g dm^{-3}. Calculate the composition of air assuming only N_2 and O_2 to be present.

The average molar mass of air can be calculated from (1.8) as

$$\bar{M} = \frac{(1.146 \text{ g dm}^{-3})(0.083\,14 \text{ dm}^3 \text{ bar K}^{-1} \text{ mol}^{-1})(300.2 \text{ K})}{0.986\,6 \text{ bar}}$$

$$= 28.99 \text{ g mol}^{-1}$$

Letting the mole fraction of N_2 be x, we obtain

$$x(28.013\,4) + (1-x)(31.998\,8) = 28.99 \qquad \text{or} \qquad x = 0.755$$

Thus $x(\text{N}_2) = 0.755$, and $x(\text{O}_2) = 0.245$.

Real Gases

1.10 CRITICAL POINT

As real gases are cooled, the nicely shaped isotherms shown in Fig. 1-1 become distorted. The isotherm that exhibits an inflection point where the tangent is vertical, i.e., $\partial P / \partial V = \partial^2 P / \partial V^2 = 0$, corresponds to the *critical temperature* (T_c). The values of P and V at the point of inflection are the *critical pressure* (P_c) and *critical volume* (V_c), respectively.

One way to determine the critical temperature and density of a material that does not associate in the liquid phase is to use the *rectilinear diameter law of Cailletet and Mathias*, which can be stated as

$$\tfrac{1}{2}(\rho_{\text{liq}} + \rho_{\text{gas}}) = A + BT + CT^2 + \cdots \tag{1.15}$$

where A, B, and C are constants. This law implies that plots of the orthobaric densities of the gas and of the liquid will intersect at the critical point, giving

$$\rho_c = A + BT_c + CT_c^2 + \cdots \tag{1.16}$$

EXAMPLE 1.11. Estimate the critical temperature for a substance having gaseous densities of 1.03×10^3, 1.07×10^3, 1.14×10^3 and 1.25×10^3 kg m^{-3} and liquid densities of 1.46×10^3, 1.43×10^3, 1.39×10^3, and 1.33×10^3 kg m^{-3} at 100, 290, 450, and 540 °C, respectively.

The plot of these densities (see Fig. 1-2) shows that the curves converge at about 560 °C $= t_c$. The point of intersection can be identified more easily by extrapolation of the $(\rho_{\text{liq}} + \rho_{\text{gas}})/2$ plot.

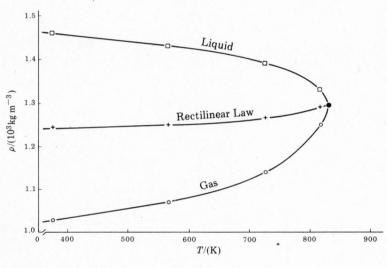

Fig. 1-2

1.11 COMPRESSIBILITY FACTOR

The *compressibility factor* (Z) is an empirical correction for the nonideal behavior of real gases that allows the simple form of the combined gas law to be retained. Thus, we write for a real gas

$$PV = ZnRT \tag{1.17}$$

The factor Z is determined by first calculating the *reduced pressure* (P_r) and the *reduced temperature* (T_r), defined as

$$P_r = P/P_c \tag{1.18}$$

and

$$T_r = T/T_c \tag{1.19}$$

Then the value of Z is read from a graph of Z plotted against P_r for various reduced temperatures (Fig. 1-3).

EXAMPLE 1.12. Calculate the volume that 1.50 mol of $(C_2H_5)_2S$ would occupy at 275 °C and 12.5 bar. $P_c = 39.6$ bar and $T_c = 283.8$ °C for $(C_2H_5)_2S$.

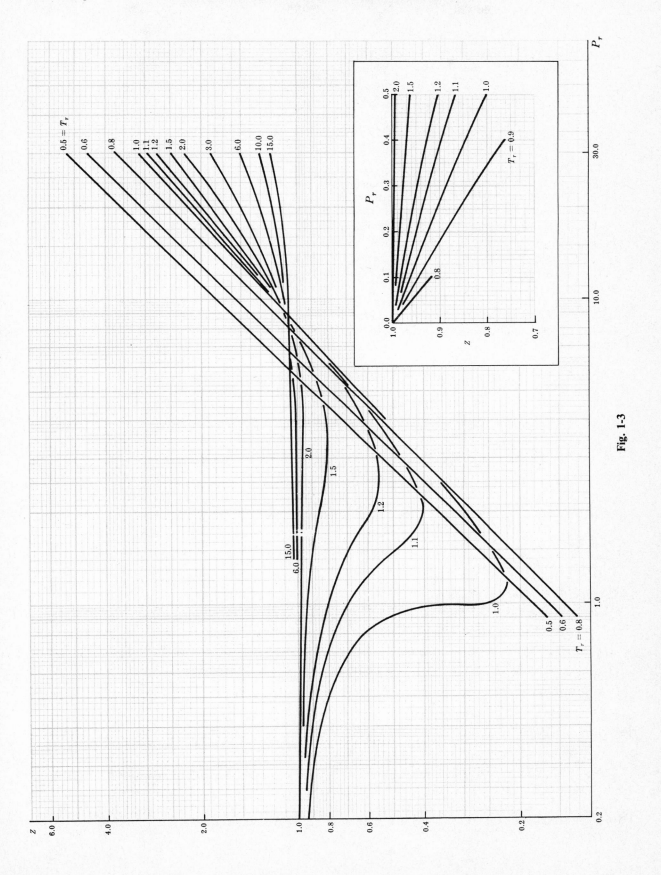

Fig. 1-3

Using (1.18) and (1.19) gives

$$P_r = \frac{12.5\ \text{bar}}{39.6\ \text{bar}} = 0.320 \qquad T_r = \frac{(275+273)\text{K}}{(283.8+273.15)\text{K}} = 0.983$$

From Fig. 1-3, $Z = 0.87$, so (1.17) gives

$$V = \frac{(0.87)(1.50\ \text{mol})(0.083\ 14\ \text{dm}^3\ \text{bar}\ \text{K}^{-1}\ \text{mol}^{-1})(548\ \text{K})}{12.5\ \text{bar}}$$

$$= 4.8\ \text{dm}^3$$

1.12 VIRIAL EQUATIONS

There are two virial equations, one describing PV as a function of $1/V$ and the second using the variable P. For 1 mol of gas, these are

$$PV_m = A_v + B_v(1/V_m) + C_v(1/V_m)^2 + \cdots \tag{1.20}$$

$$PV_m = A_p + B_pP + C_pP^2 + \cdots \tag{1.21}$$

Here V_m is the molar volume, and A_v, B_v, etc., are constants for a particular gas. The values of these virial coefficients can be determined from the van der Waals constants, statistical mechanics, or experimental data.

EXAMPLE 1.13. Evaluate A_v and A_p in (1.20) and (1.21), assuming that a real gas will approach ideality as $(1/V_m) \to 0$ and as $P \to 0$.

As the limits of zero are approached, the polynomial terms in (1.20) and (1.21) become insignificant, and both equations approach (1.6), giving $A_v = A_p = RT$.

1.13 VAN DER WAALS EQUATION

The van der Waals equation of state

$$\left(P + \frac{an^2}{V^2}\right)(V - nb) = nRT \tag{1.22}$$

corrects the ideal gas law, (1.6), for the excluded volume of the molecules (b) and the force of interaction between the molecules (a).

EXAMPLE 1.14. Using (1.6) and (1.22), calculate the volume that 1.50 mol of $(C_2H_5)_2S$ would occupy at 105 °C and 0.750 bar. Assume that $a = 19.00\ \text{dm}^6\ \text{bar}\ \text{mol}^{-2}$ and $b = 0.121\ 4\ \text{dm}^3\ \text{mol}^{-1}$.

If we assume the gas to be ideal, (1.6) gives

$$V = \frac{(1.50\ \text{mol})(0.083\ 14\ \text{dm}^3\ \text{bar}\ \text{K}^{-1}\ \text{mol}^{-1})(378\ \text{K})}{(0.750\ \text{bar})} = 62.9\ \text{dm}^3$$

On the other hand, (1.22) gives

$$\left(0.750 + \frac{(19.00)(1.50)^2}{V^2}\right)[V - (1.50)(0.121\ 4)] = (1.50)(0.083\ 14)(378)$$

or $0.750V^3 - 47.2V^2 + 42.8V - 7.79 = 0$.

Of the several ways to solve a polynomial equation for the various roots, one of the most useful is known as the *Newton–Raphson iterative process*. For the equation $f(x) = 0$, the $(n+1)$th estimate of the root is given by

$$x_{n+1} = x_n - \frac{f(x_n)}{f'(x_n)} \tag{1.23}$$

where $f(x_n)$ is the value of the function evaluated at x_n and $f'(x_n)$ is the first derivative of the function evaluated at x_n. Denoting the left-hand side of the above equation for volume as $f(V)$, we have

$$f'(V) = 2.250 V^2 - 94.4 V + 42.8$$

and (1.23) becomes

$$V_{n+1} = V_n - \frac{0.750 V_n^3 - 47.2 V_n^2 + 42.8 V_n - 7.79}{2.250 V_n^2 - 94.4 V_n + 42.8}$$

If we assume for V_1 the value predicted by (1.6), the second estimate is

$$V_2 = 62.9 - \frac{(0.750)(62.9)^3 - (47.2)(62.9)^2 + (42.8)(62.9) - 7.79}{(2.250)(62.9)^2 - (94.4)(62.9) + 42.8}$$

$$= 62.9 - \frac{2\,585}{3\,007} = 62.0 \text{ dm}^3$$

The third estimate, and the answer to three significant figures, is

$$V_3 = 62.0 - \frac{f(62.0)}{f'(62.0)} = 62.0 - \frac{-45}{2\,839} = 62.0 \text{ dm}^3$$

EXAMPLE 1.15. Critical-point data can be used to determine approximate values of the van der Waals constants. Derive these values.

Solving (1.22) for P and taking the necessary derivatives gives, for 1 mol,

$$P = \frac{RT}{V_m - b} - \frac{a}{V_m^2}$$

$$\left(\frac{\partial P}{\partial V_m}\right)_T = \frac{-RT}{(V_m - b)^2} + \frac{2a}{V_m^3} = 0$$

$$\left(\frac{\partial^2 P}{\partial V_m^2}\right)_T = \frac{2RT}{(V_m - b)^3} - \frac{6a}{V_m^4} = 0$$

Assigning the values of P, V_m, and T as P_c, V_c, and T_c and solving the three simultaneous equations gives

$$a = 3 P_c V_c^2 \qquad b = \frac{V_c}{3} \qquad R = \frac{8 P_c V_c}{3 T_c}$$

Because R is usually determined from other sources, the value of V_c can be eliminated, giving

$$a = \frac{27 R^2 T_c^2}{64 P_c} \qquad b = \frac{RT_c}{8 P_c}$$

1.14 MOLAR MASS OF A REAL GAS

For n moles of a real gas at rather low pressures, (1.21) becomes

$$PV = n(RT + B_p P)$$

Upon substituting $n = m/M$ and $\rho = m/V$, the above equation becomes

$$\frac{\rho}{P} = \frac{M/RT}{1 + (B_p P/RT)}$$

At low pressures, $[1 + (B_p P/RT)]^{-1} \approx 1 - (B_p P/RT)$ and

$$\frac{\rho}{P} = \frac{M}{RT} + \left(\frac{M}{RT}\right)\left(\frac{-B_p}{RT}\right) P \qquad\qquad (1.24)$$

Thus an isothermal plot of ρ/P against P will have a slope of $(M/RT)(-B_p/RT)$ and an intercept of M/RT.

EXAMPLE 1.16. E. Moles reports the following density-pressure data for SO_2 at 0 °C:

$P/(\text{Pa})$	101 325	50 662.5	10 132.5	1 013.25	101.325	10.132 5
$(\rho/P)/(\text{g dm}^{-3}\,\text{bar}^{-1})$	2.888 411	2.854 584	2.827 510	2.821 416	2.820 807	2.820 746

Determine the molar mass for SO_2 from these data.

A plot of ρ/P against P gives the intercept as 2.820 741 g dm^{-3} bar^{-1}; thus

$$M = (\text{intercept})(RT)$$

$$= (2.820\ 741 \text{ g dm}^{-3}\text{ bar}^{-1})(0.083\ 144\ 1 \text{ dm}^3\text{ bar K}^{-1}\text{ mol}^{-1})(273.15 \text{ K})$$

$$= 64.061\ 3 \text{ g mol}^{-1}$$

Kinetic-Molecular Theory (KMT)

1.15 KMT POSTULATES FOR GASES

The model for an ideal gas is based on the assumptions that (1) the gas consists of sufficiently many particles to allow statistical averaging to be performed; (2) the intrinsic volume of the particles is small compared to the distances between the particles and negligible relative to the volume of the container, which implies that the particles may move freely throughout the entire volume; (3) the particles are in random motion and have no mutual attractions; (4) the collisions between the particles and between the particles and the walls of the container are elastic, which implies conservation of energy and momentum; and (5) the average translational kinetic energy of the particles is proportional to the absolute temperature.

EXAMPLE 1.17. The average translational kinetic energy for a molecule ($\bar{\varepsilon}_{\text{trans}}$) is given by

$$\bar{\varepsilon}_{\text{trans}} = \tfrac{1}{2}m\overline{v^2}$$

where m is the mass of the molecule and $\overline{v^2}$ is the average of the square of the velocity. Given $\overline{v^2} = 3kT/m$, where k is *Boltzmann's constant*, calculate the ratio of the kinetic energies at 200 °C and 100 °C.

$$\frac{\bar{\varepsilon}_{\text{trans}}(200\ °C)}{\bar{\varepsilon}_{\text{trans}}(100\ °C)} = \frac{\tfrac{1}{2}m(3kT/m)_{200}}{\tfrac{1}{2}m(3kT/m)_{100}} = \frac{473 \text{ K}}{373 \text{ K}} = 1.27$$

1.16 P-V-KE RELATIONSHIPS

The KMT implies the following series of relationships for 1 mol of an ideal gas:

$$PV_m = \tfrac{1}{3}m\overline{v^2} = \tfrac{2}{3}\bar{E}_{\text{trans}} = RT \tag{1.25}$$

where $\bar{E}_{\text{trans}}$ is the average translational kinetic energy for 1 mol of gas and the other terms have been defined previously.

EXAMPLE 1.18. For oxygen molecules at 25 °C, find the *root mean square speed* (v_{rms}), which is defined by

$$v_{\text{rms}} = \sqrt{\overline{v^2}} \tag{1.26}$$

Substituting (*1.25*) into (*1.26*) gives

$$v_{\text{rms}} = \left(\frac{3RT}{M}\right)^{1/2} = \left[\frac{3(8.314 \text{ J K}^{-1}\text{ mol}^{-1})(298 \text{ K})}{32.0 \times 10^{-3} \text{ kg mol}^{-1}}\right]^{1/2} = 482 \text{ m s}^{-1}$$

Solved Problems

TEMPERATURE AND PRESSURE

1.1. Combine (*1.1*) with (*1.2*) to define the *Rankine scale* (°R), which is the absolute Fahrenheit scale.

The general form of the desired equation is $T/(\text{°R}) = t/(\text{°F}) + k$. Substituting (*1.2*) gives

$$T/(\text{°R}) = [1.800t/(\text{°C}) + 32.00] + k$$

and substituting (*1.1*) gives

$$T/(\text{°R}) = (1.800)[T/(\text{K}) - 273.15] + 32.00 + k$$

The value of k can be determined by recognizing that $0\ \text{°R} = 0\ \text{K}$, giving

$$0 = (1.800)(0 - 273.15) + 32.00 + k$$

$$k = 459.67$$

Hence $T/(\text{°R}) = t/(\text{°F}) + 459.67$.

1.2. The triple point of water is 273.16 K. Express this temperature as a Celsius temperature.

Using (*1.1*) gives

$$t/(\text{°C}) = 273.16 - 273.15 = 0.01$$

$$t = 0.01\ \text{°C}$$

1.3. The gauge pressure of the air in an automobile tire is 32 psi, and the ambient atmospheric pressure is 14.8 psi. Express the absolute pressure in the tire in bars.

The absolute pressure is found by adding the gauge pressure and the atmospheric pressure, giving

$$(14.8\ \text{psi} + 32\ \text{psi})\left(\frac{6\ 895\ \text{Pa}}{1\ \text{psi}}\right)\left(\frac{1\ \text{bar}}{10^5\ \text{Pa}}\right) = 3.2\ \text{bar}$$

LAWS FOR IDEAL GASES

1.4. A vacuum manifold was calibrated using Boyle's law. A 0.503-dm^3 flask containing dry nitrogen at 746 torr was attached to the manifold, which was at 13 mtorr. After the stopcock was opened and the system allowed to reach equilibrium, the pressure of the combined system was 273 torr. Assuming isothermal conditions, what is the volume of the manifold?

Before opening the stopcock, the original condition of the system was given by

$$P_1 V_1 = (746\ \text{torr})(0.503\ \text{dm}^3) + (13 \times 10^{-3}\ \text{torr})(V)$$

and after opening the stopcock, the condition of the system was given by

$$P_2 V_2 = (273\ \text{torr})(0.503 + V)\ \text{dm}^3$$

Equating these PV terms via (*1.3*) gives

$$(746)(0.503) + (13 \times 10^{-3})V = (273)(0.503 + V)$$

$$V = 0.872\ \text{dm}^3$$

1.5. Show for a fixed amount of an ideal gas under isothermal conditions that $(\partial V/\partial P)_{n,T} = (\text{constant})/P^2$. Using this relationship, calculate the volume change corresponding to a pressure change from 0.10 bar to 1.00 bar for 1.00 mol of gas at 500 K. The constant in (*1.3*) is 41.6 dm^3 bar for these conditions.

Solving (*1.3*) for V and differentiating with respect to P gives

$$\left(\frac{\partial V}{\partial P}\right)_{n,T} = \left(\frac{\partial(\text{constant}/P)}{\partial P}\right)_{n,T} = -\frac{\text{constant}}{P^2}$$

The volume change is found by integrating the above equation, giving

$$dV = -\frac{\text{constant}}{P^2} dP$$

$$\Delta V = -(\text{constant}) \int_{P_1}^{P_2} \frac{1}{P^2} dP = (\text{constant})\left(\frac{1}{P_2} - \frac{1}{P_1}\right)$$

$$= (41.6 \text{ dm}^3 \text{bar})\left(\frac{1}{1.00 \text{ bar}} - \frac{1}{0.10 \text{ bar}}\right) = -370 \text{ dm}^3$$

1.6. Equation (*1.4*) can be used as the basis of a thermometer. If V_{tp} represents the volume of an ideal gas in a probe at the triple point of water, 273.16 K, and if V_T represents the volume of the gas at any temperature T, then

$$T = (273.16 \text{ K})\frac{V_T}{V_{\text{tp}}}$$

Calculate the ratio of V_T to V_{tp} if the probe is immersed in boiling water at 98.3 °C.

Rearranging the given equation for the desired ratio gives

$$\frac{V_T}{V_{\text{tp}}} = \frac{T}{273.16 \text{ K}} = \frac{(98.3 + 273.15) \text{ K}}{273.16 \text{ K}} = 1.360$$

1.7. In 1702, Amonton discovered for a fixed amount of gas under constant volume (isochoric) conditions that the pressure at 0 °C increased by the fraction 1/273 for each Celsius degree that the gas was heated. Write the equation for this observation. What does this equation predict at −273 °C?

The desired equation is similar to that given in Example 1.6,

$$P = P_0\left(1 + \frac{t/(°\text{C})}{273}\right) = P[1 + (3.66 \times 10^{-3})t/(°\text{C})]$$

According to the equation, the pressure of the gas is predicted to become zero at −273 °C.

1.8. The pressure of an ideal gas as a function of height h above a reference point under isothermal conditions is given by the *barometric equation*

$$P = P_0 e^{-Mgh/RT} \qquad (1.27)$$

where g is 9.806 65 m s^{-2}. Use (*1.6*) and (*1.27*) to calculate the pressure at an altitude of 1 500 m above sea level for a 0.79-mol sample of N_2 at 24.5 dm^3 and 25 °C.

The pressure at sea level is found using (*1.6*) as

$$P_0 = \frac{nRT}{V} = \frac{(0.79 \text{ mol})(0.083 14 \text{ dm}^3 \text{ bar K}^{-1} \text{mol}^{-1})(298 \text{ K})}{24.5 \text{ dm}^3}$$

$$= 0.80 \text{ bar}$$

and (*1.27*) gives

$$P = (0.80 \text{ bar}) \exp\left(\frac{-(0.028 \text{ kg mol}^{-1})(9.8 \text{ m s}^{-2})(1 500 \text{ m})}{(8.314 \text{ J K}^{-1} \text{mol}^{-1})(298 \text{ K})}\right)$$

$$= 0.68 \text{ bar}$$

1.9. A cylinder contains 75 dm³ of N_2 at 215 psig (gauge pressure) and 25 °C. If the room pressure is 14.4 psi, what mass of N_2 could be transferred to the laboratory? Assume that the ideal gas law is valid.

The amount of N_2 originally in the cylinder is

$$n_1 = \frac{PV}{RT} = \frac{(215 + 14.4)\ \text{psi}[(6\ 895\ \text{Pa})/(1\ \text{psi})](75\ \text{dm}^3)}{(8\ 314\ \text{dm}^3\ \text{Pa}\ \text{K}^{-1}\ \text{mol}^{-1})(298\ \text{K})} = 48\ \text{mol}$$

and the amount remaining at 0 psig is

$$n_2 = \frac{(0 + 14.4)(6\ 895)(75)}{(8\ 314)(298)} = 3.0\ \text{mol}$$

The amount of N_2 transferred is

$$[(48 - 3.0)\ \text{mol}](28 \times 10^{-3}\ \text{kg mol}^{-1}) = 1.3\ \text{kg}$$

1.10. Calculate the molar mass of CO_2 assuming that a 0.308-g sample (after buoyancy corrections) at 245 torr and 25 °C occupies a volume of 0.532 dm³.

Substituting into (1.7) gives

$$M = \frac{(0.308\ \text{g})(8\ 314\ \text{dm}^3\ \text{Pa}\ \text{K}^{-1}\ \text{mol}^{-1})(298\ \text{K})}{(245\ \text{torr})[(1.013\ 25 \times 10^5\ \text{Pa})/(760\ \text{torr})](0.532\ \text{dm}^3)}$$

$$= 43.9\ \text{g mol}^{-1}$$

MIXTURES OF IDEAL GASES

1.11. A 5.0-dm³ flask containing N_2 at 5.0 bar was connected to a 4.0-dm³ flask containing He at 4.0 bar, and the gases were allowed to mix isothermally. Calculate the individual pressures and total pressure for the resulting mixture.

Using Boyle's law, (1.3), for each gas gives

$$P(\text{He}) = (4.0\ \text{bar})\frac{4.0\ \text{dm}^3}{9.0\ \text{dm}^3} = 1.8\ \text{bar} \qquad P(\text{N}_2) = 5.0\ \frac{5.0}{9.0} = 2.8\ \text{bar}$$

and (1.7) gives the total pressure as $P_t = P(\text{He}) + P(\text{N}_2) = 4.6$ bar.

1.12. Prepare a plot of the partial pressures and the total pressure of the gases as a function of the mole fraction of gas B in a binary mixture.

The partial pressure of each gas is a linear function of mole fraction; see (1.10). Thus the partial pressures are given by

$$P_B = x_B P_t \qquad \text{and} \qquad P_A = x_A P_t = (1 - x_B) P_t$$

Dalton's law, (1.9), gives the total pressure as

$$P_t = P_A + P_B = (1 - x_B) P_t + x_B P_t = P_t$$

which is independent of the composition of the mixture. The plot is shown in Fig. 1-4.

REAL GASES

1.13. The densities of liquid and gaseous CCl_4 are $0.763\ 4 \times 10^3$ and $0.359\ 7 \times 10^3$ kg m⁻³ at 280 °C and $0.866\ 6 \times 10^3$ and $0.271\ 0 \times 10^3$ kg m⁻³ at 270 °C. Find A and B in (1.15). Given $t_c = 283.2$ °C, find ρ_c and the molar volume at the critical point.

Using the density data in (1.15) gives

$$\tfrac{1}{2}[(0.763\ 4 \times 10^3) + (0.359\ 7 \times 10^3)] = A + B(553)$$

$$\tfrac{1}{2}[(0.866\ 6 \times 10^3) + (0.271\ 0 \times 10^3)] = A + B(543)$$

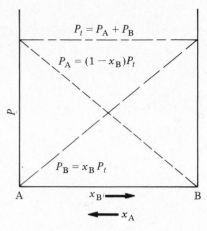

Fig. 1-4

which upon solving simultaneously gives $A = 0.960 \times 10^3 \text{ kg m}^{-3}$ and $B = -0.72 \text{ kg m}^{-3} \text{ K}^{-1}$. If we use $T_c = 556.4$ K, (1.16) gives

$$\rho_c = (0.960 \times 10^3 \text{ kg m}^{-3}) + (-0.72 \text{ kg m}^{-3} \text{ K}^{-1})(556.4 \text{ K}) = 559 \text{ kg m}^{-3}$$

The critical molar volume is given by

$$V_c = \frac{M}{\rho_c} = \frac{(153.82 \text{ g mol}^{-1})[(1 \text{ kg})/(10^3 \text{ g})]}{559 \text{ kg m}^{-3}}$$

$$= 2.75 \times 10^{-4} \text{ m}^3 \text{ mol}^{-1} = 275 \text{ cm}^3 \text{ mol}^{-1}$$

1.14. One mole of *n*-octane is confined to 20.0 dm^3 at 200. °C. Find the pressure, given that $P_c = 25.0$ bar and $T_c = 296.2$ °C.

To determine Z from Fig. 1-3, the values of both P_r and T_r must be known. Because the unknown in this problem is P, the exact value of P_r is not known at the beginning of the problem and an iterative method must be used.

If we assume an ideal gas, (1.6) gives

$$P = \frac{(1.00 \text{ mol})(0.083 \, 14 \text{ dm}^3 \text{ bar K}^{-1} \text{ mol}^{-1})(473 \text{ K})}{20.0 \text{ dm}^3} = 1.97 \text{ bar}$$

which upon substitution into (1.18) and (1.19) gives

$$P_r = \frac{1.97 \text{ bar}}{25.0 \text{ bar}} = 0.079 \qquad T_r = \frac{473 \text{ K}}{569.4 \text{ K}} = 0.831$$

The first approximation of Z from Fig. 1-3 for these values is 0.94, which upon substitution into (1.17) gives the second approximation for P as

$$P = \frac{(0.94)(1.00)(0.083 \, 14)(473)}{20.0} = 1.85 \text{ bar}$$

Substituting this value into (1.18) gives

$$P_r = \frac{1.85}{25.0} = 0.074$$

The second approximation of Z is 0.95, which gives

$$P = \frac{(0.95)(1.00)(0.083 \, 14)(473)}{20.0} = 1.87 \text{ bar}$$

This process is continued until P changes insignificantly. To three significant figures, $P = 1.87$ bar.

1.15. Solve (*1.20*) for *P*, and replace each term on the right side of (*1.21*) with this result. Find the relations between A_v and A_p, B_v and B_p, and C_v and C_p.

Rearranging (*1.20*) gives

$$P = A_v\left(\frac{1}{V_m}\right) + B_v\left(\frac{1}{V_m}\right)^2 + \cdots$$

which upon substitution into (*1.21*) gives

$$PV_m = A_p + B_p\left[A_v\left(\frac{1}{V_m}\right) + B_v\left(\frac{1}{V_m}\right)^2 + \cdots\right]$$

$$+ C_p\left[A_v\left(\frac{1}{V_m}\right) + B_v\left(\frac{1}{V_m}\right)^2 + \cdots\right]^2 + \cdots$$

$$= A_p + B_pA_v\left(\frac{1}{V_m}\right) + (B_pB_v + C_pA_v^2)\left(\frac{1}{V_m}\right)^2 + \cdots$$

Comparing each term of this result with the corresponding term in (*1.20*) shows that

$$A_v = A_p = RT \qquad\qquad B_v = B_pA_v = RTB_p$$

and

$$C_v = B_pB_v + C_pA_v^2 = B_pRTB_p + C_p(RT)^2 = RT(B_p^2 + RTC_p)$$

1.16. Discuss how B_v and C_v in (*1.20*) could be evaluated from experimental *PVT* data.

Using the result of Example 1.13, we write (*1.20*) as

$$PV_m - RT = B_v(1/V_m) + C_v(1/V_m)^2 + \cdots$$

Multiplying by V_m gives

$$V_m(PV_m - RT) = B_v + C_v(1/V_m) + \cdots$$

A plot of $V_m(PV_m - RT)$ against $1/V_m$ will have a vertical intercept of B_v and an initial slope of C_v.

1.17. Compare the pressure predicted for 1.00 mol of *n*-octane confined to 20.0 dm³ at 200 °C by the van der Waals equation ($a = 37.81$ dm⁶ bar mol⁻² and $b = 0.236\,8$ dm³ mol⁻¹) to the values found in Problem 1.14.

If we assume a van der Waals gas, (*1.22*) gives

$$P = \frac{nRT}{V - nb} - \frac{an^2}{V^2}$$

$$= \frac{(1.00)(0.083\,14)(473)}{20.0 - (1.00)(0.236\,8)} - \frac{(37.81)(1.00)^2}{(20.0)^2} = 1.99 - 0.09 = 1.90 \text{ bar}$$

There is about a 3% difference with the ideal gas law value and a 2% difference with the result of (*1.17*).

1.18. Calculate the van der Waals constants *a* and *b* for $(C_2H_5)_2S$ from the critical-point data given in Example 1.12, and compare the values to those given in Example 1.14.

From Example 1.15,

$$a = \frac{27R^2T_c^2}{64P_c} = \frac{(27)(0.083\,14 \text{ dm}^3 \text{ bar K}^{-1} \text{ mol}^{-1})^2(557.0 \text{ K})^2}{(64)(39.6 \text{ bar})}$$

$$= 22.8 \text{ dm}^6 \text{ bar mol}^{-2}$$

$$b = \frac{RT_c}{8P_c} = \frac{(0.083\,14 \text{ dm}^3 \text{ bar K}^{-1} \text{ mol}^{-1})(557.0 \text{ K})}{(8)(39.6 \text{ bar})}$$

$$= 0.146 \text{ dm}^3 \text{ mol}^{-1}$$

These results differ by about 20% from those given in Example 1.14.

1.19. Evaluate the virial coefficients A_v, B_v, and C_v for 1 mol of gas using the van der Waals equation.

For 1 mol, (1.22) becomes

$$\left(P + \frac{a}{V_m^2}\right)(V_m - b) = RT$$

which upon rearrangement gives

$$PV_m = (RT + bP) + (-a)(1/V_m) + (ab)(1/V_m)^2$$

Assuming that the bP term can be replaced by $b(RT/V_m)$ without significant error, we obtain

$$PV_m = RT + (bRT - a)(1/V_m) + (ab)(1/V_m)^2$$

Comparison with (1.20) now yields

$$A_v = RT \qquad B_v = (bRT - a) \qquad \text{and} \qquad C_v = ab$$

1.20. Any equation of state containing R and two coefficients can be written in reduced form if the expressions for R and the coefficients in terms of the critical constants are substituted into the equation and reduced variables are introduced. Determine the reduced form of the van der Waals equation of state.

Substituting a, b, and R from Example 1.15 into (1.22) for 1 mol, we find

$$\left(P + \frac{3P_c V_c^2}{V^2}\right)\left(V - \frac{V_c}{3}\right) = \left(\frac{8P_c V_c}{3T_c}\right)T$$

which upon multiplication of both sides by $3/P_c V_c$ gives

$$\left(\frac{P}{P_c} + \frac{3V_c^2}{V^2}\right)\left(\frac{3V}{V_c} - 1\right) = \frac{8T}{T_c}$$

Introducing reduced variables gives

$$\left(P_r + \frac{3}{V_r^2}\right)(3V_r - 1) = 8T_r$$

1.21. The limiting value of ρ/P as $P \to 0$ for SiF_4 at $0\,^\circ C$ is $4.583\,116\ \mathrm{g\,dm^{-3}\,bar^{-1}}$, as reported by Moles. Calculate the molar mass of SiF_4.

The intercept of (1.24) gives

$$M = (\text{intercept})(RT)$$
$$= (4.583\,116\ \mathrm{g\,dm^{-3}\,bar^{-1}})(0.083\,144\,1\ \mathrm{dm^3\,bar\,K^{-1}\,mol^{-1}})(273.15\ \mathrm{K})$$
$$= 104.086\,3\ \mathrm{g\,mol^{-1}}$$

KINETIC-MOLECULAR THEORY

1.22. Assume that an atom of neon is 0.065 nm in radius and that 1 mol of the gas occupies $22.4\ \mathrm{dm^3}$. What fraction of the volume is occupied by the atoms?

Neon is a monatomic gas, so that the desired fraction is

$$\frac{(\text{Volume of one molecule})(L)}{22.4\ \mathrm{dm^3}} = \frac{\frac{4}{3}\pi(6.5 \times 10^{-11}\ \mathrm{m})^3(6.022 \times 10^{23})}{(22.4\ \mathrm{dm^3})[(1\ \mathrm{m^3})/(1000\ \mathrm{dm^3})]} = 3.1 \times 10^{-5}$$

1.23. Calculate the ratio of $\overline{v^2}$ for Ne to that for Ar at $25\,^\circ C$.

Taking a ratio of the second and fourth terms of (1.25) for the gases gives

$$\frac{\frac{1}{3}(M_{Ne})\overline{v_{Ne}^2}}{\frac{1}{3}(M_{Ar})\overline{v_{Ar}^2}} = \frac{RT}{RT} = 1$$

which upon rearranging gives

$$\frac{\overline{v_{Ne}^2}}{\overline{v_{Ar}^2}} = \frac{M_{Ar}}{M_{Ne}} = \frac{39.984}{20.183} = 1.981\ 1$$

1.24. What is the average translational kinetic energy for 1 mol of an ideal gas at 25 °C?

Equation (1.25) gives

$$\bar{E}_{trans} = \tfrac{3}{2}RT = \tfrac{3}{2}(8.314\ \text{J K}^{-1}\ \text{mol}^{-1})(298\ \text{K}) = 3\ 716\ \text{J mol}^{-1}$$

Supplementary Problems

TEMPERATURE AND PRESSURE

1.25. There is one temperature that is common to both the Celsius and Fahrenheit scales. What is this reading?   *Ans.* $-40.0°$

1.26. The boiling temperature of pure water at exactly 1 atm (101.325 kPa) is 99.975 °C. Express this temperature in kelvins.   *Ans.*   373.125 K

1.27. The pressure in a vacuum system was measured using a thermocouple gauge calibrated in micrometers (1 μm = 10^{-3} mmHg). What is the pressure in the system expressed in pascals corresponding to a reading of 5.2 μm?   *Ans.*   0.69 Pa

1.28. Dibutyl phthalate is often used as a manometer fluid. It has a density of 1.047×10^3 kg m^{-3}. How many pascals are represented by 1 mm of this fluid?   *Ans.*   10.3 Pa

LAWS FOR IDEAL GASES

1.29. Plot the following data collected by Boyle, and evaluate the constants in (1.3).

V/(arbitrary units)	12	10	8	6	4
P/(in. Hg)	$29\frac{2}{16}$	$35\frac{5}{16}$	$44\frac{3}{16}$	$58\frac{13}{16}$	$87\frac{14}{16}$

Ans.   Plot is similar to Fig. 1-1; 352

1.30. Repeat Problem 1.11, assuming that both flasks contain N_2.   *Ans.* $P(N_2) = P_t = 4.6$ bar

1.31. Show that for a fixed amount of an ideal gas under isobaric conditions $(\partial V/\partial T)_{n,p} =$ (constant). Using this relationship, repeat the calculations of Example 1.5 using the value of the constant in (1.4) as 0.174 dm^3 K^{-1} to determine the temperature change.   *Ans.*   1 280 K

1.32. On the basis of the property of ideal gases that the temperature is directly proportional to the pressure at constant volume for a fixed amount of gas (*Amonton's Law*, see Problem 1.7), an ideal-gas thermometer containing He was standardized at an internal pressure of 305 torr at the melting point of ice. If the pressure decreased to 85 torr when the probe was placed in a Dewar flask containing boiling liquid N_2, what is the boiling point of N_2?   *Ans.* -197 °C

1.33. The probe of the gas thermometer described in Problem 1.32 is an active volume V_a, and the manometer used for pressure readings is a dead volume V_d. When the probe is placed at a low temperature T_1, a small amount of gas flows from the dead volume into the probe and the simple relationship $T_1 = T_0(P_1/P_0)$ is not valid because n has changed in the active volume. Using the relationships

$$P_0 = \frac{n_{0a}RT_0}{V_a} = \frac{n_{0d}RT_0}{V_d} \qquad P_1 = \frac{n_{1a}RT_1}{V_a} = \frac{n_{1d}RT_1}{V_d}$$

and
$$n_{0a} + n_{0d} = n_{1a} + n_{1d}$$

show that
$$T_1 = T_0 \frac{P_1}{P_0}\left[1 + \frac{V_d}{V_a}\left(\frac{P_0 - P_1}{P_0}\right)\right]^{-1}$$

1.34. Equation (1.6) can be used to calculate the value of R from experimental data. Assuming that the molar volume of most gases is 22.4 dm^3 at STP (0 °C and 1.00 atm), calculate R in units of dm^3 atm K^{-1} mol^{-1}.   *Ans.*   0.082 1 dm^3 atm K^{-1} mol^{-1}

1.35. An ideal gas at 175 K contains 5×10^{20} molecules m^{-3}. What is the pressure of this gas?   *Ans.*   1.2 Pa

1.36. The density of steam at 100 °C and 760.0 torr is 0.597 4 kg m^{-3}. Calculate the molar mass for water from these data. Explain any discrepancy from the value of 18.015 2 g mol^{-1}.

   Ans.   18.3 g mol^{-1};   the gas shows deviation from ideal behavior because it is very near the liquid state.

MIXTURES OF IDEAL GASES

1.37. Repeat Problem 1.8 for a 0.21-mol sample of O_2 under the same conditions. Does air at an altitude of 1 500 m contain a larger or smaller fraction of O_2 than air at sea level?

   Ans. $P_0(O_2) = 0.21$ bar and $x(O_2) = 0.21$ at sea level; $P(O_2) = 0.17$ bar and $x(O_2) = 0.20$ at 1 500 m;   smaller fraction of O_2

1.38. What is the difference in the density of dry air at 1 atm and 25 °C and moist air with 50% relative humidity under the same conditions? The vapor pressure of water at 25 °C is 23.7 torr. See Example 1.10 for additional data.

   Ans.   1.185 kg m^{-3} for the dry air; $P(H_2O) = 11.8$ torr, 1.178 kg m^{-3} for the wet air;   0.007 kg m^{-3} difference

1.39. The volume of air in a room is 56 m^3. Assuming the composition of the air to be $x(N_2) = 0.78$, $x(O_2) = 0.19$, and $x(CO_2) = 0.03$, find the partial volume for each of the gases.

   Ans. $V(N_2) = 44$ m^3, $V(O_2) = 11$ m^3, $V(CO_2) = 2$ m^3

1.40. The average molar mass of the vapor above $NH_4Cl(s)$ is 26.5 g mol^{-1}. Give an interpretation for this value.

   Ans.   The vapor consists of NH_3 and HCl, with $\bar{M} = 26.7$ g mol^{-1}.

REAL GASES

1.41. The densities of the gas and liquid in units of kg m^{-3} for a substance are given by

$$\rho_{gas} = 20.0 + (0.175\ 0)T + (1.500 \times 10^{-4})T^2$$

$$\rho_{liq} = 1\ 000.0 - (0.500\ 0)T - (2.000 \times 10^{-4})T^2$$

Find T_c, ρ_c, A, B, and C for this substance.

   Ans.   967 K,   329.5 kg m^{-3},   510.0 kg m^{-3}, $-0.162\ 5$ kg m^{-3} K^{-1}, -2.50×10^{-5} kg m^{-3} K^{-2}

1.42. The *Boyle temperature* for a real gas is defined as the temperature at which $dZ/dP = 0$. Show that this definition can also be written as $(1/P_c)(dZ/dP_r)$. Using Fig. 1-3, find the Boyle temperature for $P_r = 2.0$.

    Ans. $T_r = 1.1$, giving $T_{\text{Boyle}} = (1.1)\,T_c$

1.43. At low pressures, only the first two terms on the right side of (*1.21*) are important. Find the molar volume of N_2 at 100 °C and 35 bar given that $B_p = 6\ \text{cm}^3\ \text{mol}^{-1}$. Likewise, find the molar volume of Ar under the same conditions given that $B_p = -6\ \text{cm}^3\ \text{mol}^{-1}$. Given that the critical temperatures of those gases are near -150 °C and the critical pressures are near 45 bar, why is (*1.17*) difficult to use to calculate these volumes?

    Ans. $V_m(N_2) = 0.892\ \text{dm}^3\ \text{mol}^{-1}$, $V_m(\text{Ar}) = 0.880\ \text{dm}^3\ \text{mol}^{-1}$;   the values of Z are very difficult to read from Fig. 1-3.

1.44. Discuss how B_p and C_p in (*1.21*) could be evaluated from experimental PVT data.

    Ans. B_p is the vertical intercept, and C_p is the initial slope of a plot of $(PV_m - RT)/P$ against P.

1.45. Two molecules of a gas will collide when their centers are within a volume of $\frac{4}{3}\pi\sigma^3$. The excluded volume per molecule is $\frac{2}{3}\pi\sigma^3$; for a mole it is $b = \frac{2}{3}L\pi\sigma^3$. Using the diameter of argon as 0.361 nm calculate b, and compare the answer to the van der Waals value of $0.032\,19\ \text{dm}^3\ \text{mol}^{-1}$.

    Ans. $0.059\,3\ \text{dm}^3\ \text{mol}^{-1}$, about 84% larger

1.46. Transform (*1.22*) into a general polynomial as a function of volume.

    Ans. $PV^3 - n(bP + RT)V^2 + n^2 aV - n^3 ab = 0$

1.47. Substitute $1/V_m = P/RT$ into the expression for PV_m derived in Problem 1.19, and rearrange the result into the form given by (*1.21*). Determine A_p, B_p, and C_p.

    Ans. $A_p = RT$, $B_p = b - a/RT$, $C_p = ab/(RT)^2$

1.48. Assuming that $Z = PV_m/RT = (A_p + B_p P)/RT$, show that $B_p = 0$ at the Boyle temperature (see Problem 1.42). Using the results of Problem 1.47 and Example 1.15, derive an equation for T_{Boyle} in terms of T_c.   *Ans.* $T_{\text{Boyle}} = 27\,T_c/8$

1.49. Moles reported the following values of ρ/P as a function of P for CO_2 at 0 °C:

$P/(\text{Pa})$	101 325	50 662.5	10 132.5	1 013.25	101.325	10.132 5
$(\rho/P)/(\text{g dm}^{-3}\ \text{bar}^{-1})$	1.970 916	1.944 469	1.939 082	1.937 867	1.937 746	1.937 734

From these data calculate the molar mass of CO_2, and, using the accepted value of 12.011 15 u for the relative atomic mass of C, find the relative atomic mass for O.

    Ans.   Intercept = 1.937 733, 44.007 5 g mol^{-1}; 15.998 2 u

1.50. Determine the value of B_p for SO_2 from the data in Example 1.16.

    Ans.   Slope = $6.678\,9 \times 10^{-7}$ g dm^{-3} bar^{-1} Pa^{-1}, -0.538 dm^3 mol^{-1}

1.51. The *Dieterici equation*,

$$P = \frac{RT}{V_m - b}\, e^{-a/V_m RT} \tag{1.28}$$

may be expressed in a form similar to (1.21) by multiplying both sides of (1.28) by $V_m - b$; solving for PV_m; expanding the exponential as

$$e^x = 1 + x + \frac{x^2}{2} + \cdots$$

substituting the expression derived for PV_m the Pb term; collecting terms; determining A_v, B_v, and C_v by comparison with (1.22); and using the results of Problem 1.15 to find A_p, B_p, and C_p.

Ans. $A_p = RT$, $B_p = b - \dfrac{a}{RT}$, $C_p = \dfrac{a^2}{2R^3T^3}$

1.52. Evaluate the constants a and b in the Dieterici equation, (1.28), in terms of the critical point data. Determine the reduced form of this equation of state.

Ans. $a = \dfrac{4R^2T_c^2}{e^2P_c}$, $b = \dfrac{RT_c}{e^2P_c}$, $P_r = \dfrac{T_r}{2V_r - 1}\, e^{2-(2/T_rV_r)}$

1.53. Calculate the pressure for 1.00 mol of argon at 0 °C and 5.00 dm^3 using (a) the ideal gas law, (b) the van der Waals equation ($a = 1.363$ dm^6 bar mol^{-2} and $b = 0.032\,19$ dm^3 mol^{-1}), and (c) the Dieterici equation ($a = 1.75$ dm^6 bar mol^{-2} and $b = 0.035$ dm^3 mol^{-1}).

Ans.   (a) 4.54 bar,   (b) 4.52 bar,   (c) 4.50 bar

KINETIC-MOLECULAR THEORY

1.54. Calculate the ratio of v_{rms} of $^{238}UF_6$ to v_{rms} of $^{235}UF_6$ at room temperature.   Ans.   0.996

1.55. Show that for a fixed amount of gas at constant temperature (1.25) becomes Boyle's law, and for a fixed amount of gas at constant pressure (1.25) becomes Charles's law.

1.56. What is the average translational kinetic energy of a molecule of an ideal gas at 25 °C?

Ans. 6.17×10^{-21} J

Chapter 2

Translation and Transport Phenomena

Velocity and Energy Distributions of Gases

2.1 VELOCITY DISTRIBUTION

The *Maxwell relation* gives the fraction of molecules in 1 mol (dN/L) having a velocity in the x direction between v_x and $v_x + dv_x$ as

$$\frac{dN/L}{dv_x} = A\, e^{-mv_x^2/2kT} \tag{2.1}$$

where A is a constant and L is Avogadro's number. In three dimensions, the *Maxwell–Boltzmann relation* for the fraction of molecules in 1 mol having a velocity between v and $v + dv$ is

$$\frac{dN/L}{dv} = 4\pi v^2 \left(\frac{m}{2\pi kT}\right)^{3/2} e^{-mv^2/2kT} \tag{2.2}$$

The velocities of gases are often expressed using the *root mean square speed* (v_{rms}), the *average velocity* $(\bar{v})$, the *median velocity* (v_{med}), or the *most probable velocity* (v_{mp}), where

$$v_{\mathrm{rms}} = (\overline{v^2})^{1/2} = \left(\frac{3kT}{m}\right)^{1/2} = \left(\frac{3RT}{M}\right)^{1/2} \tag{2.3}$$

$$\bar{v} = \left(\frac{8kT}{\pi m}\right)^{1/2} = \left(\frac{8RT}{\pi M}\right)^{1/2} \tag{2.4}$$

$$v_{\mathrm{med}} = 1.538\, 2 \left(\frac{kT}{m}\right)^{1/2} = 1.538\, 2 \left(\frac{RT}{M}\right)^{1/2} \tag{2.5}$$

$$v_{\mathrm{mp}} = \left(\frac{2kT}{m}\right)^{1/2} = \left(\frac{2RT}{M}\right)^{1/2} \tag{2.6}$$

EXAMPLE 2.1. Evaluate the constant A in (2.1) based on the property that $\int dN = L$.

Dividing $\int dN = L$ by L gives

$$1 = \int \frac{dN}{L} = A \int_{-\infty}^{\infty} e^{-(1/2)mv_x^2/kT}\, dv_x$$

Transforming variables by letting $z = (m/2kT)^{1/2} v_x$ gives

$$1 = A \left(\frac{2kT}{m}\right)^{1/2} \int_{-\infty}^{\infty} e^{-z^2}\, dz$$

The integral is equal to $\pi^{1/2}$ (see Table 2-1), giving

$$A = (m/2\pi kT)^{1/2}$$

EXAMPLE 2.2. Derive (2.3) describing v_{rms} in terms of R, T, and M using (2.2).

With respect to the velocity distribution (2.2), the average value of any function $h(u)$ is defined as

$$\bar{h} = \int_0^{\infty} h(u)\, dN/L \tag{2.7}$$

Table 2-1

$$\int_0^\infty e^{-x^2}\,dx = \frac{\pi^{1/2}}{2}$$

$$\frac{2}{\pi^{1/2}}\int_0^z e^{-x^2}\,dx = \mathrm{erf}(z) \qquad \text{(see Fig. 2-1)}$$

$$\int_0^\infty x^n e^{-ax^2}\,dx = \begin{cases} \dfrac{(1)(3)(5)\cdots(n-1)}{2(2a)^{n/2}}\left(\dfrac{\pi}{a}\right)^{1/2} & \text{(for even } n \geq 2) \\[3mm] \dfrac{[(n-1)/2]!}{2a^{(n+1)/2}} & \text{(for odd } n \geq 1) \end{cases}$$

In particular, for $h(u) = v^2$ we have

$$\overline{v^2} = \int_0^\infty v^2\,dN/L = 4\pi\left(\frac{m}{2\pi kT}\right)^{3/2}\int_0^\infty v^4 e^{-(1/2)mv^2/kT}\,dv$$

The integral can be transformed by letting $z = (m/2kT)^{1/2}v$, which gives

$$\overline{v^2} = 4\pi\left(\frac{m}{2\pi kT}\right)^{3/2}\left(\frac{2kT}{m}\right)^{5/2}\int_0^\infty z^4 e^{-z^2}\,dz$$

The integral is equal to $3\pi^{1/2}/8$ (see Table 2-1), giving

$$\overline{v^2} = 4\pi\left(\frac{m}{2\pi kT}\right)^{3/2}\left(\frac{2kT}{m}\right)^{5/2}\frac{3\pi^{1/2}}{8} = \frac{3kT}{m}$$

So for v_{rms},

$$v_{\mathrm{rms}} = (\overline{v^2})^{1/2} = \left(\frac{3kT}{m}\right)^{1/2}$$

which, because $k = R/L$ and $m = M/L$, can also be written as

$$v_{\mathrm{rms}} = \left(\frac{3RT}{M}\right)^{1/2}$$

EXAMPLE 2.3. The fraction of molecules having a velocity $v \geq c$ is given by

$$\frac{N_{v \geq c}}{L} = \frac{2}{\pi^{1/2}}\left(\frac{c}{v_{\mathrm{mp}}}\right)e^{-(c/v_{\mathrm{mp}})^2} + 1 - \mathrm{erf}(c/v_{\mathrm{mp}}) \qquad (2.8)$$

where the definition of the *error function* $\mathrm{erf}(z)$ is given in Table 2-1. Find the fraction of neon molecules having a velocity equal to or greater than 775 m s^{-1} at 298 K.

The most probable velocity for neon at 298 K is given by (2.6) as

$$v_{\mathrm{mp}} = \left(\frac{2(8.314\ \mathrm{J\ K^{-1}\ mol^{-1}})(298\ \mathrm{K})[(1\ \mathrm{kg\ m^2\ s^{-2}})/(1\ \mathrm{J})]}{(20.18\ \mathrm{g\ mol^{-1}})[(10^{-3}\ \mathrm{kg})/(1\ \mathrm{g})]}\right)^{1/2} = 496\ \mathrm{m\ s^{-1}}$$

The argument of the error function is

$$c/v_{\mathrm{mp}} = (775\ \mathrm{m\ s^{-1}})/(491\ \mathrm{m\ s^{-1}}) = 1.58$$

which corresponds to $\mathrm{erf}(1.58) = 0.97$ (see Fig. 2-1). Substituting into (2.8) gives the fraction as

$$\frac{N_{v \geq c}}{L} = \frac{2}{\pi^{1/2}}(1.58)\,e^{-(1.58)^2} + 1 - 0.97 = 0.18$$

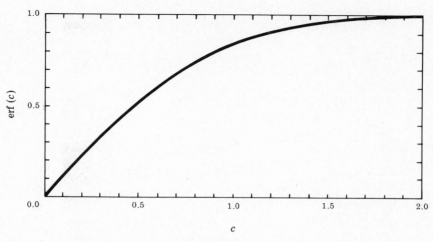

Fig. 2-1

2.2 ENERGY DISTRIBUTION

The fraction of molecules in 1 mol (dN/L) having a translational kinetic energy between $\varepsilon_{\text{trans}}$ and $\varepsilon_{\text{trans}} + d\varepsilon_{\text{trans}}$ is

$$\frac{dN/L}{d\varepsilon_{\text{trans}}} = \frac{2\varepsilon_{\text{trans}}^{1/2}}{\pi^{1/2}(kT)^{3/2}} e^{-\varepsilon_{\text{trans}}/kT} \qquad (2.9)$$

and the fraction of molecules having a translational kinetic energy $\varepsilon_{\text{trans}} \geq \varepsilon'$ is

$$\frac{N_{\varepsilon_{\text{trans}} \geq \varepsilon'}}{L} = 2\left(\frac{\varepsilon'}{\pi kT}\right)^{1/2} e^{-\varepsilon'/kT} + 1 - \text{erf}\left[\left(\frac{\varepsilon'}{kT}\right)^{1/2}\right] \qquad (2.10)$$

In the limiting case where $\varepsilon' > kT$, (2.10) becomes a special case of the *Boltzmann distribution law*, which can be used to compare the numbers of molecules having two different energies:

$$\frac{N_i}{N_j} = \frac{g_i}{g_j} e^{-(\varepsilon_i - \varepsilon_j)/kT} \qquad (2.11)$$

where g represents the *statistical weight* or *degeneracy* of the energy level (the number of states with the same energy). For these calculations, $g = 1$.

EXAMPLE 2.4.   Determine $\bar{\varepsilon}_{\text{trans}}$.

Using (2.9) with (2.7) gives

$$\bar{\varepsilon}_{\text{trans}} = \int_0^\infty \varepsilon_{\text{trans}} \frac{2\varepsilon_{\text{trans}}^{1/2}}{\pi^{1/2}(kT)^{3/2}} e^{-\varepsilon_{\text{trans}}/kT} \, d\varepsilon_{\text{trans}}$$

$$= \frac{2}{\pi^{1/2}(kT)^{3/2}} \int_0^\infty \varepsilon_{\text{trans}}^{3/2} e^{-\varepsilon_{\text{trans}}/kT} \, d\varepsilon_{\text{trans}}$$

The integral can be transformed by letting $z^2 = \varepsilon_{\text{trans}}/kT$, which gives

$$\bar{\varepsilon}_{\text{trans}} = \frac{2}{\pi^{1/2}(kT)^{3/2}} (2)(kT)^{5/2} \int_0^\infty z^4 e^{-z^2} \, dz$$

The integral is equal to $3\pi^{1/2}/8$ (see Table 2-1), giving

$$\bar{\varepsilon}_{\text{trans}} = \frac{2}{\pi^{1/2}(kT)^{3/2}} (2)(kT)^{5/2} \frac{3\pi^{1/2}}{8} = \tfrac{3}{2}kT \qquad (2.12a)$$

which can also be written in molar terms as

$$\bar{E}_{trans} = \tfrac{3}{2}RT \qquad (2.12b)$$

EXAMPLE 2.5. Compare the number of molecules with $\varepsilon_{trans} = 3.0kT$ to that with $\varepsilon_{trans} = 2.5kT$.

The ratio is given by (2.11) as

$$\frac{N_{3.0kT}}{N_{2.5kT}} = e^{-(3.0kT-2.5kT)/kT} = 0.607$$

Collision Parameters

2.3 COLLISION NUMBERS OF GASES

For a binary gaseous mixture consisting of molecules having collision diameters σ_i and σ_j, the total number of collisions per unit time per unit volume between unlike molecules (Z_{ij}) is given by

$$Z_{ij} = N_i^* N_j^* \pi \sigma_{ij}^2 \left(\frac{8kT}{\pi\mu}\right)^{1/2} \qquad (2.13)$$

where the *molecular density* (N^*), the number of molecules per unit volume, can be determined from

$$N_i^* = \frac{n_i L}{V} = \frac{LP_i}{RT} \qquad (2.14)$$

The average collision diameter (σ_{ij}) for the mixture is given by

$$\sigma_{ij} = (\sigma_i + \sigma_j)/2 \qquad (2.15)$$

and the *reduced mass* (μ) is defined as

$$\mu = \frac{m_i m_j}{m_i + m_j} \qquad (2.16)$$

The number of collisions per unit time between a given molecule of gas i with unlike molecules in a gaseous mixture (z_{ij}) is given by

$$z_{ij} = N_j^* \pi \sigma_{ij}^2 \left(\frac{8kT}{\pi\mu}\right)^{1/2} \qquad (2.17)$$

and with like molecules (z_{ii}) is given by

$$z_{ii} = N_i^* \pi \sigma_i^2 (2^{1/2}) \left(\frac{8kT}{\pi m_i}\right)^{1/2} \qquad (2.18)$$

Equation (2.18) also describes the number of collisions per unit time experienced by a single molecule in a pure gas.

The number of molecular collisions per unit time between a gas and a surface of area A (such as the wall of a container), Z_i, is given by

$$Z_i = N_i^* \left(\frac{kT}{2\pi m}\right)^{1/2} A \qquad (2.19)$$

Equation (2.19) gives the ratio of the rates of *effusion* (the rate at which molecules escape one by one through an orifice of molecular size) for a mixture of gases as

$$\frac{\text{Rate}_i}{\text{Rate}_j} = \frac{P_i/m_i^{1/2}}{P_j/m_j^{1/2}} \qquad (2.20)$$

Equation (2.20) is also valid for comparing the effusion rates of pure gases from the same container under identical temperature conditions. Equation (2.20) is *Graham's law of effusion*.

EXAMPLE 2.6.   Derive an equation describing Z_{ii}, the total rate of collisions per unit volume between like molecules, from (2.13). Using $\sigma = 0.361$ nm, calculate the total number of collisions per unit volume per unit time and the number of collisions per unit time that one molecule of argon undergoes at 25 °C and 1.00 bar.

Substituting

$$\sigma_{ij} = \frac{\sigma_i + \sigma_i}{2} = \sigma_i \qquad \mu = \frac{m_i m_i}{2m_i} = \frac{m_i}{2}$$

and $N_j^* = N_i^*$ into (2.13) gives

$$Z_{ii} = N_i^* N_i^* \pi \sigma_i^2 \left(\frac{8kT}{\pi m/2}\right)^{1/2}$$

This equation counts each collision twice, so this result is divided by 2 to give the desired equation:

$$Z_{ii} = \frac{N_i^{*2} \pi \sigma_i^2}{2^{1/2}} \left(\frac{8kT}{\pi m}\right)^{1/2} \tag{2.21}$$

Note that (2.21) also describes the collision rate per unit volume in a pure gas.

For argon at 25 °C and 1.00 bar, (2.14) gives

$$N_i^* = \frac{LP_i}{RT}$$

$$= \frac{(6.022 \times 10^{23} \text{ mol}^{-1})(1.00 \text{ bar})[(10^5 \text{ Pa})/(1 \text{ bar})][(1 \text{ kg m}^{-1} \text{ s}^{-2})/(1 \text{ Pa})]}{(8.314 \text{ J K}^{-1} \text{ mol}^{-1})[(1 \text{ kg m}^2 \text{ s}^{-2})/(1 \text{ J})](298 \text{ K})}$$

$$= 2.43 \times 10^{25} \text{ m}^{-3}$$

Thus the total number of collisions per unit volume per unit time given by (2.21) is

$$Z_{ii} = \frac{(2.43 \times 10^{25} \text{ m}^{-3})^2 (\pi)(3.61 \times 10^{-10} \text{ m})^2}{2^{1/2}}$$

$$\times \left[\frac{8(1.381 \times 10^{-23} \text{ J K}^{-1})(298 \text{ K})[(1 \text{ kg m}^2 \text{ s}^{-2})/(1 \text{ J})]}{\pi[(39.95 \times 10^{-3} \text{ kg mol}^{-1})/(6.022 \times 10^{23} \text{ mol}^{-1})]}\right]^{1/2}$$

$$= 6.79 \times 10^{34} \text{ m}^{-3} \text{ s}^{-1}$$

and the number of collisions per unit time that one molecule undergoes given by (2.18) is

$$z_{ii} = (2.43 \times 10^{25})(\pi)(3.61 \times 10^{-10})^2 2^{1/2} \left[\frac{8(1.381 \times 10^{-23})(298)}{\pi(39.95 \times 10^{-3})/(6.022 \times 10^{23})}\right]^{1/2}$$

$$= 5.59 \times 10^9 \text{ s}^{-1}$$

2.4   MEAN FREE PATH

For a binary mixture of gases, the average distance traveled by a molecule between collisions is given by

$$\lambda_i = \frac{\bar{v}_i}{z_{ii} + z_{ij}} \tag{2.22}$$

where λ is known as the *mean free path*.

EXAMPLE 2.7.   Derive an equation for the mean free path of a molecule in a pure gas. Using the results of Example 2.6, calculate λ for argon at 25 °C and 1.00 bar.

For a pure gas, $z_{ij} = 0$, and substituting (2.4) and (2.18) gives

$$\lambda_i = \frac{\bar{v}_i}{z_{ii}} = \frac{(8kT/\pi m_i)^{1/2}}{N_i^* \pi \sigma_i^2 (2^{1/2})(8kT/\pi m_i)^{1/2}} = \frac{1}{\pi(2^{1/2})\sigma_i^2 N_i^*} \tag{2.23}$$

Using $N^*(Ar) = 2.43 \times 10^{25}$ m^{-3} and σ (Ar) $= 3.61 \times 10^{-10}$ m, (2.23) gives

$$\lambda\,(Ar) = \frac{1}{\pi(2^{1/2})(3.61 \times 10^{-10}\text{ m})^2(2.43 \times 10^{25}\text{ m}^{-3})} = 7.11 \times 10^{-8}\text{ m}$$

Transport Properties

2.5 GENERAL TRANSPORT LAW

The rate of transport of a physical quantity W, such as heat (for which $W = q$), perpendicular to the direction of fluid flow is known as the *flux* or *flow* (J_W), where

$$J_W = \frac{1}{A}\left(\frac{\partial W}{\partial t}\right) \tag{2.24}$$

In (2.24), A is the area perpendicular to the flow and t is time. The *generalized force* for the transport process (X_W) is related to the negative gradient of a second physical quantity Y such as temperature, in the direction of flow (z),

$$X_W = -\left(\frac{\partial Y}{\partial z}\right) \tag{2.25}$$

The flux and the generalized force are related to each other by the *general transport law*

$$J_W = L_W X_W \tag{2.26}$$

where L_W is a proportionality constant known as the *phenomenological coefficient*. In the case where $W = q$ and $Y = T$, then, $L_W = -k_T$, where k_T is the coefficient of thermal conductivity. The general transport law for several transport properties is summarized in Table 2-2.

Table 2-2

Transport Process	$J_W = L_W X_W$	Comments
Thermal conductivity (Fourier's law)	$\dfrac{1}{A}\left(\dfrac{\partial q}{\partial t}\right) = -k_T \dfrac{\partial T}{\partial z}$	q = heat T = temperature k_T = coefficient of thermal conductivity
Viscosity (Newton's law)	$\dfrac{1}{A}\left(\dfrac{\partial p_y}{\partial t}\right) = -\eta \dfrac{\partial v_y}{\partial z}$	p_y = momentum in y direction v_y = velocity in y direction η = coefficient of viscosity
Fluid flow (Poiseuille's law)	$\dfrac{1}{A}\left(\dfrac{\partial V}{\partial t}\right) = -c \dfrac{\partial P}{\partial z}$	V = rate of volume flow ($\dot{V} = \partial V/\partial t$) P = pressure c = friction coefficient
Diffusion (Fick's first law)	$\dfrac{1}{A}\left(\dfrac{\partial N^*}{\partial t}\right) = -D \dfrac{\partial C}{\partial z}$	N^* = molecular density C = concentration D = diffusion coefficient

EXAMPLE 2.8.   The rigid, hard sphere theory for gases predicts that the coefficient of thermal conductivity will be given by

$$k_T = \frac{25}{32}\left(\frac{kT}{\pi m}\right)^{1/2}\frac{C_v}{L\sigma^2} = \frac{25\pi}{64}\,\bar{v}\,\frac{N^*}{L}\,C_v\lambda \tag{2.27}$$

where C_v is the molar heat capacity of the gas (see Chap. 3, Sec. 3.5). Calculate the value of k_T for Ar at 25 °C and 1.00 bar given that $C_v = 12.472$ J K^{-1} mol^{-1} and $\sigma = 0.361$ nm.

The value of k_T under these conditions is

$$k_T = \frac{25}{32}\left(\frac{(1.381\times 10^{-23}\ \text{J K}^{-1})(298\ \text{K})[(1\ \text{kg m}^2\ \text{s}^{-2})/(1\ \text{J})]}{\pi[(39.95\times 10^{-3}\ \text{kg mol}^{-1})/(6.022\times 10^{23}\ \text{mol}^{-1})]}\right)^{1/2}\frac{12.472\ \text{J K}^{-1}\ \text{mol}^{-1}}{(6.022\times 10^{23}\ \text{mol}^{-1})(3.61\times 10^{-10}\ \text{m})^2}$$

$$= 1.74\times 10^{-2}\ \text{J K}^{-1}\ \text{m}^{-1}\ \text{s}^{-1}$$

The experimental value is 1.75×10^{-2} J K^{-1} m^{-1} s^{-1}.

2.6 VISCOSITY

The viscosity of a fluid is related to its resistance to flow. For this transport property, (2.26) gives

$$\frac{\partial p_y}{\partial t} = -\eta A \frac{\partial v_y}{\partial z} \tag{2.28}$$

Note that because a force is the derivative of momentum with respect to time, (2.28) can be interpreted as the force necessary to move one layer of fluid with respect to an adjacent layer, where ∂v_y represents the difference in the velocity of the layers and ∂z is the distance between the layers.

If the molecules of a gas are considered to be rigid, hard spheres, η is given by

$$\eta = \frac{5\bar{v}m}{32(2^{1/2})\sigma^2} = \frac{5\pi}{32}N^*\bar{v}m\lambda \tag{2.29}$$

The cgs system unit for viscosity is the *poise* (1 P = 1 g cm^{-1} s^{-1} = 0.1 N s m^{-2}), and the SI unit is the *pascal second* (1 Pa s = 1 N s m^{-2} = 1 kg m^{-1} s^{-1} = 10 P).

The coefficient of viscosity for liquids is commonly measured with the *Ostwald viscometer* (or some revision) or with the *falling-sphere viscometer*. In the Ostwald technique, the time required for a given amount of liquid to flow at low values of Reynolds numbers (Re, see Sec. 2.7) is measured and η is calculated from

$$\eta = \frac{\pi R^4 \Delta P t}{8 V l} \tag{2.30}$$

where ΔP is the pressure drop over the length l of the tube, V is the volume of the liquid, and R is the radius of the tube. In actual practice, a comparison method is often used to bypass the determination of ΔP, V, l, and R, giving

$$\frac{\eta_1}{\eta_0} = \frac{\rho_1 t_1}{\rho_0 t_0} \tag{2.31}$$

where the subscripts are used to identify the unknown and reference liquids. The falling-sphere technique balances the force of gravitation against viscous drag, giving

$$\eta = \frac{2r_b^2(\rho_b - \rho)g}{9v} \tag{2.32}$$

where v is the velocity of the falling sphere, the subscript b refers to the dropping sphere or bead, and g is the gravitational constant. If comparison methods are used,

$$\frac{\eta_1}{\eta_0} = \frac{(\rho_b - \rho_1)t_1}{(\rho_b - \rho_0)t_0} \tag{2.33}$$

The frictional drag on one cylinder or disk by another that is rotating in a gas or liquid can also be used to measure η. Comparison methods are often used with this method also.

EXAMPLE 2.9. The coefficient of viscosity for argon at 25 °C is 222 μP. Calculate σ using (2.29), and compare this value to 4 times the atomic radius, where the atomic radius is 0.095 nm for argon.

Using (2.4), we find the average velocity

$$\bar{v} = \left(\frac{8(8.314 \text{ J K}^{-1} \text{ mol}^{-1})(298 \text{ K})}{\pi(39.95 \times 10^{-3} \text{ kg mol}^{-1})} \right)^{1/2} = 397 \text{ m s}^{-1}$$

Rearranging (2.29) gives the collision diameter as

$$\sigma = \left(\frac{5\bar{v}m}{32(2^{1/2})(\eta)} \right)^{1/2}$$

$$= \left(\frac{5(397 \text{ m s}^{-1})(39.95 \times 10^{-3} \text{ kg mol}^{-1})/(6.022 \times 10^{23} \text{ mol}^{-1})}{32(2^{1/2})(222 \times 10^{-7} \text{ Pa s})[(1 \text{ kg m}^{-1} \text{ s}^{-2})/(1 \text{ Pa})]} \right)^{1/2}$$

$$= 3.62 \times 10^{-10} \text{ m} = 0.362 \text{ nm}$$

which is about 5% lower than that predicted using the atomic radius.

EXAMPLE 2.10. Equation (2.29) predicts that η for a gas should be directly proportional to $T^{1/2}$. However, for a liquid the temperature dependence of η is given by

$$\eta = A \, e^{\Delta E(\text{viscosity})/RT} \tag{2.34}$$

where A and ΔE(viscosity) are constants for a given liquid. Glycerin has the following viscosities:

$t/(\text{°C})$	−42	−25	−10.8	0	20	30
$\eta/(\text{Pa s})$	6.710	262	35.5	12.11	1.49	0.629

Find ΔE(viscosity), and calculate η at 25 °C.

A plot of log η against $1/T$ gives a straight line with slope = 3 560 K and intercept = −11.972 (see Fig. 2-2). Then

$$\Delta E(\text{viscosity}) = (2.303 R)(\text{slope})$$

$$= (2.303)(8.314 \text{ J K}^{-1} \text{ mol}^{-1})(3 560 \text{ K})$$

$$= 6.82 \times 10^{4} \text{ J mol}^{-1} = 68.2 \text{ kJ mol}^{-1}$$

and

$$A = \log^{-1}(-11.972) = 1.07 \times 10^{-12} \text{ Pa s}$$

At 25 °C, (2.34) gives

$$\eta = (1.07 \times 10^{-12} \text{ Pa s}) \exp \left[\frac{6.82 \times 10^{4} \text{ J mol}^{-1}}{(8.314 \text{ J K}^{-1} \text{ mol}^{-1})(298 \text{ K})} \right]$$

$$= 0.98 \text{ Pa s}$$

The actual value is 0.954 Pa s.

2.7 FLUID FLOW

The flow of a fluid through a tube of radius R has associated with it a *Reynolds number* (Re) given by

$$\text{Re} = \frac{2R\bar{v}\rho}{\eta} \tag{2.35}$$

where $\bar{v}$ is the average or bulk velocity of the fluid, ρ is the density, and η is the coefficient of viscosity. If Re is greater than 4 000, the flow is *turbulent*, and if it is less than 2 100, the flow is *laminar*. In

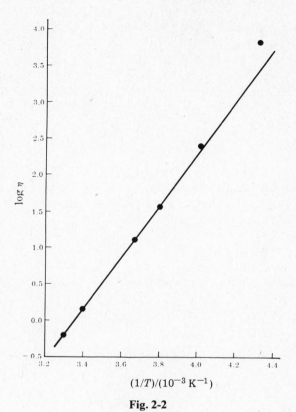

$(1/T)/(10^{-3}\,\text{K}^{-1})$

Fig. 2-2

laminar flow a velocity profile given by

$$v = \frac{\Delta P}{4\eta l}(R^2 - r^2) \qquad (2.36)$$

is observed in the tube, where ΔP is the pressure drop over a length l and r is the distance from the axis of the tube.

For an incompressible fluid such as a liquid flowing through the tube, (2.36) gives the volume flowing through a fixed cross section in time t as

$$V = \dot{V}t = \frac{\pi R^4 \Delta P t}{8\eta l} \qquad (2.37)$$

and for a compressible fluid such as a gas,

$$V = \frac{\pi R^4}{16\eta l}\left(\frac{P_i^2 - P_f^2}{P_0}\right)t \qquad (2.38)$$

where P_i is the inlet pressure, P_f is the outlet pressure, and P_0 is the pressure at which the fluid volume is measured.

EXAMPLE 2.11.   A 190-cm^3 sample of argon gas at 25 °C required 8.5 s to flow through a 1.00-m tube of 1.0-mm radius. The inlet pressure for the gas was 1.020 bar, and the outlet pressure was 1.007 bar. The volume of the gas was measured at 1.007 bar. What is η for this gas?

Rearranging (2.38) for η and substituting the given numerical values gives

$$\eta = \frac{\pi R^4}{16 Vl} \frac{P_i^2 - P_f^2}{P_0} t$$

$$= \frac{\pi\{(1.0 \text{ mm})[(10^{-3} \text{ m})/(1 \text{ mm})]\}^4[(1.020 \text{ bar})^2 - (1.007 \text{ bar})^2][(10^5 \text{ Pa})/(1 \text{ bar})](8.5 \text{ s})}{16\{(190 \text{ cm}^3)[(10^{-6} \text{ m}^3)/(1 \text{ cm}^3)]\}(1.00 \text{ m})(1.007 \text{ bar})}$$

$$= 2.3 \times 10^{-5} \text{ Pa s}$$

2.8 DIFFUSION

Diffusion involves the mixing of the molecules of substances by the random thermal motions and collisions of the molecules until the mixture attains uniform composition. The rate of diffusion in the z direction is given by (2.26) (*Fick's first law*) as

$$\frac{dN^*}{dt} = -DA \frac{\partial C}{\partial z} \qquad (2.39)$$

The changes in concentration with time and with distance are related by *Fick's second law* (the *diffusion equation*):

$$\frac{\partial C}{\partial t} = D \frac{\partial^2 C}{\partial z^2} \qquad (2.40)$$

For gaseous molecules that can be considered to be rigid, hard spheres, the diffusion coefficient of i into j is given by

$$D_{ij} = \frac{3(2^{1/2})(\pi)}{64} \lambda_i \bar{v}_i \qquad (2.41)$$

EXAMPLE 2.12. An indication of the distance that a single molecule travels in time t during diffusion is given by the *Einstein–Smoluchowski equation*, which defines the *root mean square distance* $(\overline{z^2})^{1/2}$ as

$$(\overline{z^2})^{1/2} = (2Dt)^{1/2} \qquad (2.42)$$

Given $D = 1.78 \times 10^{-5} \text{ m}^2 \text{ s}^{-1}$ for O_2 in air at 0 °C, find this distance after $t = 60.$ s.

Upon substitution of the data, (2.42) gives

$$(\overline{z^2})^{1/2} = [2(1.78 \times 10^{-5} \text{ m}^2 \text{ s}^{-1})(60. \text{ s})]^{1/2} = 0.046 \, 2 \text{ m}$$

$$= 4.62 \text{ cm}$$

Solved Problems

VELOCITY AND ENERGY DISTRIBUTIONS OF GASES

2.1.   Prepare plots of $(dN/L)/dv$ for N_2 at 100 °C and 1 000 °C for values of v up to 3 000 m s^{-1}. Indicate on the plots the values of v_{mp}, v_{med}, v_{rms}, and $\bar{v}$. Describe the distributions.

A sample calculation using (2.2) appears below for $T = 373$ K at 500. m s^{-1}.

$$\frac{dN/L}{dv} = 4\pi(500. \text{ m s}^{-1})^2 \left(\frac{(28.01 \times 10^{-3} \text{ kg mol}^{-1})[(1 \text{ mol})/(6.022 \times 10^{23})]}{2\pi(1.381 \times 10^{-23} \text{ J K}^{-1})(373 \text{ K})} \right)^{3/2}$$

$$\times \exp\left[\frac{-(28.01 \times 10^{-3})(500.)^2}{(2)(6.022 \times 10^{23})(1.381 \times 10^{-23})} \right](373)$$

$$= 1.75 \times 10^{-3} \text{ s m}^{-1}$$

This value and others are shown in Fig. 2-3. The values of v_{mp}, v_{med}, v_{rms}, and $\bar{v}$ were calculated using Eqs. (2.3)–(2.6). Both plots begin with very low fractions, increase to a maximum, and decrease exponentially. Although the maximum of the curve at 1 000 °C lies below the maximum of the curve at 100 °C, the plots cross, and the curve at 1 000 °C indicates the presence of a larger number of particles with high energy under this higher-temperature condition.

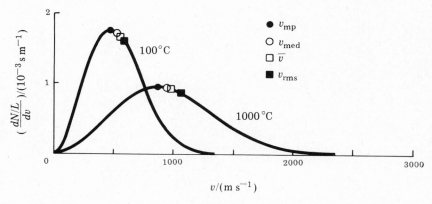

Fig. 2-3

2.2. A flask contains 1.00 mol of N_2 molecules at 373 K. How many molecules have a velocity in the range of 499–501 m s^{-1}?

The distribution function at 373 K for nitrogen (see Problem 2.1) is

$$(dN/L)/dv = 1.75 \times 10^{-3} \text{ s m}^{-1}$$

Solving for dN and substituting $dv = 2$ m s^{-1} gives

$$dN = (1.75 \times 10^{-3} \text{ s m}^{-1})(6.022 \times 10^{23} \text{ molecules})(2 \text{ m s}^{-1})$$

$$= 2.11 \times 10^{21} \text{ molecules}$$

2.3. Derive (2.6) by differentiating $(dN/L)/dv$ with respect to v, setting the result equal to zero, and solving for v_{mp}.

Performing the differentiation of (2.2) gives

$$\frac{d}{dv}\left(\frac{dN/L}{dv}\right) = 4\pi\left(\frac{m}{2\pi kT}\right)^{3/2} e^{-mv^2/2kT}\left(2v - \frac{mv^3}{kT}\right)$$

The right-hand side vanishes at $v = 0$ (but this corresponds to a minimum) and at $v = v_{mp}$, where

$$2 - \frac{mv_{mp}^2}{kT} = 0 \quad \text{or} \quad v_{mp} = \left(\frac{2kT}{m}\right)^{1/2}$$

2.4. Show that substitution of (2.5) into (2.8) for c gives $N_{v \geq c}/L = 0.5$, the value expected for the median velocity.

The value of the argument of (2.8) is

$$\frac{c}{v_{mp}} = \frac{(1.538\ 2)(RT/M)^{1/2}}{(2RT/M)^{1/2}} = 1.088$$

giving

$$\frac{N_{v \geq c}}{L} = \frac{2}{\pi^{1/2}}(1.088)\, e^{-(1.088)^2} + 1 - \text{erf}(1.088) = 0.50$$

where $\text{erf}(1.088)$ was evaluated using Fig. 2-1.

2.5. Prepare a plot of $(dN/L)/d\varepsilon_{\text{trans}}$ for N_2 at 25 °C for values of $\varepsilon_{\text{trans}}$ up to 5×10^{-20} J. Indicate on the plot the value of $\bar{\varepsilon}_{\text{trans}}$.

A sample calculation using (2.9) for 1×10^{-20} J follows.

$$\frac{dN/L}{d\varepsilon_{\text{trans}}} = \frac{2(1 \times 10^{-20}\,\text{J})^{1/2}}{\pi^{1/2}[(1.381 \times 10^{-23}\,\text{J K}^{-1})(298\,\text{K})]^{3/2}} \exp\left[\frac{-(1 \times 10^{-20}\,\text{J})}{(1.381 \times 10^{-23}\,\text{J K}^{-1})(298\,\text{K})}\right]$$

$$= 3.76 \times 10^{19}\,\text{J}^{-1}$$

This value and others are shown in Fig. 2-4. The value of $\bar{\varepsilon}_{\text{trans}}$ given by $(2.12a)$ is

$$\bar{\varepsilon}_{\text{trans}} = \tfrac{3}{2}(1.381 \times 10^{-23}\,\text{J K}^{-1})(298\,\text{K}) = 6.17 \times 10^{-21}\,\text{J}$$

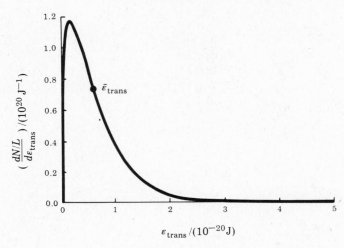

Fig. 2-4

2.6. What fraction of molecules has a translational kinetic energy greater than $\bar{\varepsilon}_{\text{trans}}$?

The argument of (2.10) is $\varepsilon'/kT = \tfrac{3}{2}kT/kT = \tfrac{3}{2}$, giving

$$\frac{N_{\varepsilon_{\text{trans}} \geq \bar{\varepsilon}_{\text{trans}}}}{L} = 2\left(\frac{\tfrac{3}{2}}{\pi}\right)^{1/2} e^{-3/2} + 1 - \text{erf}[(\tfrac{3}{2})^{1/2}] = 0.39$$

where $\text{erf}[(\tfrac{3}{2})^{1/2}]$ was evaluated using Fig. 2-1.

COLLISION PARAMETERS

2.7. Calculate the total number of collisions per unit volume per unit time in a sample of air at 25 °C and 1.00 bar. Assume that the mole fractions of N_2 and O_2 in air are 0.79 and 0.21, respectively, and that $\sigma(N_2) = 0.373$ nm and $\sigma(O_2) = 0.353$ nm.

The partial pressures of the gases are given by (1.10) as

$$P(N_2) = x(N_2)P_t = (0.79)(1.00\,\text{bar}) = 0.79\,\text{bar}$$

$$P(O_2) = (0.21)(1.00) = 0.21\,\text{bar}$$

and (2.14) gives the molecular density of the gases as

$$N^*(N_2) = \frac{LP(N_2)}{RT}$$

$$= \frac{(6.022 \times 10^{23}\ mol^{-1})(0.79\ bar)[(10^5\ Pa)/(1\ bar)][(1\ kg\ m^{-1}\ s^{-2})/(1\ Pa)]}{(8.314\ J\ K^{-1}\ mol^{-1})[(1\ kg\ m^2\ s^{-2})/(1\ J)](298\ K)}$$

$$= 1.92 \times 10^{25}\ m^{-3}$$

$$N^*(O_2) = \frac{(6.022 \times 10^{23})(0.21)(10^5)}{(8.314)(298)} = 5.1 \times 10^{24}\ m^{-3}$$

The reduced mass of the system is given by (2.16) as

$$\mu = \frac{[(28.01 \times 10^{-3}\ kg\ mol^{-1})/L][(32.00 \times 10^{-3}\ kg\ mol^{-1})/L]}{[(28.01 \times 10^{-3}\ kg\ mol^{-1})/L] + [(32.00 \times 10^{-3}\ kg\ mol^{-1})/L]}$$

$$= (14.94 \times 10^{-3}\ kg\ mol^{-1})/(6.022 \times 10^{23}\ mol^{-1}) = 2.48 \times 10^{-26}\ kg$$

and the average collision diameter is given by (2.15) as

$$\sigma(N_2, O_2) = (0.373\ nm + 0.353\ nm)/2 = 0.363\ nm$$

The number of collisions per unit volume per unit time between unlike molecules is given by (2.13) as

$$Z(N_2, O_2) = (1.92 \times 10^{25}\ m^{-3})(5.1 \times 10^{24}\ m^{-3})(\pi)(3.63 \times 10^{-10}\ m)^2$$

$$\times \left(\frac{8(1.381 \times 10^{-23}\ J\ K^{-1})(298\ K)[(1\ kg\ m^2\ s^{-2})/(1\ J)]}{\pi(2.48 \times 10^{-26}\ kg)} \right)^{1/2}$$

$$= 2.6 \times 10^{34}\ m^{-3}\ s^{-1}$$

and the number of collisions per unit volume per unit time between like molecules is given by (2.21) as

$$Z(N_2, N_2) = \frac{(1.92 \times 10^{25})^2(\pi)(3.73 \times 10^{-10})^2}{2^{1/2}} \left(\frac{8(1.381 \times 10^{-23})(298)}{\pi(28.01 \times 10^{-3})/(6.022 \times 10^{23})} \right)^{1/2}$$

$$= 5.4 \times 10^{34}\ m^{-3}\ s^{-1}$$

$$Z(O_2, O_2) = \frac{(5.1 \times 10^{24})^2(\pi)(3.53 \times 10^{-10})^2}{2^{1/2}} \left(\frac{8(1.381 \times 10^{-23})(298)}{\pi(32.00 \times 10^{-3})/(6.022 \times 10^{23})} \right)^{1/2}$$

$$= 3.2 \times 10^{33}\ m^{-3}\ s^{-1}$$

The total number of collisions in the gaseous mixture is $Z(N_2, O_2) + Z(N_2, N_2) + Z(O_2, O_2) = 8.3 \times 10^{34}\ m^{-3}\ s^{-1}$.

2.8.   Assuming an equimolar mixture of N_2 and He, how many times faster will He atoms initially leak through a pinhole in the container? Repeat the calculation for a mixture such that $x(He) = 0.25$.

For the equimolar mixture, $P(N_2) = P(He)$, and (2.20) gives

$$\frac{Rate(He)}{Rate(N_2)} = \frac{P(He)/m(He)^{1/2}}{P(N_2)/m(N_2)^{1/2}} = \frac{M(N_2)^{1/2}}{M(He)^{1/2}}$$

$$= \frac{(28.01\ g\ mol^{-1})^{1/2}}{(4.002\ 6\ g\ mol^{-1})^{1/2}} = 2.646$$

Recognizing that $x(He) = 0.25$ implies that $P(N_2) = 3P(He)$, the ratio of the rates becomes

$$\frac{Rate(He)}{Rate(N_2)} = \frac{P(He)/m(He)^{1/2}}{3P(He)/m(N_2)^{1/2}} = \frac{(28.01)^{1/2}}{3(4.002\ 6)^{1/2}} = 0.882$$

2.9. Find the rate at which O_2 molecules collide with a unit area of the surface of their container per unit time at 1.00 bar and 25 °C.

The rate is given by (2.19) as

$$\frac{Z_i}{A} = N_i^* \left(\frac{kT}{2\pi m}\right)^{1/2}$$

$$= \left(\frac{(6.022 \times 10^{23})(1.00)(10^5)}{(8.314)(298)}\right)\left(\frac{(1.381 \times 10^{-23})(298)}{2\pi(28.01 \times 10^{-3})/(6.022 \times 10^{23})}\right)^{1/2}$$

$$= 2.88 \times 10^{27} \text{ m}^{-2} \text{ s}^{-1}$$

2.10. Calculate the mean free path of an oxygen molecule in the air sample described in Problem 2.7.

For this mixture, (2.22) gives

$$\lambda(O_2) = \frac{\bar{v}(O_2)}{z(O_2, O_2) + z(O_2, N_2)}$$

The values of the collision numbers and average velocity can be found by substituting values from Problem 2.7 into (2.4), (2.17), and (2.18), giving

$$z(O_2, O_2) = (5.1 \times 10^{24})(\pi)(3.53 \times 10^{-10})^2(2^{1/2})\left(\frac{8(1.381 \times 10^{-23})(298)}{\pi(32.00 \times 10^{-3})/(6.022 \times 10^{23})}\right)^{1/2} = 1.3 \times 10^9 \text{ s}^{-1}$$

$$z(O_2, N_2) = (1.92 \times 10^{25})(\pi)(3.63 \times 10^{-10})^2\left(\frac{8(1.381 \times 10^{-23})(298)}{\pi(2.48 \times 10^{-26})}\right)^{1/2} = 5.17 \times 10^9 \text{ s}^{-1}$$

$$\bar{v}(O_2) = \left(\frac{8(1.381 \times 10^{-23})(298)}{\pi(32.00 \times 10^{-3})/(6.022 \times 10^{23})}\right)^{1/2} = 444 \text{ m s}^{-1}$$

The mean free path for an oxygen molecule is

$$\lambda(O_2) = \frac{444 \text{ m s}^{-1}}{1.3 \times 10^9 \text{ s}^{-1} + 5.17 \times 10^9 \text{ s}^{-1}} = 6.9 \times 10^{-8} \text{ m}$$

TRANSPORT PROPERTIES

2.11. The space between two surfaces is filled with argon gas at 25 °C and 1.00 bar. The area of each of the surfaces is 25 cm², and the distance between them is 3 mm. One surface is at 35 °C, and the other is at 15 °C. What is the heat flow between the two surfaces?

The thermal gradient dT/dz is defined in terms of the hotter surface being at $z = 0$ and the cooler surface being at $z = z$. Thus,

$$\frac{\partial T}{\partial z} \approx \frac{\Delta T}{\Delta z} = \frac{288 \text{ K} - 308 \text{ K}}{(3 \text{ mm})[(10^{-3} \text{ m})/(1 \text{ mm})]} = -7 \times 10^3 \text{ K m}^{-1}$$

Rearranging (2.26) and substituting $k_T = 1.75 \times 10^{-2}$ J K^{-1} m^{-1} s^{-1} (see Example 2.8) gives

$$\frac{\partial q}{\partial t} = -k_T A \frac{\partial T}{\partial z}$$

$$= -(1.75 \times 10^{-2} \text{ J K}^{-1} \text{ m}^{-1} \text{ s}^{-1})(25 \text{ cm}^2)\left(\frac{1 \text{ m}^2}{10^4 \text{ cm}^2}\right)(-7 \times 10^3 \text{ K m}^{-1})$$

$$= 0.3 \text{ J s}^{-1}$$

2.12. The average collision diameter in air is 0.363 nm. Estimate the value of the viscosity coefficient for air at 25 °C and 1.00 bar.

If we use the average velocity as 470 m s^{-1} (an average between 475 m s^{-1} for N_2 and 444 m s^{-1} for O_2) and $M = 29 \times 10^{-3} \text{ kg mol}^{-1}$ (see Example 1.10), (2.29) gives

$$\eta = \frac{5(470 \text{ m s}^{-1})(29 \times 10^{-3} \text{ kg mol}^{-1})/(6.022 \times 10^{23} \text{ mol}^{-1})}{32(2^{1/2})(3.63 \times 10^{-10} \text{ m})^2}$$

$$= 1.9 \times 10^{-5} \text{ Pa s}$$

The actual value is 1.7×10^{-5} Pa s.

2.13. An Ostwald viscometer was calibrated using water at $25 \,°\text{C}$ ($\eta = 8.9 \times 10^{-4}$ Pa s and $\rho = 1.00 \times 10^3 \text{ kg m}^{-3}$). The same viscometer was used at $-193 \,°\text{C}$ (volume changes, etc., neglected) to determine the viscosity of liquid air ($\rho = 0.92 \times 10^3 \text{ kg m}^{-3}$). Assuming $t_1/t_0 = 0.193$, find η_1.

Using (2.31) gives

$$\eta_1 = \eta_0 \frac{\rho_1}{\rho_0} \frac{t_1}{t_0} = (8.9 \times 10^{-4} \text{ Pa s}) \left(\frac{0.92 \times 10^3}{1.00 \times 10^3} \right) (0.193) = 1.6 \times 10^{-4} \text{ Pa s}$$

2.14. The times that a steel BB ($\rho_b = 7.80 \times 10^3 \text{ kg m}^{-3}$) required to drop through water and a commercial shampoo were 1 s and 7 s, respectively. Given that the densities of water and the shampoo are 1.00×10^3 and $1.03 \times 10^3 \text{ kg m}^{-3}$, respectively, find η_1/η_0.

Using (2.33) gives

$$\frac{\eta_1}{\eta_0} = \frac{(7.80 \times 10^3 - 1.03 \times 10^3)(7)}{(7.80 \times 10^3 - 1.00 \times 10^3)(1)} = 7$$

2.15. The viscosity of molten sodium is 4.50×10^{-4} Pa s at $200 \,°\text{C}$ and 2.12×10^{-4} Pa s at $600 \,°\text{C}$. Find ΔE(viscosity), and predict η at $400 \,°\text{C}$.

For two temperatures, (2.34) gives

$$\ln \left(\frac{\eta_2}{\eta_1} \right) = \frac{\Delta E \text{(viscosity)}}{R} \left(\frac{1}{T_2} - \frac{1}{T_1} \right) \qquad (2.43)$$

Solving for ΔE(viscosity) and substituting values gives

$$\Delta E \text{(viscosity)} = \frac{(8.314 \text{ J K}^{-1} \text{ mol}^{-1})[\ln(2.12 \times 10^{-4}) - \ln(4.50 \times 10^{-4})]}{1/873 \text{ K} - 1/473 \text{ K}}$$

$$= 6\,460 \text{ J mol}^{-1}$$

Using (2.43) for η_2 gives

$$\ln \eta_2 = \ln(4.50 \times 10^{-4}) + \frac{6\,460}{8.314} \left(\frac{1}{673} - \frac{1}{473} \right) = -8.19$$

which upon taking antilogarithms gives $\eta_2 = 2.8 \times 10^{-4}$ Pa s. The actual value is 2.78×10^{-4} Pa s.

2.16. Consider the flow of water through a horizontal pipe with $R = 1$ in. and $\bar{v} = 3 \text{ cm s}^{-1}$. If $\eta = 1.202$ cP at $13 \,°\text{C}$ and $\rho = 0.999\,377 \times 10^3 \text{ kg m}^{-3}$, find Re. What is v at $r = 0$, at $r = 0.5 R$, and at $r = R$? If the water is being pumped through a length of pipe equivalent to 300 ft, what pressure must the pump be able to produce for the desired flow rate?

Using (2.35) gives

$$\text{Re} = \frac{2(1 \text{ in.})[(2.54 \times 10^{-2} \text{ m})/(1 \text{ in.})](3 \times 10^{-2} \text{ m s}^{-1})(0.999\,377 \times 10^3 \text{ kg m}^{-3})}{(1.202 \times 10^{-2} \text{ P})[(0.1 \text{ Pa s})/(1 \text{ P})][(1 \text{ kg m}^{-1} \text{ s}^{-2})/(1 \text{ Pa})]}$$

$$= 1\,300$$

which means that the flow is laminar. The volume of liquid flowing in time t is given by

$$\dot{V} = V/t = \pi R^2 \bar{v} = \pi (2.54 \times 10^{-2} \text{ m})^2 (3 \times 10^{-2} \text{ m s}^{-1})$$

$$= 6.08 \times 10^{-5} \text{ m}^3 \text{ s}^{-1}$$

which combined with (2.37) gives the pumping pressure as

$$\Delta P = \frac{\dot{V} 8 \eta l}{\pi R^4}$$

$$= \frac{(6.08 \times 10^{-5} \text{ m}^3 \text{ s}^{-1})(8)(1.202 \times 10^{-2} \text{ P})[(0.1 \text{ Pa s})/(1 \text{ P})](300 \text{ ft})[(12 \text{ in.})/(1 \text{ ft})][(2.54 \times 10^{-2} \text{ m})/(1 \text{ in.})]}{\pi (2.54 \times 10^{-2} \text{ m})^4}$$

$$= 40.9 \text{ Pa}$$

Combining (2.36), (2.37), and the expression for V/t gives

$$v = 2\bar{v} \left[1 - \left(\frac{r}{R} \right)^2 \right] = (6 \text{ cm s}^{-1}) \left[1 - \left(\frac{r}{R} \right)^2 \right]$$

Hence the velocities at $r = 0$, $0.5\,R$, and R are 6, 4.5, and 0 cm s^{-1}, respectively.

2.17. The equation for the so-called evacuation method for measuring η for gases is based on combining the ideal gas law with Poiseuille's law, giving

$$\frac{1}{P} = \frac{k}{\eta} t + \frac{1}{P_0} \qquad (2.44)$$

where k is a constant for a given apparatus, P is the pressure of the gas at time t, and P_0 is the pressure of the gas at the beginning of the experiment. The slope of a plot of $1/P$ against t for dry air at 25 °C and 1.00 bar ($\eta = 1.837 \times 10^{-5}$ Pa s) was 2.50 s bar^{-1}, and for argon under the same conditions the slope was 2.02 s bar^{-1}. Calculate η for argon.

Equation (2.44) predicts that the slope of a plot of $1/P$ against t for a gas will be equal to k/η for the gas. Taking a ratio of the slopes gives

$$\frac{(k/\eta)_{\text{air}}}{(k/\eta)_{\text{Ar}}} = \frac{2.50 \text{ s bar}^{-1}}{2.02 \text{ s bar}^{-1}} = 1.24$$

Cancelling the apparatus constant from the equation gives

$$\eta_{\text{Ar}} = 1.24 \eta_{\text{air}} = (1.24)(1.837 \times 10^{-5} \text{ Pa s}) = 2.28 \times 10^{-5} \text{ Pa s}$$

2.18. For very dilute solutions of a solute i in liquid solvent j, the *Stokes–Einstein equation* gives the diffusion coefficient as

$$D_{ij}^{\infty} = \frac{2kT}{6\pi\eta_j\sigma_i} \qquad (2.45)$$

for spherical solute molecules with $r_i > r_j$. [For spherical solute molecules with $r_i \approx r_j$, the 6 in (2.45) should be replaced with a 4.]

The diffusion coefficient of water into ethanol at 25 °C is 2.4×10^{-5} cm^2 s^{-1}, and $\eta = 1.1$ cP for ethanol. Calculate $\sigma(H_2O)$ from these data.

Assuming the alcohol and water molecules to be similar in size, (2.45) gives

$$\sigma(H_2O) = \frac{2(1.381 \times 10^{-23} \text{ J K}^{-1})(298 \text{ K})[(1 \text{ kg m}^2 \text{ s}^{-2})/(1 \text{ J})]}{4\pi(0.011 \text{ P})[(0.1 \text{ Pa s})/(1 \text{ P})](2.4 \times 10^{-5} \text{ cm}^2 \text{ s}^{-1})[(10^{-4} \text{ m}^2)/(1 \text{ cm}^2)][(1 \text{ kg m}^{-1} \text{ s}^{-2})/(1 \text{ Pa})]}$$

$$= 2.5 \times 10^{-10} \text{ m} = 0.25 \text{ nm}$$

This value compares favorably with the 0.368-nm diameter calculated from gaseous viscosity measurements.

Supplementary Problems

VELOCITY AND ENERGY DISTRIBUTIONS OF GASES

2.19. Prepare plots of $(dN/L)/dv$ for He, Ne, Ar, and Kr at 298 K for values of v up to 3 000 m s^{-1}. Describe the distributions.

Ans. See Fig. 2-5; as M increases, the distribution functions become "sharper," indicating that fewer molecules have higher velocities.

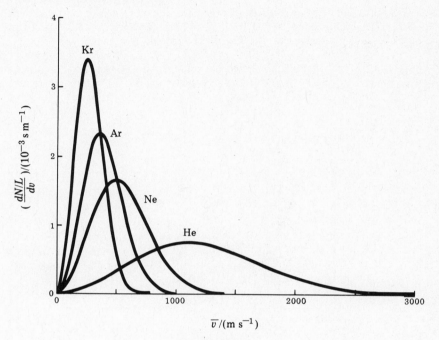

Fig. 2-5

2.20. What is the limiting slope of (*2.2*) as v approaches zero?   *Ans.* 0

2.21. Prepare a plot of $(dN/L)/dv_x$ against v_x for values from -10^4 to 10^4 m s^{-1} for v_x at 100 °C and 1 000 °C. Describe the distributions.

Ans. Both curves are "bell-shaped" curves centered at $v_x = 0$. The plot at 100 °C falls more rapidly than the plot at 1 000 °C, indicating that there are fewer molecules with high energies at lower temperatures.

2.22. Repeat the calculations in Problem 2.2 for the velocity range of (*a*) 499.9–500.1 m s^{-1}, (*b*) 499.99–500.01 m s^{-1}, and (*c*) 499.999–500.001 m s^{-1}.

Ans. (*a*) 2.11×10^{20} molecules, (*b*) 2.11×10^{19} molecules   (*c*) 2.11×10^{18} molecules

2.23. Beginning with $\bar{v} = \int v \, dN/L$, derive (*2.4*).

2.24. Find the ratio $v_{rms} : \bar{v} : v_{mp}$.

Ans. $3^{1/2} : (8/\pi)^{1/2} : 2^{1/2} = 1.000 : 0.921\,3 : 0.816\,5 = 1.225 : 1.128 : 1.000$

2.25. For $z \gg 1$, $\text{erf}(z)$ approaches $1 - 2e^{-z^2}/z\pi^{1/2}$. Using (*2.8*), derive an equation expressing $N_{v \geq c}/L$ for $c \gg v_{mp}$. Using this result, find the fraction of neon molecules having a velocity equal to or greater than 2 000 m s^{-1} at 298 K.

Ans. $N_{v \geq c}/L = 2e^{-(c/v_{mp})^2}[(c/v_{mp}) + (v_{mp}/c)]/\pi^{1/2}$; 3.12×10^{-7}

2.26. Letting $\varepsilon_{trans} = \frac{1}{2}mv^2$, derive (2.9) from (2.2). Show that $\int dN = L$ for this equation.

2.27. What is the limiting slope of (2.9) as ε_{trans} approaches zero?   *Ans.* ∞

2.28. Find the expression for the most probable energy ε_{mp}. What fraction of molecules has a translational kinetic energy greater than ε_{mp}?   *Ans.* $\varepsilon_{mp} = kT/2$;   0.79

2.29. Compare the shape of the curve in Fig. 2-4 to that in Fig. 2-3.

> *Ans.*   The energy distribution (Fig. 2-4) rises faster and falls more gently than the velocity distribution (Fig. 2-3). The limiting slopes at the origin are vertical and horizontal, respectively. Note that the energy curve does not depend on the mass of the molecule.

2.30. Prepare plots of $(dN/L)/d\varepsilon_{trans}$ for N_2 at 100 °C and 1 000 °C for values of ε_{trans} up to 5×10^{-20} J. Indicate on the plots the values of $\bar{\varepsilon}_{trans}$.   *Ans.*   See Fig. 2-6.

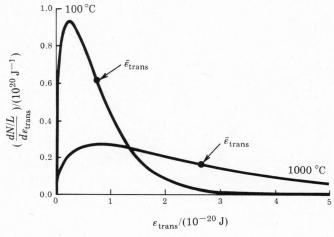

Fig. 2-6

2.31. Compare the number of molecules with $\varepsilon_{trans} = 5.0kT$ to that with $\varepsilon_{trans} = 5.5kT$. Is there a corresponding 10% increase for the 10% decrease in energy?   *Ans.*   1.65;   no, the increase is larger.

COLLISION PARAMETERS

2.32. Show that (2.17) becomes (2.18) for collisions between like molecules.

2.33. Repeat the calculations in Example 2.6 for argon at 25 °C and 0.100 bar. Compare the results.

> *Ans.* $Z_{ii} = 6.79 \times 10^{32}$ m^{-3} s^{-1}, a factor of 100 less; $z_{ii} = 5.59 \times 10^8$ s^{-1}, a factor of 10 less

2.34. Calculate the total number of collisions per unit time that one N_2 molecule makes in the air sample described in Problem 2.7. Repeat the calculations for one O_2 molecule.

> *Ans.* $z(N_2, N_2) + z(N_2, O_2) = 7.0 \times 10^9$ s^{-1}; $z(O_2, O_2) + z(O_2, N_2) = 6.5 \times 10^9$ s^{-1}

2.35. Show that under the same pressure and temperature conditions, (2.20) can be written in terms of the density of the gases as

$$\text{Rate}_i / \text{Rate}_j = (\rho_j / \rho_i)^{1/2} \qquad (2.46)$$

which is the empirical statement of Graham's law of effusion.

2.36. The occurrence of heavy water ($M = 20.0 \text{ g mol}^{-1}$) is about 1 part for 6 900 parts of regular water ($M = 18.0 \text{ g mol}^{-1}$). Compare the initial rates of effusion during a concentration step for obtaining D_2O.   *Ans.*   Rate(H_2O)/Rate(D_2O) = 7 300

2.37. The *Knudsen cell* is used to measure the vapor pressure of materials with very low vapor pressures by measuring the loss in mass of a sample of the material by effusion over a period of time. Langmuir observed a mass loss of $9.95 \times 10^{-7} \text{ g cm}^{-2} \text{ s}^{-1}$ at 3 000 K for tungsten. Derive a working equation for the Knudsen cell, and calculate the vapor pressure of W at this temperature.

Ans. $P = [(\Delta m/\Delta t)/A](2\pi RT/M)^{1/2}$; 9.19×10^{-3} Pa

2.38. Using the data of Problem 2.7, compare the mean free path of a nitrogen molecule in air to that in pure nitrogen under the same conditions.

Ans. $\lambda_{\text{air}} = 6.8 \times 10^{-8}$ m, $\lambda_{\text{pure}} = 6.66 \times 10^{-8}$ m, $\lambda_{\text{air}}/\lambda_{\text{pure}} = 1.02$

2.39. Many vacuum systems are capable of evacuating to 1.0×10^{-5} torr. Compare λ under these conditions to that at 1.00 bar.   *Ans.* 7.5×10^7 times greater

TRANSPORT PROPERTIES

2.40. Equation (*2.27*) is based on the gaseous molecules being rigid, hard spheres. Using $\sigma = 0.368$ nm and $C_v = 25.263 \text{ J K}^{-1} \text{ mol}^{-1}$ for water vapor at 25 °C, calculate k_T using (*2.27*), and compare the result to the experimental value of $1.81 \times 10^{-2} \text{ J K}^{-1} \text{ m}^{-1} \text{ s}^{-1}$. A better approximation for k_T for hard polyatomic molecules is given by

$$k_T = \frac{5}{16} \left(\frac{kT}{\pi m} \right)^{1/2} \frac{C_v + \frac{9}{4}R}{L\sigma^2} \tag{2.47}$$

Use this relationship to calculate k_T for water, and compare this result to the experimental value.

Ans. $5.07 \times 10^{-2} \text{ J K}^{-1} \text{ m}^{-1} \text{ s}^{-1}$, 180% high; $3.53 \times 10^{-2} \text{ J K}^{-1} \text{ m}^{-1} \text{ s}^{-1}$, 95% high

2.41. The molar heat capacities of gaseous argon and krypton at 1.00 bar and 25 °C are identical, and the coefficients of thermal expansion are $1.74 \times 10^{-2} \text{ J K}^{-1} \text{ m}^{-1} \text{ s}^{-1}$ and $0.87 \times 10^{-2} \text{ J K}^{-1} \text{ m}^{-1} \text{ s}^{-1}$, respectively. Calculate the ratio of the molecular collision diameters for these gases and, using $\sigma = 0.361$ nm for Ar, determine σ for Kr.

Ans. $\sigma_{\text{Kr}}/\sigma_{\text{Ar}} = 1.18$; $\sigma_{\text{Kr}} = 0.426$ nm

2.42. The coefficient of viscosity for water vapor at 150 °C and 1.01 bar is 1.445×10^{-5} Pa s. Calculate σ for a water molecule under these conditions.   *Ans.*   0.402 nm

2.43. Use (*2.27*) and (*2.29*) to predict k_T/η for a gas that consists of molecules that are rigid, hard spheres. How valid is this predicted ratio for Ar ($k_T = 1.74 \times 10^{-2} \text{ J m}^{-1} \text{ s}^{-1} \text{ K}^{-1}$, $\eta = 2.22 \times 10^{-5}$ Pa s, and $C_v = 12.472 \text{ J K}^{-1} \text{ mol}^{-1}$) and for H_2O ($k_T = 1.81 \times 10^{-2} \text{ J m}^{-1} \text{ s}^{-1} \text{ K}^{-1}$, $\eta = 1.445 \times 10^{-5}$ Pa s, and $C_v = 25.263 \text{ J K}^{-1} \text{ mol}^{-1}$) at 25 °C and 1.00 bar?

Ans. $k_T/\eta = 2.5 C_v/M$;   0.5% difference for Ar,   180% difference for H_2O

2.44. If light machinery oil has $\eta = 5 \times 10^{-2}$ Pa s and $\rho = 0.97 \times 10^3 \text{ kg m}^{-3}$ at 25 °C, how long will it take for a sample to pass through a viscometer if water under the same conditions takes 1 min?

Ans.   57.9 min, assuming $\rho = 1.00 \times 10^3 \text{ kg m}^{-3}$ and $\eta = 8.9 \times 10^{-4}$ Pa s

2.45. The viscosity of a 20 wt % aqueous solution of ethanol is 2.183×10^{-3} Pa s, and the density is $0.971\,39 \times 10^3 \text{ kg m}^{-3}$. If an olive is dropped into a tall glass of this solution, what will be the velocity once equilibrium between the gravitational and viscous forces has been established? Assume $r_b = 0.9$ cm, $\rho_b = 1.10 \times 10^3 \text{ kg m}^{-3}$, and $g = 9.8 \text{ m s}^{-2}$.   *Ans.* 10.4 m s^{-1}

2.46. If $\eta = 1.307 \times 10^{-3}$ Pa s at 10.0 °C and 5.468×10^{-4} Pa s at 50.0 °C for water, find ΔE(viscosity). At 25 °C, ΔE(vaporization) $= 41.5$ kJ mol^{-1} for water. Is the empirical relationship

$$\Delta E \text{ (viscosity)} = 0.3\Delta_{vap}U \qquad (2.48)$$

valid for water?   *Ans.* ΔE(viscosity) $= 16.56$ kJ mol^{-1};   the ratio is 0.4, reasonably valid

2.47. The viscosity of a 60 wt % aqueous solution of sucrose is 0.238 Pa s at 0 °C and 0.043 86 Pa s at 25 °C. If the density of the solution changes insignificantly over this temperature range, test the expression "as slow as molasses in January."   *Ans.* $t_0/t_{25} = 5.43$

2.48. A commercial cone-plate rotating viscometer is designed to measure η for low-viscosity liquids such as organic solvents, pharmaceuticals, and water-based systems. The maximum shear stress is designed to be 187 dyn cm^{-2}, and the shear rate is designed to be 60 s^{-1} when the cone is rotating at 300 rpm. If $\eta =$ (shear stress)/(shear rate), calculate the maximum viscosity that the viscometer can measure.

Ans.   312 P $= 0.312$ Pa s.

2.49. Repeat the calculations of Problem 2.16 at 25 °C where

$$\eta = 0.890 \ 4 \ \text{cP} \qquad \text{and} \qquad \rho = 0.997 \ 044 \times 10^3 \ \text{kg m}^{-3}$$

If the same pump is used, compare the values of v.

Ans.   Re $= 1 \ 700$, $\Delta P = 30.3$ Pa, $\bar{v}$ and v same as before;   for $\Delta P = 40.9$ Pa, $\bar{v} = 3(1.202/0.890 \ 4)$ cm s^{-1} and velocities are increased by 34.9%.

2.50. An aqueous solution of a polymer ($\eta = 13.2$ cP) is pumped through a 0.10-mm capillary tube using a pump that produces $\Delta P = 1.2 \times 10^6$ Pa. The length of the tube is 95 cm. What is the rate of volume flow of the liquid?

Ans. $\dot{V} = 3.8 \times 10^{-9}$ m^3 s$^{-1} = 3.8 \times 10^{-3}$ cm^3 s^{-1}

2.51. Starting with $\dot{V} = \int_0^R 2\pi rv \ dr$ and (2.36), derive (2.37).

2.52. Recognizing that $\partial P/\partial z = \Delta P/l$ in (2.37), evaluate L_W in (2.26) for fluid flow.   *Ans.* $c = \pi R^4/8\eta A$

2.53. The diffusion coefficient of nickel into copper is 10^{-9} cm^2 s^{-1} at 1 025 °C. How long will it take for the root mean square distance to reach 1 cm?   *Ans.* 5×10^8 s

2.54. The diffusion coefficient of sucrose in water at 25 °C is 5.2×10^{-6} cm^2 s^{-1}. What is the value of the root mean square distance after 1 day?   *Ans.*   0.95 cm

2.55. Find σ for O_2 and compare it to 0.353 nm as calculated from viscosity data, given that $D = 1.78 \times 10^{-5}$ m^2 s^{-1} at 0 °C.

Ans. $\bar{v} = 425$ m s^{-1}, $\lambda = 1.92 \times 10^{-7}$ m, $N^* = 2.68 \times 10^{25}$ m^{-3}, $\sigma = 0.209$ nm;   41% lower—note that water is not a rigid, hard sphere.

2.56. Show that (2.41) predicts *Graham's law of diffusion*,

$$\text{Rate}_i/\text{Rate}_j = (\rho_j/\rho_i)^{1/2} \qquad (2.49)$$

Chapter 3

First Law of Thermodynamics

Internal Energy and Enthalpy

3.1 INTERNAL ENERGY, U

The *internal energy* of a system is the sum of the various kinetic and potential energy contributions. These include translational, rotational, vibrational, electronic, nuclear, positional, and mass contributions. Because it is difficult to determine an absolute value of U, most calculations and experimental measurements are concerned with the change in U, ΔU, where

$$\Delta U = U_{\text{final}} - U_{\text{initial}} \tag{3.1}$$

Equation (*3.1*) expresses the basic property of any *state* (or *point*) *function*: its increment is dependent only on the final and initial states of the system (and not on the path followed between these states).

The SI derived unit for energy is the *joule* $(1\text{ J} = 1\text{ kg m}^2\text{ s}^{-2})$. Other units used to express energy are given in Table 3-1.

Table 3-1

Energy Unit	Symbol	Conversion Factor
Erg	erg	10^{-7} J*
Calorie	cal	4.184 J*
Electronvolt	eV	$1.602\,189\,2 \times 10^{-19}$ J
Liter-atmosphere	L atm	101.325 J*
Wavenumber	cm^{-1}	$1.986\,477 \times 10^{-23}$ J
Atomic mass unit	u (or amu)	$1.492\,442 \times 10^{-10}$ J

* Exact conversion factor.

3.2 ENTHALPY, H

The *enthalpy* of a system is defined as

$$H = U + PV \tag{3.2}$$

and the change in enthalpy is given by

$$\Delta H = \Delta U + \Delta(PV) \tag{3.3}$$

EXAMPLE 3.1. For processes involving gases, the $\Delta(PV)$ term in (*3.3*) can be replaced by $\Delta(nRT)$ for ideal gases and by $\Delta(ZnRT)$ for real gases. Given that $\Delta U = 7\,120$ J for heating 1.00 mol of ozone at 20 bar from 300. K to 500. K. Calculate ΔH for this process assuming ideal gas behavior and assuming real gas behavior ($Z = 0.95$ at 300. K and 0.99 at 500. K).

41

Assuming ideal gas behavior, (3.3) gives for the heating process

$$\Delta H = 7\,120\,\text{J} + nR\,\Delta T$$

$$= 7\,120\,\text{J} + (1.00\,\text{mol})(8.314\,\text{J K}^{-1}\,\text{mol}^{-1})(200.\,\text{K})$$

$$= 8\,780\,\text{J}$$

and assuming real gas behavior,

$$\Delta H = 7\,120\,\text{J} + nR\,\Delta(ZT)$$

$$= 7\,120 + (1.00)(8.314)[(0.99)(500.) - (0.95)(300.)]$$

$$= 8\,870\,\text{J}$$

[Note that less than a 1% error is introduced by assuming the gas to be ideal. For this reason, ideal gas behavior is usually assumed for all gases for most thermodynamic calculations—unless very accurate results are desired.]

EXAMPLE 3.2.   The ΔH for the formation of $NOCl(g)$ from the gaseous elements is 51.71 kJ mol^{-1} at 25 °C. If the gases are ideal, calculate ΔU.

From the reaction

$$\tfrac{1}{2}N_2(g) + \tfrac{1}{2}O_2(g) + \tfrac{1}{2}Cl_2(g) \longrightarrow NOCl(g) \qquad \Delta H = 51.71\,\text{kJ}$$

it can be seen that $\Delta(PV)$ in (3.3) can be replaced by $RT\,\Delta n_g$, where

$$\Delta n_g = \sum \nu_{i,g}$$

where $\nu_{i,g}$ represents the stoichiometric coefficients for the gaseous reactants and products in the chemical equation (positive for products and negative for reactants). Thus

$$\Delta n_g = 1 - \tfrac{1}{2} - \tfrac{1}{2} - \tfrac{1}{2} = -\tfrac{1}{2}\,\text{mol}$$

and (3.3) gives

$$\Delta U = \Delta H - \Delta(PV)$$

$$= 51.71\,\text{kJ} - (8.314 \times 10^{-3}\,\text{kJ K}^{-1}\,\text{mol}^{-1})(298\,\text{K})(-\tfrac{1}{2}\,\text{mol})$$

$$= 52.95\,\text{kJ}$$

for the reaction, or $\Delta U = 52.95\,\text{kJ mol}^{-1}$.

3.3   THERMAL ENERGY OF AN IDEAL GAS

For theoretical calculations involving ideal gases, ΔU can be equated to ΔE(thermal). The *thermal energy* represents the difference between the internal energy of the gas at some temperature T and the "rest" internal energy of the gas at 0 K:

$$E(\text{thermal}) = E_T - E_0 \tag{3.4}$$

The three major contributions to E(thermal) below 1 000 K result from the translational, rotational, and vibrational motions of the single molecules. For 1 mol of gas, each mode of molecular translational motion contributes

$$E(\text{thermal, trans}) = \tfrac{1}{2}RT \tag{3.5a}$$

each mode of rotational motion contributes

$$E(\text{thermal, rot}) = \tfrac{1}{2}RT \tag{3.5b}$$

and each mode of vibrational motion contributes

$$E(\text{thermal, vib}) = \frac{RTx}{e^x - 1} \tag{3.5c}$$

where
$$x \equiv \frac{h\nu}{kT} \tag{3.6}$$

The symbols h, ν, and k in (3.6) are Planck's constant, frequency of vibration, and Boltzmann's constant, respectively. If the vibrational data are expressed in units of cm^{-1} via the *wavenumber* ($\bar{\nu}$)

$$x = \frac{1.438\,8\,\bar{\nu}}{T} \tag{3.7a}$$

and if the vibrational data are expressed in units of K via the *vibrational characteristic temperature* (Θ_v),

$$x = \frac{\Theta_v}{T} \tag{3.7b}$$

At higher temperatures, an additional contribution to E(thermal) results from the electronic energy of the single molecule. For 1 mol of gas,

$$E(\text{thermal, elec}) = RT\frac{Q'}{Q} \tag{3.5d}$$

where
$$Q = \sum_j g_j\, e^{-\varepsilon_j/kT} \tag{3.8a}$$

$$Q' = \sum_j g_j \left(\frac{\varepsilon_j}{kT}\right) e^{-\varepsilon_j/kT} \tag{3.8b}$$

The degeneracy of each electronic energy level of energy ε_j is given by g_j. If nuclear contributions to the internal energy are included, these contributions are given by equations similar to (3.5d) and (3.8).

EXAMPLE 3.3. Derive general expressions for E(thermal) for ideal monatomic, diatomic, linear polyatomic, and nonlinear polyatomic gases.

If a molecule of the ideal gas contains Λ atoms, then the total number of degrees of freedom is given by 3Λ. Three of these are assigned to the translational motion of the molecule, leaving $3\Lambda - 3$ degrees of freedom for rotational and vibrational motion. For an ideal monatomic gas, $3(1) - 3 = 0$; thus the entire contribution to E(thermal) is from translational motion, as given by (3.5a):

$$E(\text{thermal}) = 3(\tfrac{1}{2}RT) = \tfrac{3}{2}RT \tag{3.9a}$$

For a linear molecule there are two degrees of rotational motion, leaving $3\Lambda - 5$ degrees of freedom for vibrational motion. For an ideal diatomic gas, $3(2) - 5 = 1$; thus E(thermal) is given by (3.5) as

$$E(\text{thermal}) = 3\left(\frac{1}{2}RT\right) + 2\left(\frac{1}{2}RT\right) + \frac{RTx}{e^x - 1}$$

$$= \frac{5}{2}RT + \frac{RTx}{e^x - 1} \tag{3.9b}$$

and for an ideal linear polyatomic gas,

$$E(\text{thermal}) = \frac{5}{2}RT + \sum_{i=1}^{3\Lambda-5} \frac{RTx_i}{e^{x_i} - 1} \tag{3.9c}$$

For a nonlinear molecule there are three degrees of rotational motion, leaving $3\Lambda - 6$ degrees of freedom for vibrational motion; thus,

$$E(\text{thermal}) = 3\left(\frac{1}{2}RT\right) + 3\left(\frac{1}{2}RT\right) + \sum_{i=1}^{3\Lambda-6} \frac{RTx_i}{e^{x_i} - 1}$$

$$= 3RT + \sum_{i=1}^{3\Lambda-6} \frac{RTx_i}{e^{x_i} - 1} \tag{3.9d}$$

EXAMPLE 3.4. Compare the values of ΔU for heating an ideal monatomic gas and a nonlinear triatomic gas from 25 °C to 50. °C. Assume that the triatomic gas has three vibrational frequencies near 2 000 cm^{-1}.

For the ideal monatomic gas, (3.9a) gives

$$E(\text{thermal})_{298} = \tfrac{3}{2}(8.314 \text{ J K}^{-1} \text{ mol}^{-1})(298 \text{ K}) = 3\ 716 \text{ J mol}^{-1}$$

$$E(\text{thermal})_{323} = \tfrac{3}{2}(8.314)(323) = 4\ 028 \text{ J mol}^{-1}$$

and (3.1) gives

$$\Delta U = \Delta E(\text{thermal}) = 4\ 028 - 3\ 716 = 312 \text{ J mol}^{-1}$$

For the ideal nonlinear triatomic gas, (3.7a) gives

$$x_{298} = \frac{(1.438\ 8)(2\ 000)}{298} = 9.658 \qquad x_{323} = \frac{(1.438\ 8)(2\ 000)}{323} = 8.910$$

which upon substitution into (3.9d) gives

$$E(\text{thermal})_{298} = 3(8.314)(298) + \frac{3(8.314)(298)(9.658)}{e^{9.658} - 1}$$

$$= 7\ 433 + \frac{71\ 830}{1.57 \times 10^4} = 7\ 438 \text{ J mol}^{-1}$$

$$E(\text{thermal})_{323} = 8\ 056 + 10 = 8\ 066 \text{ J mol}^{-1}$$

and (3.1) gives $\Delta U = 8\ 066 - 7\ 438 = 628 \text{ J mol}^{-1}$. The difference in the values is 316 J mol^{-1}, a factor of 2.

3.4 THERMAL ENTHALPY OF AN IDEAL GAS

The *thermal enthalpy* [H(thermal)] is defined as

$$H(\text{thermal}) = H_T - E_0 \tag{3.10}$$

where

$$H(\text{thermal}) = E(\text{thermal}) + RT \tag{3.11}$$

Heat Capacity

3.5 DEFINITIONS

The *heat capacity* (C) is defined as

$$C = \lim_{\Delta T \to 0} \frac{q}{\Delta T} \tag{3.12}$$

where q is the heat transferred corresponding to a temperature change ΔT. The *specific heat capacity* (c) is related to C by

$$c = C/m \tag{3.13}$$

For constant-volume processes, $q = \Delta U$, and (3.12) gives

$$C_V = \lim_{\Delta T \to 0} \left(\frac{\Delta U}{\Delta T}\right)_V = \left(\frac{\partial U}{\partial T}\right)_V \tag{3.14}$$

and for constant-pressure processes, $q = \Delta H$, which gives

$$C_P = \lim_{\Delta T \to 0} \left(\frac{\Delta H}{\Delta T}\right)_P = \left(\frac{\partial H}{\partial T}\right)_P \tag{3.15}$$

EXAMPLE 3.5. For the heating or cooling of an ideal gas, ΔU is a function of temperature only. Calculate ΔU for heating 2.50 mol of an ideal monatomic gas from 25 °C to 125 °C using $C_V = \frac{3}{2}R$.

Rearranging (3.14) and integrating gives

$$dU = C_V\, dT$$

$$\Delta U = \int_{T_1}^{T_2} C_V\, dT = C_V(T_2 - T_1)$$

$$= (2.50\ \text{mol})(\tfrac{3}{2})(8.314\ \text{J K}^{-1}\ \text{mol}^{-1})(398\ \text{K} - 298\ \text{K})$$

$$= 3\,120\ \text{J}$$

EXAMPLE 3.6. The pressure dependence of C_P is given by

$$\left(\frac{\partial C_P}{\partial P}\right)_T = -T\left(\frac{\partial^2 V}{\partial T^2}\right)_P \qquad (3.16)$$

Newer tabulations of C_P values are at $P = 1.00$ bar, and older tabulations of C_P were at $P = 1.00$ atm. What is the difference between these C_P values for an ideal gas?

For 1 mol of an ideal gas,

$$\left(\frac{\partial^2 V}{\partial T^2}\right)_P = \left[\frac{\partial^2}{\partial T^2}\left(\frac{RT}{P}\right)\right]_P = \left(\frac{\partial (R/P)}{\partial T}\right)_P = 0$$

giving no difference between the values of C_P.

3.6 RELATIONSHIP BETWEEN C_P AND C_V

The difference between the heat capacities can be shown to be

$$C_P - C_V = \left[P + \left(\frac{\partial U}{\partial V}\right)_T\right]\left(\frac{\partial V}{\partial T}\right)_P \qquad (3.17)$$

For an ideal gas, $V = RT/P$ and $(\partial U/\partial V)_T = 0$, so (3.17) becomes

$$C_P - C_V = R \qquad (3.18)$$

For real gases, solids, and liquids, (3.17) yields

$$C_P - C_V = \frac{\alpha^2 VT}{\kappa} \qquad (3.19)$$

where α (the *cubic expansion coefficient*) and κ (the *isothermal compressibility*) are given by

$$\alpha = \frac{1}{V}\left(\frac{\partial V}{\partial T}\right)_P = \left(\frac{\partial \ln V}{\partial T}\right)_P \qquad (3.20)$$

and

$$\kappa = \frac{-1}{V}\left(\frac{\partial V}{\partial P}\right)_T = -\left(\frac{\partial \ln V}{\partial P}\right)_T \qquad (3.21)$$

EXAMPLE 3.7. Assuming $\alpha = 2.21 \times 10^{-5}\ \text{K}^{-1}$, $\kappa = 1.32 \times 10^{-6}\ \text{bar}^{-1}$, and density of $2.702 \times 10^3\ \text{kg m}^{-3}$, find the difference between C_P and C_V for Al at 25 °C.

Substituting these values into (3.19) gives

$$C_P - C_V = \frac{(2.21 \times 10^{-5}\ \text{K}^{-1})^2[(26.98 \times 10^{-3}\ \text{kg mol}^{-1})/(2.702 \times 10^3\ \text{kg m}^{-3})](298\ \text{K})[(1\ \text{J})/(1\text{Pa m}^3)]}{(1.32 \times 10^{-6}\ \text{bar}^{-1})[(1\ \text{bar})/(10^5\ \text{Pa})]}$$

$$= 0.11\ \text{J K}^{-1}\ \text{mol}^{-1}$$

3.7 EMPIRICAL HEAT-CAPACITY EXPRESSIONS

Although theoretical values of C_V can be determined for ideal gases (see Sec. 3.8), the equations for C_V for the condensed states of matter are much more complex, and empirical relations of the types

$$C_P = a + bT + cT^2 + dT^3 \tag{3.22a}$$

$$C_P = a + bT + c'T^{-2} \tag{3.22b}$$

are often used for these phases and also to represent actual data for gases.

EXAMPLE 3.8. The values of a and b in (3.22a) for aluminum are $20.7\ \text{J K}^{-1}\ \text{mol}^{-1}$ and $0.012\ 4\ \text{J K}^{-2}\ \text{mol}^{-1}$, respectively. Calculate ΔH for heating aluminum from 25 °C to 100. °C.

Using (3.15) gives

$$\Delta H = \int_{T_1}^{T_2} (a + bT)\, dT = a(T_2 - T_1) + \tfrac{1}{2}b(T_2{}^2 - T_1{}^2)$$

$$= (20.7\ \text{J K}^{-1}\ \text{mol}^{-1})(373\ \text{K} - 298\ \text{K}) + \tfrac{1}{2}(0.012\ 4\ \text{J K}^{-2}\ \text{mol}^{-1})[(373\ \text{K})^2 - (298\ \text{K})^2]$$

$$= 1\ 860\ \text{J mol}^{-1}$$

3.8 C_V FOR IDEAL GASES

For 1 mol of an ideal gas, $C_V = (\partial E(\text{thermal})/\partial T)_V$. The contributions are

$$C_V(\text{trans}) = \tfrac{3}{2}R \tag{3.23a}$$

$$C_V(\text{rot}) = \begin{cases} R & \text{for a diatomic or linear polyatomic molecule} \\ \tfrac{3}{2}R & \text{for a nonlinear polyatomic molecule} \end{cases} \tag{3.23b}$$

$$C_V(\text{vib}) = \sum_{i=1}^{3\Lambda-5\,\text{or}\,3\Lambda-6} \frac{Rx_i{}^2 e^{x_i}}{(e^{x_i}-1)^2} \tag{3.23c}$$

$$C_V(\text{elec}) = R\left[\frac{Q''}{Q} - \left(\frac{Q'}{Q}\right)^2 \right] \tag{3.23d}$$

where x_i, Q, and Q' have been previously defined by (3.7) and (3.8) and

$$Q'' = \sum_j g_j \left(\frac{\varepsilon_j}{kT}\right)^2 e^{-\varepsilon_j/kT} \tag{3.8c}$$

If nuclear contributions are included, (3.23d) is used to determine these contributions.

EXAMPLE 3.9. The heat capacity ratio C_P/C_V for a gas was experimentally measured as 1.38. If the empirical formula is ABA, what conclusions can be made concerning the structure?

Assuming the gas to be ideal, (3.18) gives

$$C_P = C_V + R$$

which upon substitution into the desired ratio gives

$$\frac{C_V + R}{C_V} = 1.38 \quad \text{or} \quad C_V = 2.63R$$

If vibrational contributions are neglected, $C_V = 2.5R$ for a linear triatomic gas and $C_V = 3.0R$ for a nonlinear triatomic gas. Assuming the difference of $0.13R$ to be from vibrational contributions, the gas is linear.

3.9 C_V FOR CONDENSED STATES

Many metals at room temperature have an average value of $C_P = 25.9\ \text{J K}^{-1}\ \text{mol}^{-1}$. Using this in combination with (3.13) generates the *law of Dulong and Petit*, which can be used to determine approximate atomic masses for metals from values of specific heat capacity. This value, approximately equal to $3R$, is the high-temperature limit predicted by several theories.

At lower temperatures, the values of C_P drop off considerably in a nonlinear relationship. A plot of the *Einstein heat capacity relationship* for metals,

$$C_V = 3R \left(\frac{\Theta_E}{T}\right)^2 \frac{e^{\Theta_E/T}}{(e^{\Theta_E/T}-1)^2} \tag{3.24}$$

fits the experimental curves reasonably well. In this equation the *Einstein characteristic temperature* (Θ_E) describes the vibration of the atoms in the metallic crystal. Somewhat better overall agreement between experiment and theory is obtained using the *Debye heat capacity relationship* given by

$$C_V = \frac{9R}{(\Theta_D/T)^3} \int_0^{\Theta_D/T} \frac{x^4 e^x}{(e^x-1)^2}\, dx \tag{3.25}$$

where Θ_D is the *Debye characteristic temperature.*

Evaluation of (3.25) is usually done using

$$C_V = 3R\mathscr{D}(\Theta_D/T) \tag{3.26}$$

where $\mathscr{D}(\Theta_D/T)$ represents the *Debye function* (see Table 3-2). Combining the Debye result with a term for the electronic contribution as predicted by the free-electron model for metals (see Sec. 21.3) gives

$$C_V = \frac{9R}{(\Theta_D/T)^3} \int_0^{\Theta_D/T} \frac{x^4 e^x}{(e^x-1)^2}\, dx + \eta T \tag{3.27}$$

where η is a constant that can be determined as $\pi^2 Rk/2E_f$, where E_f is the *Fermi energy*.

For liquids, exact values of heat capacity cannot be predicted. However, approximations include $3R$ for the vibrational modes of the centers of mass of the molecules, $Rx^2 e^x/(e^x-1)^2$ for each intramolecular vibrational mode, and a contribution of $\frac{1}{2}R$ to R for each mode of rotational motion.

Table 3-2

$$\mathscr{D}(\Theta_D/T) \equiv \frac{3}{(\Theta_D/T)^3} \int_0^{\Theta_D/T} \frac{x^4 e^x}{(e^x-1)^2}\, dx$$

Θ_D/T	$\mathscr{D}(\Theta_D/T)$	Θ_D/T	$\mathscr{D}(\Theta_D/T)$	Θ_D/T	$\mathscr{D}(\Theta_D/T)$	Θ_D/T	$\mathscr{D}(\Theta_D/T)$
0.5	0.987 6	4.5	0.432 0	8.5	0.118 2	12.5	0.039 7
1.0	0.951 7	5.0	0.368 6	9.0	0.101 5	13.0	0.035 4
1.5	0.896 0	5.5	0.313 3	9.5	0.087 5	13.5	0.031 6
2.0	0.825 4	6.0	0.265 6	10.0	0.075 8	14.0	0.028 4
2.5	0.745 9	6.5	0.225 1	10.5	0.066 0	14.5	0.025 5
3.0	0.662 8	7.0	0.190 9	11.0	0.057 7	15.0	0.023 1
3.5	0.580 7	7.5	0.162 2	11.5	0.050 7	15.5	0.020 9
4.0	0.503 1	8.0	0.138 2	12.0	0.044 8	16.0	0.019 0

$\mathscr{D}(\Theta_D/T) = 77.927(T/\Theta_D)^3$　for　$\Theta_D/T \geq 16$

$\mathscr{D}(\Theta_D/T) = \left[1 - \frac{(\Theta_D/T)^2}{20} + \frac{(\Theta_D/T)^4}{560} - \frac{(\Theta_D/T)^6}{18\,144}\right]$　for　$\Theta_D/T \leq 0.5$

EXAMPLE 3.10. Calculate C_V for silver at 75 K using (3.24) and (3.25). The value of Θ_D is 215 K for silver. Compare the answers to the observed value of 16.9 J K^{-1} mol^{-1}.

The value of Θ_E can be calculated from Θ_D as

$$\Theta_E = (\tfrac{3}{5})^{1/2}\Theta_D = (\tfrac{3}{5})^{1/2}(215 \text{ K}) = 167 \text{ K}$$

giving
$$C_V = 3(8.314 \text{ J K}^{-1} \text{ mol}^{-1})\left(\frac{167 \text{ K}}{75 \text{ K}}\right)^2 \frac{e^{167/75}}{(e^{167/75}-1)^2}$$

$$= 16.8 \text{ J K}^{-1} \text{ mol}^{-1}$$

This value is 0.6% lower than the experimental value.

The value of Θ_D/T is $(215 \text{ K})/(75 \text{ K}) = 2.87$, and the value of $\mathscr{D}(\Theta_D/T)$ interpolated from Table 3-2 is 0.684, giving

$$C_V = (3)(8.314 \text{ J K}^{-1} \text{ mol}^{-1})(0.684) = 17.1 \text{ J K}^{-1} \text{ mol}^{-1}$$

This value is 1% higher than the experimental value.

Internal Energy, Work, and Heat

3.10 WORK, w

In thermodynamics, *work* is defined as the energy that is transferred between a system and its surroundings during a change in the state of the system and is completely convertible into some form of mechanical work in the surroundings.

Work is given by the product of an *intensity factor* X (such as force) and a *capacity factor y* (such as distance):

$$đw = X \, dy \tag{3.28}$$

where X is related to y and the potential energy E_p by

$$X = -\left(\frac{\partial E_p}{\partial y}\right) \tag{3.29}$$

The symbol $đ$ implies that work is a *path function*: Its value depends on the actual process used during a change of the system. The types of work commonly encountered in physical chemistry are summarized in Table 3-3.

The sign convention used in this book for work is that *a negative value means that the system under consideration has performed work on the surroundings and a positive value means that the surroundings have done work on the system.*

EXAMPLE 3.11. The symbol for the differential work, $đw$, used in *(3.28)* indicates that work is, in general, an inexact differential and the work involved in a process is path-dependent. To illustrate this, consider the expansion of 1 mol of an ideal gas from $0.010\,0 \text{ m}^3$ to $0.100\,0 \text{ m}^3$ at 25 °C by the following processes: (1) against a constant external pressure of 0.100 bar, (2) from $0.010\,0 \text{ m}^3$ to $0.025\,0 \text{ m}^3$ against a constant external pressure of 0.333 bar, followed by a second expansion from $0.025\,0 \text{ m}^3$ to $0.050\,0 \text{ m}^3$ against a constant pressure of 0.200 bar, followed by a third expansion from $0.050\,0 \text{ m}^3$ to $0.100\,0 \text{ m}^3$ against a constant pressure of 0.100 bar; (3) a reversible expansion.

For the first process *(3.28)* gives (see Table 3-3)

$$w = -\int_{V_1}^{V_2} P_{\text{ext}} \, dV = -P_{\text{ext}}\int_{V_1}^{V_2} dV = -P_{\text{ext}} \Delta V$$

$$= -(0.100 \text{ bar})(0.100\,0 \text{ m}^3 - 0.010\,0 \text{ m}^3)\left(\frac{10^5 \text{ Pa}}{1 \text{ bar}}\right)\left(\frac{1 \text{ J}}{1 \text{ Pa m}^3}\right) = -900 \text{ J}$$

For the second process, repeating the above calculation for each step gives

$$w = -[(0.333)(0.025\,0 - 0.010\,0) + (0.200)(0.050\,0 - 0.025\,0) + (0.100)(0.100\,0 - 0.050\,0)]$$

$$= -0.015\,0 \text{ m}^3 \text{ bar} = -1\,500 \text{ J}$$

Under reversible conditions, the external pressure and the internal pressure differ only by dP, so substituting (*1.6*) into (*3.28*) gives

$$w = -\int_{V_1}^{V_2} \frac{nRT}{V} \, dV = -nRT \ln \frac{V_2}{V_1}$$

$$= -(1.00 \text{ mol})(8.314 \text{ J K}^{-1} \text{ mol}^{-1})(298 \text{ K}) \ln \left(\frac{0.100\,0 \text{ m}^3}{0.010\,0 \text{ m}^3} \right) = -5\,705 \text{ J}$$

From the above calculations it can be seen that w is dependent on the process chosen and is greatest for a reversible process and smallest for the most irreversible process.

Table 3-3

Process	$đw = X \, dy$	Comments
Mechanical work	$đw = F_{ext} \, dl$	F_{ext} = external force l = displacement
Stretching work	$đw = kl \, dl$	kl = tension l = displacement
Gravitational work	$đw = mg \, dl$	m = mass g = gravitational constant l = displacement
Expansion work	$đw = -P_{ext} \, dV$	P_{ext} = external pressure V = volume
Surface work	$đw = \gamma \, dA$	γ = surface tension A = area
Electrochemical cell work	$đw = \Delta V \, dQ$ $= \Delta V I \, dt$	ΔV = electric potential difference Q = quantity of electricity I = electric current t = time

3.11 HEAT, q

In thermodynamics, *heat* is defined as the energy that is transferred between a system and its surroundings during a change in the state of the system and is transferred as a result of a difference in temperature between the system and its surroundings. Unless work is done, heat transfer will be directed from the point of higher temperature to the point of lower temperature.

The sign convention for heat in this book is that *a negative value represents heat flow from a system to the surroundings* (an exothermic process) *and a positive value represents heat flow from the surroundings to the system* (an endothermic process).

EXAMPLE 3.12. The heat transferred between a system and its surroundings under constant external pressure conditions is given by

$$q = nC_P \, \Delta T = mc_P \, \Delta T \tag{3.30}$$

What is q for heating 125 g of water from 15 °C to 95 °C? For liquid water, $c_P = 4.2 \text{ J K}^{-1} \text{ g}^{-1}$.

Equation (*3.30*) gives

$$q = (125 \text{ g})(4.2 \text{ J K}^{-1} \text{ g}^{-1})(368 \text{ K} - 288 \text{ K}) = 42\,000 \text{ J} = 42 \text{ kJ}$$

[Note that the positive value means that heat is absorbed by the system (the water) from the surroundings (a laboratory burner, etc.).]

3.12 STATEMENT

The *first law of thermodynamics* can be written as

$$\Delta U = q + w \tag{3.31a}$$

or, in differential form, as

$$dU = \bar{d}q + \bar{d}w \tag{3.31b}$$

For a system capable of performing only expansion work, if $dV = 0$ then $w = 0$, which upon substitution into (*3.31a*) gives

$$\Delta U = q_V \tag{3.32}$$

Thus the heat transferred under constant-volume (*isochoric*) conditions is a direct measurement of ΔU. Heat of combustion measurements for substances are made under these conditions using a "bomb" calorimeter and hence are listed as ΔU(combustion).

For a system operating under isobaric conditions, (*3.3*) becomes

$$\Delta H = \Delta U + P \, \Delta V$$

Since under these conditions $P \, \Delta V = -w$ and $\Delta U = q_P + w$,

$$\Delta H = q_P \tag{3.33}$$

Thus the heat transferred under constant-pressure conditions (such as the heat of reaction for a chemical reaction performed in a "solution" calorimeter, beaker, flask, etc.) is ΔH.

EXAMPLE 3.13. Consider a system consisting of 1 mol of a monatomic gas contained in a piston. What is the temperature change of the gas if $q = 50$ J and $w = -100$ J?

From the first law,

$$\Delta U = q + w = 50 + (-100) = -50 \text{ J mol}^{-1}$$

For a monatomic ideal gas,

$$\Delta U = \Delta U(\text{thermal}) = \tfrac{3}{2} R \, \Delta T$$

and hence

$$\Delta T = \frac{\Delta U}{\tfrac{3}{2} R} = \frac{-50 \text{ J mol}^{-1}}{\tfrac{3}{2}(8.314 \text{ J K}^{-1} \text{ mol}^{-1})} = -4.0 \text{ K}$$

The gas will decrease in temperature by 4.0 °C.

3.13 GENERAL ΔU AND ΔH CALCULATIONS

For any thermodynamic process, the internal energy change of the system is

$$\Delta U = \int_{T_1}^{T_2} C_V \, dT + \int_{V_1}^{V_2} \left(\frac{\partial U}{\partial V}\right)_T dV \tag{3.34}$$

Note that (*3.34*) gives the integral form of (*3.14*) for the special case of a constant-volume process. In general,

$$\left(\frac{\partial U}{\partial V}\right)_T = T \left(\frac{\partial P}{\partial T}\right)_V - P = \frac{\alpha}{\kappa} T - P \tag{3.35}$$

which becomes

$$\left(\frac{\partial U}{\partial V}\right)_T = 0 \qquad \text{for an ideal gas} \tag{3.36a}$$

and

$$\left(\frac{\partial U}{\partial V}\right)_T = -\left(\frac{\partial P}{\partial V}\right)_T (\mu C_p + V) - P \qquad \text{for a real gas} \tag{3.36b}$$

where μ is the Joule–Thomson coefficient (see Sec. 3.19) for the gas; and

$$\left(\frac{\partial U}{\partial V}\right)_T \approx -P \qquad \text{for condensed states of matter} \tag{3.36c}$$

Likewise, for any thermodynamic process, the enthalpy change of the system is

$$\Delta H = \int_{T_1}^{T_2} C_P\, dT + \int_{P_1}^{P_2} \left(\frac{\partial H}{\partial P}\right)_T dP \tag{3.37}$$

Note that (3.37) gives the integral form of (3.15) for the special case of a constant-pressure process. In general,

$$\left(\frac{\partial H}{\partial P}\right)_T = V(1 - \alpha T) \tag{3.38}$$

which becomes

$$\left(\frac{\partial H}{\partial P}\right)_T = 0 \qquad \text{for an ideal gas} \tag{3.39a}$$

$$\left(\frac{\partial H}{\partial P}\right)_T = -C_P\mu \qquad \text{for a real gas} \tag{3.39b}$$

and

$$\left(\frac{\partial H}{\partial P}\right)_T \approx V \qquad \text{for condensed states of matter} \tag{3.39c}$$

EXAMPLE 3.14.   Newer tabulations of "enthalpies of formation" of substances are at $P = 1.00$ bar and 25 °C and older tabulations of $\Delta_f H$ were at $P = 1.00$ atm and 25 °C. What is the difference between these $\Delta_f H$ values for a solid such as NaCl?

For 1 mol of solid, (3.37) and (3.39c) give

$$\Delta H = 0 + \int_{P_1}^{P_2} V_m\, dP = V_m\, \Delta P$$

Substituting M/ρ for V_m where $M = 58.44 \times 10^{-3}$ kg mol^{-1} and $\rho = 2.165 \times 10^3$ kg m^{-3} gives

$$\Delta H = \left(\frac{58.44 \times 10^{-3}\ \text{kg mol}^{-1}}{2.165 \times 10^3\ \text{kg m}^{-3}}\right)\left[1.00\ \text{bar} - (1.00\ \text{atm})\left(\frac{1.013\,25\ \text{bar}}{1\ \text{atm}}\right)\right]\left(\frac{10^5\ \text{Pa}}{1\ \text{bar}}\right)\left(\frac{1\ \text{J}}{1\ \text{Pa m}^3}\right)$$

$$= -0.036\ \text{J}$$

[Note that this difference is insignificant compared to the precision of most reported values of $\Delta_f H$.]

Specific Applications of the First Law

3.14   REVERSIBLE ISOTHERMAL EXPANSION OF AN IDEAL GAS

For this process, $P_{\text{ext}} = P$, giving

$$\Delta U = \Delta H = 0 \tag{3.40a}$$

$$q = -w \tag{3.40b}$$

$$w = -\int_{V_1}^{V_2} P\, dV = -nRT \ln\frac{V_2}{V_1} = nRT \ln\frac{P_2}{P_1} \tag{3.40c}$$

EXAMPLE 3.15.   Calculate ΔU, q, w, and ΔH for the reversible compression of 2.00 mol of an ideal gas from 1.00 bar to 100.0 bar at 25 °C.

Equations (3.40) give

$$\Delta U = \Delta H = 0 \qquad q = -w$$

$$w = nRT \ln \frac{P_2}{P_1}$$

$$= (2.00 \text{ mol})(8.314 \text{ J K}^{-1} \text{ mol}^{-1}) \ln (298 \text{ K}) \left(\frac{100.0 \text{ bar}}{1.00 \text{ bar}} \right)$$

$$= 22.8 \text{ kJ}$$

3.15 ISOTHERMAL ISOBARIC EXPANSION OF AN IDEAL GAS

For this irreversible process,

$$\Delta U = \Delta H = 0 \tag{3.41a}$$

$$q = -w \tag{3.41b}$$

$$w = -P_{ext} \Delta V = -P_{ext} nRT \left(\frac{1}{P_2} - \frac{1}{P_1} \right) \tag{3.41c}$$

EXAMPLE 3.16. Calculate ΔU, q, w, and ΔH for compressing 2.00 mol of an ideal gas from 1.00 bar to 100.0 bar at 25 °C, given that the external pressure is 500.0 bar.

Using ($3.41a$) gives $\Delta U = \Delta H = 0$. Using ($3.41c$) gives

$$w = -(500.0 \text{ bar})(2.00 \text{ mol})(8.314 \text{ J K}^{-1} \text{ mol}^{-1})(298 \text{ K}) \left(\frac{1}{100.0 \text{ bar}} - \frac{1}{1.00 \text{ bar}} \right)$$

$$= 2.45 \text{ MJ}$$

and ($3.41b$) gives $q = -2.45$ MJ.

3.16 ISOTHERMAL ISOBARIC PHASE CHANGE

For this reversible process,

$$q = \Delta H \tag{3.42a}$$

$$w = -\int_{V_1}^{V_2} P_{ext} \, dV = -P_{ext} \Delta V \tag{3.42b}$$

$$\Delta U = q + w \tag{3.42c}$$

EXAMPLE 3.17. Calculate q, w, and ΔU at 1.00 bar for the phase transition between two crystalline forms, s-I and s-II, of Li_2SO_4

$$Li_2SO_4(\text{s-II}) \longrightarrow Li_2SO_4(\text{s-I})$$

which takes place at 859 K. The enthalpy change is 27.2 kJ for this reaction. The densities of $Li_2SO_4(\text{s-II})$ and $Li_2SO_4(\text{s-I})$ are 2.221×10^3 and 2.07×10^3 kg m^{-3}, respectively.

For this solid–solid transition, (3.42) gives

$$q = \Delta H = 27.2 \text{ kJ}$$

$$w = -(1.00 \text{ mol})(1.00 \text{ bar}) \left[(0.109 \, 94 \text{ kg mol}^{-1}) \left(\frac{1}{2.07 \times 10^3 \text{ kg m}^{-3}} - \frac{1}{2.221 \times 10^3 \text{ kg m}^{-3}} \right) \right] \left(\frac{10^5 \text{ Pa}}{1 \text{ bar}} \right) \left(\frac{1 \text{ J}}{1 \text{ Pa m}^3} \right)$$

$$= -0.36 \text{ J}$$

$$\Delta U = q + w = 27.2 \text{ kJ} + (-0.36 \text{ J}) = 27.2 \text{ kJ}$$

3.17 REVERSIBLE ADIABATIC EXPANSION OF AN IDEAL GAS

For this process,

$$q = 0 \qquad (3.43a)$$

$$\Delta U = w = \int_{T_1}^{T_2} nC_V \, dT \qquad (3.43b)$$

$$\Delta H = \int_{T_1}^{T_2} nC_P \, dT \qquad (3.43c)$$

To perform the calculations indicated by (3.43), both the initial and final temperatures must be known. Depending on the data, one of the following relationships may be used to determine the final temperature:

$$T_1^{C_V/R} V_1 = T_2^{C_V/R} V_2 \qquad (3.44)$$

$$P_1 V_1^{\gamma} = P_2 V_2^{\gamma} \qquad (3.45)$$

$$T_2^{C_P/R} / P_2 = T_1^{C_P/R} / P_1 \qquad (3.46)$$

where the *heat capacity ratio* is defined as

$$\gamma \equiv \frac{C_P}{C_V} \qquad (3.47)$$

EXAMPLE 3.18. Calculate q, w, ΔH, and ΔU for the adiabatic and reversible compression of 1.00 mol of a monatomic ideal gas from 0.100 0 m^3 and 25 °C to 0.010 0 m^3.

Using (3.44) to determine the final temperature gives

$$298^{C_V/R}(0.100\ 0\ \text{m}^3) = T_2^{C_V/R}(0.010\ 0\ \text{m}^3)$$

Substituting $C_V = \frac{3}{2}R$ and rearranging gives

$$T_2^{3/2} = (298)^{3/2}(10.0)$$

or

$$T = 1\ 383\ \text{K}$$

Now (3.43) gives

$$q = 0$$

$$\Delta U = w = \int_{298\,\text{K}}^{1383\,\text{K}} nC_V \, dT = n(\tfrac{3}{2})R\,\Delta T$$

$$= (1.00\ \text{mol})(\tfrac{3}{2})(8.314\ \text{J K}^{-1}\ \text{mol}^{-1})(1\ 383\ \text{K} - 298\ \text{K}) = 13.6\ \text{kJ}$$

$$\Delta H = \int_{298\,\text{K}}^{1383\,\text{K}} nC_P \, dT = n(\tfrac{5}{2})R\,\Delta T = 22.6\ \text{kJ}$$

3.18 ISOBARIC ADIABATIC EXPANSION OF AN IDEAL GAS

For this irreversible process,

$$q = 0 \qquad (3.48a)$$

$$w = -\int_{V_1}^{V_2} P_{\text{ext}} \, dV = -P_{\text{ext}}\,\Delta V \qquad (3.48b)$$

$$\Delta U = \int_{T_1}^{T_2} nC_V \, dT \qquad (3.48c)$$

$$\Delta H = \int_{T_1}^{T_2} nC_P \, dT \qquad (3.48d)$$

The final temperature can be determined using either

$$T_2 = T_1 - \frac{P_{ext} \Delta V}{nC_V} \tag{3.49}$$

or

$$T_2 = T_1 \left(\frac{C_V + (P_{ext}R/P_1)}{C_V + (P_{ext}R/P_2)} \right) \tag{3.50}$$

EXAMPLE 3.19. Calculate q, w, ΔH, and ΔU for the isobaric adiabatic expansion of 1.00 mol of a monatomic ideal gas from 1.00 dm^3 and 25 °C to 10.00 dm^3 against an external pressure of 1.00 bar. (This is an irreversible expansion because of the finite difference between the applied and internal pressures.)

Using (3.49) to determine the final temperature gives

$$T_2 = 298 \text{ K} - \frac{(1.00 \text{ bar})(9.00 \text{ dm}^3)[(1 \text{ m}^3)/(10^3 \text{ dm}^3)][(10^5 \text{ Pa})/(1 \text{ bar})]}{(1.00 \text{ mol})(\frac{3}{2})(8.314 \text{ J K}^{-1} \text{ mol}^{-1})[(1 \text{ Pa m}^3)/(1 \text{ J})]}$$

$$= 226 \text{ K}$$

and using (3.48) gives

$$q = 0$$

$$\Delta U = w = (1.00 \text{ mol})(\tfrac{3}{2})(8.314 \text{ J K}^{-1} \text{ mol}^{-1})(226 \text{ K} - 298 \text{ K})$$

$$= -0.90 \text{ kJ}$$

$$\Delta H = (1.00)(\tfrac{5}{2})(8.314)(-72) = -1.50 \text{ kJ}$$

3.19 JOULE-THOMSON EFFECT

The *Joule-Thomson experiment* is a constant-enthalpy process that measures

$$(\partial T/\partial P)_H \equiv \mu \tag{3.51}$$

as real gases undergo a throttled adiabatic expansion. A positive value of μ indicates that a cooling will occur as the gas expands, because work is done at the expense of the internal energy of the gas.

EXAMPLE 3.20. The value of μ for air at 0 °C is 0.246 K bar^{-1} at 20 bar and 0.263 K bar^{-1} at 1 bar. What is the approximate cooling observed as the gas undergoes this expansion?

Rearranging (3.51) and integrating gives

$$\Delta T = \int_{P_1}^{P_2} \mu \, dP$$

If μ is assumed to be a linear function of pressure over this small pressure range, then the pressure dependence can be expressed as

$$\mu = 0.264 - (8.95 \times 10^{-4})P$$

using the data above. Substituting this relationship and integrating gives the cooling as

$$\Delta T = \int_{20}^{1} [0.264 - (8.95 \times 10^{-4} P)] \, dP$$

$$= (0.264)(1 - 20) - (4.93 \times 10^{-4})(1^2 - 20^2) = -4.82 \text{ K}$$

Solved Problems

INTERNAL ENERGY AND ENTHALPY

3.1. The heat of combustion of $H_2(g)$ to form $H_2O(l)$ under constant-pressure conditions is -285.83 kJ mol^{-1} at 25 °C. If the water is formed at 1.00 bar and has a density of 1.00×10^3 kg m^{-3}, calculate ΔU for this reaction.

For the reaction

$$H_2(g) + \tfrac{1}{2}O_2(g) \longrightarrow H_2O(l) \qquad \Delta H = -285.83 \text{ kJ}$$

the term $\Delta(PV)$ in (3.3) can be replaced by

$$\Delta(PV) = P[V(H_2O) - V(H_2) - V(O_2)]$$

$$= (1.00 \text{ bar}) \frac{(1 \text{ mol})(18.0 \times 10^{-3} \text{ kg mol}^{-1})}{1.00 \times 10^3 \text{ kg m}^{-3}} \left(\frac{10^5 \text{ Pa}}{1 \text{ bar}}\right)\left(\frac{1 \text{ J}}{1 \text{ Pa m}^3}\right)$$

$$- (1 \text{ mol } H_2)RT - (\tfrac{1}{2} \text{ mol } O_2)RT$$

$$= 1.80 \text{ J} - (\tfrac{3}{2} \text{ mol})(8.314 \text{ J K}^{-1} \text{ mol}^{-1})(298.15 \text{ K})$$

$$= -3\,716 \text{ J} = -3.716 \text{ kJ}$$

Rearranging (3.3) gives

$$\Delta U = \Delta H - \Delta(PV) = (-285.830 \text{ kJ}) - (-3.716 \text{ kJ}) = -282.114 \text{ kJ}$$

for the reaction, or $\Delta U = -282.114$ kJ mol^{-1}. Note that the PV contribution of the liquid phase is negligible compared to that of the gaseous components of the reaction.

3.2. Evaluate E(thermal) for $H_2O(g)$ at 373.12 K. The vibrational frequencies are $\bar{\nu}_1 = 3\,656.7$, $\bar{\nu}_2 = 1\,594.6$, and $\bar{\nu}_3 = 3\,755.8$ cm^{-1}.

Substituting the values of $\bar{\nu}$ into (3.7a) gives $x_1 = 14.101$, $x_2 = 6.149$, and $x_3 = 14.483$. Using (3.9d) for the nonlinear polyatomic gas gives

$$E(\text{thermal}) = 3(8.314 \text{ J K}^{-1} \text{ mol}^{-1})(373.12 \text{ K})$$

$$+ (8.314 \text{ J K}^{-1} \text{ mol}^{-1})(373.12 \text{ K})\left(\frac{14.101}{e^{14.101} - 1} + \frac{6.149}{e^{6.149} - 1} + \frac{14.483}{e^{14.483} - 1}\right)$$

$$= 9\,347 \text{ J mol}^{-1}$$

3.3. Evaluate H(thermal) for $H_2O(g)$ at 373.12 K.

Substituting the result of Problem 3.2 into (3.11) gives

$$H(\text{thermal}) = 9\,347 \text{ J mol}^{-1} + (8.314 \text{ J K}^{-1} \text{ mol}^{-1})(373.12 \text{ K}) = 12.449 \text{ kJ mol}^{-1}$$

3.4. Evaluate H(thermal) for $H_2O(g)$ at 755 K. Find ΔU and ΔH for heating H_2O from 373.12 K to 755 K.

At 755 K, the values of x for the vibrational contributions are $x_1 = 6.969$, $x_2 = 3.039$, and $x_3 = 7.157$. Using (3.9d) and (3.11) gives

$$E(\text{thermal}) = 3(8.314 \text{ J K}^{-1} \text{ mol}^{-1})(755 \text{ K}) + (8.314 \text{ J K}^{-1} \text{ mol}^{-1})(755 \text{ K})\left(\frac{6.969}{e^{6.969} - 1} + \frac{3.039}{e^{3.039} - 1} + \frac{7.157}{e^{7.157} - 1}\right)$$

$$= 19\,867 \text{ J mol}^{-1}$$

$$H(\text{thermal}) = 19\,867 \text{ J mol}^{-1} + (8.314 \text{ J K}^{-1} \text{ mol}^{-1})(755 \text{ K})$$

$$= 26\,310. \text{ J mol}^{-1}$$

Using these values and the values of E(thermal) and H(thermal) for water at 373.12 K evaluated in Problems 3.2 and 3.3 gives

$$\Delta U = \Delta E(\text{thermal}) = E(\text{thermal, 755 K}) - E(\text{thermal, 373.12 K})$$

$$= (19\,867 \text{ J mol}^{-1}) - (9\,347 \text{ J mol}^{-1}) = 10\,520. \text{ J mol}^{-1}$$

$$\Delta H = H(\text{thermal, 755 K}) - H(\text{thermal, 373.12 K})$$

$$= (26\,310. \text{ J mol}^{-1}) - (12\,499 \text{ J mol}^{-1}) = 13\,811 \text{ J mol}^{-1}$$

HEAT CAPACITY

3.5. The heat capacity of a substance is sometimes called the *molar heat capacity* (C_m) when the basis is 1 mol of that substance. What is the relationship between the molar heat capacity and the specific heat capacity?

For the basis of 1 mol of substance, the mass is equal to the molar mass M. Thus (*3.13*) gives the relationship as

$$c = C_m/M \tag{3.52}$$

3.6. The cubic expansion coefficient of water at 25 °C is $2.57 \times 10^{-4} \text{ K}^{-1}$, and the isothermal compressibility is $45.24 \times 10^{-6} \text{ bar}^{-1}$. Given that the density of water is $0.997\,075 \times 10^3 \text{ kg m}^{-3}$, calculate $C_P - C_V$.

For the liquid, (*3.19*) gives

$$C_P - C_V = \frac{\alpha^2 VT}{\kappa}$$

$$= \frac{(2.57 \times 10^{-4} \text{ K}^{-1})^2 (18.015 \times 10^{-3} \text{ kg mol}^{-1})[(1 \text{ m}^3)/(0.997\,075 \times 10^3 \text{ kg})](298 \text{ K})[(1 \text{ J})/(1 \text{ Pa m}^3)]}{(45.24 \times 10^{-6} \text{ bar}^{-1})[(1 \text{ bar})/(10^5 \text{Pa})]}$$

$$= 0.786 \text{ J K}^{-1} \text{ mol}^{-1}$$

which is about 1% of $C_P = 75.291 \text{ J K}^{-1} \text{ mol}^{-1}$.

3.7. Find ΔH for heating $H_2(g)$ from 0. °C to 100. °C, given

$$C_P = 29.066 - (0.836\,4 \times 10^{-3})T + (20.12 \times 10^{-7})T^2$$

in units of $J \text{ K}^{-1} \text{ mol}^{-1}$.

Using (*3.15*) gives

$$\Delta H = \int_{273\,\text{K}}^{373\,\text{K}} [29.066 - (0.836\,4 \times 10^{-3})T + (20.12 \times 10^{-7})T^2] \, dT$$

$$= (29.066)(373 - 273) - (0.836\,4 \times 10^{-3})(373^2 - 273^2)/2 + (20.12 \times 10^{-7})(373^3 - 273^3)/3$$

$$= 2\,900.7 \text{ J mol}^{-1}$$

3.8. Find C_V and C_P for $H_2O(g)$ at 373.12 K.

Using the results of Problem 3.2 and (*3.23*) gives

$$C_V(\text{trans}) = \tfrac{3}{2}(8.314 \text{ J K}^{-1} \text{ mol}^{-1}) = 12.471 \text{ J K}^{-1} \text{ mol}^{-1}$$

$$C_V(\text{rot}) = \tfrac{3}{2}(8.314) = 12.471 \text{ J K}^{-1} \text{ mol}^{-1}$$

$$C_V(\text{vib}) = (8.314) \left[\frac{(14.101)^2 \, e^{14.101}}{(e^{14.101} - 1)^2} + \frac{(6.149)^2 \, e^{6.149}}{(e^{6.149} - 1)^2} + \frac{(14.483)^2 \, e^{14.483}}{(e^{14.483} - 1)^2} \right]$$

$$= 0.676 \text{ J K}^{-1} \text{ mol}^{-1}$$

Summing these contributions gives $C_V = 25.618$ J K^{-1} mol^{-1}, and (3.18) gives

$$C_P = C_V + R = 25.618 + 8.314 = 33.932 \text{ J K}^{-1} \text{ mol}^{-1}$$

3.9. Calculate ΔH for heating 1 mol of an ideal diatomic gas from 0 °C to 100 °C excluding vibrational contributions. Compare the value to that determined for H_2 in Problem 3.7.

For a diatomic ideal gas, $C_V = \frac{5}{2}R$, so $C_P = \frac{7}{2}R$ and (3.15) gives

$$\Delta H = \int_{T_1}^{T_2} C_P \, dT = \int_{273\,\text{K}}^{373\,\text{K}} \tfrac{7}{2}R \, dT = \tfrac{7}{2}R(T_2 - T_1) = 2\,909.9 \text{ J mol}^{-1}$$

This answer agrees quite well with the calculated value of $2\,900.7$ J mol^{-1} that was determined for heating 1 mol of H_2.

3.10. The low-temperature limit of (3.27) can be rearranged as

$$\frac{C_V}{T} = \frac{233.8}{\Theta_D{}^3} RT^2 + \eta$$

which is in the general form of a linear equation if C_V/T is plotted against T^2. Prepare such a plot from the data given below for aluminum, assuming $C_V = C_P$ at these low temperatures, and determine the values of Θ_D and η.

$T/(\text{K})$	1	2	3	4	6	8	10
$C_P/(10^{-3} \text{ J K}^{-1} \text{ mol}^{-1})$	1.38	2.91	4.75	7.04	13.49	23.74	37.77

The slope of the straight line through the first five data is 2.59×10^{-5} J K^{-4} mol^{-1}, which gives $\Theta_D = 422$ K, and the intercept is 1.36×10^{-3} J K^{-2} mol^{-1}, which is η.

3.11. What is the high-temperature limit of C_V predicted by (3.24) for a metal? Show that at $\Theta_E/T = 2.98$, C_V is one-half this limiting value. At 77 K, the heat capacity of copper is one-half the maximum value at high temperature. Calculate Θ_E for copper.

As $T \to \infty$, $\Theta_E/T \to 0$, and the exponentials in (3.24) can be replaced by the series expansion

$$e^{\pm x} = 1 \pm x + \frac{x^2}{2!} \pm \cdots$$

giving
$$C_V = 3R \left(\frac{\Theta_E}{T}\right)^2 \frac{1 + \Theta_E/T + \cdots}{(1 + \Theta_E/T + \cdots - 1)^2} \to 3R$$

which agrees with the law of Dulong and Petit assuming $C_P \approx C_V$.

At $\Theta_E/T = 2.98$, (3.24) gives

$$C_V = 3R(2.98)^2 \frac{e^{2.98}}{(e^{2.98} - 1)^2} = (0.501)(3R)$$

The Einstein characteristic temperature for copper is

$$\Theta_E = (2.98)(77 \text{ K}) = 229 \text{ K}$$

3.12. Predict C_V for $H_2O(l)$ at 25 °C.

The vibrational contribution would be $3R = 25$ J K^{-1} mol^{-1}. The three degrees of rotational freedom each contribute between $\frac{1}{2}R$ and R for a total of 12 to 25 J K^{-1} mol^{-1}. The intramolecular vibrational frequencies for water are 3 657, 1 595, and 3 756 cm^{-1}, giving values of x equal to 17.66, 7.70, and 18.13, respectively. Substituting these values into $Rx^2e^x/(e^x-1)^2$ gives a total contribution of about 0.4 J K^{-1} mol^{-1}. Adding the contributions gives a prediction of between 37 and 50 J K^{-1} mol^{-1}. (The predicted value is quite low compared to the actual value of 75 J K^{-1} mol^{-1}.)

INTERNAL ENERGY, WORK, AND HEAT

3.13. Consider a dry cell, $\Delta V = 1.50$ V, large enough to deliver a constant current of exactly 0.010 0 A for 1 h. If this cell powered a hoist able to lift a 200-lb man, how far off the ground would the man be at the end of the hour?

Assuming no loss of work, Eq. (*3.28*) gives (see Table 3-2)

$$\int_0^{1\,h} \Delta VI\,dt = \int_{l_1}^{l_2} mg\,dl$$

Solving for the change in distance gives

$$\Delta l = \frac{\Delta VI\Delta t}{mg} = \frac{(1.50 \text{ V})(0.010\ 0 \text{ A})(3\ 600 \text{ s})[(1 \text{ kg m}^2 \text{ A}^{-1} \text{ s}^{-3}/(1\text{V})]}{(200 \text{ lb})[(0.454 \text{ kg})/(1 \text{ lb})](9.80 \text{ m s}^{-2})} = 6.07 \text{ cm}$$

3.14. Consider a 1.00-kg block of iron at 99 °C placed in contact with a 1.00-kg block of iron at 25 °C. Find the final temperature of the blocks and the amount of heat that was transferred. For iron, $c_P = 444$ J K^{-1} kg^{-1}.

Heat will be transferred from the warmer block to the cooler one until both have reached the same temperature, or

$$q_h = -q_c$$
$$m_h c_P (T_f - T_h) = -m_c c_P (T_f - T_c)$$
$$T_f = 0.5(T_h + T_c) = 0.5(372 \text{ K} + 298 \text{ K}) = 335 \text{ K} = 62 \text{ °C}$$

Using this final temperature gives

$$q_h = -q_c = (1.00 \text{ kg})(0.444 \text{ kJ K}^{-1} \text{ kg}^{-1})(335 \text{ K} - 372 \text{ K})$$
$$= -16.4 \text{ kJ}$$

3.15. Under isothermal conditions, $\Delta U = 0$ for the expansion of an ideal gas. If 100 J of work is done on the system consisting of 1 mol of an ideal gas, what amount of heat must be transferred?

The first law, (*3.31a*), gives

$$\Delta U = 0 = q + w \quad \text{or} \quad q = -w = -100 \text{ J}$$

Thus 100 J must be transferred from the system to maintain the constant temperature.

3.16. A 1.00-mol sample of argon is heated from 25 °C to 125 °C, and the corresponding pressure change is from 4.93 bar to 6.60 bar. Calculate ΔH for this process assuming ideal gas behavior. For argon, $C_P = 20.786$ J K^{-1} mol^{-1}.

For an ideal gas, (*3.39a*) gives $(\partial H/\partial P)_T = 0$, so the enthalpy change is given by (*3.37*) as

$$\Delta H = \int_{T_1}^{T_2} C_P\,dT + 0 = (1.00 \text{ mol})(20.786 \text{ J K}^{-1} \text{ mol}^{-1})(398 \text{ K} - 298 \text{ K})$$
$$= 2\ 080 \text{ J}$$

3.17. For a van der Waals gas, $\mu = [(2a/RT) - b]/C_p$. Derive an equation for $(\partial H/\partial P)_T$, and repeat the calculations of Problem 3.16 using $a = 1.363$ dm^6 bar mol^{-2} and $b = 0.032\,19$ dm^3 mol^{-1}.

Substituting the given expression for μ into (3.39b) gives

$$\left(\frac{\partial H}{\partial P}\right)_T = -C_P\mu = -C_P\left(\frac{2a/RT - b}{C_P}\right) = b - \frac{2a}{RT}$$

The average value of $(\partial H/\partial P)_T$ over this temperature change is

$$\left(\frac{\partial H}{\partial P}\right)_T = (0.032\,19 \text{ dm}^3 \text{ mol}^{-1}) - \frac{2(1.363 \text{ dm}^6 \text{ bar mol}^{-2})}{(0.083\,14 \text{ dm}^3 \text{ bar K}^{-1} \text{ mol}^{-1})(348 \text{ K})}$$

$$= -0.062\,0 \text{ dm}^3 \text{ mol}^{-1}$$

which upon substitution into (3.37) gives

$$\Delta H = \int_{T_1}^{T_2} C_P\, dT + \int_{P_1}^{P_2} (1.00 \text{ mol})(-0.062\,0 \text{ dm}^3 \text{ mol}^{-1})\, dP$$

$$= 2\,080 \text{ J} - (1.00 \text{ mol})(-0.062\,0 \text{ dm}^3 \text{ mol}^{-1})(6.60 \text{ bar} - 4.93 \text{ bar})\left(\frac{10^5 \text{ Pa}}{1 \text{ bar}}\right)\left(\frac{1 \text{ m}^3}{10^3 \text{ dm}^3}\right)\left(\frac{1 \text{ J}}{1 \text{ Pa m}^3}\right)$$

$$= 2\,070 \text{ J}$$

SPECIFIC APPLICATIONS OF THE FIRST LAW

3.18. What would be the final volume occupied by 1.00 mol of an ideal gas initially at 0 °C and 1.00 bar if $q = 1\,000$. J during a reversible isothermal expansion?

For the reversible isothermal expansion, (3.40) gives

$$q = nRT \ln \frac{V_2}{V_1}$$

Assuming $V_1 = 22.7$ dm^3 as given by (1.6), we obtain

$$1\,000. \text{ J} = (1.00 \text{ mol})(8.314 \text{ J K}^{-1} \text{ mol}^{-1})(273 \text{ K}) \ln\left(\frac{V_2}{22.7}\right)$$

$$\ln\left(\frac{V_2}{22.7}\right) = 0.441 \qquad V_2 = 35.3 \text{ dm}^3$$

3.19. For a van der Waals gas undergoing a reversible isothermal expansion, (3.40) becomes

$$\Delta U = -an^2\left(\frac{1}{V_2} - \frac{1}{V_1}\right)$$

$$\Delta H = n\left(b - \frac{2a}{RT}\right)\Delta P$$

$$w = -n\left[RT \ln\left(\frac{V_2 - nb}{V_1 - nb}\right) + an\left(\frac{1}{V_2} - \frac{1}{V_1}\right)\right]$$

$$q = nRT \ln\left(\frac{V_2 - nb}{V_1 - nb}\right)$$

Repeat the calculations of Example 3.15, and compare the results. Assume the van der Waals gas to be argon with $a = 1.363$ dm^6 bar mol^{-1}, $b = 0.032\,19$ dm^3 mol^{-1}, $V_1 = 24.7$ dm^3, and $V_2 = 0.229$ dm^3.

Substituting the values into the given equations gives

$$\Delta U = -(1.363 \text{ dm}^6 \text{ bar mol}^{-2})(2.00 \text{ mol})^2 \left(\frac{1}{0.229 \text{ dm}^3} - \frac{1}{24.7 \text{ dm}^3}\right)\left(\frac{10^5 \text{ Pa}}{1 \text{ bar}}\right)\left(\frac{1 \text{ m}^3}{10^3 \text{ dm}^3}\right)\left(\frac{1 \text{ J}}{1 \text{ Pa m}^3}\right)$$

$$= -2\,360 \text{ J}$$

$$\Delta H = 2.00 \left(0.032\,19 - \frac{2(1.363)}{(0.083\,14)(298)}\right)(100.0 - 1.00)(10^5)\left(\frac{1}{10^3}\right)$$

$$= -1\,540 \text{ J}$$

$$w = -(2.00)\left[8.314(298) \ln\left(\frac{0.229 - 2.00(0.032\,19)}{24.7 - 2.00(0.032\,19)}\right) + (1.363)(2.00)\left(\frac{1}{0.229} - \frac{1}{24.7}\right)(10^5)\left(\frac{1}{10^3}\right)\right]$$

$$= 22.6 \text{ kJ}$$

$$q = (2.00)(8.314)(298) \ln\left(\frac{0.229 - (2.00)(0.032\,19)}{24.7 - (2.00)(0.032\,19)}\right)$$

$$= -24.8 \text{ kJ}$$

The work done on the system is essentially the same for both processes. The heat transferred from the system to the surroundings was slightly larger for the van der Waals gas than for the ideal gas, which is reflected by the slight decrease that occurs in ΔU and ΔH for the real gas.

3.20. Repeat Problem 3.18 assuming an isothermal expansion against a constant pressure of 1 bar.

For the isobaric isothermal expansion, (3.41) gives $q = P(V_2 - V_1)$. Assuming $V_1 = 22.7 \text{ dm}^3$,

$$1\,000. \text{ J} = (1.00 \text{ bar})(V_2 - 22.7 \text{ dm}^3)\left(\frac{10^5 \text{ Pa}}{1 \text{ bar}}\right)\left(\frac{1 \text{ m}^3}{10^3 \text{ dm}^3}\right)\left(\frac{1 \text{ J}}{1 \text{ Pa m}^3}\right)$$

$$V_2 = 32.7 \text{ dm}^3$$

3.21. Calculate q, w, and ΔU for the conversion to steam of 1.00 mol of water at 100 °C and 1.00 atm. Pertinent data are: $\Delta H = 970.3 \text{ Btu lb}^{-1}$, 1 lb of liquid occupies 0.016 719 ft³, and 1 lb of gas occupies 26.799 ft³.

Using (3.42) gives

$$\Delta H = q = (970.3 \text{ Btu lb}^{-1})\left(\frac{1\,054.35 \text{ J}}{1 \text{ Btu}}\right)\left(\frac{1 \text{ lb}}{0.453\,59 \text{ kg}}\right)(18.015 \times 10^{-3} \text{ kg mol}^{-1})$$

$$= 40.63 \text{ kJ mol}^{-1}$$

$$w = -(1.00 \text{ atm})(26.799 \text{ ft}^3 - 0.016\,719 \text{ ft}^3)\left(\frac{28.316 \text{ dm}^3}{1 \text{ ft}^3}\right)\left(\frac{101\,325 \text{ J}}{1 \text{ atm m}^3}\right)\left(\frac{1 \text{ lb}}{0.453\,59 \text{ kg}}\right)(18.015 \times 10^{-3} \text{ kg mol}^{-1})$$

$$= -3.05 \text{ kJ mol}^{-1}$$

$$\Delta U = 40.63 + (-3.05) = 37.58 \text{ kJ mol}^{-1}$$

3.22. For a reversible adiabatic expansion of a van der Waals gas, (3.43) becomes

$$q = 0$$

$$\Delta U = w = n\left[\int_{T_1}^{T_2} C_V \, dT - an\left(\frac{1}{V_2} - \frac{1}{V_1}\right)\right]$$

$$\Delta H = \Delta U + n\left[\left(\frac{RT_2}{V_2 - nb} - \frac{an}{V_2^2}\right)V_2 - \left(\frac{RT_1}{V_1 - nb} - \frac{an}{V_1^2}\right)V_1\right]$$

where the relation between temperatures and volumes is given by

$$T_2^{C_V/R}(V_2 - nb) = T_1^{C_V/R}(V_1 - nb)$$

Repeat the calculations of Example 3.18 for 1.00 mol of argon ($a = 1.363$ dm^6 bar mol^{-1} and $b = 0.032\ 19$ dm^3 mol^{-1}), and compare results.

The final temperature is unchanged because $V - nb \approx V$. Thus,

$$q = 0$$

$$\Delta U = w = (1.00\ \text{mol})\left[13.6\ \text{kJ mol}^{-1} - (1.363\ \text{dm}^6\ \text{bar mol}^{-2})(1.00\ \text{mol})\left(\frac{1}{0.010\ 0\ \text{m}^3} - \frac{1}{0.100\ 0\ \text{m}^3}\right)\right.$$

$$\left.\times \left(\frac{1\ \text{m}^3}{10^3\ \text{dm}^3}\right)^2 \left(\frac{10^5\ \text{Pa}}{1\ \text{bar}}\right)\left(\frac{1\ \text{J}}{1\ \text{Pa m}^3}\right)\right]$$

$$= 13.6\ \text{kJ}$$

$$\Delta H = 13.6\ \text{kJ} + (1.00\ \text{mol})\left[\left(\frac{(0.083\ 14\ \text{dm}^3\ \text{bar K}^{-1}\ \text{mol}^{-1})(1\ 383\ \text{K})}{0.010\ 0\ \text{m}^3 - (1.00\ \text{mol})(3.219 \times 10^{-5}\ \text{m}^3\ \text{mol}^{-1})}\right.\right.$$

$$\left.- \frac{(1.363\ \text{dm}^6\ \text{bar mol}^{-2})(1.00\ \text{mol})}{(0.010\ 0\ \text{m}^3)^2[(10^3\ \text{dm}^3)/(1\ \text{m}^3)]}\right)(0.010\ 0\ \text{m}^3)$$

$$- \left(\frac{(0.083\ 14\ \text{dm}^3\ \text{bar K}^{-1}\ \text{mol}^{-1})(298\ \text{K})}{0.100\ 0\ \text{m}^3 - (1.00\ \text{mol})(3.219 \times 10^{-5}\ \text{m}^3\ \text{mol}^{-1})}\right.$$

$$\left.\left.- \frac{(1.363\ \text{dm}^6\ \text{bar mol}^{-2})(1.00\ \text{mol})}{(0.100\ 0\ \text{m}^3)^2[(10^3\ \text{dm}^3)/(1\ \text{m}^3)]}\right)(0.100\ 0\ \text{m}^3)\right]$$

$$\times \left(\frac{1\ \text{m}^3}{10^3\ \text{dm}^3}\right)\left(\frac{10^5\ \text{Pa}}{1\ \text{bar}}\right)\left(\frac{1\ \text{J}}{1\ \text{Pa m}^3}\right)$$

$$= 22.6\ \text{kJ}$$

At these relatively low pressures, there is very little difference between the ideal gas values and the van der Waals gas values.

3.23. The heat capacity ratio given by (3.47) can be determined for a gas using the *Clément-Désormes* method in which n_1 mol of a gas V_1, T_1, and $P_1 > P_{\text{ext}}$ is allowed to undergo a very rapid isobaric, adiabatic expansion to V_2, T_2, and $P_2 = P_{\text{ext}}$. As the expansion occurs, some of the gas escapes into the atmosphere. The remaining n_2 mol of the gas in the volume V_1 is allowed to return to T_1 and P_3. Use Eq. (3.50) to determine T_2, and use Amonton's law (see Problem 1.7) to relate P_3 to P_1 and P_2. From this last relationship, derive expressions for C_V/R and γ.

The temperature after the rapid expansion is

$$T_2 = T_1\left(\frac{C_V + P_{\text{ext}}R/P_1}{C_V + P_{\text{ext}}R/P_2}\right) = T_1\left(\frac{C_V + R(P_2/P_1)}{C_V + R}\right)$$

Amonton's law for the warming process gives

$$\frac{P_3}{T_1} = \frac{P_2}{T_2} = \frac{P_2}{\left(\dfrac{C_V + R(P_2/P_1)}{C_V + R}\right)T_1}$$

which upon rearrangement gives the final pressure in terms of the other pressures as

$$P_3 = \frac{P_2(C_V + R)}{C_V + R(P_2/P_1)}$$

Upon rearrangement,

$$\frac{C_V}{R} = \frac{P_2[1 - P_3/P_1]}{P_3 - P_2}$$

and

$$\gamma = \frac{C_P}{C_V} = \frac{C_P/R}{C_V/R} = \frac{(C_V + R)/R}{C_V/R} = \frac{C_V/R + 1}{C_V/R}$$

3.24. The Joule–Thomson coefficient for a gas can be determined from

$$\mu = \frac{V(T\alpha - 1)}{C_P} \qquad (3.53)$$

What is the value of μ for an ideal gas?

For 1 mol of an ideal gas, (3.20) gives

$$\alpha = \frac{1}{V}\left(\frac{\partial V}{\partial T}\right)_P = \frac{1}{V}\left[\frac{\partial}{\partial T}\left(\frac{RT}{P}\right)\right]_P = \frac{R}{PV} = \frac{1}{T}$$

which upon substitution into (3.53) gives

$$\mu = \frac{V[T(1/T) - 1]}{C_P} = 0$$

Supplementary Problems

INTERNAL ENERGY AND ENTHALPY

3.25. The heat of formation of FeS(α-s) is $-100.0\ \text{kJ mol}^{-1}$ at 25 °C under isobaric conditions. What is the value of ΔU? The densities of Fe, S, and FeS(α-s) at 25 °C are 7.86×10^3, 2.07×10^3, and $4.74 \times 10^3\ \text{kg m}^{-3}$, respectively. Assume all reactants and products to be at 1.00 bar.

Ans. $\Delta V = -4.05 \times 10^{-6}\ \text{m}^3$, $\Delta(PV) = -0.405\ \text{J}$, $\Delta U = -100.0\ \text{kJ mol}^{-1}$

3.26. Calculate the ratio of E(thermal) of a nonlinear polyatomic ideal gas to that of a linear polyatomic gas if only translational and rotational contributions are considered. *Ans.* 1.2

3.27. According to classical theory, each degree of freedom for a gaseous molecule has an energy of $RT/2$ associated with it (on a molar basis). Repeat the calculations of Example 3.4 using this classical approach, and compare results.

Ans. E(thermal) $= 3RT/2$, $\Delta U = 312\ \text{J mol}^{-1}$, same results for monatomic gas; E(thermal) $= 9RT/2$, $\Delta U = 935\ \text{J mol}^{-1}$, nearly 50% error for triatomic gas

3.28. Show that H(thermal) for O(g) is equal to that for H(g) at the same temperature.

Ans. H(thermal) $= \frac{5}{2}RT$ for all monatomic gases

3.29. Calculate E(thermal) and H(thermal) for CO_2(g) at 25 °C if $\bar{\nu} = 1\,342.86\ \text{cm}^{-1}$, $667.30\ \text{cm}^{-1}$ (doubly degenerate), and $2\,349.30\ \text{cm}^{-1}$ for the linear molecule.

Ans. E(thermal) $= 6\,887\ \text{J mol}^{-1}$, H(thermal) $= 9\,366\ \text{J mol}^{-1}$

3.30. Calculate E(thermal) and H(thermal) for CH_4(g) at 25 °C if $\bar{\nu} = 2\,917.0\ \text{cm}^{-1}$, $1\,533.6\ \text{cm}^{-1}$ (doubly degenerate), $3\,018.9\ \text{cm}^{-1}$ (triply degenerate), and $1\,306.2\ \text{cm}^{-1}$ (triply degenerate)

Ans. E(thermal) $= 7.540\ \text{kJ mol}^{-1}$, H(thermal) $= 10.018\ \text{kJ mol}^{-1}$

HEAT CAPACITY

3.31. The original value of the specific heat capacity of liquid water was 1.00 cal K^{-1} g^{-1}. What is the molar heat capacity of water?   *Ans.* 75.35 J K^{-1} mol^{-1}

3.32. Using $C_P = \frac{5}{2}R$ for an ideal monatomic gas, calculate ΔH for heating 2.50 mol of this gas from $25\,°C$ to $125\,°C$. Compare ΔH to the value of ΔU determined in Example 3.5.   *Ans.* $\Delta H = 5.20$ kJ, $\Delta H/\Delta U = \frac{5}{3}$

3.33. Between $100\,°C$ and $530\,°C$, $\partial^2 V/\partial T^2 = 1.90 \times 10^{-13}$ m^3 K^{-2} mol^{-1} for aluminum metal. Find the pressure needed to change C_P by 0.01 J K^{-1} mol^{-1} at 500 K, the precision to which the C_P value is known.

Ans. 1×10^8 Pa $= 1\,000$ bar

3.34. Equation (*3.19*) can be written as

$$C_P - C_V = -T\frac{(\partial V/\partial T)_P^2}{(\partial V/\partial P)_T} = -T\frac{(\partial P/\partial V)_T}{(\partial T/\partial V)_P^2}$$

for a real gas. Using (*1.22*), derive the expression for $(\partial P/\partial V)_T$ and $(\partial T/\partial V)_P$, and find $C_P - C_V$ for $(C_2H_5)_2S$ at 1.00 bar and $25\,°C$ given $a = 19.00$ dm^6 bar mol^{-2}, $b = 0.121\,4$ dm^3 mol^{-1}, and $V = 24.12$ dm^3 mol^{-1}. How different is this value from that predicted by (*3.18*)?

Ans. $(\partial P/\partial V)_T = -nRT/(V - nb)^2 + 2an^2/V^3$, $(\partial T/\partial V)_P = [P + an^2/V^2 - (2an^2/V^3)(V - nb)]/R$; $C_P - C_V = 9.493$ J K^{-1} mol^{-1},   (real gas difference)/(ideal gas difference) $= 1.142$

3.35. When working Example 3.8, a student wrote $\Delta H = a(\Delta T) + b(\Delta T)^2/2$ instead of the correct answer. How much error did the student make in the calculation of ΔH?

Ans.   Incorrect $\Delta H = 1\,590$ J mol^{-1},   15% low

3.36. When working Example 3.8, a student calculated C_P for aluminum at $63\,°C$ ($= [25\,°C + 100.\,°C]/2$) and wrote $\Delta H = C_P\,\Delta T$ instead of the correct answer. How much error did the student make in the calculation of ΔH?

Ans. 24.9 J K^{-1} mol^{-1},   incorrect $\Delta H = 1\,870$ J mol^{-1},   0.5% high

3.37. Calculate C_V and C_P for $CO_2(g)$ at $25\,°C$ using the data in Problem 3.29.

Ans. $C_V = 28.848$ J K^{-1} mol^{-1}, $C_P = 37.162$ J K^{-1} mol^{-1}

3.38. Calculate C_V and C_P for $CH_4(g)$ at $25\,°C$ using the data in Problem 3.30.

Ans. $C_V = 27.315$ J K^{-1} mol^{-1}, $C_P = 35.629$ J K^{-1} mol^{-1}

3.39. In many of the problems the vibrational contribution to E(thermal) and C_V has been neglected. Using the data in Problem 3.30, calculate E(thermal) and C_V at $25\,°C$ and $500.\,°C$, assuming the gas to be ideal. What fraction of the total contributions is the vibrational? Compare the predicted value of C_V to that calculated from $C_P - C_V = R$ if $C_P = 35.309$ J K^{-1} mol^{-1} at 298 K.

Ans. E(thermal) $= 7\,540.4$ and $26\,570.7$ J mol^{-1}, with vibrational contributions of 1.45% and 27.4% at 298 K and 773 K, respectively; $C_V = 27.311$ and 53.025 J K^{-1} mol^{-1}, with vibrational contributions of 8.67% and 53.0% at 298 K and 773 K, respectively; predicted $C_V = 26.995$ J K^{-1} mol^{-1}, which differs by 1.17% from the accepted value using (*3.18*).

3.40. The specific heat capacity of a metal is 0.46 J K^{-1} g^{-1}. What is the atomic mass of the metal?

Ans. 56 g mol^{-1}

3.41. What is the high-temperature limit of C_V for a metal predicted by (*3.26*)? What function of T at low temperatures does (*3.26*) predict that C_V will follow?

Ans. $C_V \rightarrow 3R$; C_V should follow a cubic dependence on T.

3.42. Using the low-temperature limit of (*3.26*), find an expression for the temperature at which the electronic and vibrational contributions are equal. Using $\Theta_D = 426$ K and $\eta = 1.36 \times 10^{-3}$ J K^{-2} mol^{-1}, evaluate this temperature for Al.   *Ans.* $T = (\eta \Theta_D{}^3 / 233.8\,R)^{1/2}$, 7.4 K

3.43. Prepare a plot of $C_P/3R$ against $\ln T$ for gold, given the values tabulated below and assuming that $C_P = C_V$.

$T/(K)$	$C_P/3R$	$T/(K)$	$C_P/3R$	$T/(K)$	$C_P/3R$
5	0.002	40	0.449	100	0.859
10	0.017	45	0.515	125	0.913
15	0.059	50	0.573	150	0.945
20	0.128	60	0.664	175	0.965
25	0.210	70	0.733	200	0.978
30	0.295	80	0.787	250	1.001
35	0.376	90	0.828	298	1.018

On the same graph, prepare plots of (*3.24*) and (*3.25*) using $\Theta_E = 128$ K and $\Theta_D = 165$ K. Which theoretical curve fits the experimental data better?

Ans: See Fig. 3-1;   (*3.25*) is the better fit

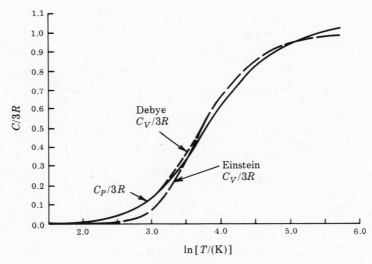

Fig. 3-1

3.44. An extension of the Dulong-Petit theory known as Kopp's rule can be used to estimate values of the heat capacity for a complex solid. If the formula of the compound contains Λ atoms, the high-temperature limit of C_V is given by

$$C_V = 3\Lambda R$$

Estimate C_V for $K_2B_8O_{13}$ at its melting point, and compare your answer to $C_P = 561$ J K^{-1} mol^{-1} assuming $C_P = C_V$.   *Ans.*   574 J K^{-1} mol^{-1},   2.3% high

3.45. Predict C_V at 25 °C for CCl_4(l) given that the intramolecular vibrational frequencies are 458, 218 (doubly degenerate), 776 (triply degenerate), and 314 cm^{-1} (triply degenerate).

Ans.   Translational = 24.9, rotational = 12.5 to 24.9, vibrational = 50.2; sum is between 87.6 and 100.0 J K^{-1} mol^{-1}

INTERNAL ENERGY, WORK, AND HEAT

3.46. If the surface tension of water is 73.05×10^{-3} N m^{-1} at 18 °C, how much of a change in area could the dry cell described in Problem 3.13 produce in 1 h?   *Ans.*   739 m^2

3.47. The potential energy of a stretched spring is given by $E_p = \frac{1}{2}kl^2$. Using (3.29), derive *Hooke's law*, which states that $F = kl$.

3.48. Repeat the calculations of Problem 3.14 if the cooler block is 1.50 kg of Ag ($c_P = 0.235$ kJ K^{-1} kg^{-1}).

   Ans.   66.3 °C, $q_h = -q_c = -14.5$ kJ

3.49. A 52.5-g sample of a yellowish metal at 99.8 °C was added to a 100.0-g sample of water at 23.2 °C. If the final temperature of the mixture was 26.7 °C, identify the metal. *Hint*: First, calculate the heat gained by the water, assuming the specific heat capacity to be 4.18×10^3 J K^{-1} kg^{-1}; second, calculate the specific heat capacity of the metal, assuming that heat was conserved in the process; third, use (3.52) to find an approximate atomic mass, assuming $C = 25.9$ J mol^{-1}; and fourth, use a periodic table.

   Ans.   1 463 J; $c = 381$ J kg^{-1} K^{-1}; $M = 0.068\ 0$ kg mol^{-1};   yellow color indicates Cu

3.50. To illustrate the path dependence of w and q, consider the initial state of a system to be an ideal gas confined to one-half of a container and the final state to be the gas confined to the entire container. Qualitatively discuss the values of q and w if the process is performed (a) irreversibly, similar to the Joule-Thomson experiment; (b) reversibly, using a piston to change the external pressure as needed.

   Ans.   (a) $q = w = 0$,   (b) $q = -w > 0$

3.51. A system receives 75 J of electrical work, delivers 274 J of expansion work, and absorbs 168 J of heat. What is the internal energy change for the system?   *Ans.* $\Delta U = -31$ J

3.52. Show that $\alpha = 1/T$ and $\kappa = 1/P$ for an ideal gas. Substitute these results into (3.35) and (3.38) to derive (3.36a) and (3.39a).

3.53. For a van der Waals gas, $\mu = (2a/RT - b)/C_p$. Derive an equation for $(\partial U/\partial V)_T$, and evaluate the volume term in (3.34).   *Ans.* $(\partial U/\partial V)_T = an^2/V^2$, $\int (\partial U/\partial V)\, dV = -an^2(1/V_2 - 1/V_1)$

3.54. Calculate ΔU for heating 1.000 mol of argon from 25.0 °C to 125.0 °C while the volume changes from 5.00 dm^3 to 6.35 dm^3 assuming ideal gas behavior and assuming argon to be a van der Waals gas ($C_V = 12.472$ J K^{-1} mol^{-1}, and $a = 1.363$ dm^6 bar mol^{-2}).   *Ans.*   1 247 J,   1 253 J

SPECIFIC APPLICATIONS OF THE FIRST LAW

3.55. Five moles of a diatomic ideal gas is allowed to expand isothermally at 25 °C from 0.020 0 to 0.100 0 m^3. Calculate q, w, ΔU, and ΔH if the expansion is performed (a) reversibly, and (b) isobarically against a constant pressure of 1.00 bar.

   Ans.   (a) $\Delta U = \Delta H = 0$, $q = -w = 19.94$ kJ;   (b) $\Delta U = \Delta H = 0$, $q = -w = 8.00$ kJ

3.56. Calculate w for the expansion described in Problem 3.55 if it is performed isobarically in four steps: (1) against 4.00 bar until the volume is 0.025 0 m^3, (2) 3.00 bar until 0.030 0 m^3, (3) 2.00 bar until 0.050 0 m^3, and (4) 1.00 bar until 0.100 0 m^3. Comment on the values of w for the processes described.

   Ans. $w = -12.50$ kJ; the reversible process described in Problem 3.55(a) generates the maximum amount of work, the pseudo-reversible process generates the secondmost amount of work, and the isobaric expansion described in Problem 3.55(b) generates the least amount of work.

3.57. Consider the reversible isothermal expansion of 1.00 mol of steam from 1.00 dm^3 to 10.00 dm^3 at 500 °C. Calculate the work assuming (a) the gas to be ideal, (b) the gas to obey the van der Waals equation with $a = 5.536$ dm^6 bar mol^{-2} and $b = 0.030\ 49$ dm^3 mol^{-1}, and (c) the gas to obey (1.17). A graphical integration of $(Z/V)\,dV$ for the third case will be necessary because the value of Z is a function of pressure. For simplicity, assume that the pressure of the steam for a given volume can be determined from the ideal gas law and that the corresponding value of Z is correct. The critical data for water are 374.1 °C and 221.2 bar.   *Ans.* (a) 14.80 kJ, (b) 14.48 kJ, (c) 14.49 kJ

3.58. Derive equations describing the reversible isothermal expansion of a solid.

Ans. $w = \Delta U = -\int P\,dV,\quad \Delta H = \int V\,dP,\quad q = 0$

3.59. Calculate q, w, and ΔU at 1.00 bar for the phase transition

$$Li_2SO_4(s\text{-}I) \rightarrow Li_2SO_4(l)$$

which occurs at 1 132 K. The enthalpy change is 7.5 kJ for this reaction. The densities of $Li_2SO_4(s\text{-}I)$ and $Li_2SO_4(l)$ are 2.07×10^3 kg m^{-3} and 2.004×10^3 kg m^{-3}, respectively.

Ans. $q = \Delta H = 7.5$ kJ, $w = -0.17$ J, $\Delta U = 7.5$ kJ

3.60. Repeat Example 3.18 for a diatomic ideal gas, excluding vibrational contributions.

Ans. $C_V = \frac{5}{2}R,\quad T_2 = 749$ K, $q = 0,\quad \Delta U = w = 9.37$ kJ, $\Delta H = 13.12$ kJ

3.61. Repeat Example 3.19 for a diatomic ideal gas, excluding vibrational contributions.

Ans. $C_V = \frac{5}{2}R,\quad T_2 = 255$ K, $q = 0,\quad w = \Delta U = -890$ J, $\Delta H = -1\,250$ J

3.62. Determine the heat capacity ratio for N_2 at room temperature using the following pressure data obtained from the Clément-Désormes method (see Problem 3.23): $P_1 = 807.6$ torr, $P_2 = 752.7$ torr, and $P_3 = 768.8$ torr.

Ans. $C_V/R = 2.29,\quad \gamma = 1.44$

3.63. Calculate μ for air at 0 °C using the following averaged parameters: $\bar{a} = 1.380$ dm^6 bar mol^{-2}, $\bar{b} = 0.037\,7$ dm^3 mol^{-1}, and $\bar{C}_p = 29.171$ J K^{-1} mol^{-1} at 1.00 bar. Compare this prediction to the values given in Example 3.20.

Ans.   0.288 K bar^{-1}; calculated value is within 10% of the experimental value at 1 bar.

3.64. At temperatures above the inversion temperature, a gas heats upon expansion. What must be done to produce liquid H_2 by expansion cooling if the inversion temperature is −80 °C?

Ans.   The gas must be cooled below this temperature before cooling will occur upon expansion.

Chapter 4

Thermochemistry

Heat of Reaction

4.1 INTRODUCTION

The *heat of reaction* is the value of ΔH or ΔU that accompanies the isothermal chemical reaction

$$\text{Reactants at } T \longrightarrow \text{products at } T$$

when carried out under constant-pressure or constant-volume conditions, respectively. Reactions with negative heats are known as *exothermic* and those with positive values are known as *endothermic*. In many cases, exothermic reactions will occur spontaneously while endothermic reactions will not. However, heat exchange is not the sole criterion for spontaneity (see Section 6.1). The value given for ΔU or ΔH beside a reaction represents the energy change for the reaction as written, and the units are simply the units of energy, e.g., kJ. Thermochemical data without an accompanying reaction usually apply to a mole of the substance in question, and the units are those of energy per mole, e.g., $kJ\,mol^{-1}$.

A complete thermochemical equation includes not only the stoichiometric and energy information, but also a description of the physical states of the substances involved. For example, (c) or (s) is used to represent the solid state, and if more information is necessary, symbols such as (α-s), (β), (rhombic), (s-II), (dia), and (graph) are used to indicate which of the several possible solids is involved. Other notation includes (l) for liquid, (g) for gas, ($c = 1$) or (1 M) for a 1-molar solution, ($m = 2.5$) for a 2.5-molal solution, and (aq) for a very dilute aqueous solution. Throughout this book molarity will be expressed in $mol\,dm^{-3}$ and molality in $mol\,kg^{-1}$. A superscript $°$ indicates that the reaction was performed under *standard pressure conditions* (1 bar).

A following subscript on U or H is used to give the absolute temperature. If no temperature is given, $T = 298.15$ K. A leading subscript r on U or H is used to identify a general heat of reaction, and certain other subscripts are used (f for formation, c for combustion, mix for mixing, etc.) for special types of reactions.

EXAMPLE 4.1. Use the thermochemical equation

$$2H_2S(g) + 3O_2(g) \longrightarrow 2H_2O(l) + 2SO_2(g) \qquad \Delta_r H^{\circ}_{298} = -1\,124.06 \text{ kJ}$$

to determine the enthalpy change for the combustion of 1.00 g of H_2S.

The enthalpy of reaction for 1.00 g of H_2S is found by interpreting the thermochemical equation on a molar basis:

$$(1.00 \text{ g } H_2S)\left(\frac{1 \text{ mol } H_2S}{34.08 \text{ g } H_2S}\right)\left(\frac{-1\,124.06 \text{ kJ}}{2 \text{ mol } H_2S}\right) = -16.5 \text{ kJ}$$

EXAMPLE 4.2. To a first approximation, a flame can be considered to be an adiabatic, isobaric process in which the heat of reaction is used to heat the product gases to the flame temperature. What would be the maximum temperature of a hydrogen-air flame?

For the reaction

$$H_2(g) + \tfrac{1}{2}O_2(g) \longrightarrow H_2O(g)$$

there is 241.82 kJ released at 25 °C for each mole of H_2 burned. In the gaseous mixture of products, for every mole of $H_2O(g)$ there are 2 mol of $N_2(g)$ from the air mixture, giving the heat capacity (in $J\ K^{-1}\ mol^{-1}$) as

$$C_P^\circ = C_P^\circ(H_2O, g) + 2C_P^\circ(N_2, g)$$

$$= (30.36 + 9.615 \times 10^{-3} T + 11.8 \times 10^{-7} T^2) + 2(27.30 + 5.230 \times 10^{-3} T - 0.04 \times 10^{-7} T^2)$$

$$= 84.96 + 2.007\ 5 \times 10^{-2} T + 1.17 \times 10^{-6} T^2$$

The ΔH° of reaction would be equal to the heat released by the cooling of the product gases from the flame temperature T_f to 25 °C; so

$$\Delta_r H^\circ = \int_{T_f}^{298\ K} C_P^\circ\, dT$$

$$-241\ 820 = 84.96(298 - T_f) + \tfrac{1}{2}(2.007\ 5 \times 10^{-2})(298^2 - T_f^2) + \tfrac{1}{3}(1.17 \times 10^{-6})(298^3 - T_f^3)$$

Rearranging gives

$$3.90 \times 10^{-7} T_f^3 + 1.003\ 8 \times 10^{-2} T_f^2 + 84.96 T_f - 268\ 040 = 0$$

and solving by (1.23) gives $T_f = 2\ 410$ K or 2 140 °C. The actual temperature would be somewhat lower because the system is not truly adiabatic and because of incomplete combustion. Commercial torches are capable of reaching temperatures near 2 000 °C.

EXAMPLE 4.3.   What would be ΔH for the combustion of benzoic acid at 25 °C given $\Delta U^\circ = -6\ 316$ cal g^{-1}?

For 1 mol of benzoic acid,

$$\Delta_r U^\circ = (-6\ 316\ \text{cal g}^{-1})[(122.13\ \text{g})/(1\ \text{mol})][(4.184\ \text{J})/(1\ \text{cal})] = -3.227\ \text{MJ mol}^{-1}$$

As demonstrated in Problem 3.1 the term $\Delta(PV)$ in (3.3) can be neglected for condensed phases, and for gases it can be replaced by $RT\ \Delta n_g$, giving

$$\Delta_r H = \Delta_r U + RT\ \Delta n_g \tag{4.1}$$

where Δn_g is determined using the coefficients of the gaseous substances only (see Example 3.2).

Thus, for the reaction

$$C_6H_5COOH(s) + \tfrac{15}{2}O_2(g) \longrightarrow 7CO_2(g) + 3H_2O(l) \qquad \Delta_r U_{298} = -3.227\ \text{MJ}$$

the value of Δn_g is $7 - \frac{15}{2} = -\frac{1}{2}$ mol. Substituting into (3.3) gives

$$\Delta_c H_{298}^\circ = (-3.227\ \text{MJ}) + (8.314\ \text{J K}^{-1}\ \text{mol}^{-1})(298\ \text{K})(-\tfrac{1}{2}\ \text{mol})[(10^{-6}\ \text{MJ})/(1\text{J})] = -3.228\ \text{MJ}$$

for the reaction, or $\Delta_c H_{298}^\circ \doteq -3.228$ MJ mol^{-1} for benzoic acid.

4.2   TEMPERATURE DEPENDENCE OF THE HEAT OF REACTION

The temperature dependence of the heat of reaction is given by

$$d(\Delta_r H) = (\Delta_r C_P)\, dT \tag{4.2a}$$

$$d(\Delta_r U) = (\Delta_r C_V)\, dT \tag{4.2b}$$

where $\Delta_r C_i$ represents the sum of the heat capacities of the products less the sum of the heat capacities of the reactants.

If the heat capacity data are given in the form (3.22) and $\Delta_r H$ is known at 25 °C, then

$$\Delta_r H_T = \Delta_r H_{298} + (\Delta a)(T - 298) + \tfrac{1}{2}(\Delta b)(T^2 - 298^2)$$

$$+ \tfrac{1}{3}(\Delta c)(T^3 - 298^3) + \tfrac{1}{4}(\Delta d)(T^4 - 298^4) \tag{4.3a}$$

$$\Delta_r H_T = \Delta_r H_{298} + (\Delta a)(T - 298) + \tfrac{1}{2}(\Delta b)(T^2 - 298^2) - (\Delta c')(T^{-1} - 298^{-1}) \tag{4.3b}$$

If J and J' are defined as

$$J \equiv \Delta_r H_{298} - (\Delta a)(298) - \tfrac{1}{2}(\Delta b)(298^2) - \tfrac{1}{3}(\Delta c)(298^3) - \tfrac{1}{4}(\Delta d)(298^4) \qquad (4.4a)$$

$$J' \equiv \Delta_r H_{298} - (\Delta a)(298) - \tfrac{1}{2}(\Delta b)(298^2) + (\Delta c')(298^{-1}) \qquad (4.4b)$$

then ($4.3a$) and ($4.3b$) become

$$\Delta_r H_T = J + (\Delta a)T + \tfrac{1}{2}(\Delta b)T^2 + \tfrac{1}{3}(\Delta c)T^3 + \tfrac{1}{4}(\Delta d)T^4 \qquad (4.5a)$$

$$\Delta_r H_T = J' + (\Delta a)T + \tfrac{1}{2}(\Delta b)T^2 - (\Delta c')T^{-1} \qquad (4.5b)$$

The above equations are valid only for the temperature regions over which the heat capacity is correctly predicted by (3.22).

EXAMPLE 4.4. The enthalpy of combustion of benzoic acid has been selected by IUPAC as the standard for calibrating calorimeters. Given that the value of $\Delta_c H^\circ_{293}$ is -771.2 kcal mol^{-1}, what is the value at 25 °C? For this small temperature interval, assume that $C^\circ_P/(\text{J K}^{-1}\,\text{mol}^{-1})$ is temperature-independent and equal to 37.1, 75.3, and 29.4 for $CO_2(g)$, $H_2O(l)$ and $O_2(g)$, respectively, and equal to 1.20 J K^{-1} g^{-1} for $C_6H_5COOH(s)$.

For the reaction written in Example 4.3, $\Delta_r C^\circ_P$ is given by

$$\Delta_r C^\circ_P = \Delta a = [7(37.1) + 3(75.3)] - [(1.20)(122.13) + \tfrac{15}{2}(29.4)] = 118.5 \text{ J K}^{-1}$$

Substituting into (4.3) gives

$$(-771.2 \text{ kcal})[(4.184 \text{ kJ})/(1 \text{ kcal})] = \Delta_c H^\circ_{298} + (118.5 \text{ J K}^{-1})(293 \text{ K} - 298 \text{ K})$$

$$\Delta_c H^\circ_{298} = (-3\,226.7 + 0.6) \text{ kJ} = -3.226\,1 \text{ MJ}$$

for the reaction, or $\Delta_c H^\circ_{298} = -3.226\,1$ MJ mol^{-1}.

4.3 CALORIMETRY

Measurements of $\Delta_r H$ are usually performed under constant-pressure conditions using a solution calorimeter such as an insulated flask or beaker, a thermos bottle or Dewar flask, a styrofoam cup, etc. Measurements of $\Delta_r U$ are usually performed under constant-volume conditions using a bomb calorimeter.

Because $\Delta_r H$ and $\Delta_r U$ are state functions under conditions of constant pressure or volume, any path can be chosen to measure their values. Consider a two-step path consisting of an adiabatic process

(Reactants & calorimeter) at T = (products & calorimeter) at T'

and a second process in which the products and calorimeter are restored to the original temperature:

(Products & calorimeter) at T' = (products & calorimeter) at T

The sum of the two processes generates the definition of the heat of reaction as given in Sec. 4.1, and $\Delta_r H$ (or $\Delta_r U$) equals the sum of the $\Delta_r H$ (or $\Delta_r U$) values for the two processes. This particular two-step process is convenient because $\Delta_r H$ (or $\Delta_r U$) for the first step is zero and $\Delta_r H$ (or $\Delta_r U$) for the second step can be easily measured or else calculated as

$$\Delta_r H = \int_{T'}^{T} C_P(\text{products \& calorimeter}) \, dT \qquad (4.6a)$$

$$\Delta_r U = \int_{T'}^{T} C_V(\text{products \& calorimeter}) \, dT \qquad (4.6b)$$

In many cases, calorimeters are designed such that $C_i(\text{products \& calorimeter})$ is equal to $C_i(\text{calorimeter})$, and $C_i(\text{calorimeter})$ is constant over reasonable temperature changes.

EXAMPLE 4.5.   For the reaction A $\longrightarrow$ B, carried out under isobaric conditions, a temperature change of $-2.7\,°C$ was observed for the calorimeter and products. To determine C_P(products & calorimeter), the electric heating circuit shown in Fig. 4-1 was used. If $\Delta V = 1.09$ V, $R_{cal} = 100.0\,\Omega$, and $R_{ref} = 10.0\,\Omega$ for the circuit, and if the calorimeter and products were heated $1.00\,°C$ by the heater in 60 s, find $\Delta_r H$.

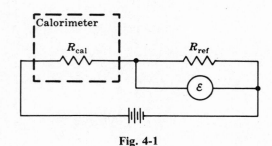

Fig. 4-1

The electrical work for the circuit is

$$w = \frac{R_{cal}}{R_{ref}^2} \int \Delta V \, dt = \frac{100.0}{10.0^2}(1.09)^2(60.0) = 71.3 \text{ V}^2 \text{ s } \Omega^{-1} = 71.3 \text{ J}$$

The heat capacity of the system is given by

$$C_P(\text{products \& calorimeter}) = \frac{w}{\Delta T} = \frac{71.3 \text{ J}}{1.00 \text{ K}} = 71.3 \text{ J K}^{-1}$$

Using (4.6a) gives

$$\Delta_r H = \int (71.3 \text{ J K}^{-1}) \, dT = (71.3 \text{ J K}^{-1})(2.7 \text{ K}) = 192 \text{ J}$$

Calculations Involving Thermochemical Equations

4.4 LAW OF HESS

The *law of Hess* states that the heat of reaction for a desired equation can be calculated by the algebraic combination of other thermochemical equations and their known heats of reaction. Because the heat of reaction is a state function, the value calculated using the path consisting of the chosen series of thermochemical equations will be valid as long as the series of equations upon summation correctly predicts the reactants and products of the desired equation.

EXAMPLE 4.6.   Combine the following thermochemical reactions

$$S(\text{rhombic}) + O_2(g) \longrightarrow SO_2(g) \qquad\qquad \Delta_r H_{298}^{\circ} = -296.830 \text{ kJ}$$

$$S(\text{monoclinic}) + O_2(g) \longrightarrow SO_2(g) \qquad\qquad \Delta_r H_{298}^{\circ} = -297.16 \text{ kJ}$$

to predict $\Delta_r H_{298}^{\circ}$ for

$$S(\text{rhombic}) \longrightarrow S(\text{monoclinic})$$

Subtracting the second equation from the first gives

$$S(\text{rhombic}) + O_2(g) \longrightarrow SO_2(g) \qquad\qquad \Delta_r H_{298}^{\circ} = -296.830 \text{ kJ}$$

$$SO_2(g) \longrightarrow S(\text{monoclinic}) + O_2(g) \qquad\qquad \Delta_r H_{298}^{\circ} = 297.16 \ \ \text{ kJ}$$

$$S(\text{rhombic}) \longrightarrow S(\text{monoclinic}) \qquad\qquad \Delta_r H_{298}^{\circ} = 0.33 \ \ \text{ kJ}$$

4.5 ENTHALPY OF FORMATION

The enthalpy of reaction for the production of 1 mol of a compound from the elements in their naturally occurring physical states is known as the *enthalpy of formation* ($\Delta_f H_T^\circ$) for the compound. The enthalpy of formation for an element in its naturally occurring physical state is zero.

If all thermochemical data to be used in predicting an enthalpy of reaction for a desired equation are enthalpies of formation, the law of Hess can be expressed as

$$\Delta_r H_T = \sum_i \nu_i \Delta_f H_{T,i}^\circ \tag{4.7}$$

where ν_i, the stoichiometric coefficients in the desired equation (ν_i is positive for products and negative for reactants), have the units of mol.

EXAMPLE 4.7. One of the first steps in the refining of sulfide ores is the process of roasting, in which the ore is heated with oxygen to form the metal oxide and $SO_2(g)$. Calculate $\Delta_r H_{298}^\circ$ for the roasting of sphalerite (ZnS), given that $\Delta_f H_{298}^\circ/(\text{kJ mol}^{-1})$ of sphalerite, ZnO(s), and $SO_2(g)$ are -205.98, -348.28, and -296.30, respectively.

The desired equation is

$$ZnS(\text{sphalerite}) + \tfrac{3}{2}O_2(g) \longrightarrow ZnO(s) + SO_2(g)$$

and substituting the data into (4.7) gives

$$\Delta_r H_{298}^\circ = \{(1\text{ mol})[\Delta_f H_{298}^\circ(\text{ZnO})] + (1\text{ mol})[\Delta_f H_{298}^\circ(\text{SO}_2)]\} - \{(1\text{ mol})[\Delta_f H_{298}^\circ(\text{ZnS})] + (\tfrac{3}{2}\text{ mol})[\Delta_f H_{298}^\circ(\text{O}_2)]\}$$

$$= [(1\text{ mol})(-348.28\text{ kJ mol}^{-1}) + (1\text{ mol})(-296.830\text{ kJ mol}^{-1})] - [(1\text{ mol})(-205.98\text{ kJ mol}^{-1}) + (\tfrac{3}{2}\text{ mol})(0)]$$

$$= -439.13\text{ kJ}$$

4.6 ENTHALPY OF COMBUSTION

The enthalpy of reaction for the oxidation of 1 mol of a compound is known as the *enthalpy of combustion* ($\Delta_c H_T^\circ$). If the substance contains C, H, O, and/or N, the products of the oxidation at 25 °C are $CO_2(g)$, $H_2O(l)$, and/or $N_2(g)$. Writing balanced equations for substances containing other elements, such as the halogens or sulfur, is difficult because a mixture of products usually occurs.

Exothermic values of $\Delta_c H_T^\circ$ are often reported in tables as positive values, and so it is necessary to change signs of these data to the usual sign convention for use in thermochemical equations.

If all thermochemical data to be used in predicting an enthalpy of reaction for a desired equation are enthalpies of combustion, the law of Hess can be expressed as

$$\Delta_r H_T^\circ = -\sum_i \nu_i \Delta_c H_{T,i}^\circ \tag{4.8}$$

where ν_i, the stoichiometric coefficients in the desired equation (ν_i is positive for products and negative for reactants), have the units of mol.

EXAMPLE 4.8. Predict $\Delta_r H_{298}^\circ$ for

$$n\text{-}C_6H_{14}(l) + CH_4(g) \longrightarrow n\text{-}C_7H_{16}(l) + H_2(g)$$

given $\Delta_c H_{298}^\circ/(\text{kJ mol}^{-1}) = -4\,141.3, -882.0, -4\,811.2$, and -286.1 for $n\text{-}C_6H_{14}$, CH_4, $n\text{-}C_7H_{16}$, and H_2, respectively.

Applying (4.8) to the desired reaction gives

$$\Delta_r H_{298}^\circ = -\{(1\text{ mol})[\Delta_c H_{298}^\circ(n-C_7H_{16})] + (1\text{ mol})[\Delta_c H_{298}^\circ(H_2)]\}$$

$$+ \{(1\text{ mol})[\Delta_c H_{298}^\circ(n-C_6H_{14})] + (1\text{ mol})[\Delta_c H_{298}^\circ(CH_4)]\}$$

$$= -[(1\text{ mol})(-4\,811.2\text{ kJ mol}^{-1}) + (1\text{ mol})(-286.1\text{ kJ mol}^{-1})]$$

$$+ [(1\text{ mol})(-4\,141.3\text{ kJ mol}^{-1}) + (1\text{ mol})(-882.0\text{ kJ mol}^{-1})]$$

$$= 74.0\text{ kJ}$$

4.7 BOND ENTHALPY

The enthalpy of reaction resulting from the breaking of a chemical bond in a gaseous molecule to form the respective gaseous molecular fragments is known as the *bond (dissociation) enthalpy*. The chemical environment of a given atom will influence the value of the bond enthalpy, so values found in tables usually represent averages over several compounds.

Bond enthalpies can be used to calculate the enthalpies of reaction by assuming the reaction to consist of two steps: (1) the decomposition of the reactants into molecular fragments and (2) the formation of the products from the fragments. Thus,

$$\Delta_r H^\circ_T = - \sum_i^{products} n_i \overline{BE}_i + \sum_j^{reactants} n_j \overline{BE}_j \qquad (4.9)$$

where $\overline{BE}$ represents the average bond enthalpy, and n_i and n_j represent the number of moles of bonds involved in the reaction. Even though bond enthalpies pertain to gaseous reactions, they are often used to predict enthalpies of reaction for condensed reactions by combining enthalpies of vaporization, sublimation, etc., or without further corrections if a cruder approximation will suffice or if the vaporization, etc., data are unknown.

Table 4-1 $\overline{BE}/(\text{kJ mol}^{-1})$

	C	N	S	O	I	Br	Cl	F	H
H—	414	389	368	464	297	368	431	569	435
F—	490	280	343	213	280	285	255	159	
Cl—	326	201	272	205	209	218	243		
Br—	272	163	209	—	176	192			
I—	218	—	—	—	151				
O—	326	230	423	142					
O=	803[a]	590[b]	523	498					
O≡	1075	—	—	—					
S—	289	—	247						
S=	582	—	—						
N—	285	159							
N=	515	473							
N≡	858	946							
C—	331								
C=	590[c]								
C≡	812								

[a] 728 if $-\overset{|}{C}=O$.

[b] 406 if $-NO_2$; 368 if $-NO_3$.

[c] 506 if alternating — and =.

EXAMPLE 4.9. The stepwise decomposition of $NH_3(g)$ and the enthalpies of reaction are given below. From these data, calculate the average bond enthalpy $\overline{BE}(N-H)$.

$$NH_3(g) \longrightarrow NH_2(g) + H(g) \qquad \Delta_r H^\circ_{298} = 449.0 \text{ kJ}$$

$$NH_2(g) \longrightarrow NH(g) + H(g) \qquad \Delta_r H^\circ_{298} = 384.6 \text{ kJ}$$

$$NH(g) \longrightarrow N(g) + H(g) \qquad \Delta_r H^\circ_{298} = 339.2 \text{ kJ}$$

Adding the three reactions and enthalpies gives

$$NH_3(g) \longrightarrow N(g) + 3H(g) \qquad \Delta_r H_{298}^\circ = 1\,172.8 \text{ kJ}$$

so the average value for one bond would be

$$\overline{BE}(N-H) = \frac{\Delta_r H_{298}^\circ}{3 \text{ mol}} = 390.9 \text{ kJ mol}^{-1}$$

EXAMPLE 4.10. Predict the enthalpy of reaction for the oxidation of $CH_3OH(l)$ using the average bond enthalpy data given in Table 4-1. The enthalpy of vaporization at 25 °C is 37.99 kJ mol^{-1} for CH_3OH and 44.011 kJ mol^{-1} for H_2O.

The desired reaction is

$$CH_3OH(l) + \tfrac{3}{2}O_2(g) \longrightarrow CO_2(g) + 2H_2O(l)$$

which can be obtained from the following reactions:

$$CH_3OH(l) \longrightarrow CH_3OH(g) \qquad \Delta_{vap} H_{298}^\circ = 37.99 \text{ kJ}$$

$$H_2O(l) \longrightarrow H_2O(g) \qquad \Delta_{vap} H_{298}^\circ = 44.011 \text{ kJ}$$

$$CH_3OH(g) + \tfrac{3}{2}O_2(g) \longrightarrow CO_2(g) + 2H_2O(g)$$

where $\Delta_r H_{298}^\circ$ for the third reaction is calculated from (4.9) as

$$\Delta_r H_{298}^\circ = -[(2 \text{ mol})\overline{BE}(C{=}O) + (4 \text{ mol})\overline{BE}(O{-}H)] + [(3 \text{ mol})\overline{BE}(C{-}H)$$

$$+ (1 \text{ mol})\overline{BE}(C{-}O) + (1 \text{ mol})\overline{BE}(O{-}H) + (\tfrac{3}{2} \text{ mol})\overline{BE}(O{=}O)]$$

$$= -[(2 \text{ mol})(803 \text{ kJ mol}^{-1}) + (4 \text{ mol})(464 \text{ kJ mol}^{-1})]$$

$$+ [(3 \text{ mol})(414 \text{ kJ mol}^{-1}) + (1 \text{ mol})(326 \text{ kJ mol}^{-1})$$

$$+ (1 \text{ mol})(464 \text{ kJ mol}^{-1}) + (\tfrac{3}{2} \text{ mol})(498 \text{ kJ mol}^{-1})]$$

$$= -683 \text{ kJ}$$

Applying the law of Hess, we obtain

$$\Delta_c H_{298}^\circ = (-683) - 2(44.011) + (37.99) = -733 \text{ kJ}$$

This value differs from the actual value at 25 °C, -726.47 kJ, by 0.9%.

4.8 THERMAL ENERGIES AND ENTHALPIES

The change in thermal enthalpy or thermal energy for a chemical reaction can be calculated by

$$\Delta_r H^\circ(\text{thermal}) = \sum_i \nu_i H^\circ(\text{thermal}, i) \qquad (4.10a)$$

$$\Delta_r E^\circ(\text{thermal}) = \sum_i \nu_i E^\circ(\text{thermal}, i) \qquad (4.10b)$$

where ν_i are the stoichiometric coefficients of the balanced equation (ν_i is positive for products and negative for reactants) and have the units of mol. The total change in enthalpy or internal energy for a reaction is given by

$$\Delta_r H_T^\circ = \Delta_r U_0^\circ + \Delta_r H^\circ(\text{thermal}) \qquad (4.11a)$$

$$\Delta_r U_T^\circ = \Delta_r U_0^\circ + \Delta_r E^\circ(\text{thermal}) \qquad (4.11b)$$

EXAMPLE 4.11. Calculate $\Delta_r H_{373.12}^\circ$ for

$$2H(g) + O(g) \longrightarrow H_2O(g)$$

given $\Delta_r H_{298}^\circ/(\text{kJ mol}^{-1}) = -926.918$. The values of $H^\circ(\text{thermal})/(\text{kJ mol}^{-1})$ at 298 K and 373.12 K are 9.924 and 12.449, respectively, for $H_2O(g)$ and 6.194 and 7.756, respectively, for both $H(g)$ and $O(g)$.

For the reaction, (4.10a) gives

$$\Delta_r H_{298}^\circ(\text{thermal}) = [(1)(9.924)] - [(2)(6.194) + (1)(6.194)] = 8.658 \text{ kJ}$$

The value of $\Delta_r U_0^\circ$ is determined by rearranging (4.11a) and substituting values of $\Delta_r H_{298}^\circ$ and $\Delta_r H_{298}^\circ(\text{thermal})$, giving

$$\Delta_r U_0^\circ = (-926.918) - (-8.658) = -918.260 \text{ kJ}$$

At the higher temperature, (4.10a) gives

$$\Delta_r H_{373.12}^\circ(\text{thermal}) = [(1)(12.449)] - [(2)(7.756) + (1)(7.756)] = -10.819 \text{ kJ}$$

and (4.11a) gives

$$\Delta_r H_{373.12}^\circ = (-918.260) + (-10.819) = -929.079 \text{ kJ}$$

Applications to Selected Chemical Reactions

4.9 IONIZATION ENERGY AND ELECTRON AFFINITY

The *ionization energy* (or *potential*) of an atom (I) is the enthalpy of reaction for

$$X(g) \longrightarrow X^+(g) + e^- \qquad I_X = \Delta_r H^\circ$$

and the *electron affinity* of an atom (E_A) is the enthalpy of reaction for

$$X(g) + e^- \longrightarrow X^-(g) \qquad E_A(X) = \Delta_r H^\circ$$

The values of I and E_A cited in many references are often values of $\Delta_r U^\circ$ and have to be converted for use with enthalpies. Often values are given at 0 K and 6.197 kJ mol^{-1} must be added to change these values to 298 K. Values given in units of eV are converted to kJ mol^{-1} using the conversion factor given in Table 3-1 and Avogadro's constant.

EXAMPLE 4.12. The *second ionization energy* of an element corresponds to the thermochemical equation

$$X^+(g) \longrightarrow X^{2+}(g) + e^-$$

Given $\Delta_f H_{298}^\circ/(\text{kJ mol}^{-1}) = 147.70$ for $Mg(g)$, 891.635 for $Mg^+(g)$, and 2 348.504 for $Mg^{2+}(g)$, calculate $I(Mg)$ and $I(Mg^+)$. Why is the second ionization energy larger than the first?

Using (4.7) for both reactions gives

$$I(Mg) = [(1 \text{ mol})(891.635 \text{ kJ mol}^{-1})] - [(1 \text{ mol})(147.70 \text{ kJ mol}^{-1})]$$

$$= 743.94 \text{ kJ}$$

$$I(Mg^+) = [(1)(2\,348.504)] - [(1)(891.635)] = 1\,456.869 \text{ kJ}$$

The second ionization energy is greater by about 700 kJ because of the additional energy needed to remove a negatively charged electron from a doubly positively charged ion.

4.10 LATTICE ENERGY

The *lattice energy* of an ionic compound, ($\Delta_{\text{lat}} H_T^\circ$), is the value of $\Delta_r H$ for

$$M_a X_b(s) \longrightarrow a M^{b+}(g) + b X^{a-}(g)$$

EXAMPLE 4.13. Calculate the lattice energy for NaCl using $\Delta_f H_{298}^\circ/(\text{kJ mol}^{-1}) = -233.13$ for $Cl^-(g)$, 609.358 for $Na^+(g)$, and -411.153 for $NaCl(s)$.

Using (4.7) for the reaction

$$NaCl(s) \longrightarrow Na^+(g) + Cl^-(g)$$

gives

$$\Delta_{lat}H^\circ_{298} = [(1\ mol)(609.358\ kJ\ mol^{-1}) + (1\ mol)(-233.13\ kJ\ mol^{-1})] - [(1\ mol)(-411.153\ kJ\ mol^{-1})]$$

$$= 787.38\ kJ$$

4.11 ENTHALPY OF NEUTRALIZATION

The enthalpy of reaction for the neutralization of an acid by a base is known as the *enthalpy of neutralization* ($\Delta_{neut}H^\circ_T$).

EXAMPLE 4.14. For the reaction between a strong acid and a strong base, i.e., those that are essentially 100% ionized or dissociated, the neutralization equation is essentially

$$H^+(aq) + OH^-(aq) \longrightarrow H_2O(l)$$

Given $\Delta_f H^\circ_{298}/(kJ\ mol^{-1}) = -285.830$, 0, and -229.994 for H_2O, H^+, and OH^-, respectively, calculate the enthalpy of neutralization for strong acids with strong bases.

Using (4.7) gives

$$\Delta_{neut}H^\circ_{298} = [(1)(-285.830)] - [(1)(0) + (1)(-229.994)] = -55.836\ kJ$$

EXAMPLE 4.15. Consider the titration of a weak acid, such as HCN, with a strong base, such as NaOH:

$$HCN(aq) + NaOH(aq) \longrightarrow NaCN(aq) + H_2O(l)$$

Given $\Delta_f H^\circ_{298}/(kJ\ mol^{-1}) = -285.830$, -89.5, -470.114, and 107.1 for H_2O, $NaCN$, $NaOH$, and HCN, respectively, calculate $\Delta_{neut}H^\circ_{298}$. Account for the difference from $-55.836\ kJ\ mol^{-1}$ as found for the strong acid-strong base case.

Substituting the data into (4.7) gives

$$\Delta_{neut}H^\circ_{298} = [(1)(-285.830) + (1)(-89.5)] - [(1)(-470.114) + (1)(107.1)] = 12.3\ kJ$$

Because HCN is a weak acid, the titration reaction may be considered to be the sum of two steps:

$$HCN(aq) \longrightarrow H^+(aq) + CN^-(aq) \qquad \Delta_r H^\circ_{298}$$

$$H^+(aq) + CN^-(aq) + NaOH(aq) \longrightarrow NaCN(aq) + H_2O(l) \qquad \Delta_{neut}H^\circ_{298} = -55.836\ kJ$$

where the ionization process required

$$\Delta_r H^\circ_{298} = (-12.3) - (-55.836) = 43.5\ kJ$$

4.12 ENTHALPY OF SOLUTION

The enthalpy of reaction for dissolving one mole of solute in n moles of solvent is known as the (*integral*) *enthalpy of solution* ($\Delta_{soln}H^\circ_T$). For a gaseous solute, $\Delta_{soln}H^\circ_T$ is the result of the solvation of the solute molecules ($\Delta_{solv}H^\circ_T$). For a solid molecular solute, the solution process can be considered to be the sum of two processes:

$$Solute(s) \longrightarrow solute(molecules) \qquad \Delta_{sub}H^\circ_T$$

$$Solute(molecules) + solvent \longrightarrow solution \qquad \Delta_{solv}H^\circ_T$$

where $\Delta_{sub}H^\circ_T$ is the enthalpy of sublimation of the solute. Thus,

$$Solute(s) + solvent \longrightarrow solution \qquad \Delta_{soln}H^\circ_T = \Delta_{sub}H^\circ_T + \Delta_{solv}H^\circ_T \qquad (4.12)$$

For a solid ionic solute, similar equations can be written involving lattice enthalpy and the solvation of the ions giving

$$\Delta_{soln}H^\circ_T = \Delta_{lat}H^\circ_T + \sum \Delta_{solv}H^\circ_T(ions) \qquad (4.13)$$

For solutions that are very dilute, enthalpies of formation of the aqueous ions can be used to predict enthalpies of reaction. The $\Delta_f H_{298}^\circ$ of $H^+(aq)$ is assumed to be zero so that individual ionic values may be defined relative to it.

EXAMPLE 4.16. From the following set of $\Delta_f H_{298}^\circ(HNO_3$ in n $H_2O)$ data, calculate the values of $\Delta_{soln} H_{298}^\circ$ for HNO_3 as a function of concentration and determine $\Delta_f H_{298}^\circ$ for the nitrate ion at infinite dilution.

$n(H_2O)/(mol)$	$HNO_3(l)$	1	2	3	4	5	7	10	15
$\Delta_f H_{298}^\circ/(kJ\ mol^{-1})$	−174.10	−187.631	−194.556	−198.568	−201.104	−202.765	−204.593	−205.819	−206.510

25	50	100	500	1 000	2 000	5 000	10 000	50 000	∞
−206.815	−206.853	−206.857	−206.974	−207.041	−207.112	−207.183	−207.229	−207.296	−207.36

The values of $\Delta_{soln} H_{298}^\circ$ corresponding to the equation

$$HNO_3(l) + nH_2O(\text{solvent}) \longrightarrow HNO_3(\text{in } nH_2O)$$

can be calculated by substituting the above data into (4.7), e.g.,

$$\Delta_{soln} H_{298}^\circ(HNO_3 \text{ in } 100\ H_2O) = [(1)\Delta_f H_{298}^\circ(HNO_3 \text{ in } 100\ H_2O)] - [(1)\Delta_f H_{298}^\circ(HNO_3)]$$

$$= (-206.857) - (-174.10) = -32.76\ kJ$$

A plot of these values, Fig. 4-2, shows the trend of $\Delta_{soln} H_{298}^\circ$ with concentration. The $\Delta_f H_{298}^\circ$, $(NO_3^-, \infty$ dil) is

$$\Delta_{soln} H_{298}^\circ(HNO_3 \text{ in } \infty\ H_2O) = (1)\Delta_f H_{298}^\circ(H^+, \infty \text{ dil}) + (1)\Delta_f H_{298}^\circ(NO_3^-, \infty \text{ dil})$$

$$-207.36 = (1)(0) + (1)\Delta_f H_{298}^\circ(NO_3^-, \infty \text{ dil})$$

$$\Delta_f H_{298}^\circ(NO_3^-, \infty \text{ dil}) = -207.36\ kJ\ mol^{-1}$$

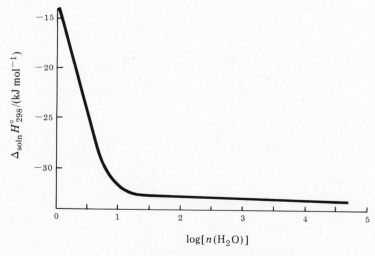

Fig. 4-2

EXAMPLE 4.17. If $\Delta_f H_{298}^\circ/(kJ\ mol^{-1}) = 105.579, -207.36, -167.159,$ and -127.068 for $Ag^+(\infty$ dil$)$, $NO_3^-(\infty$ dil$)$, $Cl^-(\infty$ dil$)$, and $AgCl(s)$, respectively, calculate the enthalpy of reaction for

$$AgNO_3(aq) + HCl(aq) \longrightarrow AgCl(s) + HNO_3(aq)$$

Substituting the data into (4.7) gives

$$\Delta_r H^\circ_{298} = [(1)\Delta_f H^\circ_{298}(AgCl, s) + (1)\Delta_f H^\circ_{298}(H^+, \infty \text{ dil})$$

$$+ (1)\Delta_f H^\circ_{298}(NO_3^-, \infty \text{ dil})] - [(1)\Delta_f H^\circ_{298}(Ag^+, \infty \text{ dil})$$

$$+ (1)\Delta_f H^\circ_{298}(NO_3^-, \infty \text{ dil}) + (1)\Delta_f H^\circ_{298}(H^+, \infty \text{ dil})$$

$$+ (1)\Delta_f H^\circ_{298}(Cl^-, \infty \text{ dil})]$$

$$= [(1 \text{ mol})(-127.068 \text{ kJ mol}^{-1})]$$

$$- [(1 \text{ mol})(105.579 \text{ kJ mol}^{-1}) + (1 \text{ mol})(-167.159 \text{ kJ mol}^{-1})]$$

$$= -65.488 \text{ kJ}$$

4.13 ENTHALPY OF DILUTION

The enthalpy of reaction for diluting 1 mol of solute in a solution of given concentration, by adding solvent to produce a solution of different concentration, is known as the (integral) enthalpy of dilution ($\Delta_{dil} H^\circ_T$).

EXAMPLE 4.18.   What is $\Delta_{dil} H^\circ_{298}$ for diluting a solution containing 1.00 mol of $AgNO_3$ in 100 mol of water by adding 400 mol of water? The enthalpies of formation are $-103.081 \text{ kJ mol}^{-1}$ and $-101.931 \text{ kJ mol}^{-1}$ for the solutions consisting of 100 and 500 mol of water, respectively, added to 1.00 mol of $AgNO_3$ at 25 °C.

Using (4.7) gives

$$\Delta_{dil} H^\circ_{298} = \{(1 \text{ mol})[\Delta_{soln} H^\circ_{298}(AgNO_3 \text{ in } 500 \text{ H}_2\text{O})]\}$$

$$- \{(1 \text{ mol})[\Delta_{soln} H^\circ_{298}(AgNO_3 \text{ in } 100 \text{ H}_2\text{O})]\}$$

$$= [(1 \text{ mol})(-101.931 \text{ kJ mol}^{-1})] - [(1 \text{ mol})(-103.081 \text{ kJ mol}^{-1})]$$

$$= 1.150 \text{ kJ}$$

Applications to Physical Changes

4.14 STATES OF MATTER

Figure 4-3 is a diagram of the various states of matter and the corresponding phase changes. For clarity, not all the phase changes involving the lower-energy crystalline forms are shown, e.g., β-solid to liquid, σ-solid to gas, α-solid to γ-solid. Because the glassy state is not a true equilibrium thermodynamic state, it has been represented by a dashed box.

To each phase transition there corresponds an energy change, e.g., s $\longrightarrow$ l, $\Delta_{fus} H^\circ_T$; l $\longrightarrow$ s, $\Delta_{crys} H^\circ_T$; s $\longrightarrow$ g, $\Delta_{sub} H^\circ_T$; l $\longrightarrow$ g, $\Delta_{vap} H^\circ_T$. Because of the conservation of energy, $\Delta_{fus} H^\circ_T = -\Delta_{crys} H^\circ_T$, etc.

EXAMPLE 4.19.   The temperature dependence for the enthalpy change corresponding to the phase transition I $\longrightarrow$ II is given by (4.2a) as

$$\Delta_{I \to II} H^\circ_{T_2} = \Delta_{I \to II} H^\circ_{T_1} + \int_{T_1}^{T_2} \Delta_{I \to II} C^\circ_P \, dT \tag{4.14}$$

where $\Delta_{I \to II} C^\circ_P$ represents the difference in the heat capacities of phase II and phase I. Find the enthalpy of vaporization for water at 25 °C given $\Delta_{vap} H_{373} = 40.656 \text{ kJ mol}^{-1}$, $C^\circ_P = 75.291 \text{ J K}^{-1} \text{ mol}^{-1}$ for $H_2O(l)$, and $C^\circ_P = 33.577 \text{ J K}^{-1} \text{ mol}^{-1}$ for $H_2O(g)$.

For the transition

$$H_2O(l) \longrightarrow H_2O(g)$$

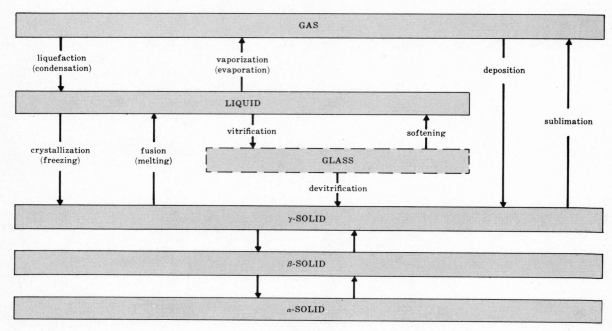

Fig. 4-3

the change in the heat capacity is

$$\Delta_{vap}C_P^\circ = \{(1\ \text{mol})[C_P^\circ(H_2O, g)]\} - \{(1\ \text{mol})[C_P^\circ(H_2O, l)]\}$$

$$= [(1\ \text{mol})(33.577\ \text{J K}^{-1}\ \text{mol}^{-1})] - [(1\ \text{mol})(75.291\ \text{J K}^{-1}\ \text{mol}^{-1})]$$

$$= -41.714\ \text{J K}^{-1}$$

which upon substitution into (*4.14*) gives

$$\Delta_{vap}H_{298}^\circ = (1\ \text{mol})\Delta_{vap}H_{373}^\circ + \int_{373\,K}^{298\,K} (-41.714\ \text{J K}^{-1})\ dT$$

$$= (1\ \text{mol})(40.656\ \text{kJ mol}^{-1}) + (-41.714\ \text{J K}^{-1})[(10^{-3}\ \text{kJ})/(1\ \text{J})](298\ \text{K} - 373\ \text{K})$$

$$= 43.785\ \text{kJ}$$

4.15 APPROXIMATE VALUES OF ENTHALPIES OF TRANSITION

For substances that are not highly associated in the liquid state or gas state, that is, hydrogen bonding or dimer formation is not present, the following approximate relationship has been observed:

$$\Delta_{vap}H_T \cong (88\ \text{J K}^{-1}\ \text{mol}^{-1})\,T_{bp} \tag{4.15}$$

where T_{bp} is the normal boiling point (1.00 atm) of the liquid. For elements,

$$\Delta_{fus}H_T^\circ \cong (9.2\ \text{J K}^{-1}\ \text{mol}^{-1})\,T_{mp} \tag{4.16}$$

where T_{mp} is the melting point of the solid.

EXAMPLE 4.20.   If the boiling point of CCl_4 is 76.7 °C, give an estimate for the enthalpy of vaporization.
Trouton's rule, (*4.15*), gives

$$\Delta_{vap}H_{349.9} = 88\,T_{bp} = (88\ \text{J K}^{-1}\ \text{mol}^{-1})(349.9\ \text{K}) = 30.8\ \text{kJ mol}^{-1}$$

The accepted value is 30.00 kJ mol^{-1}.

4.16 ENTHALPY OF HEATING

The enthalpy change in changing the temperature of a substance is given by

$$\Delta H^\circ = \sum_i^{\text{phases}} \int C_{P,i}^\circ \, dT + \sum_j^{\text{transitions}} \Delta_j H_T^\circ \qquad (4.17)$$

neglecting the last term in (3.37).

EXAMPLE 4.21. What is the enthalpy change for heating ice from $-5\,^\circ$C to steam at $105\,^\circ$C? Assume $C_P = 37.7\ \text{J K}^{-1}\ \text{mol}^{-1}$ for ice and steam, $C_P = 75.3\ \text{J K}^{-1}\ \text{mol}^{-1}$ for water, $\Delta_{\text{vap}} H_{373}^\circ = 40\,656\ \text{J mol}^{-1}$, and $\Delta_{\text{fus}} H_{273}^\circ = 6\,008\ \text{J mol}^{-1}$.

The desired process can be represented as a series of five steps, as illustrated in Fig. 4-4, for which (4.17) becomes

$$\Delta H^\circ = \Delta H^\circ(1) + \Delta H^\circ(2) + \Delta H^\circ(3) + \Delta H^\circ(4) + \Delta H^\circ(5)$$

$$= \int_{268\,\text{K}}^{273\,\text{K}} C_P^\circ \, dT + \Delta_{\text{fus}} H_{273}^\circ + \int_{273\,\text{K}}^{373\,\text{K}} C_P^\circ \, dT + \Delta_{\text{vap}} H_{373}^\circ + \int_{373\,\text{K}}^{378\,\text{K}} C_P^\circ \, dT$$

$$= (37.7)(273 - 268) + 6\,008 + (75.3)(373 - 273) + 40\,656 + (37.7)(378 - 373)$$

$$= 54.57\ \text{kJ mol}^{-1}$$

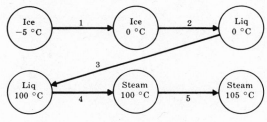

Fig. 4-4

4.17 CLAPEYRON EQUATION

For a change from phase I to phase II,

$$\frac{dP}{dT} = \frac{\Delta_{\text{I} \to \text{II}} H}{T \Delta V} \qquad (4.18)$$

where $\Delta_{\text{I} \to \text{II}} H$ is the enthalpy change for the transition and ΔV is the corresponding volume change. If the phase change involves only condensed phases, (4.18), the *Clapeyron equation*, must be used as written; but for phase transitions between a condensed phase and a gas, ΔV is essentially $V_{\text{gas}} = RT/P$, giving the *Clausius-Clapeyron equation*

$$\frac{dP}{dT} = \Delta_{\text{vap or sub}} H \frac{P}{RT^2} \qquad (4.19)$$

EXAMPLE 4.22. Determine the enthalpy of vaporization at $25\,^\circ$C for water from the following vapor pressure data:

$t/(^\circ\text{C})$	20	21	22	23	24	25	26	27	28	29	30
$P/(\text{torr})$	17.535	18.650	19.827	21.068	22.377	23.756	25.209	26.739	28.349	30.043	31.824

For small temperature intervals, ΔH is temperature-independent and (4.19) can be written as

$$\frac{dP}{P} = \left(\frac{\Delta_{\text{vap}} H}{R} \right) \frac{dT}{T^2}$$

which upon integration gives

$$\ln P = \left(\frac{-\Delta_{vap}H}{R}\right)\frac{1}{T} + k \qquad (4.20a)$$

Evaluation of the constant in $(4.20a)$ for two sets of pressure-temperature data gives

$$\ln \frac{P_2}{P_1} = \left(\frac{-\Delta_{vap}H}{R}\right)\left(\frac{1}{T_2} - \frac{1}{T_1}\right) \qquad (4.20b)$$

The form of $(4.20a)$ is such as to suggest that a plot of $\ln P$ against $1/T$ will give a straight line whose slope will be $-\Delta_{vap}H/R$, see Fig. 4-5. From Fig. 4-5 the slope at 25 °C is -5.28×10^3 K, so

$$\Delta_{vap}H^{\circ}_{298} = -(8.314 \text{ J K}^{-1} \text{ mol}^{-1})(-5.28 \times 10^3 \text{ K}) = 43.9 \text{ kJ mol}^{-1}$$

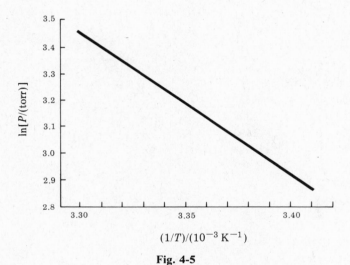

Fig. 4-5

EXAMPLE 4.23. The density of ice is 0.917×10^3 kg m^{-3} and that of water is $0.999\,8 \times 10^3$ kg m^{-3}. Express the dependence of the melting point on the pressure. Assume $\Delta_{fus}H^{\circ}$ to be pressure-independent and equal to 6.009 5 kJ mol^{-1}. At what pressure will ice melt at -1.0 °C?

Rearranging (4.18) gives

$$\frac{dT}{T} = \left(\frac{\Delta V}{\Delta_{fus}H}\right) dP$$

and integration gives

$$\ln \frac{T_2}{T_1} = \left(\frac{\Delta V}{\Delta_{fus}H}\right)(P_2 - P_1)$$

Substituting the data gives

$$\ln \frac{272.15}{273.15} = \frac{\left(\dfrac{18.015 \times 10^{-3}}{0.999\,8 \times 10^3} - \dfrac{18.015 \times 10^{-3}}{0.917 \times 10^3}\right) \text{m}^3 \text{ mol}^{-1}(P_2 - 1) \text{ bar}}{(6\,009.5 \text{ J mol}^{-1})[(1 \text{ Pa m}^3)/(1 \text{ J})][(1 \text{ bar})/(10^5 \text{ Pa})]}$$

$$-0.003\,7 = -2.71 \times 10^{-5}(P_2 - 1)$$

$$P_2 = 135 \text{ bar}$$

4.18 ENTHALPY OF FORMATION DIAGRAM

If the enthalpy of formation for a compound in its various physical states is plotted as a function of temperature, an *enthalpy of formation diagram* is generated from which it is possible to determine graphically various thermal properties.

EXAMPLE 4.24. Prepare an enthalpy of formation diagram for LiI using the data below (JANAF Thermochemical Tables):

Normal melting point	742 K	
Normal boiling point	1 449 K	

$T/(K)$	$\Delta_f H_T^\circ(s)/(\text{kcal mol}^{-1})$	$\Delta_f H_T^\circ(l)/(\text{kcal mol}^{-1})$	$\Delta_f H_T^\circ(g)/(\text{kcal mol}^{-1})$
0			−21.290
100			−21.077
200			−21.368
298	−64.550	−61.749	−21.750
300	−64.551	−61.744	−21.758
400	−66.527	−63.441	−24.122
500	−72.399	−69.111	−30.437
600	−72.192	−68.762	−30.721
700	−71.898	−68.402	−30.984
800	−71.508	−68.034	−31.234
900	−71.028	−67.667	−31.480
1 000	−70.478	−67.300	−31.721
1 100	−69.880	−66.931	−31.958
1 200	−69.252	−66.564	−32.194
1 300	−68.600	−66.194	−32.424
1 400	−67.934	−65.825	−32.653
1 500	−67.255	−65.455	−32.877

Find the enthalpy of sublimation at 298 K if a vertical line between the solid and gas lines represents this quantity. Account for the three discontinuities at about 450 K.

The data are plotted in Fig. 4-6. The dashed vertical line at 298 K intersects the solid and gas curves at −64.55 and −21.75 kcal, giving $\Delta_{sub}H_{298}^\circ = 42.80$ kcal mol^{-1} = 179.08 kJ mol^{-1}. At 450 K, the reactant Li undergoes a phase change from the solid to the liquid state and the discontinuity corresponds to the enthalpy of fusion (about 750 cal mol^{-1} = 3 100 J mol^{-1}).

Solved Problems

HEAT OF REACTION

4.1. The thermochemical equation describing the reaction of HI and Cl_2 at 1 000 K is

$$2HI(g) + Cl_2(g) \longrightarrow 2HCl(g) + I_2(g) \qquad \Delta_r H_{1000}^\circ = -175.31 \text{ kJ}$$

What will be the enthalpy change for the reaction of 2.50 kg of Cl_2 with excess HI?

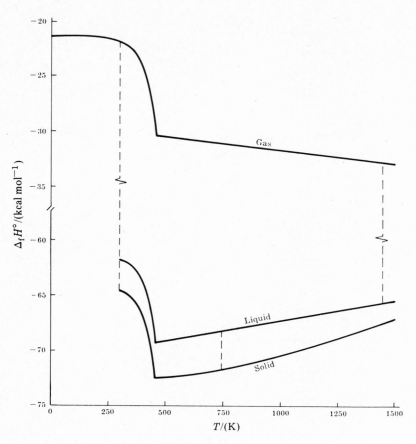

Fig. 4-6

For the given mass of Cl_2,

$$(2.50 \text{ kg } Cl_2)\left(\frac{10^3 \text{ g}}{1 \text{ kg}}\right)\left(\frac{1 \text{ mol } Cl_2}{70.91 \text{ g } Cl_2}\right)\left(\frac{-175.31 \text{ kJ}}{1 \text{ mol } Cl_2}\right)\left(\frac{1 \text{ MJ}}{10^3 \text{ kJ}}\right) = -6.18 \text{ MJ}$$

4.2. Calculate $\Delta_r H° - \Delta_r U°$ for the reaction between glycine and nitrous acid:

$$NH_2-CH_2-COOH(aq) + HONO(aq) \xrightarrow{25\,°C} HO-CH_2-COOH(aq) + N_2(g) + H_2O(l)$$

Using (*4.1*) for this reaction gives

$$\Delta_r H° - \Delta_r U° = RT \Delta n_g = (8.314 \text{ J K}^{-1} \text{ mol}^{-1})(298 \text{ K})(1 \text{ mol}) = 2\,478 \text{ J}$$

4.3. What would be the value of $\Delta_r H°$ at 1 000 K for the reaction

$$H_2(g) + Cl_2(g) \longrightarrow 2HCl(g)$$

given $\Delta_r H°_{298} = -184.614$ kJ? Assume that the heat capacities are given in J K^{-1} mol^{-1} by

$$C_P° = 29.066 - 0.836 \times 10^{-3}T + 20.17 \times 10^{-7}T^2 \quad \text{for } H_2(g)$$

$$C_P° = 31.696 + 10.144 \times 10^{-3}T - 40.38 \times 10^{-7}T^2 \quad \text{for } Cl_2(g)$$

$$C_P° = 28.166 + 1.810 \times 10^{-3}T + 15.47 \times 10^{-7}T^2 \quad \text{for } HCl(g)$$

Using the given heat capacity information,

$$\Delta a = 2(28.166) - [29.066 + 31.696] = -4.430$$

$$\Delta b = 2(1.810 \times 10^{-3}) - [-0.836 \times 10^{-3} + 10.144 \times 10^{-3}] = -5.688 \times 10^{-3}$$

$$\Delta c = 2(15.47 \times 10^{-7}) - [20.17 \times 10^{-7} + (-40.38 \times 10^{-7})] = 51.15 \times 10^{-7}$$

which upon substitution into (4.3a) gives

$$\Delta_r H^\circ_{1000} = -184.614 + (-4.430 \times 10^{-3})(1\,000 - 298)$$

$$+ \tfrac{1}{2}(-5.688 \times 10^{-6})(1\,000^2 - 298^2) + \tfrac{1}{3}(51.15 \times 10^{-10})(1\,000^3 - 298^3)$$

$$= -188.655 \text{ kJ}$$

4.4. A bomb calorimeter was constructed so that the contribution of the reaction products to the heat capacity is negligible. To calibrate this calorimeter, a 1.320-g sample of benzoic acid was oxidized, giving a temperature change of 5.88 °C. Given $\Delta_r U^\circ_{298}$ for the oxidation of benzoic acid is $-6\,316$ cal g^{-1}, calculate C_V(calorimeter).

For the combustion,

$$\Delta_r U^\circ_{298} = (1.320 \text{ g})(-6\,316 \text{ cal g}^{-1})(4.184 \text{ J cal}^{-1}) = -34.88 \text{ kJ}$$

By assuming the calorimeter constant to be temperature-independent, (4.6b) becomes

$$34.88 \text{ kJ} = \int C_V(\text{calorimeter})\, dT = C_V(\text{calorimeter})\, \Delta T$$

giving

$$C_V(\text{calorimeter}) = \frac{34.88 \text{ kJ}}{5.88 \text{ K}} = 5\,930 \text{ J K}^{-1}$$

CALCULATIONS INVOLVING THERMOCHEMICAL EQUATIONS

4.5. Using the following data

$$\text{Fe(s)} + \tfrac{1}{2}O_2(g) \longrightarrow \text{FeO(s)} \qquad\qquad \Delta_r H^\circ_{298} = -272.0 \text{ kJ}$$

$$2\text{Fe(s)} + \tfrac{3}{2}O_2(g) \longrightarrow \text{Fe}_2O_3(\text{hematite}) \qquad \Delta_r H^\circ_{298} = -824.2 \text{ kJ}$$

predict $\Delta_r H^\circ_{298}$ for the reaction

$$2\text{FeO(s)} + \tfrac{1}{2}O_2(g) \longrightarrow \text{Fe}_2O_3(\text{hematite})$$

Subtracting two times the first equation from the second gives

$$2[\text{FeO(s)} \longrightarrow \text{Fe(s)} + \tfrac{1}{2}O_2(g) \qquad \Delta_r H^\circ_{298} = 272.0 \text{ kJ}]$$

$$2\text{Fe(s)} + \tfrac{3}{2}O_2(g) \longrightarrow \text{Fe}_2O_3(\text{hematite}) \qquad \Delta_r H^\circ_{298} = -824.2 \text{ kJ}$$

$$\overline{2\text{FeO(s)} + \tfrac{1}{2}O_2(g) \longrightarrow \text{Fe}_2O_3(\text{hematite}) \qquad \Delta_r H^\circ_{298} = -280.2 \text{ kJ}}$$

4.6. Using the following data

$$\text{Li(l)} + \tfrac{1}{2}Cl_2(g) \longrightarrow \text{LiCl(l)} \qquad \Delta_f H^\circ_{883} = -386.49 \text{ kJ}$$

$$\text{Li(l)} + \tfrac{1}{2}Cl_2(g) \longrightarrow \text{LiCl(s)} \qquad \Delta_f H^\circ_{883} = -406.29 \text{ kJ}$$

predict the enthalpy of fusion for LiCl at 883 K.

For the reaction LiCl(s) $\longrightarrow$ LiCl(l), (4.7) gives

$$\Delta_{\text{fus}} H^\circ_{883} = (1)(-386.49) - (1)(-406.29) = 19.80 \text{ kJ}$$

4.7. Calculate the enthalpy of reaction for the process

$$C_2H_4(g) + HCl(g) \longrightarrow C_2H_5Cl(g)$$

given that $\Delta_f H^\circ_{298}/(\text{kJ mol}^{-1})$ of C_2H_5Cl, HCl, and C_2H_4 are -112.17, -92.307, and 52.26, respectively.

For the reaction written above, (4.7) gives

$$\Delta_r H^\circ_{298} = (1)(-112.17) - [(1)(52.26) + (1)(-92.307)] = -72.12 \text{ kJ}$$

4.8. In the formation of stalagmites and stalactites, aragonite undergoes the following reaction with a very dilute solution of carbonic acid (carbon dioxide dissolved in water):

$$CaCO_3(\text{aragonite}) + H_2CO_3(aq) \longrightarrow Ca(HCO_3)_2(aq)$$

Given that $\Delta_f H^\circ_{298}/(\text{kJ mol}^{-1}) = -1\,207.13$, -699.65, -542.83, and -691.99 for $CaCO_3$, H_2CO_3, Ca^{2+}, and HCO_3^-, respectively, calculate $\Delta_r H^\circ_{298}$.

Applying (4.7) to the reaction gives

$$\Delta_r H^\circ_{298} = [(1)(-542.83) + (2)(-691.99)] - [(1)(-1\,207.13) + (1)(-699.65)] = -20.03 \text{ kJ}$$

4.9. Consider the formation of gaseous methanol from the gaseous molecular fragments:

$$C(g) + 4H(g) + O(g) \longrightarrow CH_3OH(g)$$

Given $\Delta_f H^\circ_{298}/(\text{kJ mol}^{-1})$ for $CH_3OH(g)$ is -200.66, 716.682 for $C(g)$, 217.965 for $H(g)$, and 249.170 for $O(g)$. Using the average bond enthalpy data given in Table 4-1, calculate the bond enthalpy for $C-O$.

Substituting the data into (4.7) and (4.9) gives

$$\Delta_r H^\circ_{298} = [(1 \text{ mol})(-200.66 \text{ kJ mol}^{-1})] - [(1 \text{ mol})(716.682 \text{ kJ mol}^{-1})$$
$$+ (4 \text{ mol})(217.965 \text{ kJ mol}^{-1}) + (1 \text{ mol})(249.170 \text{ kJ mol}^{-1})]$$
$$= -2\,038.37 \text{ kJ}$$

and

$$-2\,038.37 \text{ kJ} = -[(3 \text{ mol})(414 \text{ kJ mol}^{-1}) + (1 \text{ mol})(464 \text{ kJ mol}^{-1}) + (1 \text{ mol})BE(C-O)] + [0]$$
$$BE(C-O) = 332 \text{ kJ mol}^{-1}$$

4.10. Using Table 4-1, predict the enthalpy of reaction for

$$CH_4(g) + 4F_2(g) \longrightarrow CF_4(g) + 4HF(g)$$

Application of (4.9) to the desired reaction gives

$$\Delta_r H^\circ_{298} = -[(4 \text{ mol})(490. \text{ kJ mol}^{-1}) + (4 \text{ mol})(569 \text{ kJ mol}^{-1})]$$
$$+ [(4 \text{ mol})(414 \text{ kJ mol}^{-1}) + (4 \text{ mol})(159 \text{ kJ mol}^{-1})]$$
$$= -1\,944 \text{ kJ}$$

4.11. Calculate the enthalpy of combustion of $CH_4(g)$ at 500 K, if $\Delta_c H^\circ_{298} = -802.37$ kJ for the reaction if gaseous water is formed. Spectroscopic data for $CH_4(g)$ are given in Problem 3.30, for $H_2O(g)$ in Problem 3.2, and for $CO_2(g)$ in Problem 3.29, and $\bar{\nu} = 1\,580.246 \text{ cm}^{-1}$ for $O_2(g)$.

For water at 298 K,

$$E^\circ(\text{thermal, trans}) = E^\circ(\text{thermal, rot}) = \tfrac{3}{2}(8.314)(298) = 3\,716 \text{ J mol}^{-1}$$

Using $\bar{\nu}_1$, $\bar{\nu}_2$, and $\bar{\nu}_3$ from Problem 3.2 in (3.7a), we find $x_1 = 17.65$, $x_2 = 7.699$, and $x_3 = 18.13$, so that

$$E°(\text{thermal, vib}) = (8.314)(298)\left(\frac{17.65}{e^{17.65}-1} + \frac{7.699}{e^{7.699}-1} + \frac{18.13}{e^{18.13}-1}\right)$$

$$= (8.314)(298)(3.8 \times 10^{-7} + 3.49 \times 10^{-3} + 2.4 \times 10^{-7}) = 8.65 \text{ J mol}^{-1}$$

$$E°(\text{thermal, elec}) = E°(\text{thermal, nuc}) = 0$$

The total of the contributions is $E°(\text{thermal}) = 7.441$ kJ mol^{-1}, and adding RT gives $H°(\text{thermal}) = 9.919$ kJ mol^{-1}. Performing the same operations for $O_2(g)$ gives $E°(\text{thermal}) = 6.203$ kJ mol^{-1} and $H°(\text{thermal}) = 8.681$ kJ mol^{-1}. From Problem 3.30, $H°(\text{thermal}) = 10.018$ kJ mol^{-1} for $CH_4(g)$ and from Problem 3.29, $H°(\text{thermal}) = 9.339$ kJ mol^{-1} for $CO_2(g)$.

Using (4.10a) for the reaction

$$CH_4(g) + 2O_2(g) \longrightarrow CO_2(g) + 2H_2O(g)$$

gives $\Delta_r H°_{298}(\text{thermal}) = [(1)(9.339) + (2)(9.919)] - [(1)(10.018) + (2)(8.681)] = 1.797$ kJ

Solving for $\Delta_r U°_0$ in (4.11a) gives

$$\Delta_r U°_0 = (-802.37) - (1.797) = -804.17 \text{ kJ}$$

Repeating the above calculations for $H°(\text{thermal})$ for the substances at 500 K gives 17.670 kJ mol^{-1} for $CO_2(g)$, 16.827 kJ mol^{-1} for $H_2O(g)$, 18.229 kJ mol^{-1} for $CH_4(g)$, and 14.753 kJ mol^{-1} for $O_2(g)$. Using (4.10a) gives

$$\Delta_r H°_{500}(\text{thermal}) = [(1)(17.670) + (2)(16.827)] - [(1)(18.229) + (2)(14.753)] = 3.589 \text{ kJ}$$

and (4.11a) gives

$$\Delta_c H_{500} = -804.17 + 3.589 = -800.58 \text{ kJ}$$

APPLICATIONS TO SELECTED CHEMICAL REACTIONS

4.12. Given $\Delta_f H°_{298} = 10\,087.33$ kJ mol^{-1} for $Mg^{2+}(g)$. Calculate the third ionization energy, and compare the value to those determined in Example 4.12.

For the reaction

$$Mg^{2+}(g) \longrightarrow Mg^{3+}(g) + e^-$$

using (4.7) gives

$$I(Mg^{2+}) = [(1 \text{ mol})(10\,087.33 \text{ kJ mol}^{-1})] - [(1 \text{ mol})(2\,348.504 \text{ kJ mol}^{-1})]$$

$$= 7\,738.83 \text{ kJ}$$

This value is much higher than $I(Mg)$ and $I(Mg^+)$ because the third electron that is removed is from the $2p$ subshell instead of the $3s$ subshell.

4.13. Using $\Delta_f H°_{298}/(\text{kJ mol}^{-1}) = 984$ for $O^{2-}(g)$, $1\,660.38$ for $Ba^{2+}(g)$, and -553.5 for BaO(s), calculate the lattice energy for BaO. Even though the distances between the ions in NaCl and BaO are similar (0.282 nm and 0.275 nm, respectively), there is a considerable difference between the lattice energies (see Example 4.13). Explain this difference.

For the reaction

$$BaO(s) \longrightarrow Ba^{2+}(g) + O^{2-}(g)$$

using (4.7) gives

$$\Delta_{lat} H°_{298} = [(1 \text{ mol})(1\,660.38 \text{ kJ mol}^{-1}) + (1 \text{ mol})(984 \text{ kJ mol}^{-1})]$$

$$- [(1 \text{ mol})(-553.5 \text{ kJ mol}^{-1})]$$

$$= 3\,198 \text{ kJ}$$

This value is approximately four times as large as the NaCl value because the ions being separated are divalent instead of univalent.

4.14. Commercial concentrated sulfuric acid is 17.8 M, or 0.287 mol H_2O for each mole of H_2SO_4. If this acid is diluted to 3 M (16.3 mol H_2O for each mole of H_2SO_4), what is $\Delta_{dil}H^\circ_{298}$, given that the enthalpies of formation of the dilute and concentrated solutions are -883 kJ mol^{-1} and -816 kJ mol^{-1} at 25 °C respectively? Assuming the dilution to be an adiabatic process such that the enthalpy of dilution raises the final solution from 25 °C to a higher temperature, find this final temperature if 1 mol of H_2SO_4 is contained in 400 g of final solution and the specific heat capacity is assumed to be 4.2 kJ K^{-1} kg^{-1}.

The enthalpy of dilution is

$$\Delta_{dil}H^\circ_{298} = (1)(-883) - (1)(-816) = -67 \text{ kJ}$$

which upon changing sign and substitution into (3.37) gives

$$67 \text{ kJ} = (0.400 \text{ kg})(4.2 \text{ kJ K}^{-1}\text{ kg}^{-1})(T_f - 298 \text{ K}) \qquad \text{or} \qquad T_f = 338 \text{ K}$$

APPLICATIONS TO PHYSICAL CHANGES

4.15. From the following set of data, calculate ΔH° for heating Fe from 298 K to 3 200 K:

$$\Delta H^\circ(\alpha\text{-solid} \longrightarrow \beta\text{-solid}) = 0.0 \text{ kcal mol}^{-1} \text{ at } 1\,033 \text{ K}$$

$$\Delta H^\circ(\beta\text{-solid} \longrightarrow \gamma\text{-solid}) = 0.215 \text{ kcal mol}^{-1} \text{ at } 1\,183 \text{ K}$$

$$\Delta H^\circ(\gamma\text{-solid} \longrightarrow \delta\text{-solid}) = 0.165 \text{ kcal mol}^{-1} \text{ at } 1\,673 \text{ K}$$

$$\Delta H^\circ(\delta\text{-solid} \longrightarrow \text{liq}) = 3.67 \text{ kcal mol}^{-1} \text{ at } 1\,812 \text{ K}$$

$$\Delta_{vap}H^\circ = 83.90 \text{ kcal mol}^{-1} \text{ at } 3\,160 \text{ K}$$

$$C^\circ_P/(\text{cal K}^{-1}\text{ mol}^{-1}) = 8.0 \text{ for } \alpha\text{-solid}, \ 10.8 \text{ for } \beta\text{-solid}, \ 8.5 \text{ for } \gamma\text{-solid},$$

9.5 for δ-solid, 10.7 for liquid, and 6.0 for gas

For this heating process, (4.17) gives

$$\Delta H^\circ = (8.0 \times 10^{-3})(1\,033 - 298) + (0.0) + (10.8 \times 10^{-3})(1\,183 - 1\,033) + (0.215)$$

$$+ (8.5 \times 10^{-3})(1\,673 - 1\,183) + (0.165) + (9.5 \times 10^{-3})(1\,812 - 1\,673)$$

$$+ (3.67) + (10.7 \times 10^{-3})(3\,160 - 1\,812) + (83.90) + (6.0 \times 10^{-3})(3\,200 - 3\,160)$$

$$= 115.60 \text{ kcal mol}^{-1} = 483.67 \text{ kJ mol}^{-1}$$

4.16. The sublimation pressure of CO_2(s) is 1 008.9 torr at -75 °C and 438.6 torr at -85 °C. Calculate $\Delta_{sub}H^\circ$ and the normal sublimation point.

Substituting the data into (4.20b) gives

$$\Delta_{sub}H^\circ = -\frac{(8.314 \text{ J K}^{-1}\text{ mol}^{-1}) \ln(1\,008.9/438.6)}{(1/198 \text{ K}) - (1/188 \text{ K})} = 25.65 \text{ kJ mol}^{-1}$$

Using this value and one set of the given pressure-temperature information, the predicted sublimation temperature at 1.00 atm is given by (4.20b) as

$$\frac{1}{T_2} = -\frac{(8.314) \ln(760.0/438.6)}{25.65 \times 10^3} + \frac{1}{188} = 5.14 \times 10^{-3}$$

$$T_2 = 194 \text{ K} = -78 \text{ °C}$$

4.17. Evaluate the enthalpies of fusion and vaporization at the melting and boiling points, respectively, for LiI using Fig. 4-6. At what temperature will the next discontinuity occur?

The vertical differences shown on the diagram give $\Delta_{fus}H° = 3.50$ kcal mol$^{-1} = 14.64$ kJ mol^{-1} at 742 K and $\Delta_{vap}H° = 32.90$ kcal mol$^{-1} = 137.65$ kJ mol^{-1} at 1 449 K. The curves should remain continuous until 1 590 K, at which point Li boils, giving a discontinuity of about 35 kcal mol$^{-1} = 146$ kJ mol^{-1} (the enthalpy of vaporization of Li).

Supplementary Problems

HEAT OF REACTION

4.18. Assume that the following thermochemical equation describes the combustion of gasoline:

$$C_8H_{18}(l) + \tfrac{25}{2}O_2(g) \longrightarrow 8CO_2(g) + 9H_2O(l) \qquad \Delta_rH°_{298} = -5.45 \text{ MJ}$$

What is the enthalpy released by 1.00 gal of gasoline? Assume the density of gasoline to be 0.70 g cm^{-3}.   *Ans.* -126 MJ

4.19. Repeat the calculations of Example 4.2, assuming 10% heat loss to the surroundings and 10% incomplete combustion of the hydrogen. Additional heat capacity data (in J K^{-1} mol^{-1}) are

$$C_P° = 25.72 + 12.98 \times 10^{-3}T - 38.62 \times 10^{-7}T^2 \qquad \text{for } O_2(g)$$

$$C_P° = 29.066 - 0.836\,4 \times 10^{-3}T + 20.17 \times 10^{-7}T^2 \qquad \text{for } H_2(g)$$

Ans.   Gas mixture is 0.9 mol H_2O,   0.05 mol O_2,   0.10 mol H_2, and 2.0 mol N_2;
$\Delta a = 86.12$, $\Delta b = 19.68 \times 10^{-3}$, $\Delta c = 10.6 \times 10^{-7}$;   2 040 K

4.20. Calculate $(\Delta_rH° - \Delta_rU°)$ for the explosion of TNT at 25 °C:

$$C_6H_2CH_3(NO_2)_3(s) + \tfrac{33}{4}O_2(g) \longrightarrow 7CO_2(g) + \tfrac{5}{2}H_2O(g) + 3NO_2(g)$$

Ans. $\Delta n_g = 4.25$ mol, 10.535 kJ

4.21. For the gaseous reaction

$$CO(g) + \tfrac{1}{2}O_2(g) \longrightarrow CO_2(g)$$

$\Delta_rH°_{298} = -282.987$ kJ mol^{-1}. Find $\Delta_rH°_{1000}$ if the heat capacities, in J K^{-1} mol^{-1}, are

$$C_P° = 26.86 + 6.966 \times 10^{-3}T - 8.20 \times 10^{-7}T^2 \qquad \text{for } CO(g)$$

$$C_P° = 26.00 + 43.497 \times 10^{-3}T - 148.32 \times 10^{-7}T^2 \qquad \text{for } CO_2(g)$$

$$C_P° = 25.72 + 12.98 \times 10^{-3}T - 38.6 \times 10^{-7}T^2 \qquad \text{for } O_2(g)$$

Ans. $\Delta a = -13.72$, $\Delta b = 30.04 \times 10^{-3}$, $\Delta c = -120.8 \times 10^{-7}$; -282.852 kJ

4.22. Use the following high-temperature heat capacity expressions (in J K^{-1} mol^{-1})

$$C_P° = 35.672 + 1.515 \times 10^{-3}T - 0.938 \times 10^{-7}T^2 \qquad \text{for } Cl_2(g)$$

$$C_P° = 28.999 + 5.808 \times 10^{-3}T - 6.852 \times 10^{-7}T^2 \qquad \text{for } O_2(g)$$

$$C_P° = 52.407 + 4.189 \times 10^{-3}T - 7.183 \times 10^{-7}T^2 \qquad \text{for } Cl_2O(g)$$

to predict the temperature at which the thermochemical reaction

$$Cl_2(g) + \tfrac{1}{2}O_2(g) \longrightarrow Cl_2O(g) \qquad \Delta_fH°_{298} = 87.864 \text{ kJ}$$

is most endothermic.

Ans. $\Delta a = 2.236$, $\Delta b = -0.230 \times 10^{-3}$, $\Delta c = -2.819 \times 10^{-7}$; $J = 87.210$ kJ;
$\Delta_fH°/(\text{kJ}) = 87.210 + 2.236 \times 10^{-3}T - 0.115 \times 10^{-6}T^2 - 0.940 \times 10^{-10}T^3$; $T = 2\,440$ K

4.23. Consider an adiabatic calorimeter in which 1.000 kg of water at 98.3 °C is mixed with 0.100 kg of water at 0.0 °C. What is the final temperature of the 1.100 kg of water? If the 0.100 kg of water were originally ice at 0.0 °C, what would be the final temperature of the mixture? Assume the specific heat capacity of water to be constant at 4.18 kJ K^{-1} kg^{-1} and the enthalpy of fusion to be 6.009 4 kJ mol^{-1}.

Ans.   89.4 °C;   82.1 °C

4.24. An oxygen bomb calorimeter was calibrated using a 0.325-g sample of benzoic acid ($\Delta_c U = -6\,316$ cal g^{-1}), which gave a change in temperature of 1.48 °C. What is the calorimeter constant? A 0.69-g sample of gasoline was oxidized in the calorimeter, resulting in a temperature change of 4.89 °C. What is $\Delta_c U$ for 1 g of the gasoline?

Ans. $q = 8\,58$ J,   5 803 J K^{-1}; $q = 28.4$ kJ, -41.1 kJ g^{-1}

4.25. Consider a calorimeter that measures the enthalpy of reaction by the amount of ice that is melted at 0 °C, $\Delta_{fus} H^{\circ}_{273} = 6\,009.4$ J mol^{-1}. After corrections for heat loss of the calorimeter, 0.251 g of ice melted when the following reaction was performed in the calorimeter:

$$NaNO_3(in\ 200\ H_2O) + KCl(in\ 200\ H_2O) \longrightarrow KNO_3(in\ 400\ H_2O) + NaCl(in\ 400\ H_2O)$$

Calculate the enthalpy of reaction.   *Ans.* -83.72 J

CALCULATIONS INVOLVING THERMOCHEMICAL EQUATIONS

4.26. Combine the following thermochemical equations

$$CHCl_3(l) + \tfrac{5}{4}O_2(g) \longrightarrow CO_2(g) + \tfrac{1}{2}H_2O(l) + \tfrac{3}{2}Cl_2(g) \qquad \Delta_r H^{\circ}_{298} = -401.95\ kJ$$

$$C(graph) + 2H_2(g) \longrightarrow CH_4(g) \qquad \Delta_r H^{\circ}_{298} = -74.81\ kJ$$

$$C(graph) + O_2(g) \longrightarrow CO_2(g) \qquad \Delta_r H^{\circ}_{298} = -393.509\ kJ$$

$$H_2(g) + \tfrac{1}{2}O_2(g) \longrightarrow H_2O(l) \qquad \Delta_r H^{\circ}_{298} = -285.830\ kJ$$

to predict $\Delta_r H^{\circ}_{298}$ for

$$CH_4(g) + \tfrac{3}{2}Cl_2(g) \longrightarrow CHCl_3(l) + \tfrac{3}{2}H_2(g)$$

Ans. -59.66 kJ

4.27. The enthalpy of formation of calcite, one form of $CaCO_3(s)$, is $-1\,206.92$ kJ mol^{-1} at 25 °C, and the enthalpy of formation of aragonite, another form of $CaCO_3(s)$, is $-1\,207.13$ kJ mol^{-1}. Calculate $\Delta_r H^{\circ}_{298}$ for

$$CaCO_3(calcite) \longrightarrow CaCO_3(aragonite)$$

Ans. -210 J

4.28. Calculate $\Delta_r H^{\circ}_{298}$ for the hydrogenation of benzene to cyclohexane:

$$C_6H_6(l) + 3H_2(g) \longrightarrow C_6H_{12}(l)$$

given that the enthalpies of combustion at 20 °C are $-3\,273$, -286.1, and $-3\,924$ kJ mol^{-1} for C_6H_6, H_2, and C_6H_{12}, respectively.   *Ans.* -207 kJ

4.29. From the following data for the stepwise decomposition of $CH_4(g)$, calculate $\overline{BE}(C-H)$.

$$CH_4(g) \longrightarrow CH_3(g) + H(g) \qquad \Delta H^{\circ}_{298} = 438.47\ kJ$$

$$CH_3(g) \longrightarrow CH_2(g) + H(g) \qquad \Delta H^{\circ}_{298} = 462.65\ kJ$$

$$CH_2(g) \longrightarrow CH(g) + H(g) \qquad \Delta H^{\circ}_{298} = 423.40\ kJ$$

$$CH(g) \longrightarrow C(g) + H(g) \qquad \Delta H^{\circ}_{298} = 338.85\ kJ$$

Ans.   415.84 kJ mol^{-1}

4.30. Calculate the bond enthalpy for C—I given $\Delta_f H^\circ_{298}/(kJ\ mol^{-1})$ for CH_3I is 13.0, 217.965 for H(g), 716.682 for C(g), and 106.838 for I(g), and using the average bond enthalpy data given in Table 4-1 for C—H.

Ans. $222\ kJ\ mol^{-1}$ (measured value is $232.2 \pm 12.6\ kJ\ mol^{-1}$)

4.31. Assuming benzene to consist of six C—H bonds, three C—C bonds, and three C=C bonds, (a) calculate $\Delta_r H^\circ_{298}$ for

$$6C(g) + 6H(g) \longrightarrow C_6H_6(g)$$

using the bond enthalpy data given in Table 4-1. (b) Calculate $\Delta_r H^\circ_{298}$ for the above reaction using $\Delta_f H^\circ_{298}/(kJ\ mol^{-1}) = 716.682, 217.965$, and 82.927 for C(g), H(g), and $C_6H_6(g)$, respectively. The difference between this value for the enthalpy of formation and that found in (a) is known as the *resonance energy* of the molecule; it is due to the delocalized pi bonding in the ring. Find the resonance energy.

Ans.   (a) $-5\ 247\ kJ$;   (b) $-5\ 524.955\ kJ$, $-278\ kJ$

4.32. (a) Calculate $\Delta_r H^\circ_{298}$ for the deamination of L-aspartic acid to fumaric acid:

$$HOOC—CH—CH_2—COOH \longrightarrow HOOC—CH=CH—COOH + NH_3$$
$$\overset{|}{\underset{NH_2}{}}$$

given that the enthalpies of formation at 25 °C for the acids, as reported by Burton and Krebs, are $-233.75\ kcal\ mol^{-1}$ for L-aspartic acid and $-194.13\ kcal\ mol^{-1}$ for fumaric acid, while for NH_3, $\Delta_f H^\circ_{298} = -46.11\ kJ\ mol^{-1}$. (b) Bond enthalpies are sometimes used to estimate enthalpies of reaction for condensed-phase reactions when other thermodynamic data are not available. Repeat the calculation in (a) using average bond enthalpies at 25 °C given in Table 4-1, and compare results. (In the bond enthalpy calculation, it is not necessary to break the reactant completely down to fragments, but only to consider the parts of the molecule that are changing.)

Ans.   (a) $119.66\ kJ$
         (b) $51\ kJ$, about 74% low but of correct sign and order of magnitude

4.33. Calculate the enthalpy of reaction at 1 000 K for the reaction

$$N_2(g) + 2O_2(g) \longrightarrow 2NO_2(g)$$

if $\Delta_r H^\circ_{298} = 66.36\ kJ$. Spectroscopic data are: $\bar\nu = 2\ 357.55\ cm^{-1}$ for $N_2(g)$; $1\ 580.246\ cm^{-1}$ for $O_2(g)$; and $1\ 357.8$, 756.8, and $1\ 665.5\ cm^{-1}$ for the nonlinear $NO_2(g)$ molecule.

Ans. H°(thermal, N_2) = 8.672   and   30.079 kJ mol^{-1}, H°(thermal, O_2) = 8.673   and   31.266 kJ mol^{-1},
         H°(thermal, NO_2) = 10.180 and 42.527 kJ mol^{-1}, at 298 K and 1 000 K, respectively; $\Delta_r U^\circ_0 = 72.02\ kJ$;
         $\Delta_r H^\circ_{1000} = 64.46\ kJ$

APPLICATIONS TO SELECTED CHEMICAL REACTIONS

4.34. Compare the ionization energy of a metal, such as Li, to that of a nonmetal, such as F. Use $\Delta_f H^\circ_{298}/(kJ\ mol^{-1}) = 159.37$ for Li(g), 685.783 for Li$^+$(g), 78.99 for F(g), and 1 766.44 for F$^+$(g).

Ans. $I(Li) = 526.41\ kJ\ mol^{-1}$, $I(F) = 1\ 687.45\ kJ\ mol^{-1}$; I for a metal is much less than for a nonmetal.

4.35. Calculate the ionization energy and electron affinity for oxygen using the following $\Delta_f H^\circ_{298}/(kJ\ mol^{-1})$ data: 249.170 for O(g), 1 568.770 for O$^+$(g), and 101.63 for O$^-$(g).

Ams. $I = 1\ 319.600\ kJ\ mol^{-1}$, $E_A = -147.54\ kJ\ mol^{-1}$

4.36. Using $\Delta_f H^\circ_{298}/(kJ\ mol^{-1}) = 984$ for O^{2-}(g), 2 348.504 for Mg^{2+}(g), and -601.70 for MgO(s), calculate the lattice energy for MgO. Why is this value greater than that found for BaO (see Problem 4.13)?

Ans. $3\ 934\ kJ\ mol^{-1}$;   more energy is needed to separate the ions in MgO because the distance between the ions in MgO is less than in BaO (0.211 nm and 0.275 nm, respectively).

4.37. Calculate $\Delta_{neut}H^{\circ}_{298}$ for the reaction between HCl(aq) and NaOH(aq) given $\Delta_f H^{\circ}_{298}/(kJ\ mol^{-1}) = -167.159$, -470.114, -407.27, and -285.830 for HCl, NaOH, NaCl, and H_2O, respectively, in very dilute solutions. *Ans.* $-55.83\ kJ$

4.38. Prepare a graph of the calculated values of $\Delta_{soln}H^{\circ}_{298}$ for $AgNO_3(s)$ against the number of moles of water, using the following heat of formation data,

$n(H_2O)/(mol)$	$AgNO_3(s)$	50	100	200	400	500
$\Delta_f H^{\circ}_{298}/(kJ\ mol^{-1})$	-124.39	-104.244	-103.081	-102.370	-102.002	-101.931
$n(H_2O)/(mol)$	1 000	2 000	5 000	10 000	50 000	∞
$\Delta_f H^{\circ}_{298}/(kJ\ mol^{-1})$	-101.788	-101.730	-101.709	-101.717	-101.747	-101.80

Using these data and the result of Example 4.16, calculate $\Delta_f H^{\circ}_{298}$ for $Ag^+(\infty\ dil)$.

Ans. The plot will have enthalpy of solution values ranging from $20.15\ kJ\ mol^{-1}$ at $50\ mol\ H_2O$ to $22.68\ kJ\ mol^{-1}$ at $5\ 000\ mol\ H_2O$ (a maximum) and an infinite-dilution value of $22.59\ kJ\ mol^{-1}$. The value of the ionic enthalpy of formation is $105.56\ kJ\ mol^{-1}$.

4.39. Saturated solutions of ammonium chloride are often used in sports to reduce the swelling of a sprained ankle. Calculate the temperature drop for

$$NH_4Cl(s) + 10H_2O(l) \longrightarrow NH_4Cl(in\ 10\ H_2O)$$

given that the enthalpy of formation of $NH_4Cl(in\ 10\ H_2O)$ is $-299.436\ kJ\ mol^{-1}$ and that of $NH_4Cl(s)$ is $-314.43\ kJ\ mol^{-1}$. The specific heat capacity of the solution is $3.77\ kJ\ K^{-1}\ kg^{-1}$.

Ans. $\Delta H = 14.99\ kJ$, $C = 0.881\ kJ\ K^{-1}\ mol^{-1}$; 17 K

4.40. Calculate $\Delta_{soln}H^{\circ}_{298}$ for the reactions

$$KCl(g) + \infty H_2O(l) \longrightarrow KCl(aq) \qquad KCl(s) + \infty H_2O(l) \longrightarrow KCl(aq)$$

given $\Delta_f H^{\circ}_{298}/(kJ\ mol^{-1}) = -214.14$ for KCl(g), -436.747 for KCl(s), -252.38 for $K^+(aq)$, and -167.159 for $Cl^-(aq)$. What is the enthalpy of sublimation for KCl at this temperature?

Ans. $-205.40\ kJ$, $17.21\ kJ$, $222.61\ kJ\ mol^{-1}$

APPLICATIONS TO PHYSICAL CHANGES

4.41. Consider the combustion of 1.00 mol of $CH_4(g)$ with $O_2(g)$, giving $CO_2(g)$ and H_2O at 25 °C. If the product is $H_2O(l)$, $\Delta_c H^{\circ}_{298} = -890.36\ kJ$; and if $H_2O(g)$, $\Delta_c H^{\circ}_{298} = -802.34\ kJ$. Calculate $\Delta_{vap}H^{\circ}$ for water at 25 °C. *Ans.* $44.01\ kJ\ mol^{-1}$

4.42. Krypton melts at -157.21 °C. Give an estimate of $\Delta_{fus}H^{\circ}_{115.94}$ for this element, and compare it to $1\ 636\ J\ mol^{-1}$. *Ans.* $1\ 100\ J\ mol^{-1}$, 33% low

4.43. Lozana reported a 1.57% volume increase at 1.00 atm upon melting for lithium metal at 180.54 °C. The density of the liquid is $0.515 \times 10^3\ kg\ m^{-3}$ and the enthalpy of fusion is $722.8\ cal\ mol^{-1}$. What will be the melting point at 1 000 bar?

Ans. $\Delta_{fus}H^{\circ} = 0.030\ 25\ m^3\ bar\ mol^{-1}$, $\Delta V = 0.211 \times 10^{-6}\ m^3\ mol^{-1}$, $dT = \Delta T = 3.16\ K\ mol^{-1}$; 183.70 °C

4.44. From the following vapor pressure data for Hg, calculate the enthalpy of vaporization and the normal boiling point.

$t/(°C)$	300	320	340	350	352	354	356
$P/(\text{torr})$	246.80	376.33	577.90	672.69	697.83	723.73	750.43

Ans.   Slope $= -7.16 \times 10^3$ K;   59.6 kJ mol^{-1};   629 K

4.45. One of the major components of mothballs, naphthalene, has a sublimation pressure of 1.09×10^{-2} torr at 6 °C and 5.37×10^{-2} torr at 21 °C. Using these data, calculate the enthalpy of sublimation and the sublimation pressure in a warm closet at 78 °F. If the closet is 11 ft $\times$ 3$\frac{1}{2}$ ft $\times$ 9 ft, calculate the mass of naphthalene, $C_{10}H_8$, in the gaseous state, assuming an ambient room pressure of 747 torr.

Ans.   72.4 kJ mol^{-1}; 8.47×10^{-2} torr;   0.044 6 mol = 5.72 g (ambient room pressure is irrelevant)

4.46. Prepare a graph of $\Delta_f H_T^\circ$ for $AlCl_3$ as a function of temperature from the following data (JANAF Thermochemical Tables):

$T/(\text{K})$	$\Delta_f H_T^\circ(\text{s})/(\text{kcal mol}^{-1})$	$\Delta_f H_T^\circ(\text{l})/(\text{kcal mol}^{-1})$	$\Delta_f H_T^\circ(\text{g})/(\text{kcal mol}^{-1})$
0	−168.323		−139.274
100	−169.004		−139.468
200	−168.961		−139.604
298	−168.650	−161.280	−139.700
300	−168.643	−161.258	−139.702
400	−168.200	−160.102	−139.773
500	−167.614	−159.013	−139.837
600	−166.869	−157.975	−139.911
700	−165.954	−156.977	−139.998
800	−164.873	−156.023	−140.113
900	−163.633	−155.120	−140.266
1 000	−164.760	−156.794	−142.988
1 100	−163.108	−155.899	−143.135
1 200	−161.250	−155.008	−143.281
1 300	−159.184	−154.119	−143.425
1 400	−156.912	−153.234	−143.571
1 500	−154.433	−152.352	−143.717

From the graph, determine $\Delta_{\text{fus}} H^\circ$ at 465.7 K and $\Delta_{\text{sub}} H^\circ$ at 298 K.

Ans.   8.45 kcal mol^{-1} = 35.35 kJ mol^{-1},   28.95 kcal mol^{-1} = 121.13 kJ mol^{-1}

Chapter 5

Entropy

The Second Law of Thermodynamics

5.1 STATEMENTS

Unlike the first law of thermodynamics, the second law does not have just one statement, but several: (1) a cyclic process must transfer heat from a hot to a cold reservoir if it is to convert heat into work; (2) work must be done to transfer heat from a cold to a hot reservoir; (3) no engine can operate more efficiently than a Carnot engine; (4) a perpetual-motion machine of the second kind (one that extracts heat from surroundings at T, does work on surroundings, and returns to its initial state without transferring heat to another system at a temperature lower than T) cannot exist; (5) the entropy, or randomness, of the universe is increasing; etc. Most of these statements can be reduced to the assertion that all real processes are irreversible, i.e., the system and the surroundings cannot both be restored to their original states.

Entropy changes can be used to predict the spontaneity of constant-energy processes. At fixed energy, only those processes will occur spontaneously for which there is an increase in entropy. Predicting the spontaneity of a process under combined changes of energy and entropy will be considered in Section 6.1.

5.2 THE CARNOT CYCLE

The operation of an arbitrary heat engine is represented in Fig. 5-1. A *Carnot engine* is an (idealized) heat engine that follows the cyclic process indicated in Fig. 5-2. This *Carnot cycle* consists of four steps performed on an ideal gas: (1) a reversible isothermal expansion from V_1 to V_2 at T_h, (2) a reversible adiabatic expansion from V_2 to V_3 with a temperature change from T_h to T_c, (3) a reversible isothermal compression from V_3 to V_4 at T_c, and (4) a reversible adiabatic compression from V_4 to V_1 with a temperature change from T_c to T_h.

For a Carnot engine, $q_h = q_1$ and $q_c = q_3$.

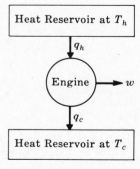

Fig. 5-1

EXAMPLE 5.1. Consider a Carnot engine operating between 500 °C and 0 °C using 1.00 mol of an ideal monatomic gas. If $V_1 = 0.010\,0\ \text{m}^3$ and $V_2 = 0.100\,0\ \text{m}^3$, calculate V_3 and V_4; q, w, and ΔU for each step; and q, w, and ΔU for the overall process. Prepare a sketch similar to Fig. 5-1 for this engine.

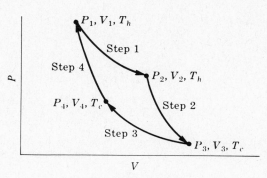

Fig. 5-2

The relationship between the temperatures is given by (*3.44*) as

$$T_h^{C_V/R} V_2 = T_c^{C_V/R} V_3 \qquad T_h^{C_V/R} V_1 = T_c^{C_V/R} V_4$$

Since $C_V = \frac{3}{2}R$ for an ideal monatomic gas,

$$V_3 = V_2 \left(\frac{T_h}{T_c}\right)^{C_V/R} = (0.100\,0\text{ m}^3)\left(\frac{773\text{ K}}{273\text{ K}}\right)^{3/2} = 0.477\text{ m}^3$$

Dividing the two volume-temperature equations written above, we obtain

$$V_4 = V_3 \frac{V_1}{V_2} = (0.477\text{ m}^3)\frac{0.010\,0\text{ m}^3}{0.100\,0\text{ m}^3} = 0.047\,7\text{ m}^3$$

The values of ΔU, q, and w for each step can be determined by using (*3.40*) and (*3.43*). For the first step:

$$\Delta U_{(1)} = 0$$

$$-q_1 = w_1 = -nRT_h \ln \frac{V_2}{V_1} = -(1.00\text{ mol})(8.314\text{ J K}^{-1}\text{ mol}^{-1})(773\text{ K}) \ln \frac{0.100\,0}{0.010\,0} = -14.80\text{ kJ}$$

For the second step:

$$q_2 = 0$$

$$\Delta U_{(2)} = w_2 = n \int_{T_h}^{T_c} C_V\,dT = n \int_{T_k}^{T_c} \frac{3}{2} R\,dT$$

$$= (1.00\text{ mol})(\tfrac{3}{2})(8.314\text{ J K}^{-1}\text{ mol}^{-1})(273\text{ K} - 773\text{ K})$$

$$= -6.24\text{ kJ}$$

For the third step:

$$\Delta U_{(3)} = 0$$

$$-q_3 = w_3 = -nRT_c \ln \frac{V_4}{V_3}$$

$$= -(1.00)(8.314)(273) \ln \frac{0.047\,7}{0.477} = 5.23\text{ kJ}$$

For the fourth step:

$$q_4 = 0$$

$$\Delta U_{(4)} = w_4 = n \int_{T_c}^{T_h} C_V\,dT = 6.24\text{ kJ}$$

For the overall cycle it can be shown that

$$\Delta U = 0$$

$$-q = w = -nR(T_h - T_c) \ln \frac{V_2}{V_1}$$

$$= -(1.00)(8.314)(733 - 273) \ln \frac{0.100\,0}{0.010\,0} = -9.57 \text{ kJ}$$

The work is equivalent to the area enclosed by the cycle shown in Fig. 5-2. The diagram for this engine is Fig. 5-3.

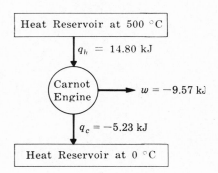

Fig. 5-3

5.3 EFFICIENCY OF A HEAT ENGINE

For an engine that operates between two heat reservoirs, the *efficiency* (expressed in %) is

$$\varepsilon \equiv 100 \frac{-w}{q_h} \tag{5.1}$$

EXAMPLE 5.2. Substitute the expressions for w and q_1 found in Example 5.1 into (5.1) to derive a general expression for the efficiency of a Carnot engine. Calculate the efficiency of the engine described in Example 5.1.

Using the expressions for w and $q_1 = q_h$ gives

$$\varepsilon_{\text{Carnot}} = 100 \frac{T_h - T_c}{T_h} \tag{5.2}$$

Substituting the numerical values into (5.1) gives

$$\varepsilon_{\text{Carnot}} = 100 \frac{-(-9.57 \text{ kJ})}{14.80 \text{ kJ}} = 64.7\%$$

or substituting the temperatures into (5.2) gives

$$\varepsilon_{\text{Carnot}} = 100 \frac{773 \text{ K} - 273 \text{ K}}{773 \text{ K}} = 64.7\%$$

5.4 REFRIGERATORS

An engine can be used as a refrigerator to pump heat from a cold to a hot reservoir, see Fig. 5-4. The *performance factor* of a refrigerator (expressed in %) is

$$\varepsilon_{\text{ref}} \equiv 100 \frac{w}{q_c} \tag{5.3a}$$

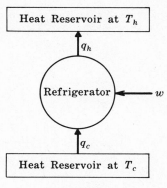

Fig. 5-4

For a Carnot refrigerator,

$$\varepsilon_{\text{ref}} = 100 \, \frac{T_h - T_c}{T_c} \qquad\qquad (5.3b)$$

EXAMPLE 5.3.   Calculate the maximum performance factor of a commercial refrigerator operating between the temperatures of $-10\,°C$ (inside temperature) and $25\,°C$ (room temperature). What minimum amount of work must be done to remove $100.\,J$ of heat from the inside of the refrigerator?

According to one of the statements of the second law, the maximum efficiency occurs in a Carnot engine. Thus $(5.3b)$ gives

$$\varepsilon_{\text{ref}} = 100 \, \frac{298\ \text{K} - 263\ \text{K}}{263\ \text{K}} = 13.3\%$$

The minimum amount of work required to remove the heat is found using $(5.3a)$ as

$$13.3 = 100 \, \frac{w}{100.\ \text{J}} \qquad \text{or} \qquad w = 13.3\ \text{J}$$

Entropy Calculations

5.5   DEFINITION OF ENTROPY

For the Carnot cycle, it can be shown that

$$\varepsilon_{\text{Carnot}} = 100 \, \frac{q_h + q_c}{q_h} = 100 \, \frac{T_h - T_c}{T_h}$$

which can be rearranged to give

$$\frac{q_c}{T_c} + \frac{q_h}{T_h} = 0$$

Any reversible cyclic process can be considered to be the sum of a large number of smaller Carnot cycles, giving

$$\sum \frac{q_{\text{rev}}}{T} = 0$$

or, as the number of subcycles becomes very large,

$$\oint \frac{dq_{\text{rev}}}{T} = 0$$

The above condition—that the integral over a cycle is zero—is the general definition of a state or point function. Calling this particular function the *entropy* (S), we have

$$dS = \frac{dq_{rev}}{T} \tag{5.4a}$$

and

$$\Delta S = \int \frac{dq_{rev}}{T} \tag{5.4b}$$

Equations (5.4) require that *the entropy change for a system be calculated using a reversible process only.* For an irreversible process, a reversible path must be defined that has the same endstates as the actual process, and (5.4) must be applied along this reversible path. For a reversible process,

$$\Delta S(\text{system}) = -\Delta S(\text{surroundings}) \tag{5.5a}$$

and

$$\Delta S(\text{universe}) = \Delta S(\text{system}) + \Delta S(\text{surroundings}) = 0 \tag{5.5b}$$

For an irreversible process,

$$\Delta S(\text{universe}) > 0 \tag{5.6}$$

The units of entropy in the older physical system are cal K^{-1} (or gibbs), which have been represented by eu; in the SI system the units are J K^{-1}.

5.6 ΔS(system) FOR HEAT TRANSFER

The entropy change for a system to which heat has been transferred reversibly and isothermally is given by (5.4b) as

$$\Delta S(\text{system}) = \frac{q_{rev}}{T} \tag{5.7}$$

A special case of (5.7) is that for a reversible adiabatic process

$$\Delta S(\text{system}) = 0 \tag{5.8}$$

For a system undergoing a phase transition from I to II (a reversible, isothermal process)

$$\Delta_{i \to II} S(\text{system}) = \frac{\Delta_{I \to II} H}{T} \tag{5.9}$$

EXAMPLE 5.4. What are the entropy changes for 1.00 mol of water undergoing reversible vaporization and fusion at 100 °C and 0 °C, respectively? Assume $\Delta_{vap} H = 40.656$ kJ mol^{-1} and $\Delta_{fus} H = 6.009$ kJ mol^{-1}. Qualitatively compare the entropy changes for the two transitions.

For the vaporization, (5.9) gives

$$\Delta_{vap} S(\text{system}) = \frac{(1.00 \text{ mol})(40.656 \text{ kJ mol}^{-1})[(10^3 \text{ J})/(1 \text{ kJ})]}{373.15 \text{ K}}$$

$$= 108.95 \text{ J K}^{-1}$$

and for the fusion,

$$\Delta_{fus} S(\text{system}) = \frac{(1.00)(6\,009)}{273.15} = 22.00 \text{ J K}^{-1}$$

The values of ΔS(system) indicate that there is a larger increase in randomness in going from the liquid state to the gaseous state than from the solid state to the liquid state.

5.7 ΔS(system) FOR VOLUME-PRESSURE-TEMPERATURE CHANGES

For a system undergoing a reversible expansion,

$$\Delta S(\text{system}) = \int_{T_1}^{T_2} \frac{C_V}{T}\, dT + \int_{V_1}^{V_2} \left(\frac{\partial P}{\partial T}\right)_V dV \qquad (5.10a)$$

$$\Delta S(\text{system}) = \int_{T_1}^{T_2} \frac{C_P}{T}\, dT - \int_{P_1}^{P_2} \left(\frac{\partial V}{\partial T}\right)_P dP \qquad (5.10b)$$

For an ideal gas, $(\partial P/\partial T)_V = R/V$ and $(\partial V/\partial T)_P = R/P$, which, upon substitution into (5.10) and integration, gives

$$\Delta S(\text{system, ideal gas}) = \int_{T_1}^{T_2} \frac{C_V}{T}\, dT + R \ln \frac{V_2}{V_1} \qquad (5.11a)$$

$$\Delta S(\text{system, ideal gas}) = \int_{T_1}^{T_2} \frac{C_P}{T}\, dT - R \ln \frac{P_2}{P_1} \qquad (5.11b)$$

For real gases, the derivatives in (5.10) are evaluated from the equations of state and substituted into (5.10) for the integration process. For condensed states, $(\partial P/\partial T)_V = \alpha/\kappa$ and $(\partial V/\partial T)_P = \alpha V$, which, upon substitution into (5.10), gives

$$\Delta S(\text{system, condensed state}) = \int_{T_1}^{T_2} \frac{C_V}{T}\, dT + \int_{V_1}^{V_2} \frac{\alpha}{\kappa}\, dV \qquad (5.12a)$$

$$\Delta S(\text{system, condensed state}) = \int_{T_1}^{T_2} \frac{C_P}{T}\, dT - \int_{P_1}^{P_2} \alpha V\, dP \qquad (5.12b)$$

EXAMPLE 5.5. Consider the reversible isothermal expansion of 1.00 mol of an ideal gas from 0.010 m³ to 0.100 0 m³ at 298 K. Calculate ΔS(system) for this process.

For the isothermal process, the first term on the right side of (5.11a) is zero, giving

$$\Delta S(\text{system}) = (1.00 \text{ mol})(8.314 \text{ J K}^{-1} \text{ mol}^{-1}) \ln \frac{0.100\,0}{0.010\,0} = 19.14 \text{ J K}^{-1}$$

EXAMPLE 5.6. Calculate ΔS(system) for the reversible heating of 1.00 mol of ethane from 298 K to 1 500 K at constant pressure. Assume

$$C_P = 5.351 + 177.669 \times 10^{-3} T - 687.01 \times 10^{-7} T^2 + 8.514 \times 10^{-9} T^3$$

in J K⁻¹ mol⁻¹.

For the isobaric process, the second term on the right side of (5.10b) is zero, giving

$$\Delta S(\text{system}) = (1.00 \text{ mol}) \int_{298\,\text{K}}^{1500\,\text{K}} \frac{C_P}{T}\, dT$$

$$= 5.351 \ln \frac{1\,500}{298} + (177.669 \times 10^{-3})(1\,500 - 298) - \left(\frac{687.01 \times 10^{-7}}{2}\right)(1\,500^2 - 298^2)$$

$$+ \left(\frac{8.514 \times 10^{-9}}{3}\right)(1\,500^3 - 298^3)$$

$$= 157.47 \text{ J K}^{-1}$$

EXAMPLE 5.7. Compare the contributions to ΔS(system) for the reversible cooling of 1.00 mol of an ideal diatomic gas from 25 °C to 0 °C at a constant pressure of 1.00 bar.

Although (5.11b) could be used to calculate the overall ΔS(system) directly, it does not give the individual contributions from the temperature decrease and from the volume decrease. However, use of (5.11a) gives both contributions. For the gas, $C_V = \frac{5}{2}R$, and

$$V_1 = \frac{nRT_1}{P} = \frac{(1 \text{ mol})(8.314 \text{ J K}^{-1} \text{ mol}^{-1})(298 \text{ K})[(1 \text{ Pa m}^3)/(1 \text{ J})]}{(1.00 \text{ bar})[(10^5 \text{ Pa})/(1 \text{ bar})]}$$

$$= 0.024\ 8 \text{ m}^3$$

$$V_2 = \frac{nRT_2}{P} = \frac{(1.00)(8.314)(298)}{(1.00)(10^5)} = 0.022\ 7 \text{ m}^3$$

Equation (5.11a) gives the contribution from the cooling as

$$\Delta S(\text{system}) = n \int_{T_1}^{T_2} \frac{C_V}{T}\, dT = nC_V \ln \frac{T_2}{T_1}$$

$$= (1.00 \text{ mol})(\tfrac{5}{2})(8.314 \text{ J K}^{-1} \text{ mol}^{-1}) \ln \frac{273 \text{ K}}{298 \text{ K}}$$

$$= -1.82 \text{ J K}^{-1}$$

and the contribution from the volume change as

$$\Delta S(\text{system}) = nR \ln \frac{V_2}{V_1}$$

$$= (1.00 \text{ mol})(8.314 \text{ J K}^{-1} \text{ mol}^{-1}) \ln \frac{0.022\ 7 \text{ m}^3}{0.024\ 8 \text{ m}^3}$$

$$= -0.74 \text{ J K}^{-1}$$

for a total of

$$\Delta S(\text{system}) = (-1.82 \text{ J K}^{-1}) + (-0.74 \text{ J K}^{-1}) = -2.56 \text{ J K}^{-1}$$

In this case, the temperature change accounts for 71% of the entropy change, and the volume change accounts for 29%.

5.8 ΔS(system) FOR ISOTHERMAL MIXING

For preparing a mixture under reversible isothermal conditions,

$$\Delta_{\text{mix}} S(\text{system}) = -R \sum_i n_i \ln x_i \qquad (5.13a)$$

where n_i is the number of moles and x_i is the mole fraction, of component i in the mixture. For 1 mol of solution, (5.13a) becomes

$$\Delta_{\text{mix}} S(\text{system}) = -R \sum_i x_i \ln x_i \qquad (5.13b)$$

EXAMPLE 5.8.   What is the entropy change for preparing a mixture containing 1.00 mol of $O_2(g)$ and 2.00 mol of $H_2(g)$, assuming no chemical reaction and reversible isothermal mixing?

For the binary mixture, (5.13) gives

$$\Delta S(\text{system}) = -(8.314 \text{ J K}^{-1} \text{ mol}^{-1})[(1.00 \text{ mol}) \ln(0.333) + (2.00 \text{ mol}) \ln(0.667)]$$

$$= 15.88 \text{ J K}^{-1}$$

5.9 ΔS(surroundings)

For the surroundings, it is assumed that the heat can be transferred reversibly and isothermally, so that

$$\Delta S(\text{surroundings}) = \frac{q(\text{surroundings})}{T} \qquad (5.14)$$

EXAMPLE 5.9. Compare the values of ΔS(surroundings) and ΔS(universe) for the process described in Example 5.5 given that the process is done reversibly and that the process is performed irreversibly against a constant external pressure of 0.100 bar.

For the surroundings, q(surroundings) $= -nRT \ln (V_2/V_1)$, which upon substitution into (5.14) gives

$$\Delta S(\text{surroundings}) = -nR \ln \frac{V_2}{V_1} = -19.14 \text{ J K}^{-1}$$

For the universe,

$$\Delta S(\text{universe}) = 19.14 + (-19.14) = 0$$

For the irreversible process, ΔS(system) must be calculated using a reversible path, so

$$\Delta S(\text{system}) = 19.14 \text{ J K}^{-1}$$

as in the reversible case. For the surroundings, (3.41b) gives q(surroundings) $= P\Delta V$, so (5.14) becomes

$$\Delta S(\text{surroundings}) = \frac{P \Delta V}{T}$$

$$= \frac{(0.100 \text{ bar})(-0.090\,0 \text{ m}^3)[(10^5 \text{ Pa})/(1 \text{ bar})][(1 \text{ J})/(1 \text{ Pa m}^3)]}{298 \text{ K}}$$

$$= -3.02 \text{ J K}^{-1}$$

which is a smaller magnitude than in the reversible case. For the universe,

$$\Delta S(\text{universe}) = 19.14 + (-3.02) = 16.12 \text{ J K}^{-1}$$

which is a positive value, implying that the entropy content of the universe is increasing for a real process.

EXAMPLE 5.10. Calculate ΔS(surroundings) and ΔS(universe) for the process described in Example 5.6, assuming that the process is performed irreversibly by placing the gas in an oven at 1 500 K.

For the irreversible process, the heat transferred from the oven is given by

$$q(\text{surroundings}) = -(1.00 \text{ mol}) \int_{298\,\text{K}}^{1500\,\text{K}} C_P \, dT$$

$$= -(5.351)(1\,500 - 298) - \left(\frac{177.669 \times 10^{-3}}{2}\right)(1\,500^2 - 298^2)$$

$$+ \left(\frac{687.01 \times 10^{-7}}{3}\right)(1\,500^3 - 298^3) - \left(\frac{8.514 \times 10^{-9}}{4}\right)(1\,500^4 - 298^4)$$

$$= 132\,497 \text{ J}$$

and applying (5.14) gives

$$\Delta S(\text{surroundings}) = \frac{-132\,497}{1\,500} = -88.33 \text{ J K}^{-1}$$

For the universe,

$$\Delta S(\text{universe}) = 157.47 + (-88.33) = 69.14 \text{ J K}^{-1}$$

The Third Law of Thermodynamics

5.10 STATEMENT

The third law of thermodynamics can be stated as "the entropy content of all perfect crystalline materials is the same at 0 K." For these materials, the value of S_0° is chosen as zero. For nonperfect crystalline materials, S_0° is greater than zero and must be determined using theoretical considerations.

The value of S_0° for crystals in which molecules may orient themselves in Ω unique ways is given by

$$S_0^{\circ} = nR \ln \Omega \qquad (5.15)$$

Because the molecules in a real crystal are not completely random in their orientation, due to weak intermolecular forces, size considerations, etc., the value of S_0° predicted by (5.15) is often slightly too large.

For a glassy material, the value of S_0° depends on the actual processes used in the production of the glass. In general, the value will lie between S_0° for the solid state and S_0° for the liquid state.

For a polymorphic material, the third law holds only for the stable crystalline state of the substance. Even though a perfect crystal is formed by a metastable polymorph at 0 K, S_0° will be positive.

For ideal solid solutions, the entropy of mixing given by (5.13) is still present at 0 K, and so $S_0^{\circ} = \Delta_{mix} S^{\circ}$.

EXAMPLE 5.11.   What is the value of S_0° for water?

A portion of an ice crystal is shown in Fig. 5-5. Each oxygen is in the center of a tetrahedron of hydrogens and a larger tetrahedron of oxygens. A given water molecule can be oriented in any of six positions if all positions are open. Because the adjacent water molecules can also be oriented in six positions, there is a 1 in 2 chance that a given position is filled; and because this can happen twice, the number of random orientations for the given molecule is cut down by a factor of $\frac{1}{2} \times \frac{1}{2} = \frac{1}{4}$. Thus $\Omega = \frac{6}{4}$ and (5.15) gives

$$S_0^{\circ} = R \ln \tfrac{6}{4} = 3.371 \text{ J K}^{-1} \text{ mol}^{-1}$$

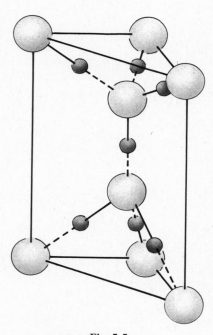

Fig. 5-5

EXAMPLE 5.12.   The stable form of Na_2SO_4 at 0 K is $Na_2SO_4(V)$. What is S_0° for $Na_2SO_4(III)$, a metastable form, if $\Delta_{V \to III} H^{\circ} = 2\,996 \text{ J mol}^{-1}$ at 298 K, $S_{298}^{\circ} - S_0^{\circ} = 154.92 \text{ J K}^{-1} \text{ mol}^{-1}$ for (III) and $S_{298}^{\circ} - S_0^{\circ} = 149.58 \text{ J K}^{-1} \text{ mol}^{-1}$ for (V)?

The value of $S_0^{\circ}(III)$ can be calculated using the series of steps shown in Fig. 5-6, where

$$\Delta S = S_0^{\circ}(III) - S_0^{\circ}(V) = S_0^{\circ}(III)$$

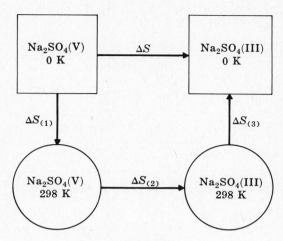

Fig. 5-6

because $S_0^\circ(V)$ is zero. Recognizing that

$$\Delta S = \Delta S_{(1)} + \Delta S_{(2)} + \Delta S_{(3)}$$

$$= [S_{298}^\circ(V) - S_0^\circ(V)] + \left[\frac{\Delta H^\circ}{T}\right] + [S_0^\circ(III) - S_{298}^\circ(III)]$$

gives

$$S_0^\circ(III) = 149.58 + \frac{2\,996}{298} + (-154.92) = 4.71 \text{ J K}^{-1} \text{ mol}^{-1}$$

5.11 VALUES OF S_T°

The contributions to S_T° for a material are given by (5.9) and (5.10b):

$$S_T^\circ = S_0^\circ + \sum_i^{\text{phases}} \int \frac{C_{P,i}^\circ}{T}\, dT + \sum_j^{\text{transitions}} \frac{\Delta_j H^\circ}{T_j} \qquad (5.16a)$$

$$= S_0^\circ + \sum_i^{\text{phases}} \int C_{P,i}^\circ\, d(\ln T) + \sum_j^{\text{transitions}} \frac{\Delta_j H^\circ}{T_j} \qquad (5.16b)$$

In (5.16) the value of S_0° is determined as in Section 5.10. The numerical value of the heat-capacity contribution is usually determined from graphical integration of a plot of $C_{P,i}^\circ / T$ against T or a plot of $C_{P,i}^\circ$ against $\ln T$, for all stable phases that the material exists in from 0 K to T.

EXAMPLE 5.13. Most compilations of low-temperature heat-capacity data for solids begin at temperatures in the range of 5 K to 25 K. Derive an equation for the contribution to S_T° for heating the solid from 0 K to the temperature T_{low} at which the first heat-capacity value $C_P^\circ(\text{low})$ is given.

The low-temperature limit of (3.27) is

$$C_P^\circ = \alpha T^3$$

which upon substitution into (5.10) gives

$$\Delta S(0 \to T_{\text{low}}) = \int_0^{T_{\text{low}}} \frac{\alpha T^3}{T}\, dT = \frac{\alpha T_{\text{low}}^3}{3} = \frac{C_P^\circ(\text{low})}{3}$$

EXAMPLE 5.14. Calculate S°_{1500} for liquid Na_2SO_4 from the following data:

$S^\circ_0(V) = 0$

$C^\circ_P(V)/(\text{cal K}^{-1} \text{ mol}^{-1})$	0.222	0.581	1.213	2.026	2.981	4.054	5.206	6.402	7.595
$T/(\text{K})$	15	20	25	30	35	40	45	50	55
$C^\circ_P(V)$	8.746	10.851	12.754	14.439	15.941	17.258	18.441	20.502	22.232
T	60	70	80	90	100	110	120	140	160
$C^\circ_P(V)$	23.793	25.217	27.656	29.744	30.603	32.819	34.710	36.672	
T	180	200	240	280	298.15	350	400	458	

$\Delta_{V \to IV} H^\circ = 75 \text{ cal mol}^{-1}$ at 458 K

$C^\circ_P(IV)/(\text{cal K}^{-1} \text{ mol}^{-1})$	36.672	37.055	37.989	38.417
$T/(\text{K})$	458	470	500	514

$\Delta_{IV \to I} H^\circ = 2\,611 \text{ cal mol}^{-1}$ at 514 K

$C^\circ_P(I)/(\text{cal K}^{-1} \text{ mol}^{-1})$	40.889	41.358	43.940	45.611	48.101
$T/(\text{K})$	514	550	750	850	950

$\Delta_{I \to \delta} H^\circ = 80 \text{ cal mol}^{-1}$ at 980 K

$C^\circ_P(\delta)/(\text{cal K}^{-1} \text{ mol}^{-1})$	48.254	50.000
$T/(\text{K})$	1\,000	1\,100

$\Delta_{\delta \to l} H^\circ = 5\,500 \text{ cal mol}^{-1}$ at 1 157 K

$C^\circ_P(l) = 47.180 \text{ cal K}^{-1} \text{ mol}^{-1}$ from 1 157 K to 1 500 K

The contribution from 0 K to 15 K is

$$\Delta S^\circ(0 \text{ K} \longrightarrow 15 \text{ K}) = \frac{0.222 \text{ cal K}^{-1} \text{ mol}^{-1}}{3} = 0.074 \text{ cal K}^{-1} \text{ mol}^{-1}$$

The remaining contribution of the heat capacities is found by graphical integration of Fig. 5-7 or Fig. 5-8, giving 103.16 cal K^{-1} mol^{-1}. The contributions from the phase changes are

$$\Delta S^\circ(V \to IV) = \frac{75}{458} = 0.164 \text{ cal K}^{-1} \text{ mol}^{-1}$$

$$\Delta S^\circ(IV \to I) = \frac{2611}{514} = 5.080 \text{ cal K}^{-1} \text{ mol}^{-1}$$

$$\Delta S^\circ(I \to \delta) = \frac{80}{980} = 0.082 \text{ cal K}^{-1} \text{ mol}^{-1}$$

$$\Delta S^\circ(\delta \to l) = \frac{5\,500}{1\,157} = 4.754 \text{ cal K}^{-1} \text{ mol}^{-1}$$

Summing these values gives $S^\circ_{1500} = 113.31$ cal K^{-1} mol^{-1} = 474.09 J K^{-1} mol^{-1}.

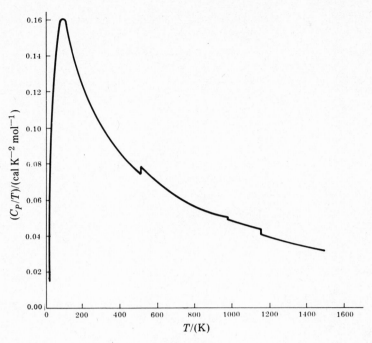

Fig. 5-7

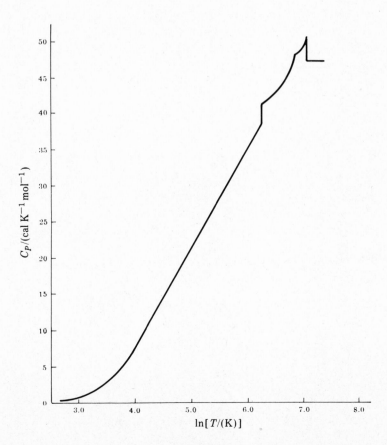

Fig. 5-8

EXAMPLE 5.15. Older thermodynamic data tables listed values of S_T at 1 atm. What is the correction factor for ideal gases to convert these values to the new standard state of 1 bar?

For the isothermal expansion from 1 atm to 1 bar, (*5.11b*) gives

$$\Delta S_{298}^{\circ}(1 \text{ atm} \longrightarrow 1 \text{ bar}) = -(8.314 \text{ J K}^{-1} \text{ mol}^{-1}) \ln \frac{1 \text{ bar}}{(1 \text{ atm})[(1.013\,25 \text{ bar})/(1 \text{ atm})]}$$

$$= 0.109\,4 \text{ J K}^{-1} \text{ mol}^{-1}$$

ΔS *for Chemical Reactions*

5.12 $\Delta_r S_T^{\circ}$ FROM THIRD LAW ENTROPIES

The value of ΔS_T° for a chemical reaction is given by

$$\Delta_r S_T^{\circ} = \sum_i \nu_i S_{T,i}^{\circ} \qquad (5.17)$$

where ν_i, the stoichiometric coefficients of the balanced equation (ν_i is positive for products and negative for reactants), have the unit of mol.

Because the entropy content of a material reflects the amount of randomness in the material, the values of $\Delta_r S_T^{\circ}$ can be checked qualitatively by observing whether any transitions between the condensed phases and the gaseous state have occurred, whether a change in the number of gaseous moles has occurred, etc.

EXAMPLE 5.16. Calculate $\Delta_f S_{298}^{\circ}$ for $CH_3OH(l)$ given that $S_{298}^{\circ}/(\text{J K}^{-1} \text{ mol}^{-1}) = 126.8$ for $CH_3OH(l)$, 5.740 for C(graph), 130.684 for $H_2(g)$, and 205.138 for $O_2(g)$.

For the reaction

$$C(\text{graph}) + 2H_2(g) + \tfrac{1}{2}O_2(g) \longrightarrow CH_3OH(l)$$

(*5.17*) gives

$$\Delta_f S_{298}^{\circ} = \{(1 \text{ mol})[S_{298}^{\circ}(CH_3OH)]\} - \{(1 \text{ mol})[S_{298}^{\circ}(C)] + (2 \text{ mol})[S_{298}^{\circ}(H_2)] + (\tfrac{1}{2} \text{ mol})[S_{298}^{\circ}(O_2)]\}$$

$$= [(1 \text{ mol})(126.8 \text{ J K}^{-1} \text{ mol}^{-1})]$$

$$- [(1 \text{ mol})(5.740 \text{ J K}^{-1} \text{ mol}^{-1}) + (2 \text{ mol})(130.684 \text{ J K}^{-1} \text{ mol}^{-1}) + (\tfrac{1}{2} \text{ mol})(205.138 \text{ J K}^{-1} \text{ mol}^{-1})]$$

$$= -242.9 \text{ J K}^{-1}$$

or $\Delta_f S_{298}^{\circ} = -242.9 \text{ J K}^{-1} \text{ mol}^{-1}$. As a qualitative check, there is 2.5 mol of gas reacting with 1 mol of solid to give 1 mol of liquid, so a large decrease in randomness would be expected.

5.13 TEMPERATURE DEPENDENCE OF $\Delta_r S_T^{\circ}$

If $\Delta_r S_{298}^{\circ}$ is known, then $\Delta_r S_T^{\circ}$ is given by

$$\Delta_r S_T^{\circ} = \Delta_r S_{298}^{\circ} + \int_{298\,K}^{T} \frac{\Delta_r C_P^{\circ}}{T} \, dT \qquad (5.18)$$

If the heat capacity data are given in the form of (*3.22*), then

$$\Delta_r S_T^{\circ} = \Delta_r S_{298}^{\circ} + \Delta a \ln \frac{T}{298} + \Delta b (T - 298) + \frac{\Delta c}{2}(T^2 - 298^2) + \frac{\Delta d}{3}(T^3 - 298^3) \qquad (5.19a)$$

$$\Delta_r S_T^{\circ} = \Delta_r S_{298}^{\circ} + \Delta a \ln \frac{T}{298} + \Delta b (T - 298) - \frac{\Delta c'}{2}\left(\frac{1}{T^2} - \frac{1}{298^2}\right) \qquad (5.19b)$$

If L and L' are defined as

$$L = \Delta_r S^\circ_{298} - \Delta a \ln(298) - \Delta b(298) - (\Delta c/2)(298)^2 - (\Delta d/3)(298)^3 \qquad (5.20a)$$

$$L' = \Delta_r S^\circ_{298} - \Delta a \ln(298) - \Delta b(298) + (\Delta c'/2)(298)^{-2} \qquad (5.20b)$$

then

$$\Delta_r S^\circ_T = L + \Delta a \ln T + \Delta b\, T + (\Delta c/2)T^2 + (\Delta d/3)T^3 \qquad (5.21a)$$

$$\Delta_r S^\circ_T = L' + \Delta a \ln T + \Delta b\, T - (\Delta c'/2)(T^{-2}) \qquad (5.21b)$$

EXAMPLE 5.17. Calculate $\Delta_r S^\circ_{1000}$ for

$$H_2(g) + Cl_2(g) \longrightarrow 2HCl(g)$$

given that S°_{298} (J K^{-1} mol^{-1}) = 130.684 for $H_2(g)$, 223.066 for $Cl_2(g)$, and 186.908 for HCl(g), and given

$$C^\circ_P = 29.066 - 0.836 \times 10^{-3}\,T + 20.17 \times 10^{-7}\,T^2 \qquad \text{for } H_2(g)$$

$$C^\circ_P = 31.696 + 10.144 \times 10^{-3}\,T - 40.38 \times 10^{-7}\,T^2 \qquad \text{for } Cl_2(g)$$

$$C^\circ_P = 28.166 + 1.810 \times 10^{-3}\,T + 15.47 \times 10^{-7}\,T^2 \qquad \text{for } HCl(g)$$

all in J K^{-1} mol^{-1}.

For the reaction, (5.17) gives

$$\Delta_r S^\circ_{298} = [(2\text{ mol})S^\circ_{298}(HCl)] - [(1\text{ mol})S^\circ_{298}(H_2) + (1\text{ mol})S^\circ_{298}(Cl_2)]$$

$$= [(2\text{ mol})(186.908\text{ J K}^{-1}\text{ mol}^{-1})]$$

$$- [(1\text{ mol})(130.684\text{ J K}^{-1}\text{ mol}^{-1}) + (1\text{ mol})(223.066\text{ J K}^{-1}\text{ mol}^{-1})]$$

$$= 20.066\text{ J K}^{-1}$$

The value of $\Delta_r C^\circ_P$ is

$$\Delta_r C^\circ_P = [(2)(28.166) - (1)(29.066) - (1)(31.696)] + [(2)(1.810) - (1)(-0.836) - (1)(10.144)] \times 10^{-3}\,T$$

$$+ [(2)(15.47) - (1)(20.17) - (1)(-40.38)] \times 10^{-7}\,T^2$$

$$= -4.430 - 5.688 \times 10^{-3}\,T + 51.15 \times 10^{-7}\,T^2$$

Substituting these values into (5.19a) gives

$$\Delta_r S^\circ_{1000} = 20.066 + (-4.430)\ln\frac{1\,000}{298} + (-5.688 \times 10^{-3})(1\,000 - 298) + \frac{51.15 \times 10^{-7}}{2}(1\,000^2 - 298^2)$$

$$= 13.040\text{ J K}^{-1}$$

5.14 PRESSURE DEPENDENCE OF $\Delta_r S^\circ_T$

Assuming the gaseous reactants and products to be ideal, and neglecting the very small pressure dependence of S°_T from the condensed states of matter, (5.11b) gives

$$\Delta_r S_T = \Delta_r S^\circ_T - R\,\Delta n_g \ln[P/(\text{bar})] \qquad (5.22)$$

where Δn_g is determined using the coefficients of the gaseous substances only (see Example 3.2).

EXAMPLE 5.18. What value of $\Delta_r S_{1000}$ would be listed in older thermodynamic tables at 1 atm for the reaction given in Example 5.17?

For the formation of HCl,

$$\Delta n_g = (2) - (1 + 1) = 0$$

which upon substitution into (5.22) gives

$$\Delta_r S_{1000} = \Delta_r S^\circ_{1000} - 0 = 13.040 \text{ J K}^{-1}$$

Solved Problems

THE SECOND LAW OF THERMODYNAMICS

5.1. Repeat the calculations of Example 5.1, using an ideal diatomic gas instead of the monatomic gas. Compare the results.

Assuming $C_V = \frac{5}{2}R$ for the diatomic gas, we obtain

$$V_3 = (0.100\,0 \text{ m}^3)\left(\frac{773 \text{ K}}{273 \text{ K}}\right)^{5/2} = 1.349 \text{ m}^3 \qquad V_4 = (1.349 \text{ m}^3)\frac{0.010\,0}{0.100\,0} = 0.134\,9 \text{ m}^3$$

The volume required for the cycle is much larger for the diatomic gas than for the monatomic gas.

For the first step,

$$\Delta U_{(1)} = 0$$

$$-q_1 = w_1 = -(1.00 \text{ mol})(8.314 \text{ J K}^{-1} \text{ mol}^{-1})(773 \text{ K}) \ln \frac{0.100\,0}{0.010\,0} = -14.80 \text{ kJ}$$

For the second step,

$$q_2 = 0$$

$$\Delta U_{(2)} = w_2 = (1.00 \text{ mol})(\tfrac{5}{2})(8.314 \text{ J K}^{-1} \text{ mol}^{-1})(273 \text{ K} - 773 \text{ K}) = -10.39 \text{ kJ}$$

For the third step,

$$\Delta U_{(3)} = 0$$

$$-q_3 = w_3 = -(1.00)(8.314)(273) \ln \frac{0.134\,5}{1.345} = 5.23 \text{ kJ}$$

For the fourth step,

$$q_4 = 0$$

$$\Delta U_{(4)} = w_4 = 10.39 \text{ kJ}$$

Overall:

$$\Delta U = 0 + (-10.39 \text{ kJ}) + 0 + 10.39 \text{ kJ} = 0$$

$$-q = w = -14.80 + 0 + (5.23) + 0 = -9.57 \text{ kJ}$$

The values determined for the individual second and fourth steps differ because of the difference in heat capacities between the diatomic and monatomic gases, but the values determined for the individual first and third steps, as well as the overall values, do not differ. The engine diagram is Fig. 5-3, as in the monatomic case.

5.2. What is the efficiency of a Carnot engine operating as a heat engine between two reservoirs at 500 K and 100 K? What is the performance factor for a Carnot engine operating as a refrigerator between the same reservoirs?

For the heat engine, (5.2) gives

$$\varepsilon_{\text{Carnot}} = 100 \frac{500 \text{ K} - 100 \text{ K}}{500 \text{ K}} = 80\%$$

and (5.3b) gives for the refrigerator

$$\varepsilon_{\text{ref}} = 100 \,\frac{500 \text{ K} - 100 \text{ K}}{100 \text{ K}} = 400\%$$

5.3.   Figure 5-9 shows a Carnot heat engine driving a Carnot refrigerator. Determine the value of q'_h.

The efficiency and performance factor of the Carnot engines are 80% and 400%, respectively, as calculated in Problem 5.2. If $q_h = 100.$ J, then from (5.1),

$$w = -\frac{\varepsilon q_h}{100} = -\frac{80(100. \text{ J})}{100} = -80. \text{ J}$$

and

$$q_c = -(100. \text{ J} - 80. \text{ J}) = -20. \text{ J}$$

If $w = 80.$ J for the refrigerator, then, from (5.3a),

$$q'_c = \frac{100w}{\varepsilon_{\text{ref}}} = \frac{100(80. \text{ J})}{400} = 20. \text{ J}$$

and

$$q'_h = -(20. \text{ J} + 80. \text{ J}) = -100. \text{ J}$$

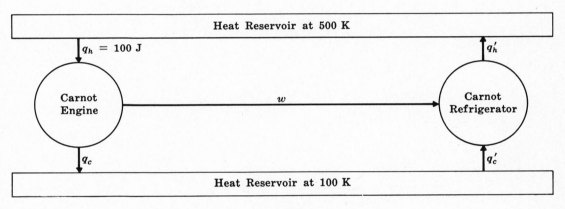

Fig. 5-9

ENTROPY CALCULATIONS

5.4.   A system at 25 °C absorbed 6.2 MJ from its surroundings without undergoing a temperature change. Calculate ΔS(system). Is this an increase or a decrease in entropy for the system?

If we assume that a reversible process is used, (5.7) gives for the system

$$\Delta S(\text{system}) = \frac{(6.2 \text{ MJ})[(10^6 \text{ J})/(1 \text{ MJ})]}{298 \text{ K}} = 21\,000 \text{ J K}^{-1}$$

Since the sign on ΔS is positive, there has been an increase in entropy for the system.

5.5.   What is the entropy change for 1.00 mol of $H_2(g)$ that has undergone a reversible expansion from 0.010 0 m³ at 100. K to 0.100 0 m³ at 600. K? For hydrogen,

$$C_P = 29.066 - 0.836 \times 10^{-3}\,T + 20.17 \times 10^{-7}\,T^2 \text{ in J K}^{-1} \text{ mol}^{-1}.$$

If we assume ideal gas behavior, (3.18) gives

$$C_V = C_P - R = 20.752 - 0.836 \times 10^{-3}\,T + 20.17 \times 10^{-7}\,T^2$$

which, upon substitution into (5.11a), gives

$$\Delta S(\text{system}) = (1.00 \text{ mol}) \int_{100\,\text{K}}^{600\,\text{K}} \frac{C_V}{T}\, dT + nR \ln \frac{V_2}{V_1}$$

$$= \left[(20.752) \ln \frac{600}{100} - (0.836 \times 10^{-3})(600 - 100) + \left(\frac{20.17 \times 10^{-7}}{2} \right)(600^2 - 100^2) \right]$$

$$+ (1.00)(8.314) \ln \frac{0.100\,0}{0.010\,0}$$

$$= 56.26 \text{ J K}^{-1}$$

5.6. Derive an equation for ΔS(system) for a van der Waals gas undergoing a reversible isothermal expansion. Compare the value calculated from this equation to that predicted by (5.11a) for 2.00 mol of argon ($b = 0.032\,19 \text{ dm}^3 \text{ mol}^{-1}$) undergoing an expansion from $0.010\,0 \text{ m}^3$ to $0.100\,0 \text{ m}^3$ at 25 °C.

For a van der Waals gas, (1.22) gives

$$P = \frac{nRT}{V - nb} - \frac{an^2}{V^2}$$

Thus,

$$\left(\frac{\partial P}{\partial T} \right)_V = \frac{nR}{V - nb}$$

which, upon substitution into (5.10a), gives

$$\Delta S(\text{system}) = 0 + \int_{V_1}^{V_2} \left(\frac{nR}{V - nb} \right) dV = nR \ln \left(\frac{V_2 - nb}{V_1 - nb} \right)$$

For the van der Waals gas,

$$\Delta S(\text{system}) = (2.00 \text{ mol})(8.314 \text{ J K}^{-1} \text{ mol}^{-1}) \ln \left(\frac{100.0 \text{ dm}^3 - (2.00 \text{ mol})(0.032\,19 \text{ dm}^3 \text{ mol}^{-1})}{10.0 \text{ dm}^3 - (2.00 \text{ mol})(0.032\,19 \text{ dm}^3 \text{ mol}^{-1})} \right)$$

$$= 38.4 \text{ J K}^{-1}$$

and for the ideal gas,

$$\Delta S(\text{system}) = (2.00)(8.314) \ln \left(\frac{100.0}{10.0} \right) = 38.3 \text{ J K}^{-1}$$

a difference of 0.3%.

5.7. Calculate ΔS(system) for the isothermal expansion of 1.00 mol of solid aluminum from 100.0 bar to 1.0 bar given $\alpha = 2.21 \times 10^{-5} \text{ K}^{-1}$ and $\rho = 2.702 \times 10^3 \text{ kg m}^{-3}$.

Assuming that the pressure change can be performed reversibly, (5.12b) gives

$$\Delta S(\text{system}) = - \int_{100.0}^{1.0} \alpha V\, dP = -\alpha V (P_2 - P_1)$$

$$= -(2.21 \times 10^{-5} \text{ K}^{-1}) \left(\frac{26.98 \times 10^{-3} \text{ kg mol}^{-1}}{2.702 \times 10^3 \text{ kg m}^{-3}} \right)(1.0 \text{ bar} - 100.0 \text{ bar}) \left(\frac{10^5 \text{ Pa}}{1 \text{ bar}} \right) \left(\frac{1 \text{ J}}{1 \text{ Pa m}^3} \right)$$

$$= 2.18 \times 10^{-3} \text{ J K}^{-1}$$

5.8. There are two different forms of molecular hydrogen—*ortho*-hydrogen, in which the nuclear spins of the atoms are parallel, and *para*-hydrogen, in which the spins are opposed. At high temperatures, an ortho-para equilibrium exists in naturally occurring hydrogen consisting of three-fourths *ortho*-hydrogen and one-fourth *para*-hydrogen. Calculate ΔS(system) for isothermally preparing 1.00 mol of this mixture under reversible conditions.

For this mixture, (5.13b) gives

$$\Delta_{mix} S(\text{system}) = -(8.314 \text{ J K}^{-1} \text{ mol}^{-1})[(0.250 \text{ mol}) \ln(0.250) + (0.750 \text{ mol}) \ln(0.750)]$$

$$= 4.68 \text{ J K}^{-1}$$

5.9. Calculate the entropy changes for the process described in Problem 5.5 if they are performed irreversibly by placing the hydrogen in an oven at 750 K and allowing it to expand against a constant external pressure of 1.00 bar.

For this irreversible process, $\Delta S(\text{system})$ will be the same. For the surroundings, $w(\text{surroundings}) = -w(\text{system})$, and the heat required is

$$q(\text{surroundings}) = -(1.00 \text{ mol}) \int_{100 \text{ K}}^{600 \text{ K}} C_V \, dT - w(\text{surroundings})$$

$$= \left[-(20.752)(600 - 100) + \left(\frac{0.836 \times 10^{-3}}{2} \right)(600^2 - 100^2) - \left(\frac{20.17 \times 10^{-7}}{3} \right)(600^3 - 100^3) \right]$$

$$+ [-(1.00 \text{ bar})(0.090 \, 0 \text{ m}^3)[(10^5 \text{ Pa})/(1 \text{ bar})][(1 \text{ J})/(1 \text{ Pa m}^3)]]$$

$$= -19 \, 374 \text{ J}$$

giving

$$\Delta S(\text{surroundings}) = \frac{-19 \, 374 \text{ J}}{750 \text{ K}} = -25.8 \text{ J K}^{-1}$$

$$\Delta S(\text{universe}) = 56.26 \text{ J K}^{-1} + (-25.8 \text{ J K}^{-1}) = 30.4 \text{ J K}^{-1}$$

5.10. Calculate the entropy changes for the transition at 368 K of 1.00 mol of sulfur from the monoclinic to the rhombic solid state, if $\Delta H = -401.7 \text{ J mol}^{-1}$ for the transition. Assume the surroundings to be an ice-water bath at 0 °C.

Assuming the phase transition at 368 K to occur reversibly, (5.9) gives

$$\Delta S(\text{system}) = \frac{(1.00 \text{ mol})(-401.7 \text{ J mol}^{-1})}{368 \text{ K}} = -1.09 \text{ J K}^{-1}$$

The ice-water bath absorbs the 401.7 J mol^{-1} at a temperature of 0 °C, so for the surroundings,

$$\Delta S(\text{surroundings}) = \frac{(1.00 \text{ mol})(401.7 \text{ J mol}^{-1})}{273 \text{ K}} = 1.47 \text{ J K}^{-1}$$

Summing these gives

$$\Delta S(\text{universe}) = (-1.09) + 1.47 = 0.38 \text{ J K}^{-1}$$

5.11. What are the values of the entropy changes for the cooling of 1.00 mol of $O_2(g)$ from 298 K to $O_2(l)$ at 90.19 K if the process is done (a) reversibly? (b) irreversibly by placing the sample in liquid hydrogen at 13.96 K? Assume $\Delta_{vap} H = 6 \, 820 \text{ J mol}^{-1}$ at 90.19 K and $C_P = \frac{7}{2} R$ for the gas.

By considering the process to be a combination of a reversible cooling at constant pressure and a reversible phase transition, (5.9) and (5.11) give for both processes

$$\Delta S(\text{system}) = \int_{298 \text{ K}}^{90.19 \text{ K}} \frac{C_P}{T} \, dT + \frac{\Delta_{vap} H}{T_{bp}}$$

$$= (1.00 \text{ mol}) (\tfrac{7}{2})(8.314 \text{ J K}^{-1} \text{ mol}^{-1}) \ln \frac{90.19 \text{ K}}{298 \text{ K}} + \frac{(1.00 \text{ mol})(-6 \, 820 \text{ J mol}^{-1})}{90.19 \text{ K}}$$

$$= -110.4 \text{ J K}^{-1}$$

For the reversible process,

$$\Delta S(\text{surroundings}) = 110.4 \text{ J K}^{-1} \qquad \text{and} \qquad \Delta S(\text{universe}) = 0$$

For the irreversible process, q for the surroundings is

$$q(\text{surroundings}) = -\int_{298\,\text{K}}^{90.19\,\text{K}} C_P \, dT + \Delta_{\text{vap}} H$$

$$= -(1.00 \text{ mol})(\tfrac{7}{2})(8.314 \text{ J K}^{-1} \text{ mol}^{-1})(90.19 \text{ K} - 298 \text{ K}) + (1.00 \text{ mol})(6\,820 \text{ J mol}^{-1})$$

$$= 12.87 \text{ kJ}$$

and (5.14) gives

$$\Delta S(\text{surroundings}) = \frac{12\,870 \text{ J}}{13.96 \text{ K}} = 922 \text{ J K}^{-1}$$

and

$$\Delta S(\text{universe}) = 922 \text{ J K}^{-1} + (-110.4 \text{ J K}^{-1}) = 812 \text{ J K}^{-1}$$

THE THIRD LAW OF THERMODYNAMICS

5.12. What is the value of S_0° for CO or NO?

In the solid state there are two possible orientations of the molecule. Using (5.15) gives

$$S_0^\circ = (8.314 \text{ J K}^{-1} \text{ mol}^{-1}) \ln 2 = 5.763 \text{ J K}^{-1} \text{ mol}^{-1}$$

5.13. For molecular oxygen, $S_{298}^\circ = 205.138 \text{ J K}^{-1} \text{ mol}^{-1}$. Find S_{1500}° for O_2 given $C_P^\circ/(\text{J K}^{-1} \text{ mol}^{-1}) = 25.72 + 12.98 \times 10^{-3} T - 38.62 \times 10^{-7} T^2$.

For O_2 at 1 500 K,

$$S_{1500}^\circ = S_{298}^\circ + \int_{298\,\text{K}}^{1500\,\text{K}} \frac{C_P}{T} \, dT$$

$$= 205.138 + (25.72) \ln \frac{1\,500}{298} + (12.98 \times 10^{-3})(1\,500 - 298)$$

$$- \frac{38.62 \times 10^{-7}}{2} (1\,500^2 - 298^2)$$

$$= 258.13 \text{ J K}^{-1} \text{ mol}^{-1}$$

ΔS FOR CHEMICAL REACTIONS

5.14. Consider the reactions

$$\tfrac{1}{2}H_2(g) + \tfrac{1}{2}Cl_2(g) \longrightarrow HCl(g)$$

$$2C(\text{graph}) + 2H_2(g) \longrightarrow C_2H_4(g)$$

$$2C(\text{graph}) + \tfrac{5}{2}H_2(g) + \tfrac{1}{2}Cl_2(g) \longrightarrow C_2H_5Cl(g)$$

and the corresponding expressions for $\Delta_f S_T^\circ$

$$\Delta_f S_T^\circ(HCl) = [(1)S_T^\circ(HCl)] - [(\tfrac{1}{2})S_T^\circ(H_2) + (\tfrac{1}{2})S_T^\circ(Cl_2)]$$

$$\Delta_f S_T^\circ(C_2H_4) = [(1)S_T^\circ(C_2H_4)] - [(2)S_T^\circ(C) + (2)S_T^\circ(H_2)]$$

$$\Delta_f S_T^\circ(C_2H_5Cl) = [(1)S_T^\circ(C_2H_5Cl)] - [(2)S_T^\circ(C) + (\tfrac{5}{2})S_T^\circ(H_2) + (\tfrac{1}{2})S_T^\circ(Cl_2)]$$

Show that $\Delta_r S_T^\circ$ for

$$HCl(g) + C_2H_4(g) \longrightarrow C_2H_5Cl(g)$$

is given by

$$\Delta_r S_T^\circ = \sum_i \nu_i \, \Delta_f S_{T,i}^\circ \qquad (5.23)$$

where ν_i are the stoichiometric coefficients.

Substituting the expressions for $\Delta_f S_T^\circ$ into (5.23) gives

$$\Delta_r S_T^\circ = [(1)S_T^\circ(C_2H_5Cl) - (2)S_T^\circ(C) - (\tfrac{5}{2})S_T^\circ(H_2) - (\tfrac{1}{2})S_T^\circ(Cl_2)]$$

$$- [(1)S_T^\circ(HCl) - (\tfrac{1}{2})S_T^\circ(H_2) - (\tfrac{1}{2})S_T^\circ(Cl_2) + (1)S_T^\circ(C_2H_4)$$

$$- (2)S_T^\circ(C) - (2)S_T^\circ(H_2)]$$

Upon cancellation of common terms and rearrangement,

$$\Delta_r S_T^\circ = [(1)S_T^\circ(C_2H_5Cl)] - [(1)S_T^\circ(HCl) + (1)S_T^\circ(C_2H_4)]$$

which is identical to the equation predicted by (5.17)

5.15. Compare the $\Delta_f S_{298}^\circ$ values for $H_2O(g)$ and $H_2O(l)$ given $S_{298}^\circ/(J\ K^{-1}\ mol^{-1}) = 130.684$ for $H_2(g)$, 205.138 for $O_2(g)$, 69.91 for $H_2O(l)$, and 188.825 for $H_2O(g)$.

Applying (5.17) to the reaction

$$H_2(g) + \tfrac{1}{2}O_2(g) \longrightarrow H_2O$$

gives $\qquad \Delta_f S_{298}^\circ = (1\ mol)[S_{298}^\circ(H_2O)] - \{(1\ mol)[S_{298}^\circ(H_2)] + (\tfrac{1}{2}\ mol)[S_{298}^\circ(O_2)]\}$

For gaseous water,

$$\Delta_f S_{298}^\circ = (1\ mol)(188.825\ J\ K^{-1}\ mol^{-1})$$

$$- [(1\ mol)(130.684\ J\ K^{-1}\ mol^{-1}) + (\tfrac{1}{2}\ mol)(205.138\ J\ K^{-1}\ mol^{-1})]$$

$$= -44.428\ J\ K^{-1}$$

or $\Delta_f S_{298}^\circ = -44.428\ J\ K^{-1}\ mol^{-1}$, indicating a slight decrease in randomness as 1.5 mol of gaseous reactants form 1 mol of gaseous product.

For forming water in the liquid state,

$$\Delta_f S_{298}^\circ = (1)(69.91) - [(1)(130.684) + (\tfrac{1}{2})(205.138)]$$

$$= -163.34\ J\ K^{-1}$$

or $\Delta_f S_{298}^\circ = -163.34\ J\ K^{-1}\ mol^{-1}$, indicating a large decrease in randomness as 1.5 mol of gaseous reactants form 1 mol of product in the liquid state.

5.16. The reaction

$$C_2H_5OH(l) + HI(g) \longrightarrow C_2H_5I(l) + H_2O(l)$$

was run at 60 °C in order to change the rate of reaction. Given $S_{298}^\circ/(J\ K^{-1}\ mol^{-1}) = 160.7$ for $C_2H_5OH(l)$, 206.594 for $HI(g)$, 211.7 for $C_2H_5I(l)$, and 69.91 for $H_2O(l)$, and given $C_P^\circ/(J\ K^{-1}\ mol^{-1}) = 111.46$ for $C_2H_5OH(l)$, 29.158 for $HI(g)$, 115.1 for $C_2H_5I(l)$, and 75.291 for $H_2O(l)$, calculate $\Delta_r S_{333}$.

Applying (5.17) to the reaction gives

$$\Delta_r S_{298}^\circ = \{(1\ mol)[S_{298}^\circ(C_2H_5I)] + (1\ mol)[S_{298}^\circ(H_2O)]\}$$

$$- \{(1\ mol)[S_{298}^\circ(C_2H_5OH)] + (1\ mol)[S_{298}^\circ(HI)]\}$$

$$= [(1\ mol)(211.7\ J\ K^{-1}\ mol^{-1}) + (1\ mol)(69.91\ J\ K^{-1}\ mol^{-1})]$$

$$- [(1\ mol)(160.7\ J\ K^{-1}\ mol^{-1}) + (1\ mol)(206.594\ J\ K^{-1}\ mol^{-1})]$$

$$= -85.7\ J\ K^{-1}$$

The value of ΔC_P° is

$$\Delta C_P^0 = [(1\ \text{mol})(115.11\ \text{J K}^{-1}\ \text{mol}^{-1}) + (1\ \text{mol})(75.291\ \text{J K}^{-1}\ \text{mol}^{-1})]$$

$$- [(1\ \text{mol})(111.46\ \text{J K}^{-1}\ \text{mol}^{-1}) + (1\ \text{mol})(29.158\ \text{J K}^{-1}\ \text{mol}^{-1})]$$

$$= 49.8\ \text{J K}^{-1}$$

and (*5.19a*) gives

$$\Delta_r S_{333}^\circ = -85.7\ \text{J K}^{-1} + (49.8\ \text{J K}^{-1})\ \ln\frac{333}{298} = -80.2\ \text{J K}^{-1}$$

5.17. The electromotive force of the electrochemical cell at 1 atm for the reaction

$$\text{Ag(s)} + \tfrac{1}{2}\text{Cl}_2\text{(g)} \longrightarrow \text{AgCl(l)}$$

as reported by Metz and Seifert is given by $E/(\text{V}) = 0.908\,1 - 0.280X + 0.110X^2$, where $X = (T - 728.2) \times 10^{-3}\text{K}$. Given

$$\Delta S = nF\frac{dE}{dT}$$

where $n = 1$ mol and $F = 96\,484.56\ \text{C mol}^{-1}$, determine $\Delta_f S_{1000}^\circ$.

Applying the chain rule of differentiation,

$$\frac{dx}{dy} = \frac{dx}{dz}\frac{dz}{dy} \qquad\qquad (5.24)$$

gives

$$\Delta_f S_T = nF\frac{dE}{dX}\frac{dX}{dT}$$

$$= (1\ \text{mol})(96\,484.56\ \text{C mol}^{-1})$$

$$\times \{[(-0.280) + (0.110)(2)X]\ \text{V K}^{-1}\}[(1\ \text{A s})/(1\ \text{C})][(1\ \text{J})/(1\ \text{V A s})]$$

$$= (-27.02 + 21.33X)\ \text{J K}^{-1}$$

At 1 000 K,

$$\Delta_f S_{1000} = -27.02 + (21.23)(1\,000 - 728.2) \times 10^{-3} = -21.25\ \text{J K}^{-1}$$

Substituting $\Delta n_g = -\tfrac{1}{2}$ mol into (*5.22*) gives

$$\Delta_f S_{1000}^\circ = -21.25\ \text{J K}^{-1} + (8.314\ \text{J K}^{-1}\ \text{mol}^{-1})(-\tfrac{1}{2}\ \text{mol})\ \ln(1.013\,25)$$

$$= -21.30\ \text{J K}^{-1}$$

Supplementary Problems

THE SECOND LAW OF THERMODYNAMICS

5.18. What is the maximum efficiency of a steam engine operating between 120 °C and 20 °C?

    Ans. $\varepsilon_{\text{Carnot}} = 25.4\%$

5.19. Equation (*5.4a*) can be rearranged into

$$đq_{\text{rev}} = T\,dS$$

which upon integration gives

$$q_{\text{rev}} = \int T \, dS$$

A graphical integration of a plot of T against S will thus give the value of q_{rev}. Prepare a plot for the Carnot cycle described in Example 5.1, and perform the graphical integration to evaluate q_{rev}, which is equal to $-w$. For this calculation it is not necessary to know the original value of S for the gas, so assume a value of zero to make the graphing and calculations easier.

Ans. $\Delta S_{(1)} = 14\,800/773 = 19.15$ J K^{-1}, $\Delta S_{(2)} = 0$, $\Delta S_{(3)} = -19.15$ J K^{-1}, $\Delta S_{(4)} = 0$ (see Fig. 5-10); $q_{\text{rev}} = -w = 9.59$ kJ.

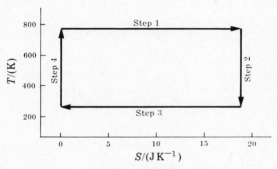

Fig. 5-10

5.20.   Prepare a plot of P against V for the Carnot engine described in Example 5.1, and perform the graphical integration to evaluate w. Assume a linear plot between V_2 and V_3 and between V_3 and V_4.

Ans.   See Fig. 5-11; -9.75 kJ

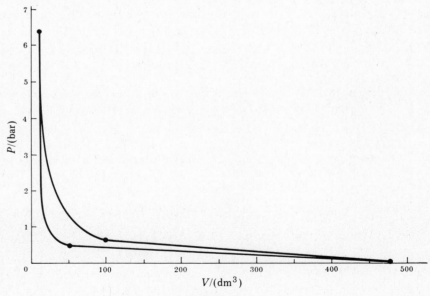

Fig. 5-11

5.21.   Repeat the calculations of Example 5.1 using an ideal nonlinear triatomic gas instead of the monatomic gas.

Ans. $C_V = 3R$, $V_3 = 2.270$ m^3, $V_4 = 0.227$ m^3; $\Delta U_{(1)} = 0$, $q_1 = -w_1 = 14.80$ kJ; $q_2 = 0$, $\Delta U_{(2)} = w_2 = -12.47$ kJ; $\Delta U_{(3)} = 0$, $q_3 = -w_3 = -5.23$ kJ; $q_4 = 0$, $\Delta U_{(4)} = w_4 = -12.47$ kJ; $\Delta U = 0$, $q = -w = 9.57$ kJ.

ENTROPY CALCULATIONS

5.22. A system absorbs 1.5 kJ at 25 °C from its surroundings and later releases 1.5 kJ to its surroundings at 75 °C. Calculate the overall entropy change for the system assuming both processes are reversible. Is this an increase or a decrease in entropy for the system?   *Ans.* $0.7 \, \text{J K}^{-1}$;   an increase

5.23. Using $(\partial V / \partial T)_P = R/P$ for an ideal gas, derive *(5.11b)* from *(5.10b)*.

5.24. Calculate ΔS(system) for the reversible isothermal expansion of 1.00 mol of steam from 100.0 bar to 1.0 bar assuming ideal gas behavior. Compare this value to that for liquid water undergoing the same pressure change ($\alpha = 2.57 \times 10^{-4} \, \text{K}^{-1}$ and $\rho = 0.97 \times 10^3 \, \text{kg m}^{-3}$).

   Ans. $-38.3 \, \text{J K}^{-1}$ for gas, $0.047 \, \text{J K}^{-1}$ for liquid; a factor of 815 larger

5.25. Calculate ΔS(system) for isothermally preparing a mixture containing 99 mol of $O_2(g)$ and 1 mol of $N_2(g)$.   *Ans.* $47 \, \text{J K}^{-1}$

5.26. What is the value of ΔS(system) if 0.500 mol of $CCl_4(l)$ is mixed with 0.500 mol of $CH_2Cl_2(l)$ at 25 °C and the final temperature of the solution is 27 °C? Assume $C_P/(\text{J K}^{-1} \, \text{mol}^{-1}) = 131.8$ for $CCl_4(l)$ and 100.0 for $CH_2Cl_2(l)$.

   Ans. $\Delta_{\text{mix}} S = 5.763 \, \text{J K}^{-1}$, $\Delta S = 0.441 \, \text{J K}^{-1}$ for heating CCl_4 and $0.334 \, \text{J K}^{-1}$
   for heating CH_2Cl_2; $6.538 \, \text{J K}^{-1}$

5.27. What are the entropy changes for the isothermal compression of 1.00 mol of an ideal gas from 1.00 bar to 5.00 bar at 25 °C if the compression is performed (*a*) reversibly? (*b*) irreversibly using an external pressure of 100.0 bar?

   Ans. ΔS(system) $= -13.38 \, \text{J K}^{-1}$; (*a*) ΔS(surroundings) $= 13.38 \, \text{J K}^{-1}$, ΔS(universe) $= 0$ for reversible process; (*b*) ΔS(surroundings) $= 664 \, \text{J K}^{-1}$, ΔS(universe) $= 651 \, \text{J K}^{-1}$ for irreversible process

5.28. Calculate the entropy changes for heating 1.00 mol of silver from 298 K to 1 500 K at constant pressure if the process is performed (*a*) reversibly; (*b*) irreversibly by placing the silver in an oven at 1 500 K. The average value of C_P for silver is $25.9 \, \text{J K}^{-1} \, \text{mol}^{-1}$.

   Ans. ΔS(system) $= 41.9 \, \text{J K}^{-1}$; (*a*) ΔS(surroundings) $= -41.9 \, \text{J K}^{-1}$, ΔS(universe) $= 0$ for reversible process; (*b*) ΔS(surroundings) $= -20.75 \, \text{J K}^{-1}$, ΔS(universe) $= 21.2 \, \text{J K}^{-1}$ for irreversible process

5.29. Calculate the entropy changes for melting 1.00 mol of sulfur from the monoclinic state at 388 K given $\Delta H = 1\,715 \, \text{J mol}^{-1}$. Assume that the sulfur is placed in a burner flame at 1 400 °C.

   Ans. ΔS(system) $= 4.42 \, \text{J K}^{-1}$, ΔS(surroundings) $= -1.03 \, \text{J K}^{-1}$, ΔS(universe) $= 3.39 \, \text{J K}^{-1}$

5.30. Calculate the entropy changes for evaporating 1.00 mol of superheated water at 110 °C and 1.01 bar, given $\Delta_{\text{vap}} H = 40\,150 \, \text{J mol}^{-1}$ at 1.433 bar and 110 °C and $(\partial V / \partial T)_P = -0.15 \times 10^{-6} \, \text{m}^3 \, \text{K}^{-1}$ for water. *Hint:* Use a cycle of pressure increase, evaporation, and pressure decrease.

   Ans. ΔS(system) $= 107.72 \, \text{J K}^{-1}$, ΔS(surroundings) $= -104.83 \, \text{J K}^{-1}$, ΔS(universe) $= 2.89 \, \text{J K}^{-1}$

THE THIRD LAW OF THERMODYNAMICS

5.31. Calculate $\Delta_{\text{mix}} S$ for preparing 1.00 mol of an ideal solid solution having the composition $x(\text{AgBr(s)}) = 0.728$ and $x(\text{AgCl(s)}) = 0.272$. The value of $S_{298}^\circ - S_{15}^\circ$ for the solution was determined as $0.0 \pm 0.1 \, \text{cal K}^{-1} \, \text{mol}^{-1}$ by Eastman and Milner. Determine S_0° for this mixture assuming that $S_{15}^\circ - S_0^\circ = 0$.

   Ans. $4.86 \, \text{J K}^{-1}$, $4.86 \, \text{J K}^{-1}$

5.32. The transition between monoclinic and rhombic sulfur is quite slow, and it is possible to supercool the monoclinic form to 0 K. Find S_0° for the metastable monoclinic form if $S_0^\circ = 0$ for the rhombic form, $S_{298}^\circ - S_0^\circ = 31.88 \, \text{J K}^{-1} \, \text{mol}^{-1}$ for the rhombic form and $S_{298}^\circ - S_0^\circ = 32.55 \, \text{J K}^{-1} \, \text{mol}^{-1}$ for the monoclinic form. Assume $C_P^\circ = 22.59$ J K^{-1} mol^{-1} for the rhombic form and 23.64 J K^{-1} mol^{-1} for the monoclinic form, between 298 K and 368.54 K. ΔH°(rhombic $\rightarrow$ monoclinic) = 401.71 J mol^{-1} at 368.54 K.

Ans.   Use the process rhombic(0 K) $\longrightarrow$ rhombic(298 K) $\longrightarrow$ rhombic(368.54 K) $\longrightarrow$ monoclinic (368.54 K) $\longrightarrow$ monoclinic(298 K) $\longrightarrow$ monoclinic(0 K) to obtain $S_{0,\text{monoclinic}}^\circ = 0.20$ J K^{-1} mol^{-1}.

5.33.   Calculate S_{298}° and S_{1500}° for Cu from the following data:

$T/$(K)	$C_P^\circ/$(J K^{-1} mol^{-1})	$T/$(K)	$C_P^\circ/$(J K^{-1} mol^{-1})
1	0.000 75	150	20.50
2	0.001 76	200	22.64
3	0.003 39	250	23.77
4	0.005 9	298	24.48
5	0.009 6	400	25.15
8	0.030 1	500	25.77
10	0.054	600	26.40
15	0.184	700	27.03
20	0.46	800	27.66
25	0.96	900	28.28
30	1.72	1 000	28.91
40	3.72	1 100	29.54
50	6.15	1 200	30.17
75	11.88	1 300	30.79
100	16.02		

$\Delta_{\text{fus}} H^\circ = 13\,055$ J mol^{-1} at 1 356 K
$C_P^\circ = 31.38$ J K^{-1} mol for the liquid from 1 356 K to 1 500 K

Ans.   33.35 J K^{-1} mol^{-1}, 87.07 J K^{-1} mol^{-1}

ΔS FOR CHEMICAL REACTIONS

5.34.   Choose the chemical equation that has the largest increase in entropy:
   (*a*) $CH_3OH(l) + \frac{3}{2}O_2(g) \rightarrow CO_2(g) + 2H_2O(l)$
   (*b*) $CaCO_3(s) \longrightarrow CaO(s) + CO_2(g)$
   (*c*) $C(s) + H_2O(g) \longrightarrow H_2(g) + CO(g)$
   (*d*) $2NH_3(g) \longrightarrow N_2(g) + 3H_2(g)$

   Ans.   (*d*)

5.35.   Calculate $\Delta_r S_{298}^\circ$ for

$$CaCO_3(\text{calcite}) \longrightarrow CaCO_3(\text{aragonite})$$

given that $S_{298}^\circ = 92.9$ J K^{-1} mol^{-1} for calcite and 88.7 J K^{-1} mol^{-1} for aragonite. Qualitatively interpret the results.

   Ans. -4.2 J K^{-1}, a small change in randomness resulting from 1 mol of a condensed phase being transformed into 1 mol of a condensed phase.

5.36.   Calculate $\Delta_r S_{298}^\circ$ for the deamination of L-aspartic acid to fumaric acid:

$$\text{HOOC—CH—CH}_2\text{—COOH} \longrightarrow \text{HOOC—CH=CH—COOH} + \text{NH}_3$$
$$\text{NH}_2$$

given that the entropies of formation for the acids, as reported by Burton and Krebs, are -194.1 cal K^{-1} mol^{-1} for L-aspartic acid and -126.2 cal K^{-1} mol^{-1} for fumaric acid, and $\Delta_f S_{298}^\circ(\text{NH}_3) = -99.38$ J K^{-1} mol^{-1}.

   Ans.   184.7 J K^{-1}, a large increase in randomness resulting from the production of 1 mol of gas and 1 mol of condensed phase from 1 mol of condensed phase.

5.37. For the reaction

$$Ag(s) + \tfrac{1}{2}Cl_2(g) \longrightarrow AgCl(l)$$

Metz and Seifert reported $\Delta C_P^\circ = 3.70 + 5.08X$ in cal K^{-1} mol^{-1}, where $X = (T - 728.2) \times 10^{-3}$ K. Given $\Delta_f S_{728}^\circ = -27.13$ J K^{-1} mol^{-1}, find $\Delta_f S_{1000}^\circ$.   *Ans.* -21.36 J K^{-1} mol^{-1}

5.38. Calculate ΔS for the flame reaction described in Example 4.2. Assume $S_{298}^\circ/(J\ K^{-1}\ mol^{-1}) = 205.138$ for $O_2(g)$, 130.684 for $H_2(g)$, and 188.825 for $H_2O(g)$.

Ans. $\Delta_r S_{298}^\circ = -44.428$ J K^{-1}, $\Delta S = 223.3$ J K^{-1} for heating;   178.9 J K^{-1}

Chapter 6

Free Energy

Free Energy

6.1 DEFINITION AND SIGNIFICANCE

The *Gibbs free energy*, or just *Gibbs energy* (G), is defined as

$$G \equiv H - TS \qquad (6.1)$$

and the *Helmholtz free energy* (A) is defined as

$$A \equiv E - TS \qquad (6.2)$$

Because absolute values for H and E are difficult to calculate, absolute values of G and A are not used. Instead, processes are analyzed in terms of changes in free energies, where

$$\Delta G = \Delta H - \Delta(TS) \qquad (6.3)$$

$$\Delta A = \Delta E - \Delta(TS) \qquad (6.4)$$

Under constant pressure and temperature conditions, ΔG will be negative for a spontaneous process, positive for a nonspontaneous process, and zero for a state of equilibrium. Similar statements hold true for the values of ΔA for processes performed under constant volume and temperature conditions.

A superscript $\circ$ attached to the symbol for the free energy implies standard pressure conditions (1 bar) for the substances involved in the process.

The differential of (6.1) is

$$dG = dE + P\,dV + V\,dP - T\,dS - S\,dT \qquad (6.5)$$

EXAMPLE 6.1. What is the value of $\Delta G^{\circ} - \Delta A^{\circ}$ for the combustion of benzoic acid at 25 °C?

Subtracting (6.4) from (6.3) and substituting (3.3) gives

$$\Delta G - \Delta A = \Delta(PV) \qquad (6.6)$$

For the reaction

$$C_6H_5COOH(s) + \tfrac{15}{2}O_2(g) = 7CO_2(g) + 3H_2O(l)$$

the contribution of the condensed phases to $\Delta(PV)$ in (6.6) is negligible (see Example 4.3), giving

$$\Delta G^{\circ} - \Delta A^{\circ} = RT\,\Delta n_g = (8.314\ \text{J K}^{-1}\ \text{mol}^{-1})(298\ \text{K})(-\tfrac{1}{2}\ \text{mol}) = -1\,239\ \text{J}$$

EXAMPLE 6.2. The standard-state emf for the Daniell-cell reaction

$$Zn(s) + Cu^{2+}(1\ M) = Zn^{2+}(1\ M) + Cu(s)$$

at 25 °C was reported by Buckbee, Surdzial, and Metz as 1.091 3 V. What is $\Delta_r G^{\circ}_{298}$? Is this reaction spontaneous?

Values of ΔG can be calculated from experimental electrochemical measurements using the relationship

$$\Delta G = -nFE \qquad (6.7)$$

where n is the number of moles of electrons in the balanced reaction, F is the Faraday constant ($9.648\,456 \times 10^4\ \text{C mol}^{-1} = 9.648\,456 \times 10^4\ \text{J mol}^{-1}\ \text{V}^{-1}$), and E is the cell potential. Substituting into (6.7) gives

$$\Delta_r G^{\circ}_{298} = -(2\ \text{mol})(96.485\ \text{kJ mol}^{-1}\ \text{V}^{-1})(1.091\,3\ \text{V}) = -210.59\ \text{kJ}$$

117

Because the reaction takes place under standard conditions, $\Delta G° = \Delta G$, and the large negative value for ΔG indicates a spontaneous reaction.

EXAMPLE 6.3. What is the significance of dG for a reversible isobaric and isothermal process? Qualitatively discuss this interpretation for the reaction considered in Example 6.2, where $\Delta_r H°_{298} = -228.53$ kJ.

Under reversible constant pressure and temperature conditions, (6.5) becomes

$$dG = P\,dV + đw_{rev} \tag{6.8}$$

Equation (6.8) implies that the free energy is the net useful work that can be obtained from a reversible process performed under these restraints.

For the Daniell cell, of the 228.53 kJ of heat released by the chemical reaction, only 210.59 kJ was available for doing useful work. The remainder of the energy was lost to PV expansion of the system, etc.

6.2 PRESSURE DEPENDENCE OF G

For reversible processes having only expansion work, (6.5) becomes

$$dG = V\,dP - S\,dT \tag{6.9}$$

which implies that

$$\left(\frac{\partial G}{\partial P}\right)_T = V \tag{6.10}$$

EXAMPLE 6.4. What is $G - G°$ for 1.00 mol of an ideal gas at 0.100 Pa and 25 °C?

For an isothermal expansion from P_1 to P_2, integration of (6.10) gives

$$\Delta G = -nRT \ln \frac{V_2}{V_1} = nRT \ln \frac{P_2}{P_1} \tag{6.11}$$

Here $P_1 = 1$ bar $= 10^5$ Pa and $P_2 = 0.100$ Pa, so that

$$\Delta G = G - G° = (1.00 \text{ mol})(8.314 \text{ J K}^{-1} \text{ mol}^{-1})(298 \text{ K}) \ln \left(\frac{0.100 \text{ Pa}}{10^5 \text{ Pa}}\right)$$

$$= -34.2 \text{ kJ}$$

EXAMPLE 6.5. Determine the correction factor that should be added to values of $\Delta_f G_{298}$ found in older thermodynamic tables (based on 1 atm as the standard reference pressure) to change these values to the newer standard reference pressure (1 bar).

The only major contribution to the correction term will be from any gases involved in the formation reaction. For each mole of gaseous substance, (6.11) gives

$$\Delta G = (8.314 \text{ J K}^{-1} \text{ mol}^{-1})(298 \text{ K}) \ln \left(\frac{1 \text{ bar}}{1.013\,25 \text{ bar}}\right) = -32.63 \text{ J mol}^{-1}$$

Thus the correction factor becomes

$$\Delta G = \Delta n_g (-32.63 \text{ J mol}^{-1})$$

where Δn_g is defined in Example 3.2.

6.3 TEMPERATURE DEPENDENCE OF G

The temperature dependence of G is given by (6.9) as

$$\left(\frac{\partial G}{\partial T}\right)_P = -S \tag{6.12}$$

Various forms of (6.12) are known as the *Gibbs–Helmholtz equations*.

EXAMPLE 6.6. Derive the Gibbs–Helmholtz equation that takes the form $[\partial(G/T)/\partial T]_P$.

Treating the numerator as a product, i.e., $G/T = GT^{-1}$, and using the product rule for differentiation, (5.24), gives

$$\left(\frac{\partial(G/T)}{\partial T}\right)_P = T^{-1}\left(\frac{\partial G}{\partial T}\right)_P + G\left(\frac{\partial T^{-1}}{\partial T}\right)_P = T^{-1}(-S) + G(-T^{-2})$$

$$= T^{-1}\frac{(G-H)}{T} - \frac{G}{T^2} = \frac{G}{T^2} - \frac{H}{T^2} - \frac{G}{T^2}$$

$$= -\frac{H}{T^2} \tag{6.13}$$

6.4 FREE ENERGY CALCULATIONS

For the large number of processes of chemical interest that occur at constant temperature, (6.3) can be simplified to

$$\Delta G = \Delta H - T\,\Delta S \tag{6.14}$$

If free energy of formation data are available for the reactants and products, $\Delta_f G_T^\circ$ can be calculated using

$$\Delta_r G_T^\circ = \sum_i \nu_i\,\Delta_f G_{T,i}^\circ \tag{6.15}$$

where ν_i represents the stoichiometric coefficients (ν_i is positive for products and negative for reactants) and has the units of mol.

EXAMPLE 6.7. For any phase transition, (5.9) gives $T\,\Delta S = \Delta H$, and so $\Delta G = 0$.

EXAMPLE 6.8. Calculate $\Delta_f G_{298}^\circ$ for gaseous and liquid water given $\Delta_f H_{298}^\circ/(\text{kJ mol}^{-1}) = -241.818$ and -285.830 and $S_{298}^\circ/(\text{J K}^{-1}\text{ mol}^{-1}) = 188.825$ and 69.91, respectively. Assume $S_{298}^\circ/(\text{J K}^{-1}\text{ mol}^{-1}) = 130.684$ for $H_2(g)$ and 205.138 for $O_2(g)$.

For the reaction

$$H_2(g) + \tfrac{1}{2}O_2(g) \longrightarrow H_2O$$

the values of $\Delta_f S_{298}^\circ$ are given by (5.17) as

$$\Delta_f S_{298}^\circ(l) = (1)[S_{298}^\circ(l)] - \{(1)[S_{298}^\circ(H_2)] + (\tfrac{1}{2})[S_{298}^\circ(O_2)]\}$$

$$= [(1\text{ mol})(69.91\text{ J K}^{-1}\text{ mol}^{-1})]$$

$$- [(1\text{ mol})(130.684\text{ J K}^{-1}\text{ mol}^{-1}) + (\tfrac{1}{2}\text{ mol})(205.138\text{ J K}^{-1}\text{ mol}^{-1})]$$

$$= -163.34\text{ J K}^{-1}$$

or $\Delta_f S_{298}^\circ(l) = -163.34\text{ J K}^{-1}\text{ mol}^{-1}$, and

$$\Delta_f S_{298}^\circ(g) = [(1)(188.825)] - [(1)(130.684) + (\tfrac{1}{2})(205.138)]$$

$$= -44.424\text{ J K}^{-1}$$

or $\Delta_f S_{298}^\circ(g) = -44.424\text{ J K}^{-1}\text{ mol}^{-1}$. Then (6.14) gives

$$\Delta_f G_{298}^\circ(l) = -285.830\text{ kJ mol}^{-1} - (298.15\text{ K})(-163.34 \times 10^{-3}\text{ kJ K}^{-1}\text{ mol}^{-1})$$

$$= -237.130\text{ kJ mol}^{-1}$$

$$\Delta_f G_{298}^\circ(g) = -241.818\text{ kJ mol}^{-1} - (298.15\text{ K})(-44.424 \times 10^{-3}\text{ kJ K}^{-1}\text{ mol}^{-1})$$

$$= -228.573\text{ kJ mol}^{-1}$$

EXAMPLE 6.9.   Use the results of Example 6.8 and (*6.11*) to predict the vapor pressure above liquid water at 25 °C.

To predict *P*, consider the reactions

$$H_2O(l) \longrightarrow H_2O(g, P)$$

$$H_2O(g, P) \longrightarrow H_2O(g, 1.00 \text{ bar})$$

For the first reaction, $\Delta G^\circ_{(1)} = 0$ (see Example 6.7); and for the second reaction, (*6.11*) gives

$$\Delta G^\circ_{(2)} = nRT \ln \frac{1.00 \text{ bar}}{P}$$

For the overall process, $H_2O(l) = H_2O(g, 1.00 \text{ bar})$,

$$\Delta G^\circ = \Delta G^\circ_{(1)} + \Delta G^\circ_{(2)}$$

For the overall reaction ΔG° is given by (*6.15*) as

$$(1)\Delta_f G^\circ(g) - (1)\Delta_f G^\circ(l) = \Delta G^\circ_{(1)} + \Delta G^\circ_{(2)}$$

$$(1)(-228\,573 \text{ J}) - (1)(-237\,130 \text{ J}) = 0 + (1.00 \text{ mol})(8.314 \text{ J K}^{-1} \text{ mol}^{-1})(298 \text{ K}) \ln(1.00/P)$$

$$\ln(1.00/P) = 3.454$$

$$P = 0.031\,6 \text{ bar}$$

6.5 TEMPERATURE DEPENDENCE OF $\Delta_r G^\circ$

The temperature depencence of $\Delta_r G^\circ$ is given by (*6.13*) as

$$\frac{1}{T_2} \Delta_r G^\circ_{T_2} = \frac{1}{T_1} \Delta_r G^\circ_{T_1} - \int_{T_1}^{T_2} \frac{\Delta_r H^\circ}{T^2} \, dT \tag{6.16}$$

If the temperature dependence of $\Delta_r H$ is described by (*4.5*), then

$$\Delta_r G_T = J + KT - \Delta a \, T \ln T - \frac{\Delta b}{2} T^2 - \frac{\Delta c}{6} T^3 - \frac{\Delta d}{12} T^4 \tag{6.17a}$$

$$\Delta_r G_T = J' + K'T - \Delta a T \ln T - \frac{\Delta b}{2} T^2 - \frac{\Delta c'}{2} T^{-1} \tag{6.17b}$$

where

$$K \equiv \frac{\Delta_r G_{298} - J}{298} + \Delta a \ln 298 + \frac{\Delta b}{2}(298) + \frac{\Delta c}{6}(298^2) + \frac{\Delta d}{12}(298^3) \tag{6.18a}$$

$$K' \equiv \frac{\Delta_r G_{298} - J'}{298} + \Delta a \ln 298 + \frac{\Delta b}{2}(298) + \frac{\Delta c'}{2}(298^{-2}) \tag{6.18b}$$

Tabulated values of the *free energy function*, $(G^\circ_T - H^\circ_0)/T$, (sometimes called the *phi* or *Planck function*) are convenient for determining $\Delta_r G^\circ_T$. The temperature dependence of $\Delta_r G^\circ_T$ is given by

$$\Delta_r G^\circ_T = T \left[\frac{\Delta(G^\circ_T - H^\circ_0)}{T} \right] + \Delta_r H^\circ_0 \tag{6.19a}$$

where the change in the free energy function for a reaction is given by

$$\frac{\Delta(G^\circ_T - H^\circ_0)}{T} = \sum_i \nu_i \left[\frac{G^\circ_T - H^\circ_0}{T} \right]_i \tag{6.20a}$$

where ν_i represents the stoichiometric coefficients (ν_i is positive for products and negative for reactants) and has the units of mol.

EXAMPLE 6.10.   For many years the standard room temperature was 20°C instead of 25 °C. What value of $\Delta G°$ would a worker report at 20 °C for the reaction

$$S(\text{monoclinic}) \longrightarrow S(\text{rhombic})$$

given $\Delta_f H°_{298}/(\text{J mol}^{-1}) = 330$ and 0, $\Delta_f G°_{298}/(\text{J mol}^{-1}) = 96$ and 0, and $C°_P/(\text{J K}^{-1}\,\text{mol}^{-1}) = 23.6$ and 22.64 for the monoclinic and rhombic forms, respectively?

At 298 K,

$$\Delta_r H° = (1\ \text{mol})(0) - (1\ \text{mol})(330\ \text{J mol}^{-1}) = -330\ \text{J}$$

$$\Delta_r G° = (1)(0) - (1)(96) = -96\ \text{J}$$

$$\Delta_r C°_P = (1)(22.64) - (1)(23.6) = -1.0\ \text{J K}^{-1}$$

Supposing $C°_P$ to be temperature-independent, we have $\Delta b = \Delta c = \Delta d = 0$ and $\Delta a = \Delta_r C°_P$ in (4.4a) and (6.18a). Hence

$$J = -330 - (-1.0)(298) = -32\ \text{J}$$

and

$$K = \frac{-96 - (-32)}{298} + (-1.0)\ln(298) = -5.9\ \text{J K}^{-1}$$

Now (6.17a) gives

$$\Delta_r G°_{293} = -32 + (-5.9)(293) - (-1.0)(293)\ln(293)$$

$$= -100\ \text{J}$$

EXAMPLE 6.11.   Because of the availability of $\Delta_f H°_{298}$ data, it is convenient to use a free energy function based on 298 K instead of 0 K. Find the relation between $(G°_T - H°_{298})/T$ and $(G°_T - H°_0)/T$, and rewrite (6.19a) and (6.20a) for the function based on 298 K.

By adding and subtracting $H°_0$ in the numerator of the function based on 298 K, we obtain

$$\frac{G°_T - H°_{298}}{T} = \frac{G°_T - H°_{298} + H°_0 - H°_0}{T} = \frac{G°_T - H°_0}{T} - \frac{H°_{298} - H°_0}{T} \tag{6.21}$$

In terms of the function based on 298 K, (6.19a) and (6.20a) become

$$\Delta_r G°_T = T\left[\frac{\Delta(G°_T - H°_{298})}{T}\right] + \Delta_r H°_{298} \tag{6.19b}$$

$$\frac{\Delta(G°_T - H°_{298})}{T} = \sum_i \nu_i \left[\frac{G°_T - H°_{298}}{T}\right]_i \tag{6.20b}$$

6.6   CHEMICAL POTENTIAL

Assuming $G = G(T, P, n_1, n_2, \ldots)$ for an open system,

$$dG = \left(\frac{\partial G}{\partial T}\right)_{P,n_i} dT + \left(\frac{\partial G}{\partial P}\right)_{T,n_i} dP + \sum_i \left(\frac{\partial G}{\partial n_i}\right)_{P,T,n_{j\neq i}} dn_i \tag{6.22}$$

where the *chemical potential* (μ_i) is defined as

$$\mu_i = \left(\frac{\partial G}{\partial n_i}\right)_{P,T,n_{j\neq i}} \tag{6.23}$$

The subscript $n_{j\neq i}$ means that the molar amounts of the other substances in the system are not changing.

EXAMPLE 6.12.   Find the relation between the chemical potentials of the liquid and vapor phases in a system at equilibrium.

For a system in which equilibrium is established, (6.22) gives

$$0 = \left(\frac{\partial G}{\partial T}\right)_{P,n_i}(0) + \left(\frac{\partial G}{\partial P}\right)_{T,n_i}(0) + \mu_{vap}\,dn_{vap} + \mu_1\,dn_1$$

Recognizing that $dn_{vap} = -dn_1$ during the dynamic equilibrium, the desired relation is found to be

$$\mu_{vap} = \mu_1$$

Activities

6.7 INTRODUCTION

For a substance undergoing an isothermal change from state 1 to state 2,

$$G_2 - G_1 = nRT \ln \frac{a_2}{a_1} \qquad (6.24)$$

where a_i is the *activity* of the substance in state i. If state 1 is the standard state where $a_1 = a° = 1$, then

$$G - G° = nRT \ln a \qquad (6.25)$$

Note that the activity is a dimensionless number. Sections 6.8–6.11 consider the determination of the activities for ideal and real gases, liquids, solids, and the solutes in electrolytic solutions. Activities of solvents and nonelectrolytic solutions are considered in Sec. 11.9.

6.8 ACTIVITIES FOR IDEAL GASES

Integration of (6.10) gives (6.25) if

$$a = P/(\text{bar}) \qquad (6.26)$$

EXAMPLE 6.13.   What is $G - G°$ for 1.00 mol of an ideal gas at 5.00 bar and 25 °C?

Using (6.25) and (6.26) gives

$$G - G° = (1.00 \text{ mol})(8.314 \text{ J K}^{-1}\text{ mol}^{-1})(298 \text{ K}) \ln 5.00 = 3\,990 \text{ J}$$

6.9 ACTIVITIES FOR REAL GASES

For real gases some appropriate equation of state may be solved for V, and the result substituted into (6.10) and integrated, giving a complicated expression for the acivity. Rather than doing this, we retain the simple form of (6.25) by letting

$$a = f/(\text{bar}) \qquad (6.27)$$

where the *fugacity* (f) is given by

$$f = \gamma P \qquad (6.28)$$

The *activity coefficient* (γ) contains the corrections for the nonideal behavior of the gas. Note that the standard state is now taken as $f° = 1$ bar (not $P° = 1$ bar). Values of the activity coefficient may be determined by a graphical integration of $V_m - (RT/P)$, where V_m is the molar volume, against P to give $RT \ln \gamma$; or $Z - 1$ against $\ln P_r$ to give $\ln \gamma$; or from Fig. 6-1. Each curve in Fig. 6-1 corresponds to a fixed value of T_r. The value of γ in a gaseous mixture is assumed to be the same as the value for the pure gas under the same conditions.

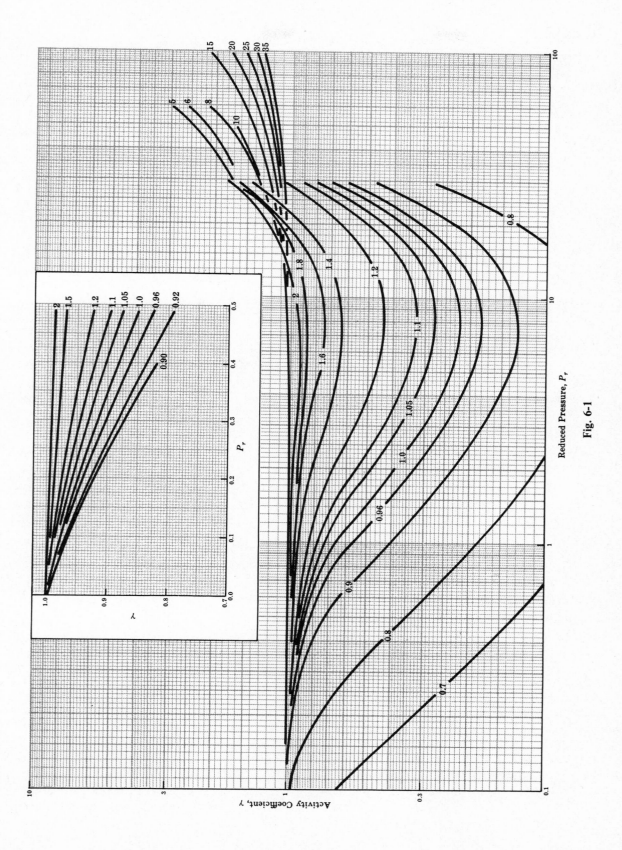

Reduced Pressure, P_r

Fig. 6-1

EXAMPLE 6.14. Determine γ and a for CO at 253 bar and 0 °C from Fig. 6-1 given that $t_c = -140.2$ °C and $P_c = 36.15$ bar.

Using Fig. 6-1 for $P_r = 253$ bar/36.15 bar $= 7.0$ and $T_r = 273$ K/133 K $= 2.05$ gives $\gamma = 0.95$. Substituting into (6.27) and (6.28) gives

$$a = f/(\text{bar}) = \gamma P/(\text{bar}) = (0.95)(253) = 240.$$

EXAMPLE 6.15. Using the following data, determine γ and a for CO at 253 bar and 0 °C by performing a graphical integration of a plot of $V_m - (RT/P)$ against P

$P/(\text{bar})$	1.01	20.3	40.5	50.7	60.8	81.1	101
$V_m/(\text{dm}^3\ \text{mol}^{-1})$	22.4	1.11	0.551	0.438	0.364	0.272	0.218
P	122	142	162	182	203	253	
V_m	0.182	0.157	0.139	0.126	0.114	0.096	

Sample calculations for the data plotted in Fig. 6-2 follow for 60.8 bar and 253 bar.

$$V_m - \frac{RT}{P} = 0.364\ \text{dm}^3\ \text{mol}^{-1} - \frac{(0.0831\ 4\ \text{dm}^3\ \text{bar}\ \text{K}^{-1}\ \text{mol}^{-1})(273\ \text{K})}{60.8\ \text{bar}} = -0.009\ \text{dm}^3\ \text{mol}^{-1}$$

$$V_m - \frac{RT}{P} = 0.096 - \frac{(0.0831\ 4)(273)}{253} = 0.006$$

Graphical integration of Fig. 6-2 gives $RT \ln \gamma = -1.171\ \text{dm}^3\ \text{bar}\ \text{mol}^{-1}$ or $\gamma = 0.949$. From (6.27) and (6.28), $a = 240$.

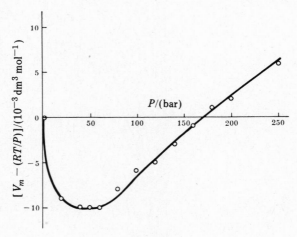

Fig. 6-2

EXAMPLE 6.16. The area under the curve of a plot of $Z - 1$ against $\ln P_r$ is equal to $\ln \gamma$. Prepare such a plot for CO at 273 K by choosing several pressures up to 253 bar, calculating P_r and $\ln P_r$, and finding values of Z from Fig. 1-3 along the $T_r = 2.05$ isotherm. Calculate γ and a.

As a sample calculation, for $P = 50$ bar

$$P_r = \frac{50\ \text{bar}}{36.15\ \text{bar}} = 1.38$$

and $\ln P_r = 0.324$. From Fig. 1-3 the value of Z is 0.98, giving $Z - 1 = -0.02$. Similar calculations yield Fig. 6-3. The graphical integration gives $\ln \gamma = -0.0466$ or $\gamma = 0.954$, and (6.27) and (6.28) give

$$a = \gamma P/(\text{bar}) = (0.954)(253) = 241$$

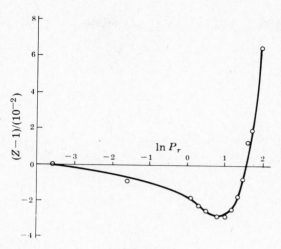

Fig. 6-3

6.10 ACTIVITIES FOR LIQUIDS AND SOLIDS

For a condensed phase under isothermal conditions

$$\ln a = \frac{1}{RT} \int_{1\,\text{bar}}^{P} V_m \, dP \tag{6.29}$$

Because the molar volume V_m is essentially a constant and much less than RT, $\ln a$ is quite small and a is usually assumed to be unity for processes involving moderate changes in pressure.

EXAMPLE 6.17.   Calculate a for water at 10 bar and 25 °C given $\kappa = 45.2 \times 10^{-6}$ bar^{-1} and $\rho = 0.997\,07 \times 10^3$ kg m^{-3}.

Assuming that $V_m = V_{m,298}(1 - \kappa P)$, (6.29) gives

$$\ln a = \frac{V_{m,298}}{RT}\left((P-1) - \frac{\kappa}{2}(P^2 - 1^2)\right)$$

$$= \frac{(18.01 \times 10^{-3}\ \text{kg mol}^{-1})/(0.997\,07 \times 10^3\ \text{kg m}^{-3})}{(8.314 \times 10^{-5}\ \text{m}^3\ \text{bar mol}^{-1}\ \text{K}^{-1})(298\ \text{K})}\left((9\ \text{bar}) - \frac{45.2 \times 10^{-6}\ \text{bar}^{-1}}{2}(99\ \text{bar}^2)\right)$$

$$= 0.006\,56$$

Taking the antilogarithm gives $a = 1.006\,6$.

6.11 ACTIVITIES FOR ELECTROLYTIC SOLUTIONS

For a solute undergoing ionization or dissociation in solution according to the equation

$$A_{\nu_+}B_{\nu_-} \longrightarrow \nu_+ A^{(z_+)+} + \nu_- B^{(z_-)-}$$

the *mean ionic activity* $(a_\pm)$ is given by

$$a_\pm = y_\pm[C_\pm/(\text{M})] \tag{6.30}$$

where $y_\pm$ is the *mean ionic activity coefficient* and $C_\pm$ is the *mean ionic concentration*. Observe that the standard state in (6.24) is now a 1 M solution. In terms of the individual ionic activities a_+ and a_-, concentrations C_+ and C_-, and activity coefficients y_+ and y_-,

$$a_\pm = (a_+^{\nu_+} a_-^{\nu_-})^{1/\nu} \tag{6.31}$$

$$y_\pm = (y_+^{\nu_+} y_-^{\nu_-})^{1/\nu} \tag{6.32}$$

$$C_\pm = (C_+^{\nu_+} C_-^{\nu_-})^{1/\nu} \tag{6.33}$$

where

$$\nu = \nu_+ + \nu_- \tag{6.34}$$

For concentrations of the electrolyte in mole fraction (x) or molality (m), we may rewrite (6.30) as $a_\pm = fx$ or $a_\pm = \gamma[m/(m)]$, where

$$f = y_\pm \left(\frac{\rho_{\text{solution}} - M_{\text{solute}}C + M_{\text{solvent}}C\nu}{\rho_{\text{solvent}}} \right) \tag{6.35}$$

$$\gamma = y_\pm \left(\frac{C}{\rho_{\text{solvent}}m \times 10^{-3}} \right) \tag{6.36}$$

Here ρ is the density (kg m^{-3}), M is the molar mass (g mol^{-1}), and C is the concentration (mol dm^{-3}).

EXAMPLE 6.18. The *Debye–Hückel theory* allows the calculation of the mean activity coefficient by

$$\log y_\pm = -\frac{z_+ z_- A I^{1/2}}{1 + aB I^{1/2}} + bI \tag{6.37}$$

where A, a, B, and b are various parameters, and the *ionic strength* (I) is given by

$$I = \frac{1}{2} \sum_i^{\text{ions}} C_i z_i^2 \tag{6.38}$$

In (6.38), z_i is the absolute value of the ionic valence, and the sum is performed over *all* ions in solution. Calculate $y_\pm$ for 1.0×10^{-4} M Al$_2$(SO$_4$)$_3$ at 25 °C.

Since 1 mol of Al$_2$(SO$_4$)$_3$ yields 2 mol of Al^{3+} and 3 mol of SO$_4^{2-}$, the ionic strength of the solution is given by (6.38) as

$$I = \tfrac{1}{2}[(2.0 \times 10^{-4}(3)^2 + (3.0 \times 10^{-4})(2)^2] = 15.0 \times 10^{-4} \text{ M}$$

For aqueous solutions at 25 °C with concentrations of $I = 10^{-2}$ M or less, (6.37) reduces to the *Debye–Hückel limiting law*:

$$\log y_\pm = -z_+ z_- (0.511\,6) I^{1/2} \tag{6.39}$$

which upon substitution of $I = 15.0 \times 10^{-4}$ gives

$$\log y_\pm = -(3)(2)(0.511\,6)(15.0 \times 10^{-4})^{1/2} = -0.119$$

$$y_\pm = 0.760$$

EXAMPLE 6.19. Although y_+ and y_- cannot be experimentally measured, the Debye-Hückel limiting law allows their calculation by

$$\log y_i = -z_i^2 (0.511\,6) I^{1/2} \tag{6.40}$$

Calculate y_+ and y_- for the Al$_2$(SO$_4$)$_3$ solution described in Example 6.18.

Substituting $I = 15.0 \times 10^{-4}$ and the appropriate valences into (6.40) gives

$$\log y_+ = -(3)^2 (0.511\,6)(15.0 \times 10^{-4})^{1/2} = -0.178$$

$$y_+ = 0.663$$

$$\log y_- = -(2)^2 (0.511\,6)(15.0 \times 10^{-4})^{1/2} = -0.079$$

$$y_- = 0.833$$

As a check, substituting these ionic activity coefficients into (6.32) gives

$$y_\pm = [(0.663)^2(0.833)^3]^{1/5} = (0.254)^{1/5} = 0.760$$

the same value as determined in Example 6.18.

6.12   REACTION QUOTIENT

A chemical equation such as

$$a\mathrm{A} + b\mathrm{B} + \cdots \longrightarrow z\mathrm{Z} + y\mathrm{Y} + \cdots$$

can be represented mathematically as

$$\sum_i \nu_i \mathrm{I}_i = 0 \qquad\qquad (6.41)$$

where ν_i represents the stoichiometric coefficients $(-a, -b, \ldots, z, y, \ldots)$ of the chemical species $\mathrm{I}_i(\mathrm{A}, \mathrm{B}, \ldots, \mathrm{Z}, \mathrm{Y}, \ldots)$ in the reaction. Note that the value of ν_i is positive for a product and negative for a reactant.

For a chemical reaction in which the reactants and/or products are not in their standard states, $\Delta_r G_T$ for the process is given by

$$\Delta_r G_T = \Delta_r G_T^\circ + RT \ln Q \qquad\qquad (6.42)$$

where Q, the *reaction quotient*, is defined as

$$Q = \prod_i a_{\mathrm{I}_i}^{\nu_i} \qquad\qquad (6.43)$$

Although the activities used in (6.43) are dimensionless, each stoichiometric coefficient (ν_i) has the unit of mol. This is important in that the overall unit for the $RT \ln Q$ term in (6.42) will be [*energy*], which matches the units on $\Delta_r G_T$ and $\Delta_r G_T^\circ$. Some authors ignore the mol term from the R completely, while others retain it throughout all calculations even though it may not be dimensionally correct in some of its applications. In keeping with the actual presence of the mol term, it will simply be inserted whenever needed during a complete dimensional setup.

EXAMPLE 6.20.   Consider the chemical reaction

$$\mathrm{Cl}_2(g) + \mathrm{F}_2(g) \longrightarrow 2\mathrm{ClF}(g) \qquad \Delta_r G_{298}^\circ = -111.88 \text{ kJ}$$

If $P(\mathrm{Cl}_2) = 16.3$ bar, $P(\mathrm{F}_2) = 5.2$ bar, and $P(\mathrm{ClF}) = 0.063$ bar, is this reaction more or less favorable under these conditions than under standard state conditions?

The reaction quotient for this equation is given by (6.43) as

$$Q = [a(\mathrm{ClF})]^2 [a(\mathrm{Cl}_2)]^{-1} [a(\mathrm{F}_2)]^{-1} = \frac{[a(\mathrm{ClF})]^2}{a(\mathrm{Cl}_2)a(\mathrm{F}_2)}$$

Assuming ideal gas behavior, (6.26) gives $a_i = P_i/(\mathrm{bar})$. Thus

$$Q = \frac{[P(\mathrm{ClF})]^2}{P(\mathrm{Cl}_2)P(\mathrm{F}_2)} = \frac{(0.063)^2}{(16.3)(5.2)} = 4.7 \times 10^{-5}$$

For the reaction under nonstandard state conditions, (6.42) gives

$$\Delta_r G_{298}^\circ = (-111.88 \text{ kJ}) + (8.314 \times 10^{-3} \text{ kJ K}^{-1} \text{ mol}^{-1})(298 \text{ K})(\text{mol}) \ln(4.7 \times 10^{-5})$$

$$= -136.6 \text{ kJ}$$

The reaction is more favorable under these nonstandard state conditions.

Thermodynamic Relations

6.13 MAXWELL RELATIONS

The *Maxwell relations* for closed reversible systems in which only PV work is considered can be derived from the following set of differential energy expressions:

$$dU = T\,dS - P\,dV \tag{6.44}$$

$$dH = T\,dS + V\,dP \tag{6.45}$$

$$dG = -S\,dT + V\,dP \tag{6.46}$$

$$dA = -S\,dT - P\,dV \tag{6.47}$$

There are several Maxwell relations because of the number of ways that the four variables—T, S, P and V—can be distributed in the following format:

$$\left(\frac{\partial W}{\partial X}\right)_Z = \pm\left(\frac{\partial Y}{\partial Z}\right)_X \tag{6.48}$$

It is possible to construct any one of the relations from (6.48) if the following rules are used: (1) the cross products XY and WZ must be work terms, i.e., PV and ST; (2) the negative sign is chosen if W and X or Z and Y are T and V.

EXAMPLE 6.21. From (6.47) derive the Maxwell relation $(\partial S/\partial V)_T = (\partial P/\partial T)_V$. Compare this result to that predicted by (6.48).

At constant V, (6.47) becomes

$$\left(\frac{\partial A}{\partial T}\right)_V = -S$$

while at constant T it becomes

$$\left(\frac{\partial A}{\partial V}\right)_T = -P$$

Differentiating the first equation with respect to V at constant T, and the second equation with respect to T at constant V, gives

$$\frac{\partial}{\partial V}\left(\left(\frac{\partial A}{\partial T}\right)_V\right)_T = -\left(\frac{\partial S}{\partial V}\right)_T \quad \text{and} \quad \frac{\partial}{\partial T}\left(\left(\frac{\partial A}{\partial V}\right)_T\right)_V = -\left(\frac{\partial P}{\partial T}\right)_V$$

But A is a point function, and one of the properties of such a function is that the order of differentiation is not important. Hence, the above second derivatives are equal, which gives the desired result.

If $W = S$ and $X = V$ in (6.48), the first rule requires that $Y = P$ and $Z = T$, giving

$$\left(\frac{\partial S}{\partial V}\right)_T = \pm\left(\frac{\partial P}{\partial T}\right)_V$$

By the second rule the plus sign is chosen, so the final result is the same as derived above.

6.14 TRANSFORMATIONS

The physical properties P, V, T, C_V, and C_P are experimentally measurable for a system. When other properties such as H, U, G, or S appear in an equation, the equation is usually transformed from those variables to the measurable variety. To make these transformations, the Maxwell relations are used, as well as the following properties of partial derivatives:

$$\left(\frac{\partial x}{\partial y}\right)_w = \left(\frac{\partial x}{\partial z}\right)_w \left(\frac{\partial z}{\partial y}\right)_w \tag{5.24}$$

$$\left(\frac{\partial x}{\partial y}\right)_z \left(\frac{\partial y}{\partial z}\right)_x \left(\frac{\partial z}{\partial x}\right)_y = -1 \tag{6.49}$$

$$\left(\frac{\partial z}{\partial x}\right)_y = \left[\left(\frac{\partial x}{\partial z}\right)_y\right]^{-1} \tag{6.50}$$

EXAMPLE 6.22. Evaluate $(\partial P/\partial S)_V$.

The transformation involves three steps: (1) inversion, (6.50); (2) chain rule, (5.24); and (3) substitution of the result for $(\partial S/\partial T)_V$ which is derived in Problem 6.21.

$$\left(\frac{\partial P}{\partial S}\right)_V = \frac{1}{(\partial S/\partial P)_V}$$

$$= \frac{1}{(\partial S/\partial T)_V (\partial T/\partial P)_V} = \frac{T}{C_V (\partial T/\partial P)_V}$$

Solved Problems

FREE ENERGY

6.1. Calculate $\Delta G° - \Delta A°$ for the reaction between glycine and nitrous acid at 25 °C:

$$NH_2-CH_2-COOH(aq) + HONO(aq) \longrightarrow HO-CH_2-COOH(aq) + N_2(g) + H_2O(l)$$

Assuming $\Delta(PV)$ for the condensed phases to be negligible, we have from (6.6):

$$\Delta G° - \Delta A° = \Delta(PV) = RT\,\Delta n_g$$

$$= (8.314\,\text{J K}^{-1}\,\text{mol}^{-1})(298\,\text{K})(1\,\text{mol}) = 2\,480\,\text{J}$$

6.2. In terms of ΔS, ΔH, and ΔG, discuss the spontaneity of the reaction $H_2O(s) \longrightarrow H_2O(l)$ in an ice-water bath at 0 °C.

For the given reaction, ΔS is positive because of the increase in randomness in going from a solid to a liquid. Thus, for a constant-enthalpy system, the spontaneous change would be for all ice to melt. However, ΔH for the process is positive (because of overcoming the intermolecular attractions), which would indicate a nonspontaneous change. Thus, if only ΔH is considered, the spontaneous change would be for the water to freeze. ΔG for the process is zero, which predicts a state of equilibrium in which the entropy change in canceled by the enthalpy change.

6.3. Calculate ΔG for the isothermal expansion of 1.00 mol of an ideal gas from 0.010 0 m³ to 0.100 0 m³ at 25 °C.

Substituting the volumes into (6.11) gives

$$\Delta G = (1.00\,\text{mol})(8.314\,\text{J K}^{-1}\,\text{mol}^{-1})(298\,\text{K}) \ln\frac{0.010\,0}{0.100\,0} = -5.70\,\text{kJ}$$

6.4. What is ΔG for changing the pressure on a piece of aluminum from 1.00 atm to 1.00 bar? For Al, $\rho = 2.702 \times 10^3\,\text{kg m}^{-3}$.

For the isothermal expansion of a condensed phase, (6.10) gives

$$\Delta G = \int_{P_1}^{P_2} V \, dP \approx V \, \Delta P$$

$$= \left(\frac{26.98 \times 10^{-3} \text{ kg mol}^{-1}}{2.702 \times 10^3 \text{ kg m}^3} \right) (1.00 \text{ bar} - 1.01 \text{ bar}) \left(\frac{10^5 \text{ Pa}}{1 \text{ bar}} \right) \left(\frac{1 \text{ J}}{1 \text{ Pa m}^3} \right)$$

$$= -0.01 \text{ J}$$

6.5. Derive an equation for the Gibbs free energy change during the reversible isothermal expansion of a van der Waals gas. Calculate ΔG for the expansion of 1.00 mol of Ar from 0.050 0 dm^3 to 25.00 dm^3 at 25 °C ($a = 1.363$ dm^6 bar mol^{-2} and $b = 0.032$ 19 dm^3 mol^{-1}). Compare this answer to that, assuming ideal gas behavior.

Solving (1.22) for P and differentiating with respect to volume gives

$$\left(\frac{\partial P}{\partial V} \right) = \frac{-nRT}{(V - nb)^2} + \frac{2an^2}{V^3}$$

Substituting this result into (6.10) gives

$$dG = V \, dP = V \left(\frac{-nRT}{(V - nb)^2} + \frac{2an^2}{V^3} \right) dV$$

which upon integration gives

$$\Delta G = \int_{V_1}^{V_2} \left(\frac{-nRTV}{(V - nb)^2} + \frac{2an^2}{V^2} \right) dV$$

$$= -nRT \ln \left(\frac{V_2 - nb}{V_1 - nb} \right) + n^2 bRT \left(\frac{1}{V_2 - nb} - \frac{1}{V_1 - nb} \right) - 2an^2 \left(\frac{1}{V_2} - \frac{1}{V_1} \right) \qquad (6.51)$$

For 1.00 mol of Ar at 298 K,

$$\Delta G = -(1.00 \text{ mol})(8.314 \text{ J K}^{-1} \text{ mol}^{-1})(298 \text{ K}) \ln \left(\frac{25.00 \text{ dm}^3 - (1.00 \text{ mol})(0.032 \text{ 19 dm}^3 \text{ mol}^{-1})}{0.050 \text{ 0 dm}^3 - (1.00 \text{ mol})(0.032 \text{ 19 dm}^3 \text{ mol}^{-1})} \right)$$

$$+ (1.00 \text{ mol})^2 (0.032 \text{ 19 dm}^3 \text{ mol}^{-1})(8.314 \text{ J K}^{-1} \text{ mol}^{-1})(298 \text{ K})$$

$$\times \left(\frac{1}{25.00 \text{ dm}^3 - (1.00 \text{ mol})(0.032 \text{ 19 dm}^3 \text{ mol}^{-1})} - \frac{1}{0.050 \text{ 0 dm}^3 - (1.00 \text{ mol})(0.032 \text{ 19 dm}^3 \text{ mol}^{-1})} \right)$$

$$- 2(1.363 \text{ dm}^6 \text{ bar mol}^{-2})(1.00 \text{ mol})^2 \left(\frac{1}{25.00 \text{ dm}^3} - \frac{1}{0.050 \text{ 0 dm}^3} \right) \left(\frac{10^5 \text{ Pa}}{1 \text{ bar}} \right) \left(\frac{1 \text{ m}^3}{10^3 \text{ dm}^3} \right) \left(\frac{1 \text{ J}}{1 \text{ Pa m}^3} \right)$$

$$= -17.0 \text{ kJ}$$

The value of ΔG assuming ideal gas behavior is given by (6.11) as

$$\Delta G = -(1.00 \text{ mol})(8.314 \text{ J K}^{-1} \text{ mol}^{-1})(298 \text{ K}) \ln \left(\frac{25.00 \text{ dm}^3}{0.050 \text{ 0 dm}^3} \right)$$

$$= -15.4 \text{ kJ}$$

a difference of 9%.

6.6. Calculate $\Delta_r G_{298}^\circ$ for the roasting of sphalerite,

$$\text{ZnS(sphalerite)} + \tfrac{3}{2}\text{O}_2(g) \longrightarrow \text{ZnO(s)} + \text{SO}_2(g)$$

given $\Delta_f G_{298}^\circ / (\text{kJ mol}^{-1}) = -201.29$ for ZnS, -318.30 for ZnO, and -300.194 for SO$_2$. Qualitatively describe the driving force of the reaction.

Equation (6.15) gives

$$\Delta_r G_{298}^\circ = [(1)\Delta_f G_{298}^\circ(ZnO) + (1)\Delta_f G_{298}^\circ(SO_2)] - [(1)\Delta_f G_{298}^\circ(ZnS) + (\tfrac{3}{2})\Delta_f G_{298}^\circ(O_2)]$$

$$= [(1)(-318.30) + (1)(-300.194)] - [(1)(-201.29) + (\tfrac{3}{2})(0)] = -417.10 \text{ kJ}$$

The change in randomness is not very large, so the driving force of the reaction is the large enthalpy change.

6.7. Find the value of ΔG_{1000}° for the reaction

$$H_2(g) + Cl_2(g) \rightarrow 2HCl(g)$$

given that $\Delta H_{298}^\circ = -184.614$ kJ, $\Delta G_{298}^\circ = -190.598$ kJ, and that the heat capacities in J K^{-1} mol^{-1} are:

$$C_P^\circ = 29.066 - 0.836 \times 10^{-3} T + 20.17 \times 10^{-7} T^2 \qquad \text{for } H_2(g)$$

$$C_P^\circ = 31.696 + 10.144 \times 10^{-3} T - 40.38 \times 10^{-7} T^2 \qquad \text{for } Cl_2(g)$$

$$C_P^\circ = 28.166 + 1.810 \times 10^{-3} T + 15.47 \times 10^{-7} T^2 \qquad \text{for } HCl(g)$$

For this reaction,

$$\Delta_r C_P^\circ = [(2)(28.166) - (1)(29.066) - (1)(31.696)]$$

$$+ [(2)(1.810) - (1)(-0.836) - (1)(10.144)] \times 10^{-3} T$$

$$+ [(2)(15.47) - (1)(20.17) - (1)(-40.38)] \times 10^{-7} T^2$$

$$= -4.430 - 5.688 \times 10^{-3} T + 51.15 \times 10^{-7} T^2$$

Using (4.4a) gives

$$J = -184\,614 - (-4.430)(298) - \tfrac{1}{2}(-5.608 \times 10^{-3})(298^2) - \tfrac{1}{3}(51.15 \times 10^{-7})(298^3)$$

$$= -183\,086 \text{ J}$$

and using (6.18a) gives

$$K = \frac{(-190\,598) - (-183\,086)}{298} + (-4.430)\ln(298) + \frac{-5.688 \times 10^{-3}}{2}(298) + \frac{51.15 \times 10^{-7}}{6}(298^2)$$

$$= -51.22 \text{ J K}^{-1}$$

Substituting into (6.17a) gives

$$\Delta_r G_{1000}^\circ = -183\,086 + (-51.22)(1\,000) - (-4.430)(1\,000)\ln 1\,000$$

$$- \frac{-5.688 \times 10^{-3}}{2}(1\,000^2) - \frac{51.15 \times 10^{-7}}{6}(1\,000^3)$$

$$= --201.71 \text{ kJ}$$

6.8. The values of $[(G_T^\circ - H_0^\circ)/T]/(\text{J K}^{-1}\text{ mol}^{-1})$ at 298 K and 500 K are -189.4 and -212.4, respectively, for $C_2H_6(g)$, -184.0 and -203.9 for $C_2H_4(g)$, and -102.2 and -116.9 for $H_2(g)$. Given $(H_{298}^\circ - H_0^\circ)/(\text{kJ mol}^{-1}) = 11.950$ for $C_2H_6(g)$, 10.565 for $C_2H_4(g)$, and 8.468 for $H_2(g)$, and given $\Delta_f H_{298}^\circ/(\text{kJ mol}^{-1}) = -84.68$ for $C_2H_6(g)$, 52.26 for $C_2H_4(g)$, and 0 for $H_2(g)$, calculate $\Delta_r G_T^\circ$ for

$$C_2H_6(g) \longrightarrow C_2H_4(g) + H_2(g)$$

at 298 K and 400 K, and compare the spontaneity at these temperatures.

Converting the values of $(G_T^\circ - H_0^\circ)/T$ to $(G_T^\circ - H_{298}^\circ)/T$ for $C_2H_6(g)$ using (6.21) gives

$$\frac{G_{298}^\circ - H_{298}^\circ}{298} = -189.4 \text{ J K}^{-1}\text{ mol}^{-1} - \frac{11\,950 \text{ J mol}^{-1}}{298 \text{ K}} = -229.5 \text{ J K}^{-1}\text{ mol}^{-1}$$

$$\frac{G_{500}^\circ - H_{298}^\circ}{500} = -212.4 - \frac{11\,950}{500} = -236.3 \text{ J K}^{-1}\text{ mol}^{-1}$$

Likewise, $[(G_T^\circ - H_{298}^\circ)/T]/(\text{J K}^{-1} \text{ mol}^{-1}) = -219.5$ and -225.0 for $C_2H_4(g)$ and -130.6 and -133.8 for $H_2(g)$ at 298 K and 500 K, respectively.

For the reaction at 298 K, $\Delta_r H_{298}^\circ$ is found by using (4.7):

$$\Delta_r H_{298}^\circ = [(1 \text{ mol})(52.26 \text{ kJ mol}^{-1}) + (1 \text{ mol})(0)] - [(1 \text{ mol})(-84.68 \text{ kJ mol}^{-1})]$$

$$= 136.94 \text{ kJ}$$

and the change in the free energy function at 298 K is given by (6.20b) as

$$\frac{\Delta(G_{298}^\circ - H_{298}^\circ)}{298} = [(1 \text{ mol})(-219.5 \text{ J K}^{-1} \text{ mol}^{-1}) + (1 \text{ mol})(-130.6 \text{ J K}^{-1} \text{ mol}^{-1})]$$

$$- [(1 \text{ mol})(-229.5 \text{ J K}^{-1} \text{ mol}^{-1})]$$

$$= -120.6 \text{ J K}^{-1}$$

which upon substitution into (6.19b) gives

$$\Delta_r G_{298}^\circ = (298 \text{ K})(-120.6 \text{ J K}^{-1})[(10^{-3} \text{ kJ})/(1 \text{ J})] + 136.94 \text{ kJ} = 101.00 \text{ kJ}$$

One of the convenient features of the free energy function is that values between known values may be determined by linear interpolation. Thus, for $C_2H_6(g)$ at 400 K,

$$\frac{G_{400}^\circ - H_{298}^\circ}{400} = -229.5 \text{ J K}^{-1} \text{ mol}^{-1}$$

$$+ \frac{400 \text{ K} - 298 \text{ K}}{500 \text{ K} - 298 \text{ K}}[-236.3 \text{ J K}^{-1} \text{ mol}^{-1} - (-229.5 \text{ J K}^{-1} \text{ mol}^{-1})]$$

$$= -232.9 \text{ J K}^{-1} \text{ mol}^{-1}$$

Likewise, $[G_{400}^\circ - H_{298}^\circ)/400]/(\text{J K}^{-1} \text{ mol}^{-1}) = -222.3$ for $C_2H_4(g)$ and -132.2 for $H_2(g)$. For the reaction at 400 K, (6.20b) gives

$$\frac{\Delta(G_{400}^\circ - H_{298}^\circ)}{400} = [(1)(-222.3) + (1)(-132.2)] - [(1)(-232.9)] = -121.6 \text{ J K}^{-1}$$

and (6.19b) gives

$$\Delta_r G_{400}^\circ = (400)(-121.6)(10^{-3}) + 136.94 = 88.3 \text{ kJ}$$

The reaction is favored slightly more under standard conditions at 400 K than at 298 K because the value of $\Delta_r G_T^\circ$ is less positive.

6.9. For the reaction

$$Ag(s) + \tfrac{1}{2}Cl_2(g) \longrightarrow AgCl(l)$$

Metz and Seifert reported

$$E^\circ/(V) = 0.907\,9 - 0.280\,X + 0.110\,X^2$$

where

$$X = (T - 728.2) \times 10^{-3} \text{ K}$$

If $n = 1$ mol, find $\Delta_f G^\circ$ as a function of X and calculate $\Delta_f G_{1000}^\circ$. Is the reaction spontaneous at this temperature?

Using (6.7) gives

$$\Delta_f G^\circ = -(1 \text{ mol})(96.485 \text{ kJ mol}^{-1} \text{ V}^{-1})(0.907\,9 - 0.280\,X + 0.110\,X^2) \text{ V}$$

$$= (-87.60 + 27.0\,X - 10.6\,X^2) \text{ kJ}$$

Evaluating X at 1 000 K gives $X = 0.271\,8$ K and

$$\Delta_f G_{1000}^\circ = -87.60 + (27.0)(0.271\,8) - (10.6)(0.271\,8)^2 = -81.04 \text{ kJ}$$

which is spontaneous under standard conditions.

6.10. Assuming ΔH in (6.16) to be constant,

$$\frac{\Delta G_2}{T_2} = \frac{\Delta G_1}{T_1} + \Delta H \left(\frac{1}{T_2} - \frac{1}{T_1} \right) \tag{6.52}$$

Show that (6.52) is equivalent to using

$$\Delta G_2 = \Delta H - T_2 \Delta S_1 \tag{6.53}$$

to predict the temperature dependence of ΔG. Use (6.53) to predict the boiling point of bromine given $\Delta_{vap} H° = 30.907$ kJ and $\Delta_{vap} S° = 93.232$ J K^{-1}.

Equation (6.52) can readily be cast into the desired form

$$\frac{\Delta G_2}{T_2} = \frac{\Delta G_1}{T_1} + \Delta H \left(\frac{1}{T_2} - \frac{1}{T_1} \right) = \frac{\Delta G_1}{T_1} - \frac{\Delta H}{T_1} + \frac{\Delta H}{T_2}$$

$$= \frac{\Delta G_i - \Delta H}{T_1} + \frac{\Delta H}{T_2} = \frac{-T_1 \Delta S_1}{T_1} + \frac{\Delta H}{T_2} = -\Delta S_1 + \frac{\Delta H}{T_2}$$

Rearranging gives

$$\Delta G_2 = \Delta H - T_2 \Delta S_1$$

For the boiling process (at 1 bar), $\Delta_{vap} G = 0$, giving

$$T_2 = \frac{\Delta_{vap} H}{\Delta_{vap} S_1} = \frac{(30.907 \text{ kJ})[(10^3 \text{ J})/(1 \text{ kJ})]}{93.232 \text{ J K}^{-1}} = 332 \text{ K} = 59 \text{ °C}$$

6.11. For the reaction

$$\tfrac{1}{2}Cl_2(g) + \tfrac{3}{2}F_2(g) \longrightarrow ClF_3(g)$$

the temperature dependence of the Gibbs free energy of formation is given by

$$\Delta_f G_T°/(\text{J}) = -158\,624 + (116.7)T + (3.363)T \ln T - (9.514 \times 10^{-3})T^2 + (16.30 \times 10^{-7})T^3$$

between 300 K and 1 500 K. At what temperature does this reaction become nonspontaneous?

To find this temperature, $\Delta_f G_T°$ is set equal to zero and the equation is solved for T.

$$(16.30 \times 10^{-7})T^3 - (9.514 \times 10^{-3})T^2 + (116.7 + 3.363 \ln T)T - (158\,624) = 0$$

The method of (1.23) can be used to solve the equation even though the equation is not exactly a simple polynomial in T. The approach used is to assume an approximate value for T (e.g., 1 000 K), calculate $\ln T$, calculate $(116.7 + 3.363 \ln T)$, solve the simple polynomial for T, and then repeat the process using the new T in the $\ln T$ term until a constant T is obtained. Using this method, $T = 1\,206$ K.

ACTIVITIES

6.12. As an indication of ideality, gases can be categorized according to molecular structure. Compare molar values at 25 °C and 1 bar of $G - G°$ for (1) a nonpolar, slightly polarizable gas such as Ar ($t_c = -122.4$ °C, $P_c = 49.4$ bar); (2) a nonpolar, easily polarizable gas such as CF$_4$ ($t_c = -47.9$ °C, $P_c = 40.3$ bar); (3) a polar, slightly polarizable gas such as CO ($t_c = -140.2$ °C, $P_c = 36.2$ bar); (4) a polar, easily polarizable gas such as CH$_3$Cl ($t_c = 143.1$ °C, $P_c = 67.5$ bar); and (5) a gas having hydrogen bonding between the molecules such as CH$_3$OH ($t_c = 239.4$ °C, $P_c = 80.2$ bar).

Using (1.18) and (1.19) for the gases gives $T_r = 1.98, 1.32, 2.24, 0.72,$ and 0.58 and $P_r = 0.020\,2, 0.024\,8, 0.027\,6, 0.013\,8,$ and $0.012\,5$ for Ar, CF$_4$, CO, CH$_3$Cl, and CH$_3$OH, respectively. Figure 6-1 gives $\gamma = 0.998, 0.997, 0.999, 0.99,$ and 0.98, respectively. The values for CH$_3$Cl and CH$_3$OH are only approximate. Using (6.27), (6.28), and (6.25) gives

$$G - G° = (1.00 \text{ mol})(8.314 \text{ J K}^{-1} \text{ mol}^{-1})(298 \text{ K}) \ln(0.998)(1.00) = -5.0 \text{ J} \qquad \text{for Ar}$$

−7.4 J for CF_4, −2.5 J for CO, −25 J for CH_3Cl, and −50 J for CH_3OH. Since $G - G°$ for an ideal gas is zero, these values indicate that gases that are easily polarizable or have hydrogen bonding within them deviate from ideality more than those that are slightly polarizable. For routine calculations where ΔG is of the order of several kJ, the assumption of ideality does not seem to present a significant problem.

6.13. For a real gas undergoing a reversible isothermal expansion, (6.24) and (6.27) give

$$\Delta G = nRT \ln\left(\frac{f_2}{f_1}\right) \qquad\qquad (6.54)$$

Consider 1.00 mol of argon at 25 °C undergoing a reversible isothermal expansion from 846 bar to 0.99 bar. Calculate ΔG for this process. For argon, $t_c = -122.4$ °C and $P_c = 49.4$ bar.

The reduced temperature is given by (1.19) as

$$T_r = (298 \text{ K})/(150.8 \text{ K}) = 1.98$$

and the reduced pressures are given by (1.18) as

$$P_r = (846 \text{ bar})/(49.4 \text{ bar}) = 17.1$$

$$P_r = (0.99 \text{ bar})/(49.4 \text{ bar}) = 0.020$$

The values of γ corresponding to these conditions are 1.13 and 0.99, respectively (see Fig. 6-1). Thus (6.54) gives

$$\Delta G = (1.00 \text{ mol})(8.314 \text{ J K}^{-1} \text{ mol}^{-1})(298 \text{ K}) \ln\left(\frac{(0.99)(0.99 \text{ bar})}{(1.13)(846 \text{ bar})}\right)$$

$$= -17.1 \text{ kJ}$$

Compare this answer to that given in Problem 6.5 for the same process.

6.14. The fugacity for a van der Waals gas is given by

$$\ln f = \ln\left(\frac{nRT}{V - nb}\right) + \frac{nb}{V - nb} - \frac{2an}{RTV} \qquad\qquad (6.55)$$

At what pressure is the standard state of argon ($a = 1.363$ dm^6 bar mol^{-2} and $b = 0.032\,19$ dm^3 mol^{-1})?

The standard state of a real gas is defined as $f = 1$ bar. Thus, for exactly 1 mol,

$$\ln(1) = \ln\left(\frac{(1)(0.083\,144\,1)(298.15)}{V - (1)(0.032\,19)}\right) + \frac{(1)(0.032\,19)}{V - (1)(0.032\,19)}$$

$$- \frac{(2)(1.363)(1)}{(0.083\,144\,1)(298.15)\,V}$$

$$0 = \ln\left(\frac{24.789}{V - 0.032\,19}\right) + \frac{0.032\,19}{V - 0.032\,19} - \frac{0.110\,0}{V}$$

Solving the equation gives $V = 24.74$ dm^3. Substitution into (1.22) gives

$$P = \frac{(1)(0.083\,144\,1)(298.15)}{24.74 - (1)(0.032\,19)} - \frac{(1.363)(1)^2}{(24.74)^2} = 1.001\,1 \text{ bar}$$

6.15. Calculate y_{Na^+} in a 0.01 M NaCl–0.01 M Na_2SO_4 mixture, assuming the Debye–Hückel limiting law to apply.

Using (6.38) gives

$$I = \tfrac{1}{2}[(C_{Na^+})(z_{Na^+})^2 + (C_{Cl^-})(z_{Cl^-})^2 + (C_{SO_4^{2-}})(z_{SO_4^{2-}})^2]$$

$$= \tfrac{1}{2}[(0.03)(1)^2 + (0.01)(1)^2 + (0.01)(2)^2] = 0.04 \text{ M}$$

Equation (6.40) gives

$$\log y_{\mathrm{Na}^+} = -(1)^2(0.511\,6)(0.04)^{1/2} = -0.102$$

and taking the antilog gives $y_{\mathrm{Na}^+} = 0.790$.

6.16. Express Q for the reactions

$$\mathrm{Ag(s)} + \tfrac{1}{2}\mathrm{Cl_2(g)} \longrightarrow \mathrm{AgCl(l)}$$

$$2\mathrm{Ag(s)} + \mathrm{Cl_2(g)} \longrightarrow 2\mathrm{AgCl(l)}$$

and find the relation between them.

Substituting into (6.43) gives for the first reaction

$$Q_1 = \frac{a(\mathrm{AgCl})}{a(\mathrm{Ag})a(\mathrm{Cl_2})^{1/2}}$$

and for the second reaction

$$Q_2 = \frac{a(\mathrm{AgCl})^2}{a(\mathrm{Ag})^2 a(\mathrm{Cl_2})} = Q_1^{\;2}$$

6.17. Calculate $\Delta_\mathrm{f}G_{298}$ for

$$\mathrm{H_2(g)} + \tfrac{1}{2}\mathrm{O_2(g)} \longrightarrow \mathrm{H_2O(l)}$$

given $\Delta_\mathrm{f}G^\circ_{298} = -237.129\ \mathrm{kJ}$, $a(\mathrm{H_2O}) = 1.00$, $P(\mathrm{H_2}) = 1.00 \times 10^{-3}$ bar, and $P(\mathrm{O_2}) = 1.00 \times 10^{-6}$ bar. Is the reaction favored under these reduced pressure conditions?

Equations (6.42) and (6.43) give

$$\Delta_\mathrm{f}G_{298} = \Delta_\mathrm{f}G^\circ_{298} + RT \ln\left(\frac{a(\mathrm{H_2O})}{[P(\mathrm{H_2})/(\mathrm{bar})][P(\mathrm{O_2})/(\mathrm{bar})]^{1/2}}\right)$$

$$= (-237.129\ \mathrm{kJ}) + (8.314\,41 \times 10^{-3}\ \mathrm{kJ\ K^{-1}\ mol^{-1}})(\mathrm{mol})(298.15\ \mathrm{K}) \ln\left(\frac{1.00}{(1.00 \times 10^{-3})(1.00 \times 10^{-6})^{1/2}}\right)$$

$$= -202.881\ \mathrm{kJ}$$

The value of ΔG is negative, so the reaction is spontaneous as written, but less so than at standard conditions.

THERMODYNAMIC RELATIONS

6.18. Derive (6.44).

Beginning with the first law of thermodynamics,

$$dU = đq + đw$$

substitute $đq = T\,dS$, which holds under reversible conditions, and $đw = -P\,dV$ to obtain

$$dU = T\,dS - P\,dV$$

6.19. Derive (6.45).

The differential of

$$H = U + PV$$

is

$$dH = dU + P\,dV + V\,dP$$

Substituting (6.44) for dU gives

$$dH = (T\,dS - P\,dV) + P\,dV + V\,dP = T\,dS + V\,dP$$

6.20. Using the format of (6.48), express the Maxwell relation that begins $(\partial S/\partial P)_T$.

Letting $W = S$ and $X = P$, the first rule says that Y must be V and Z must be T, so that the cross products are PV and ST for XY and WZ, respectively. Thus

$$\left(\frac{\partial S}{\partial P}\right)_T = \pm\left(\frac{\partial V}{\partial T}\right)_P$$

and the negative sign is chosen in view of the second rule (Y and Z are T and V).

6.21. Express $(\partial S/\partial T)_P$ and $(\partial S/\partial T)_V$ in terms of measurable properties.

Beginning with the definition of dS,

$$dS = \frac{đq}{T}$$

we recognize that at constant pressure $đq = dH = C_P\,dT$, so that

$$\frac{dS}{dT} = \left(\frac{\partial S}{\partial T}\right)_P = \frac{C_P}{T}$$

Likewise, at constant volume,

$$\left(\frac{\partial S}{\partial T}\right)_V = \frac{C_V}{T}$$

6.22. Express $(\partial P/\partial V)_S$ in terms of measurable properties.

Applying (6.49), (6.50), (5.24), and the results of Problem 6.21 to the expression gives

$$\left(\frac{\partial P}{\partial V}\right)_S = -\left(\frac{\partial S}{\partial V}\right)_P\left(\frac{\partial P}{\partial S}\right)_V = -\frac{(\partial S/\partial V)_P}{(\partial S/\partial P)_V}$$

$$= -\frac{(\partial S/\partial T)_P(\partial T/\partial V)_P}{(\partial S/\partial T)_V(\partial T/\partial P)_V}$$

$$= -\frac{C_P(\partial T/\partial V)_P}{C_V(\partial T/\partial P)_V}$$

Supplementary Problems

FREE ENERGY

6.23. What is ΔG for the reversible isothermal compression of 1.00 mol of an ideal gas from 1.00 to 5.00 bar at 25 °C? What is ΔG if the process were performed irreversibly?

Ans.   3 890 J; 3 890 J (ΔG is path-independent)

6.24. Calculate q, w, ΔU, ΔH, ΔS, ΔG, and ΔA for a reversible isothermal expansion of 1.00 mol of an ideal diatomic gas at 100.°C from 1.00 dm^3 to 10.00 dm^3. What would be the values of these parameters if the process were performed irreversibly against a constant external pressure of 1.00 bar?

Ans. $q = -w = 7\,146$ J, $\Delta U = \Delta H = 0$, $\Delta S = 19.15$ J K^{-1}, $\Delta G = \Delta A = -7\,140$ J; $q = -w = 900$ J, the rest are unchanged

6.25. Write the differential of (6.2), and determine $(\partial A/\partial V)_T$ and $(\partial A/\partial T)_V$ for reversible processes involving only expansion work.

Ans. $dA = -P\,dV - S\,dT$; $(\partial A/\partial V)_T = -P$, $(\partial A/\partial T)_V = -S$

6.26. Derive the Gibbs–Helmholtz equation that takes the form of $[\partial(G/T)/\partial(1/T)]_P$.   *Ans.*   *H*

6.27. For the freezing of water,

$$H_2O(l) \longrightarrow H_2O(s)$$

$\Delta_{crys}H°/(J) = -5\,972.2$ at $-1\,°C$, $-6\,009.5$ at $0\,°C$, and $-6\,046.7$ at $1\,°C$. Are these favorable enthalpy changes? For the same process, $\Delta_{crys}S°/(J\,K^{-1}) = -21.864\,0$ at $-1\,°C$, $-22.000\,7$ at $0\,°C$, and $-22.136\,9$ at $1\,°C$. Are these favorable entropy changes? Calculate $\Delta_{crys}G°$ at each of these temperatures, and discuss the results.

Ans.   Yes; no; -21.9 J at $-1\,°C$ (freezing process is spontaneous), 0.0 at $0\,°C$ (equilibrium), 22.1 at $1\,°C$ (melting process is spontaneous)

6.28. What is ΔG for the phase transition

$$H_2O(l) \longrightarrow H_2O(s)$$

at $-15\,°C$, given that the vapor pressure above water at this temperature is 191.5 Pa and above ice it is 165.5 Pa? Assume a three-step process consisting of an evaporation, a pressure change, and a deposition.

Ans. $\Delta G_{(1)} = \Delta G_{(3)} = 0$, $\Delta G_{(2)} = -313$ J

6.29. Calculate $\Delta_r G°_{298}$ for

$$CaCO_3(\text{calcite}) \longrightarrow CaCO_3(\text{aragonite})$$

given $\Delta_f G_{298}/(\text{kJ mol}^{-1}) = -1\,128.79$ for calcite and $-1\,127.75$ for aragonite. Which crystal form is the more stable at $25\,°C$ under standard conditions?

Ans.   1.04 kJ;   nonspontaneous, calcite more stable

6.30. Calculate $\Delta_r G°_{298}$ for

$$C_2H_4(g) + HCl(g) \longrightarrow C_2H_5Cl(g)$$

given $\Delta_f G°_{298}/(\text{kJ mol}^{-1}) = 68.15$ for $C_2H_4(g)$, -95.299 for $HCl(g)$, and -60.39 for $C_2H_5Cl(g)$. Why is $\Delta H°$ larger than $\Delta G°$?

Ans. -33.24 kJ; a decrease in entropy decreases the spontaneity

6.31. Calculate $\Delta_r G°_{298}$ for the deamination of L-aspartic acid to fumaric acid:

$$\underset{\underset{\displaystyle NH_2}{|}}{HOOC-CH-CH_2-COOH} \longrightarrow HOOC-CH=CH-COOH + NH_3$$

given that the free energies of formation of the acids, as reported by Burton and Krebs, are -174.88 kcal mol^{-1} for L-aspartic acid and -156.49 kcal mol^{-1} for fumaric acid, and $\Delta_f G°_{298} = -16.45$ kJ mol^{-1} for NH_3. Why does this reaction proceed?

Ans.   60.49 kJ; the change from the standard state may make ΔG favorable, or this reaction may be only one step in series of reactions that is overall spontaneous.

6.32. Determine the relation between K in (*6.18*) and L in (*5.20*).

Ans. $K = \Delta a - L$

6.33. For the reaction

$$\tfrac{1}{2}Cl_2(g) + \tfrac{1}{2}F_2(g) \longrightarrow ClF(g)$$

the temperature dependence of $\Delta_f G°_T$ is given by

$$\Delta_f G°_T/(J) = -50\,610 - (9.7)T + (0.741)T \ln T - (0.487 \times 10^{-3})T^2 + (0.84 \times 10^{-7})T^3$$

Using the data given in Problem 6.11, determine the temperature at which $ClF_3(g)$ is no longer the preferred product when $Cl_2(g)$ reacts with $F_2(g)$.   *Ans.*   784 K

6.34. Find the temperature at which the spontaneity of the reaction

$$2CO(g) \longrightarrow C(graph) + CO_2(g)$$

reverses. Thermodynamic data are: $\Delta_f G^\circ_{298}/(kJ\ mol^{-1})$ and $\Delta_f H^\circ_{298}/(kJ\ mol^{-1}) = -394.359$ and -393.509, respectively, for $CO_2(g)$ and -137.168 and -110.525 for $CO(g)$. In $J\ K^{-1}\ mol^{-1}$,

$$C_P = 26.00 + (43.497 \times 10^{-3})T - (148.32 \times 10^{-7})T^2 \qquad \text{for } CO_2(g)$$

$$C_P = 26.86 + (6.966 \times 10^{-3})T - (8.20 \times 10^{-7})T^2 \qquad \text{for } CO(g)$$

$$C_P = 11.18 + (10.950 \times 10^{-3})T - (4.891 \times 10^5)T^{-2} \qquad \text{for } C(graph)$$

Ans. $\Delta_r C^\circ_P/(J\ K^{-1}) = -16.54 + (40.515 \times 10^{-3})T - (131.92 \times 10^{-7})T^2 - (4.891 \times 10^5)T^{-2}$,
$\Delta_r H^\circ_{298} = -172.459$ kJ, $\Delta_r G^\circ_{298} = -120.023$ kJ, $J = -170.854$ kJ, $K = 79.43$ J K^{-1},
$\Delta_r G^\circ_T/(J) = (-170\ 854) + (79.43)T + (16.54)T \ln T - (20.258 \times 10^{-3})T^2 + (21.99 \times 10^{-7})T^3$
$+ (2.446 \times 10^5)T^{-1} = 0$;
$T = 972$ K

6.35. The tabulated values of $[(G^\circ_T - H^\circ_{298})/T]/(J\ K^{-1}\ mol^{-1})$ are -202.8 and -252.8 for $F_2(g)$, -223.0 and -274.5 for $Cl_2(g)$, and -217.8 and -268.2 for $ClF(g)$ at 298 K and 3 000 K, respectively. Given $\Delta_f G^\circ_{298} = -55.94$ kJ mol^{-1} for $ClF(g)$, find $\Delta_r H^\circ_{298}$ and $\Delta_r G^\circ_{3000}$ for

$$F_2(g) + Cl_2(g) \longrightarrow 2ClF(g)$$

Ans. $\Delta_r H^\circ_{298} = -108.96$ kJ, $\Delta_r G^\circ_{3000} = -136.3$ kJ

6.36. Given $[(G^\circ_{800} - H^\circ_{298})/800]/(J\ K^{-1}\ mol^{-1}) = -473.0$ for $WCl_6(g)$, -141.1 for $H_2(g)$, -41.7 for $W(s)$, and -197.3 for $HCl(g)$, could the reaction

$$WCl_6(g) + 3H_2(g) \longrightarrow W(s) + 6HCl(g)$$

be used at 800 K to prepare $W(s)$? Assume $\Delta_f H^\circ_{298}/(kJ\ mol^{-1}) = -513.8$ for $WCl_6(g)$ and -92.307 for $HCl(g)$.

Ans. Yes, $\Delta_r G^\circ_{800} = -303.4$ kJ

6.37. The formation of ozone in the upper atmosphere (25–30 km, 230 K) takes place as an O_2 molecule is dissociated by ultraviolet radiation and the atomic O formed reacts with a second O_2 molecule according to the reactions

$$O_2(g) + h\nu \longrightarrow 2O(g)$$

$$O(g) + O_2(g) \longrightarrow O_3(g)$$

Given $[(G^\circ_T - H^\circ_{298})/T]/(J\ K^{-1}\ mol^{-1}) = -160.95$ and -162.98 for $O(g)$, -205.03 and -207.71 for $O_2(g)$, and -238.82 and -242.29 for $O_3(g)$ at 298 K and 200 K, respectively, find $\Delta_r G^\circ_{230}$ for the second reaction. Use $\Delta_f H^\circ_{298}/(kJ\ mol^{-1}) = 249.170$ for $O(g)$ and 142.7 for $O_3(g)$.   *Ans.* -77.1 kJ

6.38. For the reaction

$$Ag(s) + \tfrac{1}{2}Br_2(g) \longrightarrow AgBr(l)$$

Metz and Seifert reported $E^\circ/(V) = 0.799\ 5 - 0.288\ Y + 0.097\ Y^2$, where $Y = (T - 705.2) \times 10^{-3}$ K. Calculate ΔG°_{1000}, and compare the value to ΔG° for the AgCl reaction considered in Problem 6.9.

Ans. $E^\circ = 0.723\ 0$ V, -69.76 kJ; less favorable

6.39. For the reaction

$$Ag(s) + \tfrac{1}{2}I_2(g) \longrightarrow AgI(l)$$

Metz and Seifert reported $E^\circ/(V) = (0.571\ 8 - 0.227\ Z + 0.220\ Z^2)V$ where $Z = (T - 831.2)10^{-3}$ K. Find expressions for $\Delta_f G^\circ$, $\Delta_f S^\circ$, $\Delta_f H^\circ$, and $\Delta_f C^\circ_P$ in terms of Z.

Ans. $\Delta_f G_T^\circ/(\text{kJ mol}^{-1}) = -nFE^\circ = -55.17 + (21.9)Z - (21.2)Z^2$,

$\Delta_f S_T^\circ/(\text{J K}^{-1}\,\text{mol}^{-1}) = -(\partial \Delta_f G_T^\circ/\partial T)_P = -21.9 + (42.4)Z$,

$\Delta_f H_T^\circ/(\text{kJ mol}^{-1}) = \Delta_f G_T^\circ + T\,\Delta_f S_T^\circ = -73.4 + (35.2)Z + (21.2)Z^2$,

$\Delta_f C_{P,T}^\circ/(\text{J K}^{-1}\,\text{mol}^{-1}) = (\partial \Delta_f H_T^\circ/\partial T)_P = 35.2 + (42.4)Z$

ACTIVITIES

6.40.   At what pressure will $G - G^\circ$ be 100. J mol^{-1} for an ideal gas at 25 °C?   *Ans.*   1.041 bar

6.41.   What is the value of $G - G^\circ$ for ethane at 101 bar and 100 °C assuming ideal-gas behavior? If $t_c = 32.2$ °C and $P_c = 48.9$ bar, find γ from Fig. 6-1, and calculate $G - G^\circ$ for the real gas. From the following data, confirm the value of γ.

$V_m^{-1}/(\text{mol dm}^{-3})$	0.5	1.0	1.5	2.0	2.5	3.0	3.5	4.0	4.5	5.0	5.5
$P/(\text{bar})$	14.62	27.64	39.30	49.68	59.06	67.56	75.35	82.58	89.42	95.99	102.70

Ans.   14.31 kJ; $T_r = 1.22$, $P_r = 2.07$ giving $\gamma = 0.69$, 13.2 kJ;
plot of $V_m - (RT/P)$ against P gives $RT \ln \gamma = -10.0$ dm^3 bar mol^{-1} or $\gamma = 0.72$

6.42.   Express I in terms of C for a 2:2 electrolyte, and calculate I for a 0.002 5 M solution of $CuSO_4$.

Ans. $I = 4C = 0.01$ M

6.43.   Calculate $y_{Cu^{2+}}$, y_{Cl^-}, $y_\pm$, and $a_\pm$ for a solution of $CuCl_2$ having $I = 0.01$ M.

Ans.   0.624, 0.889, 0.790, 0.007 90

6.44.   Calculate $\Delta_r G_{298}$ for

$$Cu^{2+}(aq) + 4NH_3(aq) \longrightarrow Cu(NH_3)_4{}^{2+}(aq)$$

given $\Delta_r G_{298}^\circ = -70.54$ kJ, $a(Cu^{2+}) = 0.010\ 0$, $a(NH_3) = 0.010\ 0$, and $a(\text{complex}) = 0.010\ 0$.
Ans. -24.89 kJ

THERMODYNAMIC RELATIONS

6.45.   Beginning with the definitions of G and A, show that (6.46) and (6.47) are valid.

6.46.   Beginning with (6.44), show that

$$\left(\frac{\partial T}{\partial V}\right)_S = -\left(\frac{\partial P}{\partial S}\right)_V$$

6.47.   Using the format of (6.48), express the Maxwell relation that begins $(\partial T/\partial P)_S$.

Ans. $\left(\dfrac{\partial T}{\partial P}\right)_S = \left(\dfrac{\partial V}{\partial S}\right)_P$

6.48.   Show that

(a) $\left(\dfrac{\partial H}{\partial P}\right)_T = -T\left(\dfrac{\partial V}{\partial T}\right)_P + V$ (b) $\left(\dfrac{\partial G}{\partial V}\right)_T = V\left(\dfrac{\partial P}{\partial V}\right)_T$

(c) $\left(\dfrac{\partial H}{\partial G}\right)_T = \dfrac{1}{V}\left[-T\left(\dfrac{\partial V}{\partial T}\right)_P + V\right]$ (d) $\left(\dfrac{\partial P}{\partial T}\right)_G = \dfrac{S}{V}$

(e) $\left(\dfrac{\partial U}{\partial V}\right)_T = T\left(\dfrac{\partial P}{\partial T}\right)_V - P$

(f) $\left(\dfrac{\partial U}{\partial P}\right)_T = -T\left(\dfrac{\partial V}{\partial T}\right)_P - P\left(\dfrac{\partial V}{\partial P}\right)_T$

(g) $\left(\dfrac{\partial H}{\partial P}\right)_T = -T^2\left(\dfrac{\partial(V/T)}{\partial T}\right)_P$

(h) $\left(\dfrac{\partial H}{\partial V}\right)_T = \left(\dfrac{\partial P}{\partial T}\right)_V\left(\dfrac{\partial(T/V)}{\partial(1/V)}\right)_P$

(i) $\left(\dfrac{\partial H}{\partial U}\right)_T = \dfrac{-V + T(\partial V/\partial T)_P}{T(\partial V/\partial T)_P + P(\partial V/\partial P)_T}$

(j) $U = -T^2\left(\dfrac{\partial(A/T)}{\partial T}\right)_V$

(k) $C_P - C_V = -T\left(\dfrac{\partial V}{\partial T}\right)_P^2\left(\dfrac{\partial P}{\partial V}\right)_T$

Chapter 7

Chemical Equilibrium

Equilibrium Constants

7.1 EQUILIBRIUM CONSTANT EXPRESSIONS

At equilibrium, $\Delta_r G = 0$ and the reaction quotient (Q) becomes the *thermodynamic equilibrium constant* (K). Equation (6.42) becomes

$$\ln K = \frac{-\Delta_r G^\circ}{RT} \qquad (7.1a)$$

$$K = e^{-\Delta_r G^\circ / RT} \qquad (7.1b)$$

Chemical equations in which equilibrium has been established are usually written using two reaction arrows facing opposite directions ($\rightleftharpoons$) to separate the reactants from the products.

Note: In using (7.1), as well as several other equations in this chapter, the unit mol needed for the correct unit analysis is generated by the stoichiometric coefficients, see Sec. 6.12 for a discussion of this point.

EXAMPLE 7.1. Consider the following oxidation-reduction reaction:

$$\text{Fe}^{2+}(\text{aq}) + \text{Ce}^{4+}(\text{aq}) \rightleftharpoons \text{Fe}^{3+}(\text{aq}) + \text{Ce}^{3+}(\text{aq})$$

Write the expression for the equilibrium constant in terms of activities of the ions and in terms of molar concentrations. Briefly describe an equilibrium mixture of these ions given $K = 10^{12}$.

Using the format of (6.43), the equilibrium constant expression is expressed as

$$K = a(\text{Fe}^{3+})a(\text{Ce}^{3+})a(\text{Fe}^{2+})^{-1}a(\text{Ce}^{4+})^{-1} = \frac{a(\text{Fe}^{3+})a(\text{Ce}^{3+})}{a(\text{Fe}^{2+})a(\text{Ce}^{4+})}$$

Substituting $a_i = y_i[i]$ for each species gives

$$K = \frac{y(\text{Fe}^{3+})y(\text{Ce}^{3+})}{y(\text{Fe}^{2+})y(\text{Ce}^{4+})} \frac{[\text{Fe}^{3+}][\text{Ce}^{3+}]}{[\text{Fe}^{2+}][\text{Ce}^{4+}]}$$

where the symbol $[i]$ represents the *equilibrium molar concentration* of chemical species *i*.

A very large value of K implies that the product of the activities of the products is much larger than the product of the activities of the reactants once equilibrium has been established. This reaction has "gone to completion."

EXAMPLE 7.2. For the chemical reaction

$$\text{Ag(s)} + \tfrac{1}{2}\text{Cl}_2(\text{g}) \longrightarrow \text{AgCl(l)}$$

Metz and Seifert reported $E^\circ = 0.839\ 8$ V at 1 000 K. If $a(\text{Ag}) = a(\text{AgCl}) = 1.00$ and $a(\text{Cl}_2) = 0.77$, find E. What is the value of K for the reaction at this temperature?

If electrochemical measurements are used to determine ΔG°, (6.42) becomes the *Nernst equation* upon substitution of (6.7):

$$E = E^\circ - \frac{RT}{nF} \ln Q \qquad (7.2)$$

and (7.1a) can be written as

$$\ln K = \frac{nFE^\circ}{RT} \tag{7.3}$$

For the given nonstandard conditions, (7.2) and (6.43) give

$$E = E^\circ - \frac{RT}{nF} \ln \frac{a(\text{AgCl})}{a(\text{Ag})[a(\text{Cl}_2)]^{1/2}}$$

$$= 0.839\,8 - \frac{(8.314\,\text{J K}^{-1}\,\text{mol}^{-1})(\text{mol})(1\,000\,\text{K})}{(1\,\text{mol})(9.648\,5 \times 10^4\,\text{J V}^{-1}\,\text{mol}^{-1})} \ln \frac{1.00}{(1.00)(0.77)^{1/2}} = 0.828\,5\,\text{V}$$

The reaction under these nonstandard conditions is less favored than under standard conditions.
    Using (7.3) gives

$$\ln K = \frac{(1\,\text{mol})(9.648\,5 \times 10^4\,\text{J V}^{-1}\,\text{mol}^{-1})(0.839\,3\,\text{V})}{(8.314\,\text{J K}^{-1}\,\text{mol}^{-1})(\text{mol})(1\,000\,\text{K})} = 9.740$$

and taking the antilog gives $K = 1.70 \times 10^4$.

EXAMPLE 7.3. The relation between equilibrium constant values at 1 atm standard pressure (K_{atm}) and those at 1 bar is

$$K = K_{\text{atm}}(1.013\,25)^{\Delta n_g} \tag{7.4}$$

where Δn_g was defined in Example 4.3. Find K_{atm} for the reaction described in Example 7.2.
    For this reaction, $\Delta n_g = -\frac{1}{2}$ mol. Thus

$$K_{\text{atm}} = \frac{K}{(1.013\,25)^{\Delta n_g}} = \frac{1.70 \times 10^4}{(1.013\,25)^{-1/2}} = 1.71 \times 10^4$$

7.2 TEMPERATURE DEPENDENCE OF K

From (6.16) and (7.1a),

$$\ln \frac{K_{T_2}}{K_{T_1}} = \frac{1}{R} \int_{T_1}^{T_2} \frac{\Delta_r H^\circ}{T^2} \, dT \tag{7.5}$$

which simplifies to

$$\ln \frac{K_{T_2}}{K_{T_1}} = -\frac{\Delta_r H^\circ}{R}\left(\frac{1}{T_2} - \frac{1}{T_1}\right) \tag{7.6}$$

for temperature intervals where $\Delta_r H^\circ$ is essentially constant. Equation (7.6) states that a plot of $\ln K$ against $1/T$ will be linear with a slope of $-\Delta_r H^\circ/R$.

EXAMPLE 7.4. Stern and Weise give $K = 3.92 \times 10^{-16}$ at 400. K and 1.69×10^{-8} at 600. K for the reaction

$$\text{BeSO}_4(\text{s}) \rightleftharpoons \text{BeO}(\text{s}) + \text{SO}_3(\text{g})$$

Assuming (7.6) to be valid over this somewhat large temperature interval, estimate $\Delta_r H^\circ_{500}$.
    Rearranging (7.5) gives

$$\Delta_r H^\circ = -R\frac{\ln(K_{T_2}/K_{T_1})}{1/T_2 - 1/T_1}$$

$$= -(8.314\,\text{J K}^{-1}\,\text{mol}^{-1})(\text{mol})\frac{\ln[(1.69 \times 10^{-8})/(3.92 \times 10^{-16})]}{(1/600 - 1/400)\,\text{K}^{-1}} = 175\,\text{kJ}$$

7.3 FREE ENERGY CURVES

The signs of the various thermodynamic properties at a given temperature can be determined from a plot of $\Delta_r G°$ against temperature.

1. For $\Delta_r G°$, the sign corresponds to the value of $\Delta_r G°$ on the plot.
2. For $(\partial \Delta_r G°/\partial T)_P$, the sign corresponds to the slope of the curve at that temperature.
3. For $\Delta_r S°$, (6.12) gives $\Delta_r S° = -(\partial \Delta_r G°/\partial T)_P$, and the sign will be opposite that of the slope.
4. For $\Delta_r H°$, (6.14) gives $\Delta_r H° = \Delta_r G° + T \Delta_r S°$, and the sign will be that of the numerically larger quantity, $\Delta_r G°$ or $T \Delta_r S°$.
5. For $(\partial \Delta_r S°/\partial T)_P$, if $\Delta_r S° = -(\partial \Delta_r G°/\partial T)_P$, then $(\partial \Delta_r S°/\partial T)_P = -(\partial^2 \Delta_r G°/\partial T^2)_P$, and the sign will be opposite that of the second derivative of $\Delta_r G°$ (which is usually zero at a point of inflection, positive at a minimum, and negative at a maximum).
6. For $(\partial \Delta_r H°/\partial T)_P$, taking the derivative of (6.14) gives $(\partial \Delta_r H°/\partial T)_P = T(\partial \Delta_r S°/\partial T)_P$, and the sign will be that of $(\partial \Delta_r S°/\partial T)_P$.
7. For $\Delta_r C_P°$, (5.10) gives $\Delta_r C_P° = T(\partial \Delta_r S°/\partial T)_P$, and the sign will be that of $(\partial \Delta_r S°/\partial T)_P$.
8. For $(\partial \Delta_r C_P°/\partial T)_P$, taking the derivative of (4.2) gives $(\partial \Delta_r C_P°/\partial T)_P = (\partial^2 \Delta_r H°/\partial T^2)_P$, and the sign will be that of the second derivative of $\Delta_r H°$ (which will be very difficult to estimate qualitatively).
9. For $\ln K$, (7.1a) gives $\ln K = -\Delta_r G°/RT$, and the sign will be opposite that of $\Delta_r G°$.
10. For $(\partial \ln K/\partial T)_P$, (7.5) gives $(\partial \ln K/\partial T)_P = \Delta_r H°/RT^2$, and the sign will be that of $\Delta_r H°$.

Equilibrium in Gases

7.4 GASEOUS EQUILIBRIUM CONSTANTS

Several different equilibrium constants are encountered when dealing with gaseous reactions. These are related to K by

$$K = K_\gamma K_p \tag{7.7}$$

$$K_p = (RT)^{\Delta n} K_c \tag{7.8}$$

and

$$K_p = P^{\Delta n} K_x \tag{7.9}$$

where K_γ is defined as

$$K_\gamma \equiv \frac{\gamma_R^r \gamma_S^s \cdots}{\gamma_L^l \gamma_M^m \cdots} \tag{7.10}$$

K_p is similarly defined in terms of partial pressures $P_i/$(bar), K_c in terms of C_i (mol dm^{-3}), and K_x in terms of mole fractions. Note that P in (7.9) is the *total* pressure of the gases, including any inert gases that may present. If the experimental conditions are not too severe, K_γ in (7.7) is often set equal to unity to simplify the calculations.

EXAMPLE 7.5. Calculate K, K_γ, K_p, K_c, and K_x at 25 °C and 1.00 bar total pressure for the reaction

$$H_2(g) + \tfrac{1}{2}O_2(g) \rightleftharpoons H_2O(g)$$

given $\Delta_f G_{298}° = -228.572$ kJ. Assume that $\gamma(H_2O) = 0.98$, $\gamma(H_2) = 1.00$, and $\gamma(O_2) = 1.00$.

Equation (7.10) gives

$$K_\gamma = \frac{\gamma(H_2O)}{\gamma(H_2)[\gamma(O_2)]^{1/2}} = \frac{0.98}{(1.00)(1.00)^{1/2}} = 0.98$$

Equation (7.1b) gives

$$K = e^{-(-228572)/(8.314)(298.15)} = 1.11 \times 10^{40}$$

and (7.7) gives

$$K_p = \frac{K}{K_\gamma} = \frac{1.11 \times 10^{40}}{0.98} = 1.13 \times 10^{40}$$

Because molarities are expressed in mol dm^{-3} and pressures are in bar, we have from (7.8)

$$K_c = K_p(RT)^{-\Delta n}$$

$$= (1.13 \times 10^{40})[(0.083\,14 \text{ dm}^3 \text{ bar K}^{-1} \text{ mol}^{-1})(298)]^{-(-1/2)} = 5.62 \times 10^{40}$$

and from (7.9),

$$K_x = K_p P^{-\Delta n} = (1.13 \times 10^{40})(1)^{-(-1/2)} = 1.13 \times 10^{40}$$

EXAMPLE 7.6. Calculate the percent conversion to PCl_5 and the partial pressure of PCl_5 at 1.00 bar total pressure and 400 K for the reaction

$$PCl_3(g) + Cl_2(g) \rightleftharpoons PCl_5(g)$$

if the original reaction mixture contained 1.00 mol of PCl_3 and 2.00 mol of Cl_2. For this reaction, $\Delta G°_{400} = -3\,533$ J.

Using (7.1b) gives

$$K = e^{-(-3533)/(8.314)(400)} = 2.89$$

Because the gases are assumed to be ideal, (7.7) gives $K = K_p = 2.89$. Letting x be the amount of $PCl_5(g)$ at equilibrium, the amount of $PCl_3(g)$ is $1.00 - x$, and that of $Cl_2(g)$ is $2.00 - x$. The total amount is

$$x + (1.00 - x) + (2.00 - x) = 3.00 - x$$

The partial pressures are given by (1.10) as

$$P(PCl_5) = \frac{xP}{3.00 - x} \qquad P(Cl_2) = \frac{(2.00 - x)P}{3.00 - x} \qquad P(PCl_3) = \frac{(1.00 - x)P}{3.00 - x}$$

and substituting into (6.43) gives

$$2.89 = \left[\frac{xP}{3.00 - x}\right]^1 \Big/ \left[\frac{(2.00 - x)P}{3.00 - x}\right]^1 \left[\frac{(1.00 - x)P}{3.00 - x}\right]^1$$

Solving for x with $P = 1.00$ bar gives $x = 0.626$ and $P(PCl_5) = 0.264$ bar.

EXAMPLE 7.7. Find K_γ for the reaction mixture described in Example 7.6, given that the actual value of x is 0.600 for real gases.

The value of K_p for the real gases is

$$K_p = \left[\frac{(0.600)(1.00)}{2.40}\right]^1 \Big/ \left[\frac{(1.40)(1.00)}{2.40}\right]^1 \left[\frac{(0.40)(1.00)}{2.40}\right]^1 = 2.57$$

and (7.7) gives

$$K_\gamma = \frac{K}{K_p} = \frac{2.89}{2.57} = 1.12$$

7.5 CALCULATIONS FOR HETEROGENEOUS SYSTEMS

Although the equilibrium constants are written as in (6.43), they are simplified by choosing the activities of the condensed phases as unity (unless large pressure changes are involved) and assuming the gases to be ideal (unless they are significantly nonideal).

EXAMPLE 7.8. Calculate the pressure of oxygen over a sample of NiO(s) at 25 °C given that $\Delta G° = 211.7$ kJ for the reaction

$$NiO(s) \rightleftharpoons Ni(s) + \tfrac{1}{2}O_2(g)$$

The equilibrium constant for the reaction is

$$K = [a(\text{Ni})]^1 [a(\text{O}_2)]^{1/2}/[a(\text{NiO})]^1$$

which simplies to

$$K = [a(\text{O}_2)]^{1/2} = [P(\text{O}_2)/(\text{bar})]^{1/2}$$

The value of K using (7.1b) is

$$K = e^{-(211700)/(8.314)(298)} = 7.8 \times 10^{-38}$$

Solving for the pressure gives

$$P(\text{O}_2) = K^2 = (7.8 \times 10^{-38})^2 = 6 \times 10^{-75} \text{ bar}$$

The assumption of ideal behavior for O_2 is valid at this very low pressure.

7.6 LE CHATELIER'S PRINCIPLE

If a stress is imposed upon a system at equilibrium, the system will change concentrations of the products and reactants in such a way as to relieve the stress. The mathematical description of the change in the system is given by combining (6.14), (7.1b), (7.7), and (7.9), and solving for the mole fraction of one of the products to give

$$x_R{}^r = \frac{x_L{}^l x_M{}^m \cdots}{x_S{}^s \cdots} K_\gamma{}^{-1} P^{-\Delta n} e^{-\Delta H°/RT} e^{\Delta S°/R} \qquad (7.11)$$

Even though a change in P will change γ and hence K_γ, the change in K_γ will be not nearly as important as the change in P for most cases.

EXAMPLE 7.9. Determine the effect on the equilibrium at 25 °C for the reaction

$$\text{H}_2(\text{g}) + \tfrac{1}{2}\text{O}_2(\text{g}) \rightleftharpoons \text{H}_2\text{O}(\text{g})$$

of (1) an increase in $x(\text{H}_2)$, (2) a decrease in $x(\text{O}_2)$, (3) an increase in total pressure, (4) an increase in temperature given that $\Delta_f H°$ is negative, and (5) forming liquid water instead of gaseous water.

According to (7.11), (1) for an increase in $x(\text{H}_2)$, the value of $x(\text{H}_2\text{O})$ increases; (2) for a decrease in $x(\text{O}_2)$, the value of $x(\text{H}_2\text{O})$ decreases; (3) for an increase in P, the term $P^{-(-1/2)}$ becomes larger, resulting in a larger value for $x(\text{H}_2\text{O})$; (4) for an increase in T, even though $\Delta_f H°$ becomes slightly more negative, the exponent becomes less positive, resulting in a decrease in $x(\text{H}_2\text{O})$; and (5) in forming liquid water, the changes in $\Delta_f S°$ and $\Delta_f H°$ are such that $\Delta_f G°$ is more negative, giving an increase in $x(\text{H}_2\text{O})$.

Equilibrium in Aqueous Solutions

7.7 MONOPROTIC ACIDS AND CONJUGATE BASES

Consider the aqueous equilibrium established by a *Brønsted–Lowry acid* (A) and its *conjugate base* (B) represented by

$$\text{A(aq)} + \text{H}_2\text{O(l)} \rightleftharpoons \text{H}_3\text{O}^+(\text{aq}) + \text{B(aq)} \qquad (7.12a)$$

Often this chemical equation is written more simply as

$$\text{A(aq)} \rightleftharpoons \text{H}^+(\text{aq}) + \text{B(aq)} \qquad (7.12b)$$

The thermodynamic equilibrium constant expression describing this equilibrium is given by

$$K_a = \frac{a(\text{H}^+)a(\text{B})}{a(\text{A})} = \frac{y(\text{H}^+)y(\text{B})}{y(\text{A})} \frac{[\text{H}^+][\text{B}]}{[\text{A}]} \qquad (7.13)$$

where K_a is known as the *acid ionization* (or *acid dissociation*) *constant*. In most calculations the activity coefficients are ignored, and (*7.13*) can be written

$$K_a = [H^+][B]/[A] \qquad (7.14)$$

The equilibrium concentrations of the acid and its conjugate base are

$$[A] = C_A - ([H^+] - [OH^-]) \qquad (7.15a)$$

$$[B] = C_B + ([H^+] - [OH^-]) \qquad (7.15b)$$

where C_A and C_B are the *formal* or *analytical concentrations* of the acid and base, respectively. Equations (*7.15*) take into account the self-ionization equilibrium of water,

$$H_2O(l) \rightleftharpoons H^+(aq) + OH^-(aq) \qquad (7.16)$$

where the *water ionization equilibrium constant* expression is given by

$$K_w = a(H^+)a(OH^-)/a(H_2O) \approx [H^+][OH^-] \qquad (7.17)$$

Substitution of (*7.15*) into (*7.14*) gives

$$K_a = \frac{[H^+](C_B + [H^+] - [OH^-])}{C_A - [H^+] + [OH^-]} \qquad (7.18)$$

Equation (*7.18*) is a complete description of the aqueous equilibrium established between an acid and its conjugate base. Special cases involving weak acids, weak bases, hydrolysis of salts, etc., will be considered in Secs. 7.8–7.10 and 7.12.

EXAMPLE 7.10.   What is the pH of a 0.010 M solution of nitric acid?

Nitric acid is a "strong" acid that is essentially 100% ionized in aqueous solution. Thus $[A] \approx 0$, and (*7.15a*) gives

$$[H^+] = C_A + [OH^-] = C_A + \frac{K_w}{[H^+]} \qquad (7.19)$$

In this case $[OH^-] \ll C_A$, and

$$[H^+] = C_A = 0.010 \text{ M}$$

giving $\qquad\qquad$ pH $= -\log[H^+] = -\log(0.010) = 2.00 \qquad (7.20)$

7.8   AQUEOUS SOLUTIONS OF WEAK ACIDS

Equation (*7.18*) can be used to describe the equilibrium solution of a weak acid by assigning $C_B = 0$, giving

$$K_a = \frac{[H^+]([H^+] - [OH^-])}{C_A - [H^+] + [OH^-]} \qquad (7.21a)$$

EXAMPLE 7.11.   Find the pH of a 0.025 M solution of benzoic acid given $K_a = 6.6 \times 10^{-5}$ for C_6H_5COOH.

Equation (*7.21a*) is a cubic equation in $[H^+]$ and could be solved using (*1.23*). However, if $[H^+] \gg [OH^-]$ (which is usually true in a solution of an acid unless $K_a \ll 10^{-15}$), then (*7.21a*) can be simplified to

$$K_a = \frac{[H^+]^2}{C_A - [H^+]} \qquad (7.21b)$$

which can be solved for $[H^+]$ using the quadratic formula or (*1.23*). Finally, if $C_A > 10^4 K_a$, (*7.21b*) can be simplified to

$$K_a = [H^+]^2/C_A \qquad (7.21c)$$

without introducing any significant error.

For the benzoic acid solution of interest, (7.21b) can be used, giving

$$6.6 \times 10^{-5} = \frac{[H^+]^2}{0.025 - [H^+]}$$

$$[H^+]^2 + (6.6 \times 10^{-5})[H^+] - (0.025)(6.6 \times 10^{-5}) = 0$$

Solving for $[H^+]$ gives 1.3×10^{-3} M. The pH of the solution is

$$pH = -\log(1.3 \times 10^{-3}) = 2.89$$

7.9 AQUEOUS SOLUTIONS OF WEAK BASES

Equation (7.18) can be used to describe the equilibrium solution of a weak base by assigning $C_A = 0$, giving

$$K_a = \frac{[H^+](C_B + [H^+] - [OH^-])}{-[H^+] + [OH^-]} \tag{7.22a}$$

When considering equilibria involving weak bases, often the classical *base ionization* (or *base dissociation*) *constant* (K_b) for the base B is substituted for K_w/K_a:

$$K_b = K_w/K_a \tag{7.23}$$

EXAMPLE 7.12. Find the pH of a 0.10 M solution of ammonia given $K_b = 1.6 \times 10^{-5}$ for NH_3.

Equation (7.22a) is a cubic equation in $[OH^-]$ and could be solved using (1.23). However, if $[OH^-] \gg [H^+]$ (which is usually true in a solution of a base), then (7.22a) can be simplified to

$$K_a = \frac{[H^+](C_B - [OH^-])}{[OH^-]} \tag{7.22b}$$

which can be solved for $[OH^-]$ using the quadratic formula or (1.23). Finally, if $C_B > 10^4 K_a$, (7.22b) can be simplified to

$$K_a = [H^+]C_B/[OH^-] \tag{7.22c}$$

For the ammonia solution being considered, (7.22c) can be used, giving

$$K_a = \frac{[H^+]C_B}{[OH^-]} = \frac{K_w}{K_b} = \frac{(K_w/[OH^-])C_B}{[OH^-]}$$

$$[OH^-]^2 = K_b C_B = (1.6 \times 10^{-5})(0.10) = 1.6 \times 10^{-6}$$

$$[OH^-] = 1.3 \times 10^{-3} \text{ M}$$

The pOH of the solution is

$$pOH = -\log[OH^-] = -\log(1.3 \times 10^{-3}) = 2.89 \tag{7.24}$$

and the pH is

$$pH = pK_w - pOH = 14.00 - 2.89 = 11.11 \tag{7.25}$$

7.10 BUFFER SOLUTIONS

A *buffer solution* resists large changes in pH upon the addition of small amounts of acid or base or upon dilution. A buffer solution is a solution containing either a weak acid and its conjugate base or a weak base and its conjugate acid.

For an acidic solution where $[H^+] \gg [OH^-]$, (7.18) becomes

$$K_a = \frac{[H^+](C_B + [H^+])}{C_A - [H^+]} \tag{7.26a}$$

which can be further simplified to

$$K_a = [H^+]C_B/C_A \tag{7.26b}$$

if C_B and $C_A \gg [H^+]$. Solving (7.26b) directly for pH gives

$$pH = pK_a + \log\left(\frac{C_B}{C_A}\right) \tag{7.27}$$

which is known as the *Henderson–Hasselbalch equation* and is useful for most calculations involving buffer solutions (provided that C_A and $C_B \gg [H^+]$).

For a basic solution where $[OH^-] \gg [H^+]$, (7.18) becomes

$$K_a = \frac{[H^+](C_B - [OH^-])}{C_A + [OH^-]} \tag{7.28}$$

which also reduces to (7.27) if C_A and $C_B \gg [OH^-]$.

EXAMPLE 7.13.   What is the pH of a buffer solution consisting of 0.20 M NH_3 and 0.10 M NH_4NO_3? For NH_3, $K_b = 1.6 \times 10^{-5}$ at 25 °C.

Under the conditions that $C(NH_3)$ and $C(NH_4^+) \gg [H^+]$, (7.27) gives

$$pH = p\left(\frac{1.00 \times 10^{-14}}{1.6 \times 10^{-5}}\right) + \log\left(\frac{0.20}{0.10}\right) = 9.51$$

where K_w/K_b was substituted for K_a.

EXAMPLE 7.14.   Determine the pH of the buffer solution described in Example 7.13 after the addition of 0.01 mol of NaOH to 1 dm^3 of the buffer.

Upon the addition of 0.01 mol OH^-, $C(NH_3) = 0.20 + 0.01 = 0.21$ M and $C(NH_4^+) = 0.10 - 0.01 = 0.09$ M. Using (7.27) gives

$$pH = p\left(\frac{1.00 \times 10^{-14}}{1.6 \times 10^{-5}}\right) + \log\left(\frac{0.21}{0.09}\right) = 9.57$$

a slight increase in pH. Note that 1 dm^3 of water would have changed from pH = 7 to pH = 12 for the same amount of NaOH.

7.11 SOLUTIONS OF AMPHOLYTES

Consider the aqueous equilibria established by the monohydrogen anion of a diprotic acid (an example of an *ampholyte*, a substance that can act as either an acid or a base):

$$HA^-(aq) \rightleftharpoons H^+(aq) + A^{2-}(aq) \qquad\qquad K_{a2} = \frac{[H^+][A^{2-}]}{[HA^-]} \tag{7.29a}$$

$$HA^-(aq) + H^+(aq) \rightleftharpoons H_2A(aq) \qquad \frac{1}{K_{a1}} = \frac{[H_2A]}{[H^+][HA^-]} \tag{7.29b}$$

The equilibrium concentration of H^+ is

$$[H^+] = [A^{2-}] + [OH^-] - [H_2A]$$

which, upon substitution of (7.17) and (7.29), gives

$$[H^+] = \left(\frac{K_{a1}(K_{a2}[HA^-] + K_w)}{K_{a1} + [HA^-]}\right)^{1/2} \tag{7.30a}$$

where $[HA^-]$ is usually assumed to be equal to the formal concentration of the salt $[C(HA^-)]$.

EXAMPLE 7.15. What is the pH of a 0.100 M solution of $NaHCO_3$? For H_2CO_3 at 25 °C, $K_{a1} = 4.5 \times 10^{-7}$ and $K_{a2} = 4.8 \times 10^{-11}$.

In this solution, we assume that $[HCO_3^-] \approx C(HCO_3^-) = 0.100$ M. Because $K_{a2}[HCO_3^-] > K_w$ and because $[HCO_3^-] > K_{a1}$, (7.30) simplifies to

$$[H^+] = (K_{a1}K_{a2})^{1/2} \tag{7.30b}$$

giving for the solution

$$[H^+] = [(4.5 \times 10^{-7})(4.8 \times 10^{-11})]^{1/2} = 4.6 \times 10^{-9} \text{ M}$$

or a pH value of 8.34.

7.12 HYDROLYSIS OF IONS

If A in (7.12) is the cation of a weak base that undergoes "hydrolysis" to form an acidic solution, then $[H^+] \gg [OH^-]$, $C_A \gg [H^+]$, and $K_a = K_w/K_b$, where K_b refers to the weak base that is formed. Equation (7.21c) becomes

$$[H^+] \approx [(K_w/K_b)C_A]^{1/2} \tag{7.21d}$$

If B in (7.12) is the anion of a weak acid that undergoes "hydrolysis" to form a basic solution, then $[OH^-] \gg [H^+]$, $C_A \gg [H^+]$, and $K_b = K_w/K_a$, where K_a refers to the weak acid that is formed. Equation (7.22c) becomes

$$[OH^-] \approx [(K_w/K_a)C_B]^{1/2} \tag{7.22d}$$

If both the anion and cation hydrolyze, (7.30a) becomes

$$[H^+] = \left(\frac{K_w K_a(C + K_b)}{K_b(C + K_a)}\right)^{1/2} \tag{7.30c}$$

where C is the concentration of the salt, K_{a1} is replaced by K_a for the weak acid involved, and K_{a2} is replaced by K_w/K_b for the weak base involved. Note that (7.30c) is valid only if $K_a K_b \gg K_w$.

EXAMPLE 7.16. Calculate the pH of the following aqueous solutions: (a) 0.010 M NH_4NO_3 given $K_b = 1.6 \times 10^{-5}$ for NH_3, (b) 0.010 M $K(CH_3COO)$ given $K_a = 1.754 \times 10^{-5}$ for CH_3COOH, (c) 0.010 M $NH_4(CH_3COO)$, and (d) 0.010 M KNO_3.

(a) For the solution of NH_4NO_3, the salt of a weak base and a strong acid, an acidic solution will be formed as the cation undergoes hydrolysis. Equation (7.21d) gives

$$[H^+] = \left(\frac{1.00 \times 10^{-14}}{1.6 \times 10^{-5}}(0.010)\right)^{1/2} = 2.5 \times 10^{-6} \text{ M}$$

or pH = 5.60.

(b) For the solution of $K(CH_3COO)$, the salt of a weak acid and a strong base, a basic solution will be formed as the anion undergoes hydrolysis. Equation (7.22d) gives

$$[OH^-] = \left(\frac{1.00 \times 10^{-14}}{1.754 \times 10^{-5}}(0.010)\right)^{1/2} = 2.4 \times 10^{-6} \text{ M}$$

or pH = 8.38.

(c) For the solution of $NH_4(CH_3COO)$ in which both ions hydrolyze, (7.30c) can be simplified, because $C \gg K_a$ and K_b, to

$$[H^+] = (K_w K_a/K_b)^{1/2} \tag{7.30d}$$

$$= \left(\frac{(1.00 \times 10^{-14})(1.754 \times 10^{-5})}{1.6 \times 10^{-5}}\right)^{1/2} = 1.05 \times 10^{-7} \text{ M}$$

or pH = 6.98.

(d) A solution of KNO_3 does not contain ions that hydrolyze. Thus pH = 7.00 for this solution.

7.13 CONSECUTIVE EQUILIBRIA

Consider the following stepwise formation reactions and equilibrium constant expressions:

$$A + B \rightleftharpoons AB \qquad K_{f1} = [AB]/[A][B]$$

$$AB + B \rightleftharpoons AB_2 \qquad K_{f2} = [AB_2]/[AB][B]$$

$$\vdots$$

$$AB_{n-1} + B \rightleftharpoons AB_n \qquad K_{fn} = [AB_n]/[AB_{n-1}][B] \qquad (7.31)$$

The equilibrium concentrations of the various species at a given value of [B] can be determined using

$$[A] = \frac{C_A}{1 + K_{f1}[B] + K_{f1}K_{f2}[B]^2 + \cdots + K_{f1}K_{f2}\cdots K_{fn}[B]^n}$$

$$[AB] = K_{f1}[B][A]$$

$$[AB_2] = K_{f1}K_{f2}[B]^2[A] \qquad (7.32)$$

$$\vdots$$

$$[AB_n] = K_{f1}K_{f2}\cdots K_{fn}[B]^n[A]$$

where C_A represents the formal concentration of substance A.

EXAMPLE 7.17. Equations (7.32) are applicable to the stepwise ionization of a polyprotic acid, where B is H^+ (or H_3O^+), A is the anion of the acid, and

$$K_{f1} = 1/K_{an} \qquad K_{f2} = 1/K_{a(n-1)} \qquad \cdots \qquad K_{fn} = 1/K_{a1}$$

Given $K_{a1} = 7.5 \times 10^{-3}$, $K_{a2} = 6.6 \times 10^{-8}$, and $K_{a3} = 1 \times 10^{-12}$ for phosphoric acid, find the equilibrium concentrations of H_3PO_4, $H_2PO_4^-$, HPO_4^{2-}, and PO_4^{3-} at pH = 6.50 in a 0.100 M H_3PO_4 solution.

The respective values of the formation constants are

$$K_{f1} = 1/K_{a3} = 1/(1 \times 10^{-12}) = 1 \times 10^{12}$$

$$K_{f2} = 1/K_{a2} = 1/(6.6 \times 10^{-8}) = 1.5 \times 10^7$$

$$K_{f3} = 1/K_{a1} = 1/(7.5 \times 10^{-3}) = 1.3 \times 10^2$$

The concentration of $[H^+]$ in the solution is found by using the antilogarithm form of (7.20):

$$[H^+] = 10^{-pH} = 10^{-6.50} = 3.2 \times 10^{-7} \text{ M}$$

and $C_A = 0.100$ M. Using (7.32) gives

$$[PO_4^{3-}] = \frac{0.100}{1 + (1 \times 10^{12})(3.2 \times 10^{-7}) + (1 \times 10^{12})(1.5 \times 10^7)(3.2 \times 10^{-7})^2 + (1 \times 10^{12})(1.5 \times 10^7)(1.3 \times 10^2)(3.2 \times 10^{-7})^3}$$

$$= 5.4 \times 10^{-8} \text{ M}$$

$$[HPO_4^{2-}] = (1 \times 10^{12})(3.2 \times 10^{-7})(5.4 \times 10^{-8}) = 1.6 \times 10^{-2} \text{ M}$$

$$[H_2PO_4^-] = (1 \times 10^{12})(1.5 \times 10^7)(3.2 \times 10^{-7})^2(5.4 \times 10^{-8}) = 8.3 \times 10^{-2} \text{ M}$$

$$[H_3PO_4] = (1 \times 10^{12})(1.5 \times 10^7)(1.3 \times 10^2)(3.2 \times 10^{-7})^3(5.4 \times 10^{-8})$$

$$= 3.2 \times 10^{-6} \text{ M}$$

This type of calculation can be repeated for various values of pH, and a plot of the concentrations or, more commonly, the fraction of concentration (given by $[AB_i]/C_A$) as a function of pH can be made as shown in Fig. 7-1.

EXAMPLE 7.18. What is the pH of a 0.100 M solution of H_3PO_4? See Example 7.17 for the values of K_a for phosphoric acid.

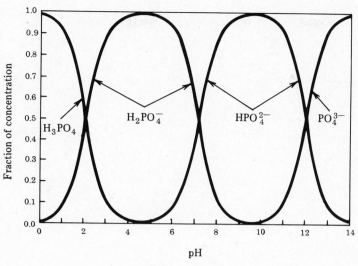

Fig. 7-1

Because $K_{a1} \gg K_{a2} \gg K_{a3}$, the stepwise ionization process can be treated as three individual reactions with essentially all the H^+ coming from the first ionization process because of the common-ion effect. Thus (*7.21b*) gives

$$7.5 \times 10^{-3} = \frac{[H^+]^2}{0.100 - [H^+]}$$

which can be solved to give $[H^+] = 2.4 \times 10^{-2}$ M, or pH = 1.62. From Fig. 7-1, we can see that additional contributions of H^+ from the ionization of $H_2PO_4^-$ or HPO_4^{2-} are not important.

7.14 SLIGHTLY SOLUBLE SALTS

For the equilibrium established by a slightly soluble salt and its ions in solution,

$$A_{\nu_+}B_{\nu_-}(s) \rightleftharpoons \nu_+ A^{(z_+)+}(aq) + \nu_- B^{(z_-)-}(aq) \qquad (7.33)$$

the thermodynamic equilibrium constant expression is given by

$$K_{sp} = (y_\pm)^\nu [A^{(z_+)+}]^{\nu_+} [B^{(z_-)-}]^{\nu_-} \qquad (7.34)$$

where K_{sp} is known as the *solubility product constant*. For very dilute solutions of pure salts, the mean ionic activity coefficient is often omitted.

EXAMPLE 7.19.   Find the concentration of Ag^+ in a solution of AgBr that (*a*) is saturated with AgBr, (*b*) contains 0.01 M NaBr (*common-ion effect*), and (*c*) contains 0.01 M HNO₃ (*diverse-ion effect*). $\Delta_r G^\circ_{298} = 70.05$ kJ for

$$AgBr(s) \rightleftharpoons Ag^+(aq) + Br^-(aq)$$

For the reaction, (*7.1b*) gives

$$K_{sp} = e^{-(70050)/(8.314)(298)} = 5.3 \times 10^{-13}$$

(*a*)   For the first solution, (*7.34*) gives

$$K_{sp} = y_\pm^2 C^2$$

where C is the concentration of either ion. Assuming $y_\pm$ to be near unity for the solution,

$$C = K_{sp}^{1/2} = (5.3 \times 10^{-13})^{1/2} = 7.3 \times 10^{-7} \text{ M}$$

As a check on the assumption concerning $y_\pm$, for a 1:1 electrolyte, (*6.38*) gives

$$I = C = 7.3 \times 10^{-7} \text{ M}$$

(excluding H^+ and OH^-), and (6.39) gives

$$\log y_{\pm} = -(1)(1)(0.511\ 6)(7.3 \times 10^{-7})^{1/2} = -4.4 \times 10^{-4}$$

Taking the antilog gives $y_{\pm} = 0.999$, which is very close to unity. Thus the Ag^+ concentration is 7.3×10^{-7} M.

(b) The effect of the common ion in the second solution will be to decrease the concentration of the AgBr. Because the total ionic strength is now appreciable, $y_{\pm}$ must be calculated. By (6.38),

$$I = \tfrac{1}{2}\{C(Na^+)[z(Na^+)]^2 + C(Br^-)[z(Br^-)]^2 + C(Ag^+)[z(Ag^+)]^2\}$$

$$= \tfrac{1}{2}[(0.01)1^2 + (0.01 + C)1^2 + (C)1^2] \approx \tfrac{1}{2}[(0.01)1^2 + (0.01)1^2] = 0.01 \text{ M}$$

and (6.39) gives

$$\log y_{\pm} = -(1)(1)(0.511\ 6)(0.01)^{1/2} = -0.051\ 16$$

Taking the antilog gives $y_{\pm} = 0.889$. The equilibrium expression becomes

$$K_{sp} = 5.3 \times 10^{-13} = (0.889)^2(C)(0.01 + C)$$

and solving for C gives 6.7×10^{-11} M. Thus the Ag^+ concentration has decreased to 6.7×10^{-11} M.

(c) For the third solution, the total ionic strength will be nearly 0.01 M, giving $y_{\pm} = 0.889$. The equilibrium expression in this case is

$$5.3 \times 10^{-13} = (0.889)^2 C^2$$

and solving for C gives 8.2×10^{-7} M, a slight increase in the solubility.

Solved Problems

EQUILIBRIUM CONSTANTS

7.1. Using the data in Examples 6.2 and 6.3 for the Daniell cell, calculate K_{298} and estimate K_{303}.

At 25 °C, ($7.1b$) gives

$$K_{298} = e^{-(-210590)/(8.314)(298)} = 8.2 \times 10^{36}$$

Assuming (7.6) to be valid,

$$\ln \frac{K_{303}}{8.2 \times 10^{36}} = -\frac{-228\ 530}{8.314}\left(\frac{1}{303} - \frac{1}{298}\right) = -1.52$$

$$K_{303} = 1.8 \times 10^{36}$$

7.2. Supposing that $\Delta_r G_T^{\circ}$ varies with temperature as shown in Fig. 7-2, prepare a table of signs for the thermodynamic properties.

Applying the rules outlined in Sec. 7.3 gives:

	$\Delta_r G^{\circ}$	$(\partial \Delta_r G^{\circ}/\partial T)_P$	$\Delta_r S^{\circ}$	$\Delta_r H^{\circ}$	$(\partial \Delta_r S^{\circ}/\partial T)_P$	$(\partial \Delta_r H^{\circ}/\partial T)_P$	$\Delta_r C_P^{\circ}$	$\ln K$	$(\partial \ln K/\partial T)_P$
T_1	−	0	0	−	−	−	−	+	−
T_2	0	+	−	−	0	0	0	0	−
T_3	+	+	−	+	+	+	+	−	+
T_4	+	0	0	+	+	+	+	−	+
T_5	+	0	0	+	−	−	−	−	+
T_6	+	+	−	+	−	−	−	−	+

For the $\Delta_r H^{\circ}$ entries at T_3 and T_6, the + value resulted from assuming $\Delta_r G^{\circ}$ to be numerically greater than $T\Delta_r S^{\circ}$.

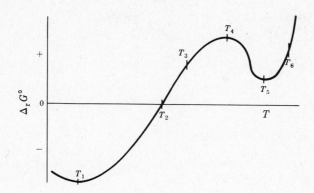

Fig. 7-2

EQUILIBRIUM IN GASES

7.3. Assuming real-gas behavior, what is K_x at 100 bar and 25 °C for the reaction

$$SO_2(g) + \tfrac{1}{2}O_2(g) \rightleftharpoons SO_3(g)$$

given $\Delta G_{298}^\circ = -70.87$ kJ, $\gamma(SO_2) = 0.059$, $\gamma(O_2) = 0.95$, and $\gamma(SO_3) = 0.022$. Given that the original reaction mixture contained 1.00 mol SO_2 and 0.50 mol_2 O_2, find $x(SO_3)$.

Equation (7.10) gives

$$K_\gamma = \frac{\gamma(SO_3)}{\gamma(SO_2)[\gamma(O_2)]^{1/2}} = \frac{0.022}{(0.059)(0.95)^{1/2}} = 0.38$$

Using (7.1b) gives

$$K = e^{-(-70870)/(8.314)(298)} = 2.6 \times 10^{12}$$

Using (7.7) gives

$$K_p = \frac{K}{K_\gamma} = \frac{2.6 \times 10^{12}}{0.38} = 6.9 \times 10^{12}$$

and using (7.9) gives

$$K_x = K_p P^{-\Delta n} = (6.9 \times 10^{12})(100)^{-(-1/2)} = 6.9 \times 10^{13}$$

If x represents the amount of SO_3 at equilibrium, then the amount of SO_2 is $1.00 - x$ and that of O_2 is $0.50 - (x/2)$. The total amount is $1.50 - (x/2)$, giving the mole fractions as

$$x_{SO_2} = \frac{1.00 - x}{1.50 - x/2} \qquad x_{O_2} = \frac{0.50 - x/2}{1.50 - x/2} \qquad x_{SO_3} = \frac{x}{1.50 - x/2}$$

Substituting into (6.43) gives

$$K_x = 6.9 \times 10^{13} = \left(\frac{x}{1.50 - x/2}\right) \bigg/ \left(\frac{1.00 - x}{1.50 - x/2}\right)\left(\frac{0.50 - x/2}{1.50 - x/2}\right)^{1/2}$$

Solving, we obtain $x = 1.00$ and $x_{SO_3} = 1.00$, to three significant figures.

7.4. What is the partial pressure of NH_3 above a sample of NH_4Cl as a result of decomposition at 25 °C given $\Delta_f G_{298}^\circ / (kJ\,mol)^{-1} = -202.87$ for $NH_4Cl(s)$, -16.45 for $NH_3(g)$, and -95.299 for $HCl(g)$?

For the reaction $NH_4Cl(s) \rightleftharpoons NH_3(g) + HCl(g)$, (6.15) gives

$$\Delta_r G_{298}^\circ = [(1)(-16.45) + (1)(-95.299)] - [(1)(-202.87)] = 91.12 \text{ kJ}$$

and (7.1b) gives

$$K = e^{-(91120)/(8.314)(298)} = 1.06 \times 10^{-16}$$

Recognizing that $a(NH_4Cl) = 1.00$ and $a(NH_3) = a(HCl) = P(HCl)$, (6.43) gives

$$1.06 \times 10^{-16} = [P(HCl)]^2 \quad \text{or} \quad P(HCl) = 1.03 \times 10^{-8} \text{ bar}$$

EQUILIBRIUM IN AQUEOUS SOLUTIONS

7.5. Using (7.15b), derive an expression for determining the pH of a solution of a strong base, and apply it to a 0.025 M solution of NaOH.

For a strong base, essentially all of the base reacts, leaving $[B] = 0$. Thus (7.15b) becomes

$$0 = C_B + [H^+] - [OH^-]$$

$$[OH^-] = C_B + [H^+] = C_B + K_w/[OH^-] \tag{7.35}$$

For the 0.025 M NaOH solution, $C_B \gg [H^+]$ and

$$[OH^-] = C_B = 0.025 \text{ M}$$

giving pOH = 1.60 and pH = 12.40.

7.6. What is the pH of a 1.0×10^{-3} M solution of hydrogen peroxide?

$$H_2O_2(aq) \rightleftharpoons H^+(aq) + HO_2^-(aq) \qquad K_a = 2.2 \times 10^{-12}$$

The low value of K_a indicates that probably (7.21a) should be used to solve for $[H^+]$. Substituting (7.17) for $[OH^-]$ and rearranging gives

$$[H^+]^3 + K_a[H^+]^2 - (K_a C_A + K_w)[H^+] - K_w K_a = 0$$

$$[H^+]^3 + (2.2 \times 10^{-12})[H^+]^2 - \{(2.2 \times 10^{-12})(1.0 \times 10^{-3}) + (1.00 \times 10^{-14})\}[H^+]$$
$$- (2.2 \times 10^{-12})(1.00 \times 10^{-14}) = 0$$

Solving gives $[H^+] = 1.10 \times 10^{-7}$ M (carrying one additional significant figure), or pH = 6.96.

7.7. Repeat the calculations of Problem 7.6 for a hydrogen peroxide solution that also contains 0.010 M NaCl.

The ionic strength of the solution is essentially that contributed by the sodium chloride, $I = 0.010$ M. Thus (6.39) gives

$$\log y_\pm = -(1)(1)(0.511\,6)(0.010)^{1/2} = -0.051$$

$$y_\pm = 0.889$$

Asssuming $y(H_2O_2) = 1$ and

$$y(H^+)y(HO_2^-) = y_\pm^2 = 0.790$$

(7.21a) becomes

$$[H^+]^3 + \left(\frac{2.2 \times 10^{-12}}{0.790}\right)[H^+]^2 - \left[\frac{(2.2 \times 10^{-12})(1.0 \times 10^{-3})}{0.790} + 1.00 \times 10^{-14}\right][H^+] - \frac{2.2 \times 10^{-26}}{0.790} = 0$$

The solution is $[H^+] = 1.13 \times 10^{-7}$ M, or pH = 6.95, a slight increase in $[H^+]$.

7.8. Assuming that (7.30b) correctly gives the equilibrium concentration of H^+ in a solution of an ampholyte, how can one quickly determine whether the solution will be acidic or basic?

If $K_1K_2 > K_w$, then $(7.30b)$ will give a value of $[H^+] > 10^{-7}$, which corresponds to an acidic solution. If $K_1K_2 < K_w$, then $(7.30b)$ will give a value of $[H^+] < 10^{-7}$, corresponding to a basic solution.

7.9. As a weak acid of concentration $C_{A,0}$ is titrated with a strong base of concentration $C_{B,0}$, use (7.18) to predict the pH (a) at the beginning of the reaction, (b) at the half-equivalence point, (c) at the equivalence point, and (d) in excess strong base.

Solving (7.18) for $[H^+]$ gives

$$[H^+] = K_a \frac{C_A - [H^+] + [OH^-]}{C_B + [H^+] - [OH^-]} \tag{7.36}$$

(a) At the beginning of the reaction, $C_B = 0$ and $[H^+] \gg [OH^-]$. Thus (7.36) simplifies to $(7.21c)$, giving

$$pH = (pK_a - \log C_A)/2$$

(b) At the half-equivalence point, $C_A \approx C_B \gg [H^+]$, and $[H^+] \gg [OH^-]$. Thus (7.36) reduces to $(7.26b)$, giving

$$pH = pK_a$$

(c) At the equivalence point, hydrolysis of the anion will occur. Thus (7.36) simplifies to $(7.22d)$, giving

$$pOH = (pK_w - pK_a - \log C_B)/2$$

$$pH = pK_w - pOH = (pK_w + pK_a + \log C_B)/2$$

where $C_B = C_{A,0}(V_A / V_{soln})$.

(d) In excess strong base, the titration mixture will be described by (7.35), which gives

$$pOH = -\log C_B \qquad\qquad pH = pK_w + \log C_B$$

where $C_B = (C_{B,0}V_B - C_{A,0}V_A)/V_{soln}$.

7.10. For the stepwise formation of the ammine complexes of Cu^{2+}, $K_{f1} = 1.0 \times 10^4$, $K_{f2} = 5 \times 10^3$, $K_{f3} = 1 \times 10^3$, $K_{f4} = 2 \times 10^2$, and $K_{f5} = 0.25$. Prepare a plot of the fraction of each ion against $\log[NH_3]$. Below what concentration of NH_3 can the formation of $Cu(NH_3)_5^{2+}$ be neglected in most systems?

As a sample calculation, consider $[NH_3] = 1.0 \times 10^{-3}$ M. The fractions of concentrations are given by (7.32) as

$$\frac{[Cu^{2+}]}{C_{Cu}} = \frac{1}{\substack{1 + (1.0\times10^4)(1.0\times10^{-3}) + (1.0\times10^4)(5\times10^3)(1.0\times10^{-3})^2 + (1.0\times10^4)(5\times10^3)(1\times10^3)(1.0\times10^{-3})^3 \\ + (1.0\times10^4)(5\times10^3)(1\times10^3)(2\times10^2)(1.0\times10^{-3})^4 + (1.0\times10^4)(5\times10^3)(1\times10^3)(2\times10^2)(0.25)(1.0\times10^{-3})^5}}$$

$$= 8.3 \times 10^{-3}$$

$$[Cu(NH_3)^{2+}]/C_{Cu} = (1.0\times10^4)(1.0\times10^{-3})(8.3\times10^{-3}) = 0.083$$

$$[Cu(NH_3)_2^{2+}]/C_{Cu} = (1.0\times10^4)(5\times10^3)(1.0\times10^{-3})^2(8.3\times10^{-3}) = 0.415$$

$$[Cu(NH_3)_3^{2+}]/C_{Cu} = (1.0\times10^4)(5\times10^3)(1\times10^3)(1.0\times10^{-3})^3(8.3\times10^{-3}) = 0.415$$

$$[Cu(NH_3)_4^{2+}]/C_{Cu} = (1.0\times10^4)(5\times10^3)(1\times10^3)(2\times10^2)(1.0\times10^{-3})^4(8.3\times10^{-3}) = 0.083$$

$$[Cu(NH_3)_5^{2+}]/C_{Cu} = (1.0\times10^4)(5\times10^3)(1\times10^3)(2\times10^2)(0.25)(1.0\times10^{-3})^5(8.3\times10^{-3}) = 2.1\times10^{-5}$$

See Fig. 7-3 for a complete plot of the various fractions. The pentaammine complex is not important for most systems in which $[NH_3] < 0.1$ M.

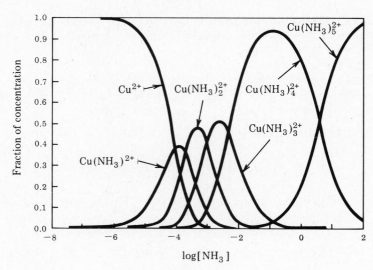

Fig. 7-3

7.11. What is the concentration of Ca^{2+} in a solution that is saturated with calcite? For the reaction

$$CaCO_3(\text{calcite}) \rightleftharpoons Ca^{2+}(aq) + CO_3^{2-}(aq)$$

$\Delta_f G_{298}^\circ = 47.40\ \text{kJ}$. Repeat the calculations if the solution is (a) 0.010 0 M Na_2CO_3 and (b) 0.010 0 M $NaNO_3$. Assume (6.39) to be valid for all solutions and that hydrolysis of the CO_3^{2-} ion is negligible.

For the reaction, (7.1b) gives $K = e^{-(47200)/(8.314)(298)} = 4.9 \times 10^{-9}$. Assuming $y_\pm$ to be unity, the first approximation of the concentration C of either ion can be found by substituting $a(CaCO_3) = 1.00$ into (7.34), giving

$$4.9 \times 10^{-9} = \frac{(y_\pm C)^2}{1.00} \qquad \text{or} \qquad C = 7.0 \times 10^{-5}\ \text{M}$$

Using this value of C, (6.38) gives

$$I = \tfrac{1}{2}[(7.0 \times 10^{-5})(2)^2 + (7.0 \times 10^{-5})(2)^2] = 28 \times 10^{-5}\ \text{M}$$

and (6.39) gives

$$\log y_\pm = -(2)(2)(0.511\ 6)(28 \times 10^{-5})^{1/2} = -0.034 \qquad \text{or} \qquad y_\pm = 0.924$$

Equation (7.34) now gives the second approximation of C as

$$4.9 \times 10^{-9} = \frac{(0.924)C(0.924)C}{1.00} \qquad \text{or} \qquad C = 7.6 \times 10^{-5}\ \text{M}$$

which corresponds to $I = 30 \times 10^{-5}$ M and $y_\pm = 0.922$. Using $y_\pm = 0.922$ gives the third estimate of C as 7.6×10^{-5} M.

For the solution containing 0.010 0 M Na_2CO_3, we assume that the contribution to I by the $CaCO_3$ will be negligible, so (6.38) and (6.39) give

$$I = \tfrac{1}{2}[(0.020\ 0)(1)^2 + (0.010\ 0)(2)^2] = 0.030\ 0\ \text{M}$$

$$\log y_\pm = -(2)(2)(0.511\ 6)(0.030\ 0)^{1/2} = -0.354\ 4$$

$$y_\pm = 0.442$$

Assuming $C(Ca^{2+}) = C$, $C(CO_3^{2-}) = 0.010\ 0 + C$, and $a(CaCO_3) = 1.00$, (7.34) gives

$$4.9 \times 10^{-9} = \frac{(0.442)C(0.442)(0.010\ 0 + C)}{1.00} \qquad \text{or} \qquad C = 2.5 \times 10^{-6}\ \text{M}$$

In solving for C, the approximation $0.010\,0 + C \approx 0.010\,0$ was made. As a check, $C = 2.72 \times 10^{-6}$ is indeed negligible compared to $0.010\,0$. Furthermore, it is negligible compared to $I = 0.030\,0$ M, and so the assumption made above is valid.

For the solution containing $0.010\,0$ M $NaNO_3$, $I = 0.010\,0$ M and $y_\pm = 0.624$. Assuming that $C = C(Ca^{2+}) = C(CO_3^{2-})$ and $a(CaCO_3) = 1.00$, (7.34) gives $C = 1.6 \times 10^{-4}$ M, a sizable increase over a pure solution of $CaCO_3$.

7.12. Repeat the calculations of Problem 7.11 for the solubility of calcite in pure water, using

$$C(Ca^{2+}) = \left[K_{sp}\left(1 + \frac{[H^+]}{K_{a2}} + \frac{[H^+]^2}{K_{a1}K_{a2}} \right) \right]^{1/2} \tag{7.37}$$

Equation (7.37) corrects for the hydrolysis of the carbonate ion, which increases the solubility of the salt. Assume that the pH of the solution is 6.0 (a typical value for natural waters) and that for H_2CO_3, $K_{a1} = 4.5 \times 10^{-7}$ and $K_{a2} = 4.8 \times 10^{-11}$.

A pH of 6.0 corresponds to $[H^+] = 1 \times 10^{-6}$ M. Substituting the data into (7.37) gives

$$C(Ca^{2+}) = \left[(4.9 \times 10^{-9})\left(1 + \frac{1 \times 10^{-6}}{4.8 \times 10^{-11}} + \frac{(1 \times 10^{-6})^2}{(4.5 \times 10^{-7})(4.8 \times 10^{-11})} \right) \right]^{1/2}$$

$$= 1.5 \times 10^{-2} \text{ M}$$

Supplementary Problems

EQUILIBRIUM CONSTANTS

7.13. For the reaction

$$2CrO_4^{2-}(aq) + 2H^+(aq) \rightleftharpoons Cr_2O_7^{2-}(aq) + H_2O(l)$$

the concentration equilibrium constant was determined by Smith and Metz as 1.3×10^{16} in solutions with $I = 0.375$ M. Given $\Delta_f G_{298}^\circ/(kJ\ mol^{-1}) = -727.75$ for CrO_4^{2-}, $-1\,301.1$ for $Cr_2O_7^{2-}$, and -237.129 for H_2O, calculate K. What is the value of K_y $[= y(Cr_2O_7^{2-})y(H_2O)/y(CrO_4^{2-})^2 y(H^+)^2]$?

Ans. $\Delta_r G_{298}^\circ = -82.7$ kJ; $K = 3 \times 10^{14}$, $K_y = 2 \times 10^{-2}$

7.14. Given $\Delta G_{298}^\circ = -237.129$ kJ for the reaction

$$H_2(g) + \tfrac{1}{2}O_2(g) \rightleftharpoons H_2O(l)$$

what is K?   *Ans.* 3.5×10^{41}

7.15. Find K, $\Delta_r H^\circ$, and $\Delta_r S^\circ$ at 25 °C for

$$Glycerol(aq) + HPO_4^{2-}(aq) = \text{DL-glycerol-1-phosphate}^{2-}(aq) + H_2O(l)$$

given $K = 0.012\,2$ at 37 °C and $\Delta_r G_{298}^\circ = 9.37$ kJ, as reported by Burton and Krebs.

Ans. $K = 0.022\,8$, $\Delta_r H^\circ = -40.07$ kJ, $\Delta_r S^\circ = -165.9$ J K^{-1}

7.16. Substitute (4.5) into (7.5) to derive an equation for the temperature dependence of $\ln K$.

Ans.   If $M = \ln K_{298} + (1/R)[J/(298) - (\Delta a)\ln(298) - (\Delta b/2)(298) - (\Delta c/6)(298)^2 - (\Delta d/12)(298)^3]$ and
$M' = \ln K_{298} + (1/R)[J'/(298) - (\Delta a)\ln(298) - (\Delta b/2)(298) - (\Delta c'/2)(298)^{-2}]$,
then $\ln K = M + (1/R)[(-J/T) + (\Delta a)\ln T + (\Delta b/2)T + (\Delta c/6)T^2 + (\Delta d/12)T^3]$
and $\ln K = M' - (1/R)[J'/T + (\Delta a)\ln T + (\Delta b/2)T + (\Delta c'/2)T^{-2}]$

7.17. From the plot of $\Delta_f G_T^\circ$ for $PbBr_2(g)$ against temperature shown in Fig. 7-4, complete the following table of signs of the thermodynamic properties:

	$\Delta_f G^\circ$	$(\partial\Delta_f G^\circ/\partial T)_P$	$\Delta_f S^\circ$	$\Delta_f H^\circ$	$(\partial\Delta_f S^\circ/\partial T)_P$	$(\partial\Delta_f H^\circ/\partial T)_P$	$\Delta_f C_P^\circ$	$\ln K$	$(\partial \ln K/\partial T)_P$
100 K				−					
1 000 K				−					
2 000 K	−			−					
4 000 K									
5 000 K				−					

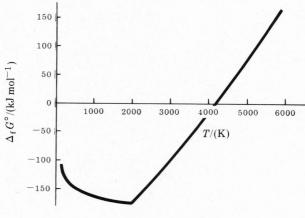

Fig. 7-4

Ans.

	$\Delta_f G^\circ$	$(\partial\Delta_f G^\circ/\partial T)_P$	$\Delta_f S^\circ$	$\Delta_f H^\circ$	$(\partial\Delta_f S^\circ/\partial T)_P$	$(\partial\Delta_f H^\circ/\partial T)_P$	ΔC_P°	$\ln K$	$(\partial \ln K/\partial T)_P$
100 K	−	−	+	−	−	−	−	+	−
1 000 K	−	−	+	−	−	−	−	+	−
2 000 K	−	−	+	−	−	−	−	+	−
4 000 K	−	+	−	−	−	−	−	+	−
5 000 K	+	+	−	−	−	−	−	−	−

EQUILIBRIUM IN GASES

7.18. Calculate the various equilibrium constants at 1 000 K and 100 bar total pressure for the reaction

$$\tfrac{3}{2}H_2(g)+\tfrac{1}{2}N_2(g) \rightleftharpoons NH_3(g)$$

given $\Delta G^\circ = 61.92$ kJ. Assume that $\gamma = 1.04$ for H_2, 1.02 for N_2, and 1.00 for NH_3.

Ans. $K_\gamma = 0.93$, $K = 5.83 \times 10^{-4}$, $K_p = 6.27 \times 10^{-4}$, $K_c = 5.21 \times 10^{-2}$, $K_x = 6.27 \times 10^{-2}$

7.19. What is the percent dissociation of $NOCl(g)$ into $NO(g)$ and $Cl_2(g)$ at 1.00 bar and 10^{-3} bar if $\Delta_f G_{298}^\circ/(\text{kJ mol}^{-1}) = 66.08$ for $NOCl(g)$ and 86.55 for $NO(g)$? Assume ideal gas behavior.

Ans. $\Delta_r G_{298}^\circ = 20.47$ kJ for 1 mol $NOCl(g)$, $K = K_p = 2.58 \times 10^{-4}$; 0.52% and 5.02%

7.20. Assuming real-gas behavior for the reaction

$$\tfrac{3}{2}H_2(g) + \tfrac{1}{2}N_2(g) \rightleftharpoons NH_3(g)$$

at 25 °C and 100.0 bar, calculate the mole fraction of NH_3 at equilibrium, given the reactants are present in stoichiometric amounts, $\gamma_{H_2} = 1.10$, $\gamma_{N_2} = 1.00$, $\gamma_{NH_3} = 0.40$, and $\Delta_r G^\circ_{298} = -16.45$ kJ.

Ans. $K = 765$, $K_\gamma = 0.345$, $K_p = 2\,220$; 0.998 1

7.21. Calculate the pressure of $CO_2(g)$ over a sample of $CaCO_3(s)$ at 1 000 K given $\Delta G^\circ_{1000} = 22.9$ kJ for the reaction

$$CaCO_3(s) \rightleftharpoons CaO(s) + CO_2(g)$$

Ans. $K = K_p = a(CO_2) = P(CO_2)/(bar) = 6.4 \times 10^{-2}$ bar

7.22. Determine the effect on the equilibrium at 25 °C for the endothermic reaction

$$CH_4(g) + HCl(g) \rightleftharpoons CH_3Cl(g) + H_2(g)$$

of (a) an increase in $x(HCl)$, (b) a decrease in $x(H_2)$, (c) an increase in pressure, and (d) an increase in temperature.

Ans. $x(CH_3Cl)$ will (a) increase, (b) increase, (c) not change, and (d) increase.

EQUILIBRIUM IN AQUEOUS SOLUTIONS

7.23. Calculate the pH of a 1.0×10^{-8} M solution of HCl.   *Ans.*   6.98

7.24. What is the pH of a 0.10 M solution of hypochlorous acid? At 25 °C, $K_a = 2.90 \times 10^{-8}$ for HClO.

Ans.   4.27

7.25. When distilled water is stored in a tank that is open to the air, the dissolved gases—mainly CO_2—change the pH from 7.00. (a) What is the pH of a saturated CO_2 solution if the solubility of CO_2 is 0.143 g in 100 g of water at 25 °C and 1.00 bar CO_2? Assume only the following reaction occurs:

$$CO_2(aq) + H_2O \rightleftharpoons H^+(aq) + HCO_3^-(aq)$$

where $\Delta_f G^\circ_{298}/(kJ\ mol^{-1}) = -385.98$ for $CO_2(aq)$, -237.129 for H_2O, and -586.77 for $HCO_3^-(aq)$. (b) If the actual concentration of CO_2 in a solution having air over it is given by

$$C(CO_2) = kP(CO_2)$$

where k is a constant, determine the pH of water in contact with air if $P(CO_2)/(bar) = 4 \times 10^{-2}$ bar.

Ans.   (a) $\Delta G^\circ_{298} = 36.34$ kJ, $K = 4.3 \times 10^{-7}$, $C(CO_2) = 0.032\,5$ M,   pH = 3.93;   (b) $C(CO_2) = 0.001\,3$ M, pH = 4.63

7.26. Compare the pH of 0.10 M solutions of methylamine ($K_b = 3.9 \times 10^{-4}$ for CH_3NH_2), dimethylamine [$K_b = 5.9 \times 10^{-4}$ for $(CH_3)_2NH$], and trimethylamine [$K_b = 6.3 \times 10^{-5}$ for $(CH_3)_3N$].

Ans.   11.78 for CH_3NH_2, 11.87 for $(CH_3)_2NH$, 11.39 for $(CH_3)_3N$; $(CH_3)_2NH$ is the most basic solution

7.27. What is the pH of a buffer solution prepared using 0.025 M benzoic acid and 0.010 M sodium benzoate? For C_6H_5COOH, $K_a = 6.6 \times 10^{-5}$.   *Ans.*   3.78

7.28. The *buffer capacity* is the number of moles of strong acid or base required to change the pH of 1 dm^3 of solution by one pH unit. To calculate the buffer capacity, the amount of strong acid or base added is divided by the resulting pH change of the solution. Calculate the capacity for adding 0.001 0 mol of NaOH (a) to 1 dm^3 of pure water, (b) to a buffer prepared by adding 0.010 0 mol of $Na(CH_3COO)$ to 1 dm^3 of 0.010 0 M CH_3COOH, and (c) to a buffer prepared by adding 0.500 mol of $Na(CH_3COO)$ to 1 dm^3 of 0.100 M CH_3COOH. $K_a = 1.76 \times 10^{-5}$ for CH_3COOH.

Ans. (a) pH = 7.00 for water, 11.00 for solution, 2.5×10^{-4}; (b) pH = 4.754 for buffer, 4.741 after, 0.077; (c) pH = 5.453 for buffer, 5.458 after, 0.020

7.29. Find the pH of a 0.010 M solution of $NaHSiO_3$ given $K_{a1} = 3.2 \times 10^{-10}$ and $K_{a2} = 1.5 \times 10^{-12}$ for metasilicic acid.

Ans. 10.55; note that K_w is not negligible compared to $K_{a2}[HSiO_3^-]$ in the numerator of (7.30a).

7.30. Consider the titration of a weak diprotic acid $(K_{a1} \gg K_{a2})$ with a strong base. Use (7.30a) to predict the pH (a) at the beginning of the reaction, (b) at the first half-equivalence point, (c) at the first equivalence point, (d) at the second half-equivalence point, and (e) at the second equivalence point.

Ans. (a) pH = $(pK_{a1} - \log C_A)/2$; (b) pH = pK_{a1}; (c) pH = $(pK_{a1} + pK_{a2})/2$; (d) pH = pK_{a2}; (e) pH = $(pK_w + pK_{a2} + \log C_B)/2$

7.31. Prepare a diagram similar to Fig. 7-1 for hydrosulfuric acid given $K_{a1} = 1.0 \times 10^{-7}$ and $K_{a2} = 3 \times 10^{-13}$ for H_2S.   *Ans.* See Fig. 7-5.

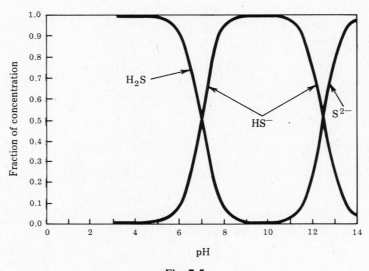

Fig. 7-5

7.32. Prepare a diagram similar to Fig. 7-3 for the ammine complexes of Ag^+ given $K_{f1} = 2.3 \times 10^3$ and $K_{f2} = 6.9 \times 10^3$.   *Ans.* See Fig. 7-6.

7.33. For the stepwise ionization of a weak diprotic acid having $K_{a1} \gg K_{a2}$, what are the approximate equilibrium concentrations of HA^- and A^{2-}?

Ans. $[HA^-] \approx [H^+]$, $[A^{2-}] \approx K_{a2}$

7.34. What are the equilibrium concentrations of H^+, H_2S, HS^-, and S^{2-} in a saturated solution of hydrogen sulfide at 1 bar pressure and 25 °C (≈ 0.1 M)? For H_2S, $K_{a1} = 1.0 \times 10^{-7}$ and $K_{a2} = 3 \times 10^{-13}$.

Ans. $[H^+] = [HS^-] = 1 \times 10^{-4}$ M, $[S^{2-}] = 3 \times 10^{-13}$ M, $[H_2S] = 0.1$ M

7.35. What is the concentration of Hg_2^{2+} in a saturated solution of Hg_2Cl_2 given $K_{sp} = 2 \times 10^{-18}$? What is the concentration of Hg_2^{2+} in a 0.500 M solution of NaCl assuming the Debye–Hückel limiting law to be valid? What is the solubility of Hg_2^{2+} in surface seawater (10 500 ppm Na^+, 1 350 ppm Mg^{2+}, 400 ppm Ca^{2+},

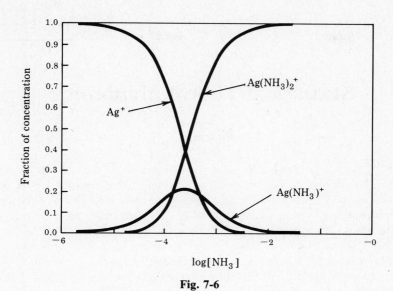

$\log[NH_3]$

Fig. 7-6

380 ppm K^+, 19 000 ppm Cl^-, and 2 700 ppm SO_4^{2-}, with $\rho = 1.02 \times 10^3$ kg m^{-3}) assuming the Debye–Hückel limiting law to be valid?

Ans. 8×10^{-7} M; $I = 0.500$, $y_\pm = 0.189$, 1.2×10^{-15}; $C/(M) = 0.46$ for Na^+, 0.06 for Mg^{2+}, 0.01 for Ca^{2+} and K^+, 0.55 for Cl^-, and 0.03 for SO_4^{2-}; $I = 0.71$ M, $y_\pm = 0.137$; 2.6×10^{-15} M

Chapter 8

Statistical Thermodynamics

Ensembles

8.1 INTRODUCTION

A *canonical ensemble* consists of N_s distinguishable systems that are identical in that they have the same number of molecules (N), volume (V), and thermodynamic state. As N_s becomes large, the time (or observable) average of any mechanical property of a system (such as P or E) becomes equal to the instantaneous ensemble average value for the property. Nonmechanical properties (such as T and S) must be calculated using known thermodynamic relationships. The ensemble acts as a single supersystem having a volume of N_sV, a number of molecules equal to N_sN, and a total energy of E_t. The number of molecules (N_i) in each of the allowed energy states in a system is called the *distribution* and is represented by ($N_1, N_2, \ldots$).

The *Boltzmann statistics* used in this chapter are obtained as the limiting case of *Fermi–Dirac statistics* (systems described by antisymmetric wave functions) and *Bose–Einstein statistics* (systems described by symmetric wave functions) when it is assumed that there are many more energy states available than molecules so that there is a very low probability of two molecules being in the same energy state. The results of Boltzmann statistics often fail for certain systems at low temperatures.

EXAMPLE 8.1. Even though the systems in a canonical ensemble have the same macroscopic thermodynamic state, they are not necessarily in the same microscopic state. Because each system in the ensemble has the same volume, each system has the same available energy states. However, a given system may be found in any quantum state consistent with the given values of N, V, and E. Determine the possible distributions among energy states for systems consisting of three simple harmonic oscillators (SHOs) such that $E_t = 9E_0$, given that the quantum-mechanical solution for energy is

$$E_v = (v + \tfrac{1}{2})h\nu \qquad (8.1)$$

where E_v is the energy of the state corresponding to the integer v, $v = 0, 1, 2, \ldots$; h is Planck's constant; and ν is the frequency of oscillation. Is $(1, 0, 1, 1, 0 \ldots)$ an acceptable answer?

The energies of the first few states are $E_0 = \tfrac{1}{2}h\nu$, $E_1 = \tfrac{3}{2}h\nu = 3E_0$, $E_2 = \tfrac{5}{2}h\nu = 5E_0$, etc. The distribution $(1, 0, 1, 1, 0, \ldots)$ means that there is one oscillator in state 0, contributing E_0 to the total energy; none in state 1; one in state 2, contributing $5E_0$; one in state 3, contributing $7E_0$; and none above state 3. The total energy would be $13E_0$, and so the proposed distribution is not acceptable. The distributions $(1, 1, 1, 0, 0, \ldots)$, $(2, 0, 0, 1, 0, \ldots)$, and $(0, 3, 0, 0, 0 \ldots)$ are the only ones that give $E_t = 9E_0$.

8.2 ENSEMBLE ENERGY STATES AND PROBABILITIES

The number of ensemble states (Ω) corresponding to a given distribution ($N_1, N_2, \ldots$) is

$$\Omega = \frac{N_s!}{\prod\limits_i N_i!} \qquad (8.2a)$$

in terms of energy states, or

$$\Omega = \frac{N_s!(\prod\limits_j g_j)^{N_j}}{\prod\limits_j N_j!} \qquad (8.2b)$$

in terms of energy levels, where g_j is the degeneracy of level j.

The conditional probability of finding a chosen system in the ith quantum state, E_i, given that the distribution is $(N_1, N_2, \ldots)$ is

$$P_i(N_1, N_2, \ldots) = \frac{N_i}{N_s} \qquad (8.3)$$

where N_i represents the number of systems in the ith quantum state under the given distribution and N_s is the total number of systems. Therefore, considering all distributions, the overall probability of finding a chosen system in state E_i is

$$P_i = \sum \left(\frac{N_i}{N_s}\right)\left(\frac{\Omega}{\sum \Omega}\right) = \frac{\sum N_i \Omega}{N_s \sum \Omega} \qquad (8.4)$$

In (8.4) the summations are extended over all distributions.

EXAMPLE 8.2. Find Ω for each of the three acceptable distributions described in Example 8.1. Prepare a table showing how these distributions are realized for three distinguishable SHOs, A, B, and C.

For $(1, 1, 1, 0, 0, \ldots)$ $(8.2a)$ gives

$$\Omega(1, 1, 1, 0, 0, \ldots) = \frac{3!}{1!\,1!\,1!\,0!\,0!\ldots} = 6$$

Similarly, $\Omega(2, 0, 0, 1, 0, \ldots) = 3$ and $\Omega(0, 3, 0, 0, 0, \ldots) = 1$. The various assignments of the SHOs are:

Energy state:	0	1	2	3	4	5	$\ldots$
Energy:	E_0	$3E_0$	$5E_0$	$7E_0$	$9E_0$	$11E_0$	$\ldots$
$(1, 1, 1, 0, 0, \ldots)$:	A	B	C				
	A	C	B				
	B	A	C				
	B	C	A				
	C	A	B				
	C	B	A				
$(2, 0, 0, 1, 0, \ldots)$:	AB			C			
	AC			B			
	BC			A			
$(0, 3, 0, 0, 0, \ldots)$:		ABC					

EXAMPLE 8.3. What is the probability of finding system B in E_0 for the acceptable SHO distribution $(1, 1, 1, 0, 0, \ldots)$? Compare the numerical answer to the actual count of the number of times system B appears in the E_0 column of the table prepared in Example 8.2 for this distribution.

Using (8.3) with $N_s = 3$ and $N_0 = 1$ gives $P_0(1, 1, 1, 0, 0, \ldots) = 1/3$. In Example 8.2, system B appears twice in the column labeled E_0 out of six entries, making the conditional probability of finding B in E_0 to be $2/6$, or $1/3$.

EXAMPLE 8.4. What is the overall probability of finding system B in E_0 for the acceptable SHO distributions given in Example 8.2? Compare the numerical answer to that predicted by actual count of the number of times system B appears in the entire E_0 column.

Substituting $\Omega(1, 1, 1, 0, 0, \ldots) = 6$, $\Omega(2, 0, 0, 1, 0, \ldots) = 3$, and $\Omega(0, 3, 0, 0, 0, \ldots) = 1$ into (8.4) gives

$$P_0 = \frac{(1)(6) + (2)(3) + (0)(1)}{3(6 + 3 + 1)} = 0.400$$

Among the 10 entries in the E_0 column of Example 8.2, system B appears 4 times, giving a probability of $4/10$ or 0.400.

EXAMPLE 8.5. In (8.2) the symbol $N!$ is defined as

$$N! = N(N-1)(N-2)\cdots(2)(1) \qquad (8.5)$$

where $1! = 0! = 1$. In many problems $N!$ or $\ln N!$ must be evaluated where N is a large number. Using *Stirling's approximation*,

$$\ln N! \approx N \ln N - N \tag{8.6}$$

evaluate $\ln L!$.

From (8.6),

$$\ln L! \approx L \ln L - L$$
$$= (6.022 \times 10^{23})(2.303) \log(6.022 \times 10^{23}) - (6.022 \times 10^{23})$$
$$= (6.022 \times 10^{23})[(2.303)(23.780) - 1] = 323.77 \times 10^{23}$$

8.3 MOST PROBABLE DISTRIBUTION

As N_s becomes large, $\bar{N}_i$, the average of N_i, approaches N_i^*, the *most probable distribution*, which is given by

$$N_i^* = \frac{N_s e^{-E_i/kT}}{Q} \tag{8.7}$$

where k is Boltzmann's constant ($k = 1.380\,662 \times 10^{-23}$ J K^{-1}), and the *partition function* (Q) is defined as

$$Q = \sum_i e^{-E_i/kT} = \sum_j g_j e^{-E_j/kT} \tag{8.8}$$

In this limiting case, (8.4) becomes

$$P_i = \frac{e^{-E_i/kT}}{Q} \tag{8.9}$$

EXAMPLE 8.6. Derive the equation for the ratio of the most probable numbers of systems at two different energies in the ensemble. Find this ratio for two states such that $E_b - E_a = kT$. If E_b and E_a represent energy levels with $g_b = 3$ and $g_a = 1$, repeat the calculations.

Taking a ratio of N_i^* given by (8.7) for the two energy states gives

$$\frac{N_b^*}{N_a^*} = \frac{N_s e^{-E_b/kT}/Q}{N_s e^{-E_a/kT}/Q} = e^{-(E_b - E_a)/kT}$$

If energy levels are used instead of energy states, a factor of g_b/g_a appears on the right side of the equation [see (2.11)], giving

$$\frac{N_b^*}{N_a^*} = \frac{g_b}{g_a} e^{-(E_b - E_a)/kT} \tag{8.10}$$

For the two energy states, $g_b = g_a = 1$, and (8.10) gives

$$N_b^*/N_a^* = e^{-(kT)/kT} = 0.368$$

and for the two energy levels, (8.10) gives

$$N_b^*/N_a^* = (3/1)\, e^{-(kT)/kT} = 1.104$$

Ideal-Gas Partition Functions

8.4   INTRODUCTION

Assume the system under consideration consists of a large number (N) of molecules, each of which may occupy a number of quantum states. The energy of the system is

$$E_i = \varepsilon_1 + \varepsilon_2 + \cdots + \varepsilon_N$$

where ε_1 is the energy of molecule 1, ε_2 is the energy of molecule 2, etc. Equation (8.8) becomes

$$Q = q_1 q_2 \cdots q_N$$

in which q_α, the *molecular partition function* for molecule α, is given by

$$q_\alpha = \sum e^{-\varepsilon_\alpha / kT} \tag{8.11}$$

The summation in (8.11) is over all the values of ε_α that correspond to the quantum states of molecule α. Because the molecules are all the same, $q_1 = q_2 = \cdots = q_N = q$ and $Q = q^N$. Correcting for the indistinguishability of the molecules, we obtain

$$Q = \frac{q^N}{N!} \tag{8.12}$$

The contributions to ε_α for an ideal gas are nuclear, electronic, vibrational, rotational, and translational (see Secs. 8.5–8.9), giving

$$\varepsilon_\alpha = \varepsilon_{\text{nuc}} + \varepsilon_{\text{elec}} + \varepsilon_{\text{vib}} + \varepsilon_{\text{rot}} + \varepsilon_{\text{trans}}$$

$$q = q_{\text{nuc}} q_{\text{elec}} q_{\text{vib}} q_{\text{rot}} q_{\text{trans}}$$

$$Q = (q_{\text{nuc}} q_{\text{elec}} q_{\text{vib}} q_{\text{rot}})^N \frac{q_{\text{trans}}^N}{N!} \tag{8.13}$$

The $N!$ correction term is traditionally associated with the translational partition function because not all gases have vibrational and rotational contributions, and the nuclear and electronic contributions are negligible for most gases. Equation (8.13) does not contain minor contributions for rotational-vibrational interactions or for intramolecular rotation energy levels.

8.5   MOLECULAR TRANSLATIONAL PARTITION FUNCTION

For an ideal gas with Λ atoms in the molecule, of the total of 3Λ degrees of freedom, three are attributed to translational motion of the molecule as a whole. The quantum-mechanical solution for a particle of mass m in a cube with side a (see Problem 14.20), gives the energy as

$$\varepsilon_{\text{trans}} = (n_x^2 + n_y^2 + n_z^2) \frac{h^2}{8ma^2} \tag{8.14}$$

where the n_i are integers and h is Planck's constant ($h = 6.626\,176 \times 10^{-34}$ J s).

EXAMPLE 8.7.   The values of $\varepsilon_{\text{trans}}$ are of the order of 10^{-42} J apart, and at room temperature a large number of states are occupied by the molecules (because $kT \approx 4 \times 10^{-21}$ J). Using the above expression for $\varepsilon_{\text{trans}}$, show that

$$q_{\text{trans}} = (2\pi mkT)^{3/2} \frac{V}{h^3} \tag{8.15}$$

where V is the volume of the cube.

Substituting (8.14) into (8.11) gives

$$q_{\text{trans}} = \sum_{n_x, n_y, n_z} \exp\left[-\frac{(n_x^2 + n_y^2 + n_z^2)h^2}{8ma^2kT}\right] = \left[\sum_{n_x=0}^{\infty} \exp\left(-\frac{n_x^2 h^2}{8ma^2kT}\right)\right]\left[\sum_{n_y=0}^{\infty} \exp\left(-\frac{n_y^2 h^2}{8ma^2kT}\right)\right]\left[\sum_{n_z=0}^{\infty} \exp\left(\frac{-n_z^2 h^2}{8ma^2kT}\right)\right]$$

Replacing the summations by integration because of the large number of states occupied gives

$$q_{trans} = \left[\int_0^\infty \exp\left(-\frac{n^2 h^2}{8ma^2 kT}\right) dn \right]^3 = \left[(2\pi mkT)^{1/2} \frac{a}{h} \right]^3 = (2\pi mkT)^{3/2} \frac{V}{h^3}$$

8.6 MOLECULAR ROTATIONAL PARTITION FUNCTION

A diatomic or linear polyatomic molecule has two rotational degrees of freedom. For the diatomic molecule, these are described by the quantum-mechanical solution for the rigid rotator (see Problem 14.22), giving

$$\varepsilon_{rot} = J(J+1)\frac{h^2}{8\pi^2 I} \tag{8.16}$$

where $J = 0, 1, 2, \ldots$, each level has a degeneracy of $2J+1$, and the moment of inertia (I) is given by

$$I = \mu r^2 \tag{8.17}$$

where r is the distance between the atoms in the diatomic molecule and the reduced mass (μ) is given by (2.16). The values of ε_{rot} are of the order of 10^{-23} J apart, and at room temperature a large number of levels are occupied by the molecules.

Substituting the above expression for ε_{rot} for a diatomic or linear polyatomic molecule into (8.11) and intergrating over J gives

$$q_{rot} = \frac{8\pi^2 IkT}{h^2 \sigma} \tag{8.18}$$

The *symmetry number* (σ) of the molecule represents the number of indistinguishable positions into which the molecule can be placed by simple rigid rotations; e.g., $\sigma = 1$ for asymmetric linear molecules such as CO and HCN and $\sigma = 2$ for symmetric linear molecules such as O_2 and CO_2. For nonlinear polyatomic molecules having three rotational degrees of freedom,

$$q_{rot} = \frac{8\pi^2 (8\pi^3 I_x I_y I_z)^{1/2} (kT)^{3/2}}{h^3 \sigma} \tag{8.19}$$

where I_x, I_y, and I_z represent the moments of inertia around three mutually perpendicular axes through the molecule and $\sigma = 2$ for C_2H_2, 3 for NH_3, 4 for C_2H_4, 12 for CH_4, etc.

8.7 MOLECULAR VIBRATIONAL PARTITION FUNCTION

The remaining degrees of freedom, $3\Lambda - 5$ for diatomic or linear polyatomic molecules or $3\Lambda - 6$ for nonlinear polyatomic molecules, are attributed to internal motion within the molecule. The vibrational energy is described by the quantum-mechanical solution for the SHO problem (see Problem 14.5):

$$\varepsilon_{vib} = \sum_{i=1}^{3\Lambda - 5 \text{ or } 3\Lambda - 6} (v_i + \tfrac{1}{2})h\nu_i \tag{8.20}$$

Here v_i and ν_i are the quantum number and fundamental frequency associated with the ith mode of vibration. The values of ε_{vib} are of the order of 10^{-20} J apart, and at room temperature most of the molecules are in the lower states.

Inserting (8.20) into (8.11) and summing over $0 \le v_1 \le \infty$, $0 \le v_2 \le \infty, \ldots$,

$$q_{vib} = \prod_{i=1}^{3\Lambda - 5 \text{ or } 3\Lambda - 6} \frac{1}{1 - e^{-x_i}} \tag{8.21}$$

where x_i is given in terms of ν_i or $\bar{\nu}_i = \nu_i/c$ by (3.6) or (3.7), if the energy levels are measured with respect to the zero vibrational state of the molecule. If the energy levels are measured with respect to the hypothetical minimum of the potential energy well, $e^{-x_i/2}$ appears in the numerator of (8.21).

The number of molecules in a given vibrational state compared to that in the ground state is given by

$$N_v = N_0 \, e^{-vx} \tag{8.22}$$

and the number in the ground state compared to the total number present is given by

$$N_0 = N(1 - e^{-x}) \tag{8.23}$$

8.8   MOLECULAR ELECTRONIC PARTITION FUNCTION

Because the electronic energy states are of the order of 10^{-19} J apart, at room temperature most of the molecules are in the ground state. The partition function for such a system is

$$q_{\text{elec}} = g_0 \tag{8.24}$$

where g_0 is the degeneracy of the ground level.

EXAMPLE 8.8.   The complete electronic partition function is given by (8.11) as

$$q_{\text{elec}} = \sum g_\alpha \, e^{-\varepsilon_\alpha/kT} \tag{8.25}$$

Show that (8.25) becomes (8.24) under the conditions mentioned above.

Choosing the reference for zero energy as the ground electronic level, (8.25) gives, upon expansion,

$$q_{\text{elec}} = g_0 \, e^{-(0)/kT} + g_1 \, e^{-\varepsilon_1/kT} + g_2 \, e^{-\varepsilon_2/kT} + \cdots$$

$$= g_0 + g_1 \exp\left[-\frac{10^{-19} \text{ J}}{(1.381 \times 10^{-23} \text{ J K}^{-1})(298 \text{ K})} \right] + \cdots$$

$$= g_0 + g_1(3 \times 10^{-11}) + \cdots \approx g_0$$

8.9   MOLECULAR NUCLEAR PARTITION FUNCTION

Because the nuclear energy states are spaced very far apart, only a negligible number of atoms in the molecule are not in the ground state at room temperature. The partition function becomes

$$q_{\text{nuc}} = g_{n,0} \tag{8.26}$$

where $g_{n,0}$ is the degeneracy of the nuclear ground state.

Application to Thermodynamics Involving Ideal Gases

8.10   GENERAL THERMODYNAMIC FUNCTIONS

In terms of the partition function,

$$A = -kT \ln Q \tag{8.27}$$

$$S = kT\left(\frac{\partial \ln Q}{\partial T}\right)_{V,N} + k \ln Q \tag{8.28}$$

$$P = kT\left(\frac{\partial \ln Q}{\partial V}\right)_{T,N} \tag{8.29}$$

$$E = kT^2\left(\frac{\partial \ln Q}{\partial T}\right)_{V,N} \tag{8.30}$$

The following sections describe the calculation of various thermodynamic properties of 1 mol of ideal gas, i.e., $N = L$. In general, the partition function is pressure-dependent, so we shall assume standard pressure conditions (denoted, as usual, by °). Also, see Sec. 3.4 for determining thermal enthalpies and Sec. 3.8 for determining heat capacities for ideal gases.

8.11 MOLAR THERMAL ENERGY

The energy content of a mole of ideal gas can be considered to be the sum of the energy contributed by the molecular ground states (E_0°) and any additional thermal energy [$E°(thermal)$] resulting from the occupation of higher energy levels by the molecules when $T > 0$. Thus,

$$E_T^\circ = E_0^\circ + E°(\text{thermal}) \tag{8.31}$$

Note that $E_0^\circ = H_0^\circ = A_0^\circ = G_0^\circ$ for ideal gases because $H°$, $A°$, and $G°$ differ from $E°$ by either RT or ST, which vanish at 0 K.

EXAMPLE 8.9. Derive expressions for $E°(thermal)$ for ideal gases.

Applying (8.30) and (8.13) to (8.15) gives

$$E_{\text{trans}}^\circ = kT^2 \left(\frac{\partial \ln q_{\text{trans}}^N}{\partial T} \right)_{V,N}$$

$$= NkT^2 \left[\frac{\partial}{\partial T} \ln \left((2\pi mkT)^{3/2} \frac{V}{h^3} \right) \right]_{V,N}$$

$$= RT^2 \left[\frac{\partial}{\partial T} \ln \left((2\pi mk)^{3/2} \frac{V}{h^3} \right) + \frac{\partial}{\partial T} \ln T^{3/2} \right]_{V,N}$$

$$= RT^2 [0 + \tfrac{3}{2}(1/T)] = \tfrac{3}{2}RT$$

$$E°(\text{thermal, trans}) = E_{\text{trans}}^\circ(T) - E_{\text{trans}}^\circ(0)$$

$$= \tfrac{3}{2}RT - \tfrac{3}{2}R(0) = \tfrac{3}{2}RT \tag{8.32a}$$

Likewise, applying (8.30) and (8.13) to (8.18) or (8.19), (8.21), (8.24), and (8.26) gives

$$E°(\text{thermal, rot}) = \begin{cases} RT & \text{for a diatomic or linear polyatomic molecule} \\ \tfrac{3}{2}RT & \text{for a nonlinear polyatomic molecule} \end{cases} \tag{8.32b}$$

$$E°(\text{thermal, vib}) = \sum_{i=1}^{3\Lambda-5 \text{ or } 3\Lambda-6} \frac{RTx_i}{e^{x_i} - 1} \tag{8.32c}$$

$$E°(\text{thermal, elec}) = 0 \tag{8.32d}$$

$$E°(\text{thermal, nuc}) = 0 \tag{8.32e}$$

8.12 MOLAR ENTROPY

Applying (8.28) and (8.13) to (8.15), (8.18) or (8.19), (8.21), (8.24), and (8.26) gives

$$S°(\text{trans}) = \frac{E°(\text{thermal, trans})}{T} + R \ln \frac{q_{\text{trans}}}{L} + R$$

$$= R \left[\frac{5}{2} + \ln \frac{(2\pi mkT)^{3/2} V°}{h^3 L} \right] \tag{8.33a}$$

$$S°(\text{rot}) = \frac{E°(\text{thermal, rot})}{T} + R \ln \frac{q_{\text{rot}}}{L}$$

$$= \begin{cases} R\left(1 + \ln \dfrac{8\pi^2 IkT}{h^2\sigma}\right) & \text{for a diatomic or linear polyatomic molecule} \\[4mm] R\left[\dfrac{3}{2} + \ln \dfrac{8\pi^2(8\pi^3 I_x I_y I_z)^{1/2}(kT)^{3/2}}{h^3\sigma}\right] & \text{for a nonlinear polyatomic molecule} \end{cases}$$

(8.33b)

$$S^\circ(\text{vib}) = \frac{E^\circ(\text{thermal, vib})}{T} + R\ln\frac{q_{\text{vib}}}{L}$$

$$= R \sum_{i=1}^{3\Lambda-5 \text{ or } 3\Lambda-6} \left[\frac{x_i}{e^{x_i}-1} - \ln(1-e^{-x_i})\right]$$

(8.33c)

$$S^\circ(\text{elec}) = R\ln g_0$$

(8.33d)

$$S^\circ(\text{nuc}) = R\ln g_{n,0}$$

(8.33e)

For most chemical reactions, $S^\circ(\text{nuc})$ is neglected.

EXAMPLE 8.10. Calculate S° for $H_2O(g)$ at 373.12 K and 1.00 bar. The moments of inertia for the molecule are such that $I_x I_y I_z = 5.765\,8 \times 10^{-141}$ kg^3 m^6. The ground-state degeneracy for the electronic contribution is 1, and $\sigma = 2$.

Assuming ideality, (1.6) gives

$$V^\circ = \frac{(8.314 \text{ m}^3 \text{ Pa K}^{-1} \text{ mol}^{-1})[(1 \text{ bar})/(10^5 \text{ Pa})](373.12 \text{ K})}{1.00 \text{ bar}}$$

$$= 3.102 \times 10^{-2} \text{ m}^3 \text{ mol}^{-1}$$

and the mass of one molecule is $m = (0.018\,01 \text{ kg mol}^{-1})/L = 2.991 \times 10^{-26}$ kg.

Using (8.33) and the results of Problem 3.2 gives

$$S^\circ(\text{trans}) = (8.314)\left\{\frac{5}{2} + \ln\frac{[(2\pi)(2.991\times 10^{-26})(1.380\,7\times 10^{-23})(373.12)]^{3/2}(3.102\times 10^{-2})}{(6.626\times 10^{-34})^3(6.022\times 10^{23})}\right\}$$

$$= 149.57 \text{ J K}^{-1} \text{ mol}^{-1}$$

$$S^\circ(\text{rot}) = (8.314)\left\{\frac{3}{2} + \ln\frac{(8\pi^2)[8\pi^3(5.765\,8\times 10^{-141})]^{1/2}[(1.380\,7\times 10^{-23})(373.12)]^{3/2}}{(6.626\times 10^{-34})^3(2)}\right\}$$

$$= 46.51 \text{ J K}^{-1} \text{ mol}^{-1}$$

$$S^\circ(\text{vib}) = (8.314)\left[\frac{14.101}{e^{14.101}-1} - \ln(1-e^{-14.101}) + \frac{6.149}{e^{6.149}-1} - \ln(1-e^{-6.149}) + \frac{14.483}{e^{14.483}-1} - \ln(1-e^{-14.483})\right]$$

$$= 0.13 \text{ J K}^{-1} \text{ mol}^{-1}$$

$$S^\circ(\text{elec}) = R\ln 1 = 0 \text{ J K}^{-1} \text{ mol}^{-1}$$

$$S^\circ(\text{nuc}) = 0 \text{ J K}^{-1} \text{ mol}^{-1}$$

The total of the contributions is $S^\circ = 196.21$ J K^{-1} mol^{-1}.

8.13 FREE ENERGY FUNCTION

The free energy is related to the partition functions by

$$G_T^\circ = E_0^\circ - kT\ln\left[\left(\frac{q_{\text{trans}}^\circ}{N}\right)^N (q_{\text{rot}}^\circ q_{\text{vib}}^\circ q_{\text{elec}}^\circ q_{\text{nuc}}^\circ)^N\right]$$

(8.34a)

or

$$\frac{G_T^\circ - E_0^\circ}{T} = -k\ln\left[\left(\frac{q_{\text{trans}}^\circ}{N}\right)^N (q_{\text{rot}}^\circ q_{\text{vib}}^\circ q_{\text{elec}}^\circ q_{\text{nuc}}^\circ)^N\right]$$

(8.34b)

Equation (8.34b) is the free energy function defined in Sec. 6.5 (where $H_0^\circ = E_0^\circ$) and can be used to calculate $\Delta_r G_T^\circ$ along with (6.19) and (6.20a).

EXAMPLE 8.11. Predict K for the isotopic exchange reaction

$$^{35}Cl_2(g) + {}^{37}Cl_2(g) \rightleftharpoons 2\,{}^{35}Cl\,{}^{37}Cl(g)$$

assuming that any differences in the translational, vibrational, electronic, and nuclear contributions are negligible and that the rotational contributions differ only with respect to σ.

For the reaction, applying (6.20a) to (8.34b) gives

$$\frac{\Delta(G_T^\circ - E_0^\circ)}{T} = 2\left\{ -k \ln\left[\left(\frac{q_{trans}^\circ}{N}\right)^N (q_{rot}^\circ q_{vib}^\circ q_{elec}^\circ q_{nuc}^\circ)^N \right]_{35\text{-}37} \right\}$$

$$-1\left\{ -k \ln\left[\left(\frac{q_{trans}^\circ}{N}\right)^N (q_{rot}^\circ q_{vib}^\circ q_{elec}^\circ q_{nuc}^\circ)^N \right]_{35\text{-}35} \right\}$$

$$-1\left\{ -k \ln\left[\left(\frac{q_{trans}^\circ}{N}\right)^N (q_{rot}^\circ q_{vib}^\circ q_{elec}^\circ q_{nuc}^\circ)^N \right]_{37\text{-}37} \right\}$$

$$= -k \ln\left[\frac{(q_{trans}^\circ/N)^{2N} (q_{vib}^\circ q_{elec}^\circ q_{nuc}^\circ)^{2N} (q_{rot,35\text{-}37}^\circ)^{2N}}{(q_{trans}^\circ/N)^{2N} (q_{vib}^\circ q_{elec}^\circ q_{nuc}^\circ)^{2N} (q_{rot,35\text{-}35}^\circ)^N (q_{rot,37\text{-}37}^\circ)^N} \right]$$

$$= -Nk \ln \frac{(q_{rot,35\text{-}37}^\circ)^2}{(q_{rot,35\text{-}35}^\circ)(q_{rot,37\text{-}37}^\circ)}$$

Now, by (8.18) and the assumption regarding the rotational contributions, q_{rot}° is proportional to $1/\sigma$. Hence,

$$\frac{\Delta(G_T^\circ - E_0^\circ)}{T} = -Nk \ln \frac{(\sigma_{35\text{-}35})(\sigma_{37\text{-}37})}{(\sigma_{35\text{-}37})^2}$$

$$= -Nk \ln \frac{(2)(2)}{(1)^2} = -Nk \ln 4 = -R \ln 4$$

Because the bond energies of the molecules are assumed to be the same, $\Delta E_0^\circ = 0$ and (6.19) gives

$$\Delta_r G^\circ = -RT \ln 4$$

and then (7.1b) gives $K = e^{\ln 4} = 4$.

Monatomic Crystals

8.14 PARTITION FUNCTIONS AND MOLAR HEAT CAPACITIES

The atoms in a crystal are localized at definite lattice points rather than being able to move freely as in a gas. Einstein assumed the vibrational motion of the atoms located at these lattice points to be given by the solution to the SHO problem; the corresponding partition function for each of the three directions is then given by a term similar to those found in (8.21). The expressions for C_V as derived from the Einstein and Debye theories were presented in Sec. 3.9.

8.15 OTHER THERMODYNAMIC PROPERTIES

In terms of the Debye theory for solids,

$$\frac{E^\circ - E_0^\circ}{3RT} = 3\left(\frac{T}{\Theta_D}\right)^3 \int_0^{\Theta_D/T} \frac{x^3 \, dx}{e^x - 1} \tag{8.35}$$

$$-\frac{A^\circ - E_0^\circ}{3RT} = 3\left(\frac{T}{\Theta_D}\right)^3 \int_0^{\Theta_D/T} x^2 \ln(1 - e^{-x}) \, dx \tag{8.36}$$

$$\frac{S^\circ}{3R} = 3\left(\frac{T}{\Theta_D}\right)^3 \int_0^{\Theta_D/T} \left[\frac{x^3}{e^x - 1} - x^2 \ln(1 - e^{-x}) \right] dx \tag{8.37}$$

The right-hand sides of (8.35)-(8.37), as well as that of (3.25) exclusive of the electronic term, have been tabulated as functions of Θ_D / T.

Solved Problems

ENSEMBLES

8.1. The quantum-mechanical solution for a particle in a one-dimensional box is

$$E_n = n^2 \frac{h^2}{8ma^2}$$

where E_n is the energy of the state corresponding to the integer n, $n = 1, 2, 3, \ldots$; h is Planck's constant; m is the mass of the particle; and a is the length of the box. (a) What are the energies of the first four states expressed in terms of E_1? (b) What are the possible distributions of three particles such that $E_t = 81 E_1$? Is $(0, 0, 0, 0, 0, 0, 0, 0, 1, 0, \ldots)$ an acceptable answer? (c) Find Ω for each of the acceptable distributions. (d) What is the probability of finding particle A in E_1 for the distribution $(1, 0, 0, 1, 0, 0, 0, 1, 0, \ldots)$? (e) What is the overall probability of finding particle A in E_1 for the acceptable distributions?

(a) $E_1, \qquad E_2 = 4E_1, \qquad E_3 = 9E_1, \qquad E_4 = 16E_1$

(b)   The distributions

$$(1, 0, 0, 1, 0, 0, 0, 1, 0, \ldots), \quad (0, 0, 0, 2, 0, 0, 1, 0, 0, \ldots), \quad (0, 0, 1, 0, 0, 2, 0, 0, 0, \ldots)$$

have $E_t = 81 E_1$. The energy of $(0, 0, 0, 0, 0, 0, 0, 0, 1, 0, \ldots)$ is $E_t = 9^2 E_1 = 81 E_1$, but because only one particle is present it does not satisfy the requirements.

(c)   Applying (8.2a) to the distributions gives

$$\Omega(1, 0, 0, 1, 0, 0, 0, 1, 0, \ldots) = \frac{3!}{1!0!0!1!0!0!0!1!0! \cdots} = 6$$

$$\Omega(0, 0, 0, 2, 0, 0, 1, 0, 0, \ldots) = \frac{3!}{0!0!0!2!0!0!1!0!0! \cdots} = 3$$

$$\Omega(0, 0, 1, 0, 0, 2, 0, 0, 0, \ldots) = \frac{3!}{0!0!1!0!0!2!0!0!0! \cdots} = 3$$

(d)   Using (8.3) with $N_s = 3$ and $N_1 = 1$ gives

$$P_1(1, 0, 0, 1, 0, 0, 0, 1, 0, \ldots) = \tfrac{1}{3}$$

(e)   Equation (8.4) gives

$$P_1 = \frac{(1)(6) + (0)(3) + (0)(3)}{(3)(6 + 3 + 3)} = \tfrac{1}{6}$$

8.2. The energy difference between the ground state and first vibrational state for O_2 is $3.139\,1 \times 10^{-20}$ J. Compare the ratios of molecules in these states at 750 K (a typical atmospheric nighttime temperature at 400 km) and 2 000 K (a typical daytime temperature).

Using (8.10) with $g_i = 1$ gives

$$\frac{N_b^*}{N_a^*} = \exp\left[-\frac{3.139\,1 \times 10^{-20}\ \text{J}}{(1.380\,7 \times 10^{-23}\ \text{J K}^{-1})(750\ \text{K})}\right] = 0.048\,2$$

at night and, similarly, $N_b^* / N_a^* = 0.320\,9$ during the day, an increase by a factor of about 6.7.

IDEAL-GAS PARTITION FUNCTIONS

8.3.   Calculate $\varepsilon_{trans}(1, 1, 2)$ for an oxygen molecule in a cubic container of side $a = 1.00$ m.

The molecular mass is given by

$$m = \frac{32.00 \times 10^{-3} \text{ kg mol}^{-1}}{6.022 \times 10^{23} \text{ mol}^{-1}} = 5.314 \times 10^{-26} \text{ kg}$$

Equation (8.14) gives

$$\varepsilon_{trans} = (1^2 + 1^2 + 2^2) \frac{(6.626 \times 10^{-34} \text{ J s})^2}{8(5.314 \times 10^{-26} \text{ kg})(1.00 \text{ m})^2} = 6.20 \times 10^{-42} \text{ J}$$

8.4.   Calculate $\epsilon_{rot}(J = 2)$ for an oxygen molecule given $r = 0.120\,74$ nm.

Using (2.16) gives

$$\mu = \frac{(16.00 \times 10^{-3}/L)(16.00 \times 10^{-3}/L)}{(16.00 \times 10^{-3}/L) + (16.00 \times 10^{-3}/L)} = 1.328 \times 10^{-26} \text{ kg}$$

and (8.17) gives

$$I = (1.328 \times 10^{-26} \text{ kg})(1.207\,4 \times 10^{-10} \text{ m})^2 = 1.936 \times 10^{-46} \text{ kg m}^2$$

which upon substitution into (8.16) gives

$$\epsilon_{rot} = (2)(2 - 1) \frac{(6.626 \times 10^{-34} \text{ J s})^2}{8\pi^2(1.936 \times 10^{-46} \text{ kg m}^2)} = 1.723 \times 10^{-22} \text{ J}$$

8.5.   Calculate ε_{vib} for an oxygen molecule if $v = 0$ and $\bar{\nu} = 1\,580.246$ cm^{-1}.

Here there is a single vibrational mode with $v_1 = v$ and $\nu_1 = c\bar{\nu}$. Thus (8.20) gives

$$\varepsilon_{vib} = (0 + \tfrac{1}{2})(6.626\,2 \times 10^{-34} \text{ J s})(2.997\,9 \times 10^{10} \text{ cm s}^{-1})(1\,580.246 \text{ cm}^{-1})$$

$$= 1.569\,6 \times 10^{-20} \text{ J}.$$

APPLICATION TO THERMODYNAMICS INVOLVING IDEAL GASES

8.6.   The expressions for the translational and rotational contributions in (8.33) can be simplified to

$$S°(\text{trans})/(\text{J K}^{-1} \text{ mol}^{-1}) = (12.471\,62) \ln M + (20.786\,0) \ln T - (9.685\,21) \qquad (8.38a)$$

$$S°(\text{rot})/(\text{J K}^{-1} \text{ mol}^{-1}) = \begin{cases} (8.314\,41) \ln[(I \times 10^{47})\,T/\sigma] - (22.377\,86) \\ 229.591\,0 + (12.471\,62) \ln T - (8.314\,41) \ln \sigma \\ + (4.157\,21) \ln(I_x I_y I_z \times 10^{114}) \end{cases} \qquad (8.38b)$$

where M is the molar mass in g mol^{-1}, I is in kg m^2, and the first form of (8.38b) is valid for a linear molecule and the second form is valid for a nonlinear molecule. Calculate $S°_{298}$ for 1 mol of O_2 using the data in Problems 8.3–8.5.

Substituting $M = 32.00$ g mol^{-1} and $I = 1.937 \times 10^{-46}$ kg m^2 into (8.38) gives

$$S°(\text{trans}) = (12.471\,62) \ln(32.00) + (20.786\,0) \ln(298.15) - 9.685\,21$$

$$= 151.968 \text{ J K}^{-1} \text{ mol}^{-1}$$

$$S°(\text{rot}) = (8.314\,41) \ln[(19.37)(298.15)/(2)] - (22.377\,86)$$

$$= 43.873 \text{ J K}^{-1} \text{ mol}^{-1}$$

The other contributions to $S°_{298}$ using $x = 7.64$ and $g_0 = 3$ are calculated from (8.33) as

$$S°(\text{vib}) = (8.314)\left[\frac{7.64}{e^{7.64} - 1} - \ln(1 - e^{-7.64})\right] = 3.5 \times 10^{-2} \text{ J K}^{-1} \text{ mol}^{-1}$$

$$S°(\text{elec}) = (8.314) \ln 3 = 9.134 \text{ J K}^{-1} \text{ mol}^{-1}$$

$$S°(\text{nuc}) = 0$$

Summing these contributions gives $S°_{298} = 205.010 \text{ J K}^{-1} \text{ mol}^{-1}$

8.7.　Calculate $\Delta_c G°_T$ at 298 K and 500. K for the burning of $CH_4(g)$ to form gaseous water given $[(G°_T - E°_0)/T]/(\text{J K}^{-1} \text{ mol}^{-1}) = -175.9$ and -191.0 for $O_2(g)$, -152.4 and -170.4 for $CH_4(g)$, -182.2 and -199.4 for $CO_2(g)$, and -155.4 and -172.7 for $H_2O(g)$, at 298 K and 500. K, respectively.

Using ($6.20a$) with $\Delta E°_0 = -804.17$ kJ (see Problem 4.11),

$$\Delta_c G°_{298} = -804.17 + (298.15)[(2)(-155.4) + (1)(-182.2)$$

$$- (2)(-175.9) - (1)(-152.4)](10^{-3}) = -800.8 \text{ kJ}$$

$$\Delta_c G°_{500} = -804.17 + (500.)[(2)(-172.7) + (1)(-199.4)$$

$$- (2)(-191.0) - (1)(-170.4)](10^{-3}) = -800.4 \text{ kJ}$$

8.8.　The relation between G and the thermodynamic partition function is

$$G = -NkT(\partial \ln Q/\partial N)_{V,T}$$

(a) Derive ($8.34a$). (b) Find a general expression for $(G°_T - E°_0)/T$ for ideal diatomic gases. (c) Evaluate $(G°_T - E°_0)/T$ for 1 mol of O_2 at 298 K given that $g_0 = 3$, $x = 7.64$, $\sigma = 2$, $I = 1.937 \times 10^{-46} \text{ kg m}^2$, and $M = 32.00 \text{ g mol}^{-1}$.

(a)　The logarithm of (8.13) is

$$\ln Q = N(\ln q_{\text{trans}} + \ln q_{\text{rot}} + \ln q_{\text{vib}} + \ln q_{\text{elec}} + \ln q_{\text{nuc}}) - \ln N!$$

and taking the derivative gives

$$\left(\frac{\partial \ln Q}{\partial N}\right)_{V,T} = \ln q_{\text{trans}} + \ln q_{\text{rot}} + \ln q_{\text{vib}} + \ln q_{\text{elec}} + \ln q_{\text{nuc}} - \left(\frac{\partial \ln N!}{\partial N}\right)_{V,T}$$

where

$$\left(\frac{\partial \ln N!}{\partial N}\right)_{V,T} = \frac{\partial(N \ln N - N)}{\partial N} = \ln N$$

upon using (8.6) for $\ln N!$. Therefore,

$$G = -NkT(\ln q_{\text{trans}} + \ln q_{\text{rot}} + \ln q_{\text{vib}} + \ln q_{\text{elec}} + \ln q_{\text{nuc}} - \ln N)$$

Evaluating G at 0 K gives $G°_0 = E°_0 = 0$, so

$$G°_T - E°_0 = -NkT[\ln(q°_{\text{trans}}/N) + \ln q°_{\text{rot}} + \ln q°_{\text{vib}} + \ln q°_{\text{elec}} + \ln q°_{\text{nuc}}]$$

which is another form of ($8.34a$).

(b) For 1 mol of an ideal diatomic gas,

$$\frac{G_T^\circ - E_0^\circ}{T} = \left\{\frac{G_T^\circ - E_0^\circ}{T}\right\}_{\text{trans}} + \left\{\frac{G_T^\circ - E_0^\circ}{T}\right\}_{\text{rot}} + \left\{\frac{G_T^\circ - E_0^\circ}{T}\right\}_{\text{vib}}$$

$$+ \left\{\frac{G_T^\circ - E_0^\circ}{T}\right\}_{\text{elec}} + \left\{\frac{G_T^\circ - E_0^\circ}{T}\right\}_{\text{nuc}}$$

where

$$\left(\frac{G_T^\circ - E_0^\circ}{T}\right)_{\text{trans}} = -R \ln \frac{q_{\text{trans}}^\circ}{L} = -R \ln \frac{(2\pi m k T)^{3/2} R T}{h^3 L}$$

$$= 30.473 \text{ J K}^{-1}\text{ mol}^{-1} - 1.5 R \ln M - 2.5 R \ln T \qquad (8.39a)$$

$$\left(\frac{G_T^\circ - E_0^\circ}{T}\right)_{\text{rot}} = -R \ln q_{\text{rot}} = -R \ln \frac{8\pi^2 I k T}{h^2 \sigma}$$

$$= 30.728 \text{ J K}^{-1}\text{ mol}^{-1} + R \ln \frac{\sigma}{T} - R \ln(I \times 10^{47}) \qquad (8.39b)$$

$$\left(\frac{G_T^\circ - E_0^\circ}{T}\right)_{\text{vib}} = R \ln(1 - e^{-x}) \qquad (8.39c)$$

$$\left(\frac{G_T^\circ - E_0^\circ}{T}\right)_{\text{elec}} = -R \ln g_0 \qquad (8.39d)$$

$$\left(\frac{G_T^\circ - E_0^\circ}{T}\right)_{\text{nuc}} = -R \ln g_{n,0} \qquad (8.39e)$$

upon substitution of (8.15), (8.18), (8.21), (8.24), and (8.26), respectively.

(c) Equations (8.39) give

$$\left(\frac{G_T^\circ - E_0^\circ}{T}\right)_{\text{trans}} = 30.473 - (1.5)(8.314) \ln 32.00 - (2.5)(8.314) \ln 298.15$$

$$= -131.17 \text{ J K}^{-1}\text{ mol}^{-1}$$

$$\left(\frac{G_T^\circ - E_0^\circ}{T}\right)_{\text{rot}} = 30.728 + (8.314) \ln(2/298.15) - (8.314) \ln(19.37)$$

$$= -35.52 \text{ J K}^{-1}\text{ mol}^{-1}$$

$$\left(\frac{G_T^\circ - E_0^\circ}{T}\right)_{\text{vib}} = (8.314) \ln(1 - e^{-7.64}) = -4.0 \times 10^{-3} \text{ J K}^{-1}\text{ mol}^{-1}$$

$$\left(\frac{G_T^\circ - E_0^\circ}{T}\right)_{\text{elec}} = -(8.314) \ln 3 = -9.13 \text{ J K}^{-1}\text{ mol}^{-1}$$

$$\left(\frac{G_T^\circ - E_0^\circ}{T}\right)_{\text{nuc}} = 0$$

Summing these gives for $(G_T^\circ - E_0^\circ)/T$ at 298 K the value $-175.82 \text{ J K}^{-1}\text{ mol}^{-1}$.

8.9.   Consider the chemical reaction A $\rightleftharpoons$ B, where the molecules of A have three equally spaced energy states (similar to electronic states) 1×10^{-22} J apart, and those of B have a triply degenerate level that is 2×10^{-22} J above the ground state of A. What is K for this reaction at 25 °C and 1 000 K?

The system is diagrammed in Fig. 8-1. Using ($8.34b$) for the reaction gives

$$\frac{\Delta(G_T^\circ - E_0^\circ)}{T} = \left\{ -k \ln\left[\left(\frac{q_{trans}^\circ}{N}\right)^N (q_{rot}^\circ q_{vib}^\circ q_{elec}^\circ q_{nuc}^\circ)^N \right]_B \right\}$$

$$- \left\{ -k \ln\left[\left(\frac{q_{trans}^\circ}{N}\right)^N (q_{rot}^\circ q_{vib}^\circ q_{elec}^\circ q_{nuc}^\circ)^N \right]_A \right\}$$

$$= -Nk \ln\left(\frac{q_{elec,B}^\circ}{q_{elec,A}^\circ}\right)$$

From (8.8),

$$q_{elec,B}^\circ = (3)\,e^{-0/kT} = 3$$

$$q_{elec,A}^\circ = (1)e^{-0/kT} + (1)\exp\left[-\frac{1\times10^{-22}}{(1.380\,7\times10^{-23})(298)}\right] + (1)\exp\left[-\frac{2\times10^{-22}}{(1.380\,7\times10^{-23})(298)}\right]$$

$$= 2.929$$

at 298 K. Recognizing that $N = L$ and $\Delta E_0^\circ = (2\times10^{-22}\text{ J})(L)$, we have from ($7.1b$):

$$K = \exp\left[-\frac{(2\times10^{-22})(6.022\times10^{23})}{(8.314)(298)}\right]\left(\frac{3}{2.929}\right) = 0.976$$

Repeating the calculations at 1 000 K gives $q_{elec,B}^\circ = 3$, $q_{elec,A}^\circ = 2.978$, and $K = 1.010$. Note that at lower temperatures the equilibrium lies on the side of the reactants, and at higher temperatures the situation is reversed.

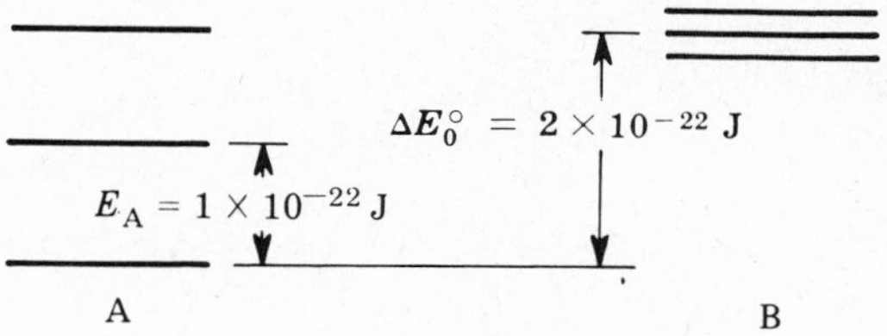

Fig. 8-1

MONATOMIC CRYSTALS

8.10. Find $E^\circ - E_0^\circ$, $A^\circ - E_0^\circ$, and S° for aluminum at 298 K.

For $\Theta_D/T = 1.43$ (see Problem 3.10), the right-hand sides of (8.35), (8.36), and (8.37) have the values 0.563 7, 0.461 9, and 1.025 5, respectively. Therefore,

$$E^\circ - E_0^\circ = (3)(8.314\times10^{-3})(298)(0.563\,7) = 4.190 \text{ kJ mol}^{-1}$$

$$A^\circ - E_0^\circ = (3)(8.314\times10^{-3})(298)(0.461\,9) = 3.433 \text{ kJ mol}^{-1}$$

$$S^\circ = (3)(8.314)(1.025\,5) = 25.58 \text{ J K}^{-1}\text{mol}^{-1}$$

Supplementary Problems

ENSEMBLES

8.11. Evaluate 11! (a) by using Stirling's approximation, (8.6); (b) by using the improved approximation

$$\ln N! = N \ln N - N + \tfrac{1}{2}\ln 2\pi N$$

and (c) by actual calculation.   *Ans.*   (a) 4.77×10^6, (b) 3.96×10^7, (c) 39 916 800

8.12. Find the overall probability of finding system A in E_1 for the acceptable distributions given in Example 8.1.

   Ans. $\Omega(1, 1, 1, 0, 0, \ldots) = 6$, $\Omega(2, 0, 0, 1, 0, \ldots) = 3$, $\Omega(0, 3, 0, 0, 0, \ldots) = 1$; $P_1 = 0.300$

8.13. Consider a system of energy states where $E_n = nE_1$. Find the possible distributions of three particles such that $E_t = 10E_1$. Determine Ω for each distribution. Find the overall probability of finding a chosen system in E_1.

   Ans. $(2, 0, 0, 0, 0, 0, 0, 1, 0, 0, \ldots)$, $(1, 1, 0, 0, 0, 0, 1, 0, 0, 0, \ldots)$, $(1, 0, 1, 0, 0, 1, 0, 0, 0, 0, \ldots)$,
            $(0, 2, 0, 0, 0, 1, 0, 0, 0, 0, \ldots)$, $(1, 0, 0, 1, 1, 0, 0, 0, 0, 0, \ldots)$, $(0, 1, 1, 0, 1, 0, 0, 0, 0, 0, \ldots)$,
            $(0, 1, 0, 2, 0, 0, 0, 0, 0, 0, \ldots)$, $(0, 0, 2, 1, 0, 0, 0, 0, 0, 0, \ldots)$; $\Omega = 3, 6, 6, 3, 6, 6, 3, 3$; 0.222

IDEAL-GAS PARTITION FUNCTIONS

8.14. Calculate $\varepsilon_{\text{trans}}(1, 1, 1)$ for an oxygen molecule in a cubic container that has $a = 1.00$ m, and determine $N_{1,1,2}^*/N_{1,1,1}^*$ at 25 °C (see Problem 8.3).

   Ans. 3.10×10^{-42} J, $e^{-7.53 \times 10^{-22}} \approx 1$

8.15. Calculate ϵ_{rot} for $J = 4$ for an oxygen molecule, and determine $N_{J=4}^*/N_{J=2}^*$ at 25 °C (see Problem 8.4).

   Ans. 5.745×10^{-22} J, $\frac{9}{5}|e^{-0.0977} = 1.632$

8.16. Calculate the number of molecules in the first three vibrational states for 1 mol of O_2 molecules at 25 °C. See Problem 8.5 for pertinent data.

   Ans. $x = 7.630$, $N_0 = 0.999\,51L$, $N_1 = 4.9 \times 10^{-4}L$, $N_2 = 2.4 \times 10^{-7}L$

8.17. Given that the first excited electronic level of O_2 is 15.72×10^{-20} J above the ground level, calculate N_1^*/N_0^* at 298 K and 1 500 K. The degeneracies are $g_0 = 3$ and $g_1 = 2$.

   Ans. $\frac{2}{3}e^{-38.2} = 1.6 \times 10^{-17}$; 3.3×10^{-4}

8.18. Derive (*8.22*) and (*8.23*), beginning with (*8.10*).

8.19. Derive (*8.18*) beginning with (*8.11*) and (*8.16*). (*Hint*: Use the technique shown in Example 8.7 to evaluate the summation.)

APPLICATION TO THERMODYNAMICS INVOLVING IDEAL GASES

8.20. Calculate S_{298}° for 1 mol of CO(g) using $r = 0.112\,81$ nm, $\bar{\nu} = 2\,169.52$ cm^{-1}, and $g_0 = 1$.

   Ans.   197.53 J K^{-1} mol^{-1}

8.21. Evaluate $(G_T^\circ - E_0^\circ)/T$ for 1 mol of N_2(g) at 1 000. K, given that $g_0 = 1$, $\bar{\nu} = 2\,357.55$ cm^{-1}, $\sigma = 2$, $r = 0.108\,758$ nm, and $M = 28.013\,4$ g mol^{-1}.

   Ans. $I = 27.5 \times 10^{-47}$ kg m^2, $x = 3.392$; $(G_T^\circ - E_0^\circ)/T = -203.50$ J K^{-1} mol^{-1}

8.22. Calculate $\Delta_r G_{1000}^\circ$ for

$$N_2(g) + 2O_2(g) \rightleftharpoons 2NO_2(g)$$

given $\Delta_r G_{298}^\circ = 102.62$ kJ and $[(G_T^\circ - E_0^\circ)/T]/(\text{J K}^{-1}\,\text{mol}^{-1}) = -162.3$ and -197.8 for N_2(g), -175.9 and -212.0 for O_2(g), and -205.7 and -251.9 for NO_2(g), at 298 K and 1 000. K, respectively.

   Ans. $[\Delta(G_T^\circ - E_0^\circ)/T]/(\text{J K}^{-1}) = 102.7$ at 298 K and 118.0 at 1 000. K, $\Delta E_0^\circ = 72.00$ kJ, $\Delta_r G_{1000}^\circ = 190.0$ kJ

8.23. Consider the equilibrium between system A having three singly degenerate levels spaced 8×10^{-22} J apart (similar to electronic levels) and system B having a triply degenerate ground level 3×10^{-22} J above the ground level of A and a doubly degenerate level 1×10^{-22} J above its own ground level. Calculate K at 10 K and 1 000 K.

Ans. $Q_A^\circ = 1.003$, $Q_B^\circ = 3.969$, $K = 0.450$; $Q_A^\circ = 2.834$, $Q_B^\circ = 4.986$, $K = 1.722$

8.24.   Predict K for the isotopic exchange reaction

$$^{16}O_2(g) + {}^{18}O(g) \rightleftharpoons {}^{16}O\,^{18}O(g) + {}^{16}O(g)$$

making assumptions similar to those in Example 8.11. What is the driving force of this reaction?

Ans.   2,   entropy increase

8.25.   Calculate (a) H°(thermal) for $Cl_2(g)$, $F_2(g)$, and $ClF(g)$ at 298 K and 1 000. K, given that $g_0 = 1$, 1, and 1; $\bar{\nu}/(cm^{-1}) = 561.1$, 923.1, and 784.39; $r/(nm) = 0.198\,6$, 0.140\,9, and 0.162\,813; $\sigma = 2$, 2, and 1; and $M/(g\ mol^{-1}) = 70.906$, 38.00, and 54.451\,4, for $Cl_2(g)$, $F_2(g)$, and $ClF(g)$, respectively. Using the values of H°(thermal) and the fact that $\Delta_f H_{298}^\circ$ of $ClF(g)$ is -54.48 kJ mol^{-1}, (b) calculate $\Delta_f E_0^\circ$ and $\Delta_f H_{1000}^\circ$. (c) Calculate S° for the gases at 298 K and 1 000. K and $\Delta_f S^\circ$ at 298 K and 1 000. K. (d) Using the values of $\Delta_f H^\circ$ and $\Delta_f S^\circ$, calculate $\Delta_f G^\circ$ at both temperatures. (e) Given $[(G_T^\circ - E_0^\circ)/T]/(J\ K^{-1}\ mol^{-1}) = -187.8$ and -231.8 for $Cl_2(g)$, -173.0 and -210.9 for $F_2(g)$, and -187.8 and -226.3 for $ClF(g)$, at 298 K and 1 000. K, respectively, calculate $\Delta_f G^\circ$ at both temperatures.

Ans.   (a) H°(thermal)/(kJ mol^{-1}) = 9.151, 8.801, and 8.890 at 298 K and 34.501, 33.085, and 33.590 at 1 000. K
      (b) $\Delta E_0^\circ = -54.39$ kJ, $\Delta_f H_{1000}^\circ = -54.60$ kJ mol^{-1}
      (c) $S^\circ/(J\ K^{-1}\ mol^{-1}) = 222.83$ and 266.18 for $Cl_2(g)$, 202.53 and 243.76 for $F_2(g)$, and 217.71 and 259.77 for $ClF(g)$, at 298 K and 1 000. K, respectively; $\Delta_f S^\circ/(J\ K^{-1}\ mol^{-1}) = 5.03$ and 4.80 at 298 K and 1 000. K, respectively
      (d) $\Delta_f G^\circ/(kJ\ mol^{-1}) = -55.98$ and -59.40 at 298 K and 1 000. K, respectively
      (e) $\Delta_f G^\circ/(kJ\ mol^{-1}) = -55.96$ and -59.34 at 298 K and 1 000. K, respectively

MONATOMIC CRYSTALS

8.26.   If the value of the right-hand side of (8.35) is 0.580\,6 and 0.858\,0 for tungsten at 298 K and 1 000 K, respectively, calculate ΔU° for heating 1.00 mol from 25 °C to 1 000 K.

Ans. E(thermal) = 4.315 kK mol^{-1} at 298 K and 21.400 kJ mol^{-1} at 1 000 K, ΔE°(thermal) = ΔU° = 17.085 kJ

Chapter 9

Electrochemistry

Oxidation-Reduction

9.1 STOICHIOMETRY

Of the several techniques used for balancing redox equations, the *ion-electron* or *half-reaction method* illustrated in Examples 9.1 and 9.2 is the best because the results are in the proper format for assigning values of potential and for performing stoichiometric calculations.

The quantity of electricity equivalent to 1 mol of electrons is 96 484.56 coulombs (1 C = 1 A s). The quantity of electricity (Q) transferred by an electric current (I) is

$$dQ = I\, dt \qquad (9.1)$$

where I is measured in amperes (A) and t is measured in seconds.

EXAMPLE 9.1. The chemical reaction in a lead storage cell of a car battery during charging involves the reduction of $PbSO_4(s)$ to $Pb(s)$ and the oxidation of $PbSO_4(s)$ to $PbO_2(s)$, both reactions occurring in the presence of $H_2SO_4(aq)$. Write the balanced equation.

The species undergoing changes in oxidation numbers are

$$PbSO_4(s) \longrightarrow Pb(s) + PbO_2(s)$$

and writing individual half-reactions gives

$$PbSO_4(s) \longrightarrow Pb(s) \qquad PbSO_4(s) \longrightarrow PbO_2(s)$$

The first half-reaction can be balanced "by inspection" by adding one $HSO_4^-(aq)$ to the right side and one $H^+(aq)$ to the left side, giving

$$PbSO_4(s) + H^+(aq) \longrightarrow Pb(s) + HSO_4^-(aq)$$

The choice of using HSO_4^- to balance the SO_4^{2-} ion is made on the basis that HSO_4^- is a weak acid and will not undergo the second ionization step to form appreciable amounts of SO_4^{2-} ion. The second equation also requires one $HSO_4^-(aq)$ on the right side, but to balance the oxygen and hydrogen atoms, $xH^+(aq)$ and $yH_2O(l)$ are added (because the reaction is occurring in neutral or acidic media):

$$PbSO_4(s) \longrightarrow PbO_2(s) + HSO_4^-(aq) + xH^+(aq) + yH_2O(l)$$

Counts of the hydrogen atoms and the oxygen atoms give

$$0 = 1 + x + 2y \qquad 4 = 2 + 4 + y$$

which yield $x = 3$ and $y = -2$. The overall half-reaction becomes

$$PbSO_4(s) + 2H_2O(l) \longrightarrow PbO_2(s) + HSO_4^-(aq) + 3H^+(aq)$$

Balancing the half-reactions electrically by adding electrons gives

$$PbSO_4(s) + H^+(aq) + 2e^- \longrightarrow Pb(s) + HSO_4^-(aq)$$

$$PbSO_4(s) + 2H_2O(l) \longrightarrow PbO_2(s) + HSO_4^-(aq) + 3H^+(aq) + 2e^-$$

Because the number of electrons required for the first half-reaction is equal to the number released in the second, these half-reactions may be added directly, and canceling one $H^+(aq)$ common to both sides we obtain

$$2PbSO_4(s) + 2H_2O(l) \longrightarrow Pb(s) + PbO_2(s) + 2HSO_4^-(aq) + 2H^+(aq)$$

for the net ionic equation. Upon checking, the equation is balanced with respect to mass and charge.

EXAMPLE 9.2. The Ni-Cd alkali cell has an electrode at which $Cd(s)$ is oxidized to $Cd(OH)_2(s)$ and an electrode at which $NiOOH(s)$ is reduced to $Ni(OH)_2(s)$. Write the balanced chemical equation describing this cell.

The species undergoing changes in oxidation states are

$$Cd(s) + NiOOH(s) \longrightarrow Cd(OH)_2(s) + Ni(OH)_2(s)$$

which gives the half-reactions

$$Cd(s) \longrightarrow Cd(OH)_2(s) \quad \text{and} \quad NiOOH(s) \longrightarrow Ni(OH)_2(s)$$

The first half-reaction can be balanced by simply adding $2OH^-(aq)$ to the left side, giving

$$Cd(s) + 2OH^-(aq) \longrightarrow Cd(OH)_2(s)$$

To balance the second half-reaction, $xOH^-(aq)$ and $yH_2O(l)$ are added (because the reaction is occurring in a basic medium):

$$NiOOH(s) + xOH^-(aq) + yH_2O(l) \longrightarrow Ni(OH)_2(s)$$

The count of H and O atoms gives the equations

$$1 + x + 2y = 2 \quad \text{and} \quad 2 + x + y = 2$$

which yield $x = -1$ and $y = 1$. The final half-reaction becomes

$$NiOOH(s) + H_2O(l) \longrightarrow Ni(OH)_2(s) + OH^-(aq)$$

Balancing electrically by adding electrons gives

$$Cd(s) + 2OH^-(aq) \longrightarrow Cd(OH)_2(s) + 2e^-$$

$$NiOOH(s) + H_2O(l) + e^- \longrightarrow Ni(OH)_2(s) + OH^-(aq)$$

Because the number of electrons in each half-reaction must be the same, the reaction involving nickel is multiplied by 2 before the two are added. Upon addition, and canceling $2OH^-(aq)$ common to both sides, we obtain

$$Cd(s) + 2NiOOH(s) + 2H_2O(l) \longrightarrow Cd(OH)_2(s) + 2Ni(OH)_2(s)$$

A check shows that the equation is balanced with respect to mass and charge.

EXAMPLE 9.3. If 10.0 A were passed through a lead storage cell for 1.50 h during a charging process, how much $PbSO_4$ would decompose?

The quantity of electricity passed through the cell is given by (*9.1*) as

$$Q = (10.0 \text{ A})(1.50 \text{ h})[(3\,600 \text{ s})/(1 \text{ h})] = 5.40 \times 10^4 \text{ C}$$

This amount of electricity corresponds to

$$\frac{5.40 \times 10^4 \text{ C}}{9.648\,5 \times 10^4 \text{ C (mol } e^-)^{-1}} = 0.560 \text{ mol } e^-$$

Recalling that $n = 2$ for each of the balanced half-reactions in Example 9.1,

$$(0.560 \text{ mol } e^-)\left(\frac{2 \text{ mol } PbSO_4}{2 \text{ mol } e^-}\right)\left(\frac{303.25 \text{ g } PbSO_4}{1 \text{ mol } PbSO_4}\right) = 170. \text{ g } PbSO_4$$

is the mass of $PbSO_4(s)$ reacting—half being oxidized and half being reduced.

9.2 GALVANIC AND ELECTROLYTIC CELLS

The relationship between the electromotive force (emf) of an electrochemical cell and the spontaneity of the chemical reaction is

$$\Delta_r G = -nFE \tag{6.7}$$

where n is the number of moles of electrons in the balanced reaction. For a negative value of $\Delta_r G$ (a positive value of E), the reaction is spontaneous and the cell will serve as a "seat of emf" or a *galvanic* (or *voltaic*) *cell*. If $\Delta_r G$ and E are zero, a state of equilibrium exists. If $\Delta_r G$ is positive (E is negative), a nonspontaneous reaction has been written, which will occur as written only if the cell is supplied energy from the surroundings, giving an *electrolytic cell*. The reverse reaction will be spontaneous. If there is a choice of reactions that may occur, the one with the most positive value of E will occur.

EXAMPLE 9.4. During the charging of the lead storage cell, the following reaction takes place:

$$2PbSO_4(s) + 2H_2O(l) \longrightarrow Pb(s) + PbO_2(s) + 2H_2SO_4(aq)$$

Given $\Delta_f G^\circ_{298}/(kJ\ mol^{-1}) = -813.14$ for $PbSO_4(s)$, -237.129 for $H_2O(l)$, 0 for $Pb(s)$, -217.33 for $PbO_2(s)$, and -744.53 for $H_2SO_4(aq)$, calculate $\Delta_r G^\circ_{298}$ and E°. Is this reaction spontaneous under standard conditions, or is an outside source of energy required for it to proceed?

Using (*6.15*) for the reaction gives

$$\Delta_r G^\circ = [(1)(0) + (1)(-217.33) + (2)(-744.53)] - [(2)(-813.14) + (2)(-237.129)]$$

$$= 394.15\ kJ$$

and using $n = 2$ (see Example 9.1) in (*6.7*) gives

$$E^\circ = \frac{-394.15\ kJ}{(2\ mol)(96.485\ kJ\ mol^{-1}\ V^{-1})} = -2.042\ 5\ V$$

The reaction is not spontaneous as written and can occur only if an external power source such as a battery charger is used. The reverse reaction would be spontaneous.

Conductivity

9.3 MOLAR CONDUCTIVITY

The measurement of the resistance to the flow of electricity in a cell is made using a bridge circuit similar to that shown in Fig. 9-1. The resistance of the cell (R) is given in terms of the other resistances that are necessary to balance the circuit by

$$R = R_3 R_1 / R_2 \tag{9.2}$$

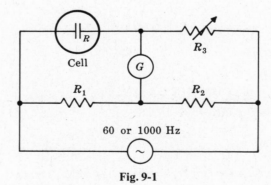

Fig. 9-1

The reciprocal of resistance ($1/R$) can be shown to be equal to

$$\frac{1}{R} = \frac{\kappa A}{l} \tag{9.3}$$

where κ is the *electrolyte conductivity* (or *specific conductance*), A is the area of an electrode surface, and l is the distance between the electrodes. The *cell constant* (A/l) is rather difficult to measure directly, so it is usually determined by measuring R for a KCl(aq) solution at a concentration for which

κ is known, using (9.3). The SI unit for conductivity is the *siemens* (S), which is equal to $1\,\Omega^{-1}$ ($= 1\,A\,V^{-1} = 1\,m^{-2}\,kg^{-1}\,s^3\,A^2$). An older unit still used is the *mho* ($= 1\,\Omega^{-1}$).

The *molar conductivity* (Λ) of an electrolytic solution is defined by

$$\Lambda = \kappa/C \qquad (9.4a)$$

where C is the concentration expressed in terms of molarity.

EXAMPLE 9.5. The units on the electrolytic conductivity are often $S\,cm^{-1}$ because the area of the electrodes and the distance between them in the conductivity cell are usually of this magnitude. Derive a practical working equation that can be used to calculate the molar conductivity in units of $S\,cm^2\,mol^{-1}$ from κ in units of $S\,cm^{-1}$ and C in units of $mol\,dm^{-3}$.

A factor of 10^3 must be introduced to convert dm^{-3} to cm^{-3} so that the units of cm^2 remain:

$$\Lambda/(S\,cm^2\,mol^{-1}) = \frac{\kappa/(S\,cm^{-1})}{C/(mol\,dm^{-3})}\,(10^3) \qquad (9.4b)$$

A plot of Λ against $C^{\frac{1}{2}}$ is nearly linear for a *strong electrolyte*, i.e., one that is highly ionized or dissociated in solution, and is highly curved for a *weak electrolyte*, i.e., one that is not highly ionized or dissociated in solution. The value of Λ at infinite dilution (Λ_0) can be considered to be the sum of the contributions of the molar conductivities of the individual ions ($\lambda_{0,i}$):

$$\Lambda_0 = v_+\lambda_{0,+} + v_-\lambda_{0,-} \qquad (9.5)$$

Because (9.5) is valid for all electrolytes, it can be shown that for the electrolytes WZ, WX, YZ, and YX

$$\Lambda_0(WZ) = \Lambda_0(WX) + \Lambda_0(YZ) - \Lambda_0(YX) \qquad (9.6)$$

For a weak electrolyte, the fraction of molecules ionized (α) is given by

$$\alpha = \Lambda/\Lambda_0 \qquad (9.7)$$

EXAMPLE 9.6. Given $\Lambda/(S\,cm^2\,mol^{-1})$ at 25 °C = 422.74, 421.36, 412.00, and 391.32 for HCl; 89.2, 88.5, 83.76, and 72.80 for $Na(CH_3COO)$; and 124.50, 123.74, 118.51, and 106.74 for NaCl at 0.000 5 M, 0.001 M, 0.01 M, and 0.1 M, respectively. Find Λ_0 for CH_3COOH. Given $\Lambda = 14.3\,S\,cm^2\,mol^{-1}$ in a 0.01 M solution of CH_3COOH, find α.

The values of $\Lambda_0/(S\,cm^2\,mol^{-1})$ for HCl, $Na(CH_3COO)$, and NaCl are found by extrapolation of a plot of Λ against $C^{1/2}$ (see Fig. 9-2), giving 426.1, 91.0, and 126.45, respectively. Using (9.6) to predict $\Lambda_0(CH_3COOH)$ gives

$$\Lambda_0(CH_3COOH) = \Lambda_0(HCl) + \Lambda_0(Na(CH_3COO)) - \Lambda_0(NaCl)$$

$$= 390.6\,S\,cm^2\,mol^{-1}$$

The percent ionization is given by (9.7) as

$$\alpha = [(14.3)/(390.6)](100) = 3.66\%$$

9.4 TRANSPORT NUMBERS

The *transport* (or *transference*) *number* (t_i) of an ion is defined as the fraction of the total current carried by that ion. For a single electrolyte in a solution,

$$t_+ + t_- = 1 \qquad (9.8)$$

There are two common experimental techniques used for determining t_i, the *Hittorf method* and the *moving boundary method*. In the former, the cell is divided into three sections, and after the passage of current the cell sections are analyzed for electrolyte content. The transport number is

$$t_i = \frac{|z_i(n_{i,0} - n_{i,f} \pm n_{i,e})|}{n_e} \qquad (9.9)$$

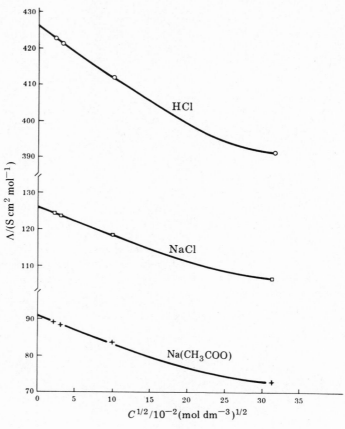

Fig. 9-2

where $n_{i,f}$ is the final number of moles present, $n_{i,0}$ is the original number, $n_{i,e}$ is the number of moles involved in the electrode reaction (the positive sign is used if the substance is generated and the negative sign is used if the substance is removed), and n_e is the number moles of electrons passed through the cell. The value of $n_{i,e}$ will be either 0 or n_e/z_i depending on whether or not inert electrodes are used.

For the moving boundary method,

$$t_i = \frac{z_i F(1\,000\,C_i)}{I}\,\frac{dV}{dt} \tag{9.10}$$

where C_i is the concentration of the ion in mol dm^{-3}, I is the current in amperes, t is the time in seconds, and V is the volume through which the moving boundary passes expressed in m^3.

EXAMPLE 9.7. Consider a hypothetical Hittorf cell having inert electrodes, in which each of the compartments contains 7 mol of electrolyte as represented by $+$ and $-$ signs in Fig. 9-3a. Construct a diagram showing the arrangement of the ions after passing 6 mol of electrons, assuming negligible migration. Construct a diagram showing the arrangement of the ions after ionic migration with $t_+ = 2t_-$.

The discharge of 6 mol e^- at the electrodes requires six of the $-$ signs in the left side of the cell to be removed at the electrode, leaving one in that compartment, and six of the $+$ signs in the right side of the cell to be removed at the electrode, leaving one in that compartment (see Fig. 9-3b). For the passing of 6 mol e^- through the solution, four $+$ signs move to the right for every two $-$ signs moving to the left across each boundary, because $t_+ = 2t_-$. This leaves 3 mol in the left side, 7 mol in the middle, and 5 mol in the right side (see Fig. 9-3c). As a check, applying (9.9) to the left portion of the cell with $n_{+,0} = 7$, $n_{+,f} = 3$, $n_{+,e} = 0$ for the cation and 6 for the anion, and

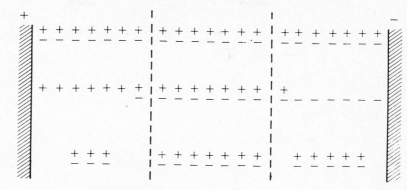

(a) Originally

(b) After discharge of 6 mol
    e^- at the electrodes

(c) After ionic migration
    with $t_+ = 2t_-$

Fig. 9-3

$n_e = 6$, gives

$$t_+ = \frac{|(1)(7-3\pm 0)|}{6} = \frac{2}{3} \qquad t_- = \frac{|(1)(7-3-6)|}{6} = \frac{1}{3}$$

so that $t_+ = 2t_-$ and (9.8) is satisfied.

9.5 IONIC MOBILITIES

The *ionic mobility* (u_i) is given by

$$u_i = \frac{l}{t(dE/dl)} \qquad (9.11)$$

where l is the distance that a moving boundary moves, t is the time, and dE/dl is the electric field strength and can be calculated by

$$\frac{dE}{dl} = \frac{I}{A\kappa} \qquad (9.12)$$

The relationship between ionic mobility and transference number is

$$t_i = \frac{|\quad z_i \nu_i u_i}{z_+\nu_+ u_+ + z_-\nu_- u_-} \qquad (9.13)$$

EXAMPLE 9.8.   The moving boundary technique was used to determine t_+ in 0.010 0 M HCl at 25 °C. A current of 3.00 mA was passed through the cell having a cross-sectional area of 3.25 cm^2 for 45.0 min, and the observed boundary moved 2.13 cm. Using these data, find u_+ and u_- given that $\Lambda = 412.00$ S cm^2 mol^{-1}.

Using $(9.4b)$ gives the electrolyte conductivity as

$$\kappa = (0.010\ 0)(412.00)(10^{-3}) = 4.12 \times 10^{-3}\ \text{S cm}^{-1} = 0.412\ \text{S m}^{-1}$$

which upon substitution into (9.12) gives

$$\frac{dE}{dl} = \frac{(3.00 \times 10^{-3}\ \text{A})[(1\ \text{V})/(1\ \text{A}\ \Omega)]}{(3.25 \times 10^{-4}\ \text{m}^2)(0.412\ \Omega^{-1}\ \text{m}^{-1})}$$

$$= 22.4\ \text{V m}^{-1}$$

Using (9.11) gives

$$u_+ = \frac{2.13 \times 10^{-2}\ \text{m}}{(45 \times 60\ \text{s})(22.4\ \text{V m}^{-1})} = 3.52 \times 10^{-7}\ \text{m}^2\ \text{V}^{-1}\ \text{s}^{-1}$$

Substituting $t_+ = 0.825$ (see Problem 9.8) and $u_+ = 3.52 \times 10^{-7}$ into (9.13) gives

$$0.825 = \frac{3.52 \times 10^{-7}}{3.52 \times 10^{-7} + u_-}$$

which upon solving gives $u_- = 7.47 \times 10^{-8} \, \text{m}^2 \, \text{V}^{-1} \, \text{s}^{-1}$.

9.6 IONIC MOLAR CONDUCTIVITY

The *ionic molar conductivity* (λ_i) is defined as

$$\nu_i \lambda_i = t_i \Lambda \qquad (9.14)$$

where

$$\nu_+ \lambda_+ + \nu_- \lambda_- = \Lambda \qquad (9.15)$$

EXAMPLE 9.9.   Find λ_+ and λ_- for the 0.010 0 M HCl solution described in Example 9.7.
With $\Lambda = 412.00 \, \text{S cm}^2 \, \text{mol}^{-1}$ and $t_+ = 0.825$, (9.14) gives

$$\lambda_+ = (0.825)(412.00) = 339.9 \, \text{S cm}^2 \, \text{mol}^{-1}$$

and (9.15) gives $\lambda_- = 72.1 \, \text{S cm}^2 \, \text{mol}^{-1}$.

Electrochemical Cells

9.7 SIGN CONVENTION AND DIAGRAMS

In a galvanic cell, the anode (site of oxidation) is negatively charged as a result of the spontaneous chemical reactions releasing electrons to the electrode. The electrons will move from the anode to the cathode in the external circuit. The cathode (site of reduction) is positively charged with respect to the anode. The positively charged cations move toward the cathode to undergo chemical reactions with the incoming electrons.

In an electrolytic cell, an external source of voltage causes the anode (site of oxidation) to be positively charged with respect to the cathode. As a result, the negatively charged anions move toward the anode to be oxidized. The cathode (site of reduction) is negatively charged, and the positively charged cations move toward the cathode to be reduced. Electrons are forced into the electrolytic cell at the cathode by the outside voltage source and leave the cell from the anode. These conventions are summarized in Fig. 9-4.

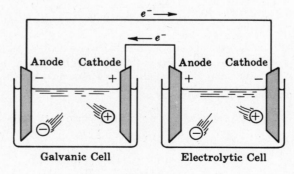

Fig. 9-4

A shorthand notation is used for describing the physical arrangement of electrochemical cells. The first terms in the notation refer to the anode reaction (oxidation), and the latter terms to the cathode reaction (reduction). A vertical bar is used to represent phase boundaries, a comma to separate different species in the same phase, and a double vertical bar to represent a salt bridge. Although far from being the most technologically advanced electrode designs, those modules illustrated in Fig. 9-5 will give an idea of what the cell will look like when constructed with this notation.

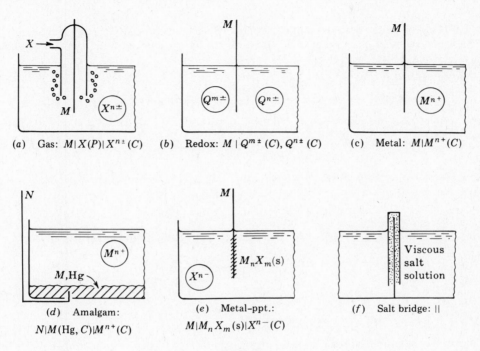

(a) Gas: $M|X(P)|X^{n\pm}(C)$ (b) Redox: $M\mid Q^{m\pm}(C), Q^{n\pm}(C)$ (c) Metal: $M|M^{n+}(C)$

(d) Amalgam:
$N|M(Hg, C)|M^{n+}(C)$ (e) Metal–ppt.:
$M|M_nX_m(s)|X^{n-}(C)$ (f) Salt bridge: ‖

Fig. 9-5

EXAMPLE 9.10. Construct a diagram using the modules shown in Fig. 9-5 for the electrochemical cell given by $Pt|Ag(s)|AgCl(s)|Cl^-(0.1\ M)\|Br^-(0.1\ M)|Br_2(1\ bar)|C(graph)|Pt.$

The Ag-AgCl anode will be represented by module (e), the salt bridge by module (f), and the Br_2 cathode by module (a). See Fig. 9-6 for the complete sketch.

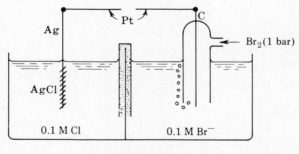

Fig. 9-6

9.8 STANDARD STATE POTENTIALS

Each reduction half-reaction (Sec. 9.1) is assigned a standard electrode potential (E^0) that is defined as the emf of a cell in which the electrode is combined with a standard hydrogen electrode $(Pt|H_2(a=1)|H^+(a=1))$ and all reacting species are in their standard states. As mentioned, these potentials are values of *electromotive force*, the limiting cell potential in which no electric current is drawn.

To use tabulated values of reduction potentials, the potential from the table is used directly for a reduction half-reaction and is used with its sign changed for an oxidation half-reaction. Values of $E°$

for an overall cell or for a different half-cell can be determined using reduction potentials (either directly or by changing $E°$ values to $\Delta G°$ values, finding the $\Delta G°$ of the process, and converting the $\Delta G°$ value back to $E°$).

EXAMPLE 9.11. To find the overall standard potential of a cell, the procedure outlined in Sec. 9.8 is used. Show that for the overall standard potential of a cell, identical results are obtained if the half-cell potentials are simply added.

For the general half-reactions

$$\text{Reactants}_1 \longrightarrow \text{products}_1 + pe^- \qquad \text{and} \qquad \text{Reactants}_2 + me^- \longrightarrow \text{products}_2$$

which have standard half-cell potentials $E_1°$ and $E_2°$, respectively, the values of $\Delta_r G°$ for the half-reactions are given by (6.7) as

$$\Delta_r G_1° = -pFE_1° \qquad \text{and} \qquad \Delta_r G_2° = -mFE_2°$$

Multiplying the oxidation half-reaction by m and the reduction half-reaction by p to eliminate electrons and adding gives the overall cell reaction, and

$$\Delta_r G° = -pmFE_1° - pmFE_2° = -pmF(E_1° + E_2°)$$

Recognizing that $n = pm$ in (6.7), the overall standard cell potential is found to be

$$E° = -\frac{-pmF(E_1° + E_2°)}{pmF} = E_1° + E_2° \tag{9.16}$$

9.9 NONSTANDARD STATE POTENTIALS

To determine the potential for nonstandard state conditions (E), $E°$ is corrected by the nonstandard state conditions using the Nernst equation,

$$E = E° - \frac{RT}{nF} \ln Q \tag{7.2}$$

At 25 °C, (7.2) becomes

$$E = E° - \frac{0.059\,157}{n} \log Q \tag{9.17}$$

EXAMPLE 9.12. What is E for the cell $\text{Ag}|\text{AgBr(s)}|\text{Br}^-(a = 0.34)$, $\text{Fe}^{3+}(a = 0.1)$, $\text{Fe}^{2+}(a = 0.02)|\text{Pt}$ given that the standard half-cell reduction potentials are 0.071 3 V for $\text{AgBr}|\text{Ag}$ and 0.771 V for $\text{Fe}^{3+}|\text{Fe}^{2+}$?

Writing the equations and half-cell potentials gives

$$\text{Ag(s)} + \text{Br}^-(a = 0.34) \longrightarrow \text{AgBr(s)} + 1e^- \qquad E° = -0.071\,3 \text{ V}$$

$$\text{Fe}^{3+}(a = 0.1) + 1e^- \longrightarrow \text{Fe}^{2+}(a = 0.02) \qquad E° = 0.771 \text{ V}$$

and adding gives

$$\text{Ag(s)} + \text{Fe}^{3+}(a = 0.1) + \text{Br}^-(a = 0.34) \longrightarrow \text{AgBr(s)} + \text{Fe}^{2+}(a = 0.02) \qquad E° = 0.700 \text{ V}$$

Observe that the half-cell with the larger reduction potential was written as the reduction reaction, making the overall potential positive (spontaneous reaction) at standard conditions. Using (6.43) for the overall reaction gives

$$Q = a(\text{AgBr})a(\text{Fe}^{2+})[a(\text{Ag})]^{-1}[a(\text{Fe}^{3+})]^{-1}[a(\text{Br}^-)]^{-1}$$

$$= (1)(0.02)/(1)(0.1)(0.34) = 0.588$$

and using (9.17) gives

$$E = 0.700 - \frac{0.059\,157}{1} \log(0.588) = 0.714 \text{ V}$$

The positive value for E implies that the reaction is spontaneous as written and even more so than at standard conditions.

9.10 CONCENTRATION CELLS AND THERMOCELLS

In either of these galvanic cells, the electrode reactions are identical except for a difference in activities or temperatures of the materials.

Because the reactions in a concentration cell are similar, $E° = 0$ and the potential arises from the nonstandard conditions of the reactants and products. Thus,

$$E = -\frac{RT}{nF} \ln Q \tag{9.18}$$

for cells that have no liquid junction or have nearly eliminated the liquid junction by using a salt bridge, and

$$E = -\frac{RT}{nF} t_i \ln Q \tag{9.19}$$

for cells with transference, where t_i is the transport number of the ion to which the electrodes are not reversible (the *spectator ion*).

The potential of a thermocell is approximately given by

$$E = -\int_{T_1}^{T_2} \frac{\partial E}{\partial T} dT \tag{9.20}$$

where $\partial E/\partial T$ is the temperature dependence of the cell potential.

Solved Problems

OXIDATION-REDUCTION

9.1. An electrochemical cell is prepared using a "quinhydrone" electrode at which hydroquinone, HOC_6H_4OH, is oxidized to quinone, OC_6H_4O, and an electrode at which $Cr_2O_7^{2-}$ is reduced to Cr^{3+}. Write the balanced net ionic equation for this cell.

Summarizing the statement of the problem gives

$$HOC_6H_4OH + Cr_2O_7^{2-} \longrightarrow OC_6H_4O + Cr^{3+}$$

from which the following half-reactions are generated:

$$HOC_6H_4OH \longrightarrow OC_6H_4O \qquad Cr_2O_7^{2-} \longrightarrow Cr^{3+}$$

The "quinhydrone" reaction is balanced with respect to C and O atoms and only needs $2H^+$ added to the right side of the reaction to complete the mass balance:

$$HOC_6H_4OH \longrightarrow OC_6H_4O + 2H^+$$

Upon adding $2e^-$, the complete half-reaction is

$$HOC_6H_4OH \longrightarrow OC_6H_4O + 2H^+ + 2e^-$$

For the reduction reaction, the Cr atoms are balanced by placing a 2 before the Cr^{3+}, and xH^+ and yH_2O are added, giving

$$Cr_2O_7^{2-} \longrightarrow 2Cr^{3+} + xH^+ + yH_2O$$

Counting H and O atoms gives

$$0 = x + 2y \quad \text{and} \quad 7 = y$$

respectively, yielding $x = -14$ and $y = 7$. The half-reaction becomes

$$Cr_2O_7^{2-} + 14H^+ \longrightarrow 2Cr^{3+} + 7H_2O$$

which requires $6e^-$ for the electrical balance:

$$Cr_2O_7^{2-} + 14H^+ + 6e^- \longrightarrow 2Cr^{3+} + 7H_2O$$

Multiplying the oxidation reaction by 3 and the reduction reaction by 1 and adding gives

$$3HOC_6H_4OH + Cr_2O_7^{2-} + 8H^+ \longrightarrow 3OC_6H_4O + 2Cr^{3+} + 7H_2O$$

after canceling common terms.

9.2. The "dry cell" or Leclanché cell involves the reaction at one electrode in which $Zn(s)$ is oxidized to $Zn(OH)_2(s)$; at the other electrode $MnO_2(s)$ is reduced to $MnOOH(s)$ in the presence of $NH_4Cl(aq)$, generating $NH_3(aq)$. Write the balanced net ionic equation for this reaction.

For the reaction

$$Zn(s) + MnO_2(s) \longrightarrow Zn(OH)_2(s) + MnOOH(s)$$

the following half-reactions can be written:

$$Zn(s) \longrightarrow Zn(OH)_2(s) \qquad MnO_2(s) \longrightarrow MnOOH(s)$$

The oxidation reaction is balanced by adding $2OH^-(aq)$ to the left side and $2e^-$ to the right side:

$$Zn(s) + 2OH^-(aq) \longrightarrow Zn(OH)_2(s) + 2e^-$$

The reduction reaction is balanced by assuming that the excess O atoms become OH^- by combining with a H^+ from the $NH_4^+(aq)$, and adding one e^-, giving

$$MnO_2(s) + NH_4^+(aq) + e^- \longrightarrow MnOOH(s) + NH_3(aq)$$

Adding the half-reactions gives

$$Zn(s) + 2MnO_2(s) + 2NH_4^+(aq) + 2OH^-(aq) \longrightarrow 2MnOOH(s) + 2NH_3(aq) + Zn(OH)_2(s)$$

9.3. What is the minimum mass of reactants for a dry cell if it is to generate 0.010 0 A for 10.0 h?

The quantity of electricity to be generated is given by (9.1) as

$$Q = (0.010\,0\ \text{A})(10.0\ \text{h})[(3\,600\ \text{s})/(1\ \text{h})] = 360\ \text{C}$$

which is equivalent to

$$\frac{360\ \text{C}}{9.648\,5 \times 10^4\ \text{C}\ (\text{mol}\ e^-)^{-1}} = 3.73 \times 10^{-3}\ \text{mol}\ e^-$$

Converting to mass of reactants (see Problem 9.2), we obtain

$$(3.73 \times 10^{-3}\ \text{mol}\ e^-)\left(\frac{1\ \text{mol Zn}}{2\ \text{mol}\ e^-}\right)\left(\frac{65.37\ \text{g Zn}}{1\ \text{mol Zn}}\right) = 0.122\ \text{g Zn}$$

$$(3.73 \times 10^{-3}\ \text{mol}\ e^-)\left(\frac{2\ \text{mol MnO}_2}{1\ \text{mol}\ e^-}\right)\left(\frac{86.94\ \text{g MnO}_2}{1\ \text{mol MnO}_2}\right) = 0.324\ \text{g MnO}_2$$

$$(3.73 \times 10^{-3}\ \text{mol}\ e^-)\left(\frac{2\ \text{mol NH}_4\text{Cl}}{2\ \text{mol}\ e^-}\right)\left(\frac{53.49\ \text{g NH}_4\text{Cl}}{1\ \text{mol NH}_4\text{Cl}}\right) = 0.200\ \text{g NH}_4\text{Cl}$$

$$(3.73 \times 10^{-3}\ \text{mol}\ e^-)\left(\frac{2\ \text{mol OH}^-}{2\ \text{mol}\ e^-}\right)\left(\frac{17.00\ \text{g OH}^-}{1\ \text{mol OH}^-}\right) = 0.063\,4\ \text{g OH}^-$$

9.4. The chemical reaction for the Daniell cell is

$$Zn(s) + Cu^{2+}(aq) \longrightarrow Zn^{2+}(aq) + Cu(s)$$

Given $\Delta_f G^\circ_{298}/(kJ\ mol^{-1}) = 0$ for $Zn(s)$ and $Cu(s)$, -147.06 for $Zn^{2+}(aq)$, and 65.49 for $Cu^{2+}(aq)$, calculate the cell potential and discuss the spontaneity of the reaction under standard conditions.

Applying (6.15) to the reaction gives

$$\Delta_r G^\circ_{298} = [(1)(-147.06) + (1)(0)] - [(1)(0) + (1)(65.49)] = -212.55\ kJ$$

and using (6.7) gives

$$E^\circ = \frac{-(-212.55\ kJ)}{(2\ mol)(96.485\ kJ\ mol^{-1}\ V^{-1})} = 1.101\ V$$

which is spontaneous as written.

9.5. During the electrolysis of a NaCl solution using inert electrodes, the following chemical reactions are possible:
At the anode,

$$Cl^-(aq) \longrightarrow \tfrac{1}{2}Cl_2(g) + 1e^-$$

$$2H_2O(l) \longrightarrow O_2(g) + 4H^+(aq) + 4e^-$$

At the cathode,

$$Na^+(aq) + 1e^- \longrightarrow Na(s)$$

$$2H^+(aq) + 2e^- \longrightarrow H_2(g)$$

The electromotive force of a cell described by the first and third equations is -4.074 V; first and fourth, $-1.359\ 7$ V; second and third, -3.943 V; and second and fourth, -1.229 V. Which reaction will proceed under standard conditions?

Choosing the combination that has the most positive value of E° gives

$$2H_2O(l) \longrightarrow O_2(g) + 2H_2(g)$$

as the reaction most favored.

CONDUCTIVITY

9.6. (a) A conductivity cell was calibrated using 0.01 M KCl ($\kappa = 1.408\ 7 \times 10^{-3}\ S\ cm^{-1}$) in the cell, and the measured resistance was 688 Ω. Find the cell constant. (b) A 0.010 0 M $AgNO_3$ solution in the same cell had a resistance of 777 Ω. What is Λ?

(a) From (9.3),

$$\frac{A}{l} = \frac{1}{\kappa R} = \frac{1}{(1.408\ 7 \times 10^{-3}\ S\ cm^{-1})(688\ \Omega)[(1\ \Omega^{-1})/(1\ S)]}$$

$$= 1.032\ cm$$

(b) Using the cell constant from (a), we have

$$\kappa = \frac{1}{R(A/l)} = \frac{1}{(777\ \Omega)(1.032\ cm)} = 1.247 \times 10^{-3}\ S\ cm^{-1}$$

The molar conductivity is given by (9.4b) as

$$\Lambda = \frac{1.247 \times 10^{-3}}{0.010\ 00}(10^3) = 124.7\ S\ cm^2\ mol^{-1}$$

9.7. Current was passed through a 0.100 M solution of KCl at 25 °C. A silver coulometer in series with the KCl cell showed that 0.613 6 g of Ag had been transferred from one electrode to the other during the electrolysis. The cathode portion weighing 117.51 g was drained and found to contain 0.566 62% KCl. The anode portion weighing 121.45 g was drained and found to contain 0.572 17% KCl. The middle portion of the Hittorf cell contained 0.742 17% KCl. Given that inert electrodes were used, find t_+.

The composition of the middle compartment is equal to the original composition of the anode and cathode compartments. In the cathode compartment there is

$$117.51 \text{ g soln} - (117.51 \text{ g soln})\left(5.666\ 2 \times 10^{-3}\ \frac{\text{g KCl}}{\text{g soln}}\right) = 117.51 \text{ g soln} - 0.665\ 8 \text{ g KCl}$$

$$= 116.84 \text{ g H}_2\text{O}$$

after the electrolysis, and for the same amount of water, there was

$$\frac{116.84}{1 - 7.421\ 7 \times 10^{-3}} = 117.71 \text{ g soln}$$

before the electrolysis containing

$$(117.71)(7.421\ 7 \times 10^{-3}) = 0.873\ 6 \text{ g KCl}$$

Thus,
$$n_{+,0} = \frac{0.873\ 6 \text{ g}}{74.56 \text{ g mol}^{-1}} = 1.172 \times 10^{-2} \text{ mol}$$

$$n_{+,f} = \frac{0.665\ 8}{74.56} = 0.893 \times 10^{-2} \text{ mol}$$

The number of moles of electrons passed is

$$n_e = \frac{(0.613\ 6 \text{ g Ag})[(1 \text{ mol } e^-)/(1 \text{ mol Ag})]}{107.868 \text{ g mol}^{-1}} = 5.688 \times 10^{-3} \text{ mol } e^-$$

Using (9.9) with $n_{+,e} = 0$ gives

$$t_+ = \frac{|(1)[(1.172 \times 10^{-2}) - (0.893 \times 10^{-2}) \pm 0]|}{5.688 \times 10^{-3}} = 0.491$$

9.8. Using the data in Example 9.8, find t_+.

Assuming dV/dt to be given by V/t, where $V = Al$, (9.10) gives

$$t_+ = \left[\frac{(1)(96\ 485)(1\ 000)(0.010\ 0)}{3.00 \times 10^{-3}}\right]\left[\frac{(3.25 \times 10^{-4})(2.13 \times 10^{-2})}{(45.0)(60)}\right] = 0.825$$

ELECTROCHEMICAL CELLS

9.9. Write the notation for the diagram shown in Fig. 9-7.

The anode compartment consists of a Pt electrode in contact with a 5% Na-Hg amalgam, which in turn is in contact with a 0.1 M solution of Na^+. The shorthand notation for the anode is

$$\text{Pt}|\text{Na}(5\% \text{ amalgam})|\text{Na}^+(0.1 \text{ M})$$

The cathode compartment consists of a Pt electrode immersed in a 1 M Cu^{2+} and 0.1 M Cu^+ solution, giving the notation $Cu^+(0.1 \text{ M}), Cu^{2+}(1 \text{ M})|\text{Pt}$. Recognizing the compartments to be separated by a salt bridge, we see that the complete cell notation becomes

$$\text{Pt}|\text{Na}(5\% \text{ amalgam})|\text{Na}^+(0.1 \text{ M})\|\text{Cu}^+(0.1 \text{ M}), \text{Cu}^{2+}(1 \text{ M})|\text{Pt}$$

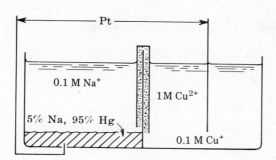

Fig. 9-7

9.10. What is $E°$ for the cell $Pt|Ag(s)|AgCl(s)\|Cl^-(a=1)|Cl_2(1\ bar)|C(graph)|Pt$ given that the standard half-cell reduction potentials are 0.222 2 V for $AgCl|Ag$ and 1.359 7 V for $Cl_2|Cl$?

The half-reaction with the smaller reduction potential is written as the oxidation, giving

$$Ag(s) + Cl^-(a=1) \longrightarrow AgCl(s) + 1e^- \qquad E° = -0.222\ 2\ V$$

and for the reduction reaction,

$$Cl_2(1\ bar) + 2e^- \longrightarrow 2Cl^-(a=1) \qquad E° = 1.359\ 7\ V$$

We multiply the first reaction by 2 and the second reaction by 1 (leaving the potentials alone) and add to obtain

$$2Ag(s) + Cl_2(1\ bar) \longrightarrow 2AgCl(s) \qquad E° = 1.137\ 5\ V$$

which is spontaneous.

9.11. Given that the standard half-cell reduction potentials are 0.521 V for $Cu^+|Cu$ and 0.337 V for $Cu^{2+}|Cu$, find the standard half-cell reduction potential for $Cu^{2+}|Cu^+$.

For the general half-reactions,

$$Reactants_1 + pe^- \longrightarrow products_1 \qquad Reactants_2 + me^- \longrightarrow products_2$$

having the standard half-cell potentials $E_1°$ and $E_2°$, respectively, the values of $\Delta G°$ are given by (6.7) as

$$\Delta_r G_1° = -pFE_1° \qquad \text{and} \qquad \Delta_r G_2° = -mFE_2°$$

If the reactions are multiplied by q and r, respectively, addition of the half-reaction equations yields the desired half-reaction

$$\Delta_r G° = -qpFE_1^0 - rmFE_2^0$$

The standard half-cell potential for the desired half-reaction is

$$E° = -\frac{-qpFE_1° - rmFE_2°}{nF} = \frac{qpE_1° + rmE_2°}{n} \qquad (9.21)$$

For the reactions involving copper,

$$Cu^+(a=1) + e^- \longrightarrow Cu(s) \qquad E° = 0.521\ V$$

$$Cu^{2+}(a=1) + 2e^- \longrightarrow Cu(s) \qquad E° = 0.337\ V$$

we can see that $p=1$ and $m=2$. Reversing the first half-reaction and adding it to the second yields the desired half-reaction:

$$Cu^{2+}(a=1) + e^- \longrightarrow Cu^+(a=1)$$

Thus $q = -1$ and $r = +1$. Substituting into (9.21) gives

$$E^\circ = \frac{(-1)(1)(0.521 \ V) + (1)(2)(0.337 \ V)}{1} = 0.153 \ V$$

9.12. The standard half-cell reduction potential for $Ag^+ | Ag$ is $0.799 \ 1 \ V$ at $25 \ ^\circ C$. Given the experimental value $K_{sp} = 1.56 \times 10^{-10}$ for AgCl, calculate the standard half-cell reduction potential for the $Ag | AgCl$ electrode.

For the desired reaction, $AgCl + 1e^- \longrightarrow Ag + Cl^-$, the value of E is given by (6.7) as

$$E^\circ = \frac{-\Delta_r G^\circ}{(1)F}$$

The needed $\Delta_r G^\circ$ can be obtained by adding the values of $\Delta_r G^\circ$ for the reactions

$$Ag^+ + 1e^- \longrightarrow Ag \qquad\qquad \Delta G^\circ = -nFE^\circ$$

$$AgCl \longrightarrow Ag^+ + Cl^- \qquad\qquad \Delta G^\circ = -RT \ln K_{sp}$$

giving $\qquad \Delta_r G^\circ = -(1 \ mol)(9.648 \ 5 \times 10^4 \ J \ mol^{-1} \ V^{-1})(0.799 \ 1 \ V)$

$$- (8.314 \ J \ K^{-1} \ mol^{-1})(298 \ K)(mol) \ln(1.56 \times 10^{-10})$$

$$= -77.10 \ kJ + 55.95 \ kJ = -21.15 \ kJ$$

The potential is

$$E^\circ = \frac{-(-21.15)}{(1)(96.485)} = -0.219 \ 2 \ V$$

9.13. What is E for the cell $Mg | Mg^{2+}(a = 10^{-3}) \| H^+(a = 10) | H_2(0.1 \ bar) | Pt$ if the standard half-cell reduction potentials are $-2.363 \ V$ for $Mg^{2+} | Mg$ and $0.000 \ V$ for $H^+ | H_2$?

For the reaction

$$Mg(s) + 2H^+(a = 10) \longrightarrow H_2(0.1 \ bar) + Mg^{2+}(a = 10^{-3})$$

$E^\circ = 2.363 \ V$ and (6.43) gives

$$Q = a(H_2)a(Mg^{2+})[a(Mg)]^{-1}[a(H^+)]^{-2} = (0.1)(10^{-3})/(1)(10)^2 = 10^{-6}$$

which upon substitution into (9.17) gives

$$E = 2.363 - \frac{0.059 \ 157}{2} \log 10^{-6} = 2.540 \ V$$

The cell is spontaneous as written, with the Mg electrode negative and the H_2 electrode positive.

9.14. What is E for the $Pb^{2+} | Pb$ half-cell at $a(Pb^{2+}) = 0.100$ given $E^\circ = -0.126 \ V$?

The reaction quotient for the half-reaction

$$Pb^{2+}(a = 0.100) + 2e^- \longrightarrow Pb(s)$$

is given by (6.43) as

$$Q = a(Pb)[a(Pb^{2+})]^{-1} = (1)/(0.100) = 10.0$$

which upon substitution into (9.17) gives

$$E = -0.126 - \frac{0.059 \ 157}{2} \log(10.0) = -0.156 \ V$$

9.15. What is the potential of the cell $C|Br_2(0.1 \text{ bar})|Br^-(0.5 \text{ M})|Br_2(1 \text{ bar})|C$ at 25 °C?

For the half-reactions

$$2Br^-(0.5 \text{ M}) \longrightarrow Br_2(0.1 \text{ bar}) + 2e^- \quad \text{and} \quad Br_2(1 \text{ bar}) + 2e^- \longrightarrow 2Br^-(0.5 \text{ M})$$

the overall reaction is $Br_2(1 \text{ bar}) \longrightarrow Br_2(0.1 \text{ bar})$, and from (9.18)

$$E = -\frac{0.059\ 157}{2} \log\left(\frac{0.1}{1}\right) = 0.029\ 579 \text{ V}$$

9.16. What is the potential at 25 °C of the cell

$$Pt|H_2(1 \text{ bar})|HCl(0.5 \text{ M})|HCl(1.0 \text{ M})|H_2(1 \text{ bar})|Pt$$

if $t_+ = 0.83$?

For this cell having a liquid junction (9.19) with $t_- = 0.17$ gives for the net reaction $HCl(1.0 \text{ M}) \longrightarrow HCl(0.5 \text{ M})$

$$E = -\frac{0.059\ 157}{1}(0.17) \log\left(\frac{0.5}{1.0}\right) = 0.003\ 0 \text{ V}$$

9.17. The potential of a neutral, saturated Weston cell is given by

$$E(t) = 1.018\ 410 - 4.93 \times 10^{-5}(t - 25) - 8.0 \times 10^{-7}(t - 25)^2 + 1 \times 10^{-8}(t - 25)^3$$

where t is the temperature between 5 °C and 50 °C. What is the approximate potential of the thermocell given below if $t_1 = 5 \text{ °C}$ and $t_2 = 50 \text{ °C}$?

$$Pt|Cd(amal)|CdSO_4, Hg_2SO_4|Hg|Pt\cdots Pt|Hg|Hg_2SO_4, CdSO_4|Cd(amal)|Pt$$
$$\quad_{t_1} \qquad\qquad\qquad\qquad\qquad\qquad\qquad\qquad\qquad\qquad\quad_{t_2}$$

The symbol $\cdots$ represents an external electrical connector between the two halves of the thermocell.

Equation (9.20) is equivalent to $E = -E(T)\big|_{T_1}^{T_2} = E(t)\big|_{t_1}^{t_2}$. Thus,

$$E = -\{(-4.93 \times 10^{-5})[(25) - (-20)] - (8.0 \times 10^{-7})[(25)^2 - (-20)^2] + (1 \times 10^{-8})[(25)^3 - (-20)^3]\}$$

$$= -(-2.22 \times 10^{-3} - 0.18 \times 10^{-3} + 0.24 \times 10^{-3}) = 2.16 \text{ mV}$$

Supplementary Problems

OXIDATION-REDUCTION

9.18. In the cells of the Edison battery, iron is oxidized to $Fe(OH)_2(s)$ in a 21% KOH solution (containing some LiOH) and $NiO_2(s)$ is reduced to $Ni(OH)_2(s)$. Write the balanced reaction for this cell.

Ans. $Fe(s) + NiO_2(s) + 2H_2O(l) \longrightarrow Fe(OH)_2(s) + Ni(OH)_2(s)$

9.19. During the electrolysis of $CdSO_4$, what volume of O_2 at 25 °C and 1.00 bar will be produced for every gram of Cd?

Ans. 1.78×10^{-2} mol e^-, 0.110 dm^3

9.20. During the electrolysis of a $CuCl_2$ solution using inert electrodes, the following chemical reactions are possible at the anode:

$$Cl^-(aq) \longrightarrow \tfrac{1}{2}Cl_2(g) + 1e^-$$

$$2H_2O(l) \longrightarrow O_2(g) + 4H^+(aq) + 4e^-$$

and at the cathode:

$$Cu^{2+}(aq) + 2e^- \longrightarrow Cu(s)$$

$$2H^+(aq) + 2e^- \longrightarrow H_2(g)$$

The electromotive force of a cell described by the first and third equations is -1.023 V; the first and fourth, $-1.359\,7$ V; the second and third, -0.892 V; and the second and fourth, -1.229 V. Which reaction will proceed under standard conditions?

Ans.   Reaction with most positive emf is $2Cu^{2+}(aq) + 2H_2O(l) \longrightarrow O_2(g) + 4H^+(aq) + 2Cu(s)$.

CONDUCTIVITY

9.21. The resistance of a conductivity cell was 702 Ω when filled with 0.010 0 M KCl ($\kappa = 1.408\,7 \times 10^{-3}$ S cm^{-1}) and 6 920 Ω when filled with 0.010 00 M CH_3COOH. Find the cell constant and Λ for the acid.

Ans.   1.011 cm,   14.29 S cm^2 mol^{-1}

9.22. The values of $\Lambda/(\text{S cm}^2 \text{ mol}^{-1})$ at 18 °C are 124.25, 118, and 106.6 for NH_4NO_3; 234, 228, and 213 for KOH; and 123.7, 118.2, and 104.8 for KNO_3, at 0.001 M, 0.01 M, and 0.1 M, respectively. (*a*) Find Λ_0 for these substances, and calculate Λ_0 for NH_3(aq). (Note that NH_3 can be written as NH_4OH for these calculations, even though molecular NH_4OH has not been shown to exist.) (*b*) Given $\Lambda/(\text{S cm}^2\text{mol}^{-1}) = 28$, 9.6, and 3.3 at 0.001 M, 0.01 M, and 0.1 M, respectively, for NH_3(aq), find α for these concentrations and comment.

Ans.   (*a*) Plot of Λ against $C^{1/2}$ gives $\Lambda_0/(\text{S cm}^2 \text{ mol}^{-1}) = 128$ for NH_4NO_3, 237 for KOH, 126 for KNO_3; $\Lambda_0 = 239$ S cm^2 mol^{-1} for NH_3(aq); (*b*) 11.7%, 4.0% and 1.4%; ionization becomes larger as the solution becomes more dilute.

9.23. From the conductivity data given below, find α, K_a, and K at each concentration of acetic acid at 25 °C.

$C/10^4(\text{mol dm}^{-3})$	0.280 14	1.113 5	1.532 1	2.184 4	10.283 1	13.634 0
$\Lambda/(\text{S cm}^2 \text{ mol}^{-1})$	210.38	127.75	112.05	96.493	48.146	42.227

Assume $\Lambda_0 = 390.13$ S cm^2 mol^{-1}.

Ans. $\alpha = 0.539\,25, 0.327\,45, 0.287\,21, 0.247\,33, 0.123\,41, 0.108\,24$; $K_\alpha = \alpha^2(C)/(1-\alpha)$, giving $K_a/10^{-5} = 1.768, 1.775, 1.773, 1.775, 1.787, 1.791$; plot of K_a against C or $C^{1/2}$ gives intercept (where $K_\gamma = 1.000$) as $K = 1.764 \times 10^{-5}$

9.24. Repeat Example 9.6 with $t_+ = 5t_-$.   *Ans.* See Fig. 9-8.

9.25. Calculate t_- for the KCl experiment described in Problem 9.7 by analyzing the anode compartment data.

Ans.   120.76 g H_2O, 121.66 g solution, 0.902 9 g KCl,
$n_{-,0} = 0.012\,11$, $n_{-,f} = 0.009\,32$, $n_{-,e} = 0.005\,688$; $t_- = 0.509$

9.26. You are to design a moving boundary experiment for students to determine t_+ for 0.010 0 M LiCl at 25 °C. The value is 0.328 9. Suppose the useful laboratory time is about 1.5 h and the current source is capable of producing 1.00 mA. To reduce the error in measurement of volume, the desired length change for the

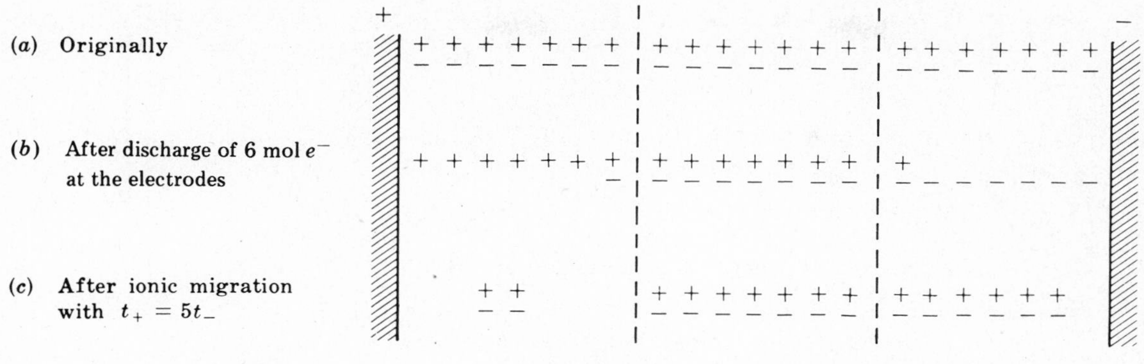

(a) Originally

(b) After discharge of 6 mol e^-
    at the electrodes

(c) After ionic migration
    with $t_+ = 5t_-$

Fig. 9-8

boundary should be 2.5 cm. If the cell is to be made from glass tubing, what size tubing should be chosen, to the nearest mm?

Ans. $A = 0.736$ cm^2, I.D. $= 9.68$ mm ≈ 10 mm

9.27.   Given $\Lambda_0/(\text{S cm}^2 \text{ mol}^{-1}) = 91.0$ for Na(CH$_3$COO) and 426.16 for HCl and given $t_+ = 0.556$ for Na(CH$_3$COO) and 0.821 for HCl, find Λ_0 for CH$_3$COOH.

Ans. $\lambda_{0,+} = 349.88$ S cm^2 mol^{-1}, $\lambda_{0,-} = 40.4$ S cm^2 mol^{-1}, $\Lambda_0 = 390.28$ S cm^2 mol^{-1}

9.28.   Calculate u_+ and u_- for 0.010 0 M LiCl given that $\Lambda = 107.32$ S cm^2 mol^{-1} using the experimental setup described in Problem 9.26.

Ans. 3.657×10^{-8} m^2 V^{-1} s^{-1}, 7.461×10^{-8} m^2 V^{-1} s^{-1}

9.29.   Consider a titration between a strong acid and a strong base. Sketch a plot of $1/R$ against volume of added base. Sketch a similar diagram for the titration of a weak acid and a strong base. Identify the equivalence points of the titrations.

Ans.   The plot for the strong acid and strong base will be nearly V-shaped (with rather steep slopes because of the high values of λ_i for H$^+$ and OH$^-$) with the equivalence point at the lowest part of the plot. In the plot for the weak acid and strong base, the slope before the neutralization equivalence point is not as great as after the equivalence point because the concentration of ions present in a weak acid is not large.

9.30.   Calculate K_{sp} for AgCl given that $\kappa = 2 \times 10^{-4}$ S m^{-1} at 25 °C and the assumption that Λ differs very little from Λ_0. Assume that $\lambda_{0,i}/(\text{S m}^2 \text{ mol}^{-1}) = 6.192 \times 10^{-3}$ for Ag$^+$ and 7.634×10^{-3} for Cl$^-$.

Ans. $\Lambda_0 = 0.013\,826$ S m^2 mol^{-1}, $C = 1.5 \times 10^{-5}$ M, $K_{sp} = 2.3 \times 10^{-10}$

9.31.   Given $\kappa = 5.7 \times 10^{-6}$ S m^{-1} for water at 25 °C, find K_w. The values of $\lambda_{0,i}/(\text{S m}^2 \text{ mol}^{-1})$ are 0.034 98 for H$^+$ and 0.019 67 for OH$^-$.

Ans. $\Lambda_0 = 0.054\,65$ S m^2 mol^{-1}, $C = 1.04 \times 10^{-7}$ M, $K_w = 1.08 \times 10^{-14}$

ELECTROCHEMICAL CELLS

9.32.   Prepare a sketch for the cell Pt|Na($a = 0.1$, amalgam)|Na$^+$($a = 0.01$)‖Cu^{2+}($a = 0.01$)|Cu(s)|Pt. Given that the standard half-cell reduction potentials at 25 °C are -2.714 V for Na$^+$|Na and 0.337 V for Cu^{2+}|Cu, calculate $E°$ and E for the cell.

Ans.   The sketch will consist of modules (d), (f), and (c) from Fig. 9-5; $E° = E = 3.051$ V for the reaction 2Na($a = 0.1$) + Cu^{2+}($a = 0.01$) $\longrightarrow$ 2Na$^+$($a = 0.01$) + Cu($a = 1$).

9.33. Calculate the potential of the cell described in Problem 9.32, assuming that the Debye–Hückel theory, (6.37), is applicable for the 0.01 M solutions of Na^+ and Cu^{2+}. Compare answers.

Ans. $\gamma(Na^+) = 0.888$, $\gamma(Cu^{2+}) = 0.442$; $E = 3.044$ V (7.5 mV lower)

9.34. What is $E°$ for the cell $Cu|Ca(s)|Ca^{2+}(a = 1)\|Fe^{3+}(a = 1)|Fe(s)|Cu$ if the standard half-cell reduction potentials are -2.866 V for $Ca^{2+}|Ca$ and -0.036 V for $Fe^{3+}|Fe$?

Ans. 2.830 V for the reaction $3\,Ca + 2\,Fe^{3+} \longrightarrow 3\,Ca^{2+} + 2\,Fe$

9.35. Given that the standard half-cell reduction potentials are 1.45 V for $ClO_3^-|Cl^-$ and 1.47 V for $ClO_3^-|Cl_2$, find $E°$ for $Cl_2|Cl^-$.   *Ans.*   1.35 V

9.36. The $E°$ for the cell $Pt|H_2(1\text{ bar})|HCl(C)|AgCl(s)|Ag(s)|Pt$ was to be determined experimentally. Write the Nernst equation for this reaction and show that

$$E + 0.118\,314 \log C(\text{HCl}) = E° - 0.118\,314 \log y_\pm$$

Because $\log y_\pm$ is a function of $C^{1/2}$ [see (6.39)] a plot of $E + 0.118\,314 \log C(\text{HCl})$ against $C^{1/2}$ will have an intercept of $E°$. Find $E°$ for the cell from the following data:

$E/(\text{V})$	0.359 8	0.389 2	0.465 0	0.579 1	0.696 1	0.814 0	0.932 2
$C/(\text{mol dm}^{-3})$	10^{-1}	5×10^{-2}	10^{-2}	10^{-3}	10^{-4}	10^{-5}	10^{-6}

Ans.   0.222 3 V

9.37. The value of K_w can be determined from emf data. Given that the standard half-reaction potentials are 0.000 0 V for $H^+|H_2$ and $-0.828\,1$ V for the half-reaction

$$2H_2O(l) + 2e^- \longrightarrow H_2(a = 1) + 2OH^-(aq)$$

find K_w from these data. Compare your answer to that obtained in Problem 9.31.

Ans. $E° = -0.828\,1$ V, $K_w = 1.00 \times 10^{-14}$; essentially the same

9.38. What is the potential of the cell $Pt|Na(10 \text{ mol } \% \text{ amalgam})|Na^+(0.1 \text{ M})|Na(5 \text{ mol } \% \text{ amalgam})|Pt$ at 25 °C?

Ans. $E° = 0$, $E = 0.017\,8$ V

9.39. The cell described in Problem 9.16 was changed to

$$Pt|H_2(1\text{ bar})|HCl(0.5 \text{ M})|AgCl|Ag\cdots Ag|AgCl|HCl(1.0 \text{ M})|H_2(1\text{ bar})|Pt$$

to eliminate the liquid junction. What is the potential of this cell at 25 °C?

Ans. $E° = 0$, $E = 0.017\,8$ V

9.40. For the thermocell $Ag|AgCl(1)|Ag$
                                                                              $t_1 \qquad t_2$

operating between 500 °C and 700 °C, Metz and Seifert reported $dE/dT = -0.378$ mV K^{-1}. Find E for this cell operating with $t_1 = 500$ °C and $t_2 = 700$ °C.   *Ans.*   75.6 mV

9.41. The potential of the Daniell cell $Zn|ZnSO_4(1 \text{ M})\|CuSO_4(1 \text{ M})|Cu$ was reported by Buckbee, Surdzial, and Metz as $E° = 1.102\,8 - 0.641 \times 10^{-3}t + 0.72 \times 10^{-5}t^2$, where t is the Celsius temperature. (a) Calculate $\Delta_r G°$, $\Delta_r S°$, and $\Delta_r H°$ from this equation at 25 °C. (b) Compare the results to the answers found in Problem 9.4. (c) The value of $\Delta_r S° = -15.6$ J K^{-1} from thermochemical tables is several times less than the cell value. Why?

Ans.   (a) $\Delta_r G° = -210.59$ kJ, $\Delta_r S° = -54.2$ J K^{-1}, $\Delta_r H° = -226.74$ kJ^{-1};
(b) $\Delta_r G°$ is 0.92% higher;
(c) A negligible error in emf from small liquid junction potential gives a significant error in $\Delta_r S°$ because the liquid junction has a different temperature coefficient than the reaction of interest.

Chapter 10

Heterogeneous Equilibria

Phase Rule

10.1 PHASES

A *phase* can be defined as a portion of the system under consideration that is submacroscopically homogeneous and is separated from other such portions by definite physical boundaries. The symbol for the number of phases present in a system will be p. There can be only one gaseous phase in a system because all gases are completely miscible.

EXAMPLE 10.1. Consider the system shown in Fig. 10-1. Find p, and describe the various phases.

The system consists of the CCl_4-rich layer, which contains small amounts of air and H_2O; the H_2O-rich layer, which contains small amounts of air and CCl_4; and the gaseous phase, which contains air, H_2O vapor, and CCl_4 vapor. Here $p = 3$.

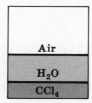

Fig. 10-1

10.2 COMPONENTS

The number of *components* in a system (c) is the minimum number of independently variable chemical species necessary to describe the composition of each phase. The establishment of chemical equilibria often reduces c.

EXAMPLE 10.2. Determine the number of components in a mixture of $H_2(g)$, $O_2(g)$, and $H_2O(g)$.

Depending on the method used to prepare the system and the final status of the system, $c = 1$, 2, or 3. If simply a mixture of gases, three components must be specified. If a mixture of three gases that have been allowed to equilibrate, only two components must be specified, because the information concerning the third can be calculated from the equilibrium constant for the reaction: $K = a(H_2O)/a(H_2)[a(O_2)]^{1/2}$. If a mixture of gases produced by the decomposition of H_2O, then only one component must be specified because the information concerning the other two can be calculated from the equilibrium constant and the known stoichiometry, where $a(H_2) = 2a(O_2)$.

10.3 DEGREES OF FREEDOM (VARIANCE)

The number of *degrees of freedom* (f) is the minimum number of intensive variables (mass-independent properties such as T, P, and concentration) that must be specified to fix the values of all remaining intensive variables. Systems with $f = 0$ are known as *invariant*, or as having no degrees of freedom; with $f = 1$, *univariant* (one degree of freedom); with $f = 2$, *divariant* (two degrees of freedom); etc.

10.4 GIBBS PHASE RULE

The number of degrees of freedom is given by

$$f = c - p + 2 \qquad (10.1)$$

EXAMPLE 10.3.   Determine f for the system described in Example 10.2.

In all cases $p = 1$, giving $f = c + 1$. For the case where $c = 3$, $f = 4$, implying that T, P, and the composition of two of the three components are required to fix the remaining variables. For $c = 2$, $f = 3$, requiring T, P, and one composition to be specified. For $c = 1$, $f = 2$, requiring only T and P to be specified. The remaining information can be calculated using the ideal gas law, etc.

Phase Diagrams for One-Component Systems

10.5 INTRODUCTION

For a one-component system (10.1) becomes

$$f = 3 - p \qquad (10.2)$$

Because p is at least 1, a maximum of two variables are needed to fix the remaining properties. Usually T and P are chosen, and the system is described by a phase diagram expressed in terms of these variables.

EXAMPLE 10.4.   Discuss the hypothetical phase diagram shown in Fig. 10-2.

Figure 10-2 consists of a plot of the vapor pressure curve of the liquid between points a and b; the sublimation pressure curves of the α-solid and the β-solid between points b and c and points c and d, respectively; the pressure dependence of the melting point of the α-solid, the β-solid, and the γ-solid between points b and e, points e and f, and points f and g, respectively; and the pressure dependence of the phase transition between the β-solid and the α-solid between points c and e, and between the β-solid and the γ-solid between points f and h. The points b, c, e, and f at which three phases are present are known as *triple points*. Point a and the vertical dashed line represent the *critical point* and the *critical isotherm*, respectively.

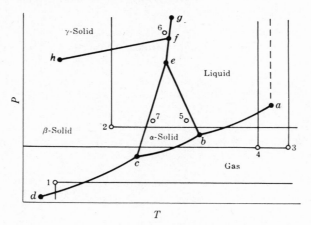

Fig. 10-2

Phase Diagrams for Two-Component Systems

10.6 INTRODUCTION

For two components, (10.1) becomes

$$f = 4 - p \qquad (10.3)$$

Because p can be as low as 1, three variables may be necessary to describe a system. Because three variables are difficult to graph, usually P is held constant on a diagram of T plotted against concentration. Useful measures of concentration are mass fraction (w_i) and mole fraction (x_i).

EXAMPLE 10.5.   Consider the hypothetical phase diagram shown in Fig. 10-3. In such a diagram certain areas will be one-phase areas and others will be two-phase areas. The compositions of the phases in equilibrium in the two-phase areas will be determined by horizontal *tie lines*. The respective masses of these two phases can be determined by

$$\frac{m_1}{m_2} = \frac{w_{B,2} - w_{B,0}}{w_{B,0} - w_{B,1}} \tag{10.4}$$

where

$$m_1 + m_2 = m_0 \tag{10.5}$$

What would be the masses of the phases for a system containing 0.050 0 kg of A and 0.050 0 kg of B in equilibrium if $w_{B,1} = 0.300$ and $w_{B,2} = 0.855$?

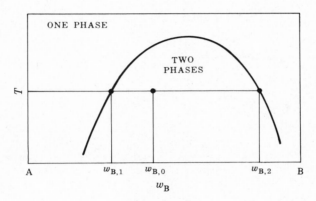

Fig. 10-3

The bulk composition of the system is

$$w_{B,0} = \frac{0.050\ 0\ \text{kg}}{0.050\ 0\ \text{kg} + 0.050\ 0\ \text{kg}} = 0.500$$

Using (10.4) gives

$$\frac{m_1}{m_2} = \frac{0.855 - 0.500}{0.500 - 0.300} = 1.78$$

Using this result with (10.5), where $m_0 = 0.100\ 0$ kg, gives $m_1 = 0.064\ 0$ kg for phase 1 and $m_2 = 0.036\ 0$ kg for phase 2.

10.7   LIQUID-LIQUID AND LIQUID-VAPOR DIAGRAMS

Typical diagrams for partially miscible liquids are shown in Fig. 10-4. The *lower consolute temperature* (point b) and the *upper consolute temperature* (point a) represent the temperatures below which or above which, respectively, only one phase will exist regardless of the composition. Above or below these temperatures, respectively, two phases might exist, depending on the composition of the mixture.

EXAMPLE 10.6.   Consider the typical liquid-vapor diagram for completely miscible liquids shown in Fig. 10-5. There are two one-phase areas and one two-phase area. Note that pressure is constant for this diagram. Determine the number of *theoretical equivalent plates* (TEP) in a distillation column necessary to separate pure B from a mixture having an original composition x_1.

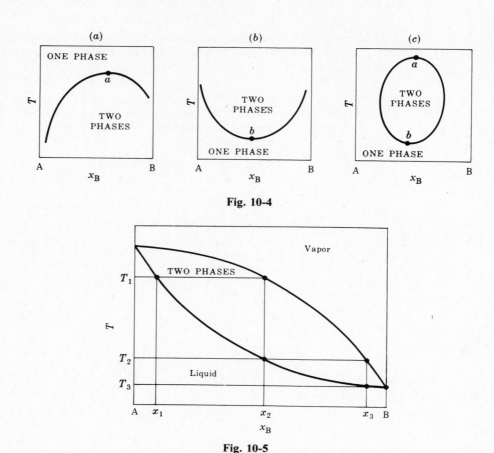

Fig. 10-4

Fig. 10-5

If a solution having composition x_1 is heated to T_1, the vapor phase in equilibrium with it has the composition x_2, somewhat richer in B. This vapor is cooled to T_2 and condenses to liquid with composition x_2. The new vapor of composition x_3 is allowed to form above the solution. The new vapor is cooled to T_3 and condensed, yet new vapor is allowed to form, etc., until the more volatile component is isolated. A TEP can be defined as a simple distillation step in which an equilibrium between the solution and vapor is established and the vapor is condensed to a liquid of different composition. The three TEPs shown in Fig. 10-5 will perform the desired separation.

10.8   SOLID-LIQUID DIAGRAMS

There are several features that may appear on these diagrams, such as partial, complete, or no mutual solubility of the solids; congruent, incongruent, or no compound formation; and solid-solid phase transitions.

Generally these diagrams are determined by cooling-curve measurements at various concentrations. The cooling curves for a pure compound or at the eutectic composition will consist of a plateau or "arrest." At other compositions the cooling curve will consist of (1) a "break" where the solid begins to solidify and the liquid changes composition and (2) an "arrest" at which the remaining liquid solidifies at the eutectic composition.

EXAMPLE 10.7.   Compare the phase diagrams shown in Figs. 10-6 and 10-7.

Figure 10-6 illustrates complete miscibility of the materials in the solid phase. There are two one-phase areas and one two-phase area. This type of diagram results from the ability of one substance to substitute freely for the other in the crystal lattice because of similarity in size of molecules (atoms or ions), charge (if any), etc. Horizontal tie lines are used in the two-phase area.

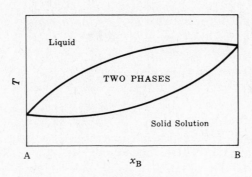

Fig. 10-6

Figure 10-7 illustrates two substances that are completely immiscible in the solid phase. There are four two-phase areas and one one-phase area. The horizontal tie lines indicate that in area 1 the two phases in equilibrium will be solid A and liquid; in area 2, solid B and liquid; and in areas 3 and 4, solid A and solid B, with one of the materials undergoing a phase transition to a second solid state.

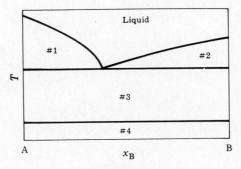

Fig. 10-7

EXAMPLE 10.8. The temperature at which pure solid i is in equilibrium with liquid having a concentration x_i is given by

$$\ln x_i = \frac{\Delta_{\text{fus}} H(i)}{R} \left(\frac{1}{T_{\text{mp}}(i)} - \frac{1}{T} \right) \tag{10.6}$$

where $T_{\text{mp}}(i)$ is the melting point of pure i. Given that $\Delta_{\text{fus}} H(\text{A})$ is $1\,500.\ \text{J mol}^{-1}$ and $t_{\text{mp}}(\text{A}) = 400.\ ^\circ\text{C}$, find the solubility of B in A at $350.\ ^\circ\text{C}$.

Using (10.6) gives

$$\ln x_{\text{A}} = \frac{1\,500\ \text{J mol}^{-1}}{8.314\ \text{J K}^{-1}\ \text{mol}^{-1}} \left(\frac{1}{673\ \text{K}} - \frac{1}{623\ \text{K}} \right) = -0.021\,5$$

or $x_{\text{A}} = 0.978\,7$. Recognizing that $x_{\text{B}} = 1 - x_{\text{A}}$ gives $x_{\text{B}} = 0.021\,3$.

EXAMPLE 10.9. Predict the eutectic temperature and composition for a binary solid-liquid system if $\Delta_{\text{fus}} H(i)/(\text{kJ mol}^{-1}) = 0.500$ and 1.000 and $t_{\text{mp}}(i)/(^\circ\text{C}) = 400.$ and $600.$, respectively, for A and B.

Substituting the data into (10.6) gives

$$\ln x_{\text{A}} = \frac{500.}{8.314} \left(\frac{1}{673} - \frac{1}{T} \right) \qquad \ln x_{\text{B}} = \frac{1\,000.}{8.314} \left(\frac{1}{873} - \frac{1}{T} \right)$$

which upon solving simultaneously with $x_{\text{A}} + x_{\text{B}} = 1$ gives $x_{\text{A}} = 0.625$ and $T = 107.5\ \text{K}$.

Phase Diagrams for Three-Component Systems

10.9   INTRODUCTION

For a three-component system (10.1) becomes

$$f = 5 - p \qquad (10.7)$$

For a system having only one phase, the phase diagram must illustrate four variables, which is difficult. For this reason P and T are fixed for a given diagram and triangular graph paper is used to illustrate the system in terms of the remaining variables—two of the three concentrations—as in Fig. 10-8. The tie lines, which are experimentally determined, must be specified on the phase diagram in the two-phase regions because they are no longer horizontal as in two-component diagrams. Tie lines are not used in the three-phase regions.

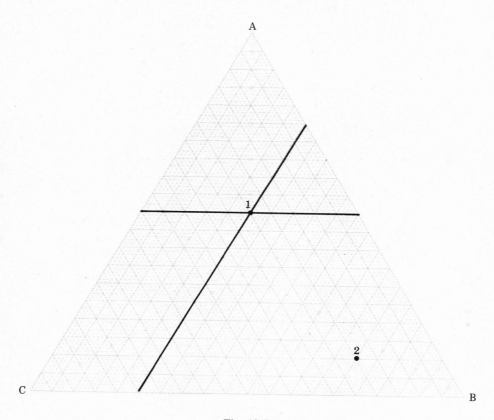

Fig. 10-8

EXAMPLE 10.10.   Construct the point $x_A = 0.50$, $x_B = x_C = 0.25$ in Fig. 10-8.

Starting with $x_A = 0.50$, the point will be located one-half the distance between side BC and vertex A. This is indicated on the graph by a heavy horizontal line at $x_A = 0.50$. Similarly, for x_B, the point will be located on a line one-fourth the distance from side AC to vertex B, which line is also drawn heavy. The intersection of these two lines (point 1) is the desired point. As a check, point 1 has $x_C = 0.25$, since the point lies one-fourth the distance from side AB to vertex C.

Solved Problems

PHASE RULE

10.1. Determine the number of components in the following systems: (a) an aqueous solution of sugar; (b) an aqueous solution of CH_3COOH; (c) an aqueous solution of KCl; (d) a mixture of $CaCO_3(s)$, $CaO(s)$, and $CO_2(g)$; and (e) $Fe(s) + H_2O(g) \rightleftharpoons FeO(s) + H_2(g)$.

(a) Here $c = 2$; the amounts of H_2O and sugar must be known to define the system.

(b) Here $c = 2$; the amounts of H_2O and CH_3COOH will define the system even though other species, such as $H^+(aq)$ and $CH_3COO^-(aq)$, are present.

(c) Here $c = 2$; the amounts of H_2O and KCl will define the system even though other species, such as $K^+(aq)$, $Cl^-(aq)$, $H^+(aq)$, and $OH^-(aq)$, are present.

(d) Here $c = 2$ or 3, depending on the method of preparation and final status of the system. If simply a mixture, all three components must be specified. If a mixture produced by the decomposition of $CaCO_3$, only two components need be specified because the third can be calculated using the stoichiometry of the reaction. The equilibrium expression cannot be used to reduce the number of components needed, because it contains only one useful term, $K = a(CO_2)$.

(e) Here $c = 3$; the chemical reaction implies that equilibrium exists, so only one gaseous component must be specified along with the two solid species, because the second may be calculated from $K = a(H_2)/a(H_2O)$.

10.2. Determine f for the systems described in Problem 10.1.

For systems (a), (b), and (c), $p = 1$ and $c = 2$ for the aqueous solutions, giving $f = 3$. The usual choice of variables is T, P, and one concentration.

For both cases of system (d), $p = 3$, giving $f = c - 1$. For the case where $c = 3$, $f = 2$, and knowing T and P is sufficient to fix the system. For $c = 2$, $f = 1$, and only T or P needs to be defined.

For system (e), $p = 3$ and $c = 3$, giving $f = 2$. Usually T and P are used to define the system, but other possibilities, such as T and the concentration of one of the gases, would work for a specific application.

PHASE DIAGRAMS FOR ONE-COMPONENT SYSTEMS

10.3. Determine f at point c, along the line between points a and b, and at point 1 for the hypothetical phase diagram shown in Fig. 10-2.

At point c, $p = 3$, which upon substitution into (10.2) gives $f = 0$; T and P are fixed. Along a line there are two phases present, in this case liquid and gas, giving $p = 2$ and $f = 1$; either T or P must be specified. At point 1, $p = 1$ and $f = 2$; both T and P must be given.

10.4. Describe the changes in the system shown in Fig. 10-2 for (a) isobarically heating from point 1, (b) isobarically heating from point 2, (c) isothermally compressing from point 3, (d) isothermally compressing from point 4, and (e) isothermally and isobarically removing heat at point b.

For the isobaric heatings, horizontal lines are drawn from the points, and for the isothermal compressions, vertical lines are drawn. The changes are:

(a) Heating β-solid until reaching the line at which sublimation occurs (isothermally) and then heating the gas.

(b) Heating β-solid until reaching the line between points c and e at which α-solid is isothermally formed, heating α-solid until reaching the line between points b and e at which isothermal melting occurs, heating the liquid until reaching the line between points a and b at which isothermal boiling occurs, and heating the gas.

(c) Pressure increases for the gas.

(d) Pressure increases for the gas until reaching the line between points a and b, at which condensation occurs and the liquid undergoes further compression.

(e) The amount of gas phase decreases as more of the condensed phases is formed.

PHASE DIAGRAMS FOR TWO-COMPONENT SYSTEMS

10.5. Consider a liquid-liquid system containing 10.00 kg of A and 5.00 kg of B at a temperature such that two phases are present, one with $w_{B,1} = 0.100$ and the other with $w_{B,2} = 0.400$. Calculate the masses of the two phases in equilibrium.

The system has a bulk composition ($w_{B,0}$) of

$$w_{B,0} = \frac{5.00 \text{ kg}}{5.00 \text{ kg} + 10.00 \text{ kg}} = 0.333$$

Using (10.4) gives

$$\frac{m_1}{m_2} = \frac{0.400 - 0.333}{0.333 - 0.100} = 0.29$$

which, solved simultaneously with (10.5), where $m_0 = 15.00$ kg, gives $m_1 = 3.4$ kg for the A-rich phase and $m_2 = 11.6$ kg for the B-rich phase.

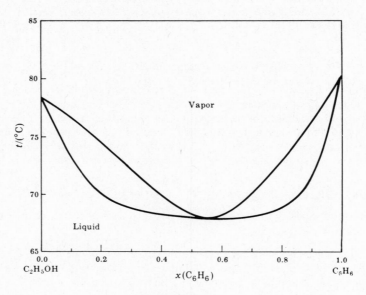

Fig. 10-9

10.6. Consider the liquid-vapor phase diagram (Fig. 10-9) for the C_6H_6/C_2H_5OH binary system. If a sample from a packed distillation tower contained $x(C_6H_6) = 0.10$ and a second sample taken from the tower the proper distance equivalent to one TEP contained $x(C_6H_6) = 0.25$, what is the operating temperature of this part of the tower? Discuss the separation of C_6H_6 from an alcohol-rich solution using distillation.

A horizontal line corresponding to $x(C_6H_6) = 0.10$ on the liquid line and $x(C_6H_6) = 0.25$ on the vapor line lies at 73 °C.

As the sketches representing the TEPs are drawn in the two-phase area from the alcohol-rich side toward the benzene-rich side, a constant-boiling mixture (*azeotrope*) is reached at 67.8 °C and $x(C_6H_6) = 0.552$, at which the liquid and vapor above it have identical composition and further separation by distillation is not possible. If the total pressure of the system is changed or a third component is added, the composition and temperature of the azeotrope will change and further separation is possible.

10.7.   Identify the phases present in the numbered areas of Fig. 10-10.

A *congruent-melting compound* is formed at $x_B = 0.667$, which means that there is 2 mol of B for every mole of A; thus the empirical formula of the compound is AB_2. Horizontal tie lines indicate that in area 1, solid A and liquid are in equilibrium; in areas 2 and 4, liquid and solid AB_2; in area 3, solid A and solid AB_2; in area 5, solid B and liquid; and in area 6, solid B and solid AB_2.

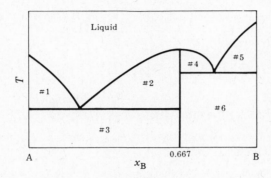

Fig. 10-10

10.8.   Identify the phases present in the numbered areas of Fig. 10-11.

An *incongruent-melting compound* is formed at $x_B = 0.250$, which means it has an empirical formula of A_3B. Horizontal tie lines indicate that in area 1, solid A and liquid are in equilibrium; in 2, solid A and solid A_3B; in 3, liquid and solid A_3B; in 4, solid B and A_3B; and in 5, solid B and liquid.

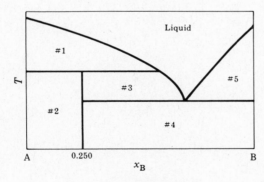

Fig. 10-11

10.9.   Figure 10-12 is a diagram for two substances that show partial miscibility with each other in the solid state. This type of diagram results from the ability of one substance to penetrate the empty spaces, etc., of the lattice of the other. There are limitations on this type of solubility, and the maximum is indicated by the curved lines enclosing areas 1 and 4. Identify the phases present in the numbered areas.

Horizontal tie lines indicate that in area 2, liquid and solid solution of B in A (saturated) are in equilibrium; in 3, liquid and solid solution of A in B (saturated); and in 5, solid solution of B in A (saturated) and solid solution of A in B (saturated). Areas 1 and 4 are single phases consisting of an unsaturated solid solution of B in A and of A in B, respectively.

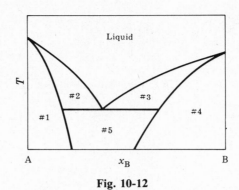

Fig. 10-12

PHASE DIAGRAMS FOR THREE-COMPONENT SYSTEMS

10.10. Describe the changes in the system shown in Fig. 10-13 as salt B is added to a solution of composition given by point 1.

    As pure B is added, the system becomes richer in B and less rich in A and C, as indicated by the line drawn from point 1 to vertex B. As B is added, it dissolves until sufficient B is present that the bulk concentration reaches point 2, at which solid B becomes a second phase in equilibrium with a solution phase of composition given by point 2. Further addition of B changes only the relative amounts of each phase, not the composition.

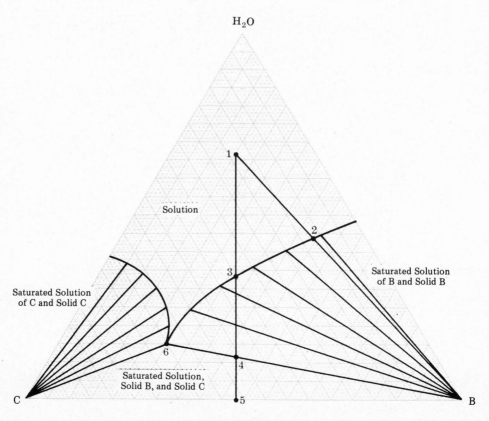

Fig. 10-13

Supplementary Problems

PHASE RULE

10.11. Consider the system in Fig. 10-14. Find p, and describe the various phases.

> *Ans.* $p = 4$;   (1) solid NaCl,   (2) aqueous solution of NaCl that contains a small amount of air,   (3) ice, and   (4) air that contains small amounts of H_2O vapor and NaCl vapor

Fig. 10-14

10.12. Determine c for the following ststems: (a) Br_2 dissolved in CCl_4; (b) a mixture of $N_2(g)$, $H_2(g)$, and $NH_3(g)$; (c) $2KClO_3(s) \rightleftharpoons 3O_2(g) + 2KCl(s)$.   *Ans.* ($a$) 2;   ($b$) 1, 2, or 3;   ($c$) 2

10.13. Determine f for the systems described in Problem 10.12.   *Ans.* (a) 3;   (b) 2, 3, or 4;   (c) 1

10.14. Can there be a "quadruple" point on a phase diagram for a one-component system?

> *Ans.*   No; $f = 1 - 4 + 2 = -1$

PHASE DIAGRAMS FOR ONE-COMPONENT SYSTEMS

10.15. Determine f at points b and e, along the line between points b and c, and at point 4 for the hypothetical phase diagram shown in Fig. 10-2.   *Ans.*   0, 0, 1, 2

10.16. Describe the changes in the system shown in Fig. 10-2 for (a) an isothermal expansion from point 1, (b) an isothermal compression from point 2, (c) an isobaric cooling from point 3, and (d) adding heat isothermally and isobarically at point e.

> *Ans.*   (a) β-solid expanding, subliming, gas expanding;   (b) β-solid compressing, changing to γ-solid, γ-solid compressing;   (c) gas cooling, condensing to α-solid, α-solid cooling and changing to β-solid, β-solid cooling;   (d) increasing the amount of liquid present compared to the amounts of α- and β-solids.

10.17. (a) Water has a solid-liquid line similar to that between the points b and e in Fig. 10-2. Describe what happens to ice if pressure is applied to point 5 isothermally. (b) What happens to dry ice, which has a solid-liquid line similar to that between points f and g in Fig. 10-2, if pressure is applied at point 6 isothermally? (c) What happens to the system in Fig. 10-2 if pressure is applied isothermally at point 7 or the system is cooled isobarically at point 7?

> *Ans.*   (a) melts;   (b) nothing;   (c) α-solid changes to β-solid

10.18. Consider the phase diagram for sulfur shown in Fig. 10-15. The diagram does not show the various transitions that occur in the liquid state as the S_8 molecules become fragmented. (a) What is the stable form of S under room conditions? (b) Suppose a sample of S were dissolved in CS_2 (bp = 46.3 °C) and the solvent evaporated. What would be the stable form of the S? (c) Suppose a sample of S were heated in boiling water for a short time and removed quickly. What would be observed? (d) What would be observed if rhombic S were heated rather quickly? (e) What would be observed if molten S at 115 °C were

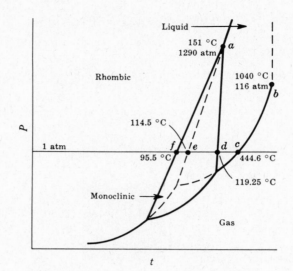

Fig. 10-15

poured into boiling water? (f) If molten S at 115 °C were allowed to cool to room temperature and allowed to remain at room temperature for a week, what change would occur?

Ans.   (a) Rhombic;   (b) rhombic;   (c) perhaps a little monoclinic would form;   (d) melt at 114.5 °C;   (e) monoclinic would form;   (f) change to rhombic

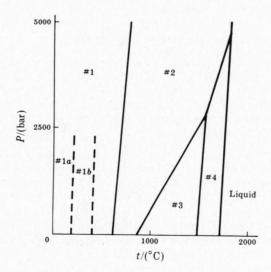

Fig. 10-16

10.19. SiO_2 is polymorphic (see Fig. 10-16). Two polymorphic crystalline forms are said to be *enantiotropic* if their interconversion takes place reversibly at a definite temperature and pressure and *monotropic* if their interconversion occurs irreversibly with the inherently unstable form going into the inherently stable form. Classify the following possible transitions $1 \rightarrow 2$, $1a \rightarrow 3$, $1b \rightarrow 4$, $2 \rightarrow 3$, $2 \rightarrow 4$, $3 \rightarrow 4$, $1a \rightarrow 1$, and $1b \rightarrow 1$.

Ans.   The last two are monotropic, the rest are enantiotropic.

PHASE DIAGRAMS FOR TWO-COMPONENT SYSTEMS

10.20. (*a*) Prepare a liquid-liquid phase diagram for the system of partially miscible liquids A and B, where w_B values at various temperatures are

$t/(°C)$	0	10	20	30	40
$w_{B,1}$	30	37	45	53	64
$w_{B,2}$	94	90	87	84	80

(*b*) What is the upper consolute temperature?

(*c*) If a mixture containing 0.060 0 kg of B and 0.040 0 kg of A were present at 25 °C, what is the composition of the two phases in equilibrium, and how much of each phase is present?

Ans. (*a*) Diagram will look like Fig. 10-4*a*; (*b*) 46 °C; (*c*) 0.070 2 kg of the phase having composition 49% B and 0.029 8 kg of the phase having composition 86% B

10.21. (*a*) Prepare a liquid-vapor phase diagram for the system of completely miscible liquids A and B from the following data:

$t/(°C)$	0	10	20	30	40	50	60
x_B of liquid	0.00	0.39	0.62	0.77	0.87	0.95	1.00
x_B of vapor	0.00	0.09	0.21	0.35	0.55	0.76	1.00

(*b*) How many TEPs would be required to separate 99% pure A from a solution originally containing $x_B = 0.80$?

Ans. (*a*) The diagram will look like the reverse of Fig. 10-5; (*b*) 3–4 plates

10.22. (*a*) Construct the liquid-vapor phase diagram for the H_2O/D_2O binary system from the following data at 1.01 bar.

$t/(°C)$	100.00	100.35	100.71	101.06	101.41
$x(D_2O)$ for liquid	0.000	0.246	0.505	0.752	1.000
$x(D_2O)$ for vapor	0.000	0.237	0.493	0.743	1.000

(*b*) The natural abundance of D_2O is 1 part in 6 900, which may be increased to $x(D_2O) = 0.15$ by an enrichment technique using H_2S. From the phase diagram, determine the approximate change in concentration that one TEP would give.

(*c*) Assuming that a technique that operates in the reverse direction of distillation exists, give an estimate of the number of TEPs necessary to prepare a solution having $x(D_2O) = 0.95$ from the enrichment product.

Ans. (*a*) The diagram has a very narrow two-phase area;
(*b*) one TEP averages $\Delta x(D_2O) = 0.01$; (*c*) 80

10.23. Identify the phase present in the liquid-vapor diagram shown in Fig. 10-17. This type of system is useful for steam distillations.

Ans.   1, vapor;   2, vapor and liquid solution of A saturated with B;   3, liquid solution of B in A;   4, vapor and liquid solution of B saturated with A;   5, liquid solution of A in B;   6, two liquid solutions of B saturated with A and liquid solution of A saturated with B

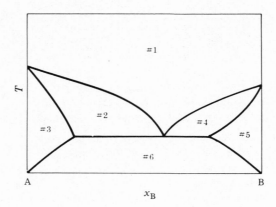

Fig. 10-17

10.24. Construct a liquid-vapor diagram for C_2H_5OH/H_2O from the following data and the fact that an azeotrope (constant-boiling mixture) is formed at $x(C_2H_5OH) = 0.89$ at 79 °C.

$t/(°C)$	100	95	90	85	80	79
$x(C_2H_5OH)$ in liquid	0.00	0.02	0.05	0.13	0.45	1.00
$x(C_2H_5OH)$ in vapor	0.00	0.18	0.33	0.47	0.68	1.00

Find the number of TEPs required to separate 190 proof alcohol (the azeotrope concentration of 95 wt %) from "corn squeezins" with $x(C_2H_5OH) = 0.10$.   *Ans.*   3 plates

10.25. (*a*)   Prepare a solid-solid phase diagram for the completely miscible substances A and B from the following data:

$t/(°C)$	60	70	80	90	100
x_B for liquid	0.00	0.19	0.42	0.65	1.00
x_B for solid	0.00	0.58	0.78	0.90	1.00

(*b*)   If 1.00 kg of a solution having $x_B = 0.50$ were just melted, what would be the composition of the liquid phase in equilibrium with it?

Ans.   (*a*) The diagram will look like Fig. 10-6;   (*b*) $x_B = 0.14$

10.26. (*a*) Construct a solid-liquid phase diagram for substances A and B that are mutually insoluble in the solid state from the following data:

x_B	0.00	0.10	0.20	0.30	0.40	0.50	0.60	0.70	0.80	0.90	1.00
$t_{break}/(°C)$		53	46		46		43	42	64	75	
$t_{arrest}/(°C)$	60	39	40	40	40	50	31	30	30	30	80

(*b*) What phases will be in equilibrium at $x_B = 0.90$ and 70 °C?

Ans. (*a*) The diagram will look like the reverse of Fig. 10-9; (*b*) compound AB formed, eutectics at $x_B = 0.30$ and 0.66 at 40 °C and 30 °C, solid B and liquid having $x_B = 0.85$

10.27. Identify the phases present in the solid-liquid diagram shown in Fig. 10-18.

Ans. 1, liquid; 2, solid solution of A_2B in A; 3, liquid and solid solution of A_2B in A (saturated); 4, liquid and solid A_2B; 5, solid A_2B and solid solution of A_2B in A (saturated); 6, liquid and solid AB; 7, solid A_2B and solid AB; 8, liquid and solid AB; 9, liquid and solid B; 10, solid AB and solid B; 11, solid A and solid A_2B; 12, solid AB and solid B

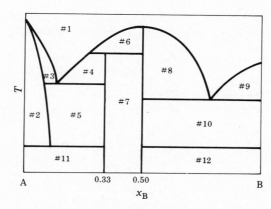

Fig. 10-18

PHASE DIAGRAMS FOR THREE-COMPONENT SYSTEMS

10.28. What are the coordinates of point 2 in Fig. 10-8?

Ans. $x_A = 0.10$, $x_B = 0.70$, $x_C = 0.20$

10.29. (*a*) What is the composition of the two phases in equilibrium if a three-component system having the bulk composition of $w_A = 0.205$, $w_B = 0.495$, and $w_C = 0.300$ obeyed the phase diagram shown in Fig. 10-19?

(*b*) What is the mass of each phase for 1.000 kg of mixture?

(*c*) What mass of A is in each phase? Hint: Use a ruler to measure $w_2 - w_0$ and $w_0 - w_1$.

Ans. (*a*) $m_2/m_1 = 2.54$

(*b*) 0.718 kg of $w_A = 0.215$, $w_B = 0.630$, and $w_C = 0.155$;
0.282 kg of $w_A = 0.180$, $w_B = 0.169$, and $w_C = 0.660$

(*c*) 0.051 kg of A in the C-rich phase and 0.154 kg of A in the B-rich phase

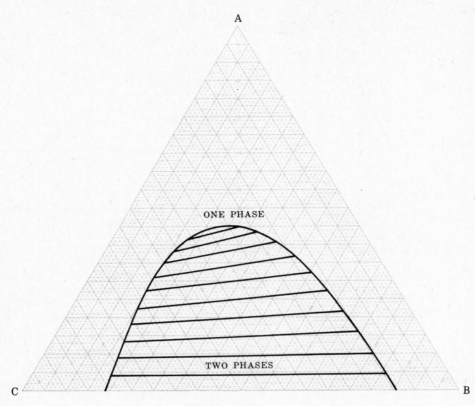

Fig. 10-19

10.30. Describe the changes in the system in Fig. 10-13 as water is evaporated from the composition given by point 1.

Ans. A solution until point 3 is reached, where two phases are in equilibrium: solid B and solution saturated with B. These phases continue until the bulk composition reaches point 4, where three phases are in equilibrium: solid A, solid B, and solution saturated with B and C having composition given by point 6. These phases continue until all the water is gone, giving solid A and solid B, a two-component system, with composition given by point 5.

Chapter 11

Solutions

Concentrations

11.1 INTRODUCTION

A *solution* is a homogeneous mixture of two or more substances. The *solvent*, indicated by the subscript 1 or A, is usually the more abundant component, although it may be a minor component chosen for convenience (e.g., if the resulting solution is a liquid, the solvent may be chosen as the liquid component). The *solutes*, indicated by the subscripts 2, 3, ... or B, C, ..., are the substances dissolved in the solvent. Because the solvent and solute in a binary mixture can be solid, liquid, or gas, there are nine possible combinations, and all are known to exist.

11.2 CONCENTRATION UNITS

The amount of solute in a solution can be expressed in terms of two systems of units. The "group A" units specify the amount of solute in a given volume of solution, such as $(g \text{ solute})(dm^3 \text{ soln})^{-1} = (kg \text{ solute})(m^3 \text{ soln})^{-1}$ and molarity (C), $(mol \text{ solute})(dm^3 \text{ soln})^{-1} = M$. "Group B" units specify the amount of solute for a given mass of solvent or solution, such as mass or weight fraction (w_2), $(g \text{ solute})(g \text{ soln})^{-1}$; $(g \text{ solute})(kg \text{ solvent})^{-1}$; molality (m), $(mol \text{ solute})(kg \text{ solvent})^{-1}$; and mole fraction (x_2). The advantage of "group A" units is the ease of solution preparation, and that of "group B" units is the temperature independence. To convert from one group to the other requires knowledge of the solution density ρ.

EXAMPLE 11.1. A solution containing 13.00 g NaOH ($M = 40.01 \text{ g mol}^{-1}$) and 87.00 g H_2O has a density of $1.142\,1 \times 10^3 \text{ kg m}^{-3}$. Find (a) $w(NaOH)$, (b) $w(H_2O)$, (c) m, (d) $x(NaOH)$, (e) $x(H_2O)$, (f) $(g \text{ solute})(dm^3 \text{ soln})^{-1}$, and (g) C.

This type of conversion problem is most easily solved by choosing a "basis" for all calculations, e.g., 100.00 g solution ($= 13.00$ g NaOH and 87.00 g H_2O).

(a)
$$w(NaOH) = \frac{m(NaOH)}{m(NaOH) + m(H_2O)} = \frac{13.00 \text{ g}}{13.00 \text{ g} + 87.00 \text{ g}} = 0.130\,0$$

(b) $w(H_2O) = 1 - 0.130\,0 = 0.870\,0$

(c) The numbers of moles of solute and solvent, respectively, are

$$n(NaOH) = \frac{13.00 \text{ g}}{40.01 \text{ g mol}^{-1}} = 0.325 \text{ mol}$$

$$n(H_2O) = \frac{87.00 \text{ g}}{18.015 \text{ g mol}^{-1}} = 4.83 \text{ mol}$$

and the mass of the solvent is 8.700×10^{-2} kg, giving a molality of

$$\frac{0.325 \text{ mol}}{8.700 \times 10^{-2} \text{ kg}} = 3.73 \text{ mol kg}^{-1}$$

(d)
$$x(NaOH) = \frac{0.325}{0.325 + 4.83} = 0.063\,1$$

213

(e) $x(H_2O) = 1.000\ 0 - 0.063\ 1 = 0.936\ 9$

(f)   The volume of solution corresponding to the basis is

$$V = \frac{(100.00\ \text{g})[(10^{-3}\ \text{kg})/(1\ \text{g})]}{1.142\ 1 \times 10^3\ \text{kg m}^{-3}} = 87.56 \times 10^{-6}\ \text{m}^3 = 87.56 \times 10^{-3}\ \text{dm}^3$$

and $(\text{g solute})(\text{dm}^3\ \text{soln})^{-1} = \dfrac{13.00\ \text{g}}{87.56 \times 10^{-3}\ \text{dm}^3} = 148.5\ \text{g NaOH} \ (\text{dm}^3\ \text{soln})^{-1}$

(g)   The molarity is

$$C = \frac{0.325\ \text{mol NaOH}}{87.56 \times 10^{-3}\ \text{dm}^3} = 3.71\ \text{M}$$

11.3   DILUTIONS

The volume of the concentrated solution (V_{conc}) required to prepare a volume of dilute solution (V_{dil}) is given by

$$V_{conc} = V_{dil} \left(\frac{C_{dil}}{C_{conc}} \right) \tag{11.1}$$

where C_i are the concentrations of the solutions expressed in "group A" units.

11.4   HENRY'S LAW

At a fixed temperature, the amount of a gas dissolved in a given quantity of solvent is proportional to the partial pressure of the gas above the solution, or, in equation form,

$$P_2 = x_2 K_2 \tag{11.2}$$

where K_2 is the *Henry's law constant.*

EXAMPLE 11.2.   Given $K_2 = 4.40 \times 10^4$ bar for a solution of $O_2(g)$ in water at 25 °C, find the solubility of oxygen under room conditions.

Under room conditions, assuming air to be 20% O_2, $P_2 = (0.20)(1.01\ \text{bar}) = 0.20$ bar. Using (11.2) gives

$$x_2 = \frac{P_2}{K_2} = \frac{0.20\ \text{bar}}{4.40 \times 10^4\ \text{bar}} = 4.5 \times 10^{-6}$$

Assuming that $n_2 + n_1 \approx n_1$ for this very dilute solution,

$$x_2 = \frac{n_2}{n_1 + n_2} \approx \frac{n_2}{n_1} = 4.5 \times 10^{-6}$$

which upon substitution of $n_1 = 5.55$ mol for 100 g of water gives $n_2 = 2.5 \times 10^{-5}$ mol, or 8.0×10^{-4} g O_2 in 100 g H_2O.

11.5   DISTRIBUTION COEFFICIENTS

At a given temperature, the concentrations of a solute distributed between two phases are related to each other by the *Nernst distribution law*

$$\frac{C_2'}{C_2} = k_d \tag{11.3}$$

where k_d is the *distribution coefficient.* It is assumed that the solute in both phases is chemically the same, i.e., no association taking place, etc.

EXAMPLE 11.3. Given that the fraction of unextracted solute remaining after n extractions is given by

$$f_n = \left(1 + k_d \frac{V'}{V}\right)^{-n}$$

where V' is the volume of the extracting liquid and V is the volume containing the unextracted solute, find n for a 99.9% extraction of a solute having $k_d = 10$ using equal volumes of solvent.

Taking logarithms and solving for n gives

$$n = \frac{-\log f_n}{\log(1 + k_d V'/V)} = \frac{-\log(0.001)}{\log[1 + (10)(1/1)]} = \frac{3}{1.042} \approx 3$$

Thermodynamic Properties of Solutions

11.6   IDEAL SOLUTIONS

An *ideal solution* is one for which

$$\Delta_{\text{soln}} H = 0 \qquad \text{and} \qquad V_{\text{soln}} = \sum_{i}^{\text{components}} V_i$$

and for which *Raoult's law*,

$$P_i = P_i^\circ x_i \tag{11.4}$$

is obeyed, where P_i is the partial pressure of component i in the vapor phase above the solution, P_i° is the vapor pressure of the pure component i at that temperature, and x_i is the mole fraction of the component in the solution.

11.7   VAPOR PRESSURE

For an ideal solution, the partial pressure of each component is given by (*11.4*), and the total pressure above the solution by

$$P_{\text{soln}} = \sum_{i}^{\text{components}} P_i \tag{11.5}$$

Those solutions with P_i and P_{soln} less than as predicted by (*11.4*) and (*11.5*) are said to show *negative deviations*, and those greater are said to show *positive deviations*.

EXAMPLE 11.4. Given $P^\circ(\text{HNO}_3) = 7\,600\,\text{Pa}$ and $P^\circ(\text{H}_2\text{O}) = 3\,167\,\text{Pa}$ at 25 °C, prepare plots of $P(\text{HNO}_3)$, $P(\text{H}_2\text{O})$, and P_{soln} against $x(\text{HNO}_3)$ assuming the solutions to be ideal. On the same graph, plot the following data for the real solutions:

$w(\text{HNO}_3)$	0.20	0.25	0.30	0.35	0.40	0.45	0.50	0.55	0.60	0.65	0.70	0.80	0.90	1.00
$P(\text{HNO}_3)/(\text{Pa})$					16	31	52	88	161	309	547	1 400	3 606	7 600
$P(\text{H}_2\text{O})/(\text{Pa})$	2 750	2 560	2 370	2 160	1 950	1 690	1 430	1 200	1 030	880	730	430	130	

Describe the system. Find $x(\text{HNO}_3)_{\text{vap}}$ at $x(\text{HNO}_3) = 0.500$ if the solution were ideal and for the real solution.

The plots of $P(\text{HNO}_3)_{\text{ideal}}$ and $P(\text{soln})_{\text{ideal}}$ against $x(\text{HNO}_3)$ are linear (see Fig. 11-1), going from 0 to $P^\circ(\text{HNO}_3)$ and from $P^\circ(\text{H}_2\text{O})$ to $P^\circ(\text{HNO}_3)$, respectively. Because the data for the real solutions are given in terms of $w(\text{HNO}_3)$, they must be converted to $x(\text{HNO}_3)$ before plotting. As a sample calculation, for the solution with $w(\text{HNO}_3) = 0.200$, on a basis of 100.0 g of solution, there are 20.0 g HNO_3 and 80.0 g H_2O or

$$n(\text{HNO}_3) = \frac{20.0\,\text{g}}{63.01\,\text{g mol}^{-1}} = 0.317\,\text{mol} \qquad n(\text{H}_2\text{O}) = \frac{80.0\,\text{g}}{18.015\,\text{g mol}^{-1}} = 4.44\,\text{mol}$$

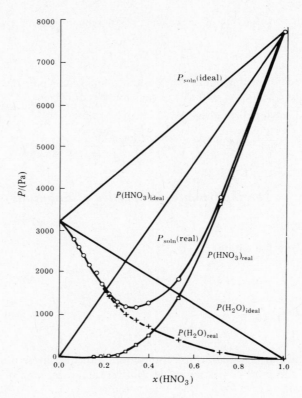

Fig. 11-1

giving
$$x(HNO_3) = \frac{0.317}{0.317 + 4.44} = 0.066\ 6$$

The plots of actual data (see Fig. 11-1) show large negative deviations from the plots of (*11.4*) and (*11.5*). For an ideal solution, using (*11.4*) and (*11.5*) at $x(HNO_3) = 0.500$ gives

$$P(HNO_3) = (7\ 600\ Pa)(0.500) = 3\ 800\ Pa$$

$$P(H_2O) = (3\ 167)(0.500) = 1\ 584\ Pa$$

$$P_{soln} = 3\ 800 + 1\ 584 = 5\ 400\ Pa$$

and the gaseous mole fractions given by (*1.10*) are

$$x(HNO_3)_{vap} = \frac{3\ 800}{5\ 400} = 0.70 \qquad x(H_2O)_{vap} = 1 - 0.70 = 0.30$$

For the real solution, the values of P_i read from Fig. 11-1 are 1 160 Pa for HNO_3 and 470 Pa for H_2O, giving $P_{soln} = 1\ 630\ Pa$ and

$$x(HNO_3)_{vap} = \frac{1\ 160}{1\ 630} = 0.71 \qquad x(H_2O)_{vap} = 1 - 0.71 = 0.29$$

The composition of the vapor for the real solution does not differ greatly from that for the ideal solution in this case, but the total pressure is much lower.

11.8 ΔS, ΔH, AND ΔG OF MIXING

For an ideal solution,

$$\Delta_{mix}S = \Delta_{soln}S = -R \sum_i^{\text{components}} n_i \ln x_i \qquad (5.13a)$$

$$\Delta_{mix}H = \Delta_{soln}H = 0 \qquad (11.6)$$

$$\Delta_{mix}G = \Delta_{soln}G = RT \sum_i^{\text{components}} n_i \ln x_i \qquad (11.7a)$$

For 1 mol of solution, $(5.13a)$ and $(11.7a)$ can be written as

$$\Delta_{soln}S = -R \sum_i^{\text{components}} x_i \ln x_i \qquad (5.13b)$$

$$\Delta_{soln}G = RT \sum_i^{\text{components}} x_i \ln x_i \qquad (11.7b)$$

EXAMPLE 11.5.   Why may real solutions show positive and negative deviations from (5.13), (11.6), and (11.7)?

Usually, negative deviations are caused by association between the solvent and solute molecules, giving a negative $\Delta_{soln}H$ and a lower value for $\Delta_{soln}S$, which results in a more negative value of $\Delta_{soln}G$. Positive deviations are usually due to the dissociation of an associated solution component, which results in a positive $\Delta_{soln}H$ and a higher value for $\Delta_{soln}S$, giving a less favorable $\Delta_{soln}G$.

11.9   ACTIVITIES AND ACTIVITY COEFFICIENTS

If a solution consists of two volatile liquids, the activity coefficient (f_i) is given by

$$f_i = \frac{P_i}{x_i P_i^\circ} \qquad (11.8)$$

and the activity (a_i) by

$$a_i = f_i x_i \qquad (11.9)$$

For a solution containing two volatile liquids for which it is not possible to vary the mole fractions of both components up to unity, or for a gaseous solute, the activity coefficient for the solvent is given by (11.8) and for the solute by

$$f_2 = \frac{P_2}{K_2 x_2} \qquad (11.10)$$

where K_2, the Henry's law constant, is the value of the intercept at $x_2 = 0$ of a plot of P_2/x_2 against x_2. Even though the values of f_2 calculated by (11.8) and (11.10) differ, as long as the same standard state is used for the substance throughout a calculation, the final result will not be changed.

If the activity of the solvent is known over a range of concentrations and the activity of the solute is known at one of these concentrations, the Gibbs–Duhem equation (11.34), in the form

$$d(\ln a_2) = -\frac{x_1}{x_2} d(\ln a_1) \qquad (11.11)$$

can be used to determine the value of the solute activity at another concentration by performing a graphical integration of a plot of x_1/x_2 against $\ln a_1$.

EXAMPLE 11.6.   The partial pressure of water over an aqueous solution of NH_3 at $22\,°C$ is $2\,300\,Pa$ for $x(NH_3) = 0.05$. The partial pressure of NH_3 is $5\,700\,Pa$ over the same solution. Calculate $f(H_2O)$, $a(H_2O)$, $f(NH_3)$, and $a(NH_3)$.

The value of $f(H_2O)$, using (11.8) for $P°(H_2O) = 2\,502$ Pa, is

$$f(H_2O) = \frac{2\,300 \text{ Pa}}{(0.95)(2\,502 \text{ Pa})} = 0.97$$

and by (11.9)

$$a(H_2O) = (0.97)(0.95) = 0.92$$

For NH_3, assuming the value of K_2 to be $96\,660$ Pa, (11.10) gives

$$f(NH_3) = \frac{5\,700}{(0.05)(96\,660)} = 1.2$$

and (11.9) gives $a(NH_3) = (1.2)(0.05) = 0.060$.

Colligative Properties of Solutions Containing Nonelectrolytic Solutes

11.10 VAPOR PRESSURE LOWERING

The vapor pressure of a solution containing a volatile solvent and a nonvolatile solute is given by

$$P_{\text{soln}} = P_1°(1 - x_2) \tag{11.12}$$

EXAMPLE 11.7. The vapor pressure of water is $3\,167.2$ Pa at $25\,°C$. What would be the vapor pressure of a solution of sucrose with $x_2 = 0.100$ and of a solution of levulose with $x_2 = 0.100$?

Equation (11.12) involves only solvent and concentration terms, and so

$$P_{\text{soln}} = (3\,167.2 \text{ Pa})(1.000 - 0.100) = 2\,850 \text{ Pa}$$

for both solutions.

11.11 BOILING POINT ELEVATION

The boiling point of a solution containing a nonvolatile solute is given by

$$T_{\text{bp,soln}} = T_{\text{bp,1}} + K_{x,\text{bp}} x_2 \tag{11.13}$$

where

$$K_{x,\text{bp}} = \frac{RT_{\text{bp,1}}^2}{\Delta_{\text{vap}} H_1} \tag{11.14}$$

or, in terms of molality,

$$T_{\text{bp,soln}} = T_{\text{bp,1}} + K_{\text{bp}} m \tag{11.15}$$

where the *ebullioscopic constant* is given by

$$K_{\text{bp}} = \frac{RT_{\text{bp,1}}^2 M_1}{1\,000 \Delta_{\text{vap}} H_1} \tag{11.16}$$

EXAMPLE 11.8. Calculate $K_{x,\text{bp}}$ for water if $\Delta_{\text{vap}} H = 40.656 \text{ kJ mol}^{-1}$ at 373.15 K. What is the boiling point of a solution of urea with $x_2 = 0.100$?

Using (11.14) gives

$$K_{x,\text{bp}} = \frac{(8.314 \times 10^{-3} \text{ kJ K}^{-1} \text{ mol}^{-1})(373.15 \text{ K})^2}{40.656 \text{ kJ mol}^{-1}} = 28.47 \text{ K}$$

and using (11.13) gives

$$T_{\text{bp}} = 373.15 \text{ K} + (28.47 \text{ K})(0.100) = 376.00 \text{ K}$$

11.12 FREEZING POINT DEPRESSION

The freezing point of a solution containing a nonvolatile solute is given by

$$T_{fp,soln} = T_{fp,1} - K_{x,fp}x_2 \qquad (11.17)$$

where

$$K_{x,fp} = \frac{RT_{fp,1}^2}{\Delta_{fus}H_1} \qquad (11.18)$$

or by

$$T_{fp,soln} = T_{fp,1} - K_{fp}m \qquad (11.19)$$

where the *cryoscopic constant* is given by

$$K_{fp} = \frac{RT_{fp,1}^2 M_1}{1\,000\Delta_{fus}H_1} \qquad (11.20)$$

11.13 OSMOTIC PRESSURE

Osmotic pressure (Π) is the external pressure required to stop the spontaneous flow of solvent from a supply of pure solvent across a semipermeable membrane into a solution. In general,

$$\Pi = -\frac{RT}{V_{m,1}} \ln \frac{P_{soln}}{P_1^\circ}$$

where $V_{m,1}$ is the molar volume of the solvent. If the solution is assumed to be ideal, this equation becomes

$$\Pi = -\frac{RT}{V_{m,1}} \ln x_1 \qquad (11.21a)$$

For ideal dilute solutions, two of the several approximations for (11.21) that are commonly used are

$$\Pi = RT\frac{x_2}{V_{m,1}} \qquad (11.21b)$$

and

$$\Pi = CRT \qquad (11.21c)$$

In (11.21), C is the molarity of the solution and R is expressed in dm^3 bar K^{-1} mol^{-1}.

EXAMPLE 11.9.   For dilute real solutions the semiempirical formula

$$\frac{\Pi}{C'} = \frac{RT}{M_2} + bC' \qquad (11.22)$$

is used to determine the molar masses of solutes, where C' is the concentration expressed in (g solute)(dm^3 soln)$^{-1}$ and b is known as the *interaction constant*. Determine the molar mass of sucrose from the following data at 20 °C:

$C'/[(\text{g solute})(dm^3 \text{ soln})^{-1}]$	103.8	50.9	40.6	30.3	20.1	10.0
$\Pi/(\text{bar})$	8.25	3.83	3.01	2.24	1.46	0.69

To determine the molar mass, the values of Π/C' are calculated—for example, for $C' = 103.8$,

$$\Pi/C' = (8.25 \text{ bar})/(103.8 \text{ g dm}^{-3}) = 7.95 \times 10^{-2} \text{ bar g}^{-1} \text{ dm}^3$$

and are plotted against C' (see Fig. 11-2). The intercept is 7.11×10^{-2} bar g^{-1} dm^3, giving

$$M_2 = \frac{(0.083\,14 \text{ dm}^3 \text{ bar K}^{-1} \text{ mol}^{-1})(293 \text{ K})}{7.11 \times 10^{-2} \text{ bar g}^{-1} \text{ dm}^3} = 343 \text{ g mol}^{-1}$$

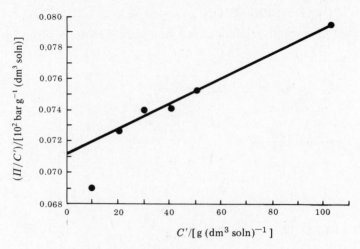

Fig. 11-2

Solutions of Electrolytes

11.14 CONDUCTIVITY

See Secs. 9.3–9.6.

11.15 COLLIGATIVE PROPERTIES OF STRONG ELECTROLYTES

For strong electrolytes, (11.12), (11.13), (11.15), (11.17), (11.19), and (11.21) become

$$P_{\text{soln}} = P_1^\circ \left(1 - \frac{n_2 i}{n_2 i + n_1} \right) \tag{11.23}$$

$$T_{\text{bp,soln}} = T_{\text{bp,1}} + K_{x,\text{bp}} \frac{n_2 i}{n_2 i + n_1} \tag{11.24}$$

$$T_{\text{bp,soln}} = T_{\text{bp,1}} + K_{\text{bp}} m i \tag{11.25}$$

$$T_{\text{fp,soln}} = T_{\text{fp,1}} - K_{x,\text{fp}} \frac{n_2 i}{n_2 i + n_1} \tag{11.26}$$

$$T_{\text{fp,soln}} = T_{\text{fp,1}} - K_{\text{fp}} m i \tag{11.27}$$

$$\Pi = -\frac{RT}{V_{\text{m,1}}} \ln \frac{n_1}{n_1 + n_2 i} \tag{11.28}$$

$$\Pi = \frac{RT}{V_{\text{m,1}}} \left(\frac{n_2 i}{n_2 i + n_1} \right) \tag{11.29}$$

$$\Pi = iCRT \tag{11.30}$$

where i is the *van't Hoff factor*. The value of i in very dilute solutions approaches the number of ions (ν) that the electrolyte forms in solution. The quantity mi is known as the *apparent molality*.

EXAMPLE 11.10. A 0.001 m solution of $Pt(NH_3)_4Cl_4$ in water had a freezing point depression of 0.005 4 K. Discuss the bonding in the compound.

Substituting the data and $K_{fp}(H_2O) = 1.860$ K mol kg^{-1} (see Problem 11.10) into (11.27) gives $i = 2.9$, which implies that the compound forms three ions upon ionization. Recognizing that the NH_3 will take preference over Cl^- in forming ligand bonds, the data imply that the structure is $[Pt(NH_3)_4Cl_2]Cl_2$.

11.16 COLLIGATIVE PROPERTIES OF WEAK ELECTROLYTES

Equations (11.23)–(11.30) are valid for weak electrolytes as well as strong electrolytes, but the value of i is related to the degree of ionization (α) for the weak electrolyte by

$$\alpha = \frac{i-1}{\nu-1} \tag{11.31}$$

Partial Molar Quantities

11.17 CONCEPT

If X is an extensive (mass-dependent) property of a system such that

$$X = X(P, T, n_1, n_2, \dots)$$

then

$$dX = \left(\frac{\partial X}{\partial T}\right)_{P,n_1,n_2,\dots} dT + \left(\frac{\partial X}{\partial P}\right)_{T,n_1,n_2,\dots} dP + \sum_{i}^{\text{components}} \left(\frac{\partial X}{\partial n_i}\right)_{T,P,n_j \neq n_i} dn_i$$

If $\bar{X}_i$ is defined as the *partial molar property*, where

$$\bar{X}_i \equiv \left(\frac{\partial X}{\partial n_i}\right)_{T,P,n_j \neq n_i} \tag{11.32}$$

then at constant P and T

$$dX = \sum_{i}^{\text{components}} \bar{X}_i \, dn_i$$

which integrates to

$$X = \sum_{i}^{\text{components}} \bar{X}_i n_i \tag{11.33}$$

It follows that for a binary solution,

$$n_1 \, d\bar{X}_1 = -n_2 \, d\bar{X}_2 \tag{11.34}$$

which is one form of the *Gibbs–Duhem equation*. If n_j in (11.32) is specified as 1 kg of component j, then $\bar{X}_i$ is known as a *partial molal quantity*.

Three important partial molar properties are the *partial molar volumes* of the components in a solution, the *partial molar heat of solution* (which is also known as the *differential heat of solution*), and the *partial molar free energy* (which is also known as the *chemical potential*). These properties can be determined by numerical differentiation of a function relating X to n_i (see Problem 11.38), from the slope of a plot of X against n_i for $n_{j \neq i} = 1$ (see Problems 11.14, 11.16, and Example 11.11), from a numerical calculation involving the *apparent molar quantity* (ϕ_i),

$$\phi_i \equiv \frac{X - n_1 X^\circ_{m,1}}{n_i} \tag{11.35}$$

where $X^\circ_{m,1}$ is the molar value for the pure solvent (see Problem 11.40), and from the "method of intercepts" (see Problems 11.17 and 11.37).

EXAMPLE 11.11. Because the total volume of a solution is an extensive property, (*11.32*)–(*11.35*) are valid with $X = V$. Determine $\bar{V}_2$ in a 0.5 m aqueous NaCl solution from the following data:

$m/(\text{mol kg}^{-1})$	0.349	0.439	0.529	0.621
$\rho/(10^3 \text{ kg m}^{-3})$	1.012 5	1.015 5	1.019 6	1.023 3

Assume a basis of 1.000 0 kg of water, so that $n_2 = (1 \text{ kg})m$. The volume of the 0.621 m solution is

$$V_{\text{soln}} = \frac{(0.621 \text{ mol})(58.45 \text{ g mol}^{-1})[(10^{-3} \text{ kg})/(1 \text{ g})] + 1.000 0 \text{ kg}}{1.023 3 \times 10^3 \text{ kg m}^{-3}}$$

$$= 1.012 7 \times 10^{-3} \text{ m}^3 = 1.012 7 \text{ dm}^3$$

Volumes for the remaining solutions were calculated as above and appear in Fig. 11-3. By (*11.32*), $\bar{V}_2 = (\partial V/\partial n_2)_{T,P,n_1}$; thus the slope of the line,

$$\frac{dV}{dm} = (1 \text{ kg})\frac{dV}{dn_2}$$

gives 0.018 1 dm³ mol⁻¹ = $\bar{V}_2$.

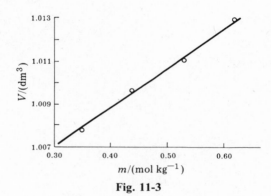

Fig. 11-3

Solved Problems

CONCENTRATIONS

11.1. Specify the solvent and solute(s) in the following solutions: (*a*) $w(\text{C}_2\text{H}_5\text{OH}) = 0.95$ and $w(\text{H}_2\text{O}) = 0.05$; (*b*) 82.3 g HCl in 100 g H_2O; (*c*) 50 g toluene and 50 g benzene; (*d*) 79 dm³ N_2, 1 dm³ CO_2, and 20 dm³ O_2; (*e*) $x(\text{Ag}) = 0.05$ in a Cu–Ag solution.

Choosing as solvent the major component or liquid component gives (*a*) ethanol as solvent and water as solute; (*b*) water as solvent and HCl as solute; (*c*) an arbitrary choice; (*d*) N_2 as solvent and O_2 and CO_2 as solutes; (*e*) Cu as solvent and Ag as solute.

11.2. How would 1.00 dm³ of a Na_2CO_3 solution such that $w(\text{Na}_2\text{CO}_3) = 0.025 0$ be prepared from $\text{Na}_2\text{CO}_3 \cdot 10\text{H}_2\text{O}$ and water? The solution density is $1.017 8 \times 10^3 \text{ kg m}^{-3}$, and the molar masses are 286.16 g mol⁻¹ for the hydrate and 106.00 g mol⁻¹ for the anhydrate.

Assuming a basis of 1 dm³ of solution or

$$(1 \text{ dm}^3)(1.017 8 \times 10^3 \text{ kg m}^{-3})[(10^{-3} \text{ m}^3/(1 \text{ dm}^3))] = 1 017.8 \text{ g soln}$$

the mass of anhydrous salt is

$$(1\,017.8 \text{ g})(0.025\,0) = 25.4 \text{ g Na}_2\text{CO}_3$$

Converting to moles gives

$$\frac{25.4 \text{ g Na}_2\text{CO}_3}{106.00 \text{ g mol}^{-1}} = 0.240 \text{ mol Na}_2\text{CO}_3$$

To obtain this amount of Na_2CO_3 from the hydrate, the mass of hydrate needed is

$$(286.16 \text{ g mol}^{-1})(0.240 \text{ mol}) = 68.7 \text{ g hydrate}$$

The preparation would consist in weighing 68.7 g of the hydrate, placing it in a 1-dm^3 (1-liter) volumetric flask, and adding sufficient water to make 1 dm^3 of solution (dilute to "the mark").

11.3. Commercial acetic acid is 17.4 M. How would 10.00 dm^3 of 3.00 M acid be prepared?

Using (*11.1*) gives

$$V_{\text{conc}} = (10.00 \text{ dm}^3)\frac{3.00 \text{ M}}{17.4 \text{ M}} = 1.72 \text{ dm}^3$$

which is diluted with sufficient water to produce 10.00 dm^3 of solution.

11.4. Given the value of K_2 in (*11.2*) is 7 120 bar for $\text{H}_2(\text{g})$ in water and 367 bar in benzene, how many times more soluble is H_2 in benzene than in water?

Taking a ratio of x_2 as expressed by (*11.2*) in the solvents gives

$$\frac{x(\text{H}_2 \text{ in C}_6\text{H}_6)}{x(\text{H}_2 \text{ in H}_2\text{O})} = \frac{K_2(\text{H}_2 \text{ in H}_2\text{O})}{K_2(\text{H}_2 \text{ in C}_6\text{H}_6)} = \frac{7\,120}{367} = 19.4$$

11.5. If the value of k_d in (*11.3*) is 410 for the distribution of I_2 between water and CS_2, find the fraction of I_2 remaining in the water phase after an amount of CS_2 equal to the water has been allowed to equilibrate with the aqueous phase.

Letting $C'_2 = n'_2/V'$ be the amount of I_2 in the CS_2 layer and $C_2 = n_2/V$ the amount of I_2 in the water layer, we have from (*11.3*)

$$\frac{n'_2/V'}{n_2/V} = 410$$

Letting $V' = V$ gives $n'_2 = 410n_2$, which upon substitution into $n_2 + n'_2 = 1.00$ gives $n_2 = 2.43 \times 10^{-3}$ of the original amount.

THERMODYNAMIC PROPERTIES OF SOLUTIONS

11.6. At 75 °C, $P(\text{HNO}_3) = 4\,670$ Pa and $P(\text{H}_2\text{O}) = 11\,500$ Pa over a nitric acid solution having $w(\text{HNO}_3) = 0.650$. Given $P°(\text{HNO}_3) = 71\,990$ Pa and $P°(\text{H}_2\text{O}) = 38\,540$ Pa, calculate P_i for the ideal solution, P_{soln} for the ideal and real solutions, and $x_{i,\text{vap}}$ for the ideal and real solutions.

A basis of 100.0 g of solution contains $(100.0 \text{ g soln})(0.650) = 65.0 \text{ g HNO}_3$. Then

$$n(\text{HNO}_3) = \frac{65.0 \text{ g HNO}_3}{63.01 \text{ g mol}^{-1}} = 1.032 \text{ mol HNO}_3$$

$$n(\text{H}_2\text{O}) = \frac{35.0 \text{ g H}_2\text{O}}{18.015 \text{ g mol}^{-1}} = 1.943 \text{ mol H}_2\text{O}$$

so that

$$x(\text{HNO}_3) = \frac{1.032}{1.032 + 1.943} = 0.347 \qquad x(\text{H}_2\text{O}) = 1.000 - 0.347 = 0.653$$

For the ideal solution, (11.4) and (11.5) give

$$P(\text{HNO}_3) = (71\,990\,\text{Pa})(0.347) = 25\,000\,\text{Pa}$$

$$P(\text{H}_2\text{O}) = (38\,540)(0.653) = 25\,200\,\text{Pa}$$

$$P_{\text{soln}} = 25\,000 + 25\,200 = 50\,200\,\text{Pa}$$

The composition of the vapor would be given by (1.10) as

$$x(\text{HNO}_3)_{\text{vap}} = \frac{25\,000\,\text{Pa}}{50\,200\,\text{Pa}} = 0.498 \qquad x(\text{H}_2\text{O})_{\text{vap}} = 1 - 0.498 = 0.502$$

For the real solution, (11.5) gives

$$P_{\text{soln}} = 4\,670 + 11\,500 = 16\,200\,\text{Pa}$$

and (11.10) gives the composition as

$$x(\text{HNO}_3)_{\text{vap}} = \frac{4\,670}{16\,200} = 0.288 \qquad x(\text{H}_2\text{O})_{\text{vap}} = 1 - 0.288 = 0.712$$

a considerable deviation from ideality.

11.7. Prepare plots of (5.13), (11.6), and (11.7b) against x_i for 1.00 mol of an ideal solution at 25 °C. To make the plot of (5.13b) more meaningful, it is usually plotted as $T\,\Delta_{\text{soln}}S$. On the same graph, plot values of $\Delta_{\text{soln}}H$ against $x(\text{CH}_3\text{OH})$ for 1.00 mol of solution, calculated from the following data for solutions of CH_3OH in C_6H_6:

$\Delta_f H^\circ_{298}/(\text{kJ mol}^{-1})$	-238.66	-238.505	-238.429	-238.358	-238.287	-237.965	-237.409	-236.957	-236.530
$n(\text{C}_6\text{H}_6)/(\text{mol})$	pure $\text{CH}_3\text{OH(l)}$	0.10	0.15	0.20	0.25	0.50	1.00	1.50	2.00
$\Delta_f H^\circ_{298}/(\text{kJ mol}^{-1})$	-236.149	-235.806	-235.174	-234.614	-234.095	-233.208	-232.471	-231.007	-229.911
$n(\text{C}_6\text{H}_6)/(\text{mol})$	2.50	3.00	4.00	5.00	6.00	8.00	10.00	15.00	20.00

Qualitatively discuss $\Delta_{\text{soln}}S$ and $\Delta_{\text{soln}}G$.

The values of $T\Delta_{\text{soln}}S$ and $\Delta_{\text{soln}}G$ were calculated at various values of x_i and plotted, see Fig. 11-4. As a sample calculation, at $x(\text{CH}_3\text{OH}) = 0.1$,

$$\Delta_{\text{soln}}S = -(8.314\,\text{J K}^{-1}\,\text{mol}^{-1})[(0.1)\ln(0.1) + (0.9)\ln(0.9)]$$

$$= -(8.314)[(-0.230) + (-0.095)] = 2.70\,\text{J K}^{-1}$$

$$T\,\Delta_{\text{soln}}S = (2.70)(298) = 805\,\text{J}$$

$$\Delta_{\text{soln}}G = (8.314)(298)[(0.1)\ln(0.1) + (0.9)\ln(0.9)] = -805\,\text{J}$$

As a sample calculation of $\Delta_{\text{soln}}H$ for 1.00 mol of solution, consider the data for 1.00 mol of CH_3OH in 5.00 mol of C_6H_6. According to (4.7),

$$\Delta_{\text{soln}}H = [(1)(-234.614)] - [(1)(-238.66)] = 4\,050\,\text{J (mol CH}_3\text{OH)}^{-1}$$

Thus, for 1 mol of solution, corresponding values are

$$\Delta_{\text{soln}}H = \frac{4\,050\,\text{J}}{6} = 675\,\text{J} \qquad \text{and} \qquad x(\text{CH}_3\text{OH}) = \frac{1}{6} = 0.167$$

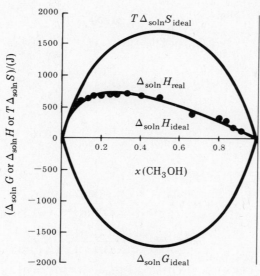

Fig. 11-4

Qualitatively, $\Delta_{soln}H$ is the endothermic result of breaking the association between the alcohol molecules as the solution is formed. As a result, $\Delta_{soln}S$ will be more positive than in the ideal case, and $\Delta_{soln}G$ will be less favorable than in the ideal case. Thus the solution will show positive deviations from ideality.

11.8. The partial pressure of ether over a solution of acetone in ether at 30 °C is 0.713 bar at $x_2 = 0.200$. The partial pressure of acetone at this same concentration is 0.120 bar. Calculate f_i given that $P_2^\circ = 0.377$ bar and $P_1^\circ = 0.861$ bar. Given $P_2 = 0.197$ bar at $x_2 = 0.400$, use the data from these two solutions to determine K_2 and calculate f_2.

Using (11.8) gives

$$f(\text{acetone}) = \frac{0.120}{(0.200)(0.377)} = 1.59$$

$$f(\text{ether}) = \frac{0.713}{(0.800)(0.861)} = 1.04$$

To estimate K_2, (11.2) is applied to each solution, giving

$$K_2 = \frac{0.120}{0.200} = 0.600 \qquad K_2 = \frac{0.197}{0.400} = 0.493$$

and the extrapolated value (see Sec. 11.9) is 0.707. Upon substitution into (11.10),

$$f(\text{acetone}) = \frac{0.120}{(0.707)(0.200)} = 0.85$$

COLLIGATIVE PROPERTIES OF SOLUTIONS CONTAINING NONELECTROLYTIC SOLUTES

11.9. The vapor pressure above a solution of 5.00 g CH_3COOH in 100.0 g H_2O [$P^\circ(H_2O) = 3\,167.2$ Pa at 25 °C] was 3 120. Pa and in 100.0 g C_6H_6 [$P^\circ(C_6H_6) = 9\,670$ Pa at 25 °C] was 9 330 Pa. Assuming that CH_3COOH is nonvolatile, use these data to discuss the intermolecular bonding in CH_3COOH.

For the solutions, (11.12) gives

$$x(CH_3COOH \text{ in } H_2O) = 1 - \frac{3\ 120.}{3\ 167.2} = 0.014\ 9$$

$$x(CH_3COOH \text{ in } C_6H_6) = 1 - \frac{9\ 330}{9\ 670} = 0.035$$

In 100.0 g of solvent, there are

$$\frac{100.0 \text{ g}}{18.015 \text{ g mol}^{-1}} = 5.551 \text{ mol } H_2O \qquad \text{and} \qquad \frac{100.0}{78.12} = 1.280 \text{ mol } C_6H_6$$

which upon substituting into (1.11) gives

$$0.014\ 9 = \frac{n(CH_3COOH \text{ in } H_2O)}{n(CH_3COOH \text{ in } H_2O) + 5.551}$$

$$0.035 = \frac{n(CH_3COOH \text{ in } C_6H_6)}{n(CH_3COOH \text{ in } C_6H_6) + 1.280}$$

and solving gives $n(CH_3COOH \text{ in } H_2O) = 0.084\ 0$ and $n(CH_3COOH \text{ in } C_6H_6) = 0.046$. For 5.00 g, the corresponding values of M are

$$M(CH_3COOH \text{ in } H_2O) = (5.00 \text{ g})/(0.084\ 0 \text{ mol}) = 59.5 \text{ g mol}^{-1}$$

$$M(CH_3COOH \text{ in } C_6H_6) = (5.00)/(0.046) = 110 \text{ g mol}^{-1}$$

The factor 2 between the molecular masses can be attributed to the formation of a dimeric species in the nonpolar solvent as two molecules orient themselves with considerable hydrogen bonding between the mutual —COOH groups.

11.10. Given that a freezing point change of 1.01 K was observed in a solution of sucrose in water such that $w(\text{sucrose}) = 0.150\ 0$, find the molar mass of sucrose. For water at 273.15 K, $\Delta_{fus}H = 6.009\ 5 \text{ kJ mol}^{-1}$.

Using (11.20) gives for water

$$K_{fp} = \frac{(8.314 \times 10^{-3} \text{ kJ K}^{-1} \text{ mol}^{-1})(273.15 \text{ K})^2(18.015 \text{ g mol}^{-1})}{[(1\ 000 \text{ g})/(1 \text{ kg})](6.009\ 5 \text{ kJ mol}^{-1})}$$

$$= 1.860 \text{ K kg mol}^{-1}$$

and (11.19) gives

$$m = \frac{T_{fp,1} - T_{fp,soln}}{K_{fp}} = \frac{1.01 \text{ K}}{1.860 \text{ K kg mol}^{-1}} = 0.543 \text{ mol (kg soln)}^{-1}$$

For 1.000 kg of solution, there are 0.150 kg solute and 0.850 kg solvent, giving

$$0.543 \text{ mol kg}^{-1} = \frac{(0.150 \text{ kg})/(M_2 \times 10^{-3} \text{ kg mol}^{-1})}{0.850 \text{ kg}}$$

which upon solving yields $M_2 = 325 \text{ g mol}^{-1}$. The correct answer, 342.30 g mol^{-1}, implies that the solution is not quite ideal.

11.11. Predict the osmotic pressure for a 0.100 m aqueous solution of sucrose at 20 °C using (11.21). A 0.100 m solution is equivalent to 0.098 M.

In a 0.100 m solution there is 0.100 mol of solute for 1.000 kg (55.56 mol) of water, giving

$$x_1 = \frac{55.56}{55.56 + 0.10} = 0.998\ 2 \qquad x_2 = \frac{0.100}{55.56 + 0.10} = 0.001\ 80$$

Using $V_{m,1} = 0.018\ 0\ dm^3\ mol^{-1}$, the most general equation, (11.21a), gives

$$\Pi = \frac{-(0.083\ 14\ dm^3\ bar\ K^{-1}\ mol^{-1})(293\ K)}{0.018\ 0\ dm^3\ mol^{-1}}\ln(0.998\ 2) = 2.44\ bar$$

The approximate equations (11.21b) and (11.21c) give, respectively,

$$\Pi = (0.083\ 14)(293)\frac{0.001\ 80}{0.018\ 0} = 2.44\ bar$$

and $$\Pi = (0.098)(0.083\ 14)(293) = 2.39\ bar$$

The observed value is 2.62 bar, indicating that even at this relatively low concentration, the solution is not quite ideal.

SOLUTIONS OF ELECTROLYTES

11.12. Prepare plots of i against m for the following aqueous solutions:

$m(acetone)/(mol\ kg^{-1})$	1.003	0.812	0.625	0.442	0.262	0.087
$(T_{fp,1} - T_{fp,soln})/K$	1.79	1.46	1.13	0.81	0.48	0.16

$m(NaNO_3)/(mol\ kg^{-1})$	0.685	0.555	0.427	0.303	0.179	0.059
$(T_{fp,1} - T_{fp,soln})/K$	2.08	1.70	1.33	0.95	0.58	0.21

$m((NH_4)_2SO_4)/(mol\ kg^{-1})$	0.441	0.357	0.275	0.195	0.115	0.038
$(T_{fp,1} - T_{fp,soln})/K$	1.63	1.35	1.07	0.78	0.48	0.17

Why is i nonintegral?

Using the value of $K_{fp} = 1.860\ K\ kg\ mol^{-1}$ (see Problem 11.10), (11.27) gives for the 1.003 m acetone solution

$$i = \frac{1.79}{(1.860)(1.003)} = 0.96$$

Performing the same calculations for the rest of the solutions generates the points shown in Fig. 11-5. As can be seen in the figure, the van't Hoff factors approach 1, 2, and 3, respectively, as molality approaches zero. The value of i represents the actual number of moles of particles present for each mole of solute added, which includes ion pairs, etc., at finite concentrations. Consequently, i will be smaller than the integer ν.

11.13. The observed freezing point depression for a 0.100 m aqueous solution of acetic acid is 0.190 °C. Find K_a at this concentration.

Using (11.27) gives $i = (0.190)/(1.860)(0.100) = 1.02$, and (11.31) gives

$$\alpha = \frac{1.02 - 1}{2 - 1} = 0.02$$

For the equilibrium

$$CH_3COOH(aq) \rightleftharpoons H^+(aq) + CH_3COO^-(aq)$$

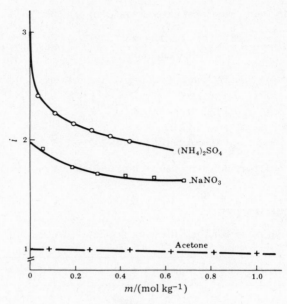

Fig. 11-5

the equilibrium constant expressed in terms of α becomes

$$K_\alpha = \frac{C(\mathrm{H}^+)\,C(\mathrm{CH_3COO}^-)}{C(\mathrm{CH_3COOH})} = \frac{\alpha C^2}{1-\alpha} = \frac{(0.02)^2(0.100)}{0.98} = 4 \times 10^{-5}$$

assuming that 0.100 m = 0.100 M. Because the values of α and K_a determined from freezing point depression measurements are precise to only one or two significant figures, values determined from conductivity measurements are usually preferred (see Example 9.6).

PARTIAL MOLAR QUANTITIES

11.14. Graphically determine $\overline{\Delta_{\mathrm{soln}}H_i}$ for water and H_2SO_4 in a 1 m solution, using the following data:

$\Delta_f H^\circ_{298}/(\mathrm{kJ\,mol}^{-1})$	−813.989	−885.983	−886.460	−886.774	−887.292	−887.636	−887.811	−888.188
$n(\mathrm{H_2O})/(\mathrm{mol})$	pure H_2SO_4(l)	30	40	50	75	100	115	150

From (*11.33*) calculate $\Delta_{\mathrm{soln}}H$ for preparing 1.00 mol of the 1 m solution.

The integral heat of solution for $n(\mathrm{H_2O})$ = 30 mol is

$$\Delta_{\mathrm{soln}}H = [(1)(-885.983)] - [(1)(-813.989)] = -71.994\ \mathrm{kJ}$$

The heat of solution for preparing the amount of solution containing 1.00 mol of water is

$$\frac{-71.994\ \mathrm{kJ}}{30\ \mathrm{mol\ H_2O}} = -2.400\ \mathrm{kJ\ (mol\ H_2O)}^{-1}$$

and the number of moles of H_2SO_4 for each mole of water is

$$n_2 = \frac{1\ \mathrm{mol\ H_2SO_4}}{30\ \mathrm{mol\ H_2O}} = 0.033\ 3\ \mathrm{(mol\ H_2SO_4)(mol\ H_2O)}^{-1}$$

The results of similar calculations for the remainder of the solutions are shown in Fig. 11-6. The slope of the plot at 1 m ($n_2 = 0.018\,0$) is

$$\overline{\Delta_{soln}H(H_2SO_4)} = -71.46 \text{ kJ (mol } H_2SO_4)^{-1}$$

The integral heat of solution, e.g., -71.994 kJ (mol $H_2SO_4)^{-1}$ for 30 mol of H_2O, is the heat of solution for preparing the amount of solution containing 1 mol of H_2SO_4. Figure 11-7 is a plot of these values, and the slope at 1 m, where $n_1 = 55.56$, is

$$\overline{\Delta_{soln}H(H_2O)} = -25.0 \text{ J (mol } H_2O)^{-1}$$

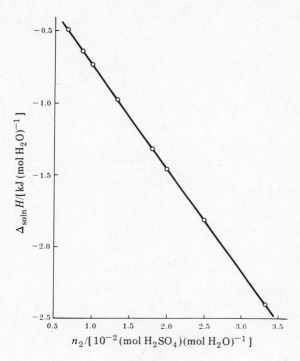

Fig. 11-6

For 1 mol of a 1 m solution, the data needed for (11.33) are

$$n_2 = \frac{1.000}{1.000 + 55.56} = 0.017\,7 \qquad \text{and} \qquad n_1 = \frac{55.56}{1.000 + 55.56} = 0.982\,3$$

giving

$$\Delta_{soln}H = n_1\overline{\Delta_{soln}H_1} + n_2\overline{\Delta_{soln}H_2}$$

$$= (0.982\,3 \text{ mol})(-0.025 \text{ kJ mol}^{-1}) + (0.017\,7 \text{ mol})(-71.46 \text{ kJ mol}^{-1})$$

$$= -1.29 \text{ J}$$

11.15. Although $\overline{\Delta_{soln}H_1}$ as shown in Fig. 11-7 changes from -25.0 to -11.4 J (mol $H_2O)^{-1}$ as the concentration changes from 1 m to 0.5 m, $\overline{\Delta_{soln}H_2}$ essentially remains constant. Show this to be true from (11.34).

Rearranging (11.34) gives

$$d\bar{X}_2 = -\frac{n_1}{n_2}\,d\bar{X}_1$$

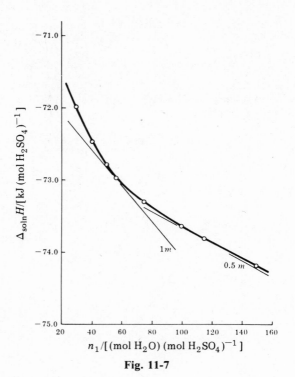

Fig. 11-7

where, in this case, $\bar{X} = \Delta_{soln}H$. Integration gives

$$\Delta \bar{X}_2 = -\int_{-25.0}^{-11.4} \frac{n_1}{n_2}\, d\bar{X}_1$$

The integration is performed graphically (see Fig. 11-8), giving $\Delta \bar{X}_2 = -1.13$ kJ $(mol\ H_2SO_4)^{-1}$. This change would not be significant on Fig. 11-6.

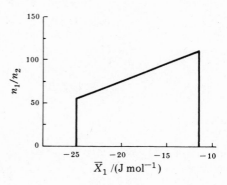

Fig. 11-8

11.16. Using the data in Problem 11.14 for the H_2SO_4/H_2O solutions, prepare a plot of the heat of solution for preparing 1 mol of the above solutions, and check the answer of -1.29 kJ for a 1 m solution. Using the value of $\overline{\Delta_{soln}H_2} = -71.46$ kJ $(mol\ H_2SO_4)^{-1}$ at 0.5 m from Fig. 11-6 and the value of $\Delta_{soln}H$ read from the new plot, determine $\overline{\Delta_{soln}H_1}$ at 0.5 m.

For $n(H_2O) = 30$, $\Delta_{soln}H$ for 1.00 mol of solution is

$$\frac{-71.994\ kJ}{30+1} = -2.322\ kJ$$

Figure 11-9 contains a plot of the data as calculated above. From the plot, $\Delta_{soln}H = -1.296$ kJ, a difference of 0.5% for the 1 m solution. For the 0.5 m solution, $\Delta_{soln}H = -0.663$ kJ. By (*11.33*)

$$-0.663 = \frac{55.56}{56.06}\overline{\Delta_{soln}H_1} + \frac{0.50}{56.06}(-71.46)$$

which upon solving gives $\overline{\Delta_{soln}H_1} = -26$ J $(mol\ H_2O)^{-1}$. The slope of the curve in Fig. 11-7 gives -11.4 J $(mol\ H_2O)^{-1}$, which is more precise.

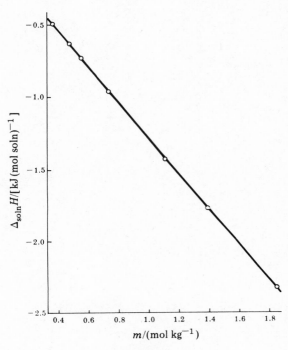

Fig. 11-9

11.17. The "method of intercepts" is a convenient method for determining both partial molar quantities from the same graph. A plot of X' against x_2, where

$$X' \equiv \frac{X}{n_1 + n_2}$$

is prepared, and a tangent is drawn to the curve at the desired concentration. The intercept of the tangent on the x_1 axis is $\bar{X}_1$ and the intercept on the x_2 axis is $\bar{X}_2$. Using the data in Problem 11.14 for the H_2SO_4/H_2O solutions, prepare a plot of $\Delta_{soln}H$ for 1.00 mol of solution against $x(H_2SO_4)$, and determine $\Delta_{soln}H_i$ for a 1 m solution.

For $n(H_2O) = 30$ mol H_2O,

$$x_2 = \frac{1}{31} = 0.032\ 3 \qquad \text{and} \qquad X' = \frac{-71.994\ \text{kJ}}{31\ \text{mol}} = -2.322\ \text{kJ}\ (mol\ soln)^{-1}$$

The results of similar calculations are plotted in Fig. 11-10. The intercepts of the tangent (superimposed on the plot of the data) give

$$\overline{\Delta_{soln}H(H_2SO_4)} = -71.1\ \text{kJ}\ (mol\ H_2SO_4)^{-1} \qquad \text{and} \qquad \overline{\Delta_{soln}H(H_2O)} = 25\ \text{J}\ (mol\ H_2O)^{-1}$$

for the 1 m solution. These values agree well with those obtained in Problem 11.14.

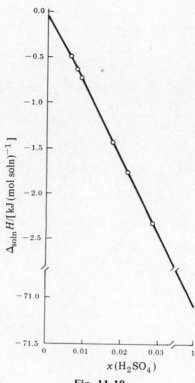

Fig. 11-10

Supplementary Problems

CONCENTRATIONS

11.18. Specify the solvent and solute in the following solutions: (*a*) 60 g ethanol and 40 g water, (*b*) 50 g ethanol and 50 g water, (*c*) 89.9 g (about 100 dm^3) $NH_3(g)$ and 100 g $H_2O(l)$, (*d*) 116.8 g NaBr and 100 g H_2O, (*e*) Pt and $H_2(g)$.

    Ans.   (*a*) The solvent is usually ethanol, and the solute is water;   (*b*) choice is arbitrary, but water might be more convenient for the solvent;   (*c*) solvent is water, and solute is NH_3;   (*d*) solvent is water, and solute is NaBr;   (*e*) solvent is Pt, and solute is H_2.

11.19. What is the concentration of a 0.509 M solution of D-mannitol $[CH_2OH(CHOH)_4CH_2OH]$ expressed in w(D-mannitol)? The molar mass of the solute is 182.17 g mol^{-1}, and the density of the solution is $1.030\ 2 \times 10^3$ kg m^{-3}. (*Hint*: Assume a basis of 1.000 dm^3 of solution.)   *Ans.* w(D-mannitol) = 0.090 0

11.20. How would 1.00 dm^3 of a $CaCl_2$ solution having $w(CaCl_2) = 0.010\ 0$ be prepared from $CaCl_2 \cdot 2H_2O$ and water? The solution density is $1.006\ 5 \times 10^3$ kg m^{-3}, and the molar masses are 110.99 g mol^{-1} for $CaCl_2$ and 147.03 g mol^{-1} for $CaCl_2 \cdot 2H_2O$.   *Ans.*   13.3 g of hydrate dilated to 1.00 dm^3

11.21. What volume of commercial $NH_3(aq)$ ($C = 7.4$ M) is required to produce 1.00 dm^3 of 3 M $NH_3(aq)$?

    Ans.   0.405 dm^3 diluted to 1.00 dm^3

11.22. The value of K_2 in (*11.2*) is 4.40×10^4 bar for $O_2(g)$ in water at 25 °C and 4.69×10^4 bar at 30 °C. What change in the solubility of O_2 at 0.20 bar in water will occur if thermal pollution of water changes the temperature by 5 K?   *Ans.* $x_2 = 4.55 \times 10^{-6}$ at 25 °C and 4.26×10^{-6} at 30 °C, a 6.4% decrease

11.23. Repeat Problem 11.5, now using two extraction samples of CS_2, each with $V'_2 = 0.5\ V$.

   Ans. $n_2 = 4.85 \times 10^{-3}$ of the original after the first extraction and $n_2 = 2.36 \times 10^{-5}$ of the original after the second equilibrium

THERMODYNAMIC PROPERTIES OF SOLUTIONS

11.24. Using the data in Example 4.16 for $\Delta_f H^\circ_{298}$ for various solutions of HNO_3, prepare a plot of $\Delta_{soln} H$ for 1.00 mol of solution against $x(HNO_3)$. Include plots for (*5.13b*) as $T\Delta_{soln} S$, (*11.6*), and (*11.7b*) on the same graph. Discuss qualitatively the results for $\Delta_{soln} H$ for the real solution and what the plots for $\Delta_{soln} S$ and $\Delta_{soln} G$ would look like.

   Ans.   The plots for the ideal solutions would look like those in Fig. 11-4. The $\Delta_{soln} H$ curve shows very negative deviations from ideality as it goes to -6.77 kJ at $x(HNO_3) = 0.500$. The very large exothermic $\Delta_{soln} H$ is the result of forming $H^+(aq)$ and $NO_3^-(aq)$ in the solution. Qualitatively, $\Delta_{soln} S$ will be less positive and $\Delta_{soln} G$ more negative than in the ideal case; thus the solution will show negative deviations from ideality.

11.25. Using the following data, prepare a plot of P_i and P_{soln} against x_B

$P_A/(\text{bar})$	0.600	0.540	0.475	0.411	0.347	0.269	0.160	0.081	0.050	0.025	0
$P_B/(\text{bar})$	0	0.013	0.025	0.042	0.088	0.179	0.350	0.467	0.543	0.622	0.700
x_B	0.0	0.1	0.2	0.3	0.4	0.5	0.6	0.7	0.8	0.9	1.0

Include plots of (*11.4*) and (*11.5*) on the graph. Describe the system. Calculate f_i at $x_B = 0.2$ using the convention given by (*11.8*). Calculate f_B using the convention given by (*11.10*) at the same concentration.

   Ans.   Both substances show negative deviations from ideality; $f_A = 0.990$, $f_B = 0.179$; plot of P_2/x_2 against x_2 has intercept $= K_B = 0.135$, $f_B = 0.926$

11.26. At $x_B = 0.3$, the partial pressure of component B in Problem 11.25 no longer agrees with the value predicted by Henry's law, but the value for component A agrees fairly well with the value predicted by Raoult's law. From points on the graph prepared for Problem 11.25, find f_A from $x_B = 0.2$ to 0.3 at 0.01 intervals. Using (*11.9*) calculate a_A for these concentrations. Prepare a plot of x_A/x_B against $\ln a_A$ and graphically integrate (*11.11*) to determine $\ln a_{B,0.3} - \ln a_{B,0.2}$. Using the value of $a_{B,0.2} = 0.185$, find $a_{B,0.3}$.

   Ans. $P_A = 0.475, 0.469, 0.463, 0.456, 0.450, 0.444, 0.438, 0.432, 0.425, 0.417,$ and 0.411 bar;
   $f_A = 0.990, 0.989, 0.989, 0.987, 0.987, 0.987, 0.986, 0.986, 0.984, 0.979,$ and 0.979;
   $a_A = 0.792, 0.781, 0.771, 0.760, 0.750, 0.740, 0.730, 0.720, 0.708, 0.695,$ and 0.685;
   coordinates of plot are $(4.00, -0.233), (3.76, -0.247), (3.55, -0.260), (3.35, -0.275), (3.17, -0.288),$
   $(3.00, -0.301), (2.85, -0.315), (2.70, -0.329), (2.57, -0.345), (2.45, -0.364), (2.33, -0.379)$; area
   corresponds to 0.442; $a_{B,0.3} = 0.288$

COLLIGATIVE PROPERTIES OF SOLUTIONS CONTAINING NONELECTROLYTIC SOLUTES

11.27. The vapor pressure of water is 3 167.2 Pa at 25 °C. If the vapor pressure over a solution of urea having $w(\text{urea}) = 0.045\ 0$ is 3 123.2 Pa, find the molar mass of urea.   *Ans.* $x_2 = 1.39 \times 10^{-2}$, $M = 60.2\ \text{g mol}^{-1}$

11.28. Calculate K_{bp} for water and the boiling point of a 0.100 m solution of urea. $\Delta_{vap} H = 40.656\ \text{kJ mol}^{-1}$ for water at 373.15 K.   *Ans.*   0.513 K m^{-1}, 373.20 K

11.29. What freezing point depression would be predicted for a 0.1 m solution of benzoic acid in benzene if $\Delta_{fus} H = 127.4\ \text{J g}^{-1}$ at 278.69 K for benzene?

   Ans. $\Delta_{fus} H = 9.951\ \text{kJ mol}^{-1}$, $K_{fp} = 5.07$ K m^{-1}; -0.507 K

11.30. The melting points and heats of fusion for Bi and Cd are 544.2 K and 11.10 kJ mol^{-1} and 594.1 K and 6.11 kJ mol^{-1}, respectively. Using (11.17) and (11.18), predict the temperature and composition of the simple eutectic formed by these metals by plotting the freezing points of solutions with various compositions for each substance against mole fraction and observing where the curves meet. The observed values are 144 °C and $w(Cd) = 0.40$.

Ans. $K_{x,fp} = 224$ K for Bi and 480 K for Cd; 134 °C at $x(Cd) = 0.612$ or $w(Cd) = 0.46$

11.31. Using (11.21), predict the osmotic pressure for a 1.000 m (0.825 M) aqueous solution of sucrose at 20 °C. Assume $x_1 = 0.018\,0$ dm^3 mol^{-1}. The observed value is 27.0 bar.

Ans. $x_1 = 0.982\,3$, $x_2 = 0.017\,7$; $\Pi = 24.2$, 24.0, and 20.1 bar

11.32. Determine the molar mass of inulin, $(C_6H_{10}O_5)_x$, from the following data at 20 °C:

$C'/[(\text{g solute})(\text{dm}^3 \text{ soln})^{-1}]$	103.8	77.1	50.9	25.2
$\Pi/(\text{bar})$	0.86	0.58	0.31	0.14

Ans.   The intercept of a plot of Π/C' against C' is 4.46×10^{-3}/bar g^{-1} dm^3, giving 5 500 g mol^{-1}

SOLUTIONS OF ELECTROLYTES

11.33. Assuming 100% dissociation of the solutes, what would be the freezing point depression for an aqueous solution that is 0.10 m in NaCl and 0.10 m in CaCl$_2$? $K_{fp} = 1.860$ K kg mol^{-1} for water.

Ans. $m = 0.50$, 0.93 K

11.34. Prepare a plot of i against m for the KH$_2$PO$_4$ solutions given below:

$m/(\text{mol kg}^{-1})$	0.037	0.074	0.112	0.227	0.306	0.427	0.552	0.683	0.771
$(T_{fp,1} - T_{fp,soln})/K$	0.13	0.25	0.37	0.74	0.97	1.31	1.64	1.94	2.14

What conclusions from the graph can be made concerning the species present at these concentrations?

Ans.   Values of i lie between 1.49 and 1.89 for the concentrations given. Intercept at $m = 0$ will be above 2, implying that H$_2$PO$_4^-$ is mainly associated.

11.35. The major solutes in surface seawater expressed in ppm by mass are: Cl$^-$, 18 980; Na$^+$, 10 560; SO$_4^{2-}$, 2 700; Mg^{2+}, 1 270; Ca^{2+}, 400; K$^+$, 380; HCO$_3^-$, 140; and Br$^-$, 65. Express these concentrations as m and predict the boiling and freezing points of a sample of seawater given $K_{bp} = 0.513$ K kg mol^{-1} and $K_{fp} = 1.860$ K kg mol^{-1}.

Ans.   965 505 ppm H$_2$O;   0.554 m Cl$^-$,   0.475 m Na$^+$,   0.029 m SO$_4^{2-}$,   0.054 m Mg^{2+};   0.010 m Ca^{2+}, 0.010 m K$^+$, 0.002 m HCO$_3^-$, 0.001 m Br$^-$; $m_i = 1.135$; 100.58 °C; -2.11 °C

11.36. Find K_a for tartaric acid if a 0.100 m solution freezes at -0.205 °C. Assume that only the first ionization is of importance and that 0.100 $m = 0.100$ M, $K_{fp} = 1.860$ K kg mol^{-1}.

Ans. $i = 1.10$, $\alpha = 0.10$, $K_a = 1.1 \times 10^{-3}$

PARTIAL MOLAR QUANTITIES

11.37. Determine $\overline{\Delta_{soln}H_i}$ for CCl_4 and C_6H_6 at $x_i = 0.500$, using the following data:

$\Delta_f H^\circ_{298}/(kJ\ mol^{-1})$	−135.44	−135.344	−135.281	−135.206	−135.156	−135.122
$n(C_6H_6)/(mol)$	pure $CCl_4(l)$	0.25	0.50	1.0	1.5	2.0

by preparing plots of $\Delta_{soln}H$ against $n(CCl_4)$ and $n(C_6H_6)$ and determining the slopes. Using (*11.33*), calculate $\Delta_{soln}H$ for preparing 1 mol of this solution. Prepare a plot of the heat of solution for preparing 1.00 mol of the above solutions against $x(C_6H_6)$, and find $\Delta_{soln}H$. From this same plot, determine $\Delta_{soln}H_i$ using the method of intercepts.

Ans. $\overline{\Delta_{soln}H_i} = 122\ J\ (mol\ C_6H_6)^{-1}$ and $111\ J\ (mol\ CCl_4)^{-1}$; $\Delta_{soln}H = 117\ J$; $\Delta_{soln}H = 115\ J$, $\overline{\Delta_{soln}H_i} = 123\ J\ (mol\ C_6H_6)^{-1}$ and $110\ J\ (mol\ CCl_4)^{-1}$

11.38. Determine $\bar{V}_2$ in a 0.5 m aqueous HCl solution from the following data:

$m/(mol\ kg^{-1})$	0.418	0.560
$\rho/(10^3\ kg\ m^{-3})$	1.005 7	1.008 1

by finding V as a function of m and using (*11.32*).

Ans.   For 1 kg of water, $V = 1.001\ 6 + (0.019\ 0)m$; $\bar{V}_2 = 0.019\ 0\ dm^3\ mol^{-1}$

11.39. The chemical potential (μ) is defined as the partial molar free energy, i.e.,

$$\mu_i \equiv \left(\frac{\partial G}{\partial n_i}\right)_{T,P,n_j \neq n_i}$$

Beginning with (*6.5*), show that

$$\left(\frac{\partial \Delta\mu_2}{\partial P}\right)_T = \Delta\bar{V}_2$$

Will the solubility of the solute increase or decrease for an increase in P if $\Delta\bar{V}_2 = \bar{V}_2 - V^\circ_2 > 0$ where V°_2 is the molar volume of the pure state?

Ans.   As P increases, $\Delta\mu_2$ increases, giving a larger μ_2 for the solution. There is a net driving force for solute to leave solution, so that solubility decreases.

11.40. Beginning with (*11.35*), show that

$$\bar{X}_2 = \phi_2 + n_2 \frac{\partial \phi_2}{\partial n_2} = \phi_2 + \frac{\partial \phi_2}{\partial(\ln n_2)}$$

Using the data in Example 11.11 for the $NaCl/H_2O$ solutions, prepare a plot of ϕ_2 against $\ln n_2$ and find $\bar{V}_2$ for a 0.5 m solution. Assume that $V^\circ_1 = 0.018\ 09\ dm^3\ mol^{-1}$.

Ans.   At 0.5 m, $\phi = 11.05$, $d\phi/d(\ln n_2) = 7.84$, $\bar{V} = 0.018\ 89\ dm^3\ mol^{-1}$

Chapter 12

Rates of Chemical Reactions

Rate Equations for Simple Reactions

12.1 REACTION RATE

A chemical equation can be written in the form

$$\sum_i \nu_i I_i \qquad (12.1)$$

where ν_i represents the stoichiometric coefficients (ν_i is positive for products and negative for reactants) and I_i represents the chemical formulas of the species involved in the reaction. The *extent of reaction* (ξ) is defined as

$$d\xi = \frac{1}{\nu_i} dn_i \qquad (12.2)$$

where n_i is the molar amount of substance i. The *rate of reaction* ($d\xi/dt$) is defined as

$$\frac{d\xi}{dt} = \frac{1}{\nu_i} \frac{dn_i}{dt} \qquad (12.3)$$

Note that the rate of reaction is independent of the choice of the reactant or product of the reaction that is being studied and is valid regardless of reaction conditions. The quantity dn_i/dt is known as the *rate of change of i*, and the quantity dC_i/dt is known as the *rate of change of the concentration of i*.

EXAMPLE 12.1. Express the rate of reaction and the rates of change of the concentrations of the reactants and products for the chemical reaction

$$Cr_2O_7^{2-}(aq) + H_2O(l) \longrightarrow 2CrO_4^{2-}(aq) + 2H^+(aq)$$

The rate of reaction is given by

$$\frac{d\xi}{dt} = \frac{1}{2} \frac{dn(CrO_4^{2-})}{dt} = \frac{1}{2} \frac{dn(H^+)}{dt}$$

$$= -\frac{dn(Cr_2O_7^{2-})}{dt} = -\frac{dn(H_2O)}{dt}$$

The rates of increase of the concentrations of the products are $dC(CrO_4^{2-})/dt$ and $dC(H^+)/dt$, and the rates of decrease of the concentrations of the reactants (including the negative sign) are $-dC(Cr_2O_7^{2-})/dt$ and $-dC(H_2O)/dt$. Because water is the solvent, it would be difficult to measure $-dC(H_2O)/dt$.

12.2 CONCENTRATION DEPENDENCE

The rate of change of the concentration of a given reactant or product in a chemical reaction must be determined experimentally. Often the mathematical relationship describing this rate will take the form of the following *rate law*:

$$\pm \frac{dC_i}{dt} = k \prod_j C_j^{x_j} \qquad (12.4)$$

where C_j represents the concentrations of the reactants (and, occasionally, the products as well) that are experimentally determined to influence the rate, x_j represents the *order of reaction* for that component, and k is the *rate constant*. The positive sign is used for products, and the negative sign is used for the reactants.

The reaction order represents the number of molecules of the substance involved in the actual reaction equation. The actual reaction equation may or may not be the same as the overall stoichiometric equation for the reaction, and so x_j does not necessarily equal ν_i for that substance. The *overall order* of the reaction is the sum of the exponents. The rate constant is a function of temperature (see Sec. 13.1) and has the units of $C^{1-\text{overall order}} t^{-1}$.

EXAMPLE 12.2. Consider the following chemical equations and associated rates of changes of concentrations:

(i) $CH_3CHO(g) \longrightarrow CH_4(g) + CO(g)$ $-\dfrac{dC(CH_3CHO)}{dt} = k[C(CH_3CHO)]^{3/2}$

(ii) $OH^- + CH_3Br \longrightarrow CH_3OH + Br^-$ $\dfrac{dC(CH_3OH)}{dt} = kC(OH^-)C(CH_3Br)$

Which of these could be *elementary reactions* (those that occur exactly as written)?

If the reaction is elementary, concentrations of all the reactants will appear in the expression for dC_i/dt, and the order of reaction for each reactant will be the same as the stoichiometric coefficient. Reaction (i) cannot be elementary, and reaction (ii) could be elementary.

12.3   ZERO-ORDER REACTIONS

Zero-order reactions are those having the differential rate equation

$$-\frac{dC}{dt} = k \tag{12.5}$$

which upon integration ($C = C_0$ at $t = 0$) gives

$$C = C_0 - kt \tag{12.6}$$

For a reaction that is zero-order, a plot of C against t will give a straight line having a slope of $-k$ and an intercept of C_0.

EXAMPLE 12.3. The *half-life* of a chemical reaction ($t_{1/2}$) is defined by the condition $C = \frac{1}{2}C_0$ at $t = t_{1/2}$. For a zero-order reaction, (12.6) gives

$$t_{1/2} = C_0/2k$$

12.4   FIRST-ORDER REACTIONS

First-order reactions are those having the differential rate equation

$$-\frac{dC}{dt} = kC \tag{12.7}$$

which upon integration gives

$$\ln C = \ln C_0 - kt \tag{12.8}$$

For a first-order reaction, a plot of $\ln C$ against t will be linear with a slope of $-k$ and an intercept of $\ln C_0$.

EXAMPLE 12.4. The liquid-phase dissociation of dicylopentadiene has been studied by Langer and Patton using gas chromatographic techniques. The technique involved measured a quantity proportional to dC/dt rather than

$-dC/dt$, so (12.8) becomes

$$\ln C' = \ln C'_0 + kt$$

where C' and C'_0 are the quantities that are proportional to the concentration. Note that k has not changed. Determine k from the following data at 190 °C:

C'	1.85	2.04	2.34	2.70	3.83	5.28
$t/(s)$	524	620	752	876	1 188	1 452

For the first-order reaction, the plot of $\ln C'$ against t is linear (see Fig. 12-1), with slope $k = 1.13 \times 10^{-3}\ \text{s}^{-1}$.

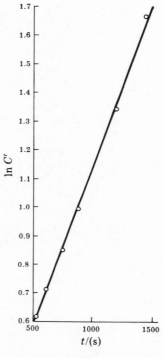

Fig. 12-1

12.5 SECOND-ORDER REACTIONS

The most general second-order reaction is one in which the stoichiometry is given by

$$a\text{A} + b\text{B} \longrightarrow \text{products}$$

where $a \neq b$ and $C_{A,0} \neq C_{B,0}$. For this reaction the integrated rate equation is

$$\frac{1}{bC_{A,0} - aC_{B,0}} \ln \left(\frac{C_{B,0}}{C_{A,0}} \frac{C_A}{C_B} \right) = kt \qquad (12.9)$$

If $a = b$ with $C_{A,0} \neq C_{B,0}$, the differential rate equation is

$$-\frac{dC_A}{dt} = -\frac{dC_B}{dt} = kC_A C_B \qquad (12.10)$$

which integrates to

$$\frac{1}{C_{A,0} - C_{B,0}} \ln\left(\frac{C_{B,0}}{C_{A,0}} \frac{C_A}{C_B}\right) = kt \qquad (12.11)$$

For (12.10) with $a = b = 1$ and $C_{A,0} = C_{B,0}$ or for the reaction

$$2A \longrightarrow \text{products}$$

the differential rate equation is

$$-\frac{dC}{dt} = kC^2 \qquad (12.12)$$

which integrates to

$$\frac{1}{C} = \frac{1}{C_0} + kt \qquad (12.13)$$

Plots of the left-hand sides of (12.9), (12.11), and (12.13) against t will be linear, with a slope in each case equal to k.

EXAMPLE 12.5. What are the dimensions of k for a second-order reaction?

Upon inspection of (12.9), (12.11), and (12.13), it can be seen that the dimensions of the left-hand sides are $[C^{-1}]$. Hence the dimensions of k are $C^{-1}t^{-1}$, in agreement with the general rule presented in Sec. 12.2.

12.6 THIRD-ORDER REACTIONS

For the reaction

$$A + B + C \longrightarrow \text{products}$$

with $C_{A,0} \neq C_{B,0} \neq C_{C,0}$, the differential rate equation is

$$-\frac{dC_A}{dt} = kC_A C_B C_C \qquad (12.14)$$

and the integrated equation is

$$\frac{\ln(C_A/C_{A,0})}{(C_{A,0} - C_{B,0})(C_{C,0} - C_{A,0})} + \frac{\ln(C_B/C_{B,0})}{(C_{A,0} - C_{B,0})(C_{B,0} - C_{C,0})} + \frac{\ln(C_C/C_{C,0})}{(C_{B,0} - C_{C,0})(C_{C,0} - C_{A,0})} = kt \quad (12.15)$$

For the case where $C_{B,0} \neq C_{A,0} = C_{C,0}$ in (12.14) or for the reaction

$$2A + B \longrightarrow \text{products}$$

with $C_{A,0} \neq C_{B,0}$ or $C_{A,0} \neq 2C_{B,0}$, the differential rate equation is

$$-\frac{dC_A}{dt} = kC_A^2 C_B \qquad (12.16)$$

and the integrated form is

$$\frac{2}{(2C_{B,0} - C_{A,0})^2}\left[\frac{2(2C_{B,0} - C_{A,0})(C_{A,0} - C_A)}{C_{A,0}C_A} + \ln\frac{C_{B,0}C_A}{C_{A,0}C_B}\right] = kt \qquad (12.17)$$

For the reaction

$$A + B \longrightarrow \text{products}$$

with $C_{A,0} \neq C_{B,0}$ where (12.16) is valid, the integrated form is

$$\frac{1}{(C_{B,0}-C_{A,0})^2}\left[\frac{(C_{B,0}-C_{A,0})(C_{A,0}-C_A)}{C_{A,0}C_A}+\ln\frac{C_{B,0}C_A}{C_{A,0}C_B}\right]=kt \qquad (12.18)$$

For the case where $C_{A,0}=C_{B,0}=C_{C,0}$ for (12.14), or $C_{A,0}=C_{B,0}$ or $C_{A,0}=2C_{B,0}$ for (12.16), or for the reaction

$$3A \longrightarrow \text{products}$$

the differential rate equation is

$$-\frac{dC}{dt}=kC^3 \qquad (12.19)$$

which integrates to

$$\frac{1}{2}\left(\frac{1}{C^2}-\frac{1}{C_0^2}\right)=kt \qquad (12.20)$$

Plots of the left-hand sides of (12.15), (12.17), (12.18), and (12.20) against t will be linear, having a slope in each case equal to k.

12.7 PSEUDO-ORDER REACTIONS

If one of the reactants is in great excess or is regenerated (a catalyst) so that its concentration is essentially constant, the concentration term in the rate equation for that component will appear as part of k unless special effort is taken to separate its contribution.

EXAMPLE 12.6. Consider the elementary reaction

$$A + H_2O + H^+(aq) \longrightarrow \text{products}$$

Write the complete rate equation and the pseudo-order rate equation if $C(H_2O) \gg C(A)$ and if H^+ is regenerated. If $k' = 1.00 \times 10^{-5}\,\text{s}^{-1}$ for the pseudo-first-order reaction, find k for the complete rate equation, given that $C(H_2O) = 55.5$ M and $C(H^+) = 0.10$ M.

The complete rate equation is

$$-\frac{dC(A)}{dt}=kC(A)C(H_2O)C(H^+)$$

If $C(H_2O)$ and $C(H^+)$ are constant, the pseudo-order equation is

$$-\frac{dC(A)}{dt}=k'C(A) \qquad \text{where} \qquad k'=kC(H_2O)C(H^+)$$

Substituting the values of concentrations gives

$$k=\frac{k'}{C(H_2O)C(H^+)}=\frac{1.00\times10^{-5}\,\text{s}^{-1}}{(55.5\ \text{M})(0.10\ \text{M})}=1.8\times10^{-6}\,\text{M}^{-2}\,\text{s}^{-1}$$

Determination of Reaction Order and Rate Constants

12.8 DIFFERENTIAL METHOD

The *differential method* determines $\pm dC_i/dt$ for substance i as a function of concentration for that substance when an excess or fixed amount of the remaining materials is used. The rate equation for the pseudo-order reaction in that component only is

$$\pm\frac{dC_i}{dt}=kC_i^n \qquad (12.21)$$

which upon taking logarithms gives

$$\log\left(\pm\frac{dC_i}{dt}\right) = \log k + n \log C_i \qquad (12.22)$$

A plot of $\log(\pm dC_i/dt)$ against $\log C_i$ will give a straight line with a slope of n for that component and an intercept of $\log k$, where k is the rate constant for the pseudo-order reaction.

EXAMPLE 12.7. The rate law for the gaseous reaction of nitric oxide and hydrogen is

$$-\frac{dP(\text{NO})}{dt} = kP(\text{NO})^x P(\text{H}_2)^y$$

In one series of experiments in which the initial pressure of hydrogen was held constant, the initial decrease in the partial pressure of NO was 200. Pa s^{-1} for $P(\text{NO})_0 = 0.479$ bar and 137 Pa s^{-1} for $P(\text{NO})_0 = 0.400$ bar. What is the order of reaction with respect to NO?

Taking a ratio of $[-dP(\text{NO})/dt]_0$ for both experiments gives

$$\frac{-[dP(\text{NO})/dt]_{0,2}}{-[dP(\text{NO})/dt]_{0,1}} = \frac{kP(\text{NO})_2{}^x P(\text{H}_2)_2{}^y}{kP(\text{NO})_1{}^x P(\text{H}_2)_1{}^y} = \left(\frac{P(\text{NO})_2}{P(\text{NO})_1}\right)^x$$

Taking logarithms, solving for x, and substituting the data gives

$$x = \frac{\log\{-[dP(\text{NO})/dt]_{0,2}/-[dP(\text{NO})/dt]_{0,1}\}}{\log\{[P(\text{NO})]_2/[P(\text{NO})]_1\}}$$

$$= \frac{\log[(200.\ \text{Pa s}^{-1})/(137\ \text{Pa s}^{-1})]}{\log[(0.479\ \text{bar})/(0.400\ \text{bar})]} = 2.10 \approx 2$$

The reaction is second-order with respect to NO.

12.9 INTEGRAL METHODS

The *graphical integral method* for determining n and k in (12.21) is a trial-and-error procedure. It begins by plotting $\log C_i$ against t, giving a linear plot only if $n = 1$ and a curved plot if $n \neq 1$; it continues by plotting C_i^{1-n} against t for various values of $n \neq 1$. The plot that turns out to be linear specifies n and has a slope given by $(n-1)k$. The *mathematical integral method* is another trial-and-error procedure, which solves for the value of k in the integrated form of the rate equation for various values of n until one value of n is found that gives the same rate constant for all the data. The integral methods usually require concentration-time data over several half-life periods in order to be accurate.

12.10 HALF-LIFE METHOD

The *half-life method* is based on measurements of the time required for one-half of the substance to disappear, as a function of $C_{i,0}$. For a first-order reaction, $t_{1/2}$ is independent of $C_{i,0}$, and for order $n \neq 1$,

$$\log t_{1/2} = \log\frac{2^{n-1}-1}{(n-1)k} - (n-1)\log C_{i,0} \qquad (12.23)$$

Thus a plot of $\log t_{1/2}$ against $\log C_{i,0}$ will give a straight line having slope $-(n-1)$ and intercept $\log[(2^{n-1}-1)/(n-1)k]$.

12.11 POWELL-PLOT METHOD

The *Powell-plot method* for determining n in (12.21) is a systematic graphical approach based on the results of (12.37). A plot of experimental values of α, the fraction of an unreacted reactant

$(\alpha = C_i / C_{i,0})$, against $\log t$ is made on the same scale as a master plot of α against $\log \phi$, where $\phi = k C_{i,0}^{n-1} t$. The experimental plot is placed over the master plot and is moved horizontally (keeping the $\log t$ and $\log \phi$ axes together) until a match is obtained between the experimental curve and one of the master curves. The order of reaction corresponds to the value of n from the matching curve on the master plot. A master Powell plot is given in Fig. 12-2.

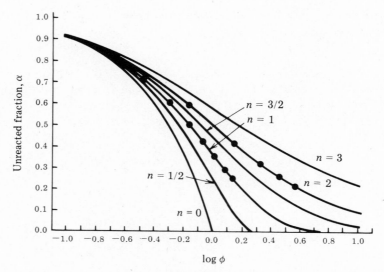

Fig. 12-2

12.12 RELAXATION METHODS

Relaxation methods allow the study of fast reactions. A reaction mixture at equilibrium is disturbed by a pressure, thermal, or electric shock and allowed to return to equilibrium. The restoration is always first-order for small displacements from equilibrium, giving

$$\Delta C_i = \Delta C_{i,0}\, e^{-t/\tau} \tag{12.24}$$

where ΔC_i is the displacement from equilibrium at time t, $\Delta C_{i,0}$ is the initial displacement from equilibrium, and τ is the *relaxation time*. Upon differentiation, (12.24) gives

$$\frac{d(\Delta C_i)}{dt} = -\Delta C_i \left(\frac{1}{\tau}\right) \tag{12.25}$$

EXAMPLE 12.8. Derive the relationship between the relaxation time and k_1 and k_2 for the reaction

$$A + B \underset{k_2}{\overset{k_1}{\rightleftharpoons}} C$$

Assuming $\tau = 2.0\ \mu s$ for $[A] = [B] = 1.0$ M and $3.3\ \mu s$ for $[A] = [B] = 0.5$ M, find k_1, k_2, and K.

The rate equation for the system is

$$\frac{dC_C}{dt} = k_1 C_A C_B - k_2 C_C$$

and at equilibrium $k_1[A][B] - k_2[C] = 0$. For a small increase in C_C, ΔC_C,

$$C_C = [C] + \Delta C_C \qquad C_B = [B] - \Delta C_C \qquad C_A = [A] - \Delta C_C$$

The rate equation for the decay of ΔC_C is

$$
\begin{aligned}
\frac{d(\Delta C_C)}{dt} &= k_1([A] - \Delta C_C)([B] - \Delta C_C) - k_2([C] + \Delta C_C) \\
&= k_1([A][B] - [B]\Delta C_C - [A]\Delta C_C + \Delta C_C{}^2) - k_2([C] + \Delta C_C) \\
&= (k_1[A][B] - k_2[C]) - (k_1[A] + k_1[B] + k_2)\Delta C_C \\
&= -(k_1[A] + k_1[B] + k_2)\Delta C_C
\end{aligned}
$$

where the results at equilibrium were substituted and the term $\Delta C_C{}^2$ was neglected. Upon comparison to (*12.15*),

$$
\frac{1}{\tau} = k_1[A] + k_1[B] + k_2 = k_2 + k_1([A] + [B])
$$

which implies that a plot of $1/\tau$ against $[A] + [B]$ will be linear with a slope of k_1 and an intercept of k_2. Rather than making a plot for two sets of data, the simultaneous equations

$$
\frac{1}{2.0 \times 10^{-6}} = k_2 + k_1(1 + 1) \qquad \frac{1}{3.3 \times 10^{-6}} = k_2 + k_1(0.5 + 0.5)
$$

can be solved, giving $k_1 = 2 \times 10^5$ and $k_2 = 1 \times 10^5$. Then

$$
K = \frac{k_1}{k_2} = 2
$$

12.13   EXPERIMENTAL PARAMETERS

In many experiments C_i is not measured, but some related property (P, pH, absorbancy, etc.) may be the basis for writing the rate equation. For a first-order reaction there is no change in the value of k, but for other orders the value and units of k must be changed.

EXAMPLE 12.9.   Consider the chromate-dichromate reaction

$$
H_2O(l) + Cr_2O_7{}^{2-}(aq) \longrightarrow 2CrO_4{}^{2-}(aq) + 2H^+(aq)
$$

How is pH is related to $C(Cr_2O_7{}^{2-})$?

From Example 12.1,

$$
\frac{1}{2} \frac{d}{dt}[C(H^+)] = -\frac{d}{dt}[C(Cr_2O_7{}^{2-})]
$$

which integrates into

$$
C(H^+) = [C(H^+)]_0 - 2\{C(Cr_2O_7{}^{2-}) - [C(Cr_2O_7{}^{2-})]_0\}
$$

By definition, pH $= -\log C(H^+)$. Hence the desired relation is

$$
\text{pH} = -\log[10^{-(\text{pH})_0} - 2\{C(Cr_2O_7{}^{2-}) - [C(Cr_2O_7{}^{2-})]_0\}]
$$

Rate Equations for Complex Reactions

12.14   DIFFERENTIAL RATE EQUATIONS

The rate equation for a series of reactions in a complex mechanism is written as a sum of the rate equations for the simple reactions making up the complex mechanism. In the expression for dC_i/dt, a term appears for each reaction in which substance i appears or disappears.

EXAMPLE 12.10. For the set of reactions

$$A+B \underset{k_{-1}}{\overset{k_1}{\rightleftharpoons}} C \qquad C+B \xrightarrow{k_2} D$$

find $-dC_A/dt$, $-dC_B/dt$, dC_C/dt, and dC_D/dt.

The substance A appears only in the first equation, but because it both reacts in and is formed by this reaction, the rate equation contains two terms:

$$-\frac{dC_A}{dt} = k_1 C_A C_B - k_{-1} C_C$$

The substance B appears in both equations, reacting and being formed in the first and reacting in the second, giving three terms for the rate expression:

$$-\frac{dC_B}{dt} = k_1 C_A C_B - k_{-1} C_C + k_2 C_C C_B$$

In the same manner,

$$\frac{dC_C}{dt} = k_1 C_A C_B - k_{-1} C_C - k_2 C_C C_B \qquad \frac{dC_D}{dt} = k_2 C_C C_B$$

12.15 STEADY-STATE APPROXIMATION

The *steady-state approximation* assumes that the concentrations of certain intermediate substances reach constant values (i.e., $dC_i/dt = 0$ for these substances).

EXAMPLE 12.11. Assuming that $dC_C/dt = 0$ for the set of reactions given in Example 12.10, find $-dC_A/dt$, $-dC_B/dt$, and dC_D/dt in terms of C_A, C_B, and C_D.

Assuming that $dC_C/dt = k_1 C_A C_B - k_{-1} C_C - k_2 C_C C_B = 0$ gives

$$C_C = \frac{k_1 C_A C_B}{k_{-1} + k_2 C_B}$$

which upon substitution into the rate expressions for the other substances gives

$$-\frac{dC_A}{dt} = k_1 C_A C_B - \frac{k_{-1} k_1 C_A C_B}{k_{-1} + k_2 C_B} = \frac{k_1 k_2 C_A C_B^2}{k_{-1} + k_2 C_B}$$

$$-\frac{dC_B}{dt} = k_1 C_A C_B - \frac{k_{-1} k_1 C_A C_B}{k_{-1} + k_2 C_B} + \frac{k_2 C_B k_1 C_A C_B}{k_{-1} + k_2 C_B} = \frac{2 k_1 k_2 C_A C_B^2}{k_{-1} + k_2 C_B}$$

$$\frac{dC_D}{dt} = \frac{k_1 k_2 C_A C_B^2}{k_{-1} + k_2 C_B}$$

EXAMPLE 12.12. If $k_2 C_B \gg k_{-1}$, what is the pseudo-order of the formation of D in Example 12.11? If $k_2 C_B \ll k_{-1}$, what is the pseudo-order?

If $k_2 C_B \gg k_{-1}$, $k_{-1} + k_2 C_B$ can be replaced by $k_2 C_B$, giving $dC_D/dt = k_1 C_A C_B$, which is pseudo-second-order. For $k_2 C_B \ll k_{-1}$, $k_{-1} + k_2 C_B$ can be replaced by k_{-1}, giving $dC_D/dt = k_3 C_A C_B^2$, a pseudo-third-order reaction, where $k_3 = k_1 k_2/k_{-1}$.

12.16 OPPOSING REACTIONS AND EQUILIBRIUM

At equilibrium, opposing reactions proceed at equal rates, yielding a net rate of change of zero for each component. It is possible to collect all concentration terms on one side and all rate constants on the other side of an equation and equate the latter side to the thermodynamic equilibrium constant K.

EXAMPLE 12.13.   Consider the opposing second-order reactions

$$A_2 + B_2 \underset{k_{-2}}{\overset{k_2}{\rightleftharpoons}} 2AB$$

Derive an expression for K in terms of k_2 and k_{-2}.

For component A at equilibrium,

$$-\frac{d}{dt} C(A_2) = k_2 C(A_2) C(B_2) - k_{-2} C_{AB}{}^2 = 0$$

To find K we solve the above equation for the ratio of product concentrations to reactant concentrations (see Sec. 7.1), obtaining

$$\frac{[C(AB)]^2}{C(A_2) C(B_2)} = \frac{k_2}{k_{-2}} = K$$

12.17 CONSECUTIVE FIRST-ORDER REACTIONS

Consider the consecutive first-order reactions

$$A \xrightarrow{k_1} B \qquad\qquad B \xrightarrow{k_2} C$$

The concentrations at time t will be

$$C_A = C_{A,0}\, e^{-k_1 t} \tag{12.26a}$$

$$C_B = \frac{k_1 C_{A,0}}{k_2 - k_1} (e^{-k_1 t} - e^{-k_2 t}) \tag{12.26b}$$

$$C_C = C_{A,0} \left(1 - \frac{k_2\, e^{-k_1 t} - k_1\, e^{-k_2 t}}{k_2 - k_1} \right) \tag{12.26c}$$

assuming that $C_{B,0} = C_{C,0} = 0$. As a check, note that $C_A + C_B + C_C = C_{A,0}$.

12.18 COMPETING (PARALLEL) REACTIONS

Quite often the reactants can form several different sets of products. The proper choice of a catalyst or temperature conditions can often alter the values of the parallel rate constants so as to favor one reaction.

EXAMPLE 12.14.   For the series of competing reactions

$$H + HO_2 \xrightarrow{k_1} H_2 + O_2 \qquad H + HO_2 \xrightarrow{k_2} 2OH \qquad H + HO_2 \xrightarrow{k_3} H_2O + O$$

Westenberg and deHass report $k_1 : k_2 : k_3 = 0.62 : 0.27 : 0.11$. Find the ratio of the products at time t.

The rate equation for the reactions is

$$-\frac{dC(H)}{dt} = k_1 C(H) C(HO_2) + k_2 C(H) C(HO_2) + k_3 C(H) C(HO_2) = k C(H) C(HO_2)$$

where $k = k_1 + k_2 + k_3$. The given ratio of rate constants implies that 62% of the reactants will form the products of the first reaction, 27% will form the products of the second reaction, etc. Thus the ratio of the products will be

$$[C(H_2) = C(O_2)] : C(OH) : [C(H_2O) = C(O)] = 0.62 : 0.54 : 0.11$$

Radioactive Decay

12.19   DECAY CONSTANT AND HALF-LIFE

Radioactive decay follows first-order kinetics (see Sec. 12.4), in which the number of nuclei undergoing decay is directly proportional to the number present. Expressing this mathematically gives

$$-\frac{dN}{dt} = \lambda N \qquad (12.27)$$

where N is the number of radioactive nuclei at time t and λ is the *decay constant*. The integrated form of (12.27) is

$$\ln \frac{N}{N_0} = -\lambda t \qquad (12.28)$$

The *activity* (A) is defined as

$$A = c\lambda N \qquad (12.29)$$

where c, the *detection coefficient*, depends on the detection instrument, geometrical arrangement of experiment, etc., and in the ideal case is unity. Substituting (12.29) into (12.28) gives

$$\ln \frac{A}{A_0} = -\lambda t \qquad (12.30)$$

The SI unit for activity is the *becquerel* ($1 \text{ Bq} = 1 \text{ s}^{-1}$). Other commonly used units are the *curie* ($1 \text{ Ci} = 3.7 \times 10^{10} \text{ Bq}$) and the *rutherford* ($1 \text{ rd} = 1 \times 10^6 \text{ Bq}$).

EXAMPLE 12.15.   Find the relationship between λ and the half-life of a radioactive nuclide.

The half-life is the time at which $A = A_0/2$. Substituting into (12.30) gives

$$\ln\left(\frac{A_0/2}{A_0}\right) = -0.693\ 15 = -\lambda t_{1/2}$$

or

$$\lambda t_{1/2} = 0.693\ 15 \qquad (12.31)$$

12.20   SUCCESSIVE DECAYS

If the "daughter nuclide" formed by the decay of the "parent nuclide" also undergoes decay, then the number of daughter nuclei present at time t is given by (see Sec. 12.17)

$$N_2 = \lambda_1 N_{1,0} \frac{e^{-\lambda_1 t} - e^{-\lambda_2 t}}{\lambda_2 - \lambda_1} + N_{2,0}\, e^{-\lambda_2 t} \qquad (12.32)$$

where the subscripts 1 and 2 refer to the parent and daughter nuclides, respectively.

Three special cases of (12.32) are often considered:

1. *Transient equilibrium*, in which the parent has a longer half-life than the daughter ($\lambda_1 < \lambda_2$) and

$$\frac{N_1}{N_2} \approx \frac{\lambda_2 - \lambda_1}{\lambda_1} \qquad (12.33)$$

2. *Secular equilibrium*, in which the activity of the parent does not decrease much during several half-lives of the daughter ($\lambda_1 \ll \lambda_2$) and

$$\frac{N_1}{N_2} \approx \frac{\lambda_2}{\lambda_1} \qquad (12.34)$$

3. *Disequilibrium*, in which case the daughter outlives the parent ($\lambda_1 > \lambda_2$) and (12.32) must be used.

12.21 RADIOACTIVE DATING

For determining the age of once-living carbon-containing materials up to 60 000 years old, $^{14}_{6}C$ dating is used. The initial activity per unit mass is assumed to be $12.6\ \text{min}^{-1}\,\text{g}^{-1}$, and the half-life is 5 730 yr.

Geological dating of potassium-bearing rocks can be done by measuring the ratio of atoms of $^{40}_{18}Ar$ to atoms of $^{40}_{19}K$ in the sample. Because $^{40}_{19}K$ undergoes two types of decay, (12.30) becomes

$$t = \frac{1}{\lambda_{EC} + \lambda_{\beta^-}} \ln \left[1 + \frac{(\lambda_{EC} + \lambda_{\beta^-}) N(^{40}Ar)}{\lambda_{EC} N(^{40}K)} \right] \tag{12.35}$$

where $\lambda_{EC} = 5.85 \times 10^{-11}\ \text{yr}^{-1}$ for electron capture and $\lambda_{\beta^-} = 4.72 \times 10^{-10}\ \text{yr}^{-1}$ for β^- decay.

If a rock contains uranium, its age can be determined by the *lead–lead technique* (assuming no loss of lead or intermediates in the decay series) using

$$\frac{N(^{207}Pb)}{N(^{206}Pb)} = (7.25 \times 10^{-3}) \frac{e^{\lambda_{235}t} - 1}{e^{\lambda_{238}t} - 1} \tag{12.36}$$

where $t_{1/2,235} = 7.1 \times 10^8$ yr and $t_{1/2,238} = 4.51 \times 10^9$ yr. This procedure is not good for ages less than 10^9 yr.

EXAMPLE 12.16. One sample of rock No. 12013 from the Apollo 12 exploration contained an extremely large amount of ^{40}Ar. If the moon rock contains $w(K) = 0.016\ 6$, and if there is $81.9 \times 10^{-7}\ \text{m}^3$ of ^{40}Ar at STP per kilogram of rock, calculate the age of the rock.

The amount of ^{40}Ar in 1 kg of rock is

$$n = (81.9 \times 10^{-7}\ \text{m}^3) \left(\frac{10^3\ \text{L}}{1\ \text{m}^3} \right) \left(\frac{1\ \text{mol}}{22.4\ \text{L}} \right) = 3.66 \times 10^{-4}\ \text{mol kg}^{-1}$$

and the number of atoms is

$$N(^{40}Ar) = (3.66 \times 10^{-4})L$$

Not all of the ^{40}Ar present in the sample originated from the ^{40}K decay. Experimentally for this sample, the $^{40}Ar/^{36}Ar$ ratio was determined as 52 700, whereas in normal Ar the ratio is 296. Applying a correction factor of

$$\frac{52\ 700 - 296}{52\ 700} = 0.994$$

to $(3.66 \times 10^{-4})L$ gives the amount of ^{40}Ar formed from the ^{40}K as

$$(0.994)(3.66 \times 10^{-4})L = 3.64 \times 10^{-4}\ L$$

The amount of potassium in the 1-kg sample is given by

$$\frac{(1.66 \times 10^{-2})(1\ 000\ \text{g})}{39.1\ \text{g (mol K)}^{-1}} = 0.425\ \text{mol K}$$

Of this, a fraction 1.18×10^{-4} is ^{40}K. Hence, the number of ^{40}K atoms per kg of rock is

$$N(^{40}K) = (0.425)(L)(1.18 \times 10^{-4}) = (5.01 \times 10^{-5})L$$

Equation (12.35) now gives

$$t = \frac{1}{5.31 \times 10^{-10}} \ln \left[1 + \frac{(5.31 \times 10^{-10})(3.64 \times 10^{-4}\ L)}{(5.85 \times 10^{-11})(5.01 \times 10^{-5}\ L)} \right] = 7.92 \times 10^9\ \text{yr}$$

Comparison of this answer to those for Problems 12.19 and 12.51 indicates that possibly a sampling error was present.

Solved Problems

RATE EQUATIONS FOR SIMPLE REACTIONS

12.1. The experimental rate of reaction for

$$2NO_2(g) + F_2(g) \rightarrow 2NO_2F(g)$$

has been found experimentally to be first-order with respect to NO_2 and first-order with respect to F_2. Express the rates of change of the partial pressures of the reactants and product in terms of the rate of reaction. Is this an elementary reaction?

The rate of reaction is given by (*12.3*) as

$$\frac{d\xi}{dt} = -\frac{1}{2}\frac{dn(NO_2)}{dt} = -\frac{dn(F_2)}{dt} = \frac{1}{2}\frac{dn(NO_2F)}{dt}$$

Substituting (*1.6*) for each gas gives

$$\left(\frac{RT}{V}\right)\left(\frac{d\xi}{dt}\right) = -\frac{1}{2}\frac{dP(NO_2)}{dt} = -\frac{dP(F_2)}{dt} = \frac{1}{2}\frac{dP(NO_2F)}{dt}$$

If the reaction were elementary, the experimental reaction order would be first-order with respect to F_2 and second-order with respect to NO_2. Thus we can conclude that the reaction is not elementary.

12.2. Consider a photochemical reaction in which one molecule of A will react for every photon of light energy absorbed: $A + h\nu \longrightarrow$ products. Assume that five photons are being absorbed each second. If $C_A = 1$ M, what will be the rate of the reaction? If $C_A = 2$ M, what will be the rate of the reaction?

If five photons are being absorbed each second, only five molecules of A can react each second, no matter what the concentration of A is (as long as it can supply the molecules as needed). The rate of reaction in both cases, pseudo-zero-order in concentration, is the same and is independent of C_A.

12.3. The concentrations of bromine at various times after flash photolysis of a bromine-SF_6 mixture with $C(Br_2)/C(SF_6) = 3.2 \times 10^{-2}$ were reported by DeGraff and Lang as

$C(Br)/(10^{-5}$ M)	2.58	1.51	1.04	0.80	0.67	0.56
$t/(\mu s)$	120	220	320	420	520	620

Given that these data are for the reaction

$$2Br \xrightarrow{k} Br_2$$

show that the reaction is pseudo-second-order, and calculate k.

This second-order reaction is described by (*12.12*) and (*12.13*). A plot of $1/C(Br)$ against t is linear (see Fig. 12-3), having a slope of 2.79×10^8 $M^{-1} s^{-1} = k$.

12.4. In the presence of an acidic solution of phenol, the iodate ion is reduced to the iodite ion by Br^-, according to the reaction

$$IO_3^- + 2Br^- + 2H^+ \longrightarrow IO_2^- + Br_2 + H_2O$$

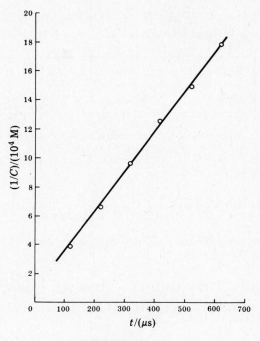

Fig. 12-3

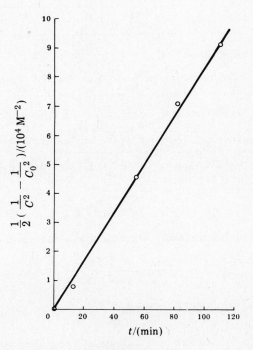

Fig. 12-4

With $[C(IO_3^-)]_0 = 5.00 \times 10^{-3}$ M and $[C(Br^-)]_0 = 1.00 \times 10^{-2}$ M, the following data are reported by Sharma and Gupta for a solution at 35 °C having $C(C_6H_5OH) = 2 \times 10^{-2}$ M:

$C(IO_3^-)/(10^{-3}$ M)	5.00	4.23	2.76	2.35	2.12
$t/(min)$	0.0	12.8	54.8	82.1	110.1

Show that these data correspond to a pseudo-third-order reaction, and find k.

Because $[C(Br^-)]_0 = 2[C(IO_3^-)]_0$, the integrated rate equation (12.20) must be used. As a sample calculation, for $t = 12.8$ min,

$$\frac{1}{2}\left(\frac{1}{C^2} - \frac{1}{C_0^2}\right) = \frac{1}{2}\left[\frac{1}{(4.23 \times 10^{-3})^2} - \frac{1}{(5.00 \times 10^{-3})^2}\right] = 0.79 \times 10^4 \text{ M}^{-2}$$

A plot of the values of the left-hand side of (12.20) against t is linear (see Fig. 12-4), with a slope of $k = 8.50 \times 10^2$ M^{-2} min^{-1} = 14.2 M^{-2} s^{-1}.

DETERMINATION OF REACTION ORDER AND RATE CONSTANTS

12.5.   Determine a and b in the rate equation

$$-\left(\frac{dC_A}{dt}\right) = kC_A{}^a C_B{}^b$$

given

$-(dC_A/dt)/(\text{M s}^{-1})$	0.05	0.10	0.20	0.40
$C_{A,0}/(\text{M})$	1	1	2	2
$C_{B,0}/(\text{M})$	1	2	1	2

Calculate k for this reaction.

To determine the order for substance B, substitute the first two entries ($C_{A,0}$ fixed) of the data into (12.22), obtaining

$$\log 0.05 = \log k_B + b \log 1 \qquad \log 0.10 = \log k_B + b \log 2$$

which upon solving gives $b = 1$. Likewise, using the first and third entries ($C_{B,0}$ fixed) gives

$$\log 0.05 = \log k_A + a \log 1 \qquad \log 0.20 = \log k_A + a \log 2$$

whence $a = 2$. Thus the rate equation is

$$-(dC_A/dt) = kC_A{}^2 C_B$$

To determine k, the rate equation is solved for k, and each set of data substituted; for example,

$$k = \frac{\text{rate}}{C_A{}^2 C_B} = \frac{0.20 \text{ M s}^{-1}}{(2 \text{ M})^2 (1 \text{ M})} = 0.05 \text{ M}^{-2} \text{ s}^{-1}$$

The average value calculated from all data is 0.05 M^{-2} s^{-1}.

12.6.   The half-life for a given reaction was halved as the initial concentration of a reactant was doubled. What is n for this component?

Substituting the data into (*12.23*) gives

$$\log t_{1/2} = \log \frac{2^{n-1}-1}{(n-1)k} - (n-1)\log C_{i,0}$$

$$\log \frac{t_{1/2}}{2} = \log \frac{2^{n-1}-1}{(n-1)k} - (n-1)\log 2C_{i,0}$$

and solving by subtraction gives $n = 2$.

12.7. Repeat the calculations of Problem 12.3 using the half-life method.

In this case, the data are not in the form of values of $t_{1/2}$ for various values of $C_{i,0}$. These can be determined, however, from a plot of C_i against t as shown in Fig. 12-5. The time required for $C(\text{Br})$ to decrease from 2.58×10^{-5} M to 1.29×10^{-5} M is $257\ \mu\text{s} - 120\ \mu\text{s} = 137\ \mu\text{s}$ and the time needed to decrease from 1.29×10^{-5} M to 0.65×10^{-5} M is $536\ \mu\text{s} - 257\ \mu\text{s} = 279\ \mu\text{s}$. The plot of $\log t_{1/2}$ against $\log C_{i,0}$ is linear for the two points and has a slope of -1.03, giving

$$n = 1 - \text{slope} = 1 - (-1.03) = 2.03 \approx 2$$

and an intercept of -8.573, giving

$$\log \left(\frac{2^{2-1}-1}{(2-1)k} \right) = -8.573$$

or $k = 3.74 \times 10^8$ M^{-1} s^{-1}. (Better agreement between values of k determined by the various methods would be obtained if the reaction had been studied over several half-life periods instead of just two periods.)

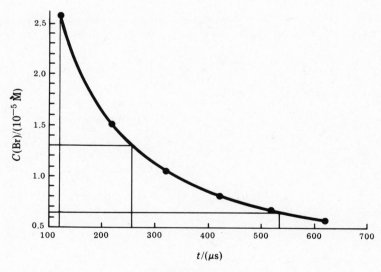

Fig. 12-5

12.8. Confirm the order of reaction using the Powell-plot method for the reaction given in Problem 12.3.

Extrapolation of Fig. 12-3 to $t = 0$ gives $C(\text{Br})_0 = 18.94 \times 10^{-5}$ M. If we use this value of C_0 to determine α, the values of α are such that not enough of the $n = 2$ curve is shown in Fig. 12-2 to make the match. However, if we redefine the data such that $[C(\text{Br})]_0' = 2.58 \times 10^{-5}$ M and $t'/(\mu\text{s}) = t/(\mu\text{s}) - 120$, the curves in Fig. 12-2 can be used. The fraction of unreacted Br atoms at $t = 220\ \mu\text{s}$ is

$$\alpha = \frac{1.51 \times 10^{-5}\ \text{M}}{2.58 \times 10^{-5}\ \text{M}} = 0.585$$

and log $t' = \log(220 \times 10^{-6} - 120 \times 10^{-6}) = -4.000$. Using this result and repeating these calculations for the remaining four data give the five solid dots shown in Fig. 12-2 on the $n = 2$ curve. The plot confirms that the reaction is second-order with respect to Br.

12.9.   Williams and Petrucci studied the system

$$\text{Ni(NCS)}^+ + \text{NCS}^- \underset{k_r}{\overset{k_f}{\rightleftharpoons}} \text{Ni(NCS)}_2$$

at 25 °C in methanol using a pressure-jump technique and obtained the following data:

$C(\text{Ni(NCS)}_2)/(M)$	0.001	0.002	0.005	0.010	0.025	0.05	0.10
$\tau/(\text{ms})$	4.08	3.74	2.63	1.84	1.31	0.88	0.67

Determine k_f, k_r, and K.

In Example 12.8 it was shown that

$$\frac{1}{\tau} = k_r + k_f([\text{Ni(NCS)}^+] + [\text{NCS}^-])$$

with $[\text{Ni(NCS)}^+] = [\text{NCS}^-]$. As for the third component,

$$[\text{Ni(NCS)}_2] = C(\text{Ni(NCS)}_2)\left(1 - \frac{\alpha}{100}\right)$$

where $C(\text{Ni(NCS)}_2)$ is the total (associated plus dissociated) concentration and α is the percent dissociation. Assuming that K will be large, and therefore that α will be small, we have

$$K = \frac{k_f}{k_r} = \frac{[\text{Ni(NCS)}_2]}{[\text{Ni(NCS)}^+]^2} \approx \frac{C(\text{Ni(NCS)}_2)}{[\text{Ni(NCS)}^+]^2}$$

and the formula for the relaxation time becomes

$$\frac{1}{\tau} = k_r + 2(k_f k_r)^{1/2}[C(\text{Ni(NCS)}_2)]^{1/2}$$

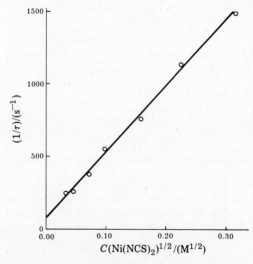

Fig. 12-6

A plot of $1/\tau$ against $[C(\text{Ni(NCS)}_2)]^{1/2}$ (see Fig. 12-6) is linear with a slope of $4\,500\ \text{M}^{-1/2}\,\text{s}^{-1}$ and an intercept of $k_r = 79\ \text{s}^{-1}$. The value of k_f is

$$k_f = \left[\frac{\text{slope}}{2k_r^{1/2}}\right]^2 = \left[\frac{4\,500}{(2)(79)^{1/2}}\right]^2 = 6.4 \times 10^4\ \text{M}^{-1}\,\text{s}^{-1}$$

and the value of K is

$$K = \frac{6.4 \times 10^4}{79} = 810$$

12.10. Consider the gaseous decomposition reaction of cyclopentene to H_2 and cyclopentadiene:

$$\text{cyclo-C}_5\text{H}_8 \longrightarrow \text{H}_2 + \text{cyclo-C}_5\text{H}_6$$

(a) How is dP/dt related to $-dC(\text{C}_5\text{H}_8)/dt$? (b) If the reaction is first order, what are the units on k? (c) Derive the first-order integrated rate equation in terms of $[P(\text{C}_5\text{H}_8)]_0$ and P.

(a)　The total pressure is given by (1.9) as

$$P = P(\text{C}_5\text{H}_8) + P(\text{H}_2) + P(\text{C}_5\text{H}_6)$$

The stoichiometry of the reaction is such that

$$P(\text{H}_2) = P(\text{C}_5\text{H}_6) = [P(\text{C}_5\text{H}_8)]_0 - P(\text{C}_5\text{H}_8)$$

Hence　　$P = P(\text{C}_5\text{H}_8) + 2\{[P(\text{C}_5\text{H}_8)]_0 - P(\text{C}_5\text{H}_8)\} = 2[P(\text{C}_5\text{H}_8)]_0 - P(\text{C}_5\text{H}_8)$

Taking derivatives gives

$$\frac{dP}{dt} = -\frac{dP(\text{C}_5\text{H}_8)}{dt}$$

(b)　The units of k are s^{-1}, independent of whether concentration or pressure is used.

(c)　For a first-order reaction

$$-\frac{d}{dt}[P(\text{C}_5\text{H}_8)] = kP(\text{C}_5\text{H}_8)$$

Then, by (a),

$$\frac{dP}{dt} = kP(\text{C}_5\text{H}_8) = k\{2[P(\text{C}_5\text{H}_8)]_0 - P\}$$

which integrates to

$$\ln\{P(\text{C}_5\text{H}_8)_0 / [2P(\text{C}_5\text{H}_8)_0 - P]\} = kt$$

RATE EQUATIONS FOR COMPLEX REACTIONS

12.11. A proposed mechanism for the reaction between $H_2(g)$ and $Br_2(g)$ is

$$\text{Br}_2 \underset{k_5}{\overset{k_1}{\rightleftharpoons}} 2\,\text{Br}$$

$$\text{Br} + \text{H}_2 \xrightarrow{k_2} \text{HBr} + \text{H}$$

$$\text{H} + \text{Br}_2 \xrightarrow{k_3} \text{HBr} + \text{Br}$$

$$\text{H} + \text{HBr} \xrightarrow{k_4} \text{H}_2 + \text{Br}$$

Write expressions for $dC(\text{HBr})/dt$, $dC(\text{H})/dt$, and $dC(\text{Br})/dt$. Assuming that $dC(\text{H})/dt = dC(\text{Br})/dt = 0$, solve for $dC(\text{HBr})/dt$ in terms of $C(\text{H}_2)$, $C(\text{Br}_2)$, and $C(\text{HBr})$. If $1 \gg (k_4/k_3)[C(\text{HBr})/C(\text{Br}_2)]$, what is the pseudo-order of the reaction?

For the four steps, the rate expressions are

$$\frac{dC(\text{HBr})}{dt} = k_2 C(\text{Br})C(\text{H}_2) + k_3 C(\text{H})C(\text{Br}_2) - k_4 C(\text{H})C(\text{HBr})$$

$$\frac{dC(\text{H})}{dt} = k_2 C(\text{Br})C(\text{H}_2) - k_3 C(\text{H})C(\text{Br}_2) - k_4 C(\text{H})C(\text{HBr}) = 0$$

$$\frac{dC(\text{Br})}{dt} = 2k_1 C(\text{Br}_2) - k_2 C(\text{Br})C(\text{H}_2) + k_3 C(\text{H})C(\text{Br}_2) + k_4 C(\text{H})C(\text{HBr}) - 2k_5[C(\text{Br})]^2 = 0$$

Solving the last two equations simultaneously gives

$$C(\text{Br}) = \left(\frac{k_1}{k_5}\right)^{1/2} [C(\text{Br}_2)]^{1/2}$$

$$C(\text{H}) = k_2 \left(\frac{k_1}{k_5}\right)^{1/2} \frac{[C(\text{Br}_2)]^{1/2}C(\text{H}_2)}{k_3 C(\text{Br}_2) + k_4 C(\text{HBr})}$$

which upon substitution into the expression for $dC(\text{HBr})/dt$ gives

$$\frac{dC(\text{HBr})}{dt} = 2k_2 \left(\frac{k_1}{k_5}\right)^{1/2} \frac{C(\text{H}_2)[C(\text{Br}_2)]^{1/2}}{1 + (k_4/k_3)[C(\text{HBr})/C(\text{Br}_2)]}$$

Upon approximating the denominator, the rate equation can be written as the following pseudo-$1\frac{1}{2}$-order reaction:

$$\frac{dC(\text{HBr})}{dt} = k' C(\text{H}_2)[C(\text{Br}_2)]^{1/2}$$

RADIOACTIVE DECAY

12.12. Calculate the ratio of N/N_0 after an hour has passed, for a material having a half-life of 47.2 s.

Using (*12.31*) gives

$$\lambda = \frac{0.693}{47.2 \text{ s}} = 1.47 \times 10^{-2} \text{ s}^{-1}$$

and (*12.28*) gives

$$\ln \frac{N}{N_0} = -(1.47 \times 10^{-2} \text{ s}^{-1})(1 \text{ h})[(3\,600 \text{ s})/(1 \text{ h})] = -52.9$$

$$\frac{N}{N_0} = 1 \times 10^{-23}$$

12.13. Given $N = 0.798\, N_0$ for a sample at the end of 4.2 day, calculate $t_{1/2}$.

Substituting the data into (*12.20*) gives

$$\lambda = -\frac{\ln 0.798}{4.2 \text{ day}} = 0.054 \text{ day}^{-1}$$

which upon substitution into (*12.31*) gives

$$t_{1/2} = \frac{0.693}{0.054} = 12.8 \text{ day}$$

12.14. $^{214}_{82}\text{Pb}$ undergoes β^- emission, forming $^{214}_{83}\text{Bi}$, which undergoes further decomposition by β^- emission to become $^{214}_{84}\text{Po}$. The half-lives of $^{214}_{82}\text{Pb}$ and $^{214}_{83}\text{Bi}$ are 26.8 min and 19.7 min, respectively. Assuming $N_0 = 100$ atoms for $^{214}_{82}\text{Pb}$, prepare a diagram showing the concentrations of $^{214}_{82}\text{Pb}$, $^{214}_{83}\text{Bi}$, and $^{214}_{84}\text{Po}$ as functions of time up to 100 min.

Substituting $t = t_{1/2}$ and $N = \frac{1}{2}N_0$ into (12.31) and solving gives

$$\lambda = \frac{\ln 2}{t_{1/2}}$$

from which

$$\lambda_1 = \frac{\ln 2}{26.8 \text{ min}} = 2.59 \times 10^{-2} \text{ min}^{-1} \qquad \lambda_2 = \frac{\ln 2}{19.7 \text{ min}} = 3.52 \times 10^{-2} \text{ min}^{-1}$$

As a sample calculation, (12.26) gives, for $t = 10$ min,

$$N_A = 100 \exp[(-2.59 \times 10^{-2})(10)] = 77 \text{ atoms}$$

$$N_B = \frac{(2.59 \times 10^{-2})(100)}{3.52 \times 10^{-2} - 2.59 \times 10^{-2}} \{\exp[(-259 \times 10^{-2})(10)] - \exp[(-3.52 \times 10^{-2})(10)]\} = 19 \text{ atoms}$$

$$N_C = 100 - 77 - 19 = 4$$

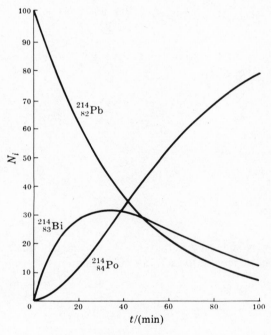

Fig. 12.7

12.15. How will the plot of the concentrations given by (12.26) look for $\lambda_1 > 100\lambda_2$ and for $100\lambda_1 < \lambda_2$?

In both cases N_A will decrease exponentially with a rate constant of λ_1; N_B will essentially be given by $N_{A,0} e^{-\lambda_2 t}$ and $(\lambda_1 N_{A,0}/\lambda_2) e^{-\lambda_1 t}$, respectively; and N_C will essentially be given by $N_{A,0}(1 - e^{-\lambda_2 t})$ and $N_{A,0}(1 - e^{-\lambda_1 t})$, respectively.

12.16. The half-life of $^{212}_{82}\text{Pb}$, which decays to $^{212}_{83}\text{Bi}$, is 10.6 h. The half-life of $^{212}_{83}\text{Bi}$ is 60.5 min. Describe the activity in a sample of $^{212}_{82}\text{Pb}$ after equilibrium has been established.

The total activity of the sample will decay with a half-life equal to 10.6 h, and the ratio of parent to daughter will be given by (12.33) as

$$\frac{N_1}{N_2} = \frac{(0.693/t_{1/2,2}) - (0.693/t_{1/2,1})}{(0.693/t_{1/2,1})} = \frac{60/60.5 - 1/10.6}{1/10.6} = 9.51$$

12.17. What mass of $^{220}_{86}$Rn having $t_{1/2} = 54.5$ s is equivalent to 1 mCi?

Using (12.31) to determine λ gives

$$\lambda = \frac{0.693}{54.5\text{ s}} = 1.27 \times 10^{-2}\text{ s}^{-1}$$

The definition of a curie gives for the millicurie

$$-\frac{dN}{dt} = 3.7 \times 10^7\text{ s}^{-1}$$

Solving (12.27) for N and substituting values gives

$$N = \frac{3.700 \times 10^7\text{ s}^{-1}}{1.27 \times 10^{-2}\text{ s}^{-1}} = 2.91 \times 10^9$$

which can be converted to mass by

$$\frac{2.91 \times 10^9}{6.022 \times 10^{23}\text{ mol}^{-1}}(220\text{ g mol}^{-1}) = 1.06 \times 10^{-15}\text{ kg}$$

12.18. A sample of wood from an Egyptian tomb gave a ^{14}C activity per unit mass of 7.3 min^{-1} g^{-1}. What is the age of the wood?

Using (12.31) gives

$$\lambda = \frac{0.693}{5\,730} = 1.21 \times 10^{-4}\text{ yr}^{-1}$$

and using (12.30) gives

$$\ln\frac{7.3}{12.6} = -0.546 = -1.21 \times 10^{-4}t$$

which upon solving gives $t = 4\,510$ yr.

12.19. A current nuclear theory suggests that ^{235}U/^{238}U was nearly unity at the time of the formation of the elements. If the current ratio is 7.25×10^{-3}, calculate the age of the elements.

Using (12.31) for the isotopes gives

$$\lambda_{235} = \frac{0.693}{7.1 \times 10^8} = 9.76 \times 10^{-10}\text{ yr}^{-1}$$

and $\lambda_{238} = 1.54 \times 10^{-10}$ yr^{-1}. The exponential form of (12.28) gives for the current ratio

$$7.25 \times 10^{-3} = (1)\exp[-(\lambda_{235} - \lambda_{238})t] = \exp(-8.22 \times 10^{-10}t)$$

Solving for t gives

$$t = -\ln\frac{7.25 \times 10^{-3}}{8.22 \times 10^{-10}} = 6.0 \times 10^9\text{ yr}$$

Supplementary Problems

RATE EQUATIONS FOR SIMPLE REACTIONS

12.20. For the reaction

$$H_2(g) + I_2(g) \longrightarrow 2HI(g)$$

the rate of reaction has been found experimentally to be first-order wth respect to H_2 and to I_2. Write the expression for the rate of reaction, and express the rates of change of the concentrations of the reactants and product in terms of the rate of reaction. Could this be an elementary reaction?

Ans. $(V^{-1})(d\xi/dt) = -dC(H_2)/dt = -dC(I_2)/dt = \frac{1}{2}dC(HI)/dt$;

Although the reaction-order data suggest that this reaction could be elementary, other data suggest that it is not.

12.21. Discuss the following overall reactions with respect to order, catalysts, etc.

(*a*) $2A \longrightarrow 4B+C$, $-\dfrac{dC_A}{dt} = kC_A$ (*b*) $A+B \longrightarrow C$, $\dfrac{dC_C}{dt} = kC_A C_B^{-1}$

(*c*) $2A+B \longrightarrow 2C+D+B$, $-\dfrac{dC_A}{dt} = kC_A C_B$ (*d*) $2A+B \longrightarrow 2C$, $\dfrac{dC_C}{dt} = kC_A C_C^{-1/2}$

(*e*) $2A+B \longrightarrow 2C$, $\dfrac{dC_C}{dt} = kC_A^2 C_B$ (*f*) $A+2B \longrightarrow C+D$, $\dfrac{dC_C}{dt} = kC_A C_B^2 C_C$

Ans.   (*a*) First-order in A, first-order overall
        (*b*) First-order in A, negative first-order in B (reactant is inhibitor), zero-order overall
        (*c*) First-order in A, first-order in B (catalyst), second-order overall
        (*d*) First-order in A, negative $\frac{1}{2}$-order in C (product is inhibitor), $\frac{1}{2}$-order overall
        (*e*) Second-order in A, first-order in B, third-order overall
        (*f*) First-order in A, second-order in B, first-order in C (product is catalyst), fourth-order overall

12.22. Find the overall order of a reaction for which the half-life and the units of k do not involve concentration.

Ans.   For a first-order reaction, $t_{1/2} = (\ln 2)/k$ and units of k are s^{-1}.

12.23. (*a*)   Integrate the rate equation

$$-\frac{dC}{dt} = kC^{1/2}$$

(*b*)   How could a group of data be checked graphically to see if they describe a half-order reaction?
(*c*)   Derive an expression for $t_{1/2}$ in terms of k and C_0.   (*d*)   What are the units of k?

Ans.   (*a*) $2(C_0^{1/2} - C^{1/2}) = kt$ (*b*) Plot of $C^{1/2}$ against t would be linear
        (*c*) $t^{1/2} = 0.586\, C_0^{1/2}/k$ (*d*) $mol^{1/2}\, dm^{-3/2}\, s^{-1}$

12.24. Consider a decomposition reaction, $A \longrightarrow$ products, that is catalyzed by a finely divided solid. If there are only enough "active sites" on the catalyst that five molecules of A can react each second, what is the rate of reaction for a 1 M solution of A? Does this change for a 2 M solution?

Ans.   5 molecules s^{-1};   no

12.25. In the region of constant Co(III) concentration, the following data were obtained by Scott and Chester for the autoxidation of toluene in acetic acid at 88 °C;

$C(C_6H_5CH_3)/(M)$	0.282	0.229	0.200	0.168	0.130
$t/(min)$	104	143	192	252	321

Determine k for this first-order reaction.   *Ans.* $5.67 \times 10^{-5}\, s^{-1}$

12.26. The reaction between ozone and CS_2 was studied at 29.3 °C with excess CS_2 to find the order of reaction with respect to O_3. Using the following data by Olszyna and Heicklen, prove that the reaction is second-order, and determine the pseudo-second-order rate constant.

$t/(min)$	0.0	0.5	1.0	2.0	3.0	4.0
$P(O_3)/(bar)$	235	139	105	69	49	39

Ans.   Plot of $1/P(O_3)$ against t is linear, slope gives $k = 8.88 \times 10^{-5}$ Pa s^{-1}

12.27. The kinetics for the reaction

$$A + B \xrightarrow{\ k\ } \text{products}$$

are known to be third-order. From the following data, determine m and n in the rate equation

$$-\frac{dC_A}{dt} = kC_A{}^m C_B{}^n$$

by calculating values of k using variations of (12.18) until a constant value is reached.

$t/(s)$	0	5	10
$C_A/(M)$	0.750	0.700	0.665
$C_B/(M)$	0.500	0.450	0.415

Ans. $k = 0.040\,0$ and $0.035\,2$ M^{-2} s^{-1} using (12.18) for $m = 2$ and $n = 1$, not very constant;
$k = 0.060\,5$ and $0.057\,9$ M^{-2} s^{-1} using (12.18) rewritten for $m = 1$ and $n = 2$, rather constant;
$k = 0.059\,2$ M^{-2} s^{-1}

12.28. The second-order reaction

$$OH^- + C_2H_5OH \longrightarrow C_2H_5O^- + H_2O$$

when studied in 0.1 M NaOH gave the pseudo-first-order rate equation

$$-\frac{d}{dt}[C(C_2H_5OH)] = kC(C_2H_5OH)$$

with $k = 3 \times 10^5$ s^{-1}. Find the rate constant for the elementary second-order reaction.

Ans. $k_2 = k/C(OH^-) = 3 \times 10^6$ M^{-1} s^{-1}

DETERMINATION OF REACTION ORDER AND RATE CONSTANTS

12.29. Starting with (12.21), show that

$$C_i^{1-n} = C_{i,0}^{1-n} - (1-n)kt \tag{12.37}$$

for $n \neq 1$. Derive (12.23) from (12.37).

12.30. Determine a, b, and c for the rate equation

$$\text{Rate} = kC_A{}^a C_B{}^b C_C{}^c$$

from the following data:

$-(dC_A/dt)/(10^5 \text{ M s}^{-1})$	5.0	5.0	2.5	14.1
$C_{A,0}/(M)$	0.010	0.010	0.010	0.020
$C_{B,0}/(M)$	0.005	0.005	0.010	0.005
$C_{C,0}/(M)$	0.010	0.015	0.010	0.010

Calculate k for this reaction.

Ans. $a = 3/2$, $b = -1$, $c = 0$, $k = 2.5 \times 10^{-4} \text{ M}^{-1/2} \text{ s}^{-1}$

12.31. For the reaction described in Example 12.8, a series of experiments in which the initial pressure of NO was held constant was performed. The initial decrease in the partial pressure of NO was 213 Pa s^{-1} for $P(H_2)_0 = 0.385$ bar and 147 Pa s^{-1} for $P(H_2)_0 = 0.273$ bar. What is the order of reaction with respect to H_2? What is the overall order of the reaction?   *Ans.* $1.08 \approx 1$, third-order overall

12.32. The thermal decomposition of 3-chloro-3-phenyldiazirine in cyclohexane at 90 °C was observed spectrophotometrically by Liu and Toriyama.

$t/(\text{min})$	0	3	6	9	12	15	18	21	24
A	1.924	1.649	1.377	1.165	0.964	0.813	0.683	0.559	0.472

where the absorbance (A) is directly proportional to C. The reaction was also studied using vapor-phase chromatography, generating the following data:

$t/(\text{min})$	0.00	3.00	6.00	9.00	10.00	15.00	25.00	30.00
Peak area ratio	2.520	2.098	1.716	1.461	1.245	1.014	0.559	0.421

where the peak area ratio is also directly proportional to the concentration. Show that these two data sets indicate the same reaction order and give the same value for k.

Ans.   Plots of $\ln A$ and $\ln$ (peak area ratio) against t are linear (first-order reaction), slopes give $k = 9.8 \times 10^{-4} \text{ s}^{-1}$ and $10.0 \times 10^{-4} \text{ s}^{-1}$.

12.33. Confirm that the absorbance data in Problem 12.32 demonstrate first-order kinetics, using both the half-life method and the Powell-plot method.

Ans. $t_{1/2} = 12.0$ min for $A_0 = 1.924$ and $t_{1/2} = 11.9$ min for $A_0 = 0.962$; $t_{1/2}$ is independent of A_0, giving $n = 1$; values of $\alpha = A/A_0$ are plotted on the $n = 1$ line of Fig. 12-2.

12.34. The half-life for a given reaction was doubled as the initial concentration of a reactant was doubled. What is n for this component?   *Ans.*   0

12.35. Express the relaxation time as a function of k_1 and k_2 for the reaction

$$A \underset{k_2}{\overset{k_1}{\rightleftharpoons}} B$$

Assuming $K = 1.0 \times 10^3$ and $\tau = 10 \, \mu\text{s}$, find k_1 and k_2.

Ans. $\tau^{-1} = k_1 + k_2$; $K = k_1/k_2$, $k_1 = 1 \times 10^5 \text{ s}^{-1}$, $k_2 = 1 \times 10^2 \text{ s}^{-1}$

12.36. Show that the relaxation time for the reaction

$$HIn^- \underset{k_2}{\overset{k_1}{\rightleftharpoons}} H^+ + In^{2-}$$

where HIn^- is bromocresol green, is given by

$$\frac{1}{\tau} = k_2[C(H^+) + C(In^{2-})] + k_1$$

From the following data reported by Warrick, Auborn, and Eyring, prepare a plot of $1/\tau$ against $C(H^+) + C(In^{2-})$, and determine k_1, k_2, and the equilibrium constant K.

$(1/\tau)/(10^6 \text{ s}^{-1})$	1.01	1.16	3.13	5.56	6.62	7.87	11.24	17.2
$[C(H^+) + C(In^{2-})]/(\mu M)$	4.30	6.91	50.94	85.70	100.5	129.1	176.0	286.5

(Concentrations have been corrected for ionic strength changes.)

Ans.   Intercept gives $k_1 = 6.75 \times 10^5 \text{ s}^{-1}$;   slope gives $k_2 = 5.76 \times 10^{10} \text{ (mol L}^{-1})^{-1} \text{ s}^{-1}$; $K = 1.17 \times 10^{-5}$

12.37. Consider the gaseous reaction between H_2 and F_2 to give HF

$$H_2(g) + F_2(g) \longrightarrow 2HF(g)$$

How is dP/dt related to $dP(HF)/dt$?

Ans.   At any time, $P = [P(H_2)]_0 + [P(F_2)]_0$, giving $dP/dt = 0$;   not related.

RATE EQUATIONS FOR COMPLEX REACTIONS

12.38. Consider the following mechanism

$$A + A \underset{k_{-2}}{\overset{k_2}{\rightleftharpoons}} A^* + A \qquad A^* \overset{k_1}{\longrightarrow} \text{products}$$

used to describe the decomposition of a gaseous molecule. (*a*) Write the differential rate equations for $-dC_A/dt$ and dC_{A^*}/dt and assuming a steady-state approximation for C_{A^*}, write $-dC_A/dt$ in terms of C_A and rate constants. (*b*) Under what conditions is this a pseudo-first-order reaction? (*c*) A pseudo-second-order reaction?

Ans.   (*a*) $-\dfrac{dC_A}{dt} = k_2 C_A{}^2 - k_{-2} C_{A^*} C_A,$ (*b*) $k_{-2} C_A \gg 1$

$\dfrac{dC_{A^*}}{dt} = k_2 C_A{}^2 - k_{-2} C_{A^*} C_A - k_1 C_{A^*},$ (*c*) $\dfrac{k_{-2} C_A}{k_1} \ll 1$

$-\dfrac{dC_A}{dt} = \dfrac{k_2 C_A{}^2}{1 + (k_{-2}/k_1) C_A}$

12.39. Consider the opposing first- and second-order reactions

If $k_2 = 1.35 \times 10^3 \text{ s}^{-1}$ and $k_{-2} = 2.41 \times 10^6 \text{ M}^{-1} \text{ s}^{-1}$, as reported by Cheung and Swinehart, determine K.

Ans. $K = k_2/k_{-2} = 5.60 \times 10^{-4}$

12.40. The proposed mechanism for the saponification of dimethyl glutarate (DMGL) by NaOH is

$$NaOH + H_3COOC(CH_2)_3COOCH_3 \xrightarrow{k_1} NaOOC(CH_2)_3COOCH_3 + CH_3OH$$

$$NaOH + NaOOC(CH_2)_3COOCH_3 \xrightarrow{k_2} NaOOC(CH_2)_3COONa + CH_3OH$$

If k_1 and k_2 are related to the probability of a Na^+ colliding with the CH_3-end of the molecule, would k_1 be greater or less than k_2?

Ans. Greater, because Na^+ can react with either end of DMGL, but with only one end of the monosubstituted salt.

12.41. A mixture of products is obtained during the thermal decomposition of cyclobutanone, as shown by the competing reactions

$$\begin{matrix} H_2C-C=O \\ |\quad\quad | \\ H_2C-CH_2 \end{matrix} \xrightarrow{k_1} C_2H_4 + H_2C=C=O \qquad\qquad \begin{matrix} H_2C-C=O \\ |\quad\quad | \\ H_2C-CH_2 \end{matrix} \xrightarrow{k_2} \begin{matrix} H_2C \\ | \quad >CH_2 + CO \\ H_2C \end{matrix}$$

Write the rate equation for $-dC(C_4H_6O)/dt$ and show that it is first-order. From the following data by McGee and Schleifer at 383 °C for $C(C_4H_6O)_0 = 6.50 \times 10^{-5}$ M, determine k_1, k_2, and the first-order rate constant for $-dC(C_4H_6O)/dt$.

$t/(min)$	0.5	1.0	3.0	6.0
$C(C_2H_4)/(10^{-5}\,M)$	0.31	0.68	1.53	2.63
$C(cyclo\text{-}C_3H_6)/(10^{-7}\,M)$	0.21	0.47	1.24	2.20

Ans. $-dC(C_4H_6O)/dt = k_1C(C_4H_6O) + k_2C(C_4H_6O) = kC(C_4H_6O)$, where $k = k_1 + k_2$;   plot of $\ln\{[C(C_4H_6O)]_0 - C(C_2H_4) - C(cyclo\text{-}C_3H_6)\}$ against t is linear, slope gives $k = 1.39 \times 10^{-3}\,s^{-1}$;   ratios of $C(C_2H_4)/C(cyclo\text{-}C_3H_6)$ give $k_1/k_2 = 120$, $k_1 = 1.38 \times 10^{-3}\,s^{-1}$, $k_2 = 1.15 \times 10^{-5}\,s^{-1}$

12.42. If the proposed mechanism for the reaction in Problem 12.32 is

$$\begin{matrix} \phi \quad\quad N \\ \diagdown \diagup \quad \| \\ C \quad\quad \| \xrightarrow{k_1} N_2 + \phi ClC: \\ \diagup \quad\quad N \\ Cl \end{matrix}$$

where the carbene can undergo further reaction, e.g.,

$$\phi ClC: + \begin{matrix} \phi \quad\quad N \\ \diagdown \diagup \quad \| \\ C \quad\quad \| \xrightarrow{k_2} \phi ClCNNCCl\phi \\ \diagup \quad\quad N \\ Cl \end{matrix}$$

$$\phi ClC: + S(solvent) \xrightarrow{k_3} \text{addition and insertion products}$$

derive the overall expression for $-dC(\phi ClCNN)/dt$, and show that in high concentration of solvent the reaction is pseudo-first order.

Ans. Using steady state,

$$C(\phi ClC:) = \frac{k_1 C(\phi ClCNN)}{k_2 C(\phi ClCNN) + k_3 C(S)}$$

and $$-\frac{d}{dt}[C(\phi ClCNN)] = k_1 C(\phi ClCNN)\left(1 + \frac{k_2 C(\phi ClCNN)}{k_2 C(\phi ClCNN) + k_3 C(S)}\right) \approx k_1 C(\phi ClCNN)$$

RADIOACTIVE DECAY

12.43. $^{214}_{83}Bi$ undergoes β^- emission (99.96%) or α emission (0.04%). Given that the half-life is 19.7 min, find λ_α and λ_β.

Ans. $\lambda = \lambda_\alpha + \lambda_\beta = 5.86 \times 10^{-4}\,s^{-1}$, $\lambda_\alpha = 2.3 \times 10^{-7}\,s^{-1}$, $\lambda_\beta = 5.86 \times 10^{-4}\,s^{-1}$

12.44. Find the amount of time, expressed in units of $t_{1/2}$, at which $A/A_0 = 0.125$.   *Ans.*   3

12.45. Compare the fractions of various radioactive Na nuclei left after 1 h, given $t_{1/2} = 0.39\,s$, 23 s, 2.602 yr, 15.0 h, 60 s, and 1 s for ^{20}Na, ^{21}Na, ^{22}Na, ^{24}Na, ^{25}Na, and ^{26}Na, respectively.

Ans. 10^{-2777}, 8×10^{-48}, 0.999 969 6, 0.954 9, 9×10^{-19}, 10^{-1083}

12.46. From the following data, prepare a plot of log A against t, and determine the half-life for the isotope.

$A/(Bq)$	86	63	45	33	24	17	13
$t/(min)$	0	2	4	6	8	10	12

Ans.   Slope is $(-\lambda/2.303)$; $t_{1/2} = 4.39$ min

12.47. $^{224}_{88}Ra$ having $t_{1/2} = 3.64$ day emits an alpha particle to form $^{220}_{86}Rn$, which has $t_{1/2} = 54.5$ s. Given that the molar volume of radon under these conditions is 35.2 dm³, what volume of radon is in secular equilibrium with 1 g of radium?   *Ans.* $2.72 \times 10^{-8}\,m^3$

12.48. What mass of ^{14}C with $t_{1/2} = 5\,730$ yr is equal to 1 Ci?   *Ans.* $1.92 \times 10^{-4}\,kg$

12.49. Find the ratio of the mass needed to generate 1 μCi of $^{226}_{88}Ra$ ($t_{1/2} = 1\,622$ yr) to that for 1 μCi of $^{222}_{86}Rn$ ($t_{1/2} = 3.825$ day).

Ans. 1.58×10^5

12.50. A chip of paint from "Leif Ericson's ship" had a ^{14}C activity per unit mass of 12.0 $min^{-1}\,g^{-1}$. Comment.

Ans. $t = 406$ yr;   fake (Ericson flourished 1 000 A.D.)

12.51. Another sample of rock No. 12013 from Apollo 12 (see Example 12.16) had an argon content of 7.17×10^{-7} (m³ STP) kg^{-1}. What is the age of this section?

Ans. $n = 3.20 \times 10^{-5}$, $N(^{40}Ar) = 2.97 \times 10^{-5}L$; $t = 3.49 \times 10^9$ yr

Chapter 13

Reaction Kinetics

Influence of Temperature

13.1 ARRHENIUS EQUATION

Over moderate temperature intervals, a plot of $\ln k$ against $1/T$ is linear, giving

$$\ln k = \ln \mathscr{A} - \frac{E_a}{R}\left(\frac{1}{T}\right) \tag{13.1a}$$

where $\mathscr{A}$ is known as the *pre-exponential factor* and E_a is known as the *activation energy*. Values of the rate constant at two different temperatures are related by

$$\ln \frac{k_2}{k_1} = -\frac{E_a}{R}\left(\frac{1}{T_2} - \frac{1}{T_1}\right) \tag{13.1b}$$

Equations (*13.1*) require that the units on E_a be those of (energy)(mol)$^{-1}$ so that the right-hand side of each equation is dimensionless. The (mol)$^{-1}$ term refers to the production of 1 mol of activated complex (see Sec. 13.2).

If a plot of E against the progress of the reaction (*reaction coordinate diagram*) is prepared, E_a represents the energy hump that must be surmounted before the products are formed. If the reverse reaction occurs as written in the equation, the energy of activation for the reverse reaction, $E_{a,r}$, is related to E_a by

$$\Delta_r U^\circ = E_a - E_{a,r} \tag{13.2}$$

where $\Delta_r U^\circ$ is the change in internal energy for the reaction (see Sec. 4.1).

EXAMPLE 13.1. The average rate constants for dicyclopentadiene dissociation in *n*-hexatriacontane on "Gas-Chrome" Q are $1.92 \times 10^{-4}\,\text{s}^{-1}$ at 170.0 °C, $4.61 \times 10^{-4}\,\text{s}^{-1}$ at 180.1 °C, $7.10 \times 10^{-4}\,\text{s}^{-1}$ at 185.2 °C, and $10.52 \times 10^{-4}\,\text{s}^{-1}$ at 189.9 °C, as reported by Langer and Patton. Prepare an *Arrhenius plot* (log k against $1/T$), and determine E_a for the decomposition. Given $U = -545\,\text{kJ}$ for the reaction

$$(C_5H_6)_2 \longrightarrow 2C_5H_6$$

prepare a reaction coordinate diagram and find $E_{a,r}$, assuming the reverse reaction to be a simple bimolecular collision as shown in the equation.

The slope of a plot of log k against $1/T$ is equal to $-7\,610\,\text{K}$ (see Fig. 13-1). Then (*13.1a*) gives

$$E_a = -(2.303)R(\text{slope}) = -(2.303)(8.314\,\text{J K}^{-1}\,\text{mol}^{-1})(-7\,610\,\text{K}) = 145.7\,\text{kJ mol}^{-1}$$

for the decomposition. The reaction coordinate diagram inserted in Fig. 13-1 shows this potential energy barrier above the reactant, with the product 545 kJ below the reactant. The value of $E_{a,r}$ is given by (*13.2*) as

$$E_{a,r} = 145.7\,\text{kJ} - (-545\,\text{kJ}) = 691\,\text{kJ}$$

for the reverse reaction.

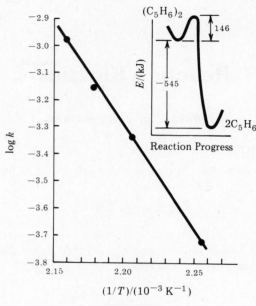

Fig. 13-1

Reaction Rate Theory

13.2   COLLISION THEORY OF BIMOLECULAR REACTIONS

According to the collison theory, the reaction rate is a function of the collision number (z_{ij} for unlike molecules and $2z_{ii}$ for like molecules) and the fraction of molecules having sufficient energy to react ($e^{-E_a/RT}$). Using (*2.17*) or (*2.18*) and (*2.14*), and changing concentrations from molecules m^{-3} to $mol\,dm^{-3}$, it can be shown that the rate of change of concentration is given by

$$-\frac{dC_i}{dt} = p\left[\frac{8\pi RT(M_1 + M_2)}{M_1 M_2}\right]^{1/2} \sigma_{12}^2(10^3 L)\, e^{-E_a/RT}\, C_1 C_2$$

$$= (2.753 \times 10^{29}) p\sigma_{12}^2\left[\frac{T(M_1 + M_2)}{M_1 M_2}\right]^{1/2} e^{-E_a/RT}\, C_1 C_2 \qquad (13.3a)$$

for unlike molecules colliding, and

$$-\frac{dC_i}{dt} = 4p\left(\frac{\pi RT}{M}\right)^{1/2} \sigma^2(10^3 L)\, e^{-E_a/RT}\, C^2$$

$$= (3.893 \times 10^{29}) p\sigma^2\left(\frac{T}{M}\right)^{1/2} e^{-E_a/RT}\, C^2 \qquad (13.3b)$$

for like molecules colliding, where σ_{12} and σ are expressed in m and the *steric factor* (p) is dependent on the relative positions of the colliding molecules.

EXAMPLE 13.2.   Upon comparison of (*13.1a*) with (*13.3*), the Arrhenius pre-exponential factor is given by the collision theory as

$$\mathscr{A} = (2.753 \times 10^{29}) p\sigma_{12}^2\left[\frac{T(M_1 + M_2)}{M_1 M_2}\right]^{1/2}$$

for unlike molecules and

$$\mathscr{A} = (3.893 \times 10^{29}) p\sigma^2 \left(\frac{T}{M}\right)^{1/2}$$

for like molecules. Although the Arrhenius theory describes $\mathscr{A}$ as a constant and the collision theory predicts a temperature dependence for $\mathscr{A}$, show that these theories are not in conflict for typical reactions over reasonable temperature intervals.

Assuming $\sigma = 0.2$ nm, $M = 100$ g mol^{-1}, and $p = 1.00$, the pre-exponential factor at 100 °C for like molecules would be

$$\mathscr{A} = (3.893 \times 10^{29})(1.00)(2 \times 10^{-10})^2 \left(\frac{373}{100}\right)^{1/2} = 3.01 \times 10^{10} \text{ dm}^3 \text{ mol}^{-1} \text{ s}^{-1}$$

and at 200 °C it would be

$$\mathscr{A} = (3.893 \times 10^{29})(1.00)(2 \times 10^{-10})^2 \left(\frac{473}{100}\right)^{1/2} = 3.39 \times 10^{10} \text{ dm}^3 \text{ mol}^{-1} \text{ s}^{-1}$$

a 12% change. Over this same temperature interval, the exponential factor increases from 9.92×10^{-22} to 2.73×10^{-17}, a factor of 2.75×10^4, assuming a typical value of $E_a = 150$ kJ mol^{-1}. Thus the temperature dependence for the exponential factor is often many orders of magnitude larger than that for the pre-exponential factor.

13.3 TRANSITION-STATE THEORY

According to the *transition-state* (or *absolute rate*) *theory*, the reactants form an activated complex under equilibrium conditions that undergoes decomposition to the products according to the mechanism

$$\underset{\text{reactants}}{A + B} \underset{}{\overset{K_c}{\rightleftharpoons}} \underset{\substack{\text{activated} \\ \text{complex}}}{(AB)^{+}} \xrightarrow{k_1} \text{products}$$

where K_c is the concentration equilibrium constant. For this process, the rate equation can be shown to be

$$-\frac{dC_i}{dt} = w\left(\frac{kT}{h}\right)\left(\frac{Q_{+}}{Q_A Q_B}\right) e^{-\Delta U_0^{\circ}/RT} C_A C_B \tag{13.4}$$

where w is the *transmission coefficient* (often neglected), ΔU_0° is the internal energy change for the formation of the activated complex, and the partition functions are given by

$$Q_i = q_{\text{trans}} q_{\text{rot}} q_{\text{vib}} \tag{13.5}$$

where q_{trans} is given by (*8.15*), q_{rot} is given by (*8.18*) or (*8.19*), and q_{vib} is given by (*8.21*), using $3\Lambda_A + 3\Lambda_B - 7$ degrees of freedom for a nonlinear activated complex and $3\Lambda_A + 3\Lambda_B - 6$ degrees of freedom for a linear activated complex and the usual values of $3\Lambda - 5$ and $3\Lambda - 6$ for linear and nonlinear reactants, respectively.

The approximate values of $\mathscr{A}$ at 25 °C, p (assuming $w = 1$), and the exponent of T in $\mathscr{A}$, given in Table 13-1, can often suggest the structure of the activated complex.

EXAMPLE 13.3. Assuming that A and B are atomic species that form a diatomic activated complex having $I = m_A m_B \sigma_{AB}^2 / m_{+}$, where $m_{+} = m_A + m_B$, show that the pre-exponential factor predicted by (*13.4*) is similar to that predicted by (*13.3a*).

For this reaction, (*13.5*) gives the partition functions as

$$Q_{+} = q_{\text{trans}} q_{\text{rot}} \qquad Q_A = Q_B = q_{\text{trans}}$$

which upon substitution of (*8.15*), with $V = 1$ m^3, and (*8.18*), with $\sigma = 1$, become

$$Q_{+} = (2\pi m_{+} kT)^{5/2} \frac{4\pi m_A m_B \sigma_{AB}^2}{m_{+}^2 h^5} \qquad Q_A = \frac{(2\pi m_A kT)^{3/2}}{h^3} \qquad Q_B = \frac{(2\pi m_B kT)^{3/2}}{h^3}$$

Table 13-1

Reactant A	+	Reactant B	=	Activated complex $(AB)^+$	Exponent of T in $\mathscr{A}$	$\mathscr{A}$ at 25 °C, $dm^3\ mol^{-1}\ s^{-1}$	Approximate p
Atom		Atom		Linear	$\frac{1}{2}$	10^{12}	1
Atom		Linear molecule		Linear	$-\frac{1}{2}$ to $\frac{1}{2}$	10^{10}	10^{-2}
Atom		Linear molecule		Nonlinear	0 to $\frac{1}{2}$	10^{11}	10^{-1}
Atom		Nonlinear molecule		Nonlinear	$-\frac{1}{2}$ to $\frac{1}{2}$	10^{10}	10^{-2}
Linear molecule		Linear molecule		Linear	$-\frac{3}{2}$ to $\frac{1}{2}$	10^{7}	10^{-4}
Linear molecule		Linear molecule		Nonlinear	-1 to $\frac{1}{2}$	10^{8}	10^{-3}
Linear molecule		Nonlinear molecule		Nonlinear	$-\frac{3}{2}$ to $\frac{1}{2}$	10^{7}	10^{-4}
Nonlinear molecule		Nonlinear molecule		Nonlinear	-2 to $\frac{1}{2}$	10^{7}	10^{-5}

Inserting these expressions into (13.4), simplifying, and changing the units from $m^3\ molecule^{-1}\ s^{-1}$ to $dm^3\ mol^{-1}\ s^{-1}$, we find

$$\mathscr{A} = w \left[\frac{8\pi RTM_+}{M_A M_B} \right]^{1/2} \sigma_{AB}^{2} (10^3 L)$$

This result is identical with (13.3a) if we set $p = w$.

13.4 THERMODYNAMIC CONSIDERATIONS

The equilibrium constant (K_c) for the formation of the transition-state complex from the reactants is given by

$$K_c = \frac{C_+}{C_A C_B} = \left(\frac{kT}{h\nu} \right) \left(\frac{Q_+}{Q_A Q_B} \right) e^{-\Delta U_0^\circ / RT} = \left(\frac{kT}{h\nu} \right) K_+ \qquad (13.6)$$

where ν is the intramolecular vibrational frequency associated with the decomposition of the complex into the products and K_+ differs from K_c by the same factor, $kT/h\nu$, that relates Q_+ and the entire Q for the activated complex. Although K_+ is rather difficult to calculate, it can be considered using classical thermodynamics; thus (5.9) and (5.39a) give

$$\Delta G_+^\circ = \Delta H_+^\circ - T\Delta S_+^\circ = -RT \ln K_+$$

Solving for K_+ and substituting into (13.6) gives

$$-\frac{dC_i}{dt} = \frac{kT}{h} e^{\Delta S_+^\circ / R} e^{-\Delta H_+^\circ / RT} C_A C_B \qquad (13.7)$$

which upon comparison with (13.1a) gives

$$\mathscr{A} = e^n \left(\frac{kT}{h} \right) e^{\Delta S_+^\circ / R} \qquad (13.8)$$

if

$$\Delta H_+^\circ = E_a - nRT \qquad (13.9)$$

where $n = 1$ for condensed-phase reactions and is the molecularity for gas-phase reactions.

EXAMPLE 13.4. The influence of ionic strength on the rate constant as predicted by the transition-state theory is known as the *primary salt effect*. Discuss the influence on k if ions A and B have like charges.

The thermodynamic equilibrium constant for the formation of the complex is given by

$$K = \frac{a_+}{a_A a_B} = \frac{C_+}{C_A C_B} \left(\frac{y_+}{y_A y_B} \right)$$

which upon solving for C_+ and substituting into the first-order reaction describing the decomposition of the complex into the products gives

$$-\frac{dC_i}{dt} = k_1 C_+ = k_1 K \frac{y_A y_B}{y_+} C_A C_B = k C_A C_B$$

For charged species in rather dilute solutions, (6.40) gives

$$\log y_A = -z_A^2 (0.511\,6) I^{1/2}$$

$$\log y_B = -z_B^2 (0.511\,6) I^{1/2}$$

$$\log y_+ = -(z_A + z_B)^2 (0.511\,6) I^{1/2}$$

which upon substitution into the logarithm of the last two terms of the rate equation gives

$$\log k = \log(k_1 K) + \log \frac{y_A y_B}{y_+} = \log(k_1 K) + (1.023\,2) z_A z_B I^{1/2} \qquad (13.10)$$

For z_A and z_B both positive or both negative, (13.10) predicts that $\log k$ will increase for an increase in I.

Catalysis

13.5 HOMOGENEOUS CATALYSIS

A *catalyst* is a substance that increases the rate of a chemical reaction without undergoing any permanent chemical change. The presence of a catalyst will not influence equilibrium conditions once they are attained. Because the catalyst may enter into the reaction, the complete rate equation should include catalytic terms. In *homogeneous catalysis*, the catalyst is found in the same physical state as the reactants.

EXAMPLE 13.5. One of the proposed mechanisms for the recombination of bromine atoms in the presence of a third body or *chaperon*, M, is

$$Br + Br \underset{k_2}{\overset{k_1}{\rightleftharpoons}} Br_2^* \qquad\qquad Br_2^* + M \overset{k_3}{\longrightarrow} Br_2 + M$$

where * indicates a highly excited molecule having an energy content of the order of the Br—Br bond dissociation energy. Why are CCl_4 and SF_6 better catalysts than Ne or other monatomic and diatomic species?

The simpler molecules have only translational—and perhaps two rotational and one vibrational—degrees of freedom available to absorb the energy from the Br_2^*, whereas the more complicated molecules have three rotational and many more vibrational degrees of freedom. Thus the energy transfer in a collision with the more complicated molecule will be more effective in increasing the rate of formation of Br_2.

EXAMPLE 13.6. Consider the following mechanism describing enzyme catalysis:

$$E + S \underset{k_{-1}}{\overset{k_1}{\rightleftharpoons}} X \underset{k_{-2}}{\overset{k_2}{\rightleftharpoons}} E + P$$

where E is the enzymatic site, S is the substrate, X is the enzyme substrate complex, and P is the product of the reaction. Derive the rate equation for this process, assuming $dC_X/dt = 0$, and discuss the results for the reaction during the initial stages.

The rate equations for this mechanism are

$$\frac{dC_X}{dt} = 0 = k_1 C_E C_S - k_{-1} C_X - k_2 C_X + k_2 C_E C_P \qquad\qquad \frac{dC_P}{dt} = k_2 C_X - k_{-2} C_E C_P$$

where $C_{E,0} = C_E + C_X$. Eliminating C_X and C_E gives

$$\frac{dC_P}{dt} = \frac{(V_S/K_S) C_S - (V_P/K_P) C_P}{1 + C_S/K_S + C_P/K_P}$$

where
$$V_S = k_2 C_{E,0} \qquad V_P = k_{-1} C_{E,0}$$

$$K_S = \frac{k_{-1} + k_2}{k_1} \qquad K_P = \frac{k_{-1} + k_2}{k_{-2}}$$

If the measurements of the reaction rate are made during the early stages of the reaction, C_P will be nearly zero, and the rate equation simplifies to give the *Michaelis–Menten equation*:

$$\frac{dC_P}{dt} = \frac{(V_S/K_S) C_S}{1 + C_S/K_S} = \frac{V_S}{1 + K_S/C_S}$$

If $C_S \ll K_S$ the reaction rate is first-order with respect to C_S, and if $C_S \gg K_S$ the reaction rate is zero-order with respect to C_S. In both cases the reaction is dependent on V_S, which is first-order with respect to $C_{E,0} \approx C_E$. The Michaelis constant for the substrate, K_S, represents the value of C_S necessary to reduce the maximum reaction rate, V_S, by a factor of 2. The rate constant for the product reaction, k_2, is known as the *turnover constant*. Casting the rate equation into the *Lineweaver–Burk linear form* gives

$$\left(\frac{dC_P}{dt} \right)^{-1} = \left(\frac{K_S}{V_S} \right) C_S^{-1} + V_S^{-1}$$

Thus a plot of $(dC_P/dt)^{-1}$ against C_S^{-1} will be linear, having an intercept of V_S^{-1} and a slope of K_S/V_S.

Photochemistry

13.6   INTRODUCTION

Only those light quanta that are absorbed by a substance will be effective in producing a photochemical change. The energy of a quantum is given by

$$E = h\nu = \frac{hc}{\lambda} \qquad\qquad (13.11)$$

where h is Planck's constant, ν is the frequency of the incident light, c is the velocity of light, and λ is the wavelength of the incident light. Many authors express the number of quanta in *einsteins*, where 1 einstein = 1 mol of quanta.

The primary step in a photochemical reaction involves one molecule being activated by one absorbed quantum of radiation. The *quantum yield* (Φ) is defined as the number of molecules of reactant consumed or product generated per quantum absorbed.

EXAMPLE 13.7.   Bridges and White propose the following mechanism for the photolysis of methanethiol:
(1) $CH_3SH + h\nu \longrightarrow CH_3S + H^*$
(2) $CH_3SH + h\nu \longrightarrow CH_3(\text{or } CH_3^*) + SH(\text{or } SH^*)$
(3) $H^* + CH_3SH \longrightarrow CH_3S + H_2$
(4a) $H^* + CH_3SH \longrightarrow CH_3 + H_2S$
(4b) $H^* + CH_3SH \longrightarrow CH_4 + SH$
(5) $CH_3(\text{or } CH_3^*) + CH_3SH \longrightarrow CH_3S + CH_4$
(6) $SH(\text{or } SH^*) + CH_3SH \longrightarrow CH_3S + H_2S$
(7) $H^* + M \longrightarrow H + M$
(8) $H + CH_3SH \longrightarrow CH_3S + H_2$
(9) $CH_3S + CH_3S \longrightarrow CH_3SSCH_3$
Assuming one quantum of light to be absorbed according to reaction (1), find the relationships among $\Phi(H_2)$, $\Phi(CH_4)$, $\Phi(H_2S)$, and $\Phi(CH_3SSCH_3)$.

The products of (1) are CH_3 and H^*. Assuming the H^* to react in (3), a second CH_3S is generated, as well as a H_2. The two CH_3S can react according to (9), giving one CH_3SSCH_3. For this path for the dissipation of H^*, $\Phi(H_2) = 1$, $\Phi(CH_3SSCH_3) = 1$, and $\Phi(H_2S) = \Phi(CH_4) = 0$.

If the H^* reacts according to (4a), a CH_3 and H_2S are formed. Letting the CH_3 react with a second CH_3SH according to (5) gives CH_4 and a second CH_3S. Again, the two CH_3S fragments react according to (9). For this path, $\Phi(CH_3SSCH_3) = 1$, $\Phi(H_2S) = \Phi(CH_4) = 1$, and $\Phi(H_2) = 0$.

Assuming the H^* to react according to (*4b*), a CH_4 and SH are formed. Assuming the latter to react according to (*6*) gives H_2S and a second CH_3S. Again, both of the CH_3S react according to (*9*). For this path, $\Phi(CH_3SSCH_3) = 1$, $\Phi(CH_4) = \Phi(H_2S) = 1$, and $\Phi(H_2) = 0$.

Finally, letting the H^* dissipate its extra energy by colliding with the inert material M according to (*7*) gives H, which in turn reacts according to (*8*), giving the second CH_3S needed for (*9*) and a H_2. For this path, $\Phi(H_2) = 1$, $\Phi(CH_3SSCH_3) = 1$, and $\Phi(CH_4) = \Phi(H_2S) = 0$.

In summary, regardless of which path was chosen for the dissipation of the H^*, $\Phi(CH_3SSCH_3) = 1$. Because the paths are competing reactions (Sec. 12.18), the exact values for the other quantum yields cannot be determined from the information given above, but the relationships $\Phi(H_2S) = \Phi(CH_4)$ and $\Phi(CH_4) + \Phi(H_2) = 1$ are valid.

EXAMPLE 13.8. For the reactions described in Example 13.7, Bridges and White reported that $\Phi(CH_4) = 0.16$ at 254 nm and 0.35 at 214 nm. Discuss these results.

The energy of the 214-nm light is given by (*13.11*) as

$$E = \frac{(6.626 \times 10^{-34}\text{ J s})(2.997\,9 \times 10^8\text{ m s}^{-1})}{(214\text{ nm})[(10^{-9}\text{ m})/(1\text{ nm})]} = 9.28 \times 10^{-19}\text{ J}$$

or 559 kJ mol^{-1}, and the energy of the 254-nm radiation is 471 kJ mol^{-1}. The energy required to break the S—H bond in reaction (*1*) is roughly 368 kJ mol^{-1}, and for the C—S bond in (*2*) it is roughly 289 kJ mol^{-1} (see Sec. 4.7). Because both wavelengths have sufficient energy to cleave both bond types, both reactions (*1*) and (*2*) are favored, and a mixture of CH_4, H_2S, and H_2 will result. As the energy is increased by changing to 214-nm light, reaction (*2*) becomes more favored than before and a larger yield of CH_4 and H_2S (see Problem 13.27) will result.

Solved Problems

INFLUENCE OF TEMPERATURE

13.1. For the reaction described in Problem 12.9, a more careful analysis of the data gave

$t/(°C)$	19.7	25	30	33.5
$k_f/(10^5\text{ M}^{-1}\text{ s}^{-1})$	0.66	1.40	2.21	3.32

Find E_a, and prepare a reaction coordinate diagram using $E_{a,r} = 135$ kJ mol^{-1}.

A plot of log k against $1/T$ is shown in Fig. 13-2. From the slope of the line (*13.1a*) gives

$$E_a = -2.303\,R(\text{slope}) = -(2.303)(8.314 \times 10^{-3}\text{ kJ K}^{-1}\text{ mol}^{-1})(-4\,470\text{ K}) = 85.6\text{ kJ mol}^{-1}$$

Equation (*13.2*) gives for the reaction

$$\Delta_r U° = 85.6 - 135 = -50.\text{ kJ}$$

The reaction coordinate diagram is shown as an insert in Fig. 13-2.

13.2. For the reaction

$$2I(g) + H_2(g) \longrightarrow 2HI(g)$$

Sullivan reported $k/(10^5\text{ M}^{-2}\text{ s}^{-1}) = 1.12$ at 417.9 K and 18.54 at 737.9 K. Determine E_a for this reaction from these data, and predict the value of k at 633.2 K.

Solving (*13.1b*) for E_a and substituting the experimental data gives

$$E_a = \frac{-R\ln(k_2/k_1)}{1/T_2 - 1/T_1} = \frac{-(8.341 \times 10^{-3}\text{ kJ K}^{-1}\text{ mol}^{-1})\ln[(1.12 \times 10^5)/(18.5 \times 10^5)]}{1/417.9\text{ K} - 1/737.9\text{ K}} = 22.5\text{ kJ mol}^{-1}$$

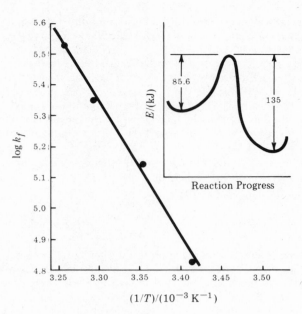

Fig. 13-2

The rate constant at 633.2 K is determined by substituting the value of E_a and one set of the experimental temperature-rate constant values into ($13.1b$) to give

$$\ln\frac{k_2}{1.12\times10^5\ \text{M}^{-2}\ \text{s}^{-1}}=\frac{-22.5\ \text{kJ mol}^{-1}}{8.314\times10^{-3}\ \text{kJ K}^{-1}\ \text{mol}^{-1}}\left(\frac{1}{633.3\ \text{K}}-\frac{1}{417.9\ \text{K}}\right)=2.20$$

$$k_2=1.0\times10^6\ \text{M}^{-2}\ \text{s}^{-1}$$

REACTION RATE THEORY

13.3.   Predict $\mathscr{A}$ for the reaction

$$\text{Cl(g)}+\text{H}_2\text{(g)}\ \longrightarrow\ \text{HCl(g)}+\text{H(g)}$$

given $M/(\text{g mol}^{-1})=35.453$ for Cl and $2.015\,94$ for H_2 and $\sigma=0.200$ nm for Cl and 0.150 nm for H_2. If the value of $\log\mathscr{A}$ is 10.08 between 250 K and 450 K, find p and interpret the result.

The pre-exponential factor at 350 K is given by ($13.3a$) as

$$\mathscr{A}=(2.753\times10^{29})p[\tfrac{1}{2}(2.00+1.50)10^{-10}]^2\left[\frac{350(35.453+2.015\,94)}{(35.453)(2.015\,94)}\right]^{1/2}=1.142\times10^{11}p$$

Comparing the predicted value of $\mathscr{A}$ to the observed value, antilog (10.08), gives

$$p=\frac{\text{antilog}(10.08)}{1.142\times10^{11}}=\frac{1.2\times10^{10}}{1.142\times10^{11}}=0.11$$

A reasonably high value for p would be expected for this reaction because the Cl can combine with either end of the H_2 molecule to eventually form the products.

13.4.   (a) Deduce the structure of the activated complex formed between Cl and H_2 in the reaction described in Problem 13.3, and confirm the entries in Table 13-1 for this configuration by calculating the approximate values of $\mathscr{A}$ and p, assuming $q_{\text{trans}}=10^{10}$, $q_{\text{rot}}=10$, and $q_{\text{vib}}=1$ for each degree of freedom. (b) If the temperature exponent is 1/2 for each translational and rotational degree of freedom and 0–1 for each vibrational degree of freedom, predict the temperature exponent in $\mathscr{A}$ (maximum value is 1/2).

(a) Problem 13.3 gave $\mathscr{A} = 1.2 \times 10^{10}$ and $p = 0.11$. For the reaction between an atom and a linear molecule, the entries in Table 13-1 are $\mathscr{A} = 10^{10}$ and $p = 10^{-2}$ for a linear complex and $\mathscr{A} = 10^{11}$ and $p = 10^{-1}$ for a nonlinear complex. Giving more weight to the value of p because the errors in its estimation tend to cancel, the predicted activated complex is nonlinear.

To confirm the table value of $\mathscr{A}$, (13.4) is applied to the reaction of an atom A with a linear molecule B to form a nonlinear activated complex $(AB)^+$. For each component, every degree of freedom gives rise to a factor in the partition function; hence (13.5) gives

$$Q_A = q_{trans}q_{rot}q_{vib} = (10^{10})^3(10)^0(1)^0 = 10^{30}$$

$$Q_B = q_{trans}q_{rot}q_{vib} = (10^{10})^3(10)^2(1)^1 = 10^{32}$$

$$Q_+ = q_{trans}q_{rot}q_{vib} = (10^{10})^3(10)^3(1)^2 = 10^{33}$$

Substituting into (13.4) and converting to $dm^3\,mol^{-1}\,s^{-1}$ gives

$$\mathscr{A} = \frac{kT}{h}(10^3 L)\frac{Q_+}{Q_A Q_B} = (4 \times 10^{39})\frac{10^{33}}{10^{30}10^{32}} = 4 \times 10^{10}$$

which is in good agreement with the table entry of 10^{11}.

The value of p for this reaction is found by comparing the estimated value of $\mathscr{A}$ to that for the reaction of rigid spheres interacting to form a diatomic activated complex. For the latter reaction,

$$Q_A = Q_B = (10^{10})^3(10)^0(1)^0 = 10^{30}$$

As for Q_+, the activated complex has, according to Sec. 13.3, $3(1) + 3(1) - 6 = 0$ degrees of vibrational freedom attributed to it, so that

$$Q_+ = (10^{10})^3(10)^2(1)^0 = 10^{32}$$

Then $$\mathscr{A} = p(4 \times 10^{39})\frac{10^{32}}{10^{30}10^{30}} = p(4 \times 10^{11})$$

from which $p = 10^{-1}$.

(b) Multiplying partition functions means adding temperature exponents. Hence, as above,

$$Q_A \propto (T^{1/2})^3(T^{1/2})^0(T^\theta)^0 = T^{3/2}$$

$$Q_B \propto (T^{1/2})^3(T^{1/2})^2(T^\theta)^1 = T^{5/2+\theta}$$

$$Q_+ \propto (T^{1/2})^3(T^{1/2})^3(T^\theta)^2 = T^{3+2\theta}$$

where θ stands for a number between 0 and 1. Then,

$$\mathscr{A} \propto T\frac{Q_+}{Q_A Q_B} \propto T\frac{T^{3+2\theta}}{T^{3/2}T^{5/2+\theta}} = T^\theta$$

Because the exponent cannot exceed $\frac{1}{2}$ (see Table 13-1), it must be between 0 and $\frac{1}{2}$.

13.5. Estimate ΔS_+° for the reaction

$$A + B \underset{}{\overset{K}{\rightleftharpoons}} (AB)^+$$

where A is an atom, B is a diatomic molecule, and $(AB)^+$ is a nonlinear triatomic molecule.

Before the formation of the complex, the system had 3 degrees of translational freedom for A and 3 degrees of translational freedom, 2 degrees of rotational freedom, and 1 degree of vibrational freedom for B, giving a total of 6 translational, 2 rotational, and 1 vibrational degrees of freedom. After the formation of the complex, the system has 3 translational, 3 rotational, and 2 vibrational degrees of freedom. Assuming numerical values similar to those calculated in Example 4.11 for the various modes, ΔS_+° will be quite negative for this process:

$$\Delta S_+^\circ = [3S^\circ(trans) + 3S^\circ(rot) + 2S^\circ(vib)] - [6S^\circ(trans) + 2S^\circ(rot) + S^\circ(vib)]$$

$$= S^\circ(rot) + S^\circ(vib) - 3S^\circ(trans) \approx 30 + 1 - 3(50) \approx -120\,J\,K^{-1}$$

13.6. Calculate $\Delta S_\ddagger^\circ$ at 350 K for the reaction described in Problem 13.3. Does this value agree with the results of Problems 13.4 and 13.5, which predicted a nonlinear complex being formed with $\Delta S_\ddagger^\circ$ estimated as -120 J K^{-1}?

Solving (13.8) for $\Delta S_\ddagger^\circ$ and substituting $\mathscr{A} = 1.2 \times 10^{10}$ gives

$$\Delta S_\ddagger^\circ = R\left[\ln\left(\frac{\mathscr{A}h}{kt}\right) - n\right] = (8.314)\left\{\ln\left[\frac{(1.2 \times 10^{10})(6.626 \times 10^{-34})}{(1.380\,7 \times 10^{-23})(350)}\right] - 2\right\}$$

$$= (8.314)[\ln(1.6 \times 10^{-3}) - 2] = -70 \text{ J K}^{-1}$$

This value is about half the estimate made in Problem 13.5, which is fair agreement.

CATALYSIS

13.7. The pseudo-first-order rate constant for the cobalt-catalyzed autooxidation of toluene in acetic acid at 87 °C was determined for several concentrations of the catalyst, Co(III), by Scott and Chester. The data are

$k/(10^{-5} \text{ s}^{-1})$	1.47	2.93	5.68
$C(\text{Co(III)})/(\text{M})$	0.053	0.084	0.1185

for $[C(\text{C}_6\text{H}_5\text{CH}_3)]_0 = 0.5 \text{ M}$. Find the order with respect to $C(\text{Co(III)})$ and the rate constant.

Assuming k to be defined in terms of the rate constant k' as

$$k = k'[C(\text{Co(III)})]^n$$

this definition can be put into linear form by taking logarithms, giving

$$\log k = \log k' + n \log C(\text{Co(III)})$$

Thus a plot of $\log k$ against $\log C(\text{Co(III)})$ will give a straight line with slope n and intercept $\log k'$. From Fig. 13-3, $n = 1.67$ (nearly 2), and then, from the data,

$$k' \approx \frac{2.93 \times 10^{-5}}{(0.084)^2} = 4.15 \times 10^{-3} \text{ M}^{-2} \text{ s}^{-1}$$

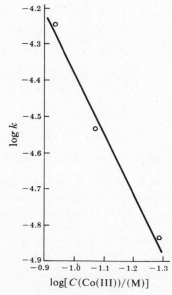

Fig. 13-3

13.8. The thermal decomposition of 3-chloro-3-phenyldiazirine in various solvents has been studied by Liu and Toriyama. Show from the data below that the catalytic solvent effect is negligible for the solvents dimethyl sulfoxide (DMSO) and diethylene glycol monoethyl ether (DEGME).

$t/(°C)$	60.0	65.0	70.0	75.0	80.0	85.0	90.0
$k_{DMSO}/(10^{-4}\,s^{-1})$	0.47		1.60	3.04	5.00		15.8
$k_{DEGME}/(10^{-4}\,s^{-1})$		0.70	1.32	2.30	4.00	6.90	

The E_a values for the reaction in the two solvents, as determined from an Arrhenius plot similar to Fig. 13-1, are 117.5 kJ mol^{-1} and 114.5 kJ mol^{-1} for DMSO and DEGME, respectively. Because these values are within 3% of each other, there is very little difference between the solvents in their catalytic effect.

13.9. Consider the following mechanism describing competitive inhibition of an enzyme catalyst in which the substrate and inhibitor are competing in binding to the enzyme:

$$E+S \underset{k_{-1}}{\overset{k_1}{\rightleftharpoons}} X \underset{k_{-2}}{\overset{k_2}{\rightleftharpoons}} E+P \qquad E+I \underset{k_{-3}}{\overset{k_3}{\rightleftharpoons}} EI$$

where I is the inhibitor and the inhibition can be described by $K_I = C_E C_I / C_{EI}$. Derive the rate equation for this process during the initial stages of the reaction, where the formation of X from E reacting with P is negligible. Assume $dC_X/dt = 0$.

The rate equations for this mechanism are

$$\frac{dC_X}{dt} = 0 = k_1 C_E C_S - k_{-1} C_X - k_2 C_X \qquad \text{and} \qquad \frac{dC_P}{dt} = k_2 C_X$$

where $C_{E,0} = C_E + C_X + C_{EI}$. Eliminating C_X, C_E, and C_{EI} gives

$$\frac{dC_P}{dt} = \frac{k_1 k_2 C_{E,0} C_S}{k_1 C_S + k_{-1} + k_{-1}(C_I/K_I) + k_2 + k_2(C_I/K_I)} = \frac{V_S C_S}{C_S + K_S[1+(C_I/K_I)]}$$

Casting the rate equation into the Lineweaver–Burk form (Example 13.6) gives

$$\left(\frac{dC_P}{dt}\right)^{-1} = \frac{K_S}{V_S}\left(1+\frac{C_I}{K_I}\right)C_S^{-1} + V_S^{-1}$$

A plot of $(dC_P/dt)^{-1}$ against C_S^{-1} will be linear, having an intercept of V_S^{-1}, as in the case of no inhibition, and a slope of $(K_S/V_S)(1+C_I/K_I)$, greater than in the case of no inhibition.

PHOTOCHEMISTRY

13.10. Shapiro and Treinin recommend the following reaction for use as an *actinometer*—a device for counting quanta:

$$HN_3 + H_2O + h\nu \longrightarrow N_2 + NH_2OH$$

Find Φ for the reaction if $I_v = 1.00 \times 10^{-7}$ mol dm^{-3} s^{-1} at 214 nm and $C(N_2) = C(NH_2OH) = 24.1 \times 10^{-5}$ M after a period of radiation of 39.38 min.

On a 1-dm^3 basis, the amount formed each second is

$$n = \frac{24.1 \times 10^{-5}\,\text{mol}}{(39.38\,\text{min})[(60\,\text{s})/(1\,\text{min})]} = 1.02 \times 10^{-7}\,\text{mol s}^{-1}$$

and

$$\Phi = \frac{1.02 \times 10^{-7}\,\text{mol s}^{-1}}{1.00 \times 10^{-7}\,\text{mol s}^{-1}} = 1.02$$

13.11. Consider the proposed mechanism for the photodimerization of A:

$$A + h\nu \xrightarrow{k_1} A^* \qquad A^* + A \xrightarrow{k_2} A_2 \qquad A^* \xrightarrow{k_3} A + h\nu'$$

Derive the expression for $\Phi(A_2)$.

Under steady-state conditions,

$$\frac{dC(A^*)}{dt} = 0 = k_1 I_v - k_2 C(A^*)C(A) - k_3 C(A^*)$$

where I_v is the intensity of the light absorbed. Upon rearrangement,

$$C(A^*) = \frac{k_1 I_v}{k_2 C(A) + k_3}$$

The formation of A_2 is given by

$$\frac{d}{dt}[C(A_2)] = k_2 C(A^*)C(A) = \frac{k_1 k_2 I_v C(A)}{k_2 C(A) + k_3}$$

and

$$\Phi(A_2) = \frac{dC(A_2)/dt}{I_v} = \frac{k_1 k_2 C(A)}{k_2 C(A) + k_3}$$

Supplementary Problems

INFLUENCE OF TEMPERATURE

13.12. Using the reaction coordinate diagrams shown in Fig. 13-4, describe the relation of k_1 to k_2 for the consecutive reactions

$$A \xrightarrow{k_1} B \qquad \text{and} \qquad B \xrightarrow{k_2} C$$

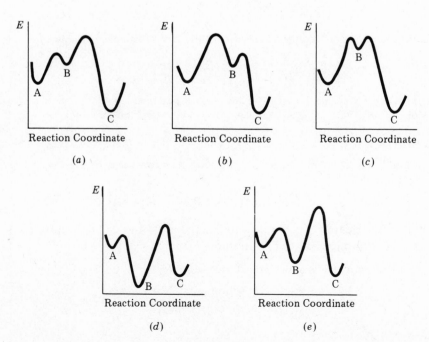

Fig. 13-4

where the overall reaction $A \longrightarrow C$ is exothermic. Describe the possibility of isolation of B in each of cases (a)–(e).

Ans. (a) k_1 about equal to k_2, fair (b) $k_1 \ll k_2$, little (c) $k_1 \ll k_2$, little
          (d) $k_1 > k_2$, excellent (e) $k_1 > k_2$, good

13.13. A rule often quoted in the laboratory is "heat the sample by $10\,°C$ and the reaction will proceed twice as fast." This is true at any given temperature only for reactions having one value of E_a. Find the value of E_a if $t_1 = 25\,°C$ and $t_2 = 35\,°C$.   *Ans.* $52.9\ \text{kJ mol}^{-1}$

13.14. The rate constants at several different temperatures for the reaction

$$CH_3CHF_2 \xrightarrow{\ k\ } CH_2CHF + HF$$

have been reported by Noble, Carmichael, and Bumgardner as

$k/(10^{-7}\ \text{s}^{-1})$	7.9	26	52	58	69	230	250	620	1 400	1 700
$t/(°C)$	429	447	460	462	463	483	487	507	521	522

Prepare an Arrhenius plot, and determine E_a. If $\Delta U°$ for this reaction is $-389\ \text{kJ}$, prepare a reaction coordinate diagram.

Ans. $E_a = 260\ \text{kJ mol}^{-1}$; plot similar to Fig. 13-1 with insert having a higher E_a barrier and smaller $\Delta_r U°$ decrease

13.15. Calculate $E_{a,1}$ and $E_{a,2}$ for the cyclobutanone decomposition described in Problem 12.41 from the following values of k_1 and k_2

$k_1/(10^{-4}\ \text{s}^{-1})$	4.6	7.2	14.5	39.8	67.5
$t/(°C)$	361	371	383	396	406

$k_2/(10^{-6}\ \text{s}^{-1})$	2.6	6.0	11.5	30.2	53.7
$t/(°C)$	360	372	383	396	406

Ans. $E_{a,1} = 215\ \text{kJ mol}^{-1}$, $E_{a,2} = 235\ \text{kJ mol}^{-1}$

13.16. Which reaction would have the greatest change of reaction rate with a change in temperature—one with a large activation energy or one with a low activation energy? Prove your answer by calculating (k_2/k_1) for reactions with $E_a/(\text{kJ mol}^{-1}) = 2.0$ and 20.0 as the temperature increases from $300.\ \text{K}$ to $500.\ \text{K}$.

Ans.   The reaction with the larger E_a value; 1.38, 24.7

REACTION RATE THEORY

13.17. Predict k and $\mathscr{A}$ for the reaction

$$Cl(g) + ICl(g) \longrightarrow Cl_2(g) + I(g)$$

given $M/(\text{g mol}^{-1}) = 35.453$ for Cl and 162.357 for ICl, $\sigma/(\text{nm}) = 0.200$ for Cl and 0.465 for ICl, and $E_a = 18.8\ \text{kJ mol}^{-1}$. If $\log \mathscr{A} = 8.7$ for this reaction between 303 and 333 K, find p and compare the value to that obtained for the reaction described in Problem 13.3.

Ans. $\mathscr{A} = 1.005 \times 10^{11}\,p\ \text{dm}^3\ \text{mol}^{-1}\ \text{s}^{-1}$ at 318 K, $k = 7.24 \times 10^7\,p\ \text{dm}^3\ \text{mol}^{-1}\ \text{s}^{-1}$;
       $p = 0.005$ (reaction occurs only at one end, therefore less)

13.18. Deduce the structure for the transition-state complex formed between Cl and ICl in the reaction described in Problem 13.17. Confirm the entries in Table 13-1 for this mechanism by calculating the approximate values of $\mathscr{A}$ and p, assuming $q_{\text{trans}} = 10^{10}$, $q_{\text{rot}} = 10$, and $q_{\text{vib}} = 1$ for each degree of freedom. If the temperature exponent is 1/2 for each translational and rotational degree of freedom, predict the temperature exponent in $\mathscr{A}$.

Ans. Linear complex; $Q_+ = 10^{32}$, $Q_A = 10^{30}$, $Q_B = 10^{32}$, $\mathscr{A} = 4 \times 10^9 \approx 10^{10}$, $p = 10^{-2}$; estimated T exponent is $-1/2$ to $3/2$, so $-1/2$ to $1/2$

13.19. Qualitatively discuss ΔS_+° for the reaction

$$M(H_2O)_6^{2+} + Y^{2-} \underset{}{\overset{K}{\rightleftharpoons}} (MY)^+ + 6H_2O$$

Ans. Negative ΔS in forming $(MY)^+$, large positive ΔS in releasing 6 mol H_2O, positive ΔS_+°.

13.20. Estimate ΔS_+° for the reaction

$$A + B \overset{K}{\rightleftharpoons} (AB)^+$$

where A is an atom, B is a diatomic molecule, and $(AB)^+$ is a linear triatomic molecule.

Ans. $\Delta S_+^{\circ} = 2S^{\circ}(\text{vib}) - 3S^{\circ}(\text{trans}) = -150 \text{ J K}^{-1}$, using values from Problem 13.5

13.21. Calculate ΔS_+° at 318 K for the reaction described in Problem 13.17. Does this value agree with that predicted by Problem 13.20?   *Ans.* 96 J K^{-1}, same order of magnitude

13.22. Equation (*13.10*) describes correctly the primary salt effect for all cases except that of $z_A = +n$ and $z_B = -n$, where the resulting charge on $(AB)^+$ is zero. Show that (*13.10*) gives the correct dependence on I, even in this case.

Ans. $\log y_A = -z_A^{\,2}(0.511\,6)I^{1/2}$, $\log y_B = -z_B^{\,2}(0.511\,6)I^{1/2}$, $y_+ \approx 1$ (by other methods),

$$\log \frac{y_A y_B}{y_{AB}} = -(z_A^{\,2} + z_B^{\,2})(0.511\,6)I^{1/2} = -1.023\,2\,I^{1/2}n^2$$

13.23. Predict the primary salt effect using (*13.10*) if $z_A = +n$ and $z_B = -m$.

Ans. $z_A z_B < 0$, so $\log k$ will decrease with increasing I.

CATALYSIS

13.24. In Problem 13.7 the second-order behavior with respect to $C(\text{Co(III)})$ fails at low concentrations of Co(III). An alternative expression for k suggested by Scott and Chester is

$$k = k_a[C(\text{Co(III)})]^{1/2}[C(\text{Co(II)})]^{-1} + k_b[C(\text{Co(III)})]^2$$

which includes the catalytic effect of Co(III) in both terms, giving an order between 1 and 2 ($n = 1.71$ in Problem 13.7), and an inhibiting effect of Co(II) in one term. Upon rearrangement,

$$kC(\text{Co(II)})[C(\text{Co(III)})]^{-1/2} = k_a + k_b C(\text{Co(II)})[C(\text{Co(III)})]^{3/2}$$

which means that a plot of $kC(\text{Co(II)})[C(\text{Co(III)})]^{-1/2}$ against $C(\text{Co(II)})[C(\text{Co(III)})]^{3/2}$ will be linear with a slope of k_b and an intercept of k_a. Prepare such a plot from the following data for $[C(C_6H_5CH_3)]_0 = 0.5$ M at 87 °C, and determine k_a and k_b.

$k/(10^{-5}\,\text{s}^{-1})$	0.455	1.47	2.93	5.68
$C(\text{Co(III)})/\text{M}$	0.017 9	0.053	0.084	0.118 5
$C(\text{Co(II)})/\text{M}$	0.044 6	0.072 0	0.103 5	0.131 5

Ans. $k_a = 1.2 \times 10^{-6}$ M$^{1/2}$ s^{-1}, $k_b = 3.74 \times 10^{-3}$ M^{-2} s^{-1}

13.25. The catalytic effect of various gases on the observed rate constants for the recombination of bromine atoms has been studied by DeGraff and Lang. From the following data, determine which chaperon gas is most effective as a catalyst.

$t/(°C)$	94	76	46	24	96	63	39	26	96	75	54	26
$k/(10^9\,M^{-2}\,s^{-1})$	1.07	1.13	1.36	1.48	4.39	6.19	6.87	8.56	9.95	12.01	14.92	22.13
Gas	Ne				SF$_6$				CCl$_4$			

Ans. $E_a = -4.4$, -8.2, and -10.5 kJ mol^{-1} for Ne, SF$_6$, and CCl$_4$, respectively; CCl$_4$ most effective in absorbing the E_a from the excited molecule

13.26. Consider the following mechanism describing noncompetitive inhibition of an enzyme catalyst in which the substrate and inhibitor bind simultaneously to the enzyme forming an inactive complex, XI

$$E+S \underset{k_{-1}}{\overset{k_1}{\rightleftharpoons}} X \underset{k_{-2}}{\overset{k_2}{\rightleftharpoons}} E+P$$

$$E+I \underset{k_{-3}}{\overset{k_3}{\rightleftharpoons}} EI$$

$$X+I \underset{k_{-4}}{\overset{k_4}{\rightleftharpoons}} XI$$

$$EI+S \underset{k_{-5}}{\overset{k_5}{\rightleftharpoons}} XI$$

where the inhibitions can be described by K_I, $K_I' = C_X C_I / C_{XI}$, and $K_S' = C_S C_{EI} / C_X$. Derive the rate equation for this process during the initial stages of the reaction where the formation of X from E reacting with P is negligible. Assume $dC_X/dt = 0$.

Ans. $\left(\dfrac{dC_P}{dt}\right)^{-1} = \dfrac{K_S}{V_S}\left(1 + \dfrac{C_I}{K_I}\right) C_S^{-1} + \left(1 + \dfrac{C_I}{K_I'}\right) V_S^{-1}$;

slope same as for competitive inhibition, intercept greater than in competitive and no-inhibition cases

PHOTOCHEMISTRY

13.27. Assuming one quantum of light to be absorbed according to reaction (2) of Example 13.7, find the relationships between $\Phi(H_2)$, $\Phi(CH_4)$, $\Phi(H_2S)$, and $\Phi(CH_3SSCH_3)$.

Ans.   Products of (2) enter into reactions (5) and (6), producing CH$_4$, H$_2$S, and 2CH$_3$S, giving $\Phi(CH_4) = \Phi(H_2S) = 1$; the 2CH$_3$S enter into reaction (9), producing CH$_3$SSCH$_3$, giving $\Phi(CH_3SSCH_3) = 1$; because no H$_2$ is formed in this path, $\Phi(H_2) = 0$.

13.28. Consider the proposed mechanism for the photodimerization of A:

$$A + h\nu \overset{k_1}{\longrightarrow} A^* \qquad A^* + A \underset{k_{-2}}{\overset{k_2}{\rightleftharpoons}} A_2 \qquad A^* \overset{k_3}{\longrightarrow} A + h\nu'$$

Is $\Phi(A_2)$ dependent on I_v?

Ans.   Using a steady-state approximation, $\Phi(A_2) = \dfrac{k_1 k_2 C(A) - k_{-2} k_3 C(A_2) I_v^{-1}}{k_2 C(A) + k_3}$; yes.

13.29. Discuss the reaction of acetone with light having $\lambda = 300$ nm. The average bond energies are 414 kJ mol^{-1} for C—H, 331 kJ mol^{-1} for C—C, and 728 kJ mol^{-1} for C=O.

Ans. $E = 399$ kJ mol^{-1}; C—C bond will break, giving CH$_3\cdot$ + CH$_3\dot{C}$=O

13.30. Find the wavelength of light necessary to photochemically break an H—H bond if the average bond energy is 414 kJ mol^{-1}. Molecular hydrogen does not absorb in this wavelength region, so the energy must be absorbed by another material and transferred to the hydrogen (*photosensitization*). Which substance, Hg(g) or Na(g), would serve as a good photosensitizing agent, given the primary absorption wavelengths are 253.651 9 nm and 330.298 8 nm, respectively?   *Ans.*   289 nm, Hg

13.31. The actinometer described in Problem 13.10 was used to find the number of quanta absorbed by a sample of HX(g). The concentrations in the actinometer after 30.0 min of light absorption were $C(N_2) =$ 43.1 × 10^{-5} M and 51.2 × 10^{-5} M, respectively, for the transmitted beam and the incident beam. (*a*) Find the quanta absorbed by the HX(g) sample given that the actinometer had a volume of 1 dm^3. (*b*) Given 0.158 × 10^{-3} mol of HX decomposed upon absorption of these quanta, find Φ. Does this value agree with that predicted by the mechanism

$$\text{HX} + h\nu \longrightarrow \text{H} + \text{X} \qquad \text{H} + \text{HX} \longrightarrow \text{H}_2 + \text{X} \qquad \text{X} + \text{X} \longrightarrow \text{X}_2 ?$$

Ans.   (*a*) 0.45 × 10^{-7} mol s^{-1};   (*b*) $\Phi = 1.95$, Φ(predicted) = 2

Chapter 14

Introduction to Quantum Mechanics

Preliminaries

14.1 ELECTROMAGNETIC RADIATION

The properties of electromagnetic radiation are currently explained in terms of a dualistic theory. One part of the theory considers the radiation to obey wave theory, with the frequency (ν), wavelength (λ), and velocity (c) of the light related by

$$\nu\lambda = c \tag{14.1}$$

The wavelength can also be represented by the *wavenumber* ($\bar{\nu}$), where

$$\bar{\nu} = \frac{1}{\lambda} \tag{14.2}$$

The second part of the dualistic theory assumes that the radiation acts as discrete packets of energy, called quanta. The wave and the corpuscular aspects are related by

$$E = h\nu \tag{14.3}$$

where E is the energy of the quantum and h is Planck's constant.

14.2 DE BROGLIE WAVELENGTH

Any particle with mass m that is moving with a velocity v has associated with it a wavelength given by

$$\lambda = \frac{h}{mv} \tag{14.4}$$

EXAMPLE 14.1. Calculate λ for an electron moving with $v = 0.1c$, using the rest mass,

$$m_e = 9.109\ 534 \times 10^{-31}\ \text{kg}$$

as the mass. In what part of the electromagnetic spectrum would this value of λ fall?

Using (14.4) gives

$$\lambda = \frac{6.626 \times 10^{-34}\ \text{J s}}{(9.11 \times 10^{-31}\ \text{kg})(0.1)(3.00 \times 10^8\ \text{m s}^{-1})} = 2.42 \times 10^{-11}\ \text{m}$$

or 0.024 2 nm, which falls into the long-γ-ray region of the spectrum (10^{-3} nm, γ-rays; 1 nm, X-rays; 10–200 nm, vacuum ultraviolet; 200–400 nm, ultraviolet; 400–700 nm, visible; 0.7–20 μm, near infrared; 20–1 000 μm, far infrared; 0.1–100 cm, microwave; and 1 m, radio).

14.3 HEISENBERG UNCERTAINTY (INDETERMINACY) PRINCIPLE

There are no restrictions on the precision of measurement of position (x) alone or on the measurement of momentum in the x direction (p_x) alone. But the product of the measurement uncertainties (Δx and Δp_x) during a simultaneous measurement must be greater than $\hbar/2$ or, in the limiting case, equal to $\hbar/2$:

$$(\Delta p_x)(\Delta x) \geq \frac{\hbar}{2} \qquad (14.5)$$

where $\hbar = h/2\pi = 1.054\,588\,7 \times 10^{-34}$ J s. Likewise, for simultaneous measurements of energy and time,

$$(\Delta E)(\Delta t) \geq \frac{\hbar}{2} \qquad (14.6)$$

14.4   RYDBERG EQUATION

Under low resolution, the emission spectra of hydrogenlike elements appear as series of lines fitting the empirical *Rydberg equation*

$$\bar{\nu} = Z^2 \mathscr{R} \left(\frac{1}{n_2{}^2} - \frac{1}{n_1{}^2} \right) \qquad (14.7)$$

where Z is the atomic number of the element, $\mathscr{R}$ is the *Rydberg constant*,

$$\mathscr{R} = (109\,737.317\,7 \text{ cm}^{-1}) \left(\frac{m_{\text{nucleus}}}{m_{\text{nucleus}} + m_e} \right) \qquad (14.8)$$

and n_i are integers, with $n_2 < n_1$.

EXAMPLE 14.2.   Find the ionization energy for atomic hydrogen.

The ionization energy is the energy required to remove the electron from the lowest energy level ($n_2 = 1$) to ∞ ($n_1 = \infty$). Using $\mathscr{R} = 109\,677.59$ cm^{-1} for H (see Problem 14.3), (14.7) and (14.3) give

$$\bar{\nu} = (1)^2 (109\,677.59 \text{ cm}^{-1}) \left(\frac{1}{1^2} - \frac{1}{\infty^2} \right) = 109\,677.59 \text{ cm}^{-1}$$

$$E = (6.626\,176 \times 10^{-34} \text{ J s})(109\,677.59 \text{ cm}^{-1})[(10^2 \text{ cm})/(1 \text{ m})](2.997\,924\,58 \times 10^8 \text{ m s}^{-1})$$

$$= 2.178\,721 \times 10^{-18} \text{ J}$$

A frequently used unit in quantum mechanics is the *electronvolt* (eV). Converting the above answer gives

$$E = \frac{2.178\,721 \times 10^{-18} \text{ J}}{(1.602\,189\,2 \times 10^{-19} \text{ J})/(1 \text{ eV})} = 13.598\,4 \text{ eV}$$

14.5   BOHR THEORY FOR HYDROGENLIKE ATOMS

The Bohr theory consists of three main postulates:
1.   The electron moves around the nucleus of charge $+Ze$ in a circular orbit of constant energy.
2.   The only orbits that are allowed are those in which the electron has an angular momentum equal to $n\hbar$, where n is an integer (quantum number).
3.   Transitions between orbits generate spectral lines, with the frequency of a line being given by $(\Delta E)/h$, where ΔE is the energy difference between the initial and final orbits.

By applying these postulates to the system shown in Fig. 14-1, it is possible to show that the radii of the orbits (r_n) are given by

$$r_n = n^2 \frac{(4\pi\varepsilon_0)\hbar^2}{\mu e^2 Z} \qquad (14.9)$$

where $e = 1.602\,189\,2 \times 10^{-19}$ C is the electronic charge, $4\pi\varepsilon_0 = 1.112\,650\,056 \times 10^{-10}$ C^2 N^{-1} m^{-2} is the permittivity constant, and

$$\mu = \frac{m_e m_{\text{nucleus}}}{m_e + m_{\text{nucleus}}} \qquad (14.10)$$

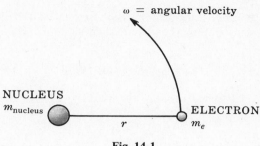

Fig. 14-1

is the reduced mass of the system. The energy of the orbit having quantum number n is

$$E_n = -\frac{\mu Z^2 e^4}{2n^2 \hbar^2 (4\pi\varepsilon_0)^2} \tag{14.11}$$

The Bohr theory explained the Rydberg equation but failed to predict the fine structure of spectra and to describe polyelectronic systems.

EXAMPLE 14.3.   Calculate the radii and energies of the first four orbits of atomic hydrogen. Prepare a sketch of these orbits to scale. Prepare an energy diagram to scale, and determine the number of transitions that can occur. Find the values of ν for these transitions.

Using $m_{\text{nucleus}} = 1.672\,648\,5 \times 10^{-27}$ kg, $m_e = 9.109\,534 \times 10^{-31}$ kg,   and $Z = 1$,   (14.10)   gives $\mu = 9.104\,575 \times 10^{-31}$ kg. Then, from (14.9) and (14.11),

$$r_n = n^2 \frac{(1.112\,650 \times 10^{-10}\,\text{C}^2\,\text{N}^{-1}\,\text{m}^{-2})(1.054\,588\,7 \times 10^{-34}\,\text{J s})^2}{(9.104\,575 \times 10^{-31}\,\text{kg})(1.602\,189\,2 \times 10^{-19}\,\text{C})^2 (1)}$$

$$= n^2 (5.294\,653 \times 10^{-11})\text{m}$$

$$E_n = -\frac{(9.104\,575 \times 10^{-31}\,\text{kg})(1)^2 (1.602\,189\,2 \times 10^{-19}\,\text{C})^4}{2n^2 (1.054\,588\,7 \times 10^{-34}\,\text{J s})^2 (1.112\,650 \times 10^{-10}\,\text{C}^2\text{N}^{-1}\,\text{m}^{-2})^2}$$

$$= -\frac{2.178\,720 \times 10^{-18}}{n^2}\,\text{J} = -\frac{109\,677.6}{n^2}\,\text{cm}^{-1}$$

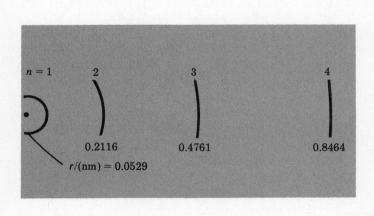

(a)

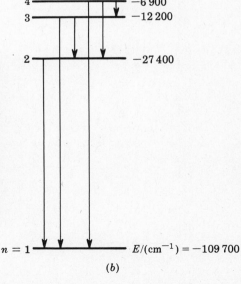

(b)

Fig. 14-2

The energy was changed from J to cm^{-1} using the conversion factor $1.986\,477 \times 10^{-23}$ J/cm^{-1}. Values corresponding to $n = 1, 2, 3,$ and 4 are indicated in Fig. 14-2. The six vertical lines shown in Fig. 14-2b represent possible transitions, with energies

$$\Delta E_{2\to 1} = (-27\,400\ \text{cm}^{-1}) - (-109\,700\ \text{cm}^{-1}) = 82\,300\ \text{cm}^{-1}$$

$\Delta E_{3\to 1} = 97\,500$ cm^{-1}; $\Delta E_{3\to 2} = 15\,200$ cm^{-1}; $\Delta E_{4\to 1} = 102\,800$ cm^{-1}; $\Delta E_{4\to 2} = 20\,500$ cm^{-1}; and $\Delta E_{4\to 3} = 5\,300$ cm^{-1}.

Postulates of Quantum Mechanics

14.6 WAVE FUNCTIONS

Any state of a dynamic system containing N particles can be described by a wave function $\psi(q_1, q_2, \ldots, q_{3N}, t)$, where the q_i are coordinates and t is time, such that $\psi^*\psi$ is proportional to the probability of finding the coordinates between q_i and $q_i + dq_i$ at t. The wave function should be continuous, single-valued, and square-integrable so that it can be in agreement with physical reality. In addition, the first and second derivatives should be continuous. The symbol ψ^* represents the *complex conjugate* of ψ, which is formed by changing the sign of the imaginary unit i in the expression for ψ.

$$\text{If} \qquad\qquad \int_{\text{all space}} \psi^*\psi\, dV = 1 \qquad\qquad (14.12)$$

where dV is the volume element of the coordinate system, the wave function is said to be *normalized*. If

$$\int_{\text{all space}} \psi_i^*\psi_j\, dV = 0 \qquad\qquad (14.13)$$

for two wave functions, the functions are said to be *orthogonal*. If ψ_i and ψ_j are normalized wave functions that satisfy (14.13), they are known as *orthonormal* wave functions.

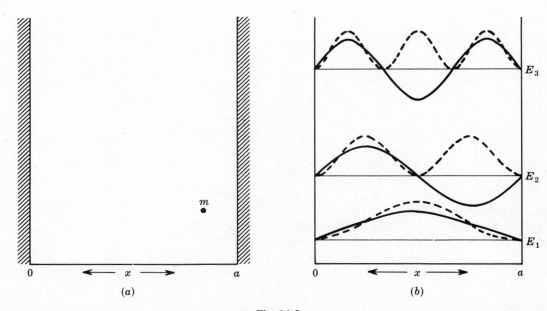

Fig. 14-3

EXAMPLE 14.4. Consider a particle having mass m located in a one-dimensional potential-energy well (box) with infinitely high walls, as shown in Fig. 14-3a. The wave function describing this system is

$$\psi_n(x) = \begin{cases} K \sin \dfrac{n\pi x}{a} & \text{for } 0 \le x \le a \\ 0 & \text{otherwise} \end{cases} \tag{14.14}$$

where K is a constant and $n = 1, 2, 3, \ldots$. (Because the potential energy is infinite—an artificial condition—outside the well, it is not possible to make the derivative of ψ continuous at the walls.) Determine $K^*K = |K|^2$.

Substituting (14.14) and $\psi_n(x)^* = K^* \sin(n\pi x/a)$ into (14.12) gives

$$1 = \int_0^a \left(K^* \sin \frac{n\pi x}{a} \right) \left(K \sin \frac{n\pi x}{a} \right) dx = K^*K \int_0^a \sin^2 \frac{n\pi x}{a} \, dx$$

The integral can be transformed into one found in most integral tables by letting $z = n\pi x/a$ and $dz = (n\pi/a)\, dx$, giving

$$1 = K^*K \left(\frac{a}{n\pi} \right) \int_0^{n\pi} \sin^2 z \, dz = K^*K \frac{a}{2}$$

Thus $K^*K = 2/a$.

EXAMPLE 14.5. Show that $\psi_n'(x) = W \cos(n\pi x/a)$ and the functions (14.14) are orthogonal.

Evaluating (14.13) gives

$$\int_0^a \psi_n^*(x)\psi_n'(x) \, dx = \int_0^a \left(K^* \sin \frac{n\pi x}{a} \right) \left(W \cos \frac{n\pi x}{a} \right) dx$$

$$= K^*W \int_0^a \sin \left(\frac{n\pi x}{a} \right) \cos \left(\frac{n\pi x}{a} \right) dx$$

The integral can be evaluated as described in Example 14.4, giving

$$\int_0^a \psi_n^*(x)\psi_n'(x) \, dx = K^*W \frac{a}{n\pi} \int_0^{n\pi} \sin z \cos z \, dz = 0$$

which according to (14.13) proves orthogonality.

14.7 OPERATORS

For every physically observable property of a system there exists a linear Hermitian operator ($\hat{o}$) that can operate on ψ. Examples of $\hat{o}$ are: position, $\hat{x} = x$; x component of linear momentum, $\hat{p}_x = (\hbar/i)(\partial/\partial x)$; x component of angular momentum, $\hat{l}_x = (\hbar/i)[y(\partial/\partial z) - z(\partial/\partial y)]$; kinetic energy along the x axis, $\hat{T}_x = -(\hbar^2/2m)(\partial^2/\partial x^2)$; and the total energy (the *Hamiltonian*) for N particles,

$$\hat{H} = \mathscr{H} = -\frac{\hbar^2}{2} \sum_{i=1}^N \frac{1}{m_i} \nabla_i^2 + U(x, y, z) \tag{14.15}$$

in which U is the total potential-energy function. The Laplacian operator is given by

$$\nabla^2 \equiv \frac{\partial^2}{\partial x^2} + \frac{\partial^2}{\partial y^2} + \frac{\partial^2}{\partial z^2} \tag{14.16a}$$

in Cartesian coordinates and in spherical polar coordinates by

$$\nabla^2 \equiv \frac{1}{r^2} \frac{\partial}{\partial r} \left(r^2 \frac{\partial}{\partial r} \right) + \frac{1}{r^2 \sin \theta} \frac{\partial}{\partial \theta} \left(\sin \theta \frac{\partial}{\partial \theta} \right) + \frac{1}{r^2 \sin^2 \theta} \left(\frac{\partial^2}{\partial \phi^2} \right) \tag{14.16b}$$

In (14.15) ∇_i^2 stands for the Laplacian with respect to the coordinates of the ith particle.

EXAMPLE 14.6.   Write the total energy operator for the system described in Example 14.4.

The Hamiltonian is found by substituting $U(x) = 0$ and ($14.16a$) written for one dimension into (14.15), giving

$$\mathcal{H} = -\frac{\hbar^2}{2m}\frac{\partial^2}{\partial x^2} \tag{14.17}$$

14.8   EIGENFUNCTIONS AND EIGENVALUES

The only possible values that an observable quantity corresponding to $\hat{o}$ can possess are given by

$$\hat{o}\psi = o\psi \tag{14.18}$$

where an allowed value o is known as an *eigenvalue* and ψ is the corresponding *eigenfunction.* In some cases, o can be continuously variable, and in others o can have only discrete values.

EXAMPLE 14.7.   The eigenvalues for position for the system described in Example 14.4 are continuously variable, but those for energy are discrete and are given by

$$E_n = \frac{n^2 h^2}{8ma^2} \tag{14.19}$$

Prepare a sketch of the energy levels E_n, and superimpose plots of $\psi_n(x)$ and $\psi_n(x)^*\psi_n(x)$ on the respective levels.

In units of $h^2/8ma^2$, the energies are given by (14.19) as $E_1 = 1$, $E_2 = 4$, $E_3 = 9$, etc. These energies are shown in Fig. 14-3b. The plots of $\psi_n(x)$ are shown as solid curves, and those of $\psi_n(x)^*\psi_n(x)$ as dashed curves.

14.9   EXPECTATION VALUES

The average of a series of measurements will be given by

$$\langle o \rangle = \frac{\displaystyle\int_{\text{all space}} \psi^*\hat{o}\psi\, dV}{\displaystyle\int_{\text{all space}} \psi^*\psi\, dV} \tag{14.20}$$

The denominator in (14.20) does not appear if normalized wave functions are used. Because of the Hermitian character of the operators,

$$\int_{\text{all space}} \psi_i^*\hat{o}\psi_j\, dV \equiv \int_{\text{all space}} \psi_j^*\hat{o}^*\psi_i\, dV \tag{14.21}$$

EXAMPLE 14.8.   Find the average position of the particle described in Example 14.4. Would this location be a good place to seek the particle if it is in energy state E_2?

Substituting $\hat{x} = x$ and (14.14) with $K^*K = 2/a$ into (14.20) gives

$$\langle x \rangle = \int_0^a \left(K^*\sin\frac{n\pi x}{a}\right)x\left(K\sin\frac{n\pi x}{a}\right)dx = \frac{2}{a}\int_0^a x\sin^2\frac{n\pi x}{a}\,dx$$

Evaluating the integral as described in Example 14.4 gives

$$\langle x \rangle = \left(\frac{2}{a}\right)\left(\frac{a}{n\pi}\right)^2\int_0^{n\pi} z\sin^2 z\,dz = \left(\frac{2}{a}\right)\left(\frac{a}{n\pi}\right)^2\left(\frac{n^2\pi^2}{4}\right) = \frac{a}{2}$$

the center of the potential-energy well. The particle will spend equal amounts of time on either side of the center because $\psi_n(x)^*\psi_n(x)$ is symmetrical about the center. However, for $n = 2$, the probability of finding the particle near the center, where $\psi_2(x)^*\psi_2(x) = 0$, is very low (see Fig. 14-3b).

EXAMPLE 14.9.   Show that (14.19) is the solution of (14.20) for the particle described in Example 14.4.

Substituting $\mathcal{H} = (-\hbar^2/2m)(\partial^2/\partial x^2)$ for $\hat{o}$ and E for o into (14.20) gives

$$E_n \equiv \langle E \rangle = \int_0^a \left(K^* \sin \frac{n\pi x}{a} \right) \left(-\frac{\hbar^2}{2m} \frac{\partial^2}{\partial x^2} \right) \left(K \sin \frac{n\pi x}{a} \right) dx$$

$$= K^* K \left(-\frac{\hbar^2}{2m} \right) \int_0^a \left(\sin \frac{n\pi x}{a} \right) \frac{\partial^2}{\partial x^2} \left(\sin \frac{n\pi x}{a} \right) dx$$

$$= \left(\frac{2}{a} \right) \left(-\frac{\hbar^2}{2m} \right) \left(\frac{n\pi}{a} \right)^2 (-1) \int_0^a \sin^2 \frac{n\pi x}{a} \, dx = \left(\frac{2}{a} \right) \left(-\frac{\hbar^2}{2m} \right) \left(\frac{n\pi}{a} \right)^2 (-1) \left(\frac{a}{2} \right)$$

$$= \frac{n^2 \hbar^2 \pi^2}{2ma^2} = \frac{n^2 h^2}{8ma^2}$$

14.10   TIME DEPENDENCE

The evolution of $\psi(q_i, t)$ is given by $i\hbar(\partial\psi/\partial t) = \mathcal{H}\psi$. Because $\mathcal{H}$ is not a function of time, $\psi(q_i, t)$ can be written as $\psi(q_i) e^{-(i/\hbar)Et}$, giving the *time-independent Schrödinger equation*

$$\mathcal{H}\psi(q_i) = E\psi(q_i) \tag{14.22}$$

for $\psi(q_i)$.

EXAMPLE 14.10.   Show that (14.14) is a solution to (14.22) for the particle described in Example 14.4.

Performing the operation $\mathcal{H}\psi_n(x)$ gives

$$\mathcal{H}\psi_n(x) = \left(-\frac{\hbar^2}{2m} \frac{\partial^2}{\partial x^2} \right) \left(K \sin \frac{n\pi x}{a} \right) = \left(-\frac{\hbar^2}{2m} \right) K \left(\frac{n\pi}{a} \right)^2 (-1) \sin \frac{n\pi x}{a}$$

$$= \left(\frac{\hbar^2}{2m} \right) \left(\frac{n\pi}{a} \right)^2 \psi_n(x) = E_n \psi_n(x)$$

14.11   THE CORRESPONDENCE PRINCIPLE

Quantum mechanics can be successfully extrapolated to macroscopic systems, generating the classical Newtonian results.

EXAMPLE 14.11.   The classical model for an ideal-gas molecule restricted to one dimension predicts that position is continuous. Show that (14.14) qualitatively predicts this result.

At room temperature a molecule with one degree of freedom has a kinetic energy given by $E = \frac{1}{2}kT = 2.06 \times 10^{-21}$ J. Assuming a molar mass of 28.0 g mol^{-1} and $a = 1$ m, (14.19) gives

$$n^2 = \frac{E(8ma^2)}{h^2} = \frac{(2.06 \times 10^{-21})(8)(28.0 \times 10^{-3}/6.022 \times 10^{23})(1)^2}{(6.626 \times 10^{-34})^2} = 1.75 \times 10^{21}$$

or $n = 4.18 \times 10^{10}$. Recognizing that the plots of $\psi_n(x)^* \psi_n(x)$ as shown in Fig. 14-3b contain n maxima, the separation between the most probable locations would be $(1 \text{ m})/(4.18 \times 10^{10}) = 0.023\,9$ nm. Because the molecular size is larger than this value, position is for all purposes continuous.

Approximation Methods

14.12   THE VARIATION METHOD

For most problems of chemical interest, the exact solution of (14.22) is not possible. However, an approximate solution corresponding to the quantum state of lowest energy, the *ground state*, can be obtained by choosing a trial wave function (ϕ) and adjusting the parameters in ϕ to give a minimum energy. Then

$$E_{\text{ground}} \leq E = \frac{\displaystyle\int_{\text{all space}} \phi^* \mathcal{H} \phi \, dV}{\displaystyle\int_{\text{all space}} \phi^* \phi \, dV} \tag{14.23}$$

that is, the best choice of ϕ provides an upper bound for the true energy of the ground state.

If ϕ is constructed by taking a linear combination of m "basis functions" (μ_i), i.e.,

$$\phi = \sum_{i=1}^{m} c_i \mu_i \tag{14.24}$$

where c_i are variable parameters, the values of c_i that give the minimum energy satisfy the following set of equations:

$$\left. \begin{array}{c} c_1(H_{11} - ES_{11}) + c_2(H_{12} - ES_{12}) + \cdots + c_m(H_{1m} - ES_{1m}) = 0 \\ c_1(H_{21} - ES_{21}) + c_2(H_{22} - ES_{22}) + \cdots + c_m(H_{2m} - ES_{2m}) = 0 \\ \vdots \\ c_1(H_{m1} - ES_{m1}) + c_2(H_{m2} - ES_{m2}) + \cdots + c_m(H_{mm} - ES_{mm}) = 0 \end{array} \right\} \tag{14.25}$$

where

$$H_{ij} = \int_{\text{all space}} \mu_i^* \mathcal{H} \mu_j \, dV \tag{14.26}$$

$$S_{ij} = \int_{\text{all space}} \mu_i^* \mu_j \, dV \tag{14.27}$$

The system (14.25) will have a nonzero solution only for those values of E that make the determinant of the coefficients vanish, i.e.,

$$\begin{vmatrix} H_{11} - ES_{11} & H_{12} - ES_{12} & \ldots & H_{1m} - ES_{1m} \\ H_{21} - ES_{21} & H_{22} - ES_{22} & \ldots & H_{2m} - ES_{2m} \\ \vdots & \vdots & & \vdots \\ H_{m1} - ES_{m1} & H_{m2} - ES_{m2} & \ldots & H_{mm} - ES_{mm} \end{vmatrix} = 0 \tag{14.28}$$

With E determined as a root of the *secular equation* (14.28), the c_i can be found by solving $m - 1$ of the equations (14.25) and imposing the condition that the wave function be normalized.

EXAMPLE 14.12. The boundary conditions in Example 14.4 are $\psi_n(0) = \psi_n(a) = 0$. Choosing $\mu_1 = x(x - a)$, which also satisfies the boundary conditions, as a simple basis function, find the estimate for E_{ground}.

For the trial function $\phi = c_1 x(x - a)$, (14.28) is

$$H_{11} - ES_{11} = 0$$

Substituting the values of H_{11} and S_{11} given by (14.26) and (14.27),

$$H_{11} = \int_0^a [x(x-a)] \left(-\frac{\hbar^2}{2m} \frac{d^2}{dx^2} \right) [x(x-a)] \, dx = \left(-\frac{\hbar^2}{2m} \right) \int_0^a x(x-a) \, dx = \frac{\hbar^2 a^3}{6m}$$

$$S_{11} = \int_0^a x^2(x-a)^2 \, dx = \frac{a^5}{30}$$

yields

$$E = \frac{H_{11}}{S_{11}} = \frac{\hbar^2 a^3/6m}{a^5/30} = \frac{5\hbar^2}{ma^2} = \frac{5h^2}{4\pi^2 ma^2}$$

A comparison with the true value, $E_{\text{ground}} = E_1 = h^2/8ma^2$, gives the percent error as

$$\frac{E - E_1}{E_1} = \frac{5h^2/4\pi^2 ma^2 - h^2/8ma^2}{h^2/8ma^2} = \frac{10}{\pi^2} - 1 = 1.3\%$$

which is rather good considering the crudeness of the trial function (a ramp function instead of a sine wave).

14.13 NONDEGENERATE PERTURBATION THEORY

If the problem for which an approximate solution is desired does not differ greatly from one for which normalized $\psi_n^{(0)}$ and $E_n^{(0)}$ are known for the corresponding $\mathscr{H}^{(0)}$, then the Hamiltonian is written as

$$\mathscr{H} = \mathscr{H}^{(0)} + \lambda \mathscr{H}^{(1)} \tag{14.29}$$

and (14.22) has the solutions

$$\psi_n = \psi_n^{(0)} + \lambda \psi_n^{(1)} + \lambda^2 \psi_n^{(1)} + \cdots \tag{14.30}$$

$$E_n = E_n^{(0)} + \lambda E_n^{(1)} + \lambda^2 E_n^{(2)} + \cdots \tag{14.31}$$

where the parameter λ is a measure of the deviation of the problem of interest from the unperturbed problem. Normally λ is set equal to 1, signifying that the perturbation is being fully applied. For first-order theory, i.e., retaining terms in λ^1, (14.30) and (14.31) become

$$\psi_n = \psi_n^{(0)} + \lambda \sum_m' \frac{H_{mn}^{(1)} \psi_m^{(0)}}{E_n^{(0)} - E_m^{(0)}} \tag{14.32}$$

and

$$E_n = E_n^{(0)} + \lambda H_{nn}^{(1)} \tag{14.33}$$

where

$$H_{mn}^{(1)} = \int_{\text{all space}} \psi_m^{(0)*} \mathscr{H}^{(1)} \psi_n^{(0)} \, dV \tag{14.34}$$

$$E_m^{(0)} = \int_{\text{all space}} \psi_m^{(0)*} \mathscr{H}^{(0)} \psi_m^{(0)} \, dV \tag{14.35}$$

and $\sum_m'$ means to sum over all values of $m \neq n$ until an insignificant addition results. The second-order estimate for the energy is

$$E_n = E_n^{(0)} + \lambda H_{nn}^{(1)} + \lambda^2 \sum_m' \frac{H_{nm}^{(1)} H_{mn}^{(1)}}{E_n^{(0)} - E_m^{(0)}} \tag{14.36}$$

EXAMPLE 14.13. Consider the particle in the one-dimensional well shown in Fig. 14-4, inside which the potential energy is given by $U(x) = C(x/a)$. What is the first-order estimate of the energy of this system using perturbation theory?

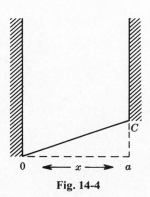

Fig. 14-4

The unperturbed problem, of which the Hamiltonian is

$$\mathcal{H}^{(0)} = \frac{-\hbar^2}{2m} \frac{d^2}{dx^2}$$

has the solution (see Examples 14.4 and 14.7)

$$\psi_n^{(0)} = K \sin \frac{n\pi x}{a}$$

with $K^*K = 2/a$, and

$$E_n^{(0)} = \frac{n^2 h^2}{8ma^2}$$

For $0 \le x \le a$, the Hamiltonian for the perturbed system is

$$\mathcal{H} = -\frac{\hbar^2}{2m} \frac{d^2}{dx^2} + C\frac{x}{a} = \mathcal{H}^{(0)} + \mathcal{H}^{(1)}$$

where $\mathcal{H}^{(1)} = C(x/a)$. Then ($14.34$) gives

$$H_{nn}^{(1)} = \int_0^a \left(K^* \sin \frac{n\pi x}{a} \right) \left(C\frac{x}{a} \right) \left(K \sin \frac{n\pi x}{a} \right) dx = \frac{K^*KC}{a} \int_0^a x \sin^2 \frac{n\pi x}{a} \, dx$$

This integral was evaluated in Example 14.8 as $a^2/4$, giving

$$H_{nn}^{(1)} = \frac{K^*KC}{a} \left(\frac{a^2}{4} \right) = \frac{C}{2}$$

Hence (14.33) gives

$$E_n = \frac{n^2 h^2}{8ma^2} + \frac{C}{2}$$

Solved Problems

PRELIMINARIES

14.1. Commercial FM radio stations operate between 88 and 108 MHz. Calculate λ, $\bar{\nu}$, and E for a radio wave having $\nu = 95.5$ MHz.

Using (14.1)–(14.3) gives

$$\lambda = \frac{2.997\,9 \times 10^8 \text{ m s}^{-1}}{95.5 \times 10^6 \text{ s}^{-1}} = 3.14 \text{ m} \qquad \bar{\nu} = \frac{1}{3.14 \text{ m}} = 0.319 \text{ m}^{-1}$$

$$E = (6.626 \times 10^{-34} \text{ J s})(95.5 \times 10^6 \text{ s}^{-1}) = 6.33 \times 10^{-26} \text{ J}$$

Note that the wavelength is about 10 ft, a multiple of the size for various quarter- and half-wave "rabbit-ear" and TV antennas.

14.2. What is the maximum precision with which the momentum can be known if the position of an electron is determined to within $\pm 1 \times 10^{-13}$ m? Will there be any problems in describing the momentum if it has a value of $\hbar/0.052\,9$ nm?

Using the limiting case of (14.5) gives

$$\Delta p_x = \frac{\hbar/2}{\Delta x} = \frac{1.055 \times 10^{-34} \text{ J s}}{(2)(1 \times 10^{-13} \text{ m})} = 5.28 \times 10^{-22} \text{ N s}$$

If the momentum is $\hbar/0.052\,9$ nm $= 1.99 \times 10^{-24}$ N s, the uncertainty in the measurement of momentum will be about 265 times as large as the momentum itself. For that reason, the concept of definite electron orbits has been replaced by "probabilities" of locating electrons.

14.3. The Balmer series of lines in the spectrum of atomic hydrogen has $n_2 = 2$ in (14.7). Calculate λ for the first six lines of this series.

Substituting $m_{\text{nucleus}} = 1.672\,648\,5 \times 10^{-27}$ kg and $m_e = 9.109\,534 \times 10^{-31}$ kg into (14.8) gives the value of the Rydberg constant for hydrogen as $\mathcal{R} = 109\,677.59$ cm^{-1}. Substituting this value of $\mathcal{R}$ with $n_1 = 3$ and $n_2 = 2$ into (14.7) gives

$$\bar{\nu} = (1)^2(109\,677.59 \text{ cm}^{-1})\left(\frac{1}{2^2} - \frac{1}{3^2}\right) = 15\,233.00 \text{ cm}^{-1}$$

and (14.2) gives

$$\lambda = \frac{1}{\bar{\nu}} = \frac{1}{15\,233.00 \text{ cm}^{-1}} = 6.564\,696 \times 10^{-5} \text{ cm} = 656.469\,6 \text{ nm}$$

Repeating the calculations for $n_1 = 4$, 5, 6, 7, and 8 gives 486.273 8, 434.173 0, 410.293 5, 397.123 6, and 389.019 0 nm, respectively. Note that the difference between successive lines becomes smaller as n_1 increases.

POSTULATES OF QUANTUM MECHANICS

14.4. If the walls of a potential-energy well have a finite height V_0, then the wave function describing a particle within the box does not go to zero at the walls but continues into the energy barrier. (a) Find an equation for the allowed energies of the system if the wave functions describing the particle in the three regions shown in Fig. 14-5a are

$$\psi_1(x) = A \exp[2m(V_0 - E)]^{1/2}(x/\hbar)$$

$$\psi_{\text{II}}(x) = B \sin[(2mE)^{1/2}(x/\hbar)] + C \cos[(2mE)^{1/2}(x/\hbar)]$$

and $$\psi_{\text{III}}(x) = D \exp\{-[2m(V_0 - E)]^{1/2}(x/\hbar)\}$$

(b) Discuss the sketches for the first three energy levels, as shown in Fig. 14-5b.

(a) Because $\psi(x)$ must be continuous at $x = 0$, $\psi_1(0) = \psi_{\text{II}}(0)$, which implies $A = C$. Furthermore, the derivatives

$$\frac{d\psi_1}{dx}\bigg|_{x=0} = A\frac{[2m(V_0 - E)]^{1/2}}{\hbar} \qquad \text{and} \qquad \frac{d\psi_{\text{II}}}{dx}\bigg|_{x=0} = B\frac{(2mE)^{1/2}}{\hbar}$$

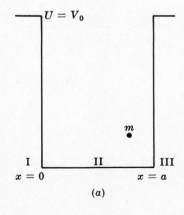
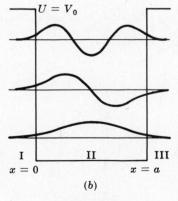
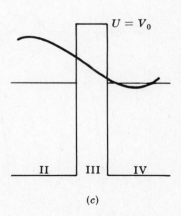

Fig. 14-5

must be equal, giving

$$A = B \left(\frac{E}{V_0 - E} \right)^{1/2}$$

Similarly, the continuity of ψ and its first derivative at $x = a$ require that

$$B \sin \frac{(2mE)^{1/2}a}{\hbar} + C \cos \frac{(2mE)^{1/2}a}{\hbar} = D \exp \left\{ -[2m(V_0 - E)]^{1/2} \left(\frac{a}{\hbar} \right) \right\}$$

and

$$B \frac{(2mE)^{1/2}}{\hbar} \cos \frac{(2mE)^{1/2}a}{\hbar} - C \frac{(2mE)^{1/2}}{\hbar} \sin \frac{(2mE)^{1/2}a}{\hbar}$$

$$= -D \frac{[2m(V_0 - E)]^{1/2}}{\hbar} \exp \left\{ -[2m(V_0 - E)]^{1/2} \left(\frac{a}{\hbar} \right) \right\}$$

In view of the relations between A and C and between A and B found above, the last two equations can be rewritten as

$$\sin \frac{(2mE)^{1/2}a}{\hbar} + \left(\frac{E}{V_0 - E} \right)^{1/2} \cos \frac{(2mE)^{1/2}a}{\hbar} = \frac{D}{B} \exp \left\{ -[2m(V_0 - E)]^{1/2} \left(\frac{a}{\hbar} \right) \right\}$$

and

$$\cos \frac{(2mE)^{1/2}a}{\hbar} - \left(\frac{E}{V_0 - E} \right)^{1/2} \sin \frac{(2mE)^{1/2}a}{\hbar} = -\frac{D}{B} \left(\frac{V_0 - E}{E} \right)^{1/2} \exp \left\{ -[2m(V_0 - E)]^{1/2} \left(\frac{a}{\hbar} \right) \right\}$$

Dividing the first equation by the second yields the desired equation for E, which, after some manipulation, can be put in the form

$$\left(\frac{E}{V_0 - E} \right)^{1/2} = \cot \frac{(2mE)^{1/2}a}{2\hbar} \qquad \text{or} \qquad -\left(\frac{E}{V_0 - E} \right)^{1/2} = \tan \frac{(2mE)^{1/2}a}{2\hbar}$$

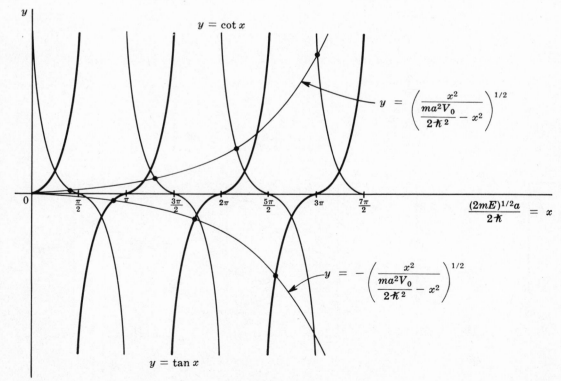

Fig. 14-6

A graphical solution is indicated in Fig. 14-6. The intersections above the x axis give the energies corresponding to wave functions that are symmetrical about the center of the well; those below the axis correspond to antisymmetric wave functions.

(b) As E approaches V_0 from below, the relative amount of "tailing off" in regions I and III increases (see Fig. 14-5b). If the width of the potential barrier for region III is small enough that $\psi_{\mathrm{III}}(x)$ has not reached zero before entering the region IV shown in Fig. 14-5c, there is a finite probability for the particle to "tunnel" from II to IV.

14.5. A simple harmonic oscillator (SHO) is characterized by two bodies of mass m_1 and m_2 separated by a distance that can change by an amount x (see Fig. 14-7). For a displacement x, it is assumed that a restoring force equal to $-kx$ is present. Write the Hamiltonian for this system.

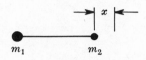

Fig. 14-7

The potential energy for the system can be determined by integrating the negative of the restoring force, giving

$$U(x) = -\int_0^x f(y)\, dy = -\int_0^x (-ky)\, dy = \frac{kx^2}{2} \tag{14.37}$$

Substituting this expression into (14.15) gives

$$\mathscr{H} = -\frac{\hbar^2}{2\mu}\frac{d^2}{dx^2} + \frac{kx^2}{2} \tag{14.38}$$

where μ is the reduced mass expressed in terms of m_1 and m_2 [see (14.10)].

14.6. The eigenvalues describing the energy states for the SHO described in Problem 14.5 are given by

$$E_v = (v + \tfrac{1}{2})h\nu_0 \tag{14.39}$$

where $v = 0, 1, 2, 3, \ldots$ and

$$\nu_0 = \frac{(k/\mu)^{1/2}}{2\pi} \tag{14.40}$$

Prepare a plot of (14.37), and sketch the energy levels given by (14.39).

A plot of (14.37) gives a parabola (see Fig. 14-8) centered about $x = 0$. The energy levels given by (14.39) are equally spaced:

$$\Delta E_v = (v + 1 + \tfrac{1}{2})h\nu_0 - (v + \tfrac{1}{2})h\nu_0 = h\nu_0$$

14.7. The normalized eigenfunctions for the SHO described in Problem 14.5 are

$$\psi_v(x) = (2^v v!)^{-1/2}(a/\pi)^{1/4}\, e^{-ax^2/2} H_v(a^{1/2}x) \tag{14.41}$$

where

$$a = 2\pi\nu_0\mu/\hbar \tag{14.42}$$

and the *Hermite polynomials* are defined as

$$H_n(z) = (-1)^n\, e^{z^2}\frac{d^n e^{-z^2}}{dz^n} \tag{14.43}$$

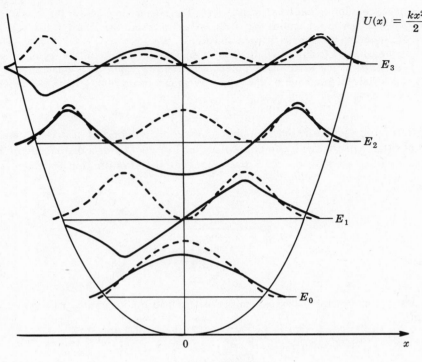

Fig. 14-8

Find the first four eigenfunctions, and sketch $\psi_v(x)$ and $\psi_v(x)^*\psi_v(x)$ on the corresponding energy levels shown in Fig. 14-8.

The first four Hermite polynomials are given by (*14.43*) as

$$H_0(z) = (-1)^0 \, e^{z^2} \frac{d^0 \, e^{-z^2}}{dz^0} = 1$$

$$H_1(z) = (-1)^1 \, e^{z^2} \frac{d \, e^{-z^2}}{dz} = 2z$$

$$H_2(z) = (-1)^2 \, e^{z^2} \frac{d^2 \, e^{-z^2}}{dz^2} = 4z^2 - 2$$

$$H_3(z) = (-1)^3 \, e^{z^2} \frac{d^3 \, e^{-z^2}}{dz^3} = 8z^3 - 12z$$

which, expressed in terms of $a^{1/2}x$ and substituted into (*14.41*), give

$$\psi_0(x) = (2^0 0!)^{-1/2}(a/\pi)^{1/4} \, e^{-ax^2/2}(1) = (a/\pi)^{1/4} \, e^{-ax^2/2}$$

$$\psi_1(x) = (2^1 1!)^{-1/2}(a/\pi)^{1/4} \, e^{-ax^2/2}(2a^{1/2}x) = (2a)^{1/2}(a/\pi)^{1/4}x \, e^{-ax^2/2}$$

$$\psi_2(x) = (2^2 2!)^{-1/2}(a/\pi)^{1/4} \, e^{-ax^2/2}(4ax^2 - 2) = (1/2)^{1/2}(a/\pi)^{1/4}(2ax^2 - 1) \, e^{-ax^2/2}$$

$$\psi_3(x) = (2^3 3!)^{-1/2}(a/\pi)^{1/4} \, e^{-ax^2/2}(8a^{3/2}x^3 - 12a^{1/2}x) = (1/3)^{1/2}(a/\pi)^{1/4}(2a^{3/2}x^3 - 3a^{1/2}x) \, e^{-ax^2/2}$$

Figure 14-8 shows these functions as solid curves and $\psi^*\psi$ as dashed curves. Note that there is a finite probability that the system lies outside the parabolic potential well.

14.8. Verify that $\psi_3(x)$ for the SHO, as given in Problem 14.7, is normalized.

To see if the wave function is normalized, the integral given by (14.12) is evaluated:

$$\int_{-\infty}^{\infty} \psi_3(x)^* \psi_3(x)\, dx = \int_{-\infty}^{\infty} \left[\left(\frac{1}{3} \right)^{1/2} \left(\frac{a}{\pi} \right)^{1/4} (2a^{3/2}x^3 - 3a^{1/2}x)\, e^{-ax^2/2} \right]^2 dx$$

$$= \left(\frac{1}{3} \right) \left(\frac{a}{\pi} \right)^{1/2} \int_{-\infty}^{\infty} e^{-ax^2}(2a^{3/2}x^3 - 3a^{1/2}x)^2\, dx$$

$$= \left(\frac{1}{3} \right) \left(\frac{a}{\pi} \right)^{1/2} \left[4a^3 \int_{-\infty}^{\infty} x^6 e^{-ax^2}\, dx - 12a^2 \int_{-\infty}^{\infty} x^4 e^{-ax^2}\, dx + 9a \int_{-\infty}^{\infty} x^2 e^{-ax^2}\, dx \right]$$

The three integrals can be evaluated using

$$\int_0^{\infty} x^{2n} e^{-ax^2}\, dx = \frac{(1)(3)(5) \cdots (2n-1)(\pi/a)^{1/2}}{2^{n+1} a^n}$$

and the fact that the three integrands are even functions. Thus,

$$\int_{-\infty}^{\infty} \psi_3(x)^* \psi_3(x)\, dx = 2 \left(\frac{1}{3} \right) \left(\frac{a}{\pi} \right)^{1/2} \left[4a^3 \frac{(1)(3)(5)(\pi/a)^{1/2}}{2^4 a^3} - 12a^2 \frac{(1)(3)(\pi/a)^{1/2}}{2^3 a^2} + 9a \frac{(1)(\pi/a)^{1/2}}{2^2 a} \right]$$

$$= 2 \left(\frac{1}{3} \right) \left(\frac{a}{\pi} \right)^{1/2} \left[(1.5) \left(\frac{\pi}{a} \right)^{1/2} \right] = 1$$

The wave function is normalized [see (14.12)].

14.9. For Hermitian operators, eigenfunctions corresponding to distinct eigenvalues are orthogonal, i.e., they obey (14.13). Verify the orthogonality of the first two eigenfunctions of the SHO, $\psi_0(x)$ and $\psi_1(x)$.

$$\int_{-\infty}^{\infty} \psi_0(x)^* \psi_1(x)\, dx = \int_{-\infty}^{\infty} \left[\left(\frac{a}{\pi} \right)^{1/4} e^{-ax^2/2} \right] \left[(2a)^{1/2} \left(\frac{a}{\pi} \right)^{1/4} x e^{-ax^2/2} \right] dx$$

$$= a \left(\frac{2}{\pi} \right)^{1/2} \int_{-\infty}^{\infty} x e^{-ax^2}\, dx$$

But the integrand is an odd function, so the value of the integral is zero.

14.10. Evaluate $\langle x \rangle$ for the SHO system in the lowest energy state.

Substituting the expression derived in Problem 14.7 for $\psi_0(x)$ into (14.20) gives

$$\langle x \rangle = \int_{-\infty}^{\infty} \left[\left(\frac{a}{\pi} \right)^{1/4} e^{-ax^2/2} \right] x \left[\left(\frac{a}{\pi} \right)^{1/4} e^{-ax^2/2} \right] dx = \left(\frac{a}{\pi} \right)^{1/2} \int_{-\infty}^{\infty} x e^{-ax^2}\, dx = 0$$

which is the center of the parabola. Thus the $v = 0$ state can be interpreted as a symmetrical oscillation around the equilibrium position.

14.11. Show that the wave function given by (14.41) for the lowest energy state of the SHO system is indeed a solution to (14.22).

Using the Hamiltonian from (14.38), E_0 from (14.39), and the expression for $\psi_0(x)$ from Problem 14.7 in (14.22) gives

$$-\frac{\hbar^2}{2\mu} \frac{d^2}{dx^2} \left[\left(\frac{a}{\pi} \right)^{1/4} e^{-ax^2/2} \right] + \frac{hx^2}{2} \left[\left(\frac{a}{\pi} \right)^{1/4} e^{-ax^2/2} \right] = \frac{1}{2} h\nu_0 \left[\left(\frac{a}{\pi} \right)^{1/4} e^{-ax^2/2} \right]$$

Taking the derivatives and simplifying gives

$$\left(-\frac{\hbar^2}{2\mu}\right)(a^2x^2 - a)\, e^{-ax^2/2} + \left(\frac{hx^2}{2}\right) e^{-ax^2/2} = \frac{h\nu_0}{2}\, e^{-ax^2/2}$$

which upon cancellation of the exponentials, substitution of (14.42) and (14.40), and simplification gives identical terms on both sides of the equation.

14.12. Figure 14-9 is a plot of $\psi^*\psi$ for the $v = 10$ state of the SHO. How does it illustrate the correspondence principle?

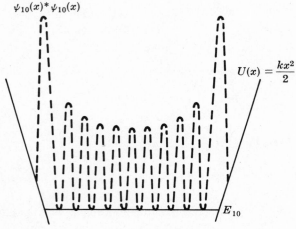

Fig. 14-9

According to the correspondence principle, at high quantum numbers the oscillator should approach a macroscopic classical oscillator in behavior, e.g., a pendulum. The figure shows that the probability distribution is concentrated near the "turning points," where the system's energy is purely potential. This indeed describes a pendulum, which spends more time slowing down, reversing direction, and accelerating near the extremes of displacement than in the center of the movement where it has its maximum speed.

APPROXIMATION METHODS

14.13. A particle is confined in a one-dimensional potential-energy well where the potential energy is given by $U(x) = U_0 \sin(p\pi x/a)$. Here p is a fixed positive integer. Use perturbation theory to obtain a first-order estimate of the energies of the system.

The given problem is not too different from that of a particle in a box with infinite walls and zero potential energy inside. Choosing this as the unperturbed problem, (14.17), (14.14), and (14.19) give

$$\mathcal{H}^{(0)} = -\frac{\hbar^2}{2m}\frac{d^2}{dx^2} \qquad \psi_n^{(0)} = K \sin\frac{n\pi x}{a} \qquad E_n^{(0)} = \frac{n^2 h^2}{8ma^2}$$

With $\lambda = 1$, (14.29) gives the perturbation term as $\mathcal{H}^{(1)} = U_0 \sin(p\pi x/a)$. Then, from (14.34)

$$H_{nn}^{(1)} = \int_0^a \left(K^* \sin\frac{n\pi x}{a}\right)\left(U_0 \sin\frac{p\pi x}{a}\right)\left(K \sin\frac{n\pi x}{a}\right) dx = \frac{2U_0}{a}\int_0^a \sin\frac{p\pi x}{a}\sin^2\frac{n\pi x}{a}\, dx$$

Employing the half-angle formula $\sin^2\alpha = \frac{1}{2} - \frac{1}{2}\cos 2\alpha$, $H_{nn}^{(1)}$ becomes

$$H_{nn}^{(1)} = \frac{U_0}{a}\int_0^a \sin\frac{p\pi x}{a}\, dx - \frac{U_0}{a}\int_0^a \sin\frac{p\pi x}{a}\cos\frac{2n\pi x}{a}\, dx$$

$$= -\frac{U_0}{p\pi}\left[\cos\frac{p\pi x}{a}\right]\Big|_0^a - \frac{U_0}{a}\int_0^a \sin\frac{p\pi x}{a}\cos\frac{2n\pi x}{a}\, dx$$

The first term is equal to zero if p is even and $2U_0/p\pi$ if p is odd. The last integral can be evaluated from tables, but three cases must be considered.

Case 1: *p is odd.* Then $2n \neq p$, and the tabulated integral

$$\int \sin \alpha x \cos \beta x \, dx = -\frac{1}{2}\left[\frac{\cos(\alpha-\beta)x}{\alpha-\beta}+\frac{\cos(\alpha+\beta)x}{\alpha+\beta}\right]$$

can be used to give

$$H_{nn}^{(1)}=\frac{2U_0}{p\pi}-\frac{U_0}{\pi}\left(\frac{1}{p-2n}+\frac{1}{p+2n}\right)=-\frac{8n^2U_0}{p(p^2-4n^2)\pi}$$

Case 2: *p is even and $n \neq p/2$.* Then the integral in case 1 still applies, giving

$$H_{nn}^{(1)}=0+0=0$$

Case 3: *p is even and $n = p/2$.* Then

$$H_{p/2,p/2}^{(1)}=0-\frac{U_0}{a}\int_0^a \sin\frac{p\pi x}{a}\cos\frac{p\pi x}{a}\,dx=-\frac{U_0}{2a}\int_0^a \sin\frac{2p\pi x}{a}\,dx=0$$

The perturbed energies are therefore

$$E_n = E_n^{(0)} + H_{nn}^{(1)} = \begin{cases} \dfrac{n^2h^2}{8ma^2}-\dfrac{8n^2U_0}{p(p^2-4n^2)\pi} & \text{if } p \text{ is odd} \\[4mm] \dfrac{n^2h^2}{8ma^2} & \text{if } p \text{ is even} \end{cases}$$

The result for even values of p might have been anticipated. When p is even, a whole number of wavelengths of the perturbing function fits into the box. The function thus has average value zero over the box and does not show up in a first-order perturbation.

Supplementary Problems

PRELIMINARIES

14.14. The visible spectrum corresponds to values of λ between 400 and 700 nm. Calculate ν, $\bar{\nu}$, and E for yellow light having $\lambda = 580$ nm.

Ans. 5.169×10^{14} Hz, $17\,240$ cm^{-1}, 3.425×10^{-19} J

14.15. The speeds of the Indy 500 racing cars are recorded to ± 0.001 mph. Assuming that the track distance is known to within ± 0.01 mi, is the uncertainty principle violated for a 3.5-ton car?

Ans. $m = 3.2 \times 10^3$ kg, $\Delta v = 4.5 \times 10^{-4}$ m s^{-1}, $\Delta p_x = m\,\Delta v = 1.4$ kg m s^{-1}, $\Delta x = 16$ m, $(\Delta p_x)(\Delta x) = 22$ J s; no

14.16. Compare the emission spectra of atomic hydrogen and deuterium. Assume $m_e = 9.109\,534 \times 10^{-31}$ kg and $m_{\text{nucleus}}/(\text{kg}) = 1.672\,648\,5 \times 10^{-27}$ for H and $3.343\,398 \times 10^{-27}$ for D.

Ans. $\mathcal{R}/(\text{cm}^{-1}) = 109\,677.59$ for H and $109\,707.43$ for D, giving slightly higher values of $\bar{\nu}$ and slightly lower values of λ for D.

14.17. Repeat Example 14.3 for He$^+$, assuming $\mu = m_e$.

Ans. $r/(\text{nm}) = 0.026\,5$, $0.105\,8$, $0.238\,1$, $0.423\,3$; $E/(\text{cm}^{-1}) = -438\,710$, $-109\,678$, $-48\,746$, $-27\,419$; $\Delta E/(\text{cm}^{-1}) = 329\,332$, $389\,964$, $60\,932$, $411\,291$, $82\,259$, $21\,327$

14.18. The element "positronium" consists of an electron moving in space around a nucleus consisting of a positron (a subatomic particle similar to the electron except possessing a positive charge). Using the Bohr theory, calculate the radii of the first four orbits of the electron, the corresponding energies, and the predicted electronic spectrum, assuming the positron to be motionless.

Ans. $\mu = m_e/2$; $r/(\text{nm}) = n^2(0.105\ 835)$, giving $r/(\text{nm}) = 0.105\ 8$, $0.423\ 3$, $0.952\ 5$, $1.693\ 4$; $E/(\text{cm}^{-1}) = (-54\ 869/n^2)$, giving $E/(\text{cm}^{-1}) = -54\ 869$, $-13\ 717$, $-6\ 097$, $-3\ 429$; $\Delta E/(\text{cm}^{-1}) = 41\ 152$, $48\ 772$, $7\ 620$, $51\ 440$, $10\ 288$, $2\ 668$

POSTULATES OF QUANTUM MECHANICS

14.19. The general form of the wave function for a free particle moving along the x axis is

$$\psi = K \cos \frac{[2m(E - U)]^{1/2}x}{\hbar}$$

where $U = \text{constant}$ and $E > U$. Write the wave function describing a particle moving over the potential-energy well shown in Fig. 14-10.

Ans. $\psi_1 = \psi_{111} = A \cos \dfrac{[2m(E - V_0)]^{1/2}x}{\hbar}$, $\psi_{11} = B \cos \dfrac{(2mE)^{1/2}x}{\hbar}$

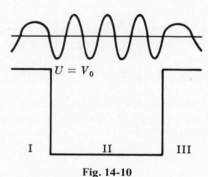

$U = V_0$

I   II   III

Fig. 14-10

14.20. The normalized wave functions and energies for a particle moving in a three-dimensional potential-energy well that is infinitely deep and has sides of lengths a, b, and c are given by

$$\psi(x, y, z) = A \sin \frac{n_x \pi x}{a} \sin \frac{n_y \pi y}{b} \sin \frac{n_z \pi z}{c} \qquad (14.44)$$

$$E = \frac{h^2}{8m}\left(\frac{n_x^2}{a^2} + \frac{n_y^2}{b^2} + \frac{n_z^2}{c^2}\right) \qquad (14.45)$$

where $A^*A = 8/abc$ and n_x, n_y, and n_z are positive integers. What is the degeneracy (the number of different quantum states having the same energy) of the energy level $E = 6h^2/8ma^2$, if the particle is in a cube?

Ans.   Triply degenerate: $n_x^2 + n_y^2 + n_z^2 = 6$ if $n_x = 2$, $n_y = 1$, $n_z = 1$ or $n_x = 1$, $n_y = 2$, $n_z = 1$ or $n_x = 1$, $n_y = 1$, $n_z = 2$.

14.21. Find $\langle z \rangle$, $\langle z^2 \rangle$, $\langle p_z \rangle$, and $\langle p_z^2 \rangle$ for a particle in a three-dimensional box using the wave function given by *(14.44)*. The operator $\widehat{p_z^2}$ is $(\hbar^2/i^2)(\partial^2/\partial z^2)$.   *Ans.* $c/2$, $c^2(2\pi^2 n_z^2 - 3)/6\pi^2 n_z^2$, 0, $\hbar^2\pi^2 n_z^2/c^2$

14.22. The *two-particle rigid rotator* undergoes rotation about its center of mass (see Fig. 14-11). Write the Hamiltonian for this system, using spherical coordinates.

Ans. $\mathcal{H} = \left(-\dfrac{\hbar^2}{2\mu}\right)\left[\dfrac{1}{r^2 \sin\theta}\dfrac{\partial}{\partial\theta}\left(\sin\theta\dfrac{\partial}{\partial\theta}\right) + \dfrac{1}{r^2 \sin^2\theta}\dfrac{\partial^2}{\partial\phi^2}\right]$ (14.46)

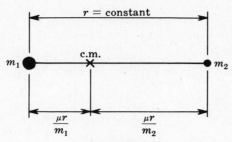

Fig. 14-11

14.23. The normalized eigenfunctions for the rigid rotator described in Problem 14.22 are given by

$$\psi_{J,m}(\theta, \phi) = (2\pi)^{-1/2}\Theta_{J,m}(\theta)\, e^{\pm im\phi} \qquad (14.47)$$

where

$$\Theta_{J,m}(\theta) = \left[\frac{(2J+1)(J-|m|)!}{2(J+|m|)!}\right]^{1/2} P_J^{|m|}(\cos\theta) \qquad (14.48)$$

for $J = 0, 1, 2, \ldots$ and $m = -J, -J+1, \ldots, J-1, J$. Here, the *associated Legendre functions*, $P_\ell^{|m|}(x)$, are defined as

$$P_\ell^{|m|}(x) = \frac{1}{2^\ell \ell!}(1-x^2)^{|m|/2}\frac{d^{\ell+|m|}}{dx^{\ell+|m|}}(x^2-1)^\ell \qquad (14.49)$$

Determine $\psi_{J,m}(\theta, \phi)$ for $J = 2$ and $m = \pm 1$, and verify that it is normalized.

Ans. $\psi_{2,\pm1}(\theta, \phi) = \left(\dfrac{15}{8\pi}\right)^{1/2}\sin\theta\cos\theta\, e^{\pm i\phi}$

14.24. The eigenvalues for the rigid rotator described in Problem 14.22 are

$$E_J = \frac{J(J+1)\hbar^2}{2I} \qquad (14.50)$$

where the moment of inertia is given by

$$I = \mu r^2 \qquad (14.51)$$

(*a*) What is the degeneracy of each level? (*b*) Prepare a plot of the energy levels expressed in units of $\hbar^2/2I$.

Ans.   (*a*) $2J+1$; (*b*) see Fig. 14-12.

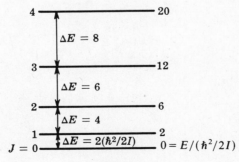

Fig. 14-12

14.25. Determine $\psi(\theta, \phi)$ for a rigid rotator using (14.47) given $J = 2$ and $m = 0$, and show that it is a solution to (14.22).

Ans. $\left(\dfrac{5}{16\pi}\right)^{1/2}(3\cos^2\theta - 1)$

APPROXIMATION METHODS

14.26. Consider a particle in a one-dimensional box $(0 \le x \le a)$ where $U(x) = U_0 \cos(n\pi x/a)$. Use first-order perturbation theory to determine E.

    Ans. $\mathcal{H}^{(1)} = U_0 \cos(n\pi x/a)$, $H_{nn}^{(1)} = 0$, $E = n^2 h^2 / 8ma^2$

14.27 Solve the secular equation

$$\begin{vmatrix} H_{aa} - E & H_{ba} - ES \\ H_{ba} - ES & H_{aa} - E \end{vmatrix} = 0$$

    Ans. $E = (H_{aa} + H_{ba})/(1 + S)$ or $(H_{aa} - H_{ba})/(1 - S)$

Chapter 15

Atomic Structure and Spectroscopy

Hydrogenlike Atoms

15.1 SYSTEM DESCRIPTION

A *hydrogenlike* atom is one in which an electron having a charge of $-e$ and mass m_e is moving around a nucleus having a charge of $+Ze$ and a mass of m_{nucleus}. Usually such a system is described using spherical coordinates (r, θ, ϕ) (see Fig. 15-1). After separating the kinetic energy terms describing the translational motion of the entire atom (a particle in a three-dimensional box), (*14.22*) becomes

$$\left(-\frac{\hbar^2}{2\mu}\right)\left[\frac{1}{r^2}\frac{\partial}{\partial r}\left(r^2\frac{\partial}{\partial r}\psi(r, \theta, \phi)\right) + \frac{1}{r^2\sin\theta}\frac{\partial}{\partial\theta}\left(\sin\theta\frac{\partial}{\partial\theta}\psi(r, \theta, \phi)\right)\right.$$

$$\left. + \frac{1}{r^2\sin^2\theta}\frac{\partial^2}{\partial\phi^2}\psi(r, \theta, \phi)\right] + \left(-\frac{Ze^2}{4\pi\varepsilon_0 r}\right)\psi(r, \theta, \phi) = E\psi(r, \theta, \phi) \qquad (15.1)$$

where μ is the reduced mass of the system and $-Ze^2/4\pi\varepsilon_0 r$ represents the potential energy of the system as a result of the coulombic attraction between the nucleus and the electron.

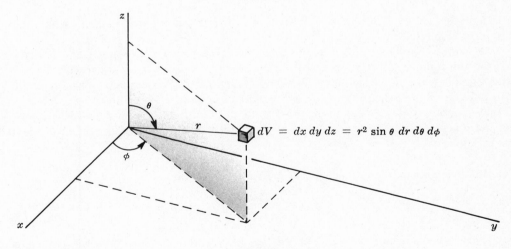

$$dV = dx\, dy\, dz = r^2 \sin\theta\, dr\, d\theta\, d\phi$$

Fig. 15-1

A wave function that satisfies (*15.1*) is

$$\psi(r, \theta, \phi) = R(r)\Theta(\theta)\Phi(\phi) \qquad (15.2)$$

where the new wave functions, $R(r)$, $\Theta(\theta)$, and $\Phi(\phi)$ are functions of only r, θ, and ϕ, respectively. Substituting (*15.2*) into (*15.1*) generates the following set of three equations:

$$\frac{1}{r^2}\frac{d}{dr}\left(r^2\frac{d}{dr}R(r)\right) - \frac{\beta}{r^2}R(r) + \frac{2\mu}{\hbar^2}\left(E + \frac{Ze^2}{4\pi\varepsilon_0 r}\right)R(r) = 0 \qquad (15.3a)$$

$$\frac{m^2}{\sin^2 \theta} \Theta(\theta) - \frac{1}{\sin \theta} \frac{d}{d\theta} \left(\sin \theta \frac{d}{d\theta} \Theta(\theta) \right) - \beta \Theta(\theta) = 0 \qquad (15.3b)$$

$$\left(-\frac{1}{\Phi(\phi)} \right) \frac{d^2}{d\phi^2} \Phi(\phi) = m^2 \qquad (15.3c)$$

where m and β are constants.

15.2 THE ANGULAR FUNCTION

The normalized solution of (15.3c) is

$$\Phi(\phi) = (2\pi)^{-1/2} e^{im\phi} \qquad (15.4)$$

where $m = 0, \pm 1, \pm 2, \ldots$. The integer m is known as the *magnetic quantum number*, because in the presence of a magnetic field, states with different values of m will have different energies (*Zeeman effect*).

The normalized solution of (15.3b) with $\beta = \ell(\ell + 1)$ is given by

$$\Theta(\theta) = \left[\frac{(2\ell+1)(\ell-|m|)!}{2(\ell+|m|)!} \right]^{1/2} P_\ell^{|m|}(\cos \theta) \qquad (15.5)$$

where $P_\ell^{|m|}(\cos \theta)$ has been defined by (14.49) and $\ell = 0, 1, 2, \ldots$. The integer ℓ is known as the *azimuthal quantum number* and describes the general shape of the wave function for an *atomic subshell*. A common notation is to substitute the letter s for $\ell = 0$, p for $\ell = 1$, d for $\ell = 2$, f for $\ell = 3$, g for $\ell = 4$, etc. The magnetic quantum number must be restricted to

$$m = 0, \pm 1, \pm 2, \ldots, \pm \ell \qquad (15.6)$$

if $\Theta(\theta)$ is to be defined by (15.5). The combination of the azimuthal and magnetic quantum numbers generates an *atomic orbital*.

The product $\Theta(\theta)\Phi(\phi)$ is known as the *angular eigenfunction* and is given the symbol $Y(\theta, \phi)$. Note that the solution for the rigid-rotator problem given by (14.47) is identical with $Y(\theta, \phi)$. Unless $m = 0$, the angular wave functions determined by (15.4) and (15.5) contain imaginary terms, which presents a difficulty in visualization of the wave function. For this reason, linear combinations of the $Y(\theta, \phi)$ that generate real wave functions are usually chosen to describe the atomic orbitals (without changing the eigenvalues of energy). It is not possible to assign one of these newly created wave functions to a given value of m.

EXAMPLE 15.1. The angular wave functions for $\ell = 2$ and $m = \pm 1$ are

$$Y(\theta, \phi) = \left(\frac{15}{8\pi} \right)^{1/2} \sin \theta \cos \theta \, e^{\pm i\phi}$$

Construct two real wave functions, $Y'(\theta, \phi)$ and $Y''(\theta, \phi)$, from these, and prepare two-dimensional plots of the new wave functions and of $Y'(\theta, \phi)^* Y'(\theta, \phi)$ and $Y''(\theta, \phi)^* Y''(\theta, \phi) = [Y''(\theta, \phi)]^2$.

In view of Euler's formula, $e^{\pm i\phi} = \cos \phi \pm i \sin \phi$, the linear combinations

$$Y'(\theta, \phi) \equiv \sqrt{2} \left[\frac{Y_+(\theta, \phi) + Y_-(\theta, \phi)}{2} \right]$$

and

$$Y''(\theta, \phi) \equiv \sqrt{2} \left[\frac{Y_+(\theta, \phi) - Y_-(\theta, \phi)}{2i} \right]$$

will be real, where the subscripts on $Y(\theta, \phi)$ indicate which sign is being used in the exponential term. (The factor $\sqrt{2}$ is merely an additional normalizing factor.) Thus,

$$Y'(\theta, \phi) = \left(\frac{15}{4\pi} \right)^{1/2} \sin \theta \cos \theta \cos \phi$$

and
$$Y''(\theta, \phi) = \left(\frac{15}{4\pi}\right)^{1/2} \sin\theta \cos\theta \sin\phi$$

Figure 15-2a is a polar plot of $|Y'(\theta, \phi)|$ in the xz plane, where $|\cos\phi| = 1$. The algebraic sign of $Y'(\theta, \phi)$ is indicated beside each lobe of the curve. Figure 15-2b shows $Y'(\theta, \phi)^* Y'(\theta, \phi)$, the scale being the same as in Fig. 15-2a. A separate figure for $Y''(\theta, \phi)$ is not needed, as $Y''(\theta, \phi)$ is just $Y'(\theta, \phi)$ rotated 90° about the z axis, i.e., $Y''(\theta, \phi)$ lies in the yz plane.

Because $Y'(\theta, \phi)$ has its peak values along 45° lines within the xz plane, and because the value of ℓ is 2, this orbital is known as the d_{xz} orbital.

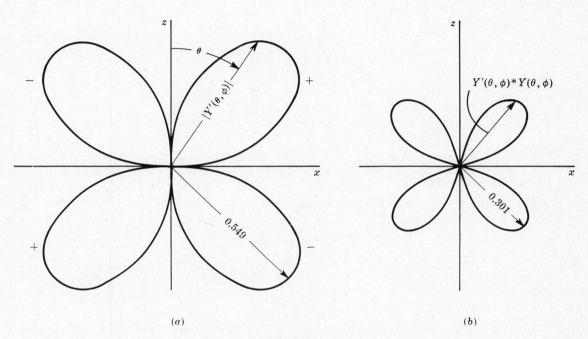

(a) $\qquad\qquad\qquad\qquad\qquad\qquad\qquad\qquad$ (b)

Fig. 15-2

15.3 THE RADIAL FUNCTION

The normalized solution of $(15.3a)$ is

$$R_{n,\ell}(r) = -\left\{\frac{(2Z/na_0)^3(n-\ell-1)!}{2n[(n+\ell)!]^3}\right\}^{1/2} e^{-\rho/2}\rho^\ell L_{n+\ell}^{2\ell+1}(\rho) \qquad (15.7)$$

where $n = 1, 2, 3, \ldots,$

$$\rho = \left(\frac{2Z}{na_0}\right)r \qquad (15.8)$$

$$a_0 = \frac{4\pi\varepsilon_0\hbar^2}{\mu e^2} \qquad (15.9)$$

and the *associated Laguerre polynomials* are defined as

$$L_q^s(x) = \frac{d^s}{dx^s}\left[e^x\frac{d^q}{dx^q}(x^q e^{-x})\right] \qquad (15.10)$$

The integer n is known as the *principal quantum number* and is related to the distance between the nucleus and the electron. Equation (15.7) restricts ℓ to the values

$$\ell = 0, 1, 2, \ldots, n-1 \tag{15.11}$$

if $R(r)$ is to be defined.

The probability of finding an electron between r and $r + dr$ is given by $R(r)^* R(r) r^2 \, dr$, where the factor r^2 appears from the volume element in the spherical coordinate system (see Problem 15.5).

EXAMPLE 15.2. Determine $R(r)$ for the 1s orbital.

Using (15.10) gives

$$L_1^1(\rho) = \frac{d}{d\rho} \left[e^\rho \frac{d}{d\rho} (\rho \, e^{-\rho}) \right] = \frac{d}{d\rho} [1 - \rho] = -1$$

for $n = 1$ and $\ell = 0$. Then (15.7) gives

$$R_{1,0}(r) = -\left(\frac{2Z}{a_0} \right)^{3/2} \left\{ \frac{0!}{2(1!)^3} \right\}^{1/2} e^{-\rho/2} \rho^0 (-1) = 2 \left(\frac{Z}{a_0} \right)^{3/2} e^{-\rho/2} = 2 \left(\frac{Z}{a_0} \right)^{3/2} e^{-Zr/a_0}$$

15.4 ELECTRON POSITION

The probability of locating an electron within a volume V_0 is given by

$$\iiint\limits_{V_0} \psi^* \psi \, dV$$

15.5 ENERGY VALUES

The energy eigenvalues for a hydrogenlike atom as determined by substituting (15.2), (15.4), (15.5), and (15.7) into (15.1) are

$$E = \frac{-\mu Z^2 e^4}{2n^2 \hbar^2 (4\pi\varepsilon_0)^2} \tag{15.12}$$

This result is identical to that of the Bohr theory, (14.11). Hence the spectrum predicted by (15.12) will be identical to that described in Secs. 14.4 and 14.5.

EXAMPLE 15.3. List the orbitals for hydrogenlike atoms in order of increasing energy.

Because (15.12) is a function of only n, all orbitals having the same value of n will have the same energy. Thus, $1s < 2s = 2p < 3s = 3p = 3d < 4s = 4p = 4d = 4f < 5s = 5p = 5d = 5f = 5g < $ etc.

Quantum Theory of Polyelectronic Atoms

15.6 ELECTRON SPIN WAVE FUNCTIONS

The coupling of the angular momenta of two or more electrons in an atom plays an important part in removing the degeneracy of the atomic orbitals having the same principal and azimuthal quantum numbers. The usual notation is α and β for the wave functions for electron spin, and $s = \pm\frac{1}{2}$ for the corresponding quantum numbers.

15.7 HAMILTONIAN OPERATOR AND WAVE FUNCTION

The Hamiltonian operator for a system containing n electrons is given by

$$\mathscr{H} = \frac{-\hbar^2}{2m_e} \sum_{i=1}^{n} \nabla_i^2 - \sum_{i=1}^{n} \frac{Ze^2}{4\pi\varepsilon_0 r_i} + \sum_{j=2}^{n} \sum_{i<j} \frac{e^2}{4\pi\varepsilon_0 r_{ij}} \qquad (15.13)$$

where r_{ij} is the distance between the ith and jth electrons. The Schrödinger equation cannot be solved exactly because of the r_{ij} term, and approximate methods must be used: neglect the r_{ij} term, self-consistent field theory, Hartree–Fock techniques, etc.

According to the *Pauli exclusion principle*, an acceptable wave function for electrons must change its algebraic sign when the coordinates (both spatial and spin) of any two particles are interchanged (antisymmetry). Such a wave function for the ground state of a polyelectronic atom is given by the *Slater determinant*

$$\psi = \frac{1}{\sqrt{n!}} \begin{vmatrix} \phi_{1s}(1)\alpha(1) & \phi_{1s}(2)\alpha(2) & \cdots & \phi_{1s}(n)\alpha(n) \\ \phi_{1s}(1)\beta(1) & \phi_{1s}(2)\beta(2) & \cdots & \phi_{1s}(n)\beta(n) \\ \phi_{2s}(1)\alpha(1) & \phi_{2s}(2)\alpha(2) & \cdots & \phi_{2s}(n)\alpha(n) \\ \vdots & \vdots & & \vdots \end{vmatrix} \qquad (15.14)$$

where $\phi_{1s}(1)\alpha(1)$ represents the product of the hydrogenlike 1s orbital containing electron number 1 and the α spin wave function for electron number 1, etc. Thus the n columns correspond to the n electrons, and the n rows to their n lowest quantum states. Interchanging the coordinates of two electrons means interchanging two columns, which reverses the sign of the determinant.

EXAMPLE 15.4.　Using (15.13) and (15.14), determine $\mathscr{H}$ and ψ for He. Verify that the expanded form of ψ is antisymmetric with respect to permutation of the electrons.

For two electrons, the Hamiltonian given by (15.13) is

$$\mathscr{H} = \frac{-\hbar^2}{2m_e}(\nabla_1^2 + \nabla_2^2) - \frac{2e^2}{4\pi\varepsilon_0}\left(\frac{1}{r_1} + \frac{1}{r_2}\right) + \frac{e^2}{4\pi\varepsilon_0 r_{12}}$$

and the determinant given by (15.14) is

$$\psi = \frac{1}{\sqrt{2!}} \begin{vmatrix} \phi_{1s}(1)\alpha(1) & \phi_{1s}(2)\alpha(2) \\ \phi_{1s}(1)\beta(1) & \phi_{1s}(2)\beta(2) \end{vmatrix}$$

which upon expansion gives

$$\psi = 2^{-1/2}[\phi_{1s}(1)\alpha(1)\phi_{1s}(2)\beta(2) - \phi_{1s}(1)\beta(1)\phi_{1s}(2)\alpha(2)]$$

To demonstrate the antisymmetric behavior of ψ, exchanging electron number 2 for electron number 1 and electron number 1 for electron number 2 gives a new wave function ψ', where

$$\psi' = 2^{-1/2}[\phi_{1s}(2)\alpha(2)\phi_{1s}(1)\beta(1) - \phi_{1s}(2)\beta(2)\phi_{1s}(1)\alpha(1)] = -\psi$$

15.8 ENERGY LEVELS

The exact ordering of the atomic subshells with respect to energy depends on Z, but in general the filling order is $1s < 2s < 2p(3) < 3s < 3p(3) < 4s < 3d(5) < 4p(3) < 5s < 4d(5) < 5p(3) < 6s < 4f(7) < 5d(5) < 6p(3) < 7s < 5f(7) = 6d(5) < 7p(3) <$ etc., where the number in parentheses represents the number of orbitals comprising the subshell. Because of spin considerations, each orbital can hold two electrons. In the accepted notation for subshells, a superscript is used to indicate the number of electrons in that subshell, e.g., $(3d)^3$ implies that there are three electrons with $n = 3$ and $\ell = 2$.

The Pauli exclusion principle can be restated as: *no two electrons can have identical values for all four quantum numbers in the same atom.* Thus, in the $(3d)^3$ subshell the three electrons must have three different combinations of $m = \pm 2, \pm 1, 0$ and $s = \pm\frac{1}{2}$.

The rule of maximum multiplicity (*Hund's first rule*) states that *the most stable state for a configuration will be the one having the most nonpaired electrons*; so, by convention, electrons with parallel spins are placed singly in the various degenerate orbitals until it is necessary to add a second electron. By convention, the orbitals having the most positive values of m and of s are lowest in energy and are used first. For example, the $(3d)^3$ configuration would correspond to $m = +2$, $s = +\frac{1}{2}$; $m = +1$, $s = +\frac{1}{2}$ and $m = 0$, $s = +\frac{1}{2}$.

EXAMPLE 15.5.   Predict the electronic configuration for the ground state of chromium, and suggest possible oxidation states for the element.

The predicted configuration for the 24 electrons in Cr would be

$$(1s)^2(2s)^2(2p)^6(3s)^2(3p)^6(4s)^2(3d)^4$$

One major exception to the rules for filling subshells is that filled and half-filled d and f subshells are definitely favored, and slight rearrangements of predicted configurations will occur to give a more energetically favorable configuration. Thus, in Cr, the attainment of the d^5 configuration changes the predicted configuration to

$$(1s)^2(2s)^2(2p)^6(3s)^2(3p)^6(4s)^1(3d)^5$$

The possible oxidation states of an atom will correspond to the number of electrons that can easily be removed—seldom more than seven—and the number of electrons needed to complete a shell—seldom more than four. The order of removing electrons is more like the reverse of the order of filling hydrogenlike atoms than that of filling polyelectron atoms. The configuration for Cr would suggest that the possible oxidation states are 0, +1 (loss of the 4s electron), +2 (loss of the 4s electron and the 3d electron that was supposed to be a 4s electron), and +6 (loss of the 4s electron and the 3d electrons). The common +3 oxidation state corresponds to the loss of the 4s and two of the 3d electrons.

Atomic Term Symbols

15.9   RUSSELL-SAUNDERS COUPLING

The coupling of the angular momenta of two atomic orbitals, known as *ℓ-ℓ coupling*, gives rise to a quantum number L, where

$$L = \ell_1 + \ell_2, \ell_1 + \ell_2 - 1, \ldots, |\ell_1 - \ell_2| \qquad (15.15)$$

The coupling of the angular momenta of two electrons, known as *s-s coupling*, gives rise to a quantum number S, where

$$S = 0 \text{ or } 1 \qquad (15.16)$$

The *Russell–Saunders coupling* between L and S gives rise to a quantum number J, where

$$J = L + S, L + S - 1, \ldots, |L - S| \qquad (15.17)$$

The atomic term symbol has the general form $^{2S+1}X_J$, where the leading superscript represents the degeneracy of the term, the subscript is the value of J as determined by (15.17), and X represents a letter symbol for the value of L; i.e., S for $L = 0$, P for $L = 1$, D for $L = 2$, F for $L = 3$, G for $L = 4$, etc.

EXAMPLE 15.6.   Determine the values of L for the interaction between a p and a d electron.

The values of ℓ for p and d electrons are 1 and 2, respectively. Using (15.15) gives

$$L = 1 + 2 = 3,$$

$$1 + 2 - 1 = 2,$$

$$1 + 2 - 2 = |1 - 2| = 1$$

which correspond to the letters F, D, and P, respectively.

15.10 POLYELECTRONIC ATOM TERM SYMBOLS

For polyelectronic systems,

$$M_L = \sum m_i \qquad \text{where } M_L = L, L-1, \ldots, -L \qquad (15.18)$$

$$M_S = \sum s_i \qquad \text{where } M_S = S, S-1, \ldots, -S \qquad (15.19)$$

As can be seen from these equations, the only contributions to M_L and M_S are from nonfilled subshells.

To determine the possible term symbols for an atom, the following steps are used:

1. Prepare a list of microstates that correspond to the possible combinations of wave functions that are permitted by the Pauli exclusion principle (see Example 15.7).
2. Determine M_L and M_S for these microstates.
3. Determine the term symbol for the microstate having the largest values of M_L and M_S, using $L = M_L$ and $S = M_S$.
4. To that same term symbol, assign all microstates whose M_L and M_S values are related to the above L and S by (15.18) and (15.19).
5. Repeat steps 3 and 4 for the largest remaining M_L-M_S combination until all microstates have been assigned.

Table 15-1 lists the results of these steps for various electronic configurations for electrons having identical values of n and ℓ.

Table 15-1

Configuration	Term Symbols
s^1	^2S
s^2	^1S
p^1 or p^5	^2P
p^2 or p^4	$^3P, {}^1D, {}^1S$
p^3	$^4S, {}^2D, {}^2P$
p^6	^1S
d^1 or d^9	^2D
d^2 or d^8	$^3F, {}^3P, {}^1G, {}^1D, {}^1S$
d^3 or d^7	$^4F, {}^4P, {}^2H, {}^2G, {}^2F, {}^2D(2), {}^2P$
d^4 or d^6	$^5D, {}^3H, {}^3G, {}^3F(2), {}^3D, {}^3P(2), {}^1I, {}^1G(2), {}^1F, {}^1D(2), {}^1S(2)$
d^5	$^6S, {}^4G, {}^4F, {}^4D, {}^4P, {}^2I, {}^2H, {}^2G(2), {}^2F(2), {}^2D(3), {}^2P, {}^2S$
d^{10}	^1S

The ground state for the atom will correspond to the term symbol having the highest value of $2S+1$, according to the rule of maximum multiplicity (*Hund's first rule*). For terms of equal multiplicity, the ground state will have the highest value of M_L (*Hund's second rule*). The value of J for a given level is usually lowest for less-than-half-filled subshells and greatest for more-than-half-filled subshells.

EXAMPLE 15.7. Determine the term symbol for C which has the configuration $\ldots (2p)^2$.

The wave functions that are used for this configuration are $\phi_{2p, m}\gamma$, where $m = \pm 1, 0$ and $\gamma = \alpha$ or β. Thus there are $3 \times 2 = 6$ choices of wave function for each electron, or $6 \times 6 = 36$ choices for the pair. Of these, 6 give the same wave function to both electrons and so are ruled out by the Pauli principle; of the remaining 30, half must be eliminated because there is no way of distinguishing between the "first" electron and the "second" electron. We are left with 15 microstates, which, using the notation $m\gamma$ instead of $\phi_{2p, m}\gamma$, can be indicated as follows: $(+1\alpha, +1\beta), (+1\alpha, 0\alpha), (+1\alpha, 0\beta), (+1\beta, 0\alpha), (+1\beta, 0\beta), (+1\alpha, -1\alpha), (+1\alpha, -1\beta), (+1\beta, -1\alpha), (+1\beta, -1\beta),$ $(0\alpha, 0\beta), (0\alpha, -1\alpha), (0\alpha, -1\beta), (0\beta, -1\alpha), (0\beta, -1\beta),$ and $(-1\alpha, -1\beta)$. The respective values of M_L and M_S for these microstates are $(2, 0), (1, 1), (1, 0), (1, 0), (1, -1), (0, 1), (0, 0), (0, 0), (0, -1), (0, 0), (-1, 1), (-1, 0), (-1, 0),$ $(-1, -1),$ and $(-2, 0)$.

Beginning with $M_L = L = 2$ and $M_S = S = 0$, the corresponding term symbol is ^1D, and that term symbol is assigned to the five microstates $(2, 0)$, $(1, 0)$, $(0, 0)$, $(-1, 0)$, and $(-2, 0)$. The value of J for this term symbol is

$$J = 2 + 0 = |2 - 0| = 2$$

giving the complete term symbol $^1D_2$.

Working with the 10 remaining microstates—$(1, 1)$, $(1, 0)$, $(1, -1)$, $(0, 1)$, $(0, 0)$, $(0, -1)$, $(0, 0)$, $(-1, 1)$, $(-1, 0)$, and $(-1, -1)$—the values of $M_L = L = 1$ and $M_S = S = 1$ define the term ^3P for nine—$(1, 1)$, $(0, 1)$, $(-1, 1)$, $(1, 0)$, $(0, 0)$, $(-1, 0)$, $(1, -1)$, $(0, -1)$, and $(-1, -1)$—leaving only the microstate $(0, 0)$. The values of J for ^3P are

$$J = 1 + 1 = 2,$$

$$1 + 1 - 1 = 1,$$

$$1 + 1 - 2 = |1 - 1| = 0$$

giving the terms $^3P_0$, $^3P_1$, and $^3P_2$. The remaining microstate, $(0, 0)$, corresponds to ^1S with $J = 0$, or $^1S_0$.

Of the possible term symbols—$^1D_2$, $^3P_0$, $^3P_1$, $^3P_2$, and $^1S_0$—Hund's first rule suggests that the ^3P state is the correct ground state for C, and because the subshell is less than half-filled the complete symbol is $^3P_0$.

The same ground-state term symbol can be derived using a method that does not require the writing of all the permitted microstates. Substituting the values of m and s for the electrons $(n = 2, \ell = 1, m = +1, s = +\frac{1}{2})$ and $(n = 2, \ell = 1, m = 0, s = +\frac{1}{2})$, into (15.18) and (15.19) gives

$$M_L = 1 + 0 = 1 \qquad M_S = \tfrac{1}{2} + \tfrac{1}{2} = 1$$

which are assumed to be L and S, respectively, neglecting signs. The value of $L = 1$ gives a P for the term, the value of S gives the multiplicity as $2S + 1 = 3$, and the values of $L = 1$ and $S = 1$ give the permitted values of J as 2, 1, and 0, using (15.17). Choosing the lowest value for J because the subshell is less than half-filled gives $^3P_0$. The reverse of this shorter technique is very useful in determining electronic configurations from experimental term symbols.

Spectra of Polyelectronic Atoms

15.11 SELECTION RULES

Only the transitions that are allowed by the selection rules

$$\Delta S = 0 \tag{15.20a}$$

$$\Delta L = 0, \pm 1, \qquad \text{but } L = 0 \nleftrightarrow L = 0 \tag{15.20b}$$

$$\Delta J = 0, \pm 1, \qquad \text{but } J = 0 \nleftrightarrow J = 0 \tag{15.20c}$$

$$\Delta \ell = \pm 1 \tag{15.20d}$$

will be observed in the spectra of complex atoms. The symbol $\nleftrightarrow$ represents a forbidden transition.

EXAMPLE 15.8. The *principal series* of sodium corresponds to electronic transitions between the 3s ground electronic state and the various p excited states. Prepare energy diagrams for the series showing both the transitions and the doublet structure that is observed.

The ground state, having a $(3s)^1$ configuration, corresponds to $\ell = 0$ and $s = +\frac{1}{2}$, giving $L = 0$ and $S = \frac{1}{2}$. The multiplicity of the ground state is $2S + 1 = 2$, and

$$J = 0 + \tfrac{1}{2} = |0 - \tfrac{1}{2}| = \tfrac{1}{2}$$

giving the term symbol for the ground state as $^2S_{1/2}$. For one electron in a p orbital, $\ell = +1$ and $s = +\frac{1}{2}$, giving $L = 1$ and $S = \frac{1}{2}$. The multiplicity is still 2, and the values of J are

$$J = 1 + \tfrac{1}{2} = \tfrac{3}{2},$$

$$1 + \tfrac{1}{2} - 1 = |1 - \tfrac{1}{2}| = \tfrac{1}{2}$$

giving the term symbols as $^2P_{3/2}$ and $^2P_{1/2}$, with the $^2P_{1/2}$ being lower in energy.

The spectra should consist of transitions between the ^{2}P states and the ^{2}S state, which are allowed by the selection rules, (15.20): (15.20a) is satisfied because the transitions are occurring between "doublets"; (15.20b) is satisfied because P $\leftrightarrow$ S is a change in L of ± 1; (15.20c) is satisfied because the $\frac{3}{2} \leftrightarrow \frac{1}{2}$ is a change of ± 1 and the $\frac{1}{2} \leftrightarrow \frac{1}{2}$ is a change of 0; and (15.20d) is satisfied because ℓ is changing by ± 1. Figure 15-3a shows the overall transitions in this series, and Fig. 15-3b shows the two-line doublet that is observed in each case.

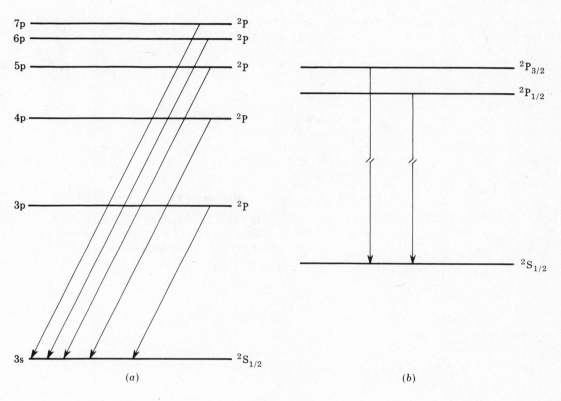

Fig. 15-3

15.12   THE NORMAL ZEEMAN EFFECT

If an atom with $S = 0$ is placed in a magnetic field parallel to the z axis of the atom, the multiplets are split into $2L + 1$ components corresponding to the quantum number M_L described in (15.18). The allowed transitions obey the additional rule

$$\Delta M_L = 0, \pm 1 \tag{15.21}$$

For atoms with $S \neq 0$, the Zeeman effect is much more complicated.

Solved Problems

HYDROGENLIKE ATOMS

15.1. The angular wave function for $\ell = 1$ and $m = 0$ is $Y(\theta, \phi) = (3/4\pi)^{1/2} \cos \theta$. Prepare two-dimensional sketches for $Y(\theta, \phi)$ and $Y(\theta, \phi)^* Y(\theta, \phi)$ in an arbitrary plane through the z axis. Describe the three-dimensional sketches.

For $\theta = 0$, the value of the wave function is

$$Y(\theta, \phi) = \left(\frac{3}{4\pi}\right)^{1/2} \cos 0 = 0.489$$

and because the function is real,

$$Y(\theta, \phi)^* Y(\theta, \phi) = (0.489)^2 = 0.239$$

Repeating the above calculations for 10° intervals in θ generates the curves shown in Fig. 15-4. The algebraic signs given in Fig. 15-4a are those of $Y(\theta, \phi)$ in the four quadrants.

Because $Y(\theta, \phi)$ lies along the z axis, and because $\ell = 1$, this orbital is known as the p_z orbital.

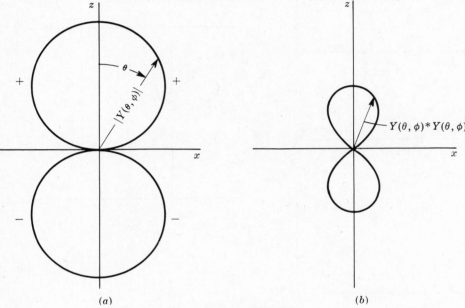

Fig. 15-4

15.2. Show that the angular wave functions for $\ell = 2$, $m = 0$ and for $\ell = 1$, $m = 1$,

$$Y(\theta, \phi) = \left(\frac{5}{16\pi}\right)^{1/2}(3\cos^2\theta - 1) \qquad \text{and} \qquad Y(\theta, \phi) = \left(\frac{3}{4\pi}\right)^{1/2}\sin\theta\cos\phi$$

are orthogonal.

To meet the criterion of (*14.13*), the angular part of $\int_{\text{all space}} \psi_i^* \psi_j \, dV$ must be shown to be zero. Performing the integration gives

$$\int_{\theta=0}^{\pi} \int_{\phi=0}^{2\pi} \left(\frac{15}{64\pi^2}\right)^{1/2}(3\cos^2\theta - 1)(\sin\theta\cos\phi)\sin\theta\,d\theta\,d\phi$$

$$= \frac{\sqrt{15}}{8\pi}\int_0^{\pi}(3\cos^2 2\theta - 1)\sin^2\theta\,d\theta \int_0^{2\pi}\cos\phi\,d\phi$$

The integral over ϕ is clearly zero. (The integral over θ also vanishes.)

15.3. Determine $R(r)$ for the 2s and 2p subshells, where $n = 2$, $\ell = 0$ and $n = 2$, $\ell = 1$, respectively.

Using (*15.10*) gives

$$L_2^1(\rho) = \frac{d}{d\rho}\left[e^\rho \frac{d^2}{d\rho^2}(\rho^2 e^{-\rho})\right] = \frac{d}{d\rho}\left[e^\rho \frac{d}{d\rho}(2\rho - \rho^2)e^{-\rho}\right] = \frac{d}{d\rho}(2 - 4\rho + \rho^2) = 2\rho - 4$$

for $n = 2$ and $\ell = 0$, and

$$L_3^3(\rho) = \frac{d^3}{d\rho^3}\left[e^\rho \frac{d^3}{d\rho^3}\left(\rho^3 e^{-\rho} \right) \right] = \frac{d^3}{d\rho^3}\left[e^\rho \frac{d^2}{d\rho^2}\left(3\rho^2 - \rho^3 \right) e^{-\rho} \right]$$

$$= \frac{d^3}{d\rho^3}\left[e^\rho \frac{d}{d\rho}\left(6\rho - 6\rho^2 + \rho^3 \right) e^{-\rho} \right] = \frac{d^3}{d\rho^3}\left(6 - 18\rho + 9\rho^2 - \rho^3 \right) = -6$$

for $n = 2$ and $\ell = 1$. Using (15.7) and (15.8) gives for these subshells

$$R_{2,0}(r) = -\left(\frac{Z}{a_0} \right)^{3/2}\left[\frac{1!}{4(2!)^3} \right]^{1/2} e^{-\rho/2}\rho^0(2\rho - 4)$$

$$= \left(\frac{Z}{a_0} \right)^{3/2}\left(\frac{1}{2} \right)^{1/2}\left(1 - \frac{Zr}{2a_0} \right) e^{-Zr/2a_0}$$

and

$$R_{2,1}(r) = -\left(\frac{Z}{a_0} \right)^{3/2}\left[\frac{0!}{4(3!)^3} \right]^{1/2} e^{-\rho/2}\rho^1(-6)$$

$$= \left(\frac{Z}{a_0} \right)^{5/2}\left(\frac{1}{2\sqrt{6}} \right)r\, e^{-Zr/2a_0}$$

15.4. Prepare plots of $a_0^{3/2}R(r)$ for the 1s, 2s, 2p, 3s, 3p, and 3d subshells for hydrogen, using values of r between 0 and $20a_0$.

The reduced mass given by (14.10) for hydrogen is

$$\mu = \frac{m_{\text{nucleus}}m_e}{m_{\text{nucleus}} + m_e} \approx m_e = 9.11 \times 10^{-31}\text{ kg}$$

The corresponding value of a_0 from (15.9) is

$$a_0 = \frac{4\pi\varepsilon_0\hbar^2}{\mu e^2} = \frac{(1.112\,65 \times 10^{-10}\text{ C}^2\text{ N}^{-1}\text{ m}^{-2})(1.055 \times 10^{-34}\text{ J s})^2}{(9.11 \times 10^{-31}\text{ kg})(1.602 \times 10^{-19}\text{ C})^2}$$

$$= 0.529 \times 10^{-10}\text{ m} = 0.052\,9\text{ nm}$$

The required values of $a_0^{3/2}R(r)$, as calculated from the wave functions determined in Example 15.2, Problem 15.20, and Problem 15.3, are plotted in Fig. 15-5.

15.5. What is the probability of finding a 1s electron for hydrogen in a sphere having $r = 0.05$ nm?

The normalization factors in (15.14), (15.15), and (15.7) have been chosen such that

$$\iiint\limits_{\text{all space}} \psi^*\psi\, dV = \int_0^\infty R^*R r^2\, dr \int_0^\pi \Theta^*\Theta \sin\theta\, d\theta \int_0^{2\pi} \Phi^*\Phi\, d\phi$$

$$= (1)(1)(1) = 1$$

This means that the (marginal) probability density functions for the r coordinate, θ coordinate, and ϕ coordinate are $R^*R r^2$, $\Theta^*\Theta \sin\theta$, and $\Phi^*\Phi$, respectively. Thus, the probability of finding a 1s electron ($R = R_{1,0}$) within 0.05 nm of the nucleus is

$$\int_0^{0.05} R_{1,0}^*R_{1,0}r^2\, dr = 4\left(\frac{1}{0.052\,9} \right)^3 \int_0^{0.05} e^{-2r/(0.0529)}r^2\, dr$$

where we have used Example 15.2 and Problem 15.4 to evaluate $R_{1,0}$. Using integration by parts (twice) or tables to evaluate the integral, we obtain

$$\int_0^{0.05} R_{1,0}^*R_{1,0}r^2\, dr = 0.295$$

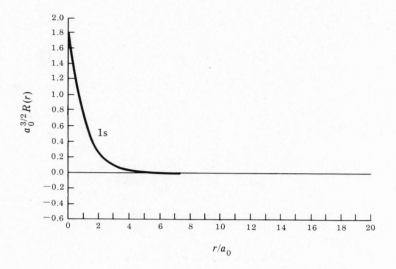

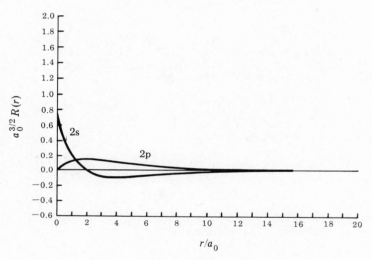

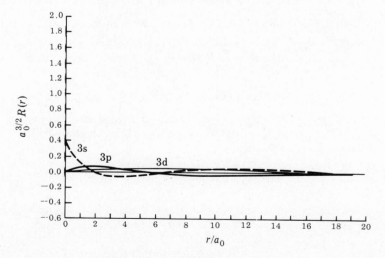

Fig. 15-5

15.6. Determine the degeneracy of the hydrogenlike energy level corresponding to $n = 2$.

For $n = 2$, (15.12) predicts that the 2s orbital and the three 2p orbitals will have the same energy in a one-electron system, giving a degeneracy of 4.

QUANTUM THEORY OF POLYELECTRONIC ATOMS

15.7. Determine $\mathscr{H}$ and ψ for Li.

For the three electrons, (15.13) gives

$$\mathscr{H} = \left(\frac{-\hbar^2}{2m_e}\right)(\nabla_1^2 + \nabla_2^2 + \nabla_3^2) - \frac{3e^2}{4\pi\varepsilon_0}\left(\frac{1}{r_1} + \frac{1}{r_2} + \frac{1}{r_3}\right) + \frac{e^2}{4\pi\varepsilon_0}\left(\frac{1}{r_{12}} + \frac{1}{r_{13}} + \frac{1}{r_{23}}\right)$$

and (15.14) gives, upon expansion,

$$\begin{aligned}
\psi = 6^{-1/2}[&\phi_{1s}(1)\alpha(1)\phi_{1s}(2)\beta(2)\phi_{2s}(3)\alpha(3) - \phi_{1s}(1)\alpha(1)\phi_{2s}(2)\alpha(2)\phi_{1s}(3)\beta(3) \\
&- \phi_{1s}(1)\beta(1)\phi_{1s}(2)\alpha(2)\phi_{2s}(3)\alpha(3) + \phi_{1s}(1)\beta(1)\phi_{2s}(2)\alpha(2)\phi_{1s}(3)\alpha(3) \\
&+ \phi_{2s}(1)\alpha(1)\phi_{1s}(2)\alpha(2)\phi_{1s}(3)\beta(3) - \phi_{2s}(1)\alpha(1)\phi_{1s}(2)\beta(2)\phi_{1s}(3)\alpha(3)]
\end{aligned}$$

15.8. Predict the electronic configurations for the ground states of Mg, P, P^{5+}, Y, Ce, O, and O^{2-}. Predict possible oxidation states for the elements.

The 12 electrons in Mg fill according to $(1s)^2(2s)^2(2p)^6(3s)^2$. In addition to 0, both 3s electrons could be lost to give a second possible oxidation state of +2.

For the 15 electrons in P, the configuration will be $(1s)^2(2s)^2(2p)^6(3s)^2(3p)^3$, giving possible oxidation states of +5, +3, and −3 as, respectively, both 3s and three 3p electrons are lost, three 3p electrons are lost, and three electrons are gained to complete the 3p subshell. In addition there is the state 0. For the 10 electrons in P^{5+}, the configuration would be $(1s)^2(2s)^2(2p)^6$.

For Y, the 39 electrons will be in the configuration

$$(1s)^2(2s)^2(2p)^6(3s)^2(3p)^6(4s)^2(3d)^{10}(4p)^6(5s)^2(4d)^1$$

In addition to 0, the oxidation states will be +2 (loss of the 5s electrons) and +3 (loss of the 5s electrons and the 4d electron). When considering the removal of electrons in transition metals, it is important to recall that the order is more like the reverse of the filling of hydrogenlike atoms. Thus the 5s electrons are removed before the 4d electron and a +1 oxidation state is not observed.

The 58 electrons in Ce will have the configuration

$$(1s)^2(2s)^2(2p)^6(3s)^2(3p)^6(4s)^2(3d)^{10}(4p)^6(5s)^2(4d)^{10}(5p)^6(6s)^2(4f)^2$$

which predicts Ce^{4+} by loss of the 4f and 6s electrons, in addition to the oxidation state of 0. The 4f and 5d orbitals are very close in energy and the following configuration is also possible for Ce:

$$(1s)^2(2s)^2(2p)^6(3s)^2(3p)^6(4s)^2(3d)^{10}(4p)^6(5s)^2(4d)^{10}(5p)^6(6s)^2(4f)^1(5d)^1$$

which predicts the common +3 oxidation state by loss of the 6s electrons and the 5d electron.

The 8 electrons in oxygen will be in $(1s)^2(2s)^2(2p)^4$, suggesting oxidation states of +6, +4, 0, and −2, where the configuration for O^{2-} is $(1s)^2(2s)^2(2p)^6$.

15.9. What are the quantum numbers for the 58th electron in Ce and the eighth electron in O?

Assuming the ... $(6s)^2(4f)^2$ configuration determined in Problem 15.8 for Ce, the last electron is the second one to enter the 4f subshell, thus $n = 4$ and $\ell = 3$. The first electron in the f subshell will be assigned $m = +3$ by convention; the second, $m = +2$; etc. The spins on the first five will be $+\frac{1}{2}$, giving the four quantum numbers as $n = 4$, $\ell = 3$, $m = +2$, $s = +\frac{1}{2}$. The eighth electron in oxygen is the fourth one to enter the 2p subshell, giving $n = 2$, $\ell = 1$, $m = +1$, $s = -\frac{1}{2}$.

ATOMIC TERM SYMBOLS

15.10. Determine the term symbols for two electrons such that $L = 2$ and $S = 1$.

The value of $L = 2$ is equivalent to a D term. The leading superscript given by

$$2S + 1 = (2)(1) + 1 = 3$$

Using (15.17) gives

$$J = 2 + 1 = 3,$$
$$2 + 1 - 1 = 2,$$
$$2 + 1 - 2 = |2 - 1| = 1$$

The term symbols are $^3D_3$, $^3D_2$, and $^3D_1$.

15.11. Determine the term symbol for the ground state of atomic O.

Assuming the ground-state electronic configuration to be that given in Problem 15.8, Table 15-1 gives ^3P, ^1D, and ^1S as possibilities for the equivalent electrons. Hund's first rule predicts that the ^3P is the preferred state. The value of L for this state is 1. Substituting $L = 1$ and $S = 1$ into (15.17) gives

$$J = 1 + 1 = 2,$$
$$1 + 1 - 1 = 1,$$
$$1 + 1 - 2 = |1 - 1| = 0$$

Choosing the largest value of J because the subshell is over half-filled, the complete term symbol is $^3P_2$.

15.12. The ground-state term of Cr is $^7S_3$. Does this term symbol correspond to the ... $(4s)^2(3d)^4$ configuration or the ... $(4s)^1(3d)^5$ configuration as presented in Example 15.5?

The d^4 configuration gives

$$M_L = L = 2 + 1 + 0 + (-1) = 2 \qquad M_S = S = \tfrac{1}{2} + \tfrac{1}{2} + \tfrac{1}{2} + \tfrac{1}{2} = 2$$

which lead to the incorrect symbol ^5D. The s^1d^5 configuration gives $M_L = L = 0$ and $M_S = S = \tfrac{6}{2}$ (assuming parallel spins), from which $2S + 1 = 7$ and $J = 0 + 3 = |0 - 3| = 3$, for a total symbol of $^7S_3$.

SPECTRA OF POLYELECTRONIC ATOMS

15.13. Is the transition between the 3s and 4s atomic orbitals allowed for Na?

The s atomic orbitals correspond to $^2S_{1/2}$ states, so that a $^2S_{1/2} \longleftrightarrow {}^2S_{1/2}$ transition is in question. Condition ($15.20a$) is met by this transition, but condition ($15.20b$) is not met, because this is an $L = 0 \longleftrightarrow L = 0$ transition. Thus this transition is not permitted.

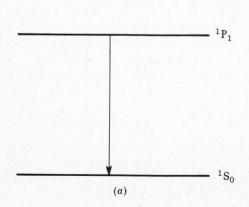
(a)

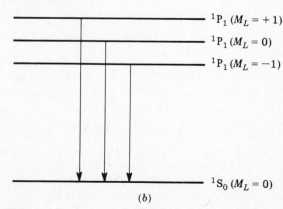
(b)

Fig. 15-6

15.14. Sketch the transitions that would occur between a $^1S_0$ and a $^1P_1$ level in the absence and in the presence of a magnetic field.

For the $^1S_0$ level, (*15.18*) gives $M_L = 0$, and for the $^1P_1$ level, $M_L = 1, 0, -1$. As shown in Fig. 15-6, what is normally a single spectra line becomes a triplet in the presence of a magnetic field.

Supplementary Problems

HYDROGENLIKE ATOMS

15.15. The angular wave function for $\ell = 0$ and $m = 0$ is $Y(\theta, \phi) = (1/4\pi)^{1/2}$. Describe plots of $Y(\theta, \phi)$ and $Y(\theta, \phi)^* Y(\theta, \phi)$ in the xz plane and the three-dimensional figure for an s orbital.

    Ans.   Wave function is independent of θ and ϕ, so in two dimensions plots are circles of radii 0.282 and 0.0796, and three-dimensional figure is spherical.

15.16. The angular wave function for the d_{z^2} orbital ($\ell = 2$ and $m = 0$) is $Y(\theta, \phi) = (5/16\pi)^{1/2}(3\cos^2\theta - 1)$. Graph in the xz plane (*a*) $Y(\theta, \phi)$ and (*b*) $Y(\theta, \phi)^* Y(\theta, \phi)$.

    Ans.   (*a*) See Fig. 15-7*a*.   (*b*) See Fig. 15-7*b*.

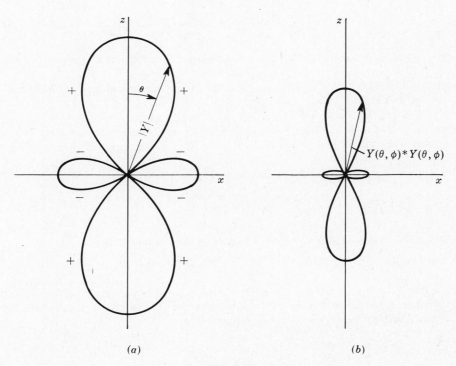

(*a*)   (*b*)

Fig. 15-7

15.17. The angular wave functions for $\ell = 1$ and $m = \pm 1$, $Y(\theta, \phi) = (3/8\pi)^{1/2} \sin\theta\, e^{\pm i\phi}$, contain imaginary terms. Construct two normalized real wave functions from these, and prepare two-dimensional plots of the new wave functions, $Y'(\theta, \phi)$ and $Y''(\theta, \phi)$, and of $Y'(\theta, \phi)^* Y'(\theta, \phi)$ and $Y''(\theta, \phi)^* Y''(\theta, \phi)$.

    Ans. $Y'(\theta, \phi) = (3/4\pi)^{1/2} \sin\theta\cos\phi$, $Y''(\theta, \phi) = (3/4\pi)^{1/2} \sin\theta\sin\phi$; plots similar to Fig. 15-4 but located along the x axis and y axis, respectively

15.18. The angular wave functions for $\ell = 2$ and $m = \pm 2$, $Y(\theta, \phi) = (15/32\pi)^{1/2} \sin^2 \theta \, e^{\pm 2i\phi}$, contain imaginary terms. From these, construct real normalized functions, $Y'(\theta, \phi)$ and $Y''(\theta, \phi)$, and prepare plots of $Y'(\theta, \phi)$, $Y''(\theta, \phi)$, $Y'(\theta, \phi)^* Y'(\theta, \phi)$, and $Y''(\theta, \phi)^* Y''(\theta, \phi)$ in the xy plane.

Ans. $Y'(\theta, \phi) = (15/16\pi)^{1/2} \sin^2 \theta \cos 2\phi$, $Y''(\theta, \phi) = (15/16\pi)^{1/2} \sin^2 \theta \sin 2\phi$; plots similar to Fig. 15-2 in size and shape, with lobes of Y' along the axes and lobes of Y'' at 45° to the axes

15.19. Show that the real angular wave functions for $\ell = 1$ and $m = \pm 1$,

$$Y'(\theta, \phi) = \left(\frac{3}{4\pi}\right)^{1/2} \sin \theta \cos \phi \qquad \text{and} \qquad Y''(\theta, \phi) = \left(\frac{3}{4\pi}\right)^{1/2} \sin \theta \sin \phi$$

are orthogonal.

15.20. Find $R(r)$ for the 3s, 3p, and 3d subshells.

Ans. $R_{3,0} = \left(\frac{Z}{a_0}\right)^{3/2} \left(\frac{2}{3\sqrt{3}}\right) e^{-Zr/3a_0} \left(1 - \frac{2Zr}{3a_0} + \frac{2Z^2r^2}{27a_0^2}\right)$

$R_{3,1} = \left(\frac{Z}{a_0}\right)^{5/2} \left(\frac{8}{27\sqrt{6}}\right) e^{-Zr/3a_0} \left(r - \frac{Zr^2}{6a_0}\right) \qquad R_{3,2} = \left(\frac{Z}{a_0}\right)^{7/2} \left(\frac{4}{81\sqrt{30}}\right) e^{-Zr/3a_0} r^2$

15.21. An alternative form for determining the associated Laguerre polynomials defined by (*15.10*) is

$$L_{n+\ell}^{2\ell+1}(x) = \sum_{i=0}^{n-\ell-1} (-1)^{i+1} \frac{[(n+\ell)!]^2}{(n-\ell-1-i)!(2\ell+1+i)!i!} x^i$$

Use this relationship to confirm the polynomials determined in Problem 15.3 for $n = 2$, $\ell = 0$ and $n = 2$, $\ell = 1$.

15.22. Prepare plots of $a_0 R(r)^* R(r) r^2$ for the 1s, 2s, 2p, 3s, 3p, and 3d subshells for hydrogen, using values of r between 0 and $20a_0$. *Ans.* See Fig. 15-8

15.23. What is the probability of finding a 1s electron for hydrogen within a sphere having $r = 0.06$ nm? Compare the answer to that found in Problem 15.5. *Ans.* 0.397, an increase of 10%

15.24. Determine the degeneracies of the various hydrogenlike levels.

Ans. $n = 1$, $\ell = 0$, $m = 0$ gives 1; $n = 2$, $\ell = 0$, $m = 0$ gives 1 and $n = 2$, $\ell = 1$, $m = \pm 1$, 0 gives 3, for a total of 4; 9; 16; etc.

15.25. Find the most probable value of r for the hydrogen 1s wave function.

Ans. $(d/dr)[R_{1,0}(r)^* R_{1,0}(r) r^2] = 0$ gives $r = a_0 = 0.052\,9$ nm

15.26. Assume that $\psi = e^{-ar}$ is a trial wave function for atomic hydrogen. Find $\int_{\text{all space}} \psi^* \mathcal{H} \psi \, dV$ and $\int_{\text{all space}} \psi^* \psi \, dV$ and calculate E. Minimize E with respect to a, solve for the constant a, and find the minimum value of E. Compare with the actual energy of the ground state.

Ans. $\int_0^\infty x^n e^{-ax} dx = \frac{n!}{a^{n+1}}$;

$\mathcal{H}\psi = \left[\frac{a\hbar^2}{2\mu}\left(\frac{2}{r} - a\right) - \frac{e^2}{4\pi\varepsilon_0 r}\right] e^{-ar}$, $\int_{\text{all space}} \psi^* \mathcal{H} \psi \, dV = \frac{\pi}{a}\left(\frac{\hbar^2}{2\mu} - \frac{e^2}{4\pi\varepsilon_0 a}\right)$;

$\int_{\text{all space}} \psi^* \psi \, dV = \frac{\pi}{a^3}$; $E = \frac{\hbar^2 a^2}{2\mu} - \frac{e^2 a}{4\pi\varepsilon_0}$; $a = \mu e^2 / 4\pi\varepsilon_0 \hbar^2$;

$E_{\text{min}} = \frac{-\mu e^4}{2\hbar^2(4\pi\varepsilon_0)^2} = E_{\text{actual}}$ (because $\psi \propto \psi_{\text{actual}}$)

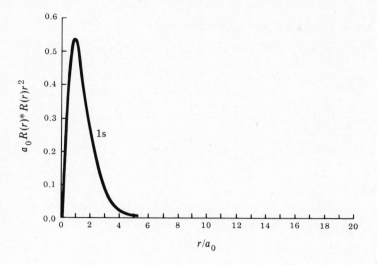

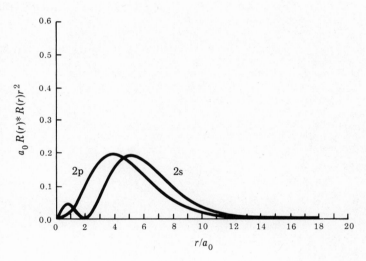

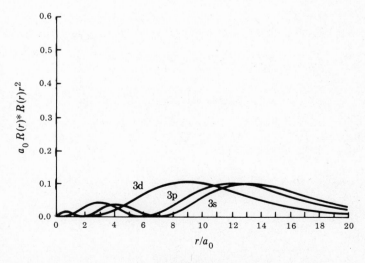

Fig. 15-8

QUANTUM THEORY OF POLYELECTRONIC ATOMS

15.27. Write the Hamiltonian and the Slater wave function for carbon.

Ans.
$$\mathcal{H} = \left(\frac{-\hbar^2}{2m_e}\right)(\nabla_1^2 + \nabla_2^2 + \nabla_3^2 + \nabla_4^2 + \nabla_5^2 + \nabla_6^2) - \frac{6e^2}{4\pi\varepsilon_0}\left(\frac{1}{r_1} + \frac{1}{r_2} + \frac{1}{r_3} + \frac{1}{r_4} + \frac{1}{r_5} + \frac{1}{r_6}\right)$$

$$+ \frac{e^2}{4\pi\varepsilon_0}\left(\frac{1}{r_{12}} + \frac{1}{r_{13}} + \frac{1}{r_{14}} + \frac{1}{r_{15}} + \frac{1}{r_{16}} + \frac{1}{r_{23}} + \frac{1}{r_{24}} + \frac{1}{r_{25}} + \frac{1}{r_{26}} + \frac{1}{r_{34}} + \frac{1}{r_{35}} + \frac{1}{r_{36}} + \frac{1}{r_{45}} + \frac{1}{r_{46}} + \frac{1}{r_{56}}\right)$$

$$\psi = \sqrt{\frac{1}{6!}} \begin{vmatrix} \phi_{1s}(1)\alpha(1) & \phi_{1s}(2)\alpha(2) & \phi_{1s}(3)\alpha(3) & \phi_{1s}(4)\alpha(4) & \phi_{1s}(5)\alpha(5) & \phi_{1s}(6)\alpha(6) \\ \phi_{1s}(1)\beta(1) & \phi_{1s}(2)\beta(2) & \phi_{1s}(3)\beta(3) & \phi_{1s}(4)\beta(4) & \phi_{1s}(5)\beta(5) & \phi_{1s}(6)\beta(6) \\ \phi_{2s}(1)\alpha(1) & \phi_{2s}(2)\alpha(2) & \phi_{2s}(3)\alpha(3) & \phi_{2s}(4)\alpha(4) & \phi_{2s}(5)\alpha(5) & \phi_{2s}(6)\alpha(6) \\ \phi_{2s}(1)\beta(1) & \phi_{2s}(2)\beta(2) & \phi_{2s}(3)\beta(3) & \phi_{2s}(4)\beta(4) & \phi_{2s}(5)\beta(5) & \phi_{2s}(6)\beta(6) \\ \phi_{2p,+1}(1)\alpha(1) & \phi_{2p,+1}(2)\alpha(2) & \phi_{2p,+1}(3)\alpha(3) & \phi_{2p,+1}(4)\alpha(4) & \phi_{2p,+1}(5)\alpha(5) & \phi_{2p,+1}(6)\alpha(6) \\ \phi_{2p,0}(1)\alpha(1) & \phi_{2p,0}(2)\alpha(2) & \phi_{2p,0}(3)\alpha(3) & \phi_{2p,0}(4)\alpha(4) & \phi_{2p,0}(5)\alpha(5) & \phi_{2p,0}(6)\alpha(6) \end{vmatrix}$$

15.28. Predict the electronic configurations for U, U^{3+}, U^{4+}, U^{5+}, U^{6+}, Cu, Cu^+, and Cu^{2+}.

Ans. $[Rn](7s)^2(6d)^1(5f)^3$ or $[Rn](7s)^2(5f)^4$, $[Rn](5f)^3$, $[Rn](5f)^2$, $[Rn](5f)$, $[Rn]$, $[Ar](4s)^1(3d)^{10}$, $[Ar](3d)^{10}$, $[Ar](3d)^9$

15.29. What are the quantum numbers for the 92nd electron in U and the 15th electron in P?

Ans. For U: $(n = 5, \ell = 3, m = +1, s = +\frac{1}{2})$ for $[Rn](7s)^2(6d)^1(5f)^3$ and
$(n = 5, \ell = 3, m = 0, s = +\frac{1}{2})$ for $[Rn](7s)^2(5f)^4$.
For P: $(n = 3, \ell = 1, m = -1, s = +\frac{1}{2})$.

15.30. Calculate the first ionization energy for He given that the ground state of He is -12.654×10^{-17} J and that (*15.12*) is valid for He^+. *Ans.* He^+ ground state is -8.715×10^{-18} J, 3.939×10^{-18} J

ATOMIC TERM SYMBOLS

15.31. Determine the values of L for the interactions of an s with an s, a p with a p, and a p with an f.

Ans. $L = 0 + 0 = |0 - 0| = 0$, giving S; $L = 1 + 1 = 2$ to $|1 - 1| = 0$, giving D, P, S;
$L = 1 + 3 = 4$ to $|1 - 3| = 2$, giving G, F, D

15.32. Determine the term symbol for two electrons such that $L = 2$ and $S = 0$. *Ans.* $^1D_2$

15.33. Determine the ground-state term for Ti using Table 15-1 for the equivalent electrons.

Ans. $[Ar](4s)^2(3d)^2$; ^3F is preferred choice; $2S + 1 = 3$ gives $S = 1$, F gives $L = 3$; $J = 3 + 1 = 4$ to $|3 - 1| = 2$, d^2 is less than half-filled; $^3F_2$

15.34. The ground state of Cu is $^2S_{1/2}$. Determine the electronic configuration from this information.

Ans. $2S + 1 = 2$ gives $S = M_S = \frac{1}{2}$, 1 unpaired electron; S term gives $L = M_L = 0$, s^1d^{10} gives $M_L = L = 0$, and s^2d^9 gives $M_L = -2$ and $L = 2$; $J = \frac{1}{2}$ agrees with S and L; structure is $(1s)^2(2s)^2(2p)^6(3s)^2(3p)^6(4s)^1(3d)^{10}$

15.35. The ground state for Nb is $^6D_{1/2}$; Pd, $^1S_0$; La, $^2D_{3/2}$; Ce, $^3H_4$; and Pr, $^4I_{9/2}$. Determine the electronic configurations.

Ans. $L = 2$, 5 unpaired electrons, ... $(4s)^2(4p)^6(5s)^1(4d)^4$;
$L = 0$, no unpaired electrons, ... $(4s)^2(4p)^6(4d)^{10}$;
$L = 2$, 1 unpaired electron, ... $(4s)^2(4p)^6(5s)^2(4d)^{10}(5p)^6(6s)^2(5d)^1$;
$L = 5$, 2 unpaired electrons, ... $(4s)^2(4p)^6(5s)^2(4d)^{10}(5p)^6(6s)^2(5d)^1(4f)^1$ or ... $(6s)^2(4f)^2$;
$L = 6$, 3 unpaired electrons, ... $(4s)^2(4p)^6(5s)^2(4d)^{10}(5p)^6(6s)^2(4f)^3$

15.36. What are the quantum numbers of the 66th electron in Dy, assuming it to be the $(4f)^{10}$ electron? Write the term symbol for the ground state of this element.

Ans. $n = 4$, $\ell = 3$, $m = +1$, $s = -\frac{1}{2}$; $M_L = 0 + (-1) + (-2) + (-3) = -6$, $L = 6$, giving I, $M_S = 4(\frac{1}{2}) = 2$, $S = 2$, $2S + 1 = 5$, $J = 8$ to 4, $^5I_8$.

15.37. Using the microstate technique shown in Example 15.7, determine the term symbols for the interaction of three equivalent p electrons.

Ans.   The 20 microstates are $(2, \frac{1}{2})$, $(1, \frac{1}{2})$, $(2, -\frac{1}{2})$, $(1, -\frac{1}{2})$, $(-1, \frac{1}{2})$, $(-1, -\frac{1}{2})$, $(1, \frac{1}{2})$, $(1, -\frac{1}{2})$, $(-1, \frac{1}{2})$, $(-2, \frac{1}{2})$, $(-1, -\frac{1}{2})$, $(-2, -\frac{1}{2})$, $(0, \frac{3}{2})$, $(0, \frac{1}{2})$, $(0, \frac{1}{2})$, $(0, -\frac{1}{2})$, $(0, \frac{1}{2})$, $(0, -\frac{1}{2})$, $(0, -\frac{1}{2})$, $(0, -\frac{3}{2})$; ^2D accounts for 10 of these, ^2P accounts for 6, and ^4S accounts for 4.

SPECTRA OF POLYELECTRONIC ATOMS

15.38. The *diffuse series* of Na corresponds to electronic transitions between the d excited states and the 3p electronic state. Prepare energy diagrams for this series showing both the transitions and compound doublet structure that are observed.   *Ans.* $^2D_{3/2}$ and $^2D_{5/2}$ to $^2P_{1/2}$ and $^2P_{3/2}$, see Fig. 15-9.

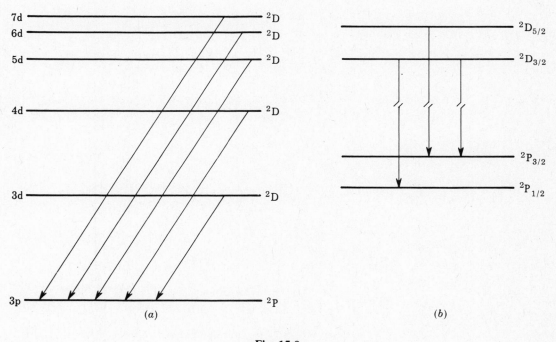

Fig. 15-9

15.39. Are the transitions between the 4p and 3p, the 4f and 3d, and the 4f and 3p atomic orbitals allowed for Na?

Ans.   No, violates (*15.20d*); yes; no, violates (*15.20b*)

15.40. How many transitions and how many spectral lines will occur for a transition between a $^1D_2$ and a $^1P_1$ level in the presence of a magnetic field?

Ans.   Nine transitions (2 to 1, 1 to 1 or 0, 0 to ±1 or 0, -1 to -1 or 0, -2 to -1), giving three lines because all similar ΔM_L values have the same energy splitting

15.41. The low-resolution spectrum of He shows that two distinct sets of principal, sharp, diffuse, and fundamental series exist. These result from orthohelium having $S = 1$ and parahelium having $S = 0$. (*a*) Are these types of helium allowed to interchange energy? (*b*) Describe the low-resolution spectra for parahelium and orthohelium.

Ans.   (*a*) No, $\Delta S = 0$ must be obeyed.   (*b*) Except that they will consist of singlets and triplets instead of doublets, the spectra will be similar to that for Na, with the lowest energy levels being the 1s and 2s, respectively

Chapter 16

Electronic Structure of Diatomic Molecules

Quantum Theory of Diatomic Molecules

16.1 HAMILTONIAN OPERATOR

The complete Hamiltonian operator for a polyatomic molecule is

$$\mathcal{H} = \frac{-\hbar^2}{2} \sum_{\alpha}^{\substack{\text{all} \\ \text{nuclei}}} \frac{1}{m_\alpha} \nabla_\alpha^2 + \frac{-\hbar^2}{2m_e} \sum_{i}^{\substack{\text{all} \\ \text{electrons}}} \nabla_i^2$$
$$+ \sum_\beta \sum_{\alpha < \beta} \frac{Z_\alpha Z_\beta e^2}{4\pi\varepsilon_0 r_{\alpha\beta}} - \sum_i \sum_\alpha \frac{Z_\alpha e^2}{4\pi\varepsilon_0 r_{i\alpha}} + \sum_j \sum_{i<j} \frac{e^2}{4\pi\varepsilon_0 r_{ij}} \qquad (16.1)$$

where the terms represent the nuclear kinetic energy, the electronic kinetic energy, the nuclear repulsion potential energy, the nuclear-electronic attraction potential energy, and the electronic repulsion potential energy, respectively. The *Born–Oppenheimer approximation* simplifies (*16.1*) by setting the nuclear kinetic energy term equal to zero and the nuclear repulsion term equal to a constant, because the nuclei are moving much more slowly than the electrons. Thus, for a two-nucleus system, the nuclear repulsion term is $Z_a Z_b e^2/4\pi\varepsilon_0 r_{ab}$, where r_{ab}, the internuclear distance, is considered constant. The electronic energy (E) is then calculated in terms of the parameter r_{ab}, whereupon the total energy of the molecule (E') is obtained as

$$E' = E + \frac{Z_a Z_b e^2}{4\pi\varepsilon_0 r_{ab}}$$

EXAMPLE 16.1. The H_2 molecule ($Z_a = Z_b = 1$) is indicated in Fig. 16-1. Write the complete Hamiltonian for this system.

Equation (*16.1*) becomes, under the Born-Oppenheimer approximation,

$$\mathcal{H} = \frac{-\hbar^2}{2m_e}(\nabla_1^2 + \nabla_2^2) + \frac{e^2}{4\pi\varepsilon_0}\left(\frac{1}{r_{ab}} - \frac{1}{r_{1a}} - \frac{1}{r_{2a}} - \frac{1}{r_{1b}} - \frac{1}{r_{2b}} + \frac{1}{r_{12}}\right)$$

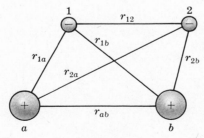

Fig. 16-1

16.2 WAVE FUNCTIONS

The wave function ϕ is usually the product of the wave functions for the individual electrons:

$$\phi = \phi_1 \phi_2 \cdots \qquad (16.2)$$

319

where the ϕ_i are produced by taking linear combinations of hydrogenlike atomic orbitals (LCAO):

$$\phi_i = a_{ia}\psi_a(i) + a_{ib}\psi_b(i) + \cdots \qquad (16.3)$$

In (16.3) the $a_{i\alpha}$ are constants and the notation $\psi_\alpha(i)$ means a hydrogenlike atomic wave function describing electron i in terms of position with respect to nucleus α. Terms in ϕ that are of the form $\psi_a(1)\psi_a(2)$ are ionic terms, and terms that are of the form $\psi_a(1)\psi_b(2)$ are covalent terms. Depending on the actual construction of ϕ, these types of terms may be weighted differently or the same. A wave function for a diatomic molecule having unshared pairs of electrons usually consists of terms involving only the bonding electrons.

EXAMPLE 16.2. Determine a trial wave function using (16.2) and (16.3) for the H_2 molecule. Assuming that the $a_{i\alpha}$ are all equal and that identical normalized hydrogenlike wave functions are chosen, show that the normalization constant is $a = (2 + 2S)^{-1}$, where $S = S_{ab} = S_{ba}$ as defined in (14.27).

Applying (16.3) to each electron and using a single-subscript notation for the a's gives

$$\phi_1 = a_1\psi_a(1) + a_2\psi_b(1) \qquad \text{and} \qquad \phi_2 = a_3\psi_a(2) + a_4\psi_b(2)$$

and (16.2) gives

$$\phi = [a_1\psi_a(1) + a_2\psi_b(1)][a_3\psi_a(2) + a_4\psi_b(2)]$$
$$= a_1 a_3 \psi_a(1)\psi_a(2) + a_2 a_4 \psi_b(1)\psi_b(2) + a_1 a_4 \psi_a(1)\psi_b(2) + a_2 a_3 \psi_b(1)\psi_a(2)$$

Assuming that identical normalized hydrogenlike wave functions are being used to describe the electrons and that

$$a_1 = a_2 = a_3 = a_4 = a$$

gives

$$\int_{\text{all space}} \phi^*\phi \, dV = \int_{\text{all space}} \phi_1^*\phi_2^*\phi_1\phi_2 \, dV$$
$$= \left(\int_{\text{all space}} \phi_1^*\phi_1 \, dV \right)\left(\int_{\text{all space}} \phi_2^*\phi_2 \, dV \right)$$
$$= \left(\int_{\text{all space}} \phi_1^*\phi_1 \, dV \right)^2$$

where

$$\int_{\text{all space}} \phi_1^*\phi_1 \, dV = \int_{\text{all space}} [a\psi_a(1) + a\psi_b(1)]^*[a\psi_a(1) + a\psi_b(1)] \, dV$$
$$= a^2 \left(\int_{\text{all space}} \psi_a^*(1)\psi_a(1) \, dV + \int_{\text{all space}} \psi_b^*(1)\psi_b(1) \, dV \right.$$
$$\left. + \int_{\text{all space}} \psi_a^*(1)\psi_b(1) \, dV + \int_{\text{all space}} \psi_b^*(1)\psi_a(1) \, dV \right)$$
$$= a^2(S_{aa} + S_{bb} + S_{ab} + S_{ba}) = a^2(1 + 1 + S + S) = a^2(2 + 2S)$$

Substituting these results into (14.12) gives

$$1 = \int_{\text{all spaces}} \phi^*\phi \, dV = \left(\int_{\text{all space}} \phi_1^*\phi_1 \, dV \right)^2 = [a^2(2 + 2S)]^2$$

which gives

$$a = (2 + 2S)^{-\frac{1}{2}}$$

Application of the Variation Method

16.3 ENERGY

The electronic contribution E (see Sec. 16.1) is calculated as a function of r_{ab} using (*14.23*) or the secular equation (*14.28*). Some integrals that are encountered in the solution to the H_2^-, H_2, and H_2^+ problems using hydrogenlike 1s atomic orbitals are

$$S = \int_{\text{all space}} \psi_a^*(1)\psi_b(1)\, dV = e^{-\rho}\left(1 + \rho + \frac{\rho^2}{3}\right) \tag{16.4a}$$

$$J = \frac{-e^2}{4\pi\varepsilon_0}\int_{\text{all space}} \psi_a^*(1)\frac{1}{r_{ib}}\psi_a(1)\, dV$$

$$= \frac{e^2}{4\pi\varepsilon_0 a_0}\left[\frac{-1}{\rho} + e^{-2\rho}\left(1 + \frac{1}{\rho}\right)\right] \tag{16.4b}$$

$$K = \frac{-e^2}{4\pi\varepsilon_0}\int_{\text{all space}} \psi_b^*(1)\frac{1}{r_{ib}}\psi_a(1)\, dV$$

$$= \frac{-e^2}{4\pi\varepsilon_0 a_0}\, e^{-\rho}(1 + \rho) \tag{16.4c}$$

$$J' = \frac{e^2}{4\pi\varepsilon_0}\int_{\text{all space}} \psi_a^*(1)\psi_b^*(2)\frac{1}{r_{12}}\psi_a(1)\psi_b(2)\, dV$$

$$= \frac{e^2}{4\pi\varepsilon_0 a_0}\left[\frac{1}{\rho} - e^{-2\rho}\left(\frac{1}{\rho} + \frac{11}{8} + \frac{3\rho}{4} + \frac{\rho^2}{6}\right)\right] \tag{16.4d}$$

$$K' = \frac{e^2}{4\pi\varepsilon_0}\int_{\text{all space}} \psi_a^*(1)\psi_b^*(2)\frac{1}{r_{12}}\psi_b(1)\psi_a(2)\, dV$$

$$= \frac{e^2}{4\pi\varepsilon_0(5a_0)}\left\{-e^{-2\rho}\left(\frac{-25}{8} + \frac{23\rho}{4} + 3\rho^2 + \frac{\rho^3}{3}\right) + \frac{6}{\rho}\left[S^2(0.577\,22 + \log\rho)\right.\right.$$

$$\left.\left. + e^{2\rho}\left(1 - \rho + \frac{\rho^2}{3}\right)^2 W(-4\rho) - 2Se^{\rho}\left(1 - \rho + \frac{\rho^2}{3}\right)W(-2\rho)\right]\right\} \tag{16.4e}$$

$$L = \frac{e^2}{4\pi\varepsilon_0}\int_{\text{all space}} \psi_a^*(1)\psi_a^*(2)\frac{1}{r_{12}}\psi_a(1)\psi_b(2)\, dV$$

$$= \frac{e^2}{4\pi\varepsilon_0 a_0}\left[e^{-\rho}\left(\rho + \frac{1}{8} + \frac{5}{16\rho}\right) + e^{-3\rho}\left(-\frac{1}{8} - \frac{5}{16\rho}\right)\right] \tag{16.4f}$$

$$C = \frac{e^2}{4\pi\varepsilon_0}\int_{\text{all space}} \psi_a^*(1)\psi_a^*(2)\frac{1}{r_{12}}\psi_a(1)\psi_a(2)\, dV$$

$$= \frac{e^2}{4\pi\varepsilon_0 a_0}\left(\frac{5}{8}\right) \tag{16.4g}$$

where
$$\rho = r_{ab}/a_0 \tag{16.5}$$

$$W(-x) = -\int_x^\infty \frac{e^{-t}}{t}\, dt \quad (x > 0) \tag{16.6}$$

$$a_0 = 0.052\,917\,706 \text{ nm}$$

EXAMPLE 16.3.   Using the Hamiltonian and trial wave function for the H_2 molecule as determined in Examples 16.1 and 16.2, find the energy of the molecule in terms of the integrals given by (*16.4*).

The electronic energy is obtained from the normalized wave function via (*14.23*):

$$E = \int_{\text{all space}} \phi^* \mathcal{H} \phi \, dV$$

$$= \int_{\text{all space}} \phi^* \left[\mathcal{H}_1 + \mathcal{H}_2 + \frac{e^2}{4\pi\varepsilon_0} \left(\frac{1}{r_{12}} \right) \right] \phi \, dV$$

$$= \int_{\text{all space}} \phi^* \mathcal{H}_i \phi \, dV + \int_{\text{all space}} \phi^* \mathcal{H}_2 \phi \, dV$$

$$+ \frac{e^2}{4\pi\varepsilon_0} \int_{\text{all space}} \phi^* \frac{1}{r_{12}} \phi \, dV$$

where

$$\mathcal{H}_i = \frac{-\hbar^2}{2m_e} \nabla_i^2 - \frac{e^2}{4\pi\varepsilon_0 r_{ia}} - \frac{e^2}{4\pi\varepsilon_0 r_{ib}}$$

Note that the r_{ab} term has been omitted from $\mathcal{H}$. Because the two electrons have identical wave functions,

$$\int_{\text{all space}} \phi^* \mathcal{H}_i \phi \, dV = \int_{\text{all space}} \phi^* \mathcal{H}_2 \phi \, dV$$

so that

$$E = 2 \int_{\text{all space}} \phi^* \mathcal{H}_1 \phi \, dV + \frac{e^2}{4\pi\varepsilon_0} \int_{\text{all space}} \phi^* \frac{1}{r_{12}} \phi \, dV$$

The first integral on the right has the value

$$\int_{\text{all space}} \phi^* \mathcal{H}_1 \phi \, dV$$

$$= (2+2S)^{-2} \int_{\text{all space}} [\psi_a(1) + \psi_b(1)]^* [\psi_a(2) + \psi_b(2)]^* \mathcal{H}_1 [\psi_a(1) + \psi_b(1)][\psi_a(2) + \psi_b(2)] \, dV$$

$$= (2+2S)^{-2} \left(\int_{\text{all space}} [\psi_a(1) + \psi_b(1)]^* \mathcal{H}_1 [\psi_a(1) + \psi_b(1)] \, dV \right)$$

$$\times \left(\int_{\text{all space}} [\psi_a(2) + \psi_b(2)]^* [\psi_a(2) + \psi_b(2)] \, dV \right)$$

The integral involving electron 2 was shown in Example 16.2 to be equal to $2+2S$, giving

$$(2+2S) \int_{\text{all space}} \phi^* \mathcal{H}_1 \phi \, dV$$

$$= \int_{\text{all space}} \left\{ [\psi_a(1) + \psi_b(1)]^* \right.$$

$$\times \left[\frac{-\hbar^2}{2m_e} \nabla_1^2 - \frac{e^2}{4\pi\varepsilon_0} \left(\frac{1}{r_{1a}} \right) - \frac{e^2}{4\pi\varepsilon_0} \left(\frac{1}{r_{1b}} \right) \right] [\psi_a(1) + \psi_b(1)] \right\} dV$$

$$= \int_{\text{all space}} \psi_a^*(1) \left[\frac{-\hbar^2}{2m_e} \nabla_1^2 - \frac{e^2}{4\pi\varepsilon_0} \left(\frac{1}{r_{1a}} \right) \right] \psi_a(1) \, dV$$

$$+ \int_{\text{all space}} \psi_b^*(1) \left[\frac{-\hbar^2}{2m_e} \nabla_1^2 - \frac{e^2}{4\pi\varepsilon_0} \left(\frac{1}{r_{1a}} \right) \right] \psi_a(1) \, dV$$

$$+ \int_{\text{all space}} \psi_a^*(1) \left[\frac{-\hbar^2}{2m_e} \nabla_1^2 - \frac{e^2}{4\pi\varepsilon_0} \left(\frac{1}{r_{1b}} \right) \right] \psi_b(1) \, dV$$

$$+ \frac{-e^2}{4\pi\varepsilon_0} \int_{\text{all space}} \psi_a^*(1) \frac{1}{r_{1a}} \psi_b(1) \, dV$$

$$+ \int_{\text{all space}} \psi_b^*(1) \left[\frac{-\hbar^2}{2m_e} \nabla_1{}^2 - \frac{e^2}{4\pi\varepsilon_0} \left(\frac{1}{r_{1b}} \right) \right] \psi_b(1)\, dV$$

$$+ \frac{-e^2}{4\pi\varepsilon_0} \int_{\text{all space}} \psi_b^*(1) \frac{1}{r_{1a}} \psi_b(1)\, dV$$

$$+ \frac{-e^2}{4\pi\varepsilon_0} \int_{\text{all space}} \psi_a^*(1) \frac{1}{r_{1b}} \psi_a(1)\, dV$$

$$+ \frac{-e^2}{4\pi\varepsilon_0} \int_{\text{all space}} \psi_b^*(1) \frac{1}{r_{1b}} \psi_a(1)\, dV$$

$$= E_H + E_H S + E_H S + K + E_H + J + J + K$$

$$= E_H(2 + 2S) + 2(J + K)$$

where the first, second, third, and fifth integrals have been evaluated by use of the Schrödinger equation. Here $E_H = -13.60 \text{ eV}$ ($= -2.179 \times 10^{-18} \text{ J}$) is the energy of the ground state of the hydrogen atom.

For the second integral in the equation for E, a similar calculation (see Problem 16.17) gives

$$\frac{e^2}{4\pi\varepsilon_0} \int_{\text{all space}} \phi^* \frac{1}{r_{12}} \phi\, dV = \frac{K' + 2L + J'/2 + C/2}{(1+S)^2}$$

Hence the electronic energy is

$$E = 2 \left(E_H + \frac{J+K}{1+S} \right) + \frac{K' + 2L + J'/2 + C/2}{(1+S)^2}$$

and the overall energy (see Sec. 16.1) is $E' = E + e^2/4\pi\varepsilon_0 r_{ab}$.

16.4 MOLECULAR ORBITALS

The variation method predicts that when two atomic orbitals are combined, two molecular orbitals are generated. One orbital has an energy lower than that of the separated atoms; hence it is known as *bonding*. The other orbital has an energy higher than that of the separated atoms; it is known as *antibonding*. Each orbital can hold a maximum of two electrons.

The molecular orbitals formed from hydrogenlike atomic orbitals for homonuclear molecules with four or fewer electrons are shown in Fig. 16-2a, and for homonuclear molecules with between four and twenty-one electrons in Fig. 16-2b. The diagram for homonuclear molecules with up to 36 electrons would be similar to Fig. 16-2b, with an additional set of higher-energy orbitals, 3s and 3p, at the top of the diagram.

For heteronuclear diatomic molecules, the molecular orbital diagrams are not symmetrical and differ for each molecule. The atomic orbitals for each atom are drawn at the approximate energy on the diagram, and the proper atomic orbitals are allowed to combine to form the molecular orbitals. Often the diagram will contain the same orbitals as shown in Fig. 16-2b.

Those molecular orbitals having a cross section perpendicular to the internuclear axis (usually the z axis) consisting of a circle are called σ *orbitals*; of two lobes (180° apart), π *orbitals*; of four lobes (90° apart), δ *orbitals*; etc. In Fig. 16-2 an asterisk indicates an antibonding orbital; in parentheses is the symbol of the atomic orbitals composing the molecular orbital. The subscripts g and u refer to the symmetry of the orbital (Sec. 16.8).

EXAMPLE 16.4. Use Fig. 16-2 to write electronic configurations for the following diatomic molecules, and discuss the bonding (if any): (a) H_2^+, H_2, and H_2^-; (b) N_2; (c) F_2.

(a) From Fig. 16-2a, the configurations are $\sigma(1s)$ for H_2^+, $\sigma(1s)^2$ for H_2, and $\sigma(1s)^2\sigma^*(1s)$ for H_2^-, indicating stable molecules having half a σ bond, a σ bond, and half a σ bond, respectively.

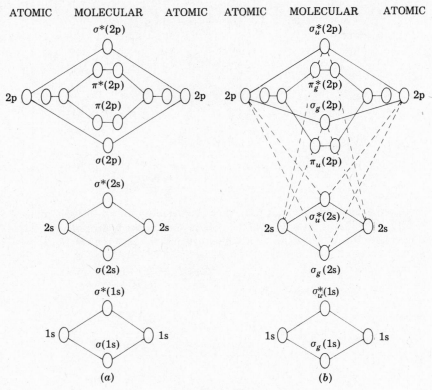

ATOMIC   MOLECULAR   ATOMIC   ATOMIC   MOLECULAR   ATOMIC

Fig. 16-2

(b)   Using Fig. 16-2b for the 14 electrons gives the configuration as

$$\sigma_g(1s)^2\sigma_u^*(1s)^2\sigma_g(2s)^2\sigma_u^*(2s)^2\pi_u(2p)^4\sigma_g(2p)^2$$

which is a triple bond consisting of a σ and two π bonds.

(c)   For the 18 electrons, Fig. 16-2b gives the configuration as a single σ bond:

$$\sigma_g(1s)^2\sigma_u^*(1s)^2\sigma_g(2s)^2\sigma_u^*(2s)^2\pi_u(2p)^4\sigma_g(2p)^2\pi_g^*(2p)^4$$

Bond Description

16.5   ELECTRONEGATIVITY

The quantitative measure of the unequal attraction of the atoms in a heteronuclear diatomic molecule for the bonding electrons is known as the *electronegativity*. Mulliken defined the electronegativity for an element (EN_i) as

$$EN_i = \tfrac{1}{2}(I_i + E_{A,i}) \tag{16.7}$$

where I_i is the ionization energy and $E_{A,i}$ is the electron affinity of the element. Pauling defined the difference in electronegativities for two elements X and Y as

$$EN_X - EN_Y = 0.102\,\Delta^{1/2} \tag{16.8}$$

where the constant 0.102 converts $\Delta^{1/2}$ from $(kJ\,mol^{-1})^{1/2}$ to $eV^{1/2}$, and

$$\Delta = BE(XY) - [BE(X_2)BE(Y_2)]^{1/2} \tag{16.9}$$

where BE_i is the bond energy (see Sec. 4.7). If $EN(H)$ is assumed to be 2.20, the Mulliken scale is related to the Pauling scale by dividing the former by $3.15\ eV^{1/2}$.

16.6 DIPOLE MOMENT

A heteronuclear molecule will be polar, with the end containing the less electronegative element having a slight positive charge and the end containing the more electronegative element having a slight negative charge. The separation of charges $\pm q$ by a distance r gives rise to a *dipole moment* (μ), where

$$\mu = qr \tag{16.10}$$

Dipole moments are often expressed in units of *debyes*, where $1\,\text{D} = 3.335\,641 \times 10^{-30}\,\text{C m}$.

16.7 IONIC CHARACTER

The *percent ionic character* (% IC) of a chemical bond can be calculated from the dipole moment by

$$\% \,\text{IC} = \frac{\mu_{\text{actual}}}{\mu_{\text{predicted}}} \times 100 \tag{16.11}$$

The actual dipole moment is experimentally measured, and the predicted dipole moment is calculated from (16.10) with $q = e$ and $r = r_{ab}$. The percent ionic character can also be estimated as

$$\% \,\text{IC} = 16(\text{EN}_i - \text{EN}_j) + 3.5(\text{EN}_i - \text{EN}_j)^2 \tag{16.12}$$

Molecular Term Symbols

16.8 CLASSIFICATION OF ELECTRONIC STATES

The component of electronic angular momentum along the z axis has the value $M_L \hbar$, where M_L is defined in (15.18). Electronic states are classified according to the value of the quantum number Λ given by

$$\Lambda = |M_L| \tag{16.13}$$

In analogy to atomic orbitals, the symbol Σ is used for $\Lambda = 0$, Π for $\Lambda = 1$, Δ for $\Lambda = 2$, Φ for $\Lambda = 3$, Γ for $\Lambda = 4$, etc. The symbol for a Σ state carries as a following superscript a minus sign if the wave function changes sign when reflected across the xz plane and a plus sign if there is no change in sign. The term symbol for a homonuclear diatomic molecule carries a following subscript of g (German: *gerade* = even) or u (*ungerade* = odd) depending on whether an inversion operation through the origin leaves the wave function unchanged or changed in sign, respectively.

A leading superscript on the symbol is $2S + 1$, where S is found by s-s coupling techniques (see Sec. 15.9). An optional numerical following subscript Ω on the term symbol represents the total angular momentum. It is found by combining the value of $\mathscr{S}$ given by

$$\mathscr{S} = S, S-1, \ldots, -S \tag{16.14}$$

with Λ, yielding

$$\Omega = |\Lambda + \mathscr{S}| \tag{16.15}$$

EXAMPLE 16.5.   Identify the term symbols that correspond to $\Lambda = 2$ and $S = 1$.

The symbol for $\Lambda = 2$ is Δ. The multiplicity is $2S + 1 = 3$, giving $^3\Delta$. Equation (16.14) gives the values of $\mathscr{S}$ as 1, 0, -1, which gives $\Omega = |2 + 1| = 3$, $|2 + 0| = 2$, and $|2 - 1| = 1$. The term symbols are $^3\Delta_3$, $^3\Delta_2$, and $^3\Delta_1$.

16.9 TERM SYMBOLS FOR ELECTRONIC CONFIGURATIONS

For various molecular orbital configurations, the major part of the term symbol can be found in Table 16-1. The *parity* or symmetry for the homonuclear diatomic molecules can be found by multiplying

Table 16-1

Configuration	Term Symbols
σ	$^2\Sigma^+$
σ^2	$^1\Sigma^+$
π	$^2\Pi$
π^2	$^1\Sigma^+, \, ^1\Delta, \, ^3\Sigma^-$
$\pi^2\sigma$	$^2\Sigma^+, \, ^2\Sigma^-, \, ^2\Delta, \, ^4\Sigma$
$\pi^2\pi$	$^2\Pi \, (3), \, ^2\Phi, \, ^4\Pi$
$\pi^2\delta$	$^2\Sigma^+, \, ^2\Sigma^-, \, ^2\Delta \, (2), \, ^2\Gamma, \, ^4\Delta$
π^3	$^2\Pi$
$\pi^3\sigma$	$^1\Pi, \, ^3\Pi$
$\pi^3\pi$	$^1\Sigma^+, \, ^1\Sigma^-, \, ^1\Delta, \, ^3\Sigma^+, \, ^3\Sigma^-, \, ^3\Delta$
$\pi^3\delta$	$^1\Pi, \, ^1\Phi, \, ^3\Pi, \, ^3\Phi$
π^4	$^1\Sigma^+$
δ	$^2\Delta$
δ^2	$^1\Sigma^+, \, ^3\Sigma^-, \, ^1\Gamma$
δ^3	$^2\Delta$
δ^4	$^1\Sigma^+$

the parities of the orbitals being used, according to the usual laws of odd and even:

$$(g)(g) = (u)(u) = g \qquad \text{and} \qquad (g)(u) = (u)(g) = u$$

The term symbol for a heteronuclear diatomic molecule does not contain the g or u subscript. Electrons in partially filled atomic orbitals that are not involved in bonding also contribute to Λ.

EXAMPLE 16.6. What is the molecular term symbol for H_2^+?

For the ground state the electronic configuration is $\sigma(1s)$, which gives rise to a $^2\Sigma^+$ symbol from Table 16-1. The parity is g, giving $^2\Sigma_g^+$.

Solved Problems

QUANTUM THEORY OF DIATOMIC MOLECULES

16.1. Determine $\mathscr{H}$ and a trial wave function for H_2^+.

The expression for $\mathscr{H}$ for the system composed of nuclei a and b and electron 1 is given by (16.1) as

$$\mathscr{H} = \frac{-\hbar^2}{2m_e} \nabla_1^2 + \frac{e^2}{4\pi\varepsilon_0} \left(\frac{1}{r_{ab}} - \frac{1}{r_{1a}} - \frac{1}{r_{1b}} \right)$$

The trial wave function is, from (16.2) and (16.3),

$$\phi = \phi_1 = a_1 \psi_a(1) + a_2 \psi_b(1)$$

where a_1 and a_2 have been written for a_{1a} and a_{1b}.

16.2. Construct a trial wave function for the bond between hydrogen and chorine in HCl.

The electrons forming the bond are a 1s electron from the hydrogen and a 3p electron from the chlorine. Assuming identical wave functions for the two electrons, (16.3) gives

$$\phi_1 = a_1 \psi_{1sH}(1) + a_2 \psi_{3pCl}(1) \qquad \phi_2 = a_1 \psi_{1sH}(2) + a_2 \psi_{3pCl}(2)$$

and then (*16.2*) gives

$$\phi = a_1{}^2[\psi_{1sH}(1)\psi_{1sH}(2)] + a_1 a_2[\psi_{1sH}(1)\psi_{3pCl}(2) + \psi_{1sH}(2)\psi_{3pCl}(1)] + a_2{}^2[\psi_{3pCl}(1)\psi_{3pCl}(2)]$$

Note that the covalent terms are of equal weight but that the ionic terms have different weights from the covalent terms and from each other.

APPLICATION OF THE VARIATION METHOD

16.3. Prepare a plot of E' against r_{ab} for the bonding orbital of the $H_2{}^+$ molecule-ion, where

$$E' = E_H + \frac{J + K}{1 + S} + \frac{e^2}{4\pi\varepsilon_0 r_{ab}}$$

Determine the dissociation energy D_e and the value of r_{ab} at which the lowest energy occurs.

Values of r_{ab} from 0.001 to 0.800 nm were used to calculate E', giving the solid curve in Fig. 16-3. A sample calculation follows for $r_{ab} = 0.100$ nm.

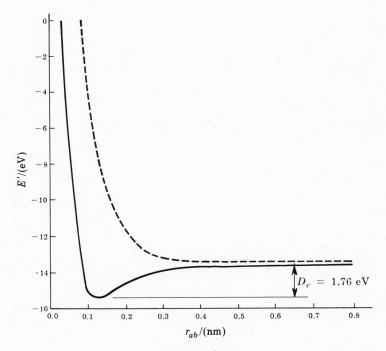

Fig. 16-3

Equation (*16.5*) gives $\rho = 0.100$ nm$/0.052\,9$ nm $= 1.89$, which upon substitution into (*16.4a*), (*16.4b*), and (*16.4c*) gives

$$S = e^{-1.89}\left[1 + 1.89 + \frac{(1.89)^2}{3}\right] = (0.151)(4.08) = 0.616$$

$$J = \frac{e^2}{4\pi\varepsilon_0 a_0}\left[\frac{-1}{1.89} + e^{-3.78}\left(1 + \frac{1}{1.89}\right)\right] = \frac{e^2}{4\pi\varepsilon_0 a_0}(-0.494)$$

$$K = \frac{-e^2}{4\pi\varepsilon_0 a_0}e^{-1.89}(1 + 1.89) = \frac{e^2}{4\pi\varepsilon_0 a_0}(-0.437)$$

The expression for E' becomes

$$E' = (-13.60 \text{ eV}) + \frac{e^2}{4\pi\varepsilon_0 a_0}\left(\frac{-0.494 - 0.437}{1 + 0.616}\right) + \frac{e^2}{4\pi\varepsilon_0 r_{ab}}$$

$$= (-13.60 \text{ eV}) + \frac{e^2}{4\pi\varepsilon_0}\left(\frac{-0.576}{a_0} + \frac{1}{r_{ab}}\right)$$

which upon substitution of numerical values of e, $4\pi\varepsilon_0$, a_0, and r_{ab} gives

$$E' = (-13.60 \text{ eV}) + \frac{(1.602\ 2 \times 10^{-19} \text{ C})^2}{1.112\ 65 \times 10^{-10} \text{ C}^2 \text{ N}^{-1} \text{ m}^{-2}}(6.241\ 45 \times 10^{18} \text{ eV N}^{-1} \text{ m}^{-1})$$

$$\times \left(\frac{-0.576}{0.529 \times 10^{-10} \text{ m}} + \frac{1}{1.00 \times 10^{-10} \text{ m}}\right)$$

$$= (-13.60) + (-15.68) + (14.40) = -14.88 \text{ eV}$$

The depth of the well is 1.76 eV ($= D_e$) at $r_{ab} = 0.132$ nm. The accepted values for these parameters are 2.791 eV ($= 4.472 \times 10^{-19}$ J) and 0.106 nm.

16.4. Qualitatively confirm that $\phi = (2 + 2S)^{-1/2}[\psi_a(1) + \psi_b(1)]$ represents the wave function for a bonding orbital in H_2^+.

The probability of finding the electron is proportional to

$$\phi^*\phi = (2 + 2S)^{-1}[\psi_a(1)^*\psi_a(1) + 2\psi_a(1)^*\psi_b(1) + \psi_b(1)^*\psi_b(1)]$$

The first and third terms in the expression for $\phi^*\phi$ are the probabilities of finding the electron around one nucleus or the other, and the second term is the probability of finding the electron between the nuclei. A qualitative plot is shown in Fig. 16-4a.

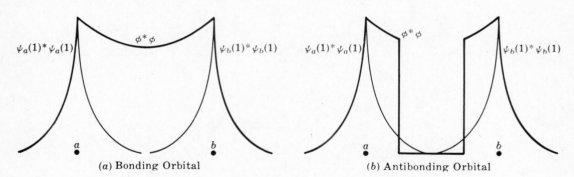

(a) Bonding Orbital   (b) Antibonding Orbital

Fig. 16-4

16.5. With $\mathscr{H}$ and the trial wave function as determined in Problem 16.1, use the variation method to obtain the wave functions and the energies of the bonding and antibonding orbitals in the H_2^+ molecule-ion.

The secular equation for the system is given by (14.28) as

$$\begin{vmatrix} H_{aa} - ES_{aa} & H_{ab} - ES_{ab} \\ H_{ba} - ES_{ba} & H_{bb} - ES_{bb} \end{vmatrix} = 0$$

In view of the symmetry of the problem, (14.26) and (14.27) give $H_{aa} = H_{bb}$, $H_{ab} = H_{ba}$, $S_{ab} = S_{ba} = S$, where S is evaluated in (16.4a). Moreover, since $\psi_a(1)$ and $\psi_b(1)$ are assumed to be normalized, $S_{aa} = S_{bb} = 1$. The secular equation thus reduces to

$$\begin{vmatrix} H_{aa} - E & H_{ab} - ES \\ H_{ab} - ES & H_{aa} - E \end{vmatrix} = 0$$

whose roots are (see Problem 14.27)

$$E_1 = \frac{H_{aa} + H_{ab}}{1 + S} \qquad E_2 = \frac{H_{aa} - H_{ab}}{1 - S}$$

Using these values of E in (14.25),

$$a_1(H_{aa} - E) + a_2(H_{ab} - ES) = 0 \qquad a_1(H_{ab} - ES) + a_2(H_{aa} - E) = 0$$

We solve for a_1/a_2, obtaining

$$\frac{a_1}{a_2} = +1 \text{ for } E = E_1 \qquad \frac{a_1}{a_2} = -1 \text{ for } E = E_2$$

Using (14.12) for E_1 gives

$$1 = \int_{\text{all space}} \phi^* \phi \, dV$$

$$= \int_{\text{all space}} [a_1\psi_a(1) + a_1\psi_b(1)]^*[a_1\psi_a(1) + a_1\psi_b(1)] \, dV = a_1^2(2 + 2S)$$

(see Example 16.2). Solving for a_1 gives $a_1 = (2 + 2S)^{-1/2}$. Likewise for E_2, where $a_1 = -a_2$,

$$1 = \int_{\text{all space}} [a_1\psi_a(1) - a_1\psi_b(1)]^*[a_1\psi_a(1) - a_1\psi_b(1)] \, dV$$

$$= a_1^2(2 - 2S)$$

which gives $a_1 = (2 - 2S)^{-1/2}$. The normalized wave functions are thus

$$\phi_1 = (2 + 2S)^{-1/2}[\psi_a(1) + \psi_b(1)] \text{ for } E_1$$
$$\phi_2 = (2 - 2S)^{-1/2}[\psi_a(1) - \psi_b(1)] \text{ for } E_2$$

All that remains is to find the values of E_1 and E_2. The evaluations of H_{aa} and H_{ab} are as follows (omitting the r_{ab} term in $\mathcal{H}$):

$$H_{aa} = \int_{\text{all space}} \psi_a^*(1) \mathcal{H} \psi_a(1) \, dV$$

$$= \int_{\text{all space}} \psi_a^*(1) \left[\frac{-\hbar^2}{2m_e} \nabla^2 - \frac{e^2}{4\pi\varepsilon_0}\left(\frac{1}{r_{1a}}\right) - \frac{e^2}{4\pi\varepsilon_0}\left(\frac{1}{r_{1b}}\right) \right] \psi_a(1) \, dV$$

$$= \int_{\text{all space}} \psi_a^*(1) \left[\frac{-\hbar^2}{2m_e} \nabla^2 - \frac{e^2}{4\pi\varepsilon_0}\left(\frac{1}{r_{1a}}\right) \right] \psi_a(1) \, dV$$

$$+ \frac{-e^2}{4\pi\varepsilon_0} \int_{\text{all space}} \psi_a^*(1) \frac{1}{r_{1b}} \psi_a(1) \, dV$$

$$= E_{\text{H}} + J$$

$$H_{ab} = \int_{\text{all space}} \psi_a^*(1) \left[\frac{-\hbar^2}{2m_e} \nabla^2 - \frac{e^2}{4\pi\varepsilon_0}\left(\frac{1}{r_{1a}}\right) - \frac{e^2}{4\pi\varepsilon_0}\left(\frac{1}{r_{1b}}\right) \right] \psi_b(1) \, dV$$

$$= \int_{\text{all space}} \psi_a^*(1) \left[\frac{-\hbar^2}{2m_e} \nabla^2 - \frac{e^2}{4\pi\varepsilon_0}\left(\frac{1}{r_{1a}}\right) \right] \psi_b(1) \, dV$$

$$+ \frac{-e^2}{4\pi\varepsilon_0} \int_{\text{all space}} \psi_a^*(1) \frac{1}{r_{1b}} \psi_b(1) \, dV$$

$$= E_{\text{H}}S + K$$

Substituting these results into the expressions for E_1 and E_2 gives

$$E_1 = E_H + \frac{J+K}{1+S} \quad \text{and} \quad E_2 = E_H + \frac{J-K}{1-S}$$

which are, respectively, the electronic energies for the bonding and antibonding orbitals. The total energies, E_1' and E_2', are obtained by adding $e^2/4\pi\varepsilon_0 r_{ab}$.

16.6. Consider the overlap of two s atomic orbitals, an s with a p, and two p atomic orbitals. Construct sketches of the newly formed molecular bonding orbitals.

As shown in Fig. 16-4, the bonding orbitals are characterized by a higher probability of finding the electrons between the nuclei than in the separated-atoms case. This is borne out by the sketches in the left half of Fig. 16-5. Thus, for an s overlapping with an s, the molecular orbital will fill in the gap between the atomic orbitals. The cross section perpendicular to the internuclear axis is circular, so a σ orbital is formed. For the overlap of an s with a p, either a σ bond or a nonbonding arrangement will occur. For the overlap of two p orbitals, a head-on overlap will produce a σ bond, and parallel overlap will produce a π bond.

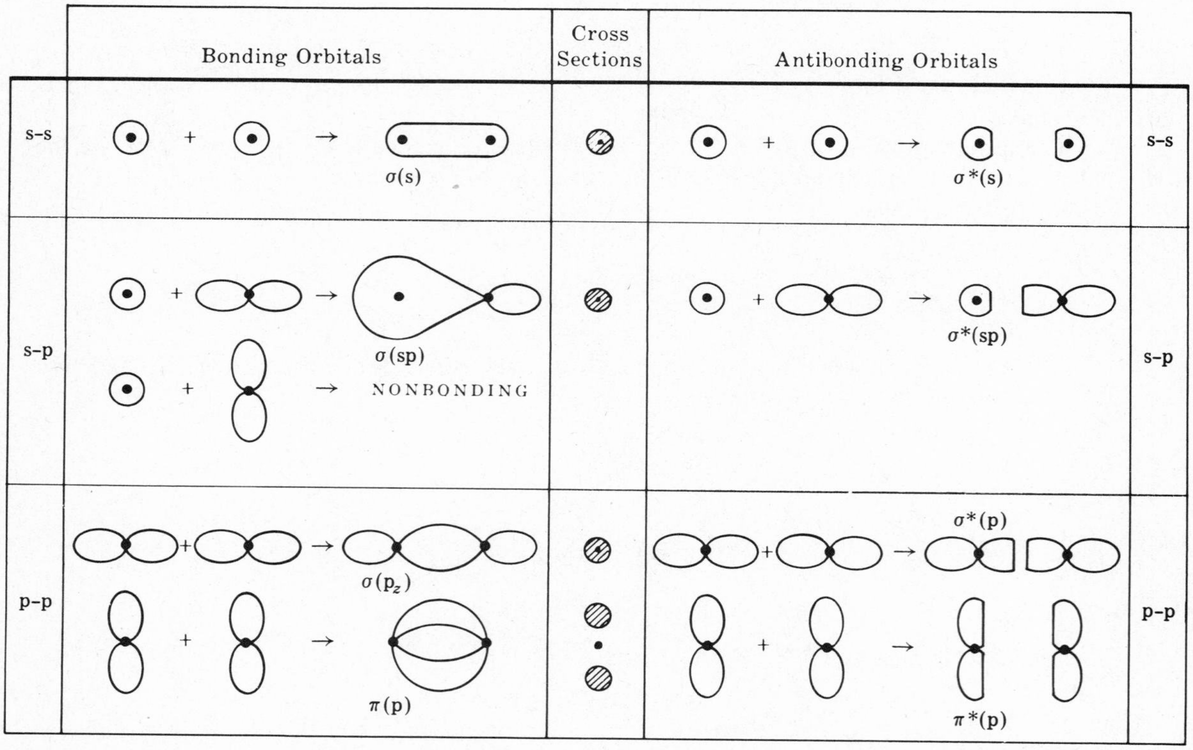

Fig. 16-5

16.7. Prepare a molecular energy diagram to represent the bonding in HCl.

Assuming that the 1s hydrogen orbital combines with the $3p_z$ chlorine orbital gives the bonding and antibonding orbitals shown in Fig. 16-6a. The remainder of the electrons on the chlorine are nonbonding.

BOND DESCRIPTION

16.8. (a)   Using $BE/(\text{kJ mol}^{-1}) = 159.0$, 192.9, and 233.5 for F_2, Br_2, and BrF, respectively, calculate the difference in electronegativities between F and Br.

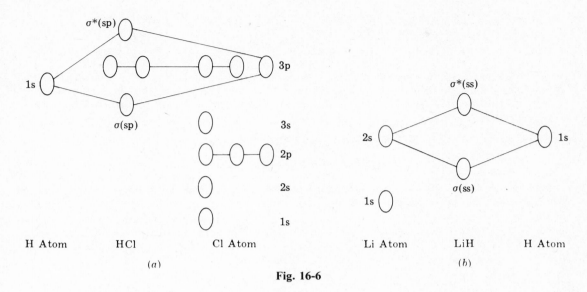

Fig. 16-6

(b)   Using $I_i/(eV) = 3.45$ and $E_{A,i}/(eV) = 17.418$ for F and 3.37 and 11.84, respectively, for Br, calculate the electronegativities from these data and the difference between the electronegativities.

(c)   The dipole moment of BrF is 1.29 D, and the bond distance is 0.175 55 nm. Calculate the percent ionic character of BrF and the difference in electronegativities.

(d)   Compare the results of (a), (b), and (c).

(a)   Using (16.9) gives

$$\Delta = 233.5 - [(192.9)(159.0)]^{1/2} = 58.4 \text{ kJ mol}^{-1}$$

and (16.8) gives

$$EN(F) - EN(Br) = (0.102)(58.4)^{1/2} = 0.78 \text{ eV}^{1/2}$$

(b)   Applying (16.7) to F and Br and converting to the scale of (a) gives

$$EN(F) = \frac{17.418 + 3.45}{(2)(3.15)} = 3.31 \text{ eV}^{1/2}$$

$$EN(Br) = \frac{11.84 + 3.37}{(2)(3.15)} = 2.41 \text{ eV}^{1/2}$$

yielding $$EN(F) - EN(Br) = 3.31 - 2.41 = 0.90 \text{ eV}^{1/2}$$

(c)   Using (16.11) gives

$$\% \text{ IC} = 100 \frac{(1.29 \text{ D})[(3.34 \times 10^{-30} \text{ C m})/(1 \text{ D})]}{(1.602\,2 \times 10^{-19} \text{ C})(1.755\,5 \times 10^{-10} \text{ m})} = 15.3\%$$

Solving the quadratic equation (16.12) gives

$$EN(F) - EN(Br) = \frac{-16 + \sqrt{(16)^2 - 4(3.5)(-15.3)}}{2(3.5)} = 0.81 \text{ eV}^{1/2}$$

(d)   The three theories predict essentially the same value for the electronegativity difference.

MOLECULAR TERM SYMBOLS

16.9.   Determine the term symbol for the ground state of (a) H_2, (b) H_2^-, (c) N_2, and (d) N_2^+.

(a)   The ground state of H_2 is $\sigma_g(1s)^2$, which gives a $^1\Sigma^+$ term (see Table 16-1) having a parity of $(g)(g) = g$, or a complete term symbol of $^1\Sigma_g^+$.

(b) H_2^- is $\sigma_g(1s)^2\sigma_u^*(1s)$ giving $^2\Sigma^+$ with $(g)(g)(u) = u$ parity, or $^2\Sigma_u^+$.

(c) N_2 is $\sigma_g(1s)^2\sigma_u^*(1s)^2\sigma_g(2s)^2\sigma_u^*(2s)^2\pi_u(2p)^4\sigma_g(2p)^2$ giving $^1\Sigma^+$ with $(g)^6(u)^8 = g$ parity, or $^1\Sigma_g^+$.

(d) N_2^+ is $\sigma_g(1s)^2\sigma_u^*(1s)^2\sigma_g(2s)^2\sigma_u^*(2s)^2\pi_u(2p)^4\sigma_g(2p)$ giving $^2\Sigma^+$ with $(g)^5(u)^8 = g$ parity, or $^2\Sigma_g^+$.

16.10. Write electronic configurations for C_2 using the orbital diagrams shown in Fig. 16-2. Which configuration predicts a paramagnetic molecule? The term for the ground state is $^1\Sigma_g^+$. Which configuration is correct?

Placing the 12 electrons in the orbitals given in Fig. 16-2a predicts the configuration as

$$\sigma(1s)^2\sigma^*(1s)^2\sigma(2s)^2\sigma^*(2s)^2\sigma(2p)^2\pi(2p)^2$$

which incorrectly gives $^3\Sigma^-$ as the term symbol (based on maximum multiplicity). Using the orbitals given in Fig. 16-2b, the configuration is

$$\sigma(1s)^2\sigma^*(1s)^2\sigma(2s)^2\sigma^*(2s)^2\pi(2p)^4$$

which correctly gives $^1\Sigma^+$ as the term symbol. The first configuration incorrectly predicts the molecule to be paramagnetic because of the two unpaired electrons in the $\pi(2p)$ orbitals.

16.11. Determine the term symbol for the ground state of (a) NO, (b) HCl, and (c) CH.

(a)   Because the atomic electronic configurations and electronegativities of N and O are not greatly different, the molecular orbital diagram for this substance will resemble that given in Fig. 16-2b. For the electronic configuration

$$\sigma(1s)^2\sigma^*(1s)^2\sigma(2s)^2\sigma^*(2s)^2\pi(2p)^4\sigma(2p)^2\pi^*(2p)^1$$

Table 16-1 gives $^2\Pi$ as the predicted term symbol for NO.

(b)   As described in Problem 16.7, the bonding in HCl involves the molecular orbitals formed by the 1s orbital of hydrogen interacting with the $3p_z$ orbital of chlorine. Two of the 3p orbitals on chlorine are not involved in bonding, giving the electronic configuration as

$$(1s_{Cl})^2(2s_{Cl})^2(2p_{Cl})^6(3s_{Cl})^2\sigma(1s, 3p_z)^2(3p_{Cl})^4$$

Because all the atomic orbitals are filled, there is no contribution to the angular momentum from this source [$M_L = (+1) + (-1) + (+1) + (-1) = 0$ for the $(3p_{Cl})^4$ electrons] and Table 16-1 gives $^1\Sigma^+$ for the σ^2 configuration.

(c)   The bonding in CH is similar to HCl in that the 1s orbital of hydrogen interacts with the $2p_z$ orbital of carbon, forming two molecular orbitals, $\sigma(1s, 2p_z)$ and $\sigma^*(1s, 2p_z)$. The two remaining 2p orbitals of the carbon are nonbonding orbitals. The electronic configuration for CH is

$$(1s_C)^2(2s_C)^2\sigma(1s, 2p_z)^2(2p_C)^1$$

The only contribution to the angular momentum is from the $2p_C$ electron, giving $M_L = +1$ and $S = +\frac{1}{2}$. Thus $\Lambda = 1$ and $S = +\frac{1}{2}$, giving $^2\Pi$ as the term for CH.

Supplementary Problems

QUANTUM THEORY OF DIATOMIC MOLECULES

16.12. Determine $\mathcal{H}$ and a trial wave function for the H_2^- molecule-ion.

Ans. $\mathcal{H} = \dfrac{-h^2}{2m_e}(\nabla_1^2 + \nabla_2^2 + \nabla_3^2) + \dfrac{e^2}{4\pi\varepsilon_0}\left(\dfrac{1}{r_{ab}} - \dfrac{1}{r_{1a}} - \dfrac{1}{r_{2a}} - \dfrac{1}{r_{3a}} - \dfrac{1}{r_{1b}} - \dfrac{1}{r_{2b}} - \dfrac{1}{r_{3b}} + \dfrac{1}{r_{12}} + \dfrac{1}{r_{13}} + \dfrac{1}{r_{23}}\right);$

$\phi = \phi_1\phi_2\phi_3$, where $\phi_1 = a_1\psi_a(1) + a_2\psi_b(1)$, $\phi_2 = a_3\psi_a(2) + a_4\psi_b(2)$, $\phi_3 = a_5\psi_a(3) + a_6\psi_b(3)$

16.13. Construct a trial wave function for the bond in ClF.

> *Ans.* $\phi = \phi_1 \phi_2$, where $\phi_1 = a_1 \psi_{2pF}(1) + a_2 \psi_{3pCl}(1)$, $\phi_2 = a_1 \psi_{2pF}(2) + a_2 \psi_{3pCl}(2)$

APPLICATION OF THE VARIATION METHOD

16.14. Prepare a plot of E' against r_{ab} for the antibonding orbital in the H_2^+ molecule-ion, where

$$E' = E_H + \frac{J - K}{1 - S} + \frac{e^2}{4\pi\varepsilon_0 r_{ab}}$$

Qualitatively discuss the plot.

> *Ans.*   See dashed curve in Fig. 16-3. The curve has no minimum, and the values of $E' > -13.60$ eV ($= -2.179 \times 10^{-18}$ J) indicate a nonbonding system.

16.15. Qualitatively demonstrate that $\phi = (2 - 2S)^{-1/2}\{\psi_a(1) - \psi_b(1)\}$ represents the wave function for an antibonding orbital.   *Ans.*   See Fig. 16-4b.

16.16. (a) The trial wave function used by Heitler and London to describe the H_2 molecule in terms of hydrogenlike wave functions is

$$\phi = c_1 \psi_a(1)\psi_b(2) + c_2 \psi_a(2)\psi_b(1)$$

Note that the ionic terms in the trial wave function determined in Example 16.2 for this same system are eliminated in this wave function. Using the variation method, find the expressions for the energies of the bonding and antibonding orbitals and evaluate c_1 and c_2.

(b) A plot of E' against r_{ab} has a minimum at 0.086 9 nm and -3.140 eV ($= -5.031 \times 10^{-19}$ J) using the Heitler–London function, and for the wave function including the ionic terms the minimum is at 0.085 nm and -2.68 eV ($= -4.29 \times 10^{-19}$ J). What can be said about the inclusion of ionic terms if the accepted values are 0.074 0 nm and -4.747 eV ($= -7.606 \times 10^{-19}$ J)?

> *Ans.*   (a) For bonding: $c_1 = c_2 = (2 + 2S^2)^{-1/2}$, $E = 2E_H + \dfrac{2(J + SK) + J' + K'}{1 + S^2}$
>
> For antibonding: $c_1 = -c_2 = (2 - 2S^2)^{-1/2}$, $E = 2E_H + \dfrac{2(J - SK) + J' - K'}{1 - S^2}$
>
> For both orbitals, $E' = E + \dfrac{e^2}{4\pi\varepsilon_0 r_{ab}}$
>
> (b) Including ionic terms with equal weight to covalent terms gives poorer results.

16.17. Show that if $\phi \propto [\psi_a(1) + \psi_b(1)][\psi_a(2) + \psi_b(2)]$ and if ϕ is normalized, then

$$\frac{e^2}{4\pi\varepsilon_0} \int_{\text{all space}} \phi^* \frac{1}{r_{12}} \phi \, dV = \frac{K' + 2L + J'/2 + C/2}{(1 + S)^2}$$

16.18. Sketch the molecular antibonding orbitals formed by the overlap of (a) two s atomic orbitals, (b) an s and a p atomic orbital, (c) two p atomic orbitals.

> *Ans.*   See right half of Fig. 16-5.

16.19. Arrange the following molecules in order of increasing bond length: O_2, O_2^+, O_2^-, and O_2^{2-}. Which molecule will have the greater bond dissociation energy: O_2 or O_2^+?

> *Ans.*   Configurations are $\ldots \pi^*(2p)^2$, $\ldots \pi^*(2p)^1$, $\ldots \pi^*(2p)^3$, and $\ldots \pi^*(2p)^4$, giving net bonding of $\sigma + \pi$, $\sigma + \frac{3}{2}\pi$, $\sigma + \frac{1}{2}\pi$ and σ; assuming bond length inversely proportional to net bonding gives $O_2^+ < O_2 < O_2^- < O_2^{2-}$. Assuming dissociation energy directly proportional to net bonding gives $O_2^+ > O_2$.

16.20. Using Fig. 16-2b, write electronic configurations for the following diatomic molecules and discuss the bonding, if any: (a) He_2^+; (b) He_2; (c) O_2; (d) Ne_2.

> *Ans.* (a) $\sigma(1s)^2\sigma^*(1s)$, giving net effect of $\frac{1}{2}$ a σ bond; (b) $\sigma(1s)^2\sigma^*(1s)^2$, giving no net bonding;
> (c) $\sigma(1s)^2\sigma^*(1s)^2\sigma(2s)^2\sigma^*(2s)^2\pi(2p)^4\sigma(2p)^2\pi^*(2p)^2$, giving a net effect of one double bond consisting of a one σ bond and two half π-bonds [the $\pi^*(2p)$ electrons have parallel spins according to the rule of maximum multiplicity];
> (d) $\sigma(1s)^2\sigma^*(1s)^2\sigma(2s)^2\sigma^*(2s)^2\pi(2p)^4\sigma(2p)^2\pi^*(2p)^4\sigma^*(2p)^2$, giving no net bonding

16.21. Prepare a molecular energy diagram to represent the bonding in LiH, neglecting any contribution from the vacant 2p atomic orbitals of Li.   *Ans.* See Fig. 16-6b.

BOND DESCRIPTION

16.22. Using $BE/(kJ\,mol^{-1}) = 431.96$, 436.0, and 242.3 for HCl, H_2, and Cl_2, respectively, calculate $EN(Cl) - EN(H)$ and infer the dipole moment given $r_{ab} = 0.127\,46$ nm. Compare with the measured value, 1.08 D.

> *Ans.* $EN(Cl) - EN(H) = 1.05\,eV^{1/2}$;   1.26 D (12% high)

16.23. For HF, HCl, HBr, and HI, $\mu/(D) = 1.82$, 1.08, 0.82, and 0.44, respectively. The values of $r_{ab}/(nm) = 0.091\,68$, $0.127\,46$, $0.141\,4$, and $0.160\,8$, respectively. Calculate the percent ionic character for these molecules, and comment.

> *Ans.*   41.3%, 17.7%, 12.1%, and 5.7%, respectively;   less ionic character as halogen gets larger

16.24. If an ionic bond is defined as one having at least 50% ionic character, what minimum difference in electronegativities will be defined by (16.12)?   *Ans.* $2.13\,eV^{1/2}$

MOLECULAR TERM SYMBOLS

16.25. Identify the term symbol that corresponds to $\Lambda = 1$ and $S = 0$.   *Ans.* $^1\Pi_1$

16.26. Using Table 16-1, determine the term symbol for the ground state of (a) He_2^+, (b) O_2, and (c) O_2^+.

> *Ans.*   (a) $^2\Sigma_u^+$;   (b) $^3\Sigma_g^-$ ($^3\Sigma^-$ has maximum multiplicity over $^1\Sigma^+$ and $^1\Delta$);   (c) $^2\Pi_g$

16.27. Using Table 16-1, determine the term symbol for the ground state of (a) LiH and (b) CN^-, NO^+, and CO.

> *Ans.*   (a) $^1\Sigma^+$;   (b) $^1\Sigma^+$ for the isoelectronic species

16.28. Using Table 16-1, determine the molecular term symbols for the molecules OH^-, OH, and OH^+.

> *Ans.* $^1\Sigma^+$, $^2\Pi$, $^3\Sigma^-$

16.29. Write the predicted electronic configuration for the ground state of BN, and determine the term symbol for this state. The actual term is $^3\Pi$. Determine the correct ground-state electronic configuration.

> *Ans.* $\sigma(1s)^2\sigma^*(1s)^2\sigma(2s)^2\sigma^*(2s)^2\pi(2p)^4$, $^1\Sigma$; $\sigma(1s)^2\sigma^*(1s)^2\sigma(2s)^2\sigma^*(2s)^2\pi(2p)^3\sigma(2p)^1$

Chapter 17

Spectroscopy of Diatomic Molecules

Rotational and Vibrational Spectra

17.1 ROTATIONAL SPECTRA

A diatomic molecule undergoing only rotational motion will be acting, to the first approximation, as a two-particle rigid rotator (see Problem 14.22) having an energy

$$E_J = B^* J(J+1) \tag{17.1}$$

where

$$B^* = \frac{\hbar^2}{2I} = \frac{\hbar^2}{2\mu r_{ab}^2} \tag{17.2}$$

[see (*14.50*) and (*14.51*)]. Most spectroscopists express energy in units of cm^{-1} ($1\ cm^{-1} = 1.986\,477 \times 10^{-23}$ J) and write (*17.1*) and (*17.2*) as

$$F(J) = B_e J(J+1) \tag{17.3}$$

and

$$B_e = \frac{\hbar \times 10^{-2}}{4\pi c \mu r_{ab}^2} = \frac{B^* \times 10^{-2}}{hc} = \frac{2.799\,3 \times 10^{-46}}{\mu r_{ab}^2} \tag{17.4}$$

where B_e will be given in cm^{-1} and the normal SI units of kg and m are used for μ and r_{ab}, respectively, in (*17.4*).

If the molecule has a permanent dipole moment, it can absorb or emit microwave radiation corresponding to the rotational transitions that obey the selection rule $\Delta J = \pm 1$. For these transitions the wave numbers are

$$\bar{\nu} = F(J) - F(J-1) = 2B_e J \tag{17.5a}$$

Thus the spectrum will consist of a series of evenly spaced lines separated by

$$\Delta \bar{\nu} = 2B_e(J+1) - 2B_e J = 2B_e \tag{17.5b}$$

Because the energy differences between the rotational levels are quite small, several levels are populated under normal temperatures, giving rise to several intense lines. The degeneracy of any level is $2J + 1$.

EXAMPLE 17.1. The microwave spectrum of CN shows a series of lines separated by $3.797\,8\ cm^{-1}$. Find the internuclear distance in the molecule.

Substituting $B_e = 3.797\,8/2 = 1.898\,9\ cm^{-1}$ into (*17.4*) gives

$$\mu r_{ab}^2 = \frac{2.799\,3 \times 10^{-46}}{1.898\,9} = 1.474\,2 \times 10^{-46}\ kg\ m^2$$

The reduced mass of the molecule is found from the molar masses to be

$$\mu = \frac{M_C M_N}{(M_C + M_N)L} = \frac{(12.011)(14.006\,7)}{(26.018)(6.022\,045 \times 10^{23})}$$

$$= 1.073\,7 \times 10^{-23}\ g = 1.073\,7 \times 10^{-26}\ kg$$

and so

$$r_{ab} = \left(\frac{1.474\,2 \times 10^{-46}}{1.073\,7 \times 10^{-26}}\right)^{1/2} = 1.171\,7 \times 10^{-10}\ m = 0.117\,17\ nm$$

17.2 VIBRATIONAL SPECTRA

The diatomic molecule undergoing only vibrational motion will be acting, to the first approximation, as a simple harmonic oscillator (see Problem 14.5) having an energy

$$E_v = \omega^*(v + \tfrac{1}{2}) \tag{17.6}$$

where

$$\omega^* = h\nu_0 = \hbar \left(\frac{k}{\mu}\right)^{1/2} \tag{17.7}$$

[see (14.39)]. If the energy is expressed in units of cm^{-1}, (17.6) and (17.7) become

$$G(v) = \omega_e(v + \tfrac{1}{2}) \tag{17.8}$$

and

$$\omega_e = \frac{10^{-2}}{2\pi c}\left(\frac{k}{\mu}\right)^{1/2} = (5.308\,8 \times 10^{-12})\left(\frac{k}{\mu}\right)^{1/2} \tag{17.9}$$

where ω_e will be given in cm^{-1} and the normal SI units of kg and $N\,m^{-1}$ are used for μ and k, respectively, in (17.9).

Vibrational transitions can occur only if $\Delta v = \pm 1$ and only in molecules having an oscillating dipole moment. Under these conditions,

$$\bar{\nu} = G(v) - G(v-1) = \omega_e \tag{17.10}$$

which means that the spectrum will consist of one line having wave number ω_e. Because the energy difference between vibrational levels is quite high, most of the molecules will be in the $v = 0$ level. Thus the major contribution to the spectrum will be the transition between the $v = 0$ and $v = 1$ levels.

The value of k, the force constant for the chemical bond, is such that ω_e will be in the infrared region of the spectrum. Homonuclear diatomic molecules will not be infrared-active, i.e., the vibrational spectrum is not observable, while heteronuclear diatomic molecules will show an infrared spectrum. Vibrational data for homonuclear diatomics can be obtained from the Raman spectrum.

17.3 ANHARMONIC OSCILLATOR

The potential-energy well describing an electronic state (solid curve in Fig. 16-3) does not look like that for a simple harmonic oscillator (Fig. 14-8), except at very low values of v. Experimentally it is found that

$$G(v) = \omega_e(v + \tfrac{1}{2}) - \omega_e x_e(v + \tfrac{1}{2})^2 + \omega_e y_e(v + \tfrac{1}{2})^3 + \cdots \tag{17.11}$$

where x_e and y_e are molecular constants that are usually given only in the combinations $\omega_e x_e$ and $\omega_e y_e$. In (17.11) the cubic and higher terms are usually neglected. The separation between levels v and $v+1$ is given by

$$\Delta G(v) = \omega_e - \omega_e x_e(2v + 2) \tag{17.12}$$

EXAMPLE 17.2. The dissociation energy predicted by equations similar to (17.11) is denoted D_e, whereas the experimental dissociation energy is denoted D_0. What is the relation between these dissociation energies?

Because D_e is measured from the bottom of the potential well to the top, while D_0 is measured from the ground level ($v = 0$ and $J = 0$) to the top, $D_e = D_0 + \omega_e/2$.

17.4 VIBRATIONAL–ROTATIONAL SPECTRA

Because a real molecule is simultaneously undergoing both rotational and vibrational motion, the energy will be

$$T(v, J) = \omega_e(v + \tfrac{1}{2}) - \omega_e x_e(v + \tfrac{1}{2})^2 + \cdots + B_e J(J+1)$$

$$- \bar{D}_e J^2(J+1)^2 + \cdots - \alpha_e(v + \tfrac{1}{2})J(J+1) \tag{17.13}$$

where the first term represents the harmonic vibration contribution [see (17.8)], the second term represents the anharmonic vibration contribution [see (17.11)], the third term represents the rotational contribution [see (17.3)], the fourth term represents the centrifugal stretching of the chemical bond, and the fifth term accounts for rotational–vibrational interaction. Usually the terms involving the centrifugal distortion constant $\bar{D}_e$ and the vibrational-rotation coupling constant α_e are small for low values of J.

The selection rules for the vibrational-rotational transitions in heteronuclear diatomic molecules with $\Lambda = 0$ are $\Delta J = \pm 1$ and $\Delta v = \pm 1$. Because of the order-of-magnitude difference between the rotation and vibrational contributions, the spectrum will appear as a series of bands corresponding to different values of v, which under high resolution give lines corresponding to different values of J. As mentioned in Sec. 17.2, homonuclear diatomic molecules will be infrared-inactive, while heteronuclear diatomic molecules will show an infrared spectrum.

The rotational fine structure having $\Delta J = +1$ is known as the R branch, and that having $\Delta J = -1$ is known as the P branch. If the upper state is designated with a prime and the lower state with a double prime, it can be shown that

$$\bar{\nu}_R = \bar{\nu}_0 + (2B'_e - 3\alpha_e) + (3B'_e - B''_e - 4\alpha_e)J + (B'_e - B''_e - \alpha_e)J^2 \qquad (17.14)$$

where $J = 0, 1, 2, 3, \ldots,$ and

$$\bar{\nu}_P = \bar{\nu}_0 - (B'_e + B''_e - 2\alpha_e)J + (B'_e - B''_e - \alpha_e)J^2 \qquad (17.15)$$

where $J = 1, 2, 3, \ldots,$ and the wave number of the "forbidden" transition between $v' = 0$, $J' = 0$ and $v'' = 0$, $J'' = 0$ is given by

$$\bar{\nu}_0 = \omega_e - 2\omega_e x_e \qquad (17.16)$$

if $\bar{D}_e$ is assumed to be zero. If $B'_e = B''_e$, then

$$\bar{\nu}_R = \bar{\nu}_0 + (2B_e - 3\alpha_e) + (2B_e - 4\alpha_e)J - \alpha_e J^2 \qquad (17.17)$$

and

$$\bar{\nu}_P = \bar{\nu}_0 - (2B_e - 2\alpha_e)J - \alpha_e J^2 \qquad (17.18)$$

If m is defined as $J + 1$ for the R branch and as $-J$ for the P branch, (17.17) and (17.18) can be written as

$$\bar{\nu} = \bar{\nu}_0 + (2B_e - 2\alpha_e)m - \alpha_e m^2 \qquad (17.19)$$

and the separation between the lines will be given by

$$\Delta\bar{\nu} = (2B_e - 3\alpha_e) - 2\alpha_e m \qquad (17.20)$$

The selection rules for vibrational–rotational interactions with $\Lambda \neq 0$ are $\Delta J = 0, \pm 1$ and $\Delta v = \pm 1$. The rotational fine structure having $\Delta J = 0$ is known as the Q $branch$, given by

$$\bar{\nu}_Q = \bar{\nu}_0 + (B'_e - B''_e)J + (B'_e - B''_e)J^2 \qquad (17.21)$$

where $J = 0, 1, 2, 3, \ldots,$ if α_e and $\bar{D}_e$ are assumed to be zero.

17.5 THE RAMAN EFFECT

Various rotational and vibrational transitions can occur within the ground electronic state as a result of the scattering of incident radiation on a sample, provided the energy $\bar{\nu}_{\text{incident}}$ is greater than the vibrational energies and less than the electronic energies. Information concerning homonuclear diatomic molecules can be obtained because a permanent dipole is not required as in infrared spectroscopy. Because of the population of vibrational levels, the most intense contributions to the Raman spectrum will be $\Delta v = \pm 1$ and, in particular, $v = 0 \longrightarrow v = 1$.

The spectrum will contain a band centered at $\bar{\nu}_{\text{incident}}$ resulting from Rayleigh scattering, a very weak *anti-Stokes band* centered at $\bar{\nu}_{\text{incident}} + \bar{\nu}_0$, and an intense *Stokes band* centered at $\bar{\nu}_{\text{incident}} - \bar{\nu}_0$.

Under high resolution these bands give rotational information corresponding to the selection rules $\Delta J = 0, \pm 2$. The lines in the Q branch (usually unresolved), corresponding to $\Delta J = 0$, are separated from the center line by

$$\Delta \bar{\nu}_\text{Q} = (B'_e - B''_e)J + (B'_e - B''_e)J^2 \qquad \text{where } J = 0, 1, 2, 3, \ldots \qquad (17.22)$$

those in the S branch, corresponding to $\Delta J = +2$, by

$$\Delta \bar{\nu}_\text{S} = 6B'_e + (5B'_e - B''_e)J + (B'_e - B''_e)J^2 \qquad \text{where } J = 0, 1, 2, 3, \ldots \qquad (17.23)$$

and those in the O branch, corresponding to $\Delta J = -2$, by

$$\Delta \bar{\nu}_\text{O} = 2B'_e - (3B'_e + B''_e)J + (B'_e - B''_e)J^2 \qquad \text{where } J = 2, 3, 4, \ldots \qquad (17.24)$$

Equations (17.22)–(17.24) do not include α_e or $\bar{D}_e$ terms.

EXAMPLE 17.3.   Describe the rotational fine structure in a Raman spectrum if $B'_e = B''_e = B_e$.

Substituing B_e for B'_e and B''_e in (17.22) gives $\Delta \bar{\nu}_\text{Q} = 0$. Thus the Q branch would be predicted to consist of one line located at the center. Similarly, (17.23) becomes

$$\Delta \bar{\nu}_\text{S} = 6B_e + (5B_e - B_e)J + (B_e - B_e)J^2 = 6B_e + 4B_e J$$

which predicts that the first line ($J = 0$) of the S branch would be $6B_e$ above the center line and the rest of the lines would be separated by $4B_e$. For the O branch, (17.24) becomes

$$\Delta \bar{\nu}_\text{O} = 2B_e - (3B_e + B_e)J + (B_e - B_e)J^2 = 2B_e - 4B_e J$$

which predicts that the first line ($J = 2$) would be $6B_e$ below the center line and the rest of the lines would be separated by $4B_e$.

Electronic Spectra

17.6   SELECTION RULES

The selection rules describing interacting electronic states are very complicated. However, the selection rule for the vibrational transitions between permitted electronic states is quite simple: Δv is arbitrary. The vibrational transitions will also include the rotational structure, but unless a very high resolution instrument is used, meaningful data are difficult to obtain; they are best found from infrared or Raman spectra.

EXAMPLE 17.4.   The intensities of the observed bands willl depend on the value of $\psi(r_{ab})^*\psi(r_{ab})$ at the same value of r_{ab} for the two levels in the transition. Consider the interacting electronic states shown in Fig. 17-1a. Will the interactions indicated by vertical lines from a'' to a' and to b', and from b'' to a' and to b', be observed?

The transition from a'' to a' will not be observed, because the positions of $\psi^*\psi$ are such that there is very little overlap. The line from a'' to b' will be observed, because the $\psi^*\psi$'s indicate high populations of both levels at this value of r_{ab}. The line from b'' to both a' and b' passes through high population values of these levels, so both of these transitions will be observed.

17.7   DESLANDRES TABLE

The analysis of a discharge spectrum is complicated by the trial-and-error assignment of the values of $v' \longrightarrow v''$ to the observed bands. Fortunately, three trends in the data reduce the chances for error: (1) values of $\bar{\nu}$ for *sequences* (bands having the same value of Δv) are nearly constant but decrease slightly as v' and v'' increase, (2) values of $\bar{\nu}$ for *progressions from the upper state* (bands having the same value of v') decrease as v'' increases, and (3) values of $\bar{\nu}$ for *progressions to the lower state* (bands

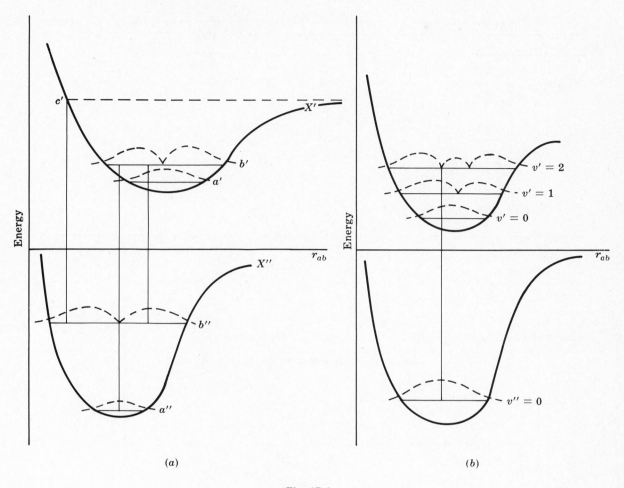

(a)   (b)

Fig. 17-1

having the same value of v'') increase as v' increases. Once the correct assignments are made, (17.12) can be used to determine ω_e and $\omega_e x_e$ for each state by plotting the differences between the values of $\bar{\nu}$ for the various progressions against $2v+2$, giving ω_e as the intercept and $\omega_e x_e$ as the slope. The value of $\bar{\nu}_{00}$, for the $v'=0 \longrightarrow v''=0$ transition, is usually calculated from

$$\bar{\nu}_{00} = \bar{\nu} - [(\omega_e' - \omega_e' x_e')v' - \omega_e' x_e'(v')^2] + [(\omega_e'' - \omega_e'' x_e'')v'' - \omega_e'' x_e''(v'')^2] \qquad (17.25)$$

using several of the observed frequencies rather than relying on only one observed value.

EXAMPLE 17.5.   Prepare a Deslandres table summarizing the (hypothetical) transitions shown in Fig. 17.2.

The table consists of entries of $\bar{\nu}$ corresponding to the values of v' and v'' (see Table 17-1). Note that the three trends discussed above are present: (1) values of $\bar{\nu}$ along diagonals, which correspond to sequences ($0 \longrightarrow 0$, $1 \longrightarrow 1$, etc.), decrease as v' and v'' increase; (2) values of $\bar{\nu}$ in rows, which correspond to progressions from the upper state ($0 \longrightarrow 0$, $0 \longrightarrow 1$, $0 \longrightarrow 2$, etc.), decrease as v'' increases; and (3) values of $\bar{\nu}$ in columns, which correspond to progressions to the lower state ($0 \longrightarrow 1$, $1 \longrightarrow 1$, $2 \longrightarrow 1$, etc.), increase as v' increases. The differences between the entries for the first two rows give the separation between the $v'=0$ and $v'=1$ levels as 300 cm^{-1}; the second two rows give the separation between the $v'=1$ and $v'=2$ levels as 200 cm^{-1}; etc. The differences between the first two columns give the separation between the $v''=0$ and $v''=1$ levels as $1\,000$ cm^{-1}; the second two columns give the separation between the $v''=1$ and $v''=2$ levels as 750 cm^{-1}, etc. For a more realistic set of data, see Problem 17.8.

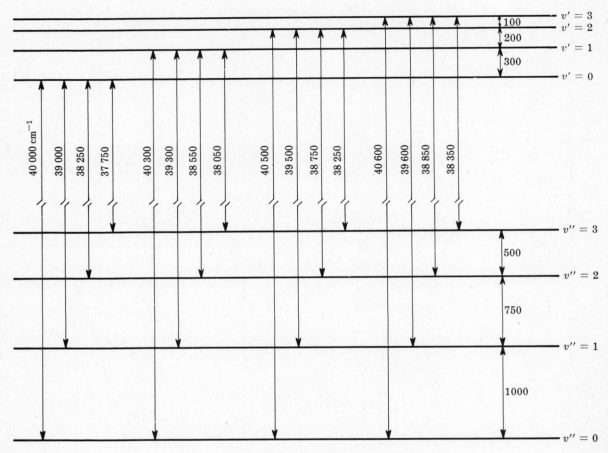

Fig. 17-2

Table 17-1

Upper State	Lower State							Average Separation in Upper State
	$v'' = 0$		$v'' = 1$		$v'' = 2$		$v'' = 3$	
$v' = 0$	40 000	1 000	39 000	750	38 250	500	37 750	
	300		300		300		300	300
1	40 300	1 000	39 300	750	38 550	500	38 050	
	200		200		200		200	200
2	40 500	1 000	39 500	750	38 750	500	38 250	
	100		100		100		100	100
3	40 600	1 000	39 600	750	38 850	500	38 350	
Average Separation in Lower State	1 000		750		500			

Solved Problems

ROTATIONAL AND VIBRATIONAL SPECTRA

17.1. The bond length in CN^+ is 0.129 nm. Predict the position of the first four lines in the microwave spectrum.

Substituting $\mu = 1.073\ 7 \times 10^{-26}$ kg (see Example 17.1) into (17.4) gives

$$B_e = \frac{2.799\ 3 \times 10^{-46}}{(1.073\ 7 \times 10^{-26})(1.29 \times 10^{-10})^2} = 1.567 \text{ cm}^{-1}$$

The first four lines predicted by (17.5) are

$$\bar{\nu}_1 = 2(1.567 \text{ cm}^{-1})(1) = 3.133 \text{ cm}^{-1}$$

$\bar{\nu}_2 = 6.267$ cm^{-1}, $\bar{\nu}_3 = 9.400$ cm^{-1}, and $\bar{\nu}_4 = 12.534$ cm^{-1}.

17.2. Compare the force constants for the bond strengths in CN and CN^+ given that $\omega_e/(\text{cm}^{-1}) = 2\ 068.61$ and $1\ 580$, respectively. The reduced mass for both molecules is $1.073\ 7 \times 10^{-26}$ kg.

Upon rearrangement, (17.9) gives

$$k = \frac{\omega_e^2 \mu}{(5.308\ 8 \times 10^{-12})^2}$$

which upon substitution of the data for the molecules gives

$$k_{\text{CN}} = \frac{(2\ 068.61)^2(1.073\ 7 \times 10^{-26})}{(5.308\ 8 \times 10^{-12})^2} = 16.302 \times 10^2 \text{ N m}^{-1}$$

and $k_{\text{CN}^+} = 9.51 \times 10^2$ N m^{-1}. There is one fewer bonding electron in CN^+ than in CN; thus the weaker bond predicted by the values of k is correct.

17.3. For BH, $\omega_e x_e = 49$ cm^{-1} and $\omega_e = 2\ 368$ cm^{-1}. Find the energy of the first three vibrational levels with respect to the bottom of the potential well, and determine the separation between these levels.

Using (17.11) gives

$$G(0) = (2\ 368)(0 + \tfrac{1}{2}) - (49)(0 + \tfrac{1}{2})^2 = 1\ 172 \text{ cm}^{-1}$$

$G(1) = 3\ 442$ cm^{-1}, and $G(2) = 5\ 614$ cm^{-1}. We then have for the separation between the $v = 0$ and $v = 1$ levels

$$\Delta G(0\text{-}1) = 3\ 442 - 1\ 172 = 2\ 270 \text{ cm}^{-1}$$

and for the separation between the $v = 1$ and $v = 2$ levels, $\Delta G(1\text{-}2) = 2\ 172$ cm^{-1}.

17.4. The contributions of the harmonic oscillator $[G(v)]$ and the rigid rotator $[F(J)]$ are the largest terms in (17.13). For lack of complete data, $T(v, J)$ may be estimated fairly well by $T = G(v) + F(J)$. Compare values of T and $T(v, J)$ for $v = 0$, $J = 2$ and for $v = 1$, $J = 10$, given $\omega_e = 2\ 068.1$ cm^{-1}, $\omega_e x_e = 13.114$ cm^{-1}, $B_e = 1.898\ 9$ cm^{-1}, and $\alpha_e = 0.017\ 2$ cm^{-1}.

Using (17.13) gives

$$T(0, 2) = (2\ 068.1)(0 + \tfrac{1}{2}) - (13.114)(0 + \tfrac{1}{2})^2 + (1.898\ 9)(2)(2 + 1)$$

$$+ 0 - (0.017\ 2)(0 + \tfrac{1}{2})(2)(2 + 1)$$

$$= 1\ 042.1 \text{ cm}^{-1}$$

Using (*17.3*) and (*17.8*), which ignores the anharmonicity, rotational-vibrational coupling, and centrifugal distortion, gives

$$T = (1.898\ 9)(2)(2+1) + (2\ 068.1)(0+\tfrac{1}{2}) = 1\ 045.4\ \text{cm}^{-1}$$

which differs by 3.3 cm^{-1}, or 0.32%. Similarly, $T(1, 10) = 3\ 278.7$ cm^{-1} and $T = 3\ 311.0$ cm^{-1}, a difference of 32.3 cm^{-1}, or 0.99%.

17.5. Using the Boltzmann distribution given in Problem 17.10, determine J_{max} as a function of temperature, where J_{max} is the quantum number of the most intensely populated rotational state. Using the data from Problem 17.1, find J_{max} for CN$^+$ at 298 K and at 1 000 K.

Substituting (*17.1*) into the expression for N_J found in Problem 17.10 gives

$$N_J = (2J+1)N_0 \exp[-B^* J(J+1)/kT]$$

For a maximum (treating J as a continuous variable),

$$\frac{\partial N_J}{\partial J} = 2N_0 \exp\left[\frac{-B^* J(J+1)}{kT}\right] + (2J+1)N_0\left(\frac{-B^*}{kT}\right)(2J+1)\exp\left[\frac{-B^* J(J+1)}{kT}\right] = 0$$

Solving for J_{max} gives

$$J_{\text{max}} = \frac{(2kT/B^*)^{1/2} - 1}{2}$$

where the right-hand side is to be rounded off to the nearest integral value.

In terms of B_e, the above result becomes

$$J_{\text{max}} = \frac{(2kT/10^2 B_e hc)^{1/2} - 1}{2}$$

which upon substitution of the data for CN$^+$ gives

$$J_{\text{max}} = \frac{1}{2}\left\{\left[\frac{2(1.380\ 7 \times 10^{-23}\ \text{J K}^{-1})T}{[(10^2\ \text{cm})/(1\ \text{m})](1.566\ \text{cm}^{-1})(6.626 \times 10^{-34}\ \text{J s})(2.997\ 9 \times 10^8\ \text{m s}^{-1})}\right]^{1/2} - 1\right\}$$

$$= \frac{(0.887\ 7\ T)^{1/2} - 1}{2}$$

At 298 K, J_{max} is 8, and at 1 000 K the value is 14.

ELECTRONIC SPECTRA

17.6. Consider the interacting electronic states shown in Fig. 17-1*b*. Will the $v = 0 \longrightarrow 0$, $v = 0 \longrightarrow 1$, and $v = 0 \longrightarrow 2$ absorptions be observed?

Because the values of $\psi^*\psi$ are large, large, and small, respectively, these transitions will be strong, strong, and very weak, respectively.

17.7. What will happen if the transition between b'' and c' in Fig. 17-1*a* takes place?

Because the upper state c' is above D_e, the molecule will dissociate.

17.8. Using the following values for $\lambda/(\text{nm})$ obtained by Tilford and Simons for the $^1\Pi \longrightarrow {}^1\Sigma^+$ electronic transition of $^{12}\text{C}^{16}\text{O}$, determine ω_e and $\omega_e x_e$ for both states and find $\bar{\nu}_{00}$; 111.510, 113.034, 113.969, 116.115, 117.315, 118.599, 119.967, 121.421, 121.691, 122.965, 124.604, 124.665, 126.293, 126.341, 128.022, 128.183, 129.856, 130.137, 131.803, 132.210, 133.874, 134.413, 136.066, 136.756, 138.400, 139.246, 140.885, 141.897, 143.529, 144.726, 146.347, 147.746, 149.358, 150.965, 152.576, 154.431, 156.021, and 159.716.

Table 17-2

Upper State	Lower State v'' = 0		v'' = 1	Average Separation in Upper State	2(v' + 1)
v' = 0	64 754.0	2 142.9	62 611.1		
	1 486.6		1 482.8	1 484.7	2
1	66 240.6	2 146.7	64 093.9		
	1 443.3		1 447.2	1 445.3	4
2	67 683.9	2 142.8	65 541.1		
	1 412.0		1 412.1	1 412.1	6
3	69 095.9	2 142.7	66 953.2		
	1 377.8		1 377.5	1 377.7	8
4	70 473.7	2 143.0	68 330.7		
	1 341.8		1 341.6	1 341.7	10
5	71 815.5	2 143.2	69 672.3		
	1 307.3		1 307.6	1 307.5	12
6	73 122.8	2 142.9	70 979.9		
	1 274.5		1 274.4	1 274.5	14
7	74 397.3	2 143.0	72 254.3		
	1 239.8		1 239.4	1 239.6	16
8	75 637.1	2 143.4	73 493.7		
	1 205.1		1 203.4	1 204.3	18
9	76 842.2	2 145.1	74 697.1		
	1 171.1		1 173.7	1 172.4	20
10	78 013.3	2 142.5	75 870.8		
	1 137.6		1 137.6	1 137.6	22
11	79 150.9	2 142.5	77 008.4		
	1 103.6		1 103.2	1 103.4	24
12	80 254.5	2 142.9	78 111.6		
	1 069.5		1 069.4	1 069.5	26
13	81 324.0	2 143.0	79 181.0		
	1 033.9		1 034.0	1 034.0	28
14	82 357.9	2 142.9	80 215.0		
	998.3			998.3	30
15	83 356.2				
	961.8			961.8	32
16	84 318.0	2 142.7	82 175.3		
	922.3			922.3	34
17	85 240.3				
	880.8			880.8	36
18	86 121.1				
19					
20	87 743.2				
	726.1			726.1	42
21	88 469.3				
22					
23	89 677.7				
Average Separation in Lower State	2 143.3				

The values of $\bar{\nu}(\text{cm}^{-1})$ corresponding to the data are, by (14.2), 89 677.7, 88 469.3, 87 743.2, 86 121.1, 85 240.3, 84 318.0, 83 356.2, 82 357.9, 82 175.3, 81 324.0, 80 215.0, 80 254.5, 79 181.0, 79 150.9, 78 111.6, 78 013.3, 77 008.4, 76 842.2, 75 870.8, 75 637.1, 74.697.1, 74 397.3, 73 493.7, 73 122.8, 72 254.3, 71 815.5, 70 979.9, 70 473.3, 69 672.3, 69 095.9, 68 330.7, 67 683.9, 66 953.2, 66 240.6, 65 541.1, 64 754.0, 64 093.9, and 62 611.1. If the lower-energy data are analyzed by taking differences, values near 2 140 and 1 500 cm^{-1} appear several times, and a similar analysis of the high-energy data gives values near 2 140 and 1 000 cm^{-1}. Assuming that the 2 140 cm^{-1} value corresponds to the $v'' = 1 \longrightarrow v'' = 0$ change (because it appears in both the high- and low-energy emissions) and that the 1 500 to 1 000 cm^{-1} values correspond to changes in v' (which decrease as v' increases), the Deslandres table (Table 17-2) can be constructed.

To find ω_e and $\omega_e x_e$ for the $^1\Pi$ state, values of the average energy difference between the v' states were plotted against $2(v+1)$ (see Fig. 17-3). Equation (17.12) indicates that the intercept of this line is $\omega'_e = 1\,516$ cm^{-1} and the slope is $\omega'_e x'_e = 17.3$ cm^{-1}. Using the value of the $0 \longrightarrow 0$ assigned transition as $\bar{\nu}_{00}$ gives $\bar{\nu}_{00} = 64\,754.0$ cm^{-1}. Because only one $\Delta G(v'')$ was obtained, ω''_e and $\omega''_e x''_e$ cannot be determined, but for $v'' = 0$, (17.12) indicates that ω''_e should be about 2 143 cm^{-1} if $\omega''_e x''_e$ is not too large.

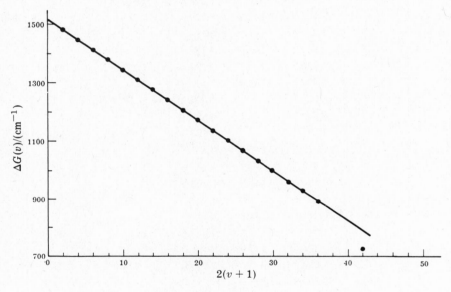

Fig. 17-3

Supplementary Problems

ROTATIONAL AND VIBRATIONAL SPECTRA

17.9.   If it is assumed that $r_{ab} = 0.127\,5$ nm for H^{35}Cl, D^{35}Cl, H^{37}Cl, and D^{37}Cl, what will be the respective spacings for the rotational spectra? Assume atomic masses of 1.007 825, 2.014 0, 34.968 85, and 36.959 for H, D, ^{35}Cl, and ^{37}Cl, respectively. The change in the spacings of the rotational lines resulting from isotopic substitition is known as the *isotopic shift effect*. In which spectra will this effect be greater: (*a*) H^{35}Cl with H^{37}Cl, or D^{35}Cl with D^{37}Cl? (*b*) H^{35}Cl with D^{35}Cl, or H^{37}Cl with D^{37}Cl?

Ans.   (*a*)   HCl change is 0.028 cm^{-1}, DCl change is 0.032 cm^{-1}; greater effect in D^{35}Cl with D^{37}Cl.
          (*b*)   Both changes are 10.280 cm^{-1}; same effect.

17.10.   The Boltzmann distribution [see (2.11)] describing the population of the various rotational states is

$$N_J / N_0 = g_J \, e^{-E_J/kT} / g_0 \, e^{-E_0/kT}$$

where g_J and g_0 are the degeneracies of the levels.

(a) Show that $N_J = (2J+1) N_0 \, e^{-E_J/kT}$.

(b) If $N_0 = 100$, find N_J for $J = 1$ to $J = 15$ at 25 °C for $D^{35}Cl$, given $I = 5.08 \times 10^{-47}$ kg m^2.

(c) The intensity of a peak for a given transition is proportional to the number of molecules in that state. Calculate the theoretical relative intensities of the first 15 peaks by dividing N_J by the maximum value of N_J.

(d) Compare the theoretical values to the experimental values calculated from observed intensities of 55, 66, 73, 75, 75, 73, 70, 64, 58, 49, 40, 31, 24, 18, 11, and 7 by dividing the observed intensity by the maximum observed intensity.

Ans. (b) $E_J/(J) = 1.095 \times 10^{-22} J(J+1)$;
  $N_J = 100, 285, 426, 508, 530, 494, 426, 338, 250, 173, 113, 69, 39, 22, 10, 5$
  (c) 0.19, 0.54, 0.80, 0.96, 1, 0.93, 0.80, 0.64, 0.47, 0.33, 0.21, 0.13, 0.07, 0.04, 0.02, 0.01
  (d) 0.73, 0.88, 0.97, 1, 1, 0.97, 0.93, 0.85, 0.77, 0.65, 0.53, 0.41, 0.32, 0.24, 0.15, 0.09; same general shape with same maximum peaks, theoretical drops off faster from the maximum than does the experimental

17.11. Compare the force constants for the bond strengths in $H^{35}Cl$ and $D^{35}Cl$ given $\omega_e/(cm^{-1}) = 2\,888$ and $2\,092$, respectively. The reduced masses for these molecules are $1.626\,6 \times 10^{-27}$ kg and $3.162\,2 \times 10^{-27}$ kg, respectively.

Ans. 4.81×10^2 N m^{-1} and 4.91×10^2 N m^{-1}; essentially the same

17.12. What is the energy separation between the 23rd and 24th vibrational levels for BH, assuming the system to be described (a) by the anharmonic oscillator model (see Problem 17.3 for pertinent data)? (b) by the harmonic oscillator model? Ans. (a) 16 cm^{-1}, (b) 2 368 cm^{-1}

17.13. The *Morse potential*, given by

$$V(r) = D_e [1 - e^{-a(r-r_e)}]^2$$

generates the first two terms of (*17.11*) if

$$\omega_e = 10^{-1} \, a \left(\frac{\hbar D_e}{\pi c \mu} \right)^{1/2} \quad \text{and} \quad \omega_e x_e = 10^{-2} \frac{\hbar a^2}{4 \pi c \mu}$$

Here, D_e, ω_e, and $\omega_e x_e$ are in cm^{-1}, while the other quantities are in normal SI units. Given $r_e = 0.074\,17$ nm, $D_e = 38\,318$ cm^{-1}, $\omega_e = 4\,405.3$ cm^{-1}, and $\omega_e x_e = 125.325$ cm^{-1} for H_2, plot $U(r) \equiv V(r) - D_e$ against r, showing the first 11 vibrational levels.

Ans. Potential well like solid curve in Fig. 16-3, with minimum at 0.074 17 nm and $-38\,318$ cm^{-1}; vibrational lines at $-36\,147, -31\,922, -28\,088, -24\,435, -21\,032, -17\,880, -14\,979, -12\,328, -9\,928, -7\,778, -5\,879$ cm^{-1}

17.14. The ground state for NO is $^2\Pi$. (a) In addition to the P and R branches of the infrared spectrum, what else will be observed? (b) Given $B'_e = B''_e$ for the molecule, describe the spectrum.

Ans. (a) Q branch will be present. (b) (*17.20*) suggests that the R branch lines get closer together as J increases and the P branch lines get farther apart as J increases; (*17.21*) suggests that the Q branch is one line at $\bar{\nu}_0$.

17.15. What will be the separation of the lines for $J = 10$ in the P and R branches of HN given that the spectroscopic data are $\omega_e = 3\,315$ cm^{-1}, $\omega_e x_e = 94.7$ cm^{-1}, $B_e = 16.668\,4$ cm^{-1}, and $\alpha_e = 0.646$ cm^{-1}?

Ans. $m = -10$ for P branch, giving $\Delta\bar{\nu} = 44.32$ cm^{-1}; $m = 11$ for R branch, giving $\Delta\bar{\nu} = 17.19$ cm^{-1}

17.16. The high-resolution infrared spectrum of $D^{35}Cl$ shows the following fine structure: $\bar{\nu}/(cm^{-1}) = 2\,244.7$, 2 238.7, 2 232.0, 2 225.0, 2 218.5, 2 211.0, 2 203.5, 2 195.5, 2 187.2, 2 179.0, 2 170.2, 2 161.5, 2 152.5, 2 143.0, 2 133.2, 2 123.2, 2 113.2, 2 103.0, 2 081.5, 2 070.2, 2 059.2, 2 047.7, 2 035.7, 2 024.0, 2 012.0, 1 997.2, 1 985.0, 1 972.2, 1 959.7, 1 946.7, 1 933.2, 1 920.2, 1 906.5, 1 892.7, 1 879.0, and 1 864.2. Assign values of m to these lines. Prepare a plot of $\Delta \bar{\nu}$ against m as suggested by (17.20) to obtain α_e and B_e. Use (17.19) for values of m from -5 to $+5$ to determine an average value of $\bar{\nu}_0$. If $\mu = 3.162\,2 \times 10^{-27}$ kg, find r_e.

Ans. $\quad \alpha_e = 0.121$ cm^{-1}, $B_e = 5.46$ cm^{-1}, $\bar{\nu}_0 = 2\,092.3$ cm^{-1}, $r_e = 0.127\,4$ nm

17.17. If $B'_e = B''_e = B_e$, where will the first lines of the O and S branches be located with respect to $\bar{\nu}_0$?

Ans. The first line of the O branch is $J = 2$, which is $-6B_e$ from $\bar{\nu}_0$, and the first line of the S branch is $J = 0$, which is $6B_e$ from $\bar{\nu}_0$.

17.18. Assume that the rotational spectrum is described by

$$F(J) = B_e J(J+1) - \bar{D}_e J^2 (J+1)^2$$

where the $\bar{D}_e$ term corrects for a centrifugal distortion effect. Show that

$$\bar{\nu} = 2B_e J - 4\bar{D}_e J^3$$

Given $\bar{D}_e = 10^{-4} B_e$ and $J = 4$, find $\bar{\nu}$ for CN^+ using the data in Problem 17.1.

Ans. $\quad 12.488$ cm^{-1}

17.19. If the electronic potential-energy well is approximated by an empirical function $V(r)$, then the values for ω_e and $\omega_e x_e$ in (17.11) can be determined from (17.9) with

$$k = 10^2 \, hc V''(r_e)$$

and from $\qquad \omega_e x_e = \dfrac{1}{24} \left\{ 5 \left[\dfrac{V'''(r_e)}{V''(r_e)} \right]^2 - 3 \dfrac{V^{(iv)}(r_e)}{V''(r_e)} \right\} \dfrac{10^{-2} \, \hbar}{4\pi c \mu}$

where $V'(r) = dV(r)/dr$, etc. Show that the expressions for ω_e and $\omega_e x_e$ given in Problem 17.13 indeed follow from the Morse potential.

Ans. $\quad V^{(n)}(r) = 2a^n D_e (-1)^{n-1} [e^{-a(r-r_e)} - 2^{n-1} e^{-2a(r-r_e)}]$

ELECTRONIC SPECTRA

17.20. (a) Describe the absorption spectrum for the system shown in Fig. 17-4a.
   (b) Describe the spectrum for the system shown in Fig. 17-4b for $v'' = 1$ to $v' = 0, 1, 2$.
   (c) Describe the spectrum for the system shown in Fig. 17-4b for $v'' = 1$ to $v' = 3$.

Ans. (a) No bands (continuum); (b) normal bands; (c) continuum if crossing over to X''' occurs

17.21. Using the following $\lambda/$(nm) data reported by Tilford and Simmons for the $^1\Pi \longrightarrow {}^1\Sigma^+$ electronic transition of $^{13}C^{16}O$, determine ω'_e and $\omega'_e x'_e$: 147.868, 144.905, 142.125, 139.515, 137.061, 134.751, 132.576, 130.525, 128.595, 126.764, 125.040, 123.412, 121.874, 120.424, and 119.057. All data are for a $v'' = 0$ progression, and the $v' = 1$ and $v' = 0$ lines are not resolvable from the spectrum of $^{12}C^{16}O$.

Ans. $\quad 1\,418$ cm^{-1}, 15.8 cm^{-1}

17.22. The following bands were observed as part of the discharge spectrum of N_2: $\bar{\nu}/(\text{cm}^{-1}) = 31\,678.65, 29\,698.27,$ $28\,600.03, 28\,270.95, 27\,991.60, 26\,959.99, 26\,638.25, 26\,297.10, 25\,700.99, 25\,361.40, 25\,028.16,$ and $24\,642.07$. Assign these bands in sequences between v' and v'', prepare a Deslandres table, determine the average vibrational level separation in each state, find ω_e and $\omega_e x_e$ for both states, and calculate $\bar{\nu}_{00}$.

Ans. Band assignments are $v' \longrightarrow v''$ as follows: $1 \longrightarrow 0$, $0 \longrightarrow 0$, $2 \longrightarrow 3$, $1 \longrightarrow 2$, $0 \longrightarrow 1$, $2 \longrightarrow 4$, $1 \longrightarrow 3$, $0 \longrightarrow 2$, $3 \longrightarrow 6$, $2 \longrightarrow 5$, $1 \longrightarrow 4$, $0 \longrightarrow 3$. Separations are: $v' = 0$ to 1, $1\,983.47$ cm^{-1}; $v' = 1$ to 2, $1\,946.81$ cm^{-1}; $v'' = 0$ to 1, $1\,706.67$ cm^{-1}; $v'' = 1$ to 2, $1\,694.50$ cm^{-1}; $v'' = 2$ to 3, $1\,643.87$ cm^{-1}; $v'' = 3$ to 4, $1\,625.07$ cm^{-1}; and $v'' = 4$ to 5, $1\,598.59$ cm^{-1}. Plots of $\Delta \bar{\nu}$ against $2(v+1)$ give intercepts $\omega'_e = 2\,020.13$ cm^{-1} and $\omega''_e = 1\,739$ cm^{-1} and slopes $\omega'_e x'_e = 18.33$ cm^{-1} and $\omega''_e x''_e = 14.2$ cm^{-1}. Average value of $\bar{\nu}_{00}$ is $29\,676.6$ cm^{-1}.

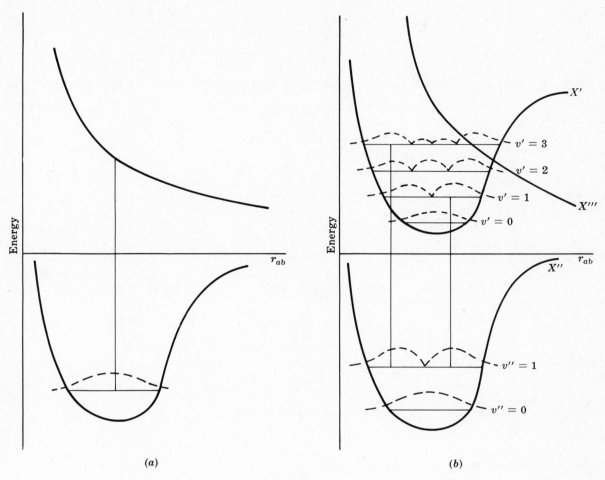

(a)   (b)

Fig. 17-4

Chapter 18

Electronic Structure of Polyatomic Molecules

Hybridization

18.1 ANGULAR WAVE FUNCTIONS

The concept of localized bonds around an atom in its ground electronic state, as presented for diatomic molecules, often fails to predict the correct formula or geometrical structure for a polyatomic molecule. This difficulty can be overcome by assuming that localized bonds are formed between atoms that have been excited to the extent of allowing the partially filled and unfilled atomic orbitals to mix to form the proper number of *hybrid atomic orbitals* having the correct geometrical arrangement.

Although various combinations of s, p, d, etc., wave functions could be used to construct the wave functions for the hybrid orbitals, only the s and p angular wave functions are necessary to describe most bonding between the representative elements. These functions are given by (*15.4*), (*15.5*), and (*14.49*) as

$$Y_s = \left(\frac{1}{4\pi}\right)^{1/2} \qquad Y_{p_z} = \left(\frac{3}{4\pi}\right)^{1/2} \cos\theta,$$

$$Y_{p_x} = \left(\frac{3}{4\pi}\right)^{1/2} \sin\theta\cos\phi \qquad Y_{p_y} = \left(\frac{3}{4\pi}\right)^{1/2} \sin\theta\sin\phi$$

The angular part of the ith hybrid orbital is then expressed as

$$X_i = a_i Y_s + b_i Y_{p_z} + c_i Y_{p_x} + d_i Y_{p_y} \tag{18.1}$$

where, by requirements of normalization and orthogonality,

$$a_i^2 + b_i^2 + c_i^2 + d_i^2 = 1 \tag{18.2a}$$

$$a_i a_j + b_i b_j + c_i c_j + d_i d_j = 0 \qquad (i \neq j) \tag{18.2b}$$

Additional restraints on the coefficients are determined by the amounts of s and p "character" desired in the hybrid orbital; e.g., $a_i^2 = b_i^2$ for sp.

EXAMPLE 18.1. The bond angles between hydrogens in H_2O, H_2S, H_2Se, and H_2Te are 104.45°, 92.2°, 91.0°, and 89.5°, respectively. What can be said about the importance of sp^3 hybridization, which predicts bonding angles of 109°28′, in describing all these molecules?

The importance decreases as the period number increases. The concept of hybridization approaches the notion of localized bonds around atoms in the ground states as the period number increases.

EXAMPLE 18.2. Give a molecular orbital description for NH_x.

The molecular orbital diagram in Fig. 18-1a predicts that three equivalent N—H bonds will be formed between the 2p atomic orbitals on the N and the 1s atomic orbitals on the H's. Even though the correct formula is predicted, these bonds would form H—N—H angles of 90°, which does not agree with the experimental value of 106.67°.

The molecular orbital diagram in Fig. 18-1b shows the N atom undergoing sp^3 hybridization before interacting with H atomic orbitals. Again three equivalent N—H bonds will be formed, but the H—N—H bond angles will be predicted as 109°28′ (see Problem 18.2), which agrees with the experimental value fairly well.

In diagrams like Fig. 18-1, the notation $n(\ )$ is used for *nonbonding* atomic orbitals.

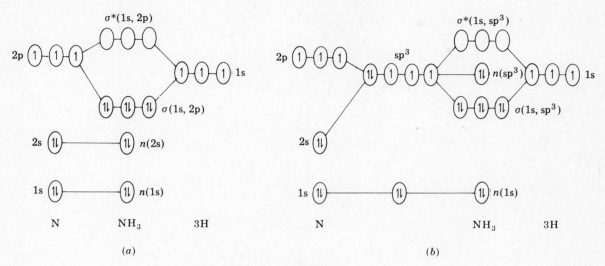

Fig. 18-1

18.2 RELATIVE BOND STRENGTH

The *relative bond strength* (rbs) of an orbital can be defined as

$$\text{rbs} \equiv \frac{\max X_i}{Y_s} = \frac{\max X_i}{(1/4\pi)^{1/2}} \tag{18.3}$$

where the numerator is the maximum value of X_i for that orbital. For an unmixed s atomic orbital, $X_i = Y_s$ and rbs = 1.

EXAMPLE 18.3. Determine the angular wave functions describing sp hybridization. Determine the relative bond strengths of these wave functions and the angle between the orbitals. Assume the bonding axis to be the x axis; i.e., contributions from p_y and p_z need not be considered.

From (18.1),

$$X_1 = a_1 Y_s + b_1 Y_{p_x} = a_1 \left(\frac{1}{4\pi}\right)^{1/2} + b_1 \left(\frac{3}{4\pi}\right)^{1/2} \sin\theta \cos\phi$$

For sp hybridization, the ratio of "p character" to "s character" must be unity; thus,

$$\frac{\displaystyle\int_{\text{all space}} b_1 Y_{p_x}^* b_1 Y_{p_x}\, dV}{\displaystyle\int_{\text{all space}} a_1 Y_s^* a_1 Y_s\, dV} = \frac{b_1^2 \displaystyle\int_{\text{all space}} Y_{p_x}^* Y_{p_x}\, dV}{a_1^2 \displaystyle\int_{\text{all space}} Y_s^* Y_s\, dV} = \frac{b_1^2}{a_1^2} = 1$$

or $a_1^2 = b_1^2$. A second relation between a_1 and b_1 is given by (18.2a) as

$$a_1^2 + b_1^2 = 1$$

Solving simultaneously gives $a_1 = \pm 2^{-1/2}$ and $b_1 = \pm 2^{-1/2}$. The positive value for a_1 is always chosen when determining the maximum value for X_1. The expression for X_1 becomes

$$X_1 = \left(\frac{1}{8\pi}\right)^{1/2}(1 \pm 3^{1/2}\sin\theta\cos\phi)$$

By inspection, the maximum value of X_1 occurs at $\theta = 90°$ and $\phi = 0°$ (along the positive x axis) if the plus sign is chosen or at $\theta = 90°$ and $\phi = 180°$ (along the negative x axis) if the minus sign is chosen. Equation (18.3) then gives

$$\text{rbs} = \frac{(1/8\pi)^{1/2}[1 \pm 3^{1/2}(1)(\pm 1)]}{(1/4\pi)^{1/2}} = 1.932$$

The coefficients for X_2 are found in a similar manner, using

$$\frac{\int_{\text{all space}} b_2 Y_{p_x}^* b_2 Y_{p_x} \, dV}{\int_{\text{all space}} a_2 Y_s^* a_2 Y_s \, dV} = \frac{b_2^2}{a_2^2} = 1 \qquad a_2^2 + b_2^2 = 1$$

which give $a_2 = \pm 2^{-1/2}$ and $b_2 = \pm 2^{-1/2}$. Similarly, the positive value of a_2 is chosen to determine the maximum of X_2, but the value of b_2 for these choices of a_1, a_2, and b_1 is restricted by (18.2b) as

$$b_2 = -\frac{a_1 a_2}{b_1} = -\frac{(2^{-1/2})(2^{-1/2})}{\pm 2^{-1/2}} = \mp 2^{-1/2}$$

giving

$$X_2 = \left(\frac{1}{8\pi}\right)^{1/2}(1 \mp 3^{1/2} \sin\theta \cos\phi)$$

The maximum for X_2 occurs at $\theta = 90°$ and $\phi = 180°$ (along the negative x axis) if the minus sign is chosen, or at $\theta = 90°$ and $\phi = 0°$ (along the positive x axis) if the plus sign is chosen. Hence,

$$\text{rbs} = \frac{(1/8\pi)^{1/2}[1 \mp 3^{1/2}(1)(\mp 1)]}{(1/4\pi)^{1/2}} = 1.932$$

The identity $\cos(\phi + 180°) = -\cos\phi$ shows that X_1 and X_2 are identical wave functions $180°$ apart. Usually X_1 is written using only the positive sign between the terms, and X_2 using the negative sign.

Localized Multiple Bonds

18.3 MOLECULAR ORBITAL THEORY

Molecular orbital theory predicts that for every two atomic orbitals used, two molecular orbitals will be generated. For head-on overlap, these are σ and σ^* orbitals; for parallel overlap, they are π and π^* orbitals. See Fig. 16-5.

EXAMPLE 18.4. Prepare a molecular orbital energy diagram similar to Fig. 18-1 representing the bonding in C_2H_2, and predict the types of electronic transitions that could occur.

The molecular orbitals shown in Fig. 18-2a are generated by using carbon atoms in their ground state and hydrogen atoms. If the carbon atoms are assumed to be sp hybridized, the molecular orbitals shown in Fig. 18-2b are formed. The dashed circles in Fig. 18-2b represent the original carbon orbitals used in hybridization. The hybridization steps in the diagram (see Fig. 18-1b) have been omitted for clarity.

Both diagrams predict a $\sigma^2 \pi^4$ configuration for the triple bond. The two lowest electronic transitions that can take place will be from the uppermost filled orbital, π, to the nearest unfilled orbitals, π^* and σ^*, giving the predicted spectrum as $\pi \longrightarrow \pi^*$ and $\pi \longrightarrow \sigma^*$ (weak). The $\sigma \longrightarrow \sigma^*$ transitions for the C—H bonds could also be observed (weak). Both diagrams predict a linear molecule.

Conjugated Bonds

18.4 CHAIN MOLECULES

For conjugated chain molecules, the *free-electron molecular orbital treatment* (FEMO) assumes the π electrons to be moving in a one-dimensional box (see Examples 14.4 and 14.7). No more than two electrons are allowed in a molecular orbital, whose energy is given by (14.19), so in the ground state the π electrons fill the lowest $n_\pi/2$ orbitals, where n_π is the number of π electrons. For this simple theory it can be shown that the transition from the highest occupied to the lowest vacant molecular

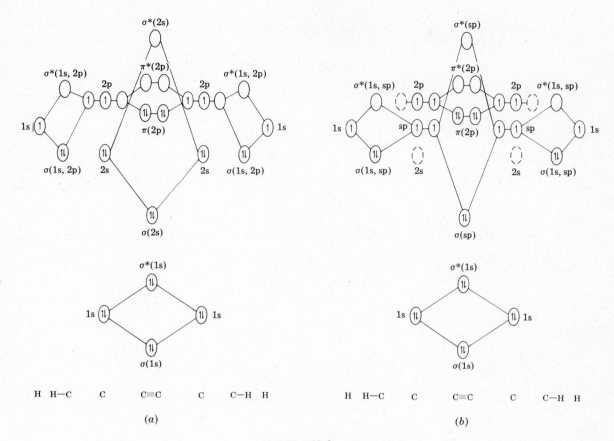

(a)   (b)

Fig. 18-2

orbital has an energy given (in cm^{-1}) by

$$\bar{\nu} = \frac{h(n_\pi + 1)}{8 m_e c a^2} \qquad (18.4)$$

where a is the length of the carbon chain. Better agreement between (18.4) and experiment is realized if a is defined as the zigzag distance along the carbon chain and allowance is made for the molecular orbitals to extend past the end carbons. This gives

$$\bar{\nu} = \frac{153\,000\ \text{cm}^{-1}}{n_c + 1} \qquad (18.5)$$

where $n_\pi = n_c$, the number of conjugated carbon atoms in the chain. An improved FEMO theory uses a sinusoidal potential-energy function with a minimum at the center of each double bond, giving

$$\bar{\nu} = \frac{153\,000\ \text{cm}^{-1}}{n_c + 1} + (16\,000\ \text{cm}^{-1})\left(1 - \frac{1}{n_c}\right) \qquad (18.6)$$

For polymethine ions having the formula

$$\text{R}\overset{+}{\text{N}}=\text{CH}(-\text{CH}=\text{CH})_k-\overset{..}{\text{N}}\text{R}$$

(18.4) can be shown to give

$$\bar{\nu} = \frac{(2k+5)(155\,000\ \text{cm}^{-1})}{(2k+4)^2} \qquad (18.7)$$

The *Hückel molecular orbital method* (HMO) assumes that the π bonding for n_c conjugated carbons is described by a secular equation of the form

$$
\begin{vmatrix}
x & 1 & 0 & 0 & \cdots & 0 & 0 & 0 \\
1 & x & 1 & 0 & \cdots & 0 & 0 & 0 \\
0 & 1 & x & 1 & \cdots & 0 & 0 & 0 \\
\vdots & & & & & & & \vdots \\
0 & 0 & 0 & 0 & \cdots & 1 & x & 1 \\
0 & 0 & 0 & 0 & \cdots & 0 & 1 & x
\end{vmatrix} = 0
\tag{18.8}
$$

The determinant has n_c rows and columns, and

$$
x = \frac{\alpha - E}{\beta}
\tag{18.9}
$$

In obtaining (18.8) from (14.28), it is assumed that the overlap integrals are $S_{jj} = 1$ and $S_{jk} = 0$; the coulombic integrals are $H_{jj} = \alpha$; and the resonance integrals are, for atoms bonded together, $H_{jk} = \beta$, and for nonbonded atoms, $H_{jk} = 0$. The solution to (18.8) is

$$
x = -2 \cos\left(\frac{j\pi}{n_c + 1}\right) \qquad \text{where } j = 1, 2, \ldots, n_c,
\tag{18.10}
$$

and, from (18.9),

$$
E_j = \alpha + 2\beta \cos\left(\frac{j\pi}{n_c + 1}\right)
\tag{18.11}
$$

Because β is negative, E_1 will be the lowest energy level, and the predicted absorption will be at

$$
\bar{\nu} = \frac{-4\beta}{kc} \sin\left(\frac{\pi}{2n_c + 2}\right)
\tag{18.12}
$$

where k has the same significance as in (18.7).

EXAMPLE 18.5. Calculate the energy levels for the conjugated bonding in 1,3-butadiene. Prepare an energy diagram showing the ground and first excited states.

Using (18.11) with $n_c = 4$ gives

$$
E_1 = \alpha + 2\beta \cos\frac{\pi}{5} = \alpha + 2\beta \cos 36° = \alpha + 1.618\,\beta
$$

$$
E_2 = \alpha + 2\beta \cos 72° = \alpha + 0.618\,\beta
$$

$$
E_3 = \alpha - 0.618\,\beta
$$

$$
E_4 = \alpha - 1.618\,\beta
$$

If the midline of Fig. 18-3 is α, then E_1 and E_2 lie below α, and E_3 and E_4 lie above. In the ground state the four electrons occupy the lowest two orbitals (see Fig. 18-3a), and one electron is promoted to E_3 for the first excited state (see Fig. 18-3b).

18.5 CYCLIC MOLECULES

The FEMO theory for a cyclic molecule assumes that π electrons have energies given by

$$
E_n = \frac{h^2 n^2}{2m_e a^2}
\tag{18.13}
$$

where a is the circumference of a circle drawn through the conjugated nuclei and $n = 0, \pm 1, \pm 2, \ldots$. Note that the energy levels above the ground state are doubly degenerate.

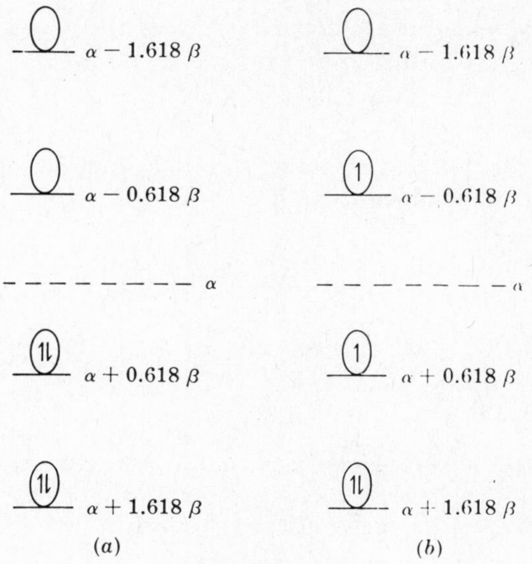

(a) (b)

Fig. 18-3

The HMO theory for a cyclic conjugated molecule uses a secular equation of the form

$$\begin{vmatrix} x & 1 & 0 & 0 & \ldots & 0 & 0 & 1 \\ 1 & x & 1 & 0 & \ldots & 0 & 0 & 0 \\ 0 & 1 & x & 1 & \ldots & 0 & 0 & 0 \\ \vdots & & & & & & & \vdots \\ 0 & 0 & 0 & 0 & \ldots & 1 & x & 1 \\ 1 & 0 & 0 & 0 & \ldots & 0 & 1 & x \end{vmatrix} = 0 \qquad (18.14)$$

whose solutions are

$$E_k = \alpha + 2\beta \cos\left(\frac{2\pi k}{n_c}\right) \qquad (18.15)$$

where $k = 0, 1, 2, \ldots, n_c - 1$. Because of the cyclic nature of the cosine term, values of E_k as given by (18.15) may be degenerate. A mnemonic device is available to determine the energies of the HMOs for C_nH_n. A vertex of the regular polygon of $n = n_c$ sides inscribed in a circle of radius 2β is placed at the bottom of the sketch, and an energy state is drawn beside it. Additional energy states are drawn corresponding to the other corners of the polygon, creating n energy states. The value of α in the diagram corresponds to the center of the polygon, and the value of $\alpha + 2\beta$ corresponds to the lowest energy level.

To have the extra stability associated with aromatic compounds, the number of π electrons must be

$$n_\pi = 4m + 2 \qquad (18.16)$$

where $m = 0, 1, 2, 3, \ldots$. If $n_\pi = 4m \pm 1$, the compound is a free radical with a singlet ground state; and if $n_\pi = 4m$, the compound is a diradical with a triplet ground state.

18.6 BOND ORDER AND LENGTH

The *total bond order* (P_{rs}) between two atoms, r and s, including a bond order of 1 for the σ bond, is given by

$$P_{rs} = 1 + p_{rs} \qquad (18.17)$$

where the π-electron bond order (p_{rs}) is given by

$$p_{rs} = \sum_j \frac{n_j}{2} (C_{jr}^* C_{js} + C_{js}^* C_{jr}) \qquad (18.18)$$

where n_j represents the number of electrons in the jth orbital and the sum is over all the π molecular orbitals. The values of C_{jr} are determined from

$$C_{jr} = \left(\frac{2}{n_c+1}\right)^{1/2} \sin\left(\frac{jr\pi}{n_c+1}\right) \qquad (18.19a)$$

for chain molecules and from

$$C_{jr} = \left(\frac{1}{n_c}\right)^{1/2} e^{i2\pi jr/n_c} \qquad (18.19b)$$

for cyclic molecules. The bond length (in nm) is given by

$$r_{rs} = 0.170\,7 - 0.018\,6 P_{rs} \qquad (18.20)$$

Coordination Compounds

18.7 VALENCE BOND THEORY

The *valence bond theory* assumed that coordinate-covalent bonds are formed between the ligands and the central atom, with the central atom making available the proper number of orbitals (equal to the coordination number). *Inner-orbital complexes* are those in which incompletely filled d orbitals are used for bonding, forming *inert complexes* (those that undergo slow ligand substitution); *outer-orbital complexes* use outer d orbitals for bonding, forming *labile complexes* (those that undergo rapid ligand substitution). The shape of the complex is determined by the orbitals used: sp^3 is tetrahedral, sp^2d is square planar, sp^3d or spd^3 is trigonal bipyramidal, sp^2d^2 or sd^4 is square pyramidal, and sp^3d^2 is octahedral. The major contribution of the valence bond theory is in predicting shapes of molecules and the number of unpaired electrons.

EXAMPLE 18.6. The ligands F^- and CN^- are quite different in their abilities to form labile and inert complexes. F^- is a weak ligand and CN^- a strong ligand, forming outer- and inner-orbital complexes, respectively. Prepare a diagram showing the configuration of CoF_6^{3-} and $Co(CN)_6^{3-}$, and predict the number of unpaired electrons.

The electronic configuration of Co^{3+} is such that the outer orbitals can be illustrated as in Fig. 18-4a(1). The six pairs of bonding electrons from the weak ligand simply fill the 4s, 4p(3), and 4d(2) orbitals, producing an outer-orbital complex [see Fig. 18-4a(2)], with four unpaired electrons on the Co. The six pairs of bonding electrons and the strong ligand pair up the 3d electrons of Co^{3+} and fill the 3d(2), 4s, and 4p(3) orbitals, producing an inner-orbital complex [see Fig. 18-4a(3)] with no unpaired electrons. In both cases sp^3d^2 orbitals are being used by the ligands, giving an octahedral shape.

18.8 CRYSTAL FIELD THEORY

The basis of this theory is that the degeneracy of the d orbitals on the central atom is removed as the ligands are placed on the molecule. Figure 18-5 gives the crystal field splitting for several geometric configurations. Though the actual separations between levels depend on the strength of the ligand, the separations between the orbitals and the spherical field ion always have the relative magnitudes indicated by the vertical numbers in Fig. 18-5. If a strong ligand is used, the magnitude of the CF splitting will be large and the lower-lying orbitals will fill first, giving rise to "low-spin" complexes. If a weak ligand is used, the magnitude of the splitting will be small, giving rise to "high-spin" complexes because all orbitals will fill with parallel spins before a second electron is placed in an orbital.

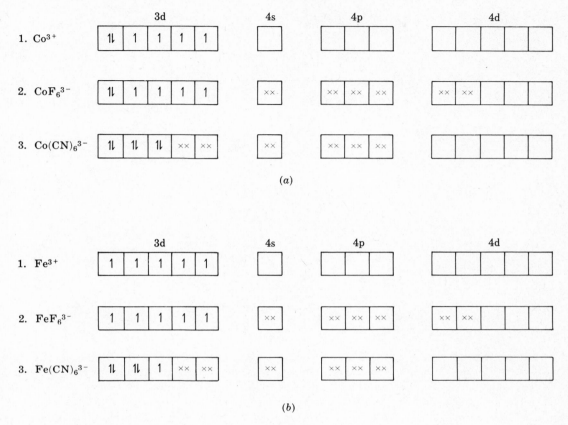

Fig. 18-4

The differences between these energy levels lie near the visible region of the spectrum, and many of these compounds are highly colored. For a d^1 configuration in an octahedral field, the CF splitting is shown in Fig. 18-6. The ^2D on the vertical axis is the Russell–Saunders term for the free ion, and at the right are the permitted components in the octahedral field—namely, E_g (doubly degenerate) and T_{2g} (triply degenerate). The allowed absorption transition is $^2T_{2g} \longrightarrow \ ^2E_g$. The corresponding diagram for a d^6 configuration is similar to Fig. 18-6 except that the terms are ^5D, ^5E_g, and $^5T_{2g}$.

EXAMPLE 18.7. $Cu(NH_3)_4 \cdot 2H_2O^{2+}$ absorbs at the wavelength 600 nm (orange), and $Cu(H_2O)_4 \cdot 2H_2O^{2+}$ absorbs at 800 nm (red). Compare the relative strengths of the ligands, and predict the colors of the solutions containing these ions.

Because energy is inversely proportional to λ [see (*14.1*) and (*14.3*)], the NH_3 has separated the energy levels on Cu^{2+} more than the H_2O has, making it the stronger ligand. A substance that absorbs orange light will appear as blue-purple, and red light absorption will result in a blue solution.

18.9 MOLECULAR ORBITAL THEORY FOR COMPLEXES

The molecular orbitals shown in Fig. 18-7 for an octahedral complex can be constructed by combining atomic orbitals of the central atom and the bonding orbitals of the ligands. Note that MO theory allows for π bonding using another energy diagram. The energy separation between the nonbonding T_{2g} and antibonding E_g^* levels will determine whether high- or low-spin complexes will result.

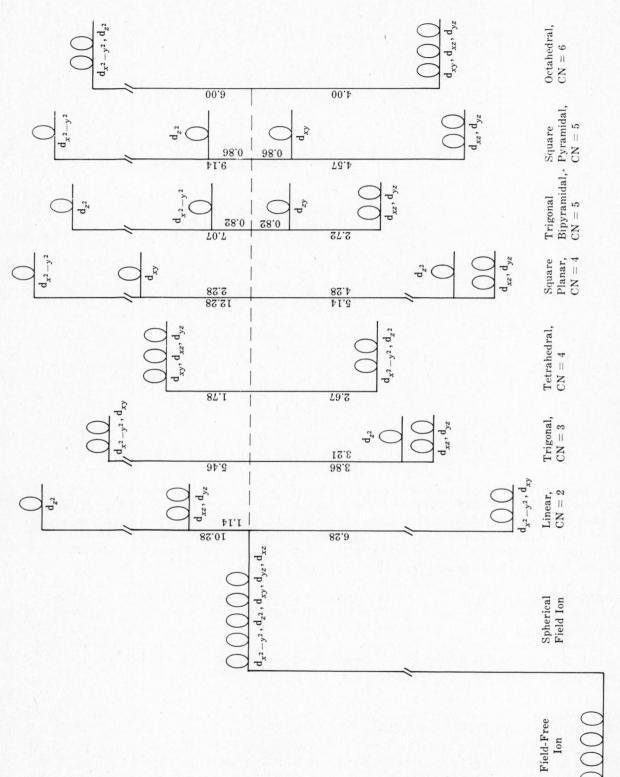

Fig. 18-5

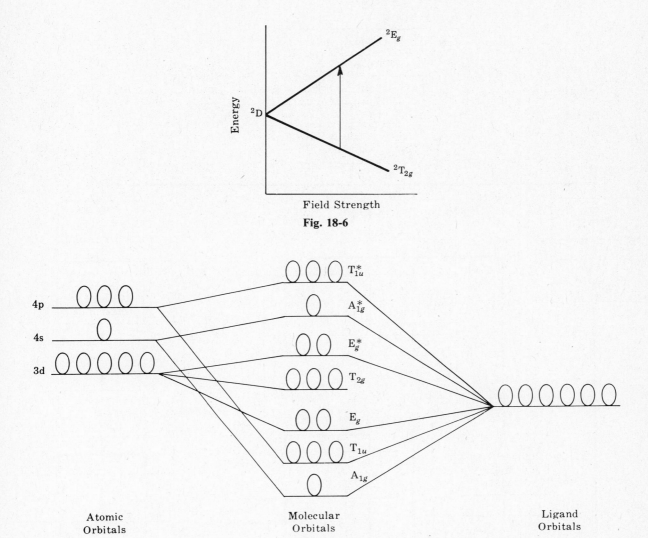

Fig. 18-6

Fig. 18-7

Spatial Relationships

18.10 INTRODUCTION

The spatial arrangement of atoms and electrons in many covalently bonded molecules or ions can be predicted correctly using the scheme in Fig. 18-8. In order to find the molecular geometry, the number of available electrons (AE) and the number of electrons needed to construct a molecule containing only single bonds (NE) are calculated. Depending on the relationship of AE to NE, one of the three major procedures shown in Fig. 18-8 will determine the correct Lewis structure(s) (Sec. 18.11), the structure number (SN_X) for each atom, the correct three-dimensional sketch, and the molecular geometry.

In calculating AE, only the valence electrons are considered, not the lower-lying kernel electrons. Thus,

$$AE = \sum_X (\text{group number})_X - (\text{charge on species}) \qquad (18.21)$$

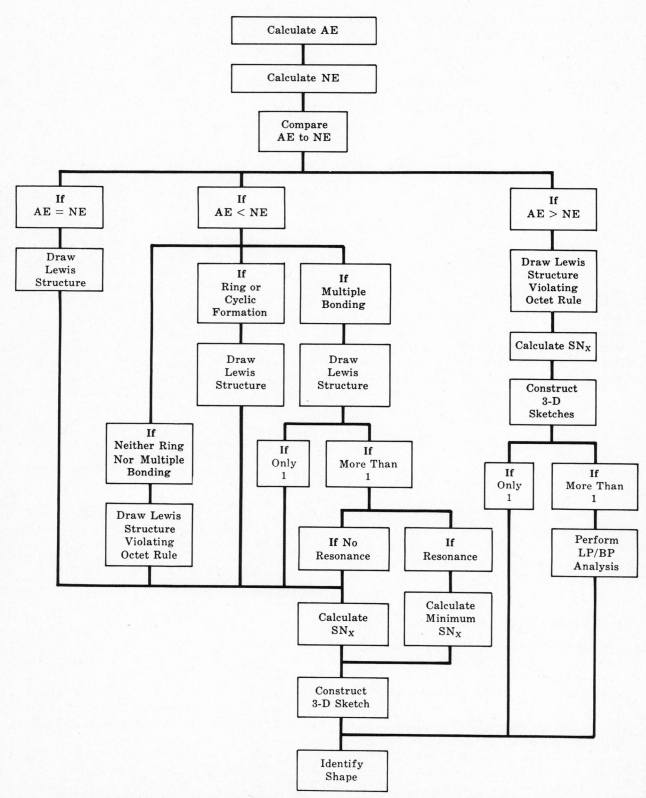

Fig. 18-8 (After C. R. Metz, "Molecular Geometry and Bonding," STRC-352, *Modular Laboratory Program*, Chemical Education Resources, Inc., Palmyra, PA, 1988.)

where the summation of the periodic table group numbers for atoms X is carried out for every atom present. The group number for transition metals in complexes is rather loosely interpreted as the oxidation state for the metal rather than the actual group number.

Assuming that each hydrogen present in the species needs two valence electrons and that each nonhydrogen needs eight valence electrons, the number of electrons needed is given by

$$NE = 2(\text{number of H atoms}) + 8(\text{number of non-H atoms}) - 2(\text{number of atoms} - 1) \qquad (18.22)$$

where the last term corrects for the number of shared electrons assuming that the entire molecule (or ion) contains only single bonds.

18.11 LEWIS STRUCTURES

The *Lewis structure* for the species under consideration is drawn as an aid in determining the structure number for each atom in the species. The *kernel* of the atom (the nucleus and inner electrons) is represented by the elemental symbol for the atom. Dots are used to represent the electrons in the outermost orbital, and dashes are used to represent pairs of bonding electrons.

If $AE = NE$, the species is said to be *saturated*. In such a species, all bonds are single bonds and all "octets" are completely filled.

If $AE < NE$, an electron deficiency exists in the species. There are three ways in which an electron deficiency can be satisfied—(1) formation of a cyclic or ring structure, (2) multiple bonding, or (3) violation of the "octet" rule for one or more atoms in the bond—with the proper choice depending on the information known for the species. Some elements, particularly C, N, O, and S, will form double bonds to eliminate a two-electron deficiency and triple bonds to eliminate a four-electron deficiency. These multiple bonds are formed using hybridized orbitals to create a σ bond, and unhybridized atomic orbitals with parallel overlap to create one or two π bonds, between the atoms.

For some substances, more than one Lewis structure can be drawn that satisfies the data for AE and NE. To determine whether the various structures all contribute to the molecular geometry, the *formal charge* (FC_X) given by

$$FC_X = (\text{group number})_X - [2(LP) + \tfrac{1}{2}(\text{number of shared electrons})]_X \qquad (18.23)$$

is calculated for each atom in the proposed structures, where LP is the number of lone or unshared pairs of electrons around atom X. The structures for which the sum of the absolute values of FC_X is small are usually the important structures to consider, and those with higher formal charge are discarded. When more than one Lewis structure having the same skeleton is acceptable based on values of FC_X, the contributing forms are drawn with double-headed arrows placed between them to indicate that the bonding and spatial arrangement is somewhere between the limiting structures (*resonance*).

If $AE > NE$, the Lewis structure is drawn showing only saturated bonding, and the "octet" rule for one or more of the atoms is violated.

18.12 STRUCTURE NUMBER AND SHAPE

The structure number for atom X (SN_X) is given by

$$SN_X = (\text{number of bonds})_X + LP_X \qquad (18.24)$$

An unpaired electron or a multiple bond in the Lewis structure counts as one bond in (18.24).

The value of SN_X corresponds to a particular geometrical arrangement of the electron pairs around atom X as shown in Table 18-1. After a three-dimensional sketch has been prepared, the shape of the species is determined by the arrangement of the atoms in the molecule.

In those cases where $AE < NE$ and resonance has been determined to exist, SN_X is calculated for each atom in each structure and the minimum value is used. The three-dimensional sketch is then drawn using these minimum SN_X's, and the remaining electrons are assumed to be in the parallel overlapping orbitals forming extended π bonding.

Table 18-1

Structure Number SN_X	Geometric Shape for X	Hybridization	Sketch of Shape	Molecular Shape Possibilities	
1	Spherical	—			
2	Linear	sp		LP = 0, BP = 2 LP = 1, BP = 1	linear linear
3	Triangular	sp^2		LP = 0, BP = 3 LP = 1, BP = 2 LP = 2, BP = 1	triangular bent (angle ~ 120°) linear
4	Tetrahedral	sp^3		LP = 0, BP = 4 LP = 1, BP = 3 LP = 2, BP = 2 LP = 3, BP = 1	tetrahedral trigonal pyramidal bent (angle ~ 109°) linear
4	Square planar	dsp^2 or sp^2d		LP = 0, BP = 4	square planar
5	Trigonal bipyramidal	dsp^3 or sp^3d		LP = 0, BP = 5 LP = 1, BP = 4 LP = 2, BP = 3 LP = 3, BP = 2 LP = 4, BP = 1	trigonal bipyramidal seesaw T-shaped linear linear
6	Octahedral	d^2sp^3 or sp^3d^2		LP = 0, BP = 6 LP = 1, BP = 5 LP = 2, BP = 4	octahedral square pyramidal square planar
7	Pentagonal bipyramidal	...			
8	Cubic	...			
8	Square antiprismal	...			

Source: After C. R. Metz, "Molecular Geometry and Bonding," STRC-352, *Modular Laboratory Program*, Chemical Education Resources, Inc., Palmyra, PA, 1988.

In those cases where AE > NE and it is possible to have more than one three-dimensional isomer, an analysis of the lone-lone pair, lone-bonded pair, and bonded-bonded pair interactions of electrons for angles up to and including 90° can often predict the correct shape. This qualitative analysis is based on the assumption that LP-LP interactions are less favored than LP-BP interactions, which, in turn, are less favored than BP-BP interactions.

Solved Problems

HYBRIDIZATION

18.1. Give a molecular orbital description for H_xO.

The MO diagram shown in Fig. 18-9a predicts two equivalent H—O bonds 90° apart if direct overlap of atomic hydrogen and oxygen orbitals occurs. Although predicting the correct formula of the compound, the predicted bond angle does not agree with the experimental value of 104.45°. If the oxygen is assumed to undergo sp^3 hybridization before interacting with the hydrogens (see Fig. 18-9b), two equivalent bonds are again predicted but at an approximate angle of 109°28′, which agrees fairly well with the experimental value.

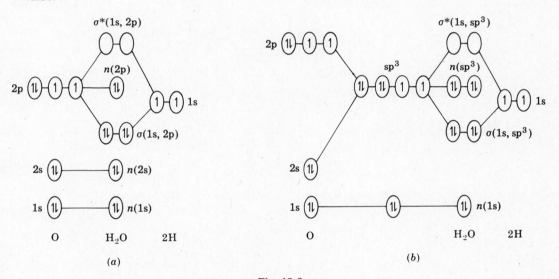

Fig. 18-9

18.2. Determine the angular wave functions describing the four hybrid orbitals representing sp^3 hybridization. Determine the relative bond strengths of these wave functions and the angles between the orbitals.

Assuming the first hybrid orbital to lie along the z axis, (18.1) gives

$$X_1 = a_1 Y_s + b_1 Y_{p_z} = a_1 \left(\frac{1}{4\pi}\right)^{1/2} + b_1 \left(\frac{3}{4\pi}\right)^{1/2} \cos\theta$$

because Y_{p_x} and Y_{p_y} are zero along this axis. Recognizing that the ratio of "p character" to "s character" requires that

$$\frac{\displaystyle\int_{\text{all space}} b_1 Y_{p_z}^* b_1 Y_{p_z}\, dV}{\displaystyle\int_{\text{all space}} a_1 Y_s^* a_1 Y_s\, dV} = \frac{b_1^2}{a_1^2} = 3$$

or $b_1^2 = 3a_1^2$, and that normalization, ($18.2a$), requires that

$$a_1^2 + b_1^2 = 1$$

the values of the coefficients are $a_1 = \pm\frac{1}{2}$ and $b_1 = \pm 3^{1/2}/2$. Choosing the positive value for a_1 so that the maximum of X_1 can be considered,

$$X_1 = \frac{1}{2}\left(\frac{1}{4\pi}\right)^{1/2}(1 \pm 3\cos\theta)$$

The maximum of X_1 will lie along the z axis (where $\theta = 0°$ or $180°$), giving

$$\text{rbs} = \frac{(1/2)(1/4\pi)^{1/2}[1 \pm 3(\pm 1)]}{(1/4\pi)^{1/2}} = 2.000$$

It is possible to retain the $\pm$ sign in X_1 as in Example 18.3, but traditionally only the positive sign is used (i.e., $b_1 = +3^{1/2}/2$), thus assigning X_1 to lie along the positive z axis. This assignment is then reflected in the choice of other coefficients, and the extraneous coefficients are discarded.

Choosing the second hybrid orbital to lie in the xz plane, (18.1) gives

$$X_2 = a_2 Y_s + b_2 Y_{p_z} + c_2 Y_{p_x} = a_2\left(\frac{1}{4\pi}\right)^{1/2} + b_2\left(\frac{3}{4\pi}\right)^{1/2}\cos\theta + c_2\left(\frac{3}{4\pi}\right)^{1/2}\sin\theta\cos\phi$$

The amount of p character requires that

$$\frac{\displaystyle\int_{\text{all space}} (b_2 Y_{p_z} + c_2 Y_{p_x})^*(b_2 Y_{p_z} + c_2 Y_{p_x})\,dV}{\displaystyle\int_{\text{all space}} a_2 Y_s^* a_2 Y_s\,dV} = 3$$

which because of the orthogonality of Y_{p_z} and Y_{p_x} simplifies to

$$\frac{b_2^2 + c_2^2}{a_2^2} = 3$$

Combining this result with ($18.2a$) gives

$$a_2^2 + 3a_2^2 = 1$$

or $a_2 = \pm\frac{1}{2}$ and $b_2^2 + c_2^2 = \frac{3}{4}$. Because X_1 and X_2 are orthogonal, ($18.2b$) gives

$$b_2 = -\frac{a_1 a_2}{b_1} = -\frac{(\frac{1}{2})(\frac{1}{2})}{+3^{1/2}/2} = -\frac{1}{2(3)^{1/2}}$$

and hence

$$c_2 = \pm[\tfrac{3}{4} - b_2^2]^{1/2} = \pm(\tfrac{2}{3})^{1/2}$$

The second wave function is then

$$X_2 = \left(\frac{1}{4\pi}\right)^{1/2}(\tfrac{1}{2} - \tfrac{1}{2}\cos\theta \pm 2^{1/2}\sin\theta\cos\phi)$$

The orientation of the maximum of the orbital can be found from

$$0 = \frac{\partial X_2}{\partial \theta} = \left(\frac{1}{4\pi}\right)^{1/2}(0 + \tfrac{1}{2}\sin\theta \pm 2^{1/2}\cos\theta\cos\phi)$$

which upon rearrangement gives

$$\tan\theta = \mp 2^{3/2}\cos\phi$$

For $\phi = 0°$ (toward the positive x axis), $\tan\theta = \mp 2^{3/2}$, giving $\theta = \mp 70°32'$ or $\pm 109°28'$; and for $\phi = 180°$ (toward the negative x axis), $\tan\theta = \pm 2^{3/2}$, giving $\theta = \pm 70°32'$ or $\mp 109°28'$. The negative values for θ are discarded because $0° \leq \theta \leq 180°$, and the values of $70°32'$ are discarded because they are too near X_1. Thus $\theta = 109°28'$. Depending on the choice of the sign for c_2, the maximum will lie at an angle of $109°28'$ from the maximum of X_1, at either $\phi = 0$ or $\phi = 180°$. Normally c_2 is chosen as positive, giving

$$X_2 = \left(\frac{1}{4\pi}\right)^{1/2}\left(\frac{1}{2} - \frac{1}{2}\cos\theta + 2^{1/2}\sin\theta\cos\phi\right)$$

which has an rbs of

$$\text{rbs} = \frac{(1/4\pi)^{1/2}[(1/2) - (1/2)\cos 109°28' + 2^{1/2}\sin 109°28'\cos 0°]}{(1/4\pi)^{1/2}} = 2.000$$

Likewise it can be shown that

$$X_3 = \left(\frac{1}{4\pi}\right)^{1/2}\left[\frac{1}{2} - \frac{1}{2}\cos\theta + \left(\frac{3}{2}\right)^{1/2}\sin\theta\cos\phi - \left(\frac{1}{2}\right)^{1/2}\sin\theta\sin\phi\right]$$

$$X_4 = \left(\frac{1}{4\pi}\right)^{1/2}\left[\frac{1}{2} - \frac{1}{2}\cos\theta - \left(\frac{3}{2}\right)^{1/2}\sin\theta\cos\phi - \left(\frac{1}{2}\right)^{1/2}\sin\theta\sin\phi\right]$$

each with an rbs of 2.000.

18.3. Prepare a plot in the xz plane of $X = (1/2)(1/4\pi)^{1/2}(1 + 3\cos\theta)$, the sp^3 orbital along the z axis.

Figure 18-10a is the polar plot of $|X|$ as a function of θ. On the larger lobe X is positive; on the smaller it is negative.

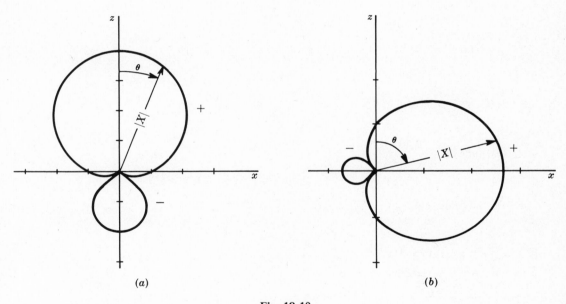

(a)   (b)

Fig. 18-10

LOCALIZED MULTIPLE BONDS

18.4. Prepare a molecular orbital energy diagram representing the bonding in C_2H_4, and predict the types of electronic transitions that could occur.

See Fig. 18-11 for diagrams based (a) on a ground-state carbon atom and (b) on an sp^2 hybridized carbon atom. The most probable transitions are $\pi \longrightarrow \pi^*$ and $\pi \longrightarrow \sigma^*$ (weak) within the C=C double bond and $\sigma \longrightarrow \sigma^*$ (weak) within the C—H bond. Figure 18-11a incorrectly predicts H—C—C angles of 90°, and Fig. 18-11b correctly predicts H—C—C angles of 120°.

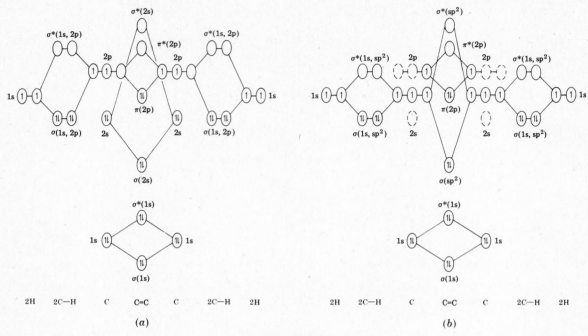

Fig. 18-11

CONJUGATED BONDS

18.5. Calculate $\bar{\nu}$ for the absorption band of $H—(CH=CH—)_k—H$ for $k = 1$ and 10, using the FEMO and improved FEMO theories.

For $H—CH=CH—H$, (18.5) gives

$$\bar{\nu} = \frac{153\,000 \text{ cm}^{-1}}{2+1} = 51\,000 \text{ cm}^{-1}$$

and (18.6) gives

$$\bar{\nu} = \frac{153\,000 \text{ cm}^{-1}}{2+1} + (16\,000 \text{ cm}^{-1})\left(1 - \frac{1}{2}\right) = 59\,000 \text{ cm}^{-1}$$

The observed value is $61\,500 \text{ cm}^{-1}$. Likewise for $H—(CH=CH—)_{10}—H$, (18.5) gives $7\,300 \text{ cm}^{-1}$ and (18.6) gives $22\,500 \text{ cm}^{-1}$, the latter agreeing quite well with the experimental value of $22\,400 \text{ cm}^{-1}$.

18.6. Calculate the energy levels for the conjugated bonding in benzene, and prepare an energy diagram showing the ground state.

Equation (18.14) becomes, for six atoms,

$$\begin{vmatrix} x & 1 & 0 & 0 & 0 & 1 \\ 1 & x & 1 & 0 & 0 & 0 \\ 0 & 1 & x & 1 & 0 & 0 \\ 0 & 0 & 1 & x & 1 & 0 \\ 0 & 0 & 0 & 1 & x & 1 \\ 1 & 0 & 0 & 0 & 1 & x \end{vmatrix} = 0$$

and using (18.15) gives

$$E_0 = \alpha + 2\beta \cos(0) = \alpha + 2\beta \qquad E_1 = \alpha + 2\beta \cos(2\pi/6) = \alpha + \beta$$

$$E_2 = \alpha - \beta \qquad E_3 = \alpha - 2\beta \qquad E_4 = \alpha - \beta \qquad E_5 = \alpha + \beta$$

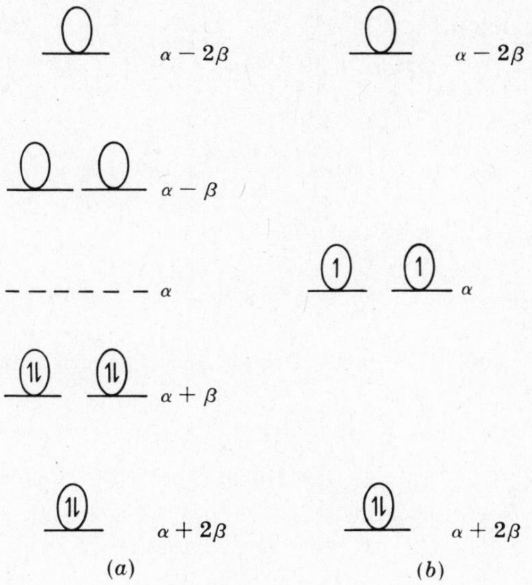

Fig. 18-12 energy diagram with levels labeled (top to bottom): $\alpha - 2\beta$, $\alpha - \beta$, α, $\alpha + \beta$, $\alpha + 2\beta$ for columns (a) and (b).

Fig. 18-12

These are sketched in Fig. 18-12a. Because E_1 and E_5, as well as E_2 and E_4, are degenerate, only four energy levels are present.

18.7. Determine the energy levels in C_6H_6 for the conjugated bonding using the mnemonic device, and compare the results to those found in Problem 18.6.

Placing the hexagon on a vertex and drawing energy levels parallel to each atom gives the system shown in Fig. 18-13a. The spacings are the same as in Fig. 18-12a.

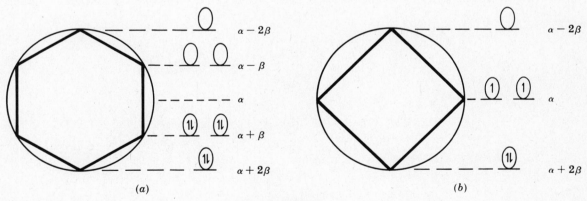

Fig. 18-13

18.8. Predict the aromaticity of C_6H_6.

Benzene satisfies (18.16) with $m = 1$, so C_6H_6 is predicted to have extra stability resulting from "resonance energy" associated with aromatic compounds.

18.9. Find P_{rs} and r_{rs} for the 1—2 and 3—4 bonds in 1,3-butadiene.

Using (*18.19a*) gives for atom 1

$$C_{11} = \left(\frac{2}{5}\right)^{1/2} \sin\left[\frac{(1)(1)\pi}{5}\right] = (0.633)\sin(36°) = 0.371$$

$$C_{21} = (0.633)\sin\left[\frac{(2)(1)\pi}{5}\right] = 0.601$$

$C_{31} = 0.601$, and $C_{41} = 0.371$. Likewise for atom 2,

$$C_{12} = (0.633)\sin\left[\frac{(1)(2)\pi}{5}\right] = 0.601$$

$C_{22} = 0.371$, $C_{32} = -0.371$, and $C_{42} = -0.601$. Using the electronic distribution shown in Fig. 18.3*a*, (*18.18*) gives

$$p_{12} = \tfrac{2}{2}[(0.371)(0.601) + (0.601)(0.371)] + \tfrac{2}{2}[(0.601)(0.371) + (0.371)(0.601)]$$

$$+ \tfrac{0}{2}[(0.601)(-0.371) + (-0.371)(0.601)] + \tfrac{0}{2}[(0.371)(-0.601) + (-0.601)(0.371)]$$

$$= 0.893$$

and (*18.17*) gives

$$P_{12} = 1 + 0.893 = 1.893$$

The bond length for this value of P_{12} is given by (*18.20*) as

$$r_{12} = 0.017\,07 - (0.018\,6)(1.893) = 0.135\,5 \text{ nm}$$

The observed value is 0.134 nm.

COORDINATION COMPOUNDS

18.10. Prepare sketches of the electronic configurations of CoF_6^{3-}, $Co(CN)_6^{3-}$, and $CoCl_4^{2-}$ (tetrahedral) using the crystal field splittings shown in Fig. 18-5.

The six electrons in Co^{3+} will be arranged as shown in Fig. 18-14 for the weak octahedral complex formed by the F^- and as shown in Fig. 18-14*b* for the strong octahedral complex formed by the CN^-. The seven d electrons in Co^{2+} will be arranged as shown in Fig. 18-14*c* for the weak tetrahedral complex.

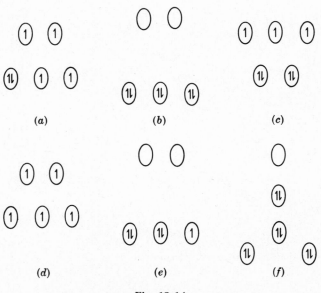

Fig. 18-14

18.11. $Fe(H_2O)_6^{2+}$ is a high-spin complex that absorbs light at about $1\,000$ nm, corresponding to a transition between the $^5T_{2g}$ and ^5E_g levels. What is the energy between these levels? Predict the color of this ion.

Using (*14.2*) gives

$$\bar{\nu} = \frac{1}{1\,000 \times 10^{-9}\,\text{m}}\,[(10^{-2}\,\text{cm}^{-1})/(1\,\text{m})] = 10\,000\,\text{cm}^{-1}$$

The $1\,000$-nm absorption peak is in the very near infrared region, and part of the absorption peak falls into the red visible region. Thus the ion will absorb some red, giving a pale blue-green color.

18.12. Sketch the molecular orbital diagram for CoF_6^{3-}.

F^-, a weak ligand, will not separate the T_{2g} and E_g^* levels sufficiently to produce electron pairing, so the high-spin complex shown in Fig. 18-15*a* results.

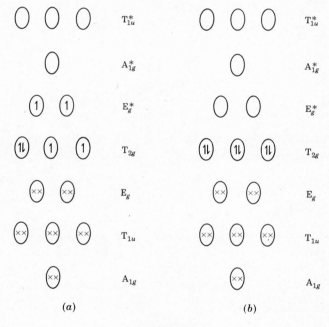

(*a*)   (*b*)

Fig. 18-15

SPATIAL RELATIONSHIPS

18.13. Determine the shapes of the following species: (*a*) $n\text{-}C_8H_{18}$, (*b*) BCl_3, (*c*) S_8, (*d*) C_2H_4, (*e*) C_2H_2, (*f*) HCNO, (*g*) N_2O, (*h*) $CrCl_6^{3-}$, and (*i*) XeF_4.

The Lewis diagrams are given in Fig. 18-16, and the three-dimensional sketches are given in Fig. 18-17. The information used to determine the structures is given below.

(a) $n\text{-}C_8H_{18}$:
$$AE = [(8)(4) + (18)(1)] - (0) = 50$$
$$NE = 2(18) + 8(8) - 2(26 - 1) = 50$$
$$AE = NE$$
$$SN_H = (1) + (0) = 1$$
$$SN_C = (4) + (0) = 4$$

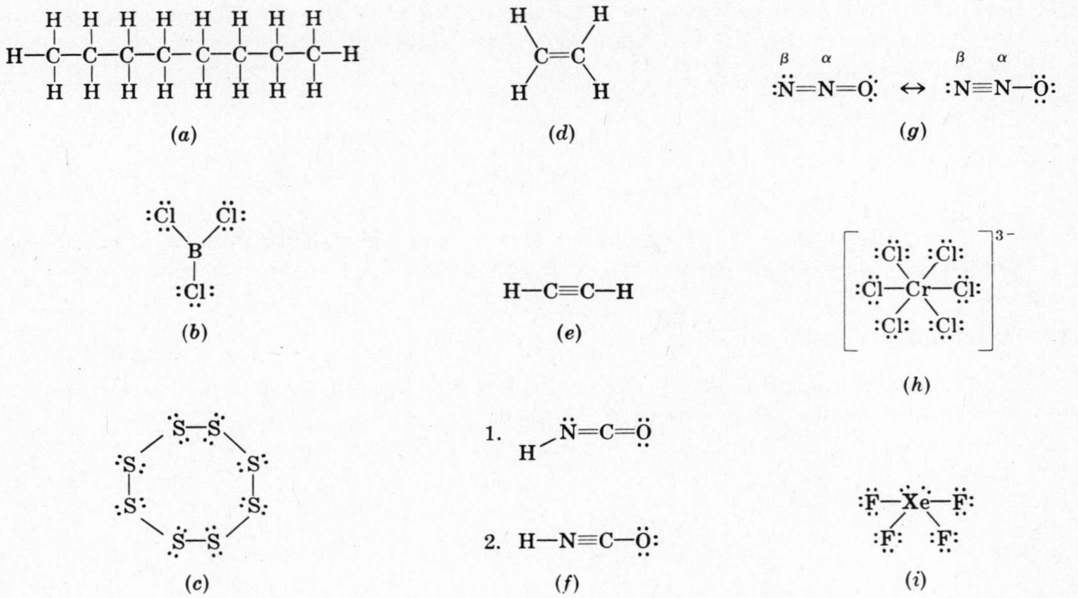

Fig. 18-16

There are four bonded pairs of electrons located tetrahedrally around each C atom, giving the entire molecule many shapes depending on rotation of the tetrahedrons with respect to each other.

(b) BCl_3:
$$AE = [(3) + (3)(7)] - (0) = 24$$
$$NE = 2(0) + 8(4) - 2(4 - 1) = 26$$
$$AE < NE$$

BCl_3 is experimentally known to not contain significant multiple bonding and not to be cyclic; it is known to have three equivalent B—F bonds. On the basis of the electronegativities of F and B, the electron deficiency is assigned to the B.

$$SN_B = (3) + (0) = 3   SN_{Cl} = (1) + (3) = 4$$

This is a triangular molecule.

(c) S_8:
$$AE = [(8)(6)] - (0) = 48$$
$$NE = 2(0) + 8(8) - 2(8 - 1) = 50$$
$$AE < NE$$

As only one type of bonding is present, multiple bonding and branching are not present.

$$SN_S = (2) + (2) = 4$$

This is a "crown" with four atoms in one plane and four in another plane.

(d) C_2H_4:
$$AE = [(2)(4) + (4)(1)] - (0) = 12$$
$$NE = 2(4) + 8(2) - 2(6 - 1) = 14$$
$$AE < NE$$

A triple bond between the C's eliminates the deficiency of four electrons.

$$SN_H = (1) + (0) = 1$$
$$SN_C = (3) + (0) = 3$$

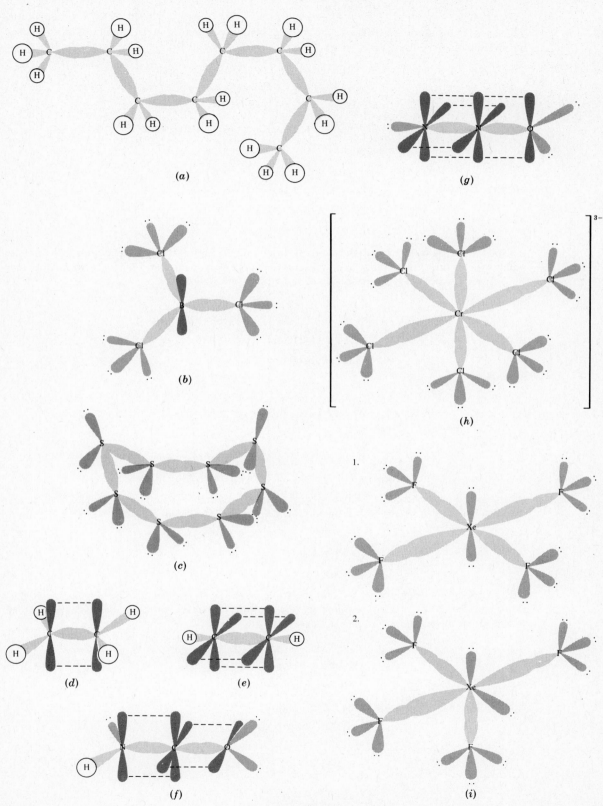

Fig. 18-17

This is a planar molecule because of the required parallel overlap of the p atomic orbitals used for the π bonding.

(e) C_2H_2:

$$AE = [(2)(4) + (2)(1)] - (0) = 10$$

$$NE = 2(2) + 8(2) - 2(4 - 1) = 14$$

$$AE < NE$$

A triple bond between the C's eliminates the deficiency of four electrons.

$$SN_H = (1) + (0) = 1 \qquad SN_C = (2) + (0) = 2$$

This is a linear molecule.

(f) HCNO:

$$AE = [(1) + (5) + (4) + (6)] - (0) = 16$$

$$NE = 2(1) + 8(3) - 2(4 - 1) = 20$$

$$AE < NE$$

A triple bond between the C and N (based on additional formal charge calculations, C—O triple bonds are very little favored) or two double bonds eliminates the deficiency of four electrons. Structure 1 is favored, as shown by the following formal charge calculations:

Structure 1

$$FC_H = (1) - [2(0) + \tfrac{1}{2}(2)] = 0 \qquad FC_N = (5) - [2(1) + \tfrac{1}{2}(6)] = 0$$

$$FC_C = (4) - [2(0) + \tfrac{1}{2}(8)] = 0 \qquad FC_O = (6) - [2(2) + \tfrac{1}{2}(4)] = 0$$

$$\text{Sum of absolute values} = 0$$

Structure 2

$$FC_H = (1) - [2(0) + \tfrac{1}{2}(2)] = 0 \qquad FC_N = (5) - [2(0) + \tfrac{1}{2}(8)] = 1$$

$$FC_C = (4) - [2(0) + \tfrac{1}{2}(8)] = 0 \qquad FC_O = (6) - [2(3) + \tfrac{1}{2}(2)] = -1$$

$$\text{Sum of absolute values} = 2$$

$$SN_H = (1) + (0) = 1 \qquad SN_N = (2) + (1) = 3$$

$$SN_C = (2) + (0) = 2 \qquad SN_O = (1) + (2) = 3$$

The preferred Lewis structure shown in Fig. 18-16f(1) predicts a nonlinear molecule. The atomic ordering HCNO can be eliminated by additional formal charge calculations.

(g) N_2O:

$$AE = [(2)(5) + (6)] - (0) = 16$$

$$NE = 2(0) + 8(3) - 2(3 - 1) = 20$$

$$AE < NE$$

Either a triple bond between the N's (again, triple bonds using O's are not favored) or two double bonds eliminates the deficiency of four electrons. Both structures have the same formal charge content and therefore contribute as resonance forms.

$$\text{Minimum } SN_{\alpha N} = (1) + (1) = 2$$

$$\text{Minimum } SN_{\beta N} = (2) + (0) = 2$$

$$\text{Minimum } SN_O = (1) + (2) = 3$$

This is a linear molecule with extended π bonding.

(h) $CrCl_6^{3-}$:

$$AE = [(3) + (6)(7)] - (-3) = 48$$

$$NE = 2(0) + 8(7) - 2(7 - 1) = 44$$

$$AE > NE$$

The "octet" rule is to be violated for the central Cr atom.

$$SN_{Cr} = (6) - (0) = 6 \qquad SN_{Cl} = (1) + (3) = 4$$

This is an octahedral molecule.

(i) XeF_4:

$$AE = [(8) + (4)(7)] - (0) = 36$$

$$NE = 2(0) + 8(5) - 2(5 - 1) = 32$$

$$AE > NE$$

The "octet" rule is to be violated for Xe.

$$SN_{Xe} = (4) + (2) = 6 \qquad SN_F = (1) + (3) = 4$$

There are two corresponding three-dimensional figures (see Fig. 18-17i). In the first structure there are 8 LP-BP and 4 BP-BP interactions at 90° on the central atom, and in the second structure there are 1 LP-LP, 6 LP-BP, and 5 BP-BP interactions at 90° on the central atom. On the basis of the LP-LP interaction, the first structure, square planar, is favored.

Supplementary Problems

HYBRIDIZATION

18.14.   Give a molecular orbital description for CH_x.

   Ans.   Figure 18-18a predicts CH_2 with H—C—H angle of 90°; Fig. 18-18b predicts CH_4 with one C—H bond of different energy and angles of 90° and 125°; Fig. 18-18c predicts (correctly) tetrahedral CH_4 with four equivalent bonds.

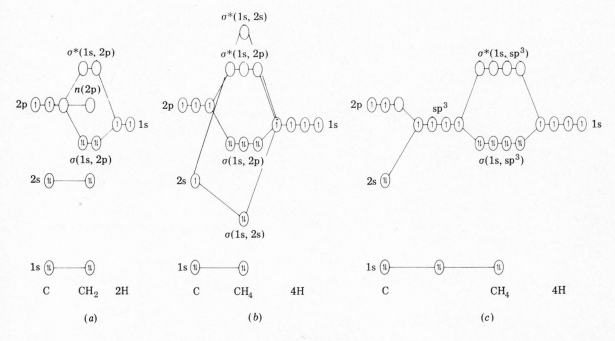

Fig. 18-18

18.15.   Represent sp^2 hybridization by means of three hybrid orbitals in the xy plane. Determine the relative bond strengths and the angles between the orbitals.

   Ans. $a_1 = +(1/3)^{1/2}, \; b_1 = +(2/3)^{1/2}, \; c_1 = 0; \; a_2 = +(1/3)^{1/2}, \; b_2 = -(1/6)^{1/2}, \; c_2 = +(1/2)^{1/2};$
   $a_3 = +(1/3)^{1/2}, \; b_3 = -(1/6)^{1/2}, \; c_3 = -(1/2)^{1/2}; \; 1.992; \; 120°$

18.16. Besides the set found in Problem 18.2, the following set of hybrid atomic orbitals can be written to describe the sp^3 orientation:

$$X_1 = \tfrac{1}{2}(Y_s + Y_{p_x} + Y_{p_y} + Y_{p_z}) \qquad X_3 = \tfrac{1}{2}(Y_s - Y_{p_x} + Y_{p_y} - Y_{p_z})$$
$$X_2 = \tfrac{1}{2}(Y_s + Y_{p_x} - Y_{p_y} - Y_{p_z}) \qquad X_4 = \tfrac{1}{2}(Y_s - Y_{p_x} - Y_{p_y} + Y_{p_z})$$

Show that these wave functions predict a relative bond strength of 2.000. Show that X_1 is normalized. Show that X_1 and X_2 are orthogonal.

18.17. Prepare a plot in the xz plane of $X = (1/8\pi)^{1/2}[1 + 3^{1/2}\sin\theta\cos\phi]$, the sp orbital along the positive x axis, and compare the plot to Fig. 18-10a.   *Ans.*   See Fig. 18-10b.

LOCALIZED MULTIPLE BONDS

18.18. Prepare molecular orbital energy diagrams for the carbonyl group

$$\diagdown\!\!\diagup C = \ddot{O} :$$

assuming the bonding to consist of a carbon atom that is sp^2 hybridized with (*a*) an O atom in its ground state and (*b*) an sp^2-hybridized O atom. If $\pi \longrightarrow \pi^*$, $n \longrightarrow \pi^*$, and $n \longrightarrow \sigma^*$ transitions are observed, which assumption correctly describes the bonding? Assume that the only contributions to the spectrum are from electrons making angles with the bonding axis of from 0° to 90°.

     Ans.   Nonhybridized oxygen, because $n(2p)$ is 90° from bonding axis and $n(sp^2)$ are 120° from bonding axis (see Fig. 18-19).

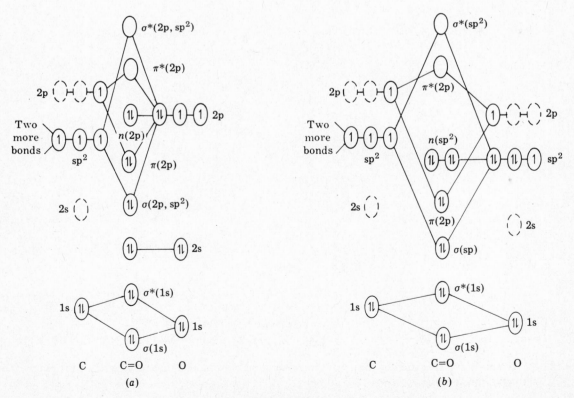

Fig. 18-19

CONJUGATED BONDS

18.19. An important compound for the sight process in animals is β-carotene, which has the formula

Predict the color of this compound.

Ans. $n_c = 22$ gives $\bar{\nu} = 21\,900$ cm^{-1} using the improved FEMO theory, $\lambda = 456.6$ nm; orange (coloring in carrots)

18.20. The three polymethine dyes 1,1'-diethyl-2,2'-cyanine iodide, 1,1'-diethyl-2,2'carbocyanine iodide, and 1,1'-diethyl-2,2'-dicarbocyanine iodide have the structural formulas

where $x = 0$, 1, and 2, respectively. (a) Predict the frequency of light at which the maximum absorbancy should occur. (b) Predict the color of the solutions of the substances if the actual frequencies are 520, 601, and 701 nm, respectively.

Ans.   (a) For the shortest path (direct movement between nitrogens) $k = x + 1$ in (*18.7*), giving 30 100, 21 800, and 17 100 cm^{-1}, which are too high; $k = x + 3$ for the intermediate path (movement between one nitrogen and around one inner ring), giving 17 100, 14 000, and 11 900 cm^{-1}, which are a little low; $k = x + 5$ for the longest path (movement between the nitrogens via both outer rings), giving 11 900, 10 300, and 9 100 cm^{-1}, which are too low.

   (b) Red-orange, red-violet, and green.

18.21. Assume that the FEMO theory correctly predicts the energies of the conjugated π electrons in benzene. Predict the values of λ at which absorption will occur for the transition between $n = 1$ and $n = 2$, which corresponds to the $(\alpha + \beta) \longrightarrow (\alpha - \beta)$ transition. Assume that the C—C bond length is 0.139 7 nm.

Ans. $a = 6(0.139\,7)$, $\lambda = 2m_e a^2 c/3h = 193$ nm; $a = 2\pi(0.139\,7)$, $\lambda = 212$ nm; observed is 203.8 nm.

18.22. The total energy of the six electrons in Fig. 18-13a is $2(\alpha + 2\beta) + 4(\alpha + \beta) = 6\alpha + 8\beta$. If π orbitals were used, the energy would be $6(\alpha + \beta) = 6\alpha + 6\beta$. The difference between these descriptions is the "resonance energy" of the molecule (see Problem 4.31). Evaluate β. (This value is only an estimate because of additional energy required to make all C—C bond lengths the same and because of the assumptions made in the HMO theory.)   *Ans.* -72.0 kJ mol^{-1}

18.23. Calculate the energy levels for the conjugated bonding in cyclobutadiene using (*18.15*), and prepare an energy diagram showing the ground state for C_4H_4. Repeat the calculations using the mnemonic device (Sec. 18.5), and compare results. Discuss the aromaticity of C_4H_4 using the $4m + 2$ rule.

Ans. $E_0 = \alpha + 2\beta$, $E_1 = E_3 = \alpha$, $E_2 = \alpha - 2\beta$, see Fig. 18-12b; same results, see Fig. 18-13b; diradical, which is very reactive

18.24. Using the mnemonic device (Sec. 18.5), sketch the ground-state configurations of $C_5H_5^+$, $C_5H_5\cdot$, and $C_5H_5^-$. Which is the most stable according to the $4m + 2$ rule?   *Ans.*   See Fig. 18-20. $C_5H_5^-$.

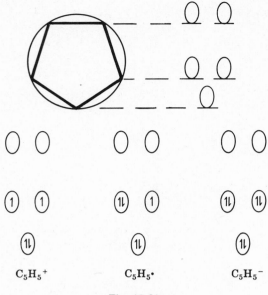

$C_5H_5^+$ $C_5H_5\cdot$ $C_5H_5^-$

Fig. 18-20

18.25. The bond order in naphthalene

between the 1 and 2 carbons is 1.725; between 2 and 3, 1.603; between 1 and 9, 1.555; and between 9 and 10, 1.518. Determine these bond lengths.   *Ans.* 0.138 6, 0.140 9, 0.141 8, and 0.142 5 nm

18.26. Find P_{rs} and r_{rs} for the 2—3 bond in 1,3-butadiene.

Ans. $C_{13} = C_{43} = 0.601$, $C_{23} = C_{33} = -0.371$, $p_{23} = 0.447$, $P_{rs} = 1.447$; 0.143 8 nm

COORDINATION COMPOUNDS

18.27. Predict the configurations, the number of unpaired electrons, and the shapes of the molecules for FeF_6^{3-} and $Fe(CN)_6^{3-}$.   *Ans.* See Fig. 18-4b; 5 and 1; octahedral for both.

18.28. A coordination compound having a ligancy of 4 can be either square planar or tetrahedral. Using the CF splitting shown in Fig. 18-5, predict the configuration for $Ni(CN)_4^{2-}$ for both structures. If $Ni(CN)_4^{2-}$ is diamagnetic, which structure is correct?   *Ans.* See Fig. 18-21; square planar.

18.29. Prepare sketches of the electronic configurations of FeF_6^{3-}, $Fe(CN)_6^{3-}$, and $Ni(CN)_4^{2-}$ (square planar) using the crystal field splittings shown in Fig. 18-5.   *Ans.* See Fig. 18-14d, e, and f.

18.30. $Ti(H_2O)_6^{3+}$ absorbs light at 490 nm (blue-green), corresponding to the $^2T_{2g} \longrightarrow {}^2E_g$ transition shown in Fig. 18-6. Calculate the energy between these levels, and predict the color of the solution containing this ion.

Ans. $20\,400$ cm^{-1}; passes blue and red, giving a red-violet solution

18.31. Is it possible for a sample of $Pt(NH_3)_2Cl_2$ to be polar and another sample to be nonpolar?

Ans.   Yes, if the compound is square planar, the cis isomer will be polar and the trans isomer will be nonpolar.

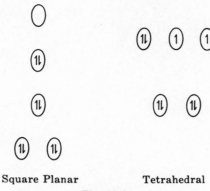

Square Planar   Tetrahedral
Fig. 18-21

18.32. Describe the bonding in a substance having the empirical formula $AgCN_2H_3$. The description should be consistent with the following properties: (1) The van't Hoff factor is 2; (2) AgCl will not precipitate if the substance is added to an aqueous solution of Cl^-; and (3) if Zn is added to a solution of this substance, 2 mol of Ag and 1 mol of $Zn(CN)_4^{2-}$ are formed for each mole of Zn added.

Ans. $Ag(NH_3)_2^+Ag(CN)_2^-$, first ion is linear with Ag—N bonds, and second ion is linear with C—Ag bonds.

18.33. Sketch the molecular orbital diagram for $Co(CN)_6^{3-}$.   *Ans.*   See the low-spin complex in Fig. 18-15*b*.

SPATIAL RELATIONSHIPS

18.34. Identify the shapes of the following species: (*a*) H_2O_2; (*b*) $SnCl_2$; (*c*) Al_2Cl_6; (*d*) $COCl_2$; (*e*) cyanogen, NCCN; (*f*) CO_2; (*g*) NO_3^-; (*h*) PCl_5; (*i*) $SeBr_4$.

Ans.   Values of AE and NE are (*a*) 14, 14; (*b*) 18, 20; (*c*) 48, 50; (*d*) 24, 26; (*e*) 18, 26; (*f*) 16, 20; (*g*) 24, 26; (*h*) 40, 38; (*i*) 34, 32. Lewis diagrams are given in Fig. 18.22. SN_X and shapes are (*a*) $SN_H = 1$, $SN_O = 4$, nonlinear; (*b*) $SN_{Sn} = 3$, $SN_{Cl} = 4$, bent; (*c*) $SN_{Cl} = 4$, $SN_{Al} = 4$, Al's and 4 Cl's planar, with shared Cl's above and below the plane, or Al's and shared Cl's planar, with end Cl's above and below the plane; (*d*) $SN_C = 3$, $SN_O = 3$, $SN_{Cl} = 4$, planar and Y-shaped, (*e*) $SN_C = 2$, $SN_N = 2$, linear; (*f*) $SN_C = 2$, $SN_O = 3$, linear; (*g*) $SN_N = 3$, $SN_O = 3$ (minimum), planar and Y-shaped; (*h*) $SN_P = 5$, $SN_{Cl} = 4$, triangular bipyramidal; (*i*) $SN_{Se} = 5$, $SN_{Br} = 4$, seesaw.

18.35. Determine the change in the molecular geometry caused by adding a proton to NH_3 to form NH_4^+.

Ans.   Trigonal pyramidal to tetrahedral

18.36. Determine the molecular geometry for BrF_3 and ICl_4^-.   *Ans.*   T-shaped, square planar

18.37. The molecule B_2H_6 has a deficiency of two electrons from that needed for normal bonding. Experimentally the molecule is known to have two types of B—H bonding, but neither can be a multiple bond. Four of the six H's can be replaced by methyl groups without breaking the molecule down. Write a Lewis structure consistent with these data, and describe the geometrical shape of the molecule.

Ans.   See Fig. 18-23 and see answer to Problem 18.34(*c*).

18.38. (*a*) Determine the molecular geometry of $NO_2(g)$, assuming that the unpaired electron remains in an unhybridized atomic orbital or, alternatively, occupies a hybrid orbital. (*b*) Which structure predicts the molecule to be paramagnetic? (*c*) Which structure predicts the molecule to be polar? (*d*) If three nondegenerate vibrational frequencies are observed, which structure is correct? (*e*) Because an unpaired electron system is very reactive, NO_2 dimerizes, forming N_2O_4. Describe the molecular geometry of N_2O_4.

Ans.   (*a*) Linear and bent; (*b*) both; (*c*) bent; (*d*) bent; (*e*) N's bonded with very little change in N—O arrangements

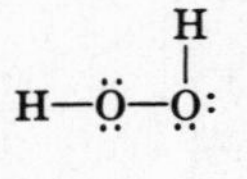

H
|
H—O̤—O̤:

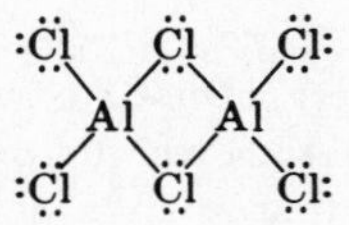

(a)

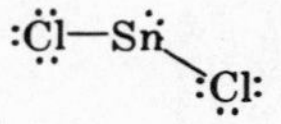

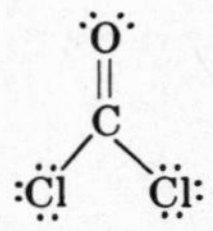

1. :Ö=C=Ö:

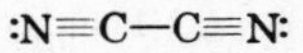

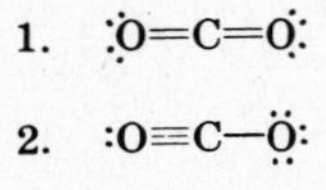

2. :O≡C—Ö:

(f)

:C̤l—Sṅ
        :C̤l:

(b)

:C̤l   C̤l   C̤l:
    \ / \ /
     Al   Al
    / \ / \
:C̤l   C̤l   C̤l:

(c)

$$\left[\begin{array}{c} :\ddot{O}\cdot \\ N \\ :\ddot{O} \quad \ddot{O}: \end{array} \right] \leftrightarrow \left[\begin{array}{c} :\ddot{O}: \\ N \\ :\ddot{O} \quad \ddot{O}: \end{array} \right]^{-} \leftrightarrow \left[\begin{array}{c} :\ddot{O}: \\ N \\ :\ddot{O} \quad \ddot{O}\cdot \end{array} \right]^{-}$$

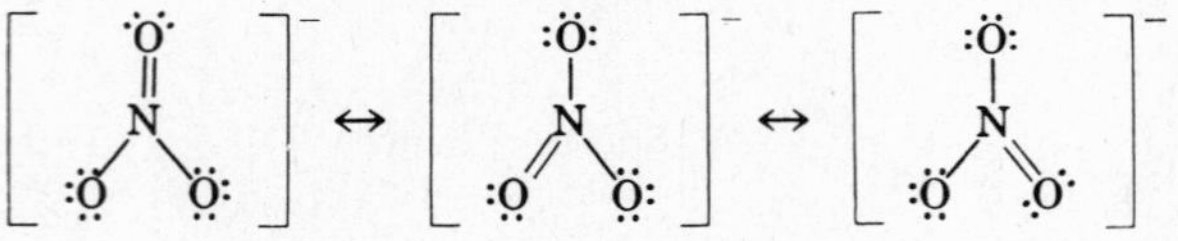

(g)

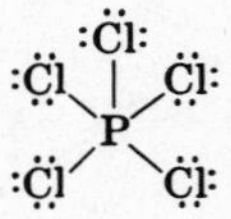

:Ö·
‖
C
/ \
:C̤l   C̤l:

(d)

    :C̤l:
:C̤l | C̤l:
   \ | /
      P
   / | \
:C̤l   C̤l:

(h)

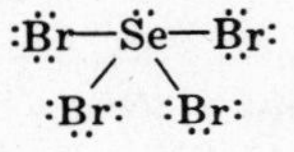

:N≡C—C≡N:

(e)

:Br̤—Se—Br̤:
:Br̤: :Br̤:

(i)

Fig. 18-22

H   H   H
 \ | /
   B   B
 / | \
H   H   H

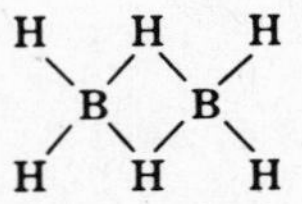

Fig. 18-23

18.39. Determine the molecular geometry for $Cu(H_2O)_6^{2+}$. Two of the waters of hydration (trans to each other) are farther from the Cu^{2+} than the other four, changing the formula to $Cu(H_2O)_4 \cdot 2H_2O^{2+}$ or $Cu(H_2O)_4^{2+}$. What is the molecular geometry for the $Cu(H_2O)_4 \cdot 2H_2O^{2+}$ and $Cu(H_2O)_4^{2+}$ ions?

Ans.   Octahedral; distorted octahedral or square bipyramidal; square planar

Chapter 19

Spectroscopy of Polyatomic Molecules

Rotational Spectra

19.1 MOMENTS OF INERTIA FOR A RIGID MOLECULE

The *principal moments of inertia* (A, B, and C) for a molecule containing n atoms are the solutions for I of the determinantal equation

$$\begin{vmatrix} I_{xx} - I & -I_{xy} & -I_{xz} \\ -I_{xy} & I_{yy} - I & -I_{yz} \\ -I_{xz} & -I_{yz} & I_{zz} - I \end{vmatrix} = 0 \qquad (19.1)$$

where by convention $C \geq B \geq A$ and

$$I_{xx} = \sum_{i=1}^{n} m_i(y_i^2 + z_i^2) \qquad (19.2a)$$
$$\vdots$$
$$I_{xy} = \sum_{i=1}^{n} m_i x_i y_i \qquad (19.2d)$$
$$\vdots$$

Calculations using (19.1) are greatly simplified if the Cartesian coordinate system used in (19.2) has its origin at the center of mass of the molecule. The coordinates of the center of mass of the molecule with respect to an arbitrary Cartesian coordinate system are

$$x_{cm} = \frac{\sum_{i=1}^{n} m_i x_i}{\sum_{i=1}^{n} m_i} \qquad y_{cm} = \frac{\sum_{i=1}^{n} m_i y_i}{\sum_{i=1}^{n} m_i} \qquad z_{cm} = \frac{\sum_{i=1}^{n} m_i z_i}{\sum_{i=1}^{n} m_i} \qquad (19.3)$$

The location of the principal axes of inertia can often be simplified by determining the symmetry elements present in a molecule (see Chap. 20). Usually a principal axis coincides with a higher-order axis of rotation, a mirror plane contains two principal axes and is perpendicular to the third, and a center of symmetry coincides with the origin of the principal axes.

Depending on the values of the principal moments of inertia, molecules are divided into three general classes: (1) *spherical tops*, where $A = B = C$; (2) *symmetrical tops*, where $A < B = C$ (prolate) or $A = B < C$ (oblate); and (3) *asymmetrical tops*, where $A < B < C$.

EXAMPLE 19.1. Determine the principal moments of inertia for CH_4, and classify this molecule. The C—H bond length is 0.109 1 nm and the H—C—H bond angle is 109°28'.

The center of mass is located at the carbon atom. In the coordinate system shown in Fig. 19-1a, the co-ordinates (in nm) of the atoms are $(0, 0, 0)$ for C, $(0, 0, 0.109\ 1)$ for H_1, $(0, 0, 102\ 8, -0.036\ 4)$ for H_2,

377

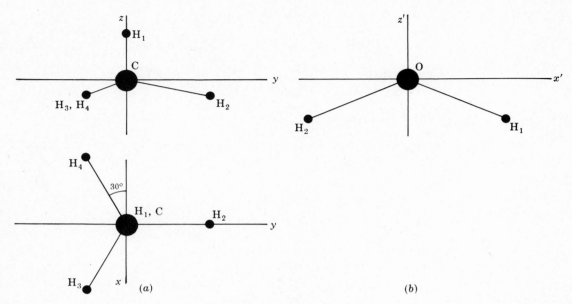

Fig. 19-1

(0.089 0, −0.051 4, −0.036 4) for H_3, and (−0.089 0, −0.051 4, −0.036 4) for H_4, giving

$$I_{xx} = m_C[(0)^2 + (0)^2] + m_H[(0)^2 + (0.109\ 1)^2]$$
$$+ m_H[(0.102\ 8)^2 + (−0.036\ 4)^2] + m_H[(−0.051\ 4)^2 + (−0.036\ 4)^2]$$
$$+ m_H[(−0.051\ 4)^2 + (−0.036\ 4)^2] = 0.031\ 72 m_H$$

$$I_{xy} = m_C[(0)(0)] + m_H[(0)(0) + (0)(0.102\ 8) + (0.089\ 0)(−0.051\ 4) + (−0.089\ 0)(−0.051\ 4)] = 0$$

$I_{yy} = I_{zz} = 0.031\ 72 m_H$, and $I_{xz} = I_{yz} = 0$, in units of nm^2.

Using (*19.1*) gives

$$\begin{vmatrix} 0.031\ 72 m_H − I & 0 & 0 \\ 0 & 0.031\ 72 m_H − I & 0 \\ 0 & 0 & 0.031\ 72 m_H − I \end{vmatrix} = 0$$

and solving gives $I = A = B = C = 0.031\ 72 m_H\ nm^2 = 5.309 \times 10^{-47}\ kg\ m^2$, a spherical top molecule.

19.2 SPHERICAL TOP MOLECULES

The rotational energy levels for a spherical top molecule are given by

$$E_J = \frac{J(J+1)\hbar^2}{2I} \tag{19.4}$$

where $J = 0, 1, 2, \ldots$. Because of the symmetry of the molecule, the rotational levels must be analyzed from Raman spectra and the infrared-active transitions.

19.3 SYMMETRICAL TOP MOLECULES

The rotational energy levels for an oblate symmetrical top are given by

$$E_{J,K} = \frac{J(J+1)\hbar^2}{2B} − K^2\hbar^2\left(\frac{1}{2B} − \frac{1}{2C}\right) \tag{19.5}$$

where $K = 0, \pm 1, \pm 2, \ldots, \pm J$. Equation (19.5) can be rewritten as

$$F(J, K) = B'J(J+1) - (B' - C')K^2 \qquad (19.6)$$

Likewise, for a prolate top,

$$F(J, K) = B'J(J+1) + (A' - B')K^2 \qquad (19.7)$$

In (19.6) and (19.7) the rotational constants A', B', and C' are defined in units of s^{-1} as

$$A' = \frac{h}{8\pi^2 A} \qquad B' = \frac{h}{8\pi^2 B} \qquad C' = \frac{h}{8\pi^2 C} \qquad (19.8a)$$

and in units of cm^{-1} as

$$A' = \frac{h \times 10^{-2}}{8\pi^2 Ac} \qquad B' = \frac{h \times 10^{-2}}{8\pi^2 Bc} \qquad C' = \frac{h \times 10^{-2}}{8\pi^2 Cc} \qquad (19.8b)$$

The linear molecule is a special case of a prolate top with $A = 0$ and $B = C$ and will have a rotational spectrum only if it has a permanent dipole moment. The energy levels will be given by (19.7), where $K = 0$, and the energy diagram is similar to that given for the diatomic molecule (Fig. 14-12), with energy spacing (in Hz) of

$$\nu = 2B'(J+1) \qquad (19.9)$$

19.4 ASYMMETRICAL TOP MOLECULES

The spectra and energy spacings for the rotational motion of an asymmetrical top molecule are too complex for treatment in this book.

Vibrational Spectra

19.5 DEGREES OF FREEDOM

A linear polyatomic molecule containing N atoms will have $3N - 5$ internal degrees of freedom, and a nonlinear molecule will have $3N - 6$ degrees. These degrees of freedom can be assigned to normal models of vibration and internal rotation (see Sec. 20.22).

19.6 INFRARED SPECTRA

Lines corresponding to

$$\bar{\nu} = G'(v_1, v_2, \ldots) - G''(v_1, v_2, \ldots) \qquad (19.10)$$

where $\Delta v_i = 0, \pm 1, \pm 2, \ldots$, will be observed in the infrared only if the molecule has a permanent dipole moment or if the mode of vibration under consideration induces a dipole moment in the molecule. *Fundamental levels* are those in which all v_i are zero except one, which is equal to unity. *Overtone levels* are those in which all v_i are zero except one, which is greater than unity. *Combination levels* are those in which various combinations of v_i exist.

Rotational branches are present in the infrared spectra, but the analysis of the data is quite complicated.

EXAMPLE 19.2. A shorthand notation for the transitions uses a mathematical combination of the fundamentals corresponding to the transition. An advantage of this notation is that it is numerically nearly equal to the observed wave number. For example, the (000) $\longrightarrow$ (010) transition in a molecule having three vibrational frequencies is represented by $\bar{\nu}_2$, and the $(011) \longrightarrow (110)$ transition by $\bar{\nu}_1 - \bar{\nu}_3$. Find similar expressions for the $(000) \longrightarrow (002)$, $(000) \longrightarrow (211)$, and $(110) \longrightarrow (002)$ transitions.

The notations for the transitions are $2\bar{\nu}_3$, $2\bar{\nu}_1 + \bar{\nu}_2 + \bar{\nu}_3$, and $2\bar{\nu}_3 - \bar{\nu}_1 - \bar{\nu}_2$, respectively.

Electron Magnetic Properties

19.7   MAGNETIC SUSCEPTIBILITY

The observed effects of a magnetic field on a material are related to the magnetic induction or flux density (B) given by

$$B = \mu H \qquad (19.11)$$

where H is the magnetic field strength and μ is the magnetic permeability. The SI units for H are $A\,m^{-1}$; for B, they are T (tesla), where $1\,T = 1\,kg\,s^{-2}\,A^{-1}$; and for μ, they are $H\,m^{-1}$ (henry per meter), where $1\,H = 1\,m^2\,kg\,s^{-2}\,A^{-2}$. The magnetic permeability of a material is defined with reference to the permeability of a vacuum (μ_0) as

$$\mu = \mu_0(1 + \chi_v) \qquad (19.12)$$

where $\mu_0 = 4\pi \times 10^{-7}\,H\,m^{-1}$ and χ_v is the dimensionless parameter known as the *magnetic susceptibility* (or *volume susceptibility* or *susceptibility per unit volume*). Reported values of χ_v having the units of $H\,m^{-1}$ are actually values of $\mu_0\chi_v$. Older commonly used units include the oersted for H [$1\,Oe = (1\,000/4\pi)\,A\,m^{-1}$] and the gauss for B $(1\,G = 10^{-4}\,T)$.

Materials with negative values of χ_v are called *diamagnetic*, and B in the material will be less than H in a vacuum. A diamagnetic material in a nonuniform field will move, if possible, toward the weakest region of the field, and an elongated sample will have a tendency to orient itself at right angles to the field. *Paramagnetic* materials have values of χ_v greater than zero, and B in these materials will be greater than H in a vacuum. A paramagnetic material in a nonuniform field will move, if possible, toward the strongest region of the field, and an elongated sample will have a tendency to orient itself parallel to the field. *Ferromagnetic* materials have values of χ_v that are about 1 000 times the normal values for paramagnetic materials.

Although not included within the SI system, chemists still retain the quantities known as the *mass magnetic susceptibility* or *magnetic susceptibility per gram* (χ_g) given by

$$\chi_g = (\chi_v/\rho) \times 10^{-3} \qquad (19.13)$$

where ρ is the density expressed in $kg\,m^{-3}$ and χ_g has the units $m^3\,g^{-1}$, and the *molar magnetic susceptibility* or *magnetic susceptibility per mole* (χ_m) given by

$$\chi_m = \chi_g M \qquad (19.14)$$

where M is the molar mass expressed in $g\,mol^{-1}$ and χ_m has the units of $m^3\,mol^{-1}$.

The value of χ_m is the sum of two terms, the *molar diamagnetic susceptibility* $(\chi_{m,d})$ and the *molar paramagnetic susceptibility* $(\chi_{m,p})$:

$$\chi_m = \chi_{m,d} + \chi_{m,p} \qquad (19.15)$$

The value of $\chi_{m,d}$ for many ions is of the order of 10^{-11} to $10^{-12}\,m^3\,mol^{-1}$, and it is negligible compared to $\chi_{m,p}$ for paramagnetic materials. For n unpaired electrons in a paramagnetic species and a temperature T,

$$n = (1 + 7.998 \times 10^6\,T\chi_{m,p})^{1/2} - 1 \qquad (19.16)$$

Magnetic susceptibility measurements for a solute in a solution are usually made with a *Gouy balance*. The apparent mass of the sample (in g) is measured in the presence of the magnetic field of the earth (H_e) only, giving m'_e, and in the presence of an applied field of strength H_f, giving m'_f. It can be shown that

$$\chi_v = \frac{2g \times 10^{-3}(m_f' - m_e')}{A(H_f^2 - H_e^2)\mu_0} + \chi_{v,\text{air}} \qquad (19.17)$$

where g is the gravitational constant (in m s^{-2}) and A is the sample cross-sectional area (in m^2). Usually $\chi_{v,\text{air}}$ is neglected in (19.17), and the Gouy balance constant, $2g \times 10^{-3}/A(H_f^2 - H_e^2)\mu_0$, is determined by standardization using $NiCl_2(aq)$. The molar susceptibility of the solute is related to χ_v by

$$\chi_m = \frac{\chi_v \times 10^{-3} + (7.20 \times 10^{-13})(\rho - CM)}{C} \qquad (19.18)$$

where C is the concentration in mol dm^{-3}.

19.8 ELECTRON SPIN (MAGNETIC OR PARAMAGNETIC) RESONANCE

In the presence of an external field, the degeneracy of the spin wave functions for an unpaired electron is removed, with the state having $s = -\frac{1}{2}$ taken, by convention, to be the lower energy state and the $s = +\frac{1}{2}$ state to be the higher state. Each of these states is split into $n+1$ components for each n equivalent nuclei having a nuclear spin present in the molecule. By convention, the nuclear spin states with positive spin are assumed to be lower than those having negative spin in the $s = -\frac{1}{2}$ state, and vice versa in the $s = +\frac{1}{2}$ state. The selection rule for allowed transitions is $\Delta M_I = 0$, where $M_I = I$, $I-1, \ldots, -I+1, -I$, and $I = n(1/2)$.

The intensities of the resonance peaks (as well as the NMR splittings, see Sec. 19.11) can be shown to be proportional to the number of combinations of nuclear spins that will create the value of M_I. An easy way to determine the relative intensities is to use Pascal's triangle of binomial coefficients, given by

$$
\begin{array}{cccccccccccccc}
n & & & & & & \textit{Relative intensities} & & & & & & & \\
0 & & & & & & & 1 & & & & & & \\
1 & & & & & & 1 & & 1 & & & & & \\
2 & & & & & 1 & & 2 & & 1 & & & & \\
3 & & & & 1 & & 3 & & 3 & & 1 & & & \\
4 & & & 1 & & 4 & & 6 & & 4 & & 1 & & \\
5 & & 1 & & 5 & & 10 & & 10 & & 5 & & 1 & \\
6 & 1 & & 6 & & 15 & & 20 & & 15 & & 6 & & 1 \\
\end{array}
\qquad (19.19)
$$

where the coefficients in additional rows can be determined by adding the coefficients to the right and left of the desired coefficient in the previous row.

EXAMPLE 19.3. Describe the ESR spectrum of $C_6H_6{}^-$.

In the presence of the field, the spin wave functions for the unpaired electron will split into two components are shown in Fig. 19-2. The six equivalent protons give $I = 6(\frac{1}{2}) = 3$, so that both levels are split into seven sublevels corresponding to $M_I = 3, 2, 1, 0, -1, -2$, and -3. The selection rule $\Delta M_I = 0$ gives seven equally spaced peaks having relative intensities of 1, 6, 15, 20, 15, 6, and 1 according to (19.19). The ^{12}C nuclei have no net nuclear spin and do not contribute to the spectrum.

Nuclear Magnetic Resonance

19.9 INTRODUCTION

The energy levels of an isolated nuclear magnetic moment in an applied magnetic field of flux density B_0 are given by

$$E = -g_N \mu_N B_0 M_I \qquad (19.20)$$

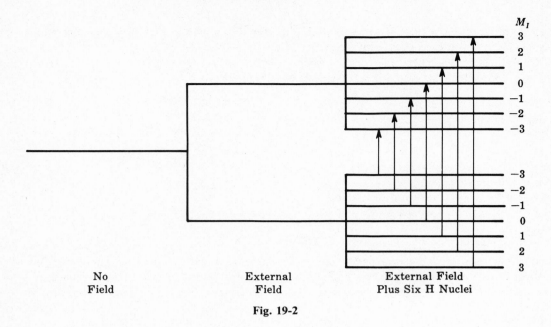

Fig. 19-2

where g_N is the nuclear "g factor" ($g_N = 5.585\,6$ for ^{1}H), μ_N is the "nuclear magneton" ($\mu_N = 5.050\,824 \times 10^{-27}$ J T^{-1}), and $M_I = I, I-1, \ldots, -I$, where I is the nuclear spin ($I = \frac{1}{2}$ for ^{1}H). The allowed transitions obey the selection rule $\Delta M_I = \pm 1$, giving for a proton

$$\nu = g_N \mu_N B_0 / h \qquad (19.21)$$

Sections 19.9 and 19.10 will be restricted to proton magnetic resonance, although other nuclei having nonzero spins are also used to identify molecular structure, e.g., ^{11}B, ^{13}C, ^{19}F.

EXAMPLE 19.4.   Assuming a room temperature of 25 °C, find the ratio of protons having $-\frac{1}{2}$ spin to the number having $+\frac{1}{2}$ spin in a 14 100-G magnetic field (60 MHz).

Using (19.20) gives

$$\Delta E = -(5.585\,6)(5.051 \times 10^{-27} \text{ J T}^{-1})(14\,100 \text{ G})[(10^{-4}\text{ T})/(1 \text{ G})][(-\tfrac{1}{2}) - (\tfrac{1}{2})]$$

$$= 3.978 \times 10^{-26} \text{ J}$$

The Boltzmann distrbution law, (2.11), then gives

$$\frac{N_-}{N_+} = \exp\left[-\frac{3.978 \times 10^{-26} \text{ J}}{(1.380\,6 \times 10^{-23} \text{ J K}^{-1})(298 \text{ K})} \right] = 0.999\,990\,33$$

Thus, in a mole of protons, the number of protons in each level is very nearly 3.01×10^{23}. Because of the very small value of ΔE, very sensitive equipment and very low temperatures are required to obtain strong signals.

19.10   CHEMICAL SHIFTS

The magnetic flux density (B_i) at the nucleus of an atom in a chemical environment i is given by

$$B_i = B_0(1 - \sigma_i) \qquad (19.22)$$

where the *screening* (*shielding*) *constant* (σ_i) is the result of the electronic interaction of the chemical environment with the magnetic field. The frequencies of the NMR lines are given by

$$\nu = \frac{(g_N)_j(1 - \sigma_i)\mu_N B_0}{h} \qquad (19.23)$$

Thus the different values of g_N corresponding to different nuclear species j will produce well-separated frequencies that undergo small shifts as a result of the respective chemical environments i.

Chemical shifts occurring in organic proton NMR work are expressed relative to the absorption of a reference material such as TMS [tetramethylsilane, $Si(CH_3)_4$] by

$$\delta_i/(ppm) = (\sigma_{ref} - \sigma_i) \times 10^6 \qquad (19.24)$$

or, in terms of an alternative scale, by

$$\tau = 10.000 - \delta_i \qquad (19.25)$$

The area under each absorption peak in the spectrum is directly proportional to the number of protons having that type of chemical environment.

EXAMPLE 19.5.   Predict the chemical shifts in the NMR spectrum for CH_3OH given that $\sigma(OH) > \sigma(CH_3)$ in a CCl_4 solution.

There will be two peaks observed , one for the proton in the OH-type environment, having a relative area of 1, and the second for the protons in the CH_3-type environment, having a relative area of 3. Because $\sigma(OH) > \sigma(CH_3)$, the shift will be farther away from the TMS reference for the OH than for the CH_3.

19.11   SPIN–SPIN SPLITTINGS

Under high resolution, the chemical shift peak for one type of nuclei often shows splittings resulting from magnetic fields from adjacent nuclei. For a molecule having m protons of type A and n protons of type X, generally a set of $n + 1$ lines centered about the frequency for proton type A will appear with relative intensities as given by (19.19), and a set of $m + 1$ lines centered about the frequency for the proton type X will appear with similar relative intensities.

Unless the spin–spin coupling constants are small compared to the chemical shifts, spectra of systems containing several nuclei are complex. Because chemical shifts are proportional to B_0 and spin–spin splittings are independent of B_0, a technique of simplifying the spectra is to use higher-frequency spectrometers. Changes in the coupling effected by using two radio-frequency magnetic fields superimposed on the sample (*decoupling*) or by isotopic substitution often aid in the identification of a spin–spin interaction.

EXAMPLE 19.6.   Predict the splittings in the spectrum for CH_3OH.

The methyl group containing three protons will split the alcohol peak into four components, centered at the original frequency, with relative intensities of 1, 3, 3, and 1. The alcohol proton will split the methyl peak into two components of equal relative intensity, centered at the original methyl frequency.

Solved Problems

ROTATIONAL SPECTRA

19.1.   Using the arbitrary Cartesian axes (x', y', z') in Fig. 19-1b, determine the center of mass for water. Find the coordinates of the atoms with respect to the center of mass, and calculate A, B, and C. Classify the molecule. The O—H bond length is 0.095 84 nm, and the H—O—H bond angle is 104.45°.

The coordinates of the atoms shown in Fig. 19-1b are (in nm)

$$x'(H_1) = (0.095\ 84) \sin \frac{104.45°}{2} = 0.075\ 75$$

$$y'(H_1) = 0$$

$$z'(H_1) = -(0.095\ 84)\cos\frac{104.45°}{2} = -0.058\ 71$$

$$x'(H_2) = -0.075\ 75 \qquad y'(H_2) = 0 \qquad z'(H_2) = -0.058\ 71$$

and
$$x'(O) = 0 \qquad y'(O) = 0 \qquad z'(O) = 0$$

Substituting these into (*19.3*) gives

$$x'_{cm} = \frac{m_H(0.075\ 75) + m_H(-0.075\ 75) + m_O(0)}{m_H + m_H + m_O} = 0$$

$$y'_{cm} = \frac{m_H(0) + m_H(0) + m_O(0)}{m_H + m_H + m_O} = 0$$

$$z'_{cm} = \frac{m_H(-0.058\ 71) + m_H(-0.058\ 71) + m_O(0)}{m_H + m_H + m_O}$$

$$= \frac{2(0.058\ 71)(1.007\ 825/L)}{2(1.007\ 825/L) + 15.994\ 91/L} = -0.006\ 57$$

The coordinates (x, y, z) of the atoms relative to the center of mass are (in nm)

$$x(H_1) = x'(H_1) - x'_{cm} = 0.075\ 75$$

$$y(H_1) = y'(H_1) - y'_{cm} = 0$$

$$z(H_1) = z'(H_1) - z'_{cm} = -0.052\ 14$$

$$x(H_2) = -0.075\ 75 \qquad y(H_2) = 0 \qquad z(H_2) = -0.052\ 14$$

and
$$x(O) = 0 \qquad y(O) = 0 \qquad z(O) = 0.006\ 57$$

Substituting the latter set of coordinates into (*19.2*) gives

$$I_{xx} = \frac{1.007\ 825}{L}[(0)^2 + (-0.052\ 14)^2] + \frac{1.007\ 825}{L}[(0)^2 + (-0.052\ 14)^2] + \frac{15.994\ 91}{L}[(0)^2 + (0.006\ 57)^2]$$

$$= \frac{0.061\ 68}{L}\ \text{g nm}^2 = 1.025 \times 10^{-47}\ \text{kg m}^2$$

$$I_{yy} = \frac{0.177\ 36}{L} = 2.945 \times 10^{-47}\ \text{kg m}^2 \qquad I_{zz} = \frac{0.115\ 66}{L} = 1.921 \times 10^{-47}\ \text{kg m}^2$$

$$I_{xy} = m_H(0.075\ 75)(0) + m_H(-0.075\ 75)(0) + m_O(0)(0) = 0$$

$$I_{xz} = 0 \qquad I_{yz} = 0$$

Because the off-diagonal terms in (*19.1*) are zero, the principal moments of inertia are just I_{xx}, I_{yy}, and I_{zz}, namely,

$$A = I_{xx} \qquad B = I_{zz} \qquad C = I_{yy}$$

The molecule is an asymmetrical top.

19.2. Sketch the energy levels for a prolate top molecule with $A' = 5B'$.

For the molecule, (*19.7*) gives

$$F(J, K) = B'J(J+1) + 4B'K^2$$

The first few allowed terms are $F(0, 0)$, $F(1, 0)$, $F(1, \pm 1)$, $F(2, 0)$, $F(2, \pm 1)$, and $F(2, \pm 2)$, which have values of

$$F(0, 0) = B'(0)(0+1) + 4B'(0)^2 = 0$$

$$F(1, 0) = B'(1)(1+1) + 4B'(0)^2 = 2B'$$

$$F(1, \pm 1) = B'(1)(1+1) + 4B'(\pm 1)^2 = 6B'$$

$F(2, 0) = 6B'$, $F(2, \pm 1) = 10B'$, and $F(2, \pm 2) = 22B'$. These are plotted in Fig. 19-3a along with a few other low terms.

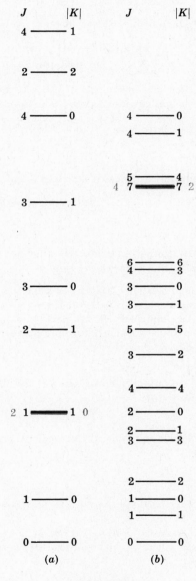

Fig. 19-3

19.3. The rotational spectrum of HCN shows an absorption at 88 631.62 MHz, and that of DCN at 72 414.61 MHz. Calculate the H—C and C—N bond lengths from these data.

Combining (*19.9*) with (*19.8*) gives

$$B = \frac{h(J+1)}{4\pi^2 \nu}$$

Thus, for the $J = 0 \longrightarrow J = 1$ transition,

$$B_{HCN} = \frac{(6.626\,176 \times 10^{-34}\,\text{J s})(0+1)}{4\pi^2(8.863\,162 \times 10^{10}\,\text{Hz})} = 1.893\,715 \times 10^{-46}\,\text{kg m}^2$$

$$B_{DCN} = \frac{(6.626\,176 \times 10^{-34}\,\text{J s})(0+1)}{4\pi^2(7.241\,461 \times 10^{10}\,\text{Hz})} = 2.317\,806 \times 10^{-46}\,\text{kg m}^2$$

For a linear triatomic molecule it can be shown that

$$B = \frac{m_1 m_2 r_{12}^2 + m_1 m_3 r_{13}^2 + m_2 m_3 r_{23}^2}{m_1 + m_2 + m_3}$$

where $r_{13} = r_{12} + r_{23}$. Assuming that the substitution of D for H does not significantly change the H—C bond length, solving the simultaneous equations gives

$$r_{HC} = 0.106\,8\,\text{nm} \qquad \text{and} \qquad r_{CN} = 0.115\,6\,\text{nm}$$

VIBRATIONAL SPECTRA

19.4. Determine the number of internal degrees of freedom for linear and bent triatomic molecules.

For the linear molecule $3N - 5 = 4$, and for the bent molecule $3N - 6 = 3$. All of these are vibrational degrees of freedom.

19.5. Describe the structure of a triatomic molecule AB_2 that has three vibrational frequencies, two in the infrared and the third in the Raman spectrum.

The molecule will have four vibrational frequencies if linear and three if bent. If the molecule were BAB bent or BBA bent or linear, all three or four frequencies would be in the infrared, which is not the case; so these possibilities can be eliminated. The last possibility, linear BAB, will show the two infrared-active transitions for the three asymmetric modes (two are degenerate) and one Raman-active transition for the symmetric stretching mode. See Fig. 20.19.

19.6. The fundamental vibrational frequencies of H_2O are 3 657.05, 1 594.59, and 3 755.79 cm^{-1}. Predict the energies of the absorption bands for (000) $\longrightarrow$ (002), (020), (200), (021), (120), (121), and (111).

For the transition between (000) and (002), the predicted energy change is

$$\bar{\nu} = 2\bar{\nu}_3 = 2(3\,755.79) = 7\,511.58\,\text{cm}^{-1}$$

Likewise, for (000) $\longrightarrow$ (121),

$$\bar{\nu} = \bar{\nu}_1 + 2\bar{\nu}_2 + \bar{\nu}_3 = 3\,657.05 + 2(1\,594.59) + 3\,755.79 = 10\,602.02\,\text{cm}^{-1}$$

The remaining energy differences are 7 511.58, 3 189.18, 7 314.10, 6 944.97, 6 846.23, 10 602.02, and 9 007.43 cm^{-1}, respectively. These values will be high because of the anharmonicity of the potential-energy well (see Sec. 17.3).

ELECTRON MAGNETIC PROPERTIES

19.7. A Gouy balance was calibrated at 25 °C using $NiCl_2(aq)$ such that $(2g \times 10^{-3})/A(H_f^2 - H_e^2)\mu_0 = 1.05 \times 10^{-4}\,\text{g}^{-1}$. A sample of 0.521 M $MnSO_4$ solution had an apparent mass of 10.216 4 g in the magnetic field and 10.148 0 g with the field removed. The density of the solution was $1.070 \times 10^3\,\text{kg m}^{-3}$ and $M = 151.00\,\text{g mol}^{-1}$. Calculate the number of unpaired electrons on a Mn(II) ion.

If $\chi_{v,\text{air}}$ is neglected, (19.17) gives

$$\chi_v = (1.05 \times 10^{-4})(10.216\,4 - 10.148\,0) = 7.18 \times 10^{-6}$$

which upon substitution into (19.18) gives

$$\chi_m = \frac{(7.18 \times 10^{-6})(10^{-3}) + (7.20 \times 10^{-13})[(1.070 \times 10^3) - (0.521)(151.00)]}{0.521}$$

$$= 1.515 \times 10^{-8} \text{ m}^3 \text{ mol}^{-1}$$

Assuming $\chi_{m,d} \ll \chi_{m,p}$, $\chi_{m,p} = 1.515 \times 10^{-8}$, and ($19.16$) gives

$$n = [1 + (7.998 \times 10^6)(298)(1.515 \times 10^{-8})]^{1/2} - 1 = 5.09$$

Because n should be an integer, the answer is assumed to be 5.

19.8. The ESR spectrum of a radical having the empirical formula $C_3H_7 \cdot$ showed 14 absorption peaks with relative intensities of 1, 1, 6, 6, 15, 15, 20, 20, 15, 15, 6, 6, 1, and 1. Is this an n-propyl or an isopropyl radical?

An n-propyl radical

$$CH_3—CH_2—CH_2 \cdot$$

would show three peaks of intensities 1, 2, and 1, each split into three peaks of intensities 1, 2, and 1, each in turn split into four peaks of intensities 1, 3, 3, and 1, giving a total of 36 peaks, which does not agree with the experimental data. The isopropyl radical

$$CH_3—\dot{C}H—CH_3$$

would show two peaks of intensities 1 and 1, each split into seven peaks of intensities 1, 6, 15, 20, 15, 6, and 1, giving a total of 14 peaks with intensities that fit the data.

NUCLEAR MAGNETIC RESONANCE

19.9. A low-resolution NMR spectrum of a compound having the empirical formula $C_2H_3Cl_3$ showed two peaks, one area being twice the other. Another substance having the same formula showed only one NMR peak. Identify these substances.

The substance showing two peaks has two types of protons, giving the structure as

```
      H   H
      |   |
  H —C — C — Cl
      |   |
      Cl  Cl
```

which agrees with the areas under the peaks. For the substance showing only one peak, only one type of proton is present, as indicated by the structure

```
      H   Cl
      |   |
  H —C — C — Cl
      |   |
      H   Cl
```

19.10. Describe the splittings in the spectrum for 1,1,1-trichloroethane.

As can be seen from the structure given in Problem 19.9, there are no protons on the carbon atom adjacent to the CH_3 group, so no splitting will occur.

Supplementary Problems

ROTATIONAL SPECTRA

19.11. The molecules C_6H_6 and CH_3Br are both symmetric tops. Classify these as oblate or prolate.

Ans. C_6H_6 is planar, giving $A = B < C$, oblate;
CH_3Br is trigonal pyramidal, giving $A < B = C$, prolate.

19.12. Determine the principal moments of inertia for H_2CO, and classify this molecule. The C—H bond length is 0.112 nm, the C=O bond length is 0.121 nm, and the H—C—H angle is 118°.

Ans. $A = 3.09 \times 10^{-47}$ kg m^2, $B = 21.70 \times 10^{-47}$ kg m^2, $C = 24.79 \times 10^{-47}$ kg m^2; asymmetric top

19.13. Sketch the energy levels for an oblate top molecule with $B' = 5C'$.

Ans.   See Fig. 19-3b for plot of $F(J, K) = B'J(J+1) - \frac{4}{5}B'K^2$.

19.14. The absorption spectrum of ^{16}OCS shows a peak at 24 325.92 MHz. Calculate B for this molecule. Given $B = 147.03 \times 10^{-47}$ kg m^2 for ^{18}OCS, find r_{CO} and r_{CS}. Assume atomic masses of 15.994 91 for ^{16}O, 17.999 2 for ^{18}O, 12.011 15 for C, and 32.064 for S.

Ans. $B = 137.996 \times 10^{-47}$ kg m^2; $r_{CO} = 0.116\,4$ nm, $r_{CS} = 0.155\,8$ nm

VIBRATIONAL SPECTRA

19.15. Determine the number of vibrational degrees of fredom for CH_4.   *Ans.* 9

19.16. Suppose the molecule in Problem 19.5 showed three strong infrared absorption bands. What would be the structure of AB_2?   *Ans.*   ABB bent or BAB bent.

19.17. The fundamental vibrational frequencies of SO_2 are 1 151.38, 517.69, and 1 361.76 cm^{-1}. Account for the absorption bands at 1 875.55, 2 295.88, and 2 499.55 cm^{-1}.

Ans. $\bar{\nu}_2 + \bar{\nu}_3$, $2\bar{\nu}_1$, and $\bar{\nu}_1 + \bar{\nu}_3$

ELECTRON MAGNETIC PROPERTIES

19.18 A 0.102 7 M solution of $KMnO_4$ had an apparent mass of 9.805 9 g in a magnetic field and 9.809 9 g out of the field when placed in the Gouy balance described in Problem 19.7. Calculate the number of unpaired electrons for Mn(VII).

Ans.   The sample is diamagnetic (for paramagnetic materials, $m'_f > m'_e$) and $n = 0$.

19.19 A 0.507 0 M solution of $K_3Fe(CN)_6$ had an apparent mass of 8.049 8 g both in and out of the field when placed in the Gouy balance described in Problem 19.7. The density of the solution was 1.085×10^3 kg m^{-3}, and the molar mass of the solute is 329.25 g mol^{-1}. Is CN^- a strong or a weak ligand?

Ans. $\chi_v = 0$, $\chi_m = 1.304 \times 10^{-9}$ m^3 mol^{-1}, $n \approx 1$. A low-spin complex is formed by a strong ligand.

19.20. Describe the ESR spectrum of $CH_3 \cdot$.

Ans.   Four equally spaced peaks of relative itensities 1, 3, 3, and 1

$(M_I = \frac{3}{2} \longrightarrow \frac{3}{2}, \frac{1}{2} \longrightarrow \frac{1}{2}, -\frac{1}{2} \longrightarrow -\frac{1}{2}, \text{ and } -\frac{3}{2} \longrightarrow -\frac{3}{2})$

19.21. An ESR spectrum for an organic radical containing two carbons consisted of 12 lines having relative intensities of 1, 2, 3, 1, 6, 3, 3, 6, 1, 3, 2, and 1. What is the radical?

Ans. $CH_3CH_2 \cdot$, because CH_3 splitting gives four peaks with intensities $1:3:3:1$ and CH_2 splitting gives $3(4) = 12$ peaks with intensities $(1:2:1)(1:3:3:1)$.

19.22. The ESR spectrum of the naphthalene anion consists of 25 lines. Show that this corresponds to the structure given in Problem 18.25. What are the relative intensities of these peaks?

Ans.   H's on the four α carbons give 5 peaks and those on the four β carbons give 5 peaks, 25 peaks total; four at intensity 1, eight at 4, four at 6, four at 16, four at 24, and one at 36.

NUCLEAR MAGNETIC RESONANCE

19.23. Given that magnetic field strength is directly proportional to applied frequency, repeat Example 19.4 at 100 MHz and liquid nitrogen temperatures ($-195\,°C$).

Ans. $e^{-6.16 \times 10^{-5}} = 0.999\,938\,4$

19.24. Predict the major components of the NMR spectrum for CH_3CHO given that $\sigma(CHO) > \sigma(CH_3)$ in a CCl_4 solution.

Ans.   Two groups of peaks, CHO farther from TMS with area 1 and CH_3 nearer to TMS with area 3

19.25. Discuss the splitting in the NMR spectrum peaks described in Problem 19.24.

Ans. CH_3 is doublet (1 and 1 relative intensities), and CHO is quartet (1, 3, 3, and 1 relative intensities)

19.26. A compound having the empirical formula $C_4H_{10}O$ gave an NMR spectrum consisting of two groups of lines with relative areas 3 and 2. Another substance with the same formula gave an NMR spectrum consisting of two lines with relative areas 9 and 1. Identify these substances.

Ans.   Diethyl ether, t-butyl alcohol

19.27. Describe the splittings in the spectrum for 1,1,2-trichloroethane.

Ans.   Triplet (1, 2, 1 intensities) and doublet (1 and 1 intensities)

19.28. In ethanol, $\sigma(CH_2)$ is between $\sigma(OH)$ and $\sigma(CH_3)$. How many lines will be observed for each peak in the NMR spectrum?

Ans. CH_3 peak is 3, OH peak is 3, and CH_2 peak is 8.

Chapter 20

Symmetry and Group Theory

Symmetry Operations and Elements

20.1 INTRODUCTION

Several types of *symmetry elements* are used to describe the symmetry present in molecules and crystals. These include the identity element, axes of proper rotation, a center of inversion, mirror planes, and axes of improper rotation (rotoreflection and rotoinversion)—although not all of these elements are necessarily present in a given case. Associated with each of these elements is a *symmetry operation* that transforms the molecule or crystal into a configuration indistinguishable from the original configuration. In crystallography, the additional operation of translation is permitted, which upon combination with the operations of proper rotation or reflection across a mirror plane generates the operations corresponding to the screw axis and glide plane elements, respectively.

There are two systems of symbols for representing symmetry elements: the *Schönflies system*, which is used primarily for molecular geometry and group theory, and the *Hermann–Mauguin system*, which is used in crystallography. The basic symmetry operation corresponding to a symmetry element is represented by putting a caret over the Schönflies symbol for that element; a superscript $1, 2, 3, \ldots$ is added to denote the symmetry operation equivalent to $1, 2, 3, \ldots$ repetitions of the basic operation (the superscript 1 is seldom used). An operation is called *distinct* if the equivalent position generated for the molecule or crystal cannot be generated in another way that is less complicated.

EXAMPLE 20.1. Describe the symmetry elements found in the letter M.

There are four operations that generate a configuration that is indistinguishable from the original letter: (1) If the letter is rotated by 180° out of and into the plane of the paper around the axis shown in Fig. 20-1*a*, an equivalent letter is formed—thus an axis of proper rotation is present (see Sec. 20-3); (2) if a plane is constructed perpendicular to the plane of the letter as shown in Fig. 20-1*b* and the various parts of the letter "reflected," an equivalent letter is formed—thus a mirror plane is present (see Sec. 20.5); (3) the plane of the letter (Fig. 20-1*c*) is, by the same reasoning, a mirror plane; (4) if the letter is simply left alone, an equivalent letter is certainly obtained—thus the identity element is present (see Sec. 20.2).

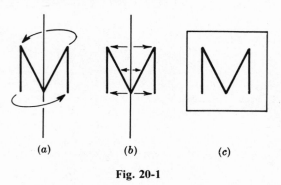

(a)   (b)   (c)

Fig. 20-1

20.2 IDENTITY

The identity element is always present in a molecule or crystal by virtue of the motif's being itself. The corresponding symmetry operation is the operation in which nothing is moved. The Schönflies symbols E and $\hat{E}$ represent the element and the operation, respectively, and the Hermann–Mauguin symbol for the element is 1. Although trivial in concept, the identity operation is important as a limit for the repetitions of a given operation (see Example 20.3).

EXAMPLE 20.2   To help in understanding an operation or to illustrate the relationship of crystal faces or atoms in a molecule around a symmetry element, perspective sketches and/or orthographic projections can be prepared. In a perspective sketch a rhombus represents a plane perpendicular to the page and a line represents an axis that is perpendicular to this plane (and lies in the plane of the page). The crystal face or atom is represented by a motif, usually a point or an asymmetrical figure like a 2, that is drawn in a general position above the plane and off the axis. The perspective sketch in Fig. 20-2a illustrates the identity operation.

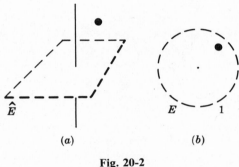

$$(a) \qquad\qquad (b)$$

Fig. 20-2

The orthographic projection is essentially a top view of a perspective sketch with the axis shown as a dot and the plane as a dashed circle. A point or motif above the plane is indicated by a solid dot or solid motif, and below the plane by a cross or dashed motif. Figure 20-2b is an orthographic projection illustrating both the result of the identity operation and the distribution of points (in this case, one point) around the identity element.

20.3   AXIS OF PROPER ROTATION

The basic operation corresponding to the element C_n (Schönflies) or n (Hermann–Mauguin) consists in rotating the motif around the axis (by convention, counterclockwise as viewed in the orthographic projection) by an angle of $2\pi/n$ and generating the motif in the new location. The various operations are represented in the Schönflies notation by $\hat{C}_n^k$ and generate a total of n equivalent crystal faces or positions of atoms around the element. The axis for which n is largest is chosen as the z axis of a Cartesian coordinate system and is known as the *principal rotation axis*.

EXAMPLE 20.3.   Make perspective and orthographic drawings for $\hat{C}_4^k$. Identify the distinct operations. Prepare similar diagrams illustrating the complete set of equivalent positions for the C_4 (or 4) element.

The basic operation consists in rotating the motif around the axis by an angle of $2\pi/4 = 90°$ from the original point. The required sketches for $1 \le k \le 4$ are shown in Fig. 20-3. No new locations are generated for values of $k > 4$, so these values of k need not be considered. Because $\hat{C}_4^4 = \hat{C}_1 = \hat{E}$ and $\hat{C}_4^2 = \hat{C}_1^2$, these operations are not distinct; but the operations $\hat{C}_4^1$ and $\hat{C}_4^3$ generate points in locations that cannot be generated in a less complex manner and are distinct.

The symmetry element C_4 or 4 will have about it all four of the points generated by the above operations (see Fig. 20-3e). Note that the planar shapes ●, ◖, ▲, ■, etc., are used to describe the appropriate axes in the figures.

20.4   CENTER OF SYMMETRY AND INVERSION

The basic inversion operation corresponding to i (Schönflies) or $\bar{1}$ (Hermann–Mauguin) consists in projecting a motif equidistant through a center of symmetry located at the point of intersection

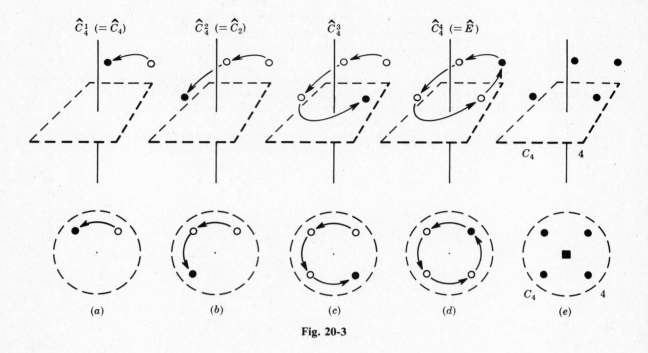

Fig. 20-3

between the plane and the axis used in the orthographic and perspective drawings (see Fig. 20-4*a*). If this element appears in a molecule or crystal, the two equivalent points will be related as shown in Fig. 20-4*b*.

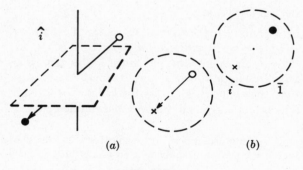

Fig. 20-4

20.5 MIRROR PLANE

The basic reflection operation corresponding to the element σ (Schönflies) or m (Hermann-Mauguin) consists in creating a mirror image of a motif equidistant from and perpendicular to a plane. If the mirror plane is perpendicular to or contains the principal axis of rotation, the respective Schönflies symbols are σ_h and σ_v. The drawings usually indicate the σ_h plane by a solid figure (see Fig. 20-5*a*). If the mirror plane is perpendicular to or contains any *n*-fold axis of rotation, the respective Hermann-Mauguin symbols are n/m and m. The Schönflies symbol σ_d means a σ_v that bisects the angle formed by two C_2 axes that lie in a plane perpendicular to the axis. Figure 20-5*b* shows the operation $\hat{\sigma}_v$, and Fig. 20-5*c* shows the distribution of points about the σ_h ($=1/m$) and σ_v ($=m$) elements.

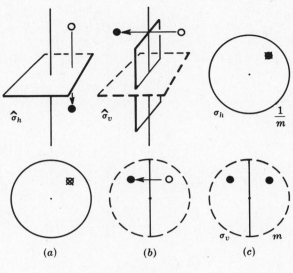

Fig. 20-5

20.6 ROTOREFLECTION

The basic operation corresponding to the element S_n (Schönflies) or $\tilde{n}$ (Hermann–Mauguin, pronounced "tilday en") consists in a rotation about an axis through $2\pi/n$ (i.e., $\hat{C}_n$) and a reflection through a plane perpendicular to the axis (i.e., $\hat{\sigma}_h$), in either order. For some values of n, the operation $\hat{S}_n^k$ will not produce the same result as $\hat{E}$ until $k = 2n$. Rotoreflection operations are much used in describing molecular geometry, whereas a second method of generating the same distribution of points in space—rotoinversion—is used extensively by crystallographers. Unshaded geometrical figures are used to represent S_n axes. Combination figures may appear; e.g., ◪, which indicates coincident S_4 and C_2 axes.

EXAMPLE 20.4. Prepare perspective and orthographic drawings for $\hat{S}_1^k$, $\hat{S}_2^k$, and $\hat{S}_4^k$. Prepare orthographic projections for the equivalent positions found around the elements S_1, S_2, and S_4.

The basic operation for S_1 consists in rotating the point 360° and reflecting it through the orthographic plane (see Fig. 20-6a). The operation performed twice, $\hat{S}_1^2$, generates the original point (see Fig. 20-6b). Because $\hat{S}_1 = \hat{\sigma}_h$ and $\hat{S}_1^2 = \hat{E}$, these operations are not distinct.

The diagrams in Fig. 20-6c illustrate $\hat{S}_2^1$, $\hat{S}_2^2$ and $\hat{S}_4^1$, $\hat{S}_4^2$, $\hat{S}_4^3$, $\hat{S}_4^4$. Figure 20-6d illustrates the complete set of equivalent faces present in a crystal, or atoms in a molecule, that contains the S_1, S_2, or S_4 element.

20.7 ROTOINVERSION

The basic operation corresponding to $\bar{n}$ (Hermann–Mauguin only, pronounced "n bar" or "bar n") consists in a rotation about an axis through $2\pi/n$ (i.e., $\hat{C}_n$) and an inversion through the point of intersection between the plane and the axis used in the orthographic and perspective drawings (i.e., $\hat{i}$), in either order. For a system containing the $\bar{n}$ element, there will be n or $2n$ points alternating above and below the orthographic projection plane.

EXAMPLE 20.5. Prepare drawings for $\bar{1}$, $\bar{2}$, and $\bar{4}$ to determine the equivalences between $\bar{n}$ and $\tilde{n}$, where $n = 1$, 2, and 4.

The basic operations corresponding to $\bar{1}$ and $\bar{2}$ are shown in Fig. 20-7a, and the complete sets of points for the three rotoinversion elements are shown in Fig. 20-7b. Upon comparison of these latter figures with Fig. 20-6d, it is obvious that $\bar{1} = \tilde{2}$, $\bar{2} = \tilde{1}$, and $\bar{4} = \tilde{4}$.

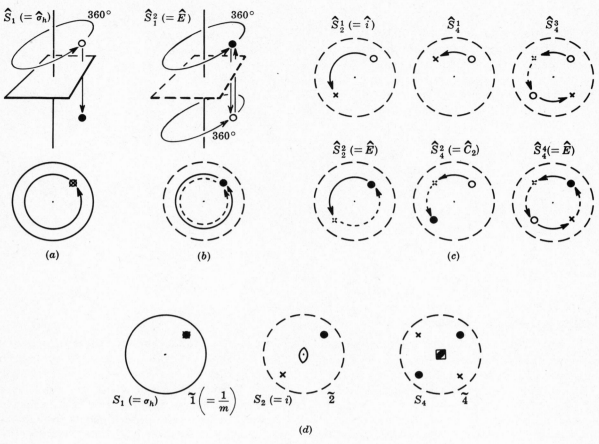

Fig. 20-6

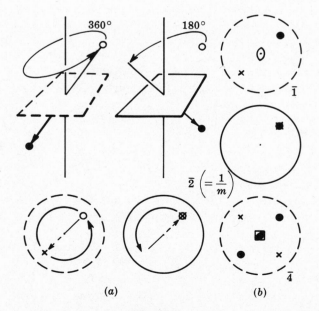

Fig. 20-7

20.8 TRANSLATION

The translational operation is the movement of a motif along a straight line and the construction of its image at a distance t from the original position (see Fig. 20-8a). If the operation is continued a large number of times, a one-dimensional array of points (commonly called a *row*) is generated. Just as the various rotation, reflection, and inversion operations are performed until the result of $\hat{E}$ is generated, the various operations involving translation are continued until a motif in a position similar to the original at some multiple of the distance t is generated. The distance t is chosen (anticipating the discussion of unit cells given in Sec. 22.1) such that it contains the points necessary to generate the entire row of points by placing several of these line segments end to end (see Fig. 20-8b).

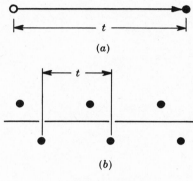

(a)

(b)

Fig. 20-8

20.9 SCREW AXIS

The basic operation corresponding to the element n_k (Hermann–Mauguin only), where k is an integer, consists in a rotation by $2\pi/n$ about an axis (by convention, clockwise as viewed in the orthographic projection) and a translation by kt/n, in either order. This operation is coninued n times until a motif is generated in a position that is identical with the original except that it has been translated by a multiple of t. The permitted screw axis elements in crystals are 2_1; 3_1, 3_2; 4_1, 4_2, 4_3; and 6_1, 6_2, 6_3, 6_4, 6_5. The planar symbols associated with rotational axes in the drawings are modified with "tails" for screw axes to indicate the effective direction of rotation.

EXAMPLE 20.6　Prepare perspective and orthographic drawings for 4_1 and 4_2.

The basic operation associated with 4_1 is a rotation by $2\pi/4 = 90°$ and a translation by $t/4$. By continuing the operation four times, a motif is generated in a position similar to the original at a distance t; thus the set of points shown in Fig. 20-9a are generated. If the orthographic plane is placed at t, the projection shown in Fig. 20-9a results, where the fractions represent the distances in fractions of t between the points and the plane.

The basic operation corresponding to 4_2 is rotation by $90°$ and translation by $2t/4 = t/2$. This operation must be continued four times, covering a distance $2t$, before generating a motif in a similar position to the original (see Fig. 20-9b). Because all points must be contained within a distance t (i.e., must keep within the unit cell), the positions generated by the various operations are transformed to positions within the permitted length by subtraction of suitable integral multiples of t. This is shown by the double-headed arrows in Fig. 20-9b. The rather messy working diagram is usually simplified to that shown on the right of Fig. 20-9b, which apparently illustrates the operation 2_1 performed on a pair of points. Assuming that the orthographic plane passes through $t/2$, the projection shown in Fig. 20-9b results.

20.10 GLIDE PLANES

The basic operation corresponding to the element t/k consists in a translation by t/k parallel to a line and a reflection across a plane that contains the translation axis and is normal to the perpendicular

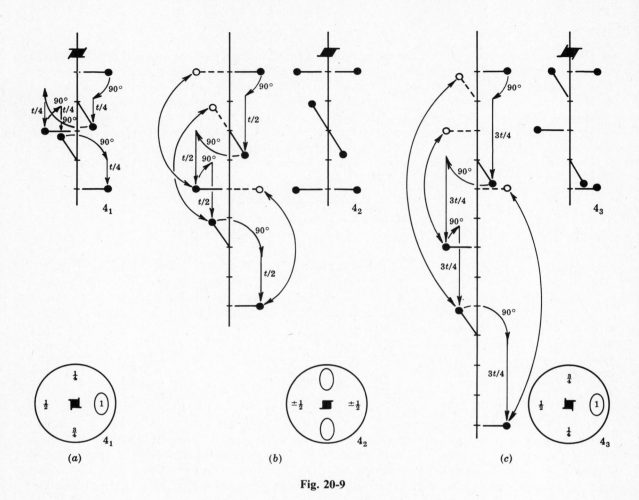

Fig. 20-9

from the motif to the translation axis. The operation is continued until a motif is generated in a position similar to the original at some multiple of t. The three classifications of glide planes permitted in crystals are (1) an *axial glide* along an axis of the crystal, $t/2$; (2) a *diagonal glide*, which occurs in the plane containing two axes of the crystal, $t_1/2 + t_2/2$; and (3) *a diamond glide*, which is a three-dimensional operation, $t_1/4 + t_2/4$.

EXAMPLE 20.7.   Find the relationship between $t/2$ and 2_1.

The perspective sketches of the location of all points around these elements are shown in Fig. 20-10. The symmetry operations used to generate these points are shown in dashed lines. Because the two elements have identical points, $t/2 = 2_1$.

Point Groups

20.11   CONCEPT

Usually more than one symmetry element is present in a molecule or crystal. The set of all symmetry elements (or their representative operations) present in a physical system is called a *group*. If the set includes no elements involving translation (namely, t, n_k, and t/k), a *point group* is formed. For crystals, only the rotational values $n = 1, 2, 3, 4$, and 6 are permitted, and 32 *crystallographic point groups* are generated. If all symmetry operations possible in crystals are considered, 230 *space groups* result. The

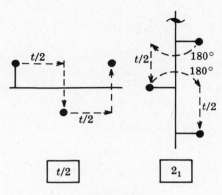

Fig. 20-10

Schönflies symbol for a group is a script or boldface letter similar to the notation for the symmetry elements, and the Hermann–Mauguin representation is simply a combination of symbols (up to three) for the symmetry elements present in that group.

A number and a symbol in a listing of the various elements in a group represents the number of elements of that type that are related by other symmetry operations in the group, and if these elements need to be identified, alphanumeric subscripts are used. Similar elements that are not related by other symmetry operations in the group are listed separately and are differentiated by primes or parentheses containing orientation information.

EXAMPLE 20.8.   Consider a molecule of water (nonlinear). Determine the symmetry elements present.

Figure 20.11 shows that the elements E, C_2, and two types of σ_v are present.

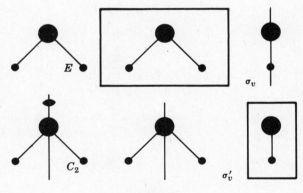

Fig. 20-11

20.12 MATHEMATICAL PROPERTIES OF A POINT GROUP

1. If $\hat{A}$ and $\hat{B}$ are symmetry operations in the group, then $\hat{A} \times \hat{B} = \hat{F}$, where $\hat{F}$ is also a symmetry operation in the group. The product $\hat{A} \times \hat{B}$ means that operation $\hat{B}$ is performed and then operation $\hat{A}$ is performed on the result. The order in which two operations are performed is important because usually $\hat{A} \times \hat{B} \neq \hat{B} \times \hat{A}$.
2. The group contains an identity operation $\hat{E}$ such that $\hat{A} \times \hat{E} = \hat{E} \times \hat{A} = \hat{A}$ for every $\hat{A}$ in the group.
3. For every operation $\hat{A}$ in the group there exists an inverse operation $\hat{A}^{-1}$ in the group, such that $\hat{A}^{-1} \times \hat{A} = \hat{A} \times \hat{A}^{-1} = \hat{E}$.
4. The associative law, $\hat{A} \times (\hat{B} \times \hat{C}) = (\hat{A} \times \hat{B}) \times \hat{C}$, holds.

EXAMPLE 20.9.   Prepare the multiplication table for the group of distinct symmetry operations in water: $\hat{E}$, $\hat{C}_2$, $\hat{\sigma}_v$, and $\hat{\sigma}_v'$. Find $\hat{C}_2^{-1}$.

The first step is to make projections of the known distinct operations (see Fig. 20-12). Always place the motif in a general position in these projections, so that a complete set of results can be obtained. The second step is to construct orthographic projections for the various multiplications, except for those involving $\hat{E}$ (see Fig. 20-13). The third step is to prepare a table with the operations $\hat{A}$ listed at the left, the operations $\hat{B}$ listed at the top, and the products $\hat{F}$ listed within the table (see Table 20-1). By inspection of Table 20-1, $\hat{C}_2^{-1} = \hat{C}_2$ because $\hat{C}_2 \times \hat{C}_2 = \hat{E}$. The group specified by Table 20-1 is designated $\mathscr{C}_{2v}$ in the Schönflies system.

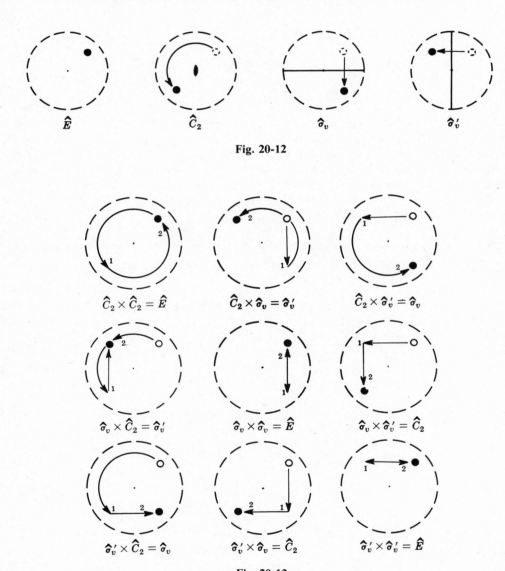

Fig. 20-12

Fig. 20-13

20.13   DETERMINATION OF A POINT GROUP

Once the symmetry elements have been identified, the point group to which a molecule belongs can be determined using the flow chart given in Fig. 20-14. The symmetry elements contained in the point groups are summarized in Table 20-2. In a macroscopic crystal, screw axes and glide planes

Table 20-1

$\mathscr{C}_{2v}$	$\hat{E}$	$\hat{C}_2$	$\hat{\sigma}_v$	$\hat{\sigma}_v' = \hat{B}$
$\hat{A} = \hat{E}$	$\hat{E}$	$\hat{C}_2$	$\hat{\sigma}_v$	$\hat{\sigma}_v'$
$\hat{C}_2$	$\hat{C}_2$	$\hat{E}$	$\hat{\sigma}_v'$	$\hat{\sigma}_v$
$\hat{\sigma}_v$	$\hat{\sigma}_v$	$\hat{\sigma}_v'$	$\hat{E}$	$\hat{C}_2$
$\hat{\sigma}_v'$	$\hat{\sigma}_v'$	$\hat{\sigma}_v$	$\hat{C}_2$	$\hat{E}$

appear as axes of proper rotation and mirror planes, respectively, and Fig. 20-14 can be used to determine which of the 32 possible crystallographic point groups pertains. This assignment will agree with that for the unit cell of the crystal (Sec. 22.1) if the macroscopic crystal displays all the symmetry elements present in the unit cell.

EXAMPLE 20.10. Determine the point group for water (see Fig. 20-11).

Using Fig. 20-14, the following analysis can be made: (1) Are there ∞ C_∞ axes present? no; (2) is there a pentagonal dodecahedron or icosahedron present? no; (3) are there four C_3 axes at $50°44'$? no; (4) is there at least one C_n where $n \geq 2$? yes, C_2; (5) is there an S_{2n} present? no; (6) are there n C_2 axes perpendicular to C_n? no; (7) are there any σ_h planes present? no; (8) are there n σ_v planes present? yes, 2, therefore $\mathscr{C}_{2v}$.

Representation of Groups

20.14 MATRIX EXPRESSIONS FOR OPERATIONS

If the distinct operations of a group are considered to form the point group, many results can be derived abstractly that are universally applicable to any molecule or crystalline unit cell that is found in that group. For example, all nonlinear molecules having the formula AB_2 (e.g., NO_2, H_2O, SO_2) can be shown to have identical modes of intramolecular vibration by using group theory.

Each operation contained in a point group can be expressed in matrix form (see Table 20-3) such that the matrix will serve as well as the original operator in performing coordinate transformations, producing a valid multiplication table for the group, etc.

EXAMPLE 20.11. Consider the symmetry elements for the $\mathscr{C}_{2v}$ point group displayed in Fig. 20-15. Show that in the given coordinate system the matrix expression for $\hat{\sigma}_v$ is

$$\begin{bmatrix} 1 & 0 & 0 \\ 0 & -1 & 0 \\ 0 & 0 & 1 \end{bmatrix}$$

A point whose coordinates are x_1, y_1, z_1 is taken by $\hat{\sigma}_v$ into the point x_2, y_2, z_2, where

$$x_2 = \hat{\sigma}_v \times x_1 = x_1 \qquad y_2 = \hat{\sigma}_v \times y_1 = -y_1 \qquad z_2 = \hat{\sigma}_v \times z_1 = z_1$$

This can be stated alternatively as

$$x_2 = 1x_1 + 0y_1 + 0z_1 \qquad y_2 = 0x_1 - 1y_1 + 0z_1 \qquad z_2 = 0x_1 + 0y_1 + 1z_1$$

or in matrix form as

$$\begin{bmatrix} x_2 \\ y_2 \\ z_2 \end{bmatrix} = \begin{bmatrix} 1 & 0 & 0 \\ 0 & -1 & 0 \\ 0 & 0 & 1 \end{bmatrix} \begin{bmatrix} x_1 \\ y_1 \\ z_1 \end{bmatrix}$$

which is the desired result. Note that the matrix can also be obtained by substituting $\beta = 0$ in Table 20-3.

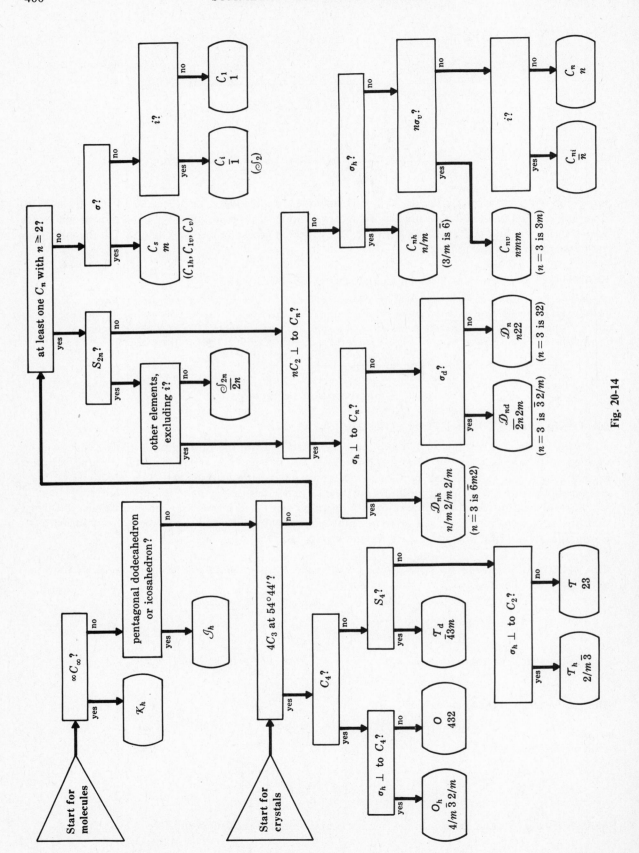

Fig. 20-14

Table 20-2

Point Group	Special Comments	E	Axes of Rotation	σ_h	σ_v	i
$\mathscr{K}_h$	Sphere	√	∞C_∞	√	$\infty\sigma_d$	√
$\mathscr{I}_h$	Regular pentagonal dodecahedron (12 pentagons) or icosahedron (20 triangles)	√	$6C_5\,(S_{10})$ $15C_2\,(S_4)$		$15\sigma_d$	√
$\mathscr{O}_h$ $4/m\,\bar{3}\,2/m$	Octahedron or cube	√	$4C_3\,(S_6)$ $3C_4\,(S_4)$ $6C_2$	√	$6\sigma_d$	√
$\mathscr{O}$ 432		√	$4C_3$ $3C_4$ $6C_2$			
$\mathscr{T}_d$ $\bar{4}3m$	Tetrahedron	√	$4C_3$ $3C_2\,(S_4)$		$6\sigma_d$	
$\mathscr{T}_h$ $2/m\,\bar{3}$		√	$4C_3\,(S_6)$ $3C_2$	√		√
$\mathscr{T}$ 23		√	$4C_3$ $3C_2$			
$\mathscr{S}_{2n}$ $\overline{2n}$	$n=2,3,4,\ldots$ $\mathscr{S}_1$ is $\mathscr{C}_s$, $\mathscr{S}_2$ is $\mathscr{C}_i$ $\mathscr{S}_n$ is $\mathscr{C}_{nh}$ if n odd	√	$S_{2n}\,(C_n)$			If n odd
$\mathscr{D}_{nh}$ $n/m\,2/m\,2/m$	$n=2,3,4,\ldots$ $\mathscr{D}_{1h}$ is $\mathscr{C}_{2v}$	√	$C_n\,(S_n)$ nC_2	√	$n\sigma_v$	If n even
$\mathscr{D}_{nd}$ $\overline{2n}\,2m$	$n=2,3,4,\ldots$ D_{1d} is $\mathscr{C}_{2h}$	√	$C_n\,(S_{2n})$ nC_2		$n\sigma_d$	If n odd
$\mathscr{D}_n$ $n22$	$n=2,3,4,\ldots$ $\mathscr{D}_1$ is $\mathscr{C}_2$	√	C_n nC_2			
$\mathscr{C}_{nh}$ n/m	$n=2,3,4,\ldots$ $\mathscr{C}_{1h}$ is $\mathscr{C}_s$	√	$C_n\,(S_n)$	√		If n even
$\mathscr{C}_{nv}$ nmm	$n=2,3,4,\ldots$ $\mathscr{C}_{1v}$ is $\mathscr{C}_s$	√	C_n		$n\sigma_v$	
$\mathscr{C}_{ni}$ $\bar{n}$	$n=3,5,7,\ldots$ $\mathscr{C}_{1i}$ is $\mathscr{C}_i$ For even values of n, $\mathscr{C}_{ni}$ is $\mathscr{C}_{nh}$ if $n/2$ is odd and $\mathscr{S}_{2n}$ if $n/2$ is even.	√	$C_n\,(S_{2n})$			√
$\mathscr{C}_n$ n	$n=2,3,4,\ldots$	√	C_n			
$\mathscr{C}_s$ m		√		√		
$\mathscr{C}_i$ $\bar{1}$		√				√
$\mathscr{C}_1$ 1		√				

Table 20-3

$\hat{E} \longrightarrow \begin{bmatrix} 1 & 0 & 0 \\ 0 & 1 & 0 \\ 0 & 0 & 1 \end{bmatrix}$	$\hat{C}_n(z)^m \longrightarrow \begin{bmatrix} \cos(2\pi m/n) & -\sin(2\pi m/n) & 0 \\ \sin(2\pi m/n) & \cos(2\pi m/n) & 0 \\ 0 & 0 & 1 \end{bmatrix}$
$\hat{i} \longrightarrow \begin{bmatrix} -1 & 0 & 0 \\ 0 & -1 & 0 \\ 0 & 0 & -1 \end{bmatrix}$	$\hat{S}_n(z)^m \longrightarrow \begin{bmatrix} \cos(2\pi m/n) & -\sin(2\pi m/n) & 0 \\ \sin(2\pi m/n) & \cos(2\pi m/n) & 0 \\ 0 & 0 & -1 \end{bmatrix}$ for odd m
$\hat{\sigma}_h \longrightarrow \begin{bmatrix} 1 & 0 & 0 \\ 0 & 1 & 0 \\ 0 & 0 & -1 \end{bmatrix}$	$\hat{\sigma}_v \longrightarrow \begin{bmatrix} \cos 2\beta & \sin 2\beta & 0 \\ \sin 2\beta & -\cos 2\beta & 0 \\ 0 & 0 & 1 \end{bmatrix}$ β = angle between σ_v and the x axis

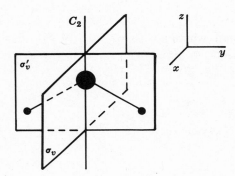

Fig. 20-15

EXAMPLE 20.12. Using matrix expressions for $\hat{C}_2$ and $\hat{E}$, show that $\hat{C}_2 \times \hat{C}_2 = \hat{E}$ as given in Table 20-1 for the $\mathscr{C}_{2v}$ point group.

Substituting $n = 2$ and $m = 1$ into the general expression given in Table 20-3 for $\hat{C}_n^m$ gives

$$\hat{C}_2 \longrightarrow \begin{bmatrix} \cos \pi & -\sin \pi & 0 \\ \sin \pi & \cos \pi & 0 \\ 0 & 0 & 1 \end{bmatrix} = \begin{bmatrix} -1 & 0 & 0 \\ 0 & -1 & 0 \\ 0 & 0 & 1 \end{bmatrix}$$

Thus,
$$\hat{C}_2 \times \hat{C}_2 = \begin{bmatrix} -1 & 0 & 0 \\ 0 & -1 & 0 \\ 0 & 0 & 1 \end{bmatrix}\begin{bmatrix} -1 & 0 & 0 \\ 0 & -1 & 0 \\ 0 & 0 & 1 \end{bmatrix} = \begin{bmatrix} 1 & 0 & 0 \\ 0 & 1 & 0 \\ 0 & 0 & 1 \end{bmatrix} = \hat{E}$$

20.15   REPRESENTATIONS

A *representation* (V) for a point group is any set of square matrices that multiply as the symmetry operations of the group.

A *reducible representation* contains matrices that can all be partitioned into the same block-diagonal form consisting of a series of submatrices that lie along the main diagonal, exclude only zero elements, and can be treated as single elements in obtaining a matrix product. Usually these submatrices are the matrices for the *irreducible representations* (V_i) that cannot be partitioned in the above fashion. In that case the representation V is said to be the *direct sum* (common notations are +, + and ⊕) of the irreducible representations V_i. For example, if a reducible representation V consists of the matrices

$[A]$, $[B]$, and $[C]$, each of which can be partitioned into the block-diagonal form

$$[X] = \begin{bmatrix} x_{11} & x_{12} & 0 & 0 \\ x_{21} & x_{22} & 0 & 0 \\ 0 & 0 & x_{33} & 0 \\ 0 & 0 & 0 & x_{44} \end{bmatrix} = \begin{bmatrix} x_1 & 0 & 0 \\ 0 & x_2 & 0 \\ 0 & 0 & x_3 \end{bmatrix}$$

where $[x_1] = \begin{bmatrix} x_{11} & x_{12} \\ x_{21} & x_{22} \end{bmatrix}$, $[x_2] = [x_{33}]$, and $[x_3] = [x_{44}]$, then

$$[A] \times [B] = \begin{bmatrix} a_1 & 0 & 0 \\ 0 & a_2 & 0 \\ 0 & 0 & a_3 \end{bmatrix} \times \begin{bmatrix} b_1 & 0 & 0 \\ 0 & b_2 & 0 \\ 0 & 0 & b_3 \end{bmatrix} = \begin{bmatrix} c_1 & 0 & 0 \\ 0 & c_2 & 0 \\ 0 & 0 & c_3 \end{bmatrix} = [C]$$

where $[a_1] \times [b_1] = [c_1]$, $[a_2] \times [b_2] = [c_2]$, $[a_3] \times [b_3] = [c_3]$, and

$$V = V_1 \oplus V_2 \oplus V_3$$

where V_i is the irreducible representation consisting of the matrices $[a_i]$, $[b_i]$, and $[c_i]$.

The number of nonequivalent irreducible representations for a group is equal to the number of *classes of operations* in the group, where a class can be loosely defined as a subset of operations that are closely related (e.g., in $\mathscr{C}_{3v}$ the classes are $\hat{E}$; $\hat{C}_3$ and $\hat{C}_3^2$; and $\hat{\sigma}_{v1}$, $\hat{\sigma}_{v2}$, and $\hat{\sigma}_{v3}$). If l_i represents the dimension of the ith irreducible representation (number of rows or columns in the square matrices), then

$$\sum_{V_i} l_i^2 = h \tag{20.1}$$

where h is known as the *order* of the group and is equal to the number of distinct operations within the group.

EXAMPLE 20.13. Show that the representation for $\mathscr{C}_{2v}$ derived from Table 20-3 is reducible. Find the irreducible representations for this group.

The representation consisting of the expressions given by Table 20-3 is

$$V: \quad \hat{E} \longrightarrow \begin{bmatrix} 1 & 0 & 0 \\ 0 & 1 & 0 \\ 0 & 0 & 1 \end{bmatrix} \quad \hat{C}_2 \longrightarrow \begin{bmatrix} -1 & 0 & 0 \\ 0 & -1 & 0 \\ 0 & 0 & 1 \end{bmatrix}$$

$$\hat{\sigma}_v \longrightarrow \begin{bmatrix} 1 & 0 & 0 \\ 0 & -1 & 0 \\ 0 & 0 & 1 \end{bmatrix} \quad \hat{\sigma}'_v \longrightarrow \begin{bmatrix} -1 & 0 & 0 \\ 0 & 1 & 0 \\ 0 & 0 & 1 \end{bmatrix}$$

and has dimension 3. If this representation were one of the irreducible ones, its contribution to the left-hand side of (20.1) would be such that the sum over all the irreducible representations would be equal to 4, the number of distinct operations. But the contribution from this representation alone is $3^2 = 9$, so it must be reducible.

Three of the irreducible representations of this group can be found by converting all of the above matrices to the block-diagonal form as shown above. Each row of blocks provides an irreducible representation. In this case the blocks are 1 by 1 matrices. Labeling these representations using subscripts corresponding to entries in the character table for this group (see Table 20-4), we have $V = V_3 \oplus V_4 \oplus V_1$, where

$$V_3: \quad \hat{E} \longrightarrow [1] \quad \hat{C}_2 \longrightarrow [-1] \quad \hat{\sigma}_v \longrightarrow [1] \quad \hat{\sigma}'_v \longrightarrow [-1]$$

$$V_4: \quad \hat{E} \longrightarrow [1] \quad \hat{C}_2 \longrightarrow [-1] \quad \hat{\sigma}_v \longrightarrow [-1] \quad \hat{\sigma}'_v \longrightarrow [1]$$

$$V_1: \quad \hat{E} \longrightarrow [1] \quad \hat{C}_2 \longrightarrow [1] \quad \hat{\sigma}_v \longrightarrow [1] \quad \hat{\sigma}'_v \longrightarrow [1]$$

For the dimension of the fourth irreducible representation, V_2, (20.1) gives

$$1^2 + l_2^2 + 1^2 + 1^2 = 4$$

Table 20-4

$\mathscr{C}_{2v}$ **Representation**	$\hat{E}$	$\hat{C}_2$	$\hat{\sigma}_v$	$\hat{\sigma}_v'$	**Translation and Rotation**
$V_1 = A_1$	1	1	1	1	z
$V_2 = A_2$	1	1	-1	-1	R_z
$V_3 = B_1$	1	-1	1	-1	x, R_y
$V_4 = B_2$	1	-1	-1	1	y, R_x

or $l_2 = 1$. Using methods not discussed here, it can be shown that

$$V_2: \qquad \hat{E} \longrightarrow [1] \qquad \hat{C}_2 \longrightarrow [1] \qquad \hat{\sigma}_v \longrightarrow [-1] \qquad \hat{\sigma}_v' \longrightarrow [-1]$$

20.16 CHARACTER

Many applications of group theory can be treated thoroughly using the *characters* of the matrices making up the irreducible representations rather than the set of entire matrices. The *character* (or *trace*) for the operation $\hat{R}$, $\chi(V, \hat{R})$, is defined as

$$\chi(V, \hat{R}) = \sum r_{jj} \tag{20.2}$$

where the r_{jj} are the diagonal elements of the matrix corresponding to $\hat{R}$ in the representation V. Some properties of the characters of irreducible representations that are of interest include

$$\sum_{\hat{R}} \chi(V_i, \hat{R})\chi(V_j, \hat{R}) = h\delta_{ij} \tag{20.3}$$

where $\delta_{ij} = 0$ if $i \neq j$ and $\delta_{ij} = 1$ if $i = j$;

$$\sum_{V_i} [\chi(V_i, \hat{E})]^2 = h \tag{20.4}$$

and if $\hat{R}_1$ and $\hat{R}_2$ belong to the same class,

$$\chi(V_i, \hat{R}_1) = \chi(V_i, \hat{R}_2) \tag{20.5}$$

EXAMPLE 20.14. The characters of the irreducible representations of a group can be used to determine the number of times the irreducible representation V_i occurs in a reducible representation V. If v_i represents thus number,

$$v_i = \frac{1}{h}\sum_{\hat{R}} \chi(V, \hat{R})\chi(V_i, \hat{R}) \tag{20.6}$$

Show that the representation for $\mathscr{C}_{2v}$ given in Example 20.13 is the direct sum of V_1, V_3, and V_4.

Using (20.2) the characters of the reducible representation are

$$\chi(V, \hat{E}) = [1]+[1]+[1] = 3 \qquad \chi(V, \hat{C}_2) = [-1]+[-1]+[1] = -1$$

$$\chi(V, \hat{\sigma}_v) = [1]+[-1]+[1] = 1 \qquad \chi(V, \hat{\sigma}_v') = [-1]+[1]+[1] = 1$$

and those of the irreducible representations are

$$\chi(V_1, \hat{E}) = 1 \qquad \chi(V_1, \hat{C}_2) = 1 \qquad \chi(V_1, \hat{\sigma}_v) = 1 \qquad \chi(V_1, \hat{\sigma}_v') = 1$$

$$\chi(V_2, \hat{E}) = 1 \qquad \chi(V_2, \hat{C}_2) = 1 \qquad \chi(V_2, \hat{\sigma}_v) = -1 \qquad \chi(V_2, \hat{\sigma}_v') = -1$$

$$\chi(V_3, \hat{E}) = 1 \qquad \chi(V_3, \hat{C}_2) = -1 \qquad \chi(V_3, \hat{\sigma}_v) = 1 \qquad \chi(V_3, \hat{\sigma}_v') = -1$$

$$\chi(V_4, \hat{E}) = 1 \qquad \chi(V_4, \hat{C}_2) = -1 \qquad \chi(V_4, \hat{\sigma}_v) = -1 \qquad \chi(V_4, \hat{\sigma}_v') = 1$$

Applying (20.6) gives

$$v_1 = \frac{1}{h}[\chi(V, \hat{E})\chi(V_1, \hat{E}) + \chi(V, \hat{C}_2)\chi(V_1, \hat{C}_2) + \chi(V, \hat{\sigma}_v)\chi(V_1, \hat{\sigma}_v) + \chi(V, \hat{\sigma}'_v)\chi(V_1, \hat{\sigma}'_v)]$$

$$= \tfrac{1}{4}[(3)(1) + (-1)(1) + (1)(1) + (1)(1)] = 1$$

$$v_2 = \tfrac{1}{4}[(3)(1) + (-1)(1) + (1)(-1) + (1)(-1)] = 0$$

$$v_3 = \tfrac{1}{4}[(3)(1) + (-1)(-1) + (1)(1) + (1)(-1)] = 1$$

$$v_4 = \tfrac{1}{4}[(3)(1) + (-1)(-1) + (1)(-1) + (1)(1)] = 1$$

which means that V can be reduced to

$$V = 1V_1 \oplus 0V_2 \oplus 1V_3 \oplus 1V_4 = V_1 \oplus V_3 \oplus V_4$$

20.17   CHARACTER TABLES

The characters of the symmetry operations for each nonequivalent irreducible representation can be found in a character table for the group, e.g., Table 20-4. In many tables the first column lists the designations of the representations in the notation developed by Mulliken (see Table 20-5). The first column may include additional information such as equivalent spectroscopic designations.

The second group of columns gives the values of $\chi(V_i, \hat{R})$, although the entries are usually made according to classes of operations instead of individual operations. The number of operations in a given class is given by an integer before the symbol for that class and must be considered when performing summations of characters, e.g., in (20.3) and (20.6). The symbol $\hat{C}_\infty^\phi$ or $\hat{S}_\infty^\phi$ represents a rotation of ϕ about a C_∞ or S_∞ axis.

Many character tables have third and fourth columns that assign molecular rotational and translational modes and various vector products to give V_i.

Applications of Group Theory to Molecular Properties

20.18   OPTICAL ACTIVITY

The S_n operation transforms a molecule or crystalline unit cell into its mirror image, and thus any potential optical activity vanishes. Because $\hat{S}_1 = \hat{\sigma}$ and $\hat{S}_2 = \hat{i}$, the presence of a mirror plane or center of symmetry also renders a system inactive.

EXAMPLE 20.15.   Discuss the optical activity of SiO_2.

The molecular structure of SiO_2 in the gaseous, liquid, or true solution phase is such that it contains a mirror plane and is optically inactive. However, SiO_4 tetrahedra are formed in the various solid forms of SiO_2, and, depending on their arrangement, the substance may be active. The crystalline arrangement for SiO_2 as cristobalite is $4/m\,\bar{3}\,2/m$ and the crystal cannot be optically active; as tridymite, it is $6/m\,2/m\,2/m$ and the crystal cannot be optically active; and as α-quartz, the arrangement is $3\,2$ and the crystal is observed to be optically active.

20.19   DIPOLE MOMENT

The dipole moment of a molecule must lie along the principal axis of rotation, and the molecule cannot contain more than one axis of this order. If a mirror plane is present, the moment will lie in this plane or along the line of intersection of several planes. The presence of an inversion center eliminates the presence of a dipole moment for the molecule.

Table 20-5   Mulliken Designation for Irreducible Representations

Designation	Interpretation
Capital letter	
A	One-dimensional and symmetric to rotation of $2\pi/n$ about the principal C_n axis as indicated by $\chi(V_i, \hat{C}_n) = 1$
B	One-dimensional and antisymmetric to rotation of $2\pi/n$ about the principal C_n axis as indicated by $\chi(V_i, \hat{C}_n) = -1$
E	Two-dimensional
T or F	Three-dimensional
G	Four-dimensional
H	Five-dimensional
Numeral subscript	
1	Symmetric to rotation about a secondary C_2 axis as indicated by $\chi(V_i, \hat{C}_2) = 1$
2	Antisymmetric to rotation about a secondary C_2 axis as indicated by $\chi(V_i, \hat{C}_2) = -1$ (if a secondary C_2 axis does not exist, used for σ_v if present)
Letter subscript	
g	Symmetric to inversion as indicated by $\chi(V_i, \hat{i}) = 1$
u	Antisymmetric to inversion as indicated by $\chi(V_i, \hat{i}) = -1$
Superscript	
'	Symmetric to σ_h as indicated by $\chi(V_i, \hat{\sigma}_h) = 1$
"	Antisymmetric to σ_h as indicated by $\chi(V_i, \hat{\sigma}_h) = -1$

20.20   MOLECULAR TRANSLATIONAL MOTION

The matrix expressions given in Table 20-3 are applicable to the translational motion of a rigid molecule, and by using (20.6) it is possible to find the sum of the irreducible representations that make up V_{trans}. Each of the x, y, and z components of the translational motion can be assigned to a specific V_i in the sum by equating the results of the various symmetry operations on the component to the results of the transformations effected by the matrices making up V_i. Because these assignments are a property of the point group and are valid for all molecules in the group, they are usually given in the character table.

An entry of x, y in a table has a different meaning than an entry of (x, y): the presence of parentheses indicates that x and y together give a two-dimensional irreducible representation, whereas the former notation means that both x and y are represented by the same one-dimensional irreducible representation.

EXAMPLE 20.16.   Determine which irreducible representation corresponds to translation along the y axis for a molecule in the $\mathscr{C}_{2v}$ point group.

In Example 20.14 it was shown that $V_{\text{trans}} = V_1 \oplus V_3 \oplus V_4$. To find which of these V_i corresponds to the y component of the translational motion, the four operations of the group are performed on y, giving

$$\hat{E} \times y = (+1)y \qquad \hat{C}_2 \times y = (-1)y \qquad \hat{\sigma}_v \times y = (-1)y \qquad \hat{\sigma}'_v \times y = (+1)y$$

which are the same as the transformations contained in V_4.

20.21 MOLECULAR ROTATIONAL MOTION

Each of the components of rotational motion—R_x, R_y, and R_z for a nonlinear molecule, and R_x and R_y for a linear molecule (R_z is used for optional orientations)—can be assigned to a specific V_i in the character table. The simplest technique for finding these assignments is based on the values of the characters in the V_i for the corresponding translational motion: $\chi(V_{R_i}, \hat{R})$ will be given by $\chi(V_i, \hat{R})$ for $\hat{R} = \hat{E}$ and $\hat{C}_n$, and by $-\chi(V_i, \hat{R})$ for $\hat{R} = \hat{S}_n$, $\hat{\sigma}$, and $\hat{\imath}$. Because these assignments are a property of the point group and are valid for all molecules in the group, they are usually given in the character table.

EXAMPLE 20.17. Determine which irreducible representation corresponds to R_x, the rotation component about the x axis, in the $\mathscr{C}_{2v}$ point group.

Applying the above rules to the characters given for V_3, to which the x component of translation is assigned, gives

$$\chi(V_{R_x}, \hat{E}) = \chi(V_3, \hat{E}) = 1 \qquad\qquad \chi(V_{R_x}, \hat{\sigma}_v) = -\chi(V_3, \hat{\sigma}_v) = -1$$

$$\chi(V_{R_x}, \hat{C}_2) = \chi(V_3, \hat{C}_2) = -1 \qquad\qquad \chi(V_{R_x}, \hat{\sigma}'_v) = -\chi(V_3, \hat{\sigma}'_v) = -(-1) = 1$$

which corresponds to V_4 in Table 20-4.

20.22 VIBRATIONAL MOTION FOR POLYATOMIC MOLECULES

The symmetry and degeneracy of the normal modes of intramolecular vibration can be determined for a polyatomic molecule from the irreducible representations of its symmetry group. Rather than working with the complete matrices, it is necessary only to consider the characters of these matrices, where

$$\chi(V_{3\Lambda}, \hat{E}) = 3\Lambda_{\hat{E}} \tag{20.7a}$$

$$\chi(V_{3\Lambda}, \hat{C}_n^m) = \Lambda_{\hat{C}_n^m}\left(1 + 2\cos\frac{2\pi m}{n}\right) \tag{20.7b}$$

$$\chi(V_{3\Lambda}, \hat{S}_n^m) = \Lambda_{\hat{S}_n^m}\left(-1 + 2\cos\frac{2\pi m}{n}\right) \qquad (m\text{ odd}) \tag{20.7c}$$

$$\chi(V_{3\Lambda}, \hat{\imath}) = -3\Lambda_{\hat{\imath}} \tag{20.7d}$$

$$\chi(V_{3\Lambda}, \hat{\sigma}) = \Lambda_{\hat{\sigma}} \tag{20.7e}$$

where $\Lambda_{\hat{R}}$ represents the number of unmoved nuclei for the operation $\hat{R}$ (e.g., $\Lambda_{\hat{E}} = \Lambda = $ number of nuclei in the molecule, $\Lambda_{\hat{\imath}} = 0$ or 1, etc.). Once the $\chi(V_{3\Lambda}, \hat{R})$ have been determined using (20.7), (20.6) is used to determine a sum of irreducible representations from which the six (for nonlinear) or five (for linear) translational and rotational contributions are subtracted, leaving the sum of irreducible representations that describe the internal motion of the molecule. For simple molecules, these degrees of freedom will be assigned to vibrational motion, V_{vib}. Because this final sum depends on the number of atoms present in the molecule, rather than being a property of the group, it must be determined for each molecule and will not appear in a character table.

EXAMPLE 20.18. Describe the vibrational motion in water.

For a molecule in $\mathscr{C}_{2v}$, equations (20.7) give

$$\chi(V_{3\Lambda}, \hat{E}) = 3(3) = 9 \qquad\qquad \chi(V_{3\Lambda}, \hat{C}_2) = (1)\left(1 + 2\cos\frac{2\pi(1)}{2}\right) = -1$$

$$\chi(V_{3\Lambda}, \hat{\sigma}_v) = 1 \qquad\qquad\qquad \chi(V_{3\Lambda}, \hat{\sigma}'_v) = 3$$

Using (20.6) gives

$$v_1 = \tfrac{1}{4}[(9)(1) + (-1)(1) + (1)(1) + (3)(1)] = 3$$

$$v_2 = \tfrac{1}{4}[(9)(1) + (-1)(1) + (1)(-1) + (3)(-1)] = 1$$

$$v_3 = \tfrac{1}{4}[(9)(1) + (-1)(-1) + (1)(1) + (3)(-1)] = 2$$

$$v_4 = \tfrac{1}{4}[(9)(1) + (-1)(-1) + (1)(-1) + (3)(1)] = 3$$

or $V_{3\Lambda} = 3V_1 \oplus V_2 \oplus 2V_3 \oplus 3V_4$. Subtracting V_1 for z, V_2 for R_z, $2V_3$ for x and R_y, and $2V_4$ for y and R_x gives

$$V_{\text{vib}} = 2V_1 \oplus V_4$$

which means that the three modes of vibration for water are singly degenerate, with two of them being completely symmetrical and one being symmetric with respect to $\hat{E}$ and $\hat{\sigma}'_v$ but antisymmetric with respect to $\hat{C}_2$ and $\hat{\sigma}_v$. The normal modes are illustrated in Fig. 20-16.

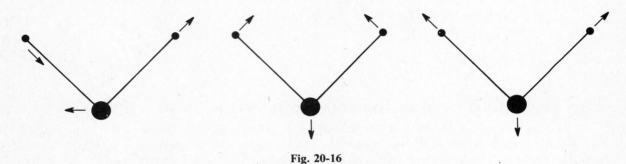

Fig. 20-16

EXAMPLE 20.19. The Mulliken convention for numbering the vibrational frequencies is to assign them according to the appearance of the corresponding irreducible representations in the character table. For modes of the same symmetry, the assignment is made according to decreasing frequency. An exception to this convention is made for linear triatomic molecules, where v_2 has traditionally been assigned to the doubly degenerate irreducible representation. Using this convention, assign the molecular vibrational frequencies for H_2O.

The results of Example 20.18 gave $V_{\text{vib}} = 2V_1 \oplus V_4$, so v_1 would be assigned to the higher frequency corresponding to V_1 (symmetric stretch), v_2 to the lower frequency correspnding to V_1 (bend), and v_3 to V_4 (asymmetric stretch).

Solved Problems

SYMMETRY ELEMENTS AND OPERATIONS

20.1. Diatomic and linear polyatomic molecules contain a C_∞ axis of rotation. Construct a perspective sketch for the points around this element.

The notation C_∞ is defined as an axis of proper rotation for which an infinite number of points are related to the original point by various angles ϕ, where $0 \le \phi \le 2\pi$. If the original (finite-sized) point is placed in a general position off the axis of rotation, a torus is generated that has no importance in discussing molecular and crystallographic symmetry. An alternative way of indicating a C_∞ axis is to place the point on the axis (see Fig. 20-17).

20.2. Is $\hat{\sigma}^k$ distinct for all k?

As can be seen in Fig. 20-5, $\hat{\sigma}^1$ generates a point that cannot be generated in a less complicated manner and is therefore distinct; $\hat{\sigma}^2$ generates the original point—the result of $\hat{E}$—and is therefore not distinct; and $\hat{\sigma}^k$ for $k > 2$ generates one of the above points and is therefore not distinct.

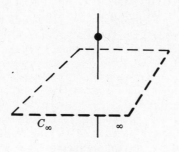

Fig. 20-17

20.3. Prepare an orthographic projection illustrating the points around the S_3 symmetry element.

The projection is shown in Fig. 20-18. The points are the images of the original point under the operations $\hat{S}_3^k$, $1 \le k \le 6$.

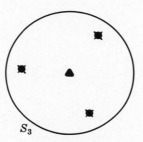

Fig. 20-18

20.4. Prepare drawings for the points around the 4_3 element.

With the fundamental operation of rotation by 90° and translation by $3t/4$, the sketches shown in Fig. 20-9c result. After translating all points to similar positions within the distance t (see Example 20.6) and constructing the simplified diagram, it appears that 4_1 and 4_3 are *enantiomorphic* operations (i.e., they correspond to right- and left-hand screws of the same pitch).

20.5. Identify the symmetry elements present in (a) an ice-cream cone, (b) a donut, (c) a football, (d) a crescent, (e) the letter A.

From inspection of the figures mentioned: (a) C_∞, $\infty\sigma_v$, E; (b) C_∞, ∞C_2, $\infty\sigma_v$, σ_h, i, E; (c) C_∞, ∞C_2, $\infty\sigma_v$, σ_h, i, E; (d) C_2, σ_v, σ_v', E; (e) C_2, σ_v, σ_v', E

POINT GROUPS

20.6. Consider a molecule of ammonia (trigonal pyramid with the N at the apex). Determine the symmetry elements present.

The elements are E, C_3, and $3\sigma_v$.

20.7. Prepare a multiplication table of distinct operations in ammonia: $\hat{E}$, $\hat{C}_3$, $\hat{C}_3^2$, $\hat{\sigma}_{v1}$, $\hat{\sigma}_{v2}$, and $\hat{\sigma}_{v3}$.

See Table 20-6.

20.8. Determine the point group for (a) NH_3, (b) CH_4, and (c) C_6H_6.

The following analyses can be made using Fig. 20-14:

(a) NH$_3$:   (1) Are there ∞ C_∞ axes present? no; (2) is there a pentagonal dodecahedron or icosahedron present? no; (3) are there four C_3 axes at 54°44'? no; (4) is there at least one C_n where $n \geq 2$? yes, C_3; (5) is there an S_{2n} present? no; (6) are there n C_2 axes perpendicular to C_n? no; (7) are there any σ_h planes present? no; (8) are there n σ_v planes present? yes, 3, therefore $\mathscr{C}_{3v}$.

(b) CH$_4$:   (1) Are there ∞ C_∞ axes present? no; (2) is there a pentagonal dodecahedron or icosahedron present? no; (3) are there four C_3 axes at 54°44'? yes; (4) is there a C_4 axis present? no; (5) is there an S_4 axis present? yes, therefore $\mathscr{T}_d$.

(c) C$_6$H$_6$:   (1) Are there ∞ C_∞ axes present, no; (2) is there a pentagonal dodecahedron or icosahedron present? no; (3) are there four C_3 axes at 54°44'? no; (4) is there at least one C_n where $n \geq 2$? yes, C_6; (5) is there an S_{2n} present? no; (6) are there n C_2 axes present perpendicular to C_n? yes, 6; (7) are there any σ_h planes perpendicular to C_n? yes, therefore $\mathscr{D}_{6h}$.

Table 20-6

$\mathscr{C}_{3v}$	$\hat{E}$	$\hat{C}_3$	$\hat{C}_3^2$	$\hat{\sigma}_{v1}$	$\hat{\sigma}_{v2}$	$\hat{\sigma}_{v3} = \hat{B}$
$\hat{A} = \hat{E}$	$\hat{E}$	$\hat{C}_3$	$\hat{C}_3^2$	$\hat{\sigma}_{v1}$	$\hat{\sigma}_{v2}$	$\hat{\sigma}_{v3}$
$\hat{C}_3$	$\hat{C}_3$	$\hat{C}_3^2$	$\hat{E}$	$\hat{\sigma}_{v3}$	$\hat{\sigma}_{v1}$	$\hat{\sigma}_{v2}$
$\hat{C}_3^2$	$\hat{C}_3^2$	$\hat{E}$	$\hat{C}_3$	$\hat{\sigma}_{v2}$	$\hat{\sigma}_{v3}$	$\hat{\sigma}_{v1}$
$\hat{\sigma}_{v1}$	$\hat{\sigma}_{v1}$	$\hat{\sigma}_{v2}$	$\hat{\sigma}_{v3}$	$\hat{E}$	$\hat{C}_3$	$\hat{C}_3^2$
$\hat{\sigma}_{v2}$	$\hat{\sigma}_{v2}$	$\hat{\sigma}_{v3}$	$\hat{\sigma}_{v1}$	$\hat{C}_3^2$	$\hat{E}$	$\hat{C}_3$
$\hat{\sigma}_{v3}$	$\hat{\sigma}_{v3}$	$\hat{\sigma}_{v1}$	$\hat{\sigma}_{v2}$	$\hat{C}_3$	$\hat{C}_3^2$	$\hat{E}$

REPRESENTATION OF GROUPS

20.9. Considering the symmetry elements shown in Fig. 20-15 for the $\mathscr{C}_{2v}$ point group, show that $\hat{C}_2$ can be expressed as

$$\begin{bmatrix} -1 & 0 & 0 \\ 0 & -1 & 0 \\ 0 & 0 & 1 \end{bmatrix}$$

Assuming x_2, y_2, and z_2 to be the result of $\hat{C}_2$ operating on x_1, y_1, and z_1, then

$$x_2 = \hat{C}_2 \times x_1 = -x_1 \qquad y_2 = \hat{C}_2 \times y_1 = -y_1 \qquad z_2 = \hat{C}_2 \times z_1 = z_1$$

which can be restated as

$$x_2 = -1x_1 + 0y_1 + 0z_1$$
$$y_2 = 0x_1 - 1y_1 + 0z_1$$
$$z_2 = 0x_1 + 0y_1 + 1z_1$$

or in matrix form as

$$\begin{bmatrix} x_2 \\ y_2 \\ z_2 \end{bmatrix} = \begin{bmatrix} -1 & 0 & 0 \\ 0 & -1 & 0 \\ 0 & 0 & 1 \end{bmatrix} \begin{bmatrix} x_1 \\ y_1 \\ z_1 \end{bmatrix}$$

which is the desired result.

Another method would be to substitute $n = 2$ and $m = 1$ into the expression for $\hat{C}_2(z)^m$ in Table 20-3.

20.10. Confirm that $\hat{C}_2 \times \hat{\sigma}_v = \hat{\sigma}_v'$ for the $\mathscr{C}_{2v}$ point group using the respective matrix expressions.

Using $\hat{C}_2 \longrightarrow \begin{bmatrix} -1 & 0 & 0 \\ 0 & -1 & 0 \\ 0 & 0 & 1 \end{bmatrix}$ (see Problem 20.9) and $\hat{\sigma}_v \longrightarrow \begin{bmatrix} 1 & 0 & 0 \\ 0 & -1 & 0 \\ 0 & 0 & 1 \end{bmatrix}$ (see

Example 20.11), the product is

$$\begin{bmatrix} -1 & 0 & 0 \\ 0 & -1 & 0 \\ 0 & 0 & 1 \end{bmatrix} \begin{bmatrix} 1 & 0 & 0 \\ 0 & -1 & 0 \\ 0 & 0 & 1 \end{bmatrix} = \begin{bmatrix} -1 & 0 & 0 \\ 0 & 1 & 0 \\ 0 & 0 & 1 \end{bmatrix}$$

which is the same as $\hat{\sigma}'_v$, see Problem 20.32.

20.11. Demonstrate that the elements of the irreducible representation V_2 for the $\mathscr{C}_{2v}$ point group multiply as the symmetry operations of the group.

A selection of the 16 multiplications using

$$\hat{E} \longrightarrow [1] \qquad \hat{C}_2 \longrightarrow [1] \qquad \hat{\sigma}_v \longrightarrow [-1] \qquad \hat{\sigma}'_v \longrightarrow [-1]$$

is

$$\begin{array}{cccc}
\hat{E} \times \hat{E} = \hat{E} & \hat{E} \times \hat{\sigma}'_v = \hat{\sigma}_v & \hat{\sigma}_u \times \hat{C}_2 = \hat{\sigma}'_v & \hat{\sigma}'_v \times \hat{\sigma}_v = \hat{C}_2 \\
\downarrow \quad \downarrow \quad \downarrow & \downarrow \quad \downarrow \quad \downarrow & \downarrow \quad \downarrow \quad \downarrow & \downarrow \quad \downarrow \quad \downarrow \\
[1] \times [1] = [1] & [1] \times [-1] = [-1] & [-1] \times [1] = [-1] & [-1] \times [-1] = [1]
\end{array}$$

20.12. Find the characters of V_2 for the $\mathscr{C}_{2v}$ point group, given the incomplete character table

$\mathscr{C}_{2v}$	$\hat{E}$	$\hat{C}_2$	$\hat{\sigma}_v$	$\hat{\sigma}'_v$
V_1	1	1	1	1
V_3	1	-1	1	-1
V_4	1	-1	-1	1

as determined from the irreducible representations found in Example 20.13.

The order of V_2 was determined as one-dimensional, which means $\chi(V_2, \hat{E}) = 1$. Applying (20.3) to V_1 and V_2 gives

$$0 = \chi(V_1, \hat{E})\chi(V_2, \hat{E}) + \chi(V_1, \hat{C}_2)\chi(V_2, \hat{C}_2) + \chi(V_1, \hat{\sigma}_v)\chi(V_2, \hat{\sigma}_v) + \chi(V_1, \hat{\sigma}'_v)\chi(V_2, \hat{\sigma}'_v)$$

$$= (1)(1) + (1)\chi(V_2, \hat{C}_2) + (1)\chi(V_2, \hat{\sigma}_v) + (1)\chi(V_2, \hat{\sigma}'_v)$$

and applying it to V_2 and V_3 and to V_2 and V_4 gives

$$0 = (1)(1) + (-1)\chi(V_2, \hat{C}_2) + (1)\chi(V_2, \hat{\sigma}_v) + (-1)\chi(V_2, \hat{\sigma}'_v)$$

$$0 = (1)(1) + (-1)\chi(V_2, \hat{C}_2) + (-1)\chi(V_2, \hat{\sigma}_v) + (1)\chi(V_2, \hat{\sigma}'_v)$$

Solving simultaneously gives $\chi(V_2, \hat{C}_2) = 1$, $\chi(V_2, \hat{\sigma}_v) = -1$, and $\chi(V_2, \hat{\sigma}'_v) = -1$, which agrees with the entry in Table 20-4.

20.13. Demonstrate that the irreducible representations for $\mathscr{C}_{2v}$ obey (20.3) when $i = j$ and (20.4).

For $i = j = 1$, (20.3) gives for V_1

$$\chi(V_1, \hat{E})\chi(V_1, \hat{E}) + \chi(V_1, \hat{C}_2)\chi(V_1, \hat{C}_2) + \chi(V_1, \hat{\sigma}_v)\chi(V_1, \hat{\sigma}_v) + \chi(V_1, \hat{\sigma}'_v)\chi(V_1, \hat{\sigma}'_v)$$

$$= (1)(1) + (1)(1) + (1)(1) + (1)(1) = 4$$

and similarly for V_2, V_3, and V_4. Substituting the values of $\chi(V_i, \hat{E})$ from Table 20-4 into (20.4) gives

$$(1)^2 + (1)^2 + (1)^2 + (1)^2 = 4$$

20.14. Determine the Mulliken designations for the irreducible representations of $\mathscr{C}_{2v}$.

Upon comparing the entries in Table 20-4 with the code given in Table 20-5, the letter A should be chosen for V_1 because $\chi(V_1, \hat{E}) = 1$ and $\chi(V_1, \hat{C}_2) = 1$, and the subscript 1 should be chosen because $\chi(V_1, \hat{\sigma}_v) = 1$; hence $V_1 = A_1$. Likewise, V_2 is designated as A_2 because $\chi(V_2, \hat{E}) = 1$, $\chi(V_2, \hat{C}_2) = 1$, and $\chi(V_2, \hat{\sigma}_v) = -1$; V_3 as B_1 because $\chi(V_3, \hat{E}) = 1$, $\chi(V_3, \hat{C}_2) = -1$, and $\chi(V_3, \hat{\sigma}_v) = 1$; and V_4 as B_2 because $\chi(V_4, \hat{E}) = 1$, $\chi(V_4, \hat{C}_2) = -1$, and $\chi(V_4, \hat{\sigma}_v) = -1$.

APPLICATIONS OF GROUP THEORY TO MOLECULAR PROPERTIES

20.15. Which of the following two carbohydrates would be potentially optically active in the gas phase or in solution?

Even though there are conformations of I that would be optically active, the low energy necessary for rotation about the C—C single bonds would permit I to exist in essentially equal numbers in the various forms and the ensemble would be inactive. Because no $\hat{S}_n$ operation is allowed by II, it is potentially active in both phases.

20.16. Discuss the dipole moments of CH_4, CH_3Cl, CH_2Cl_2, $CHCl_3$, and CCl_4 using symmetry arguments.

Both CH_4 and CCl_4 belong to $\mathscr{T}_d$, which contains four C_3 axes, and thus they have no dipole moment. Both CH_3Cl and $CHCl_3$ belong to $\mathscr{C}_{3v}$, which does not contain additional C_3 axes, a center of inversion, or mirror planes other than σ_v, and thus they have a dipole moment that lies along the C_3 axis (and the intersection of the $3\sigma_v$). The molecule CH_2Cl_2 belongs to $\mathscr{C}_{2v}$, which contains only E, C_2, and two types of σ_v, and thus it has a dipole moment that lies along the C_2 axis (and the intersection of the σ_v planes).

20.17. Determine which irreducible representations correspond to the x, y, and z components of translation for the $\mathscr{C}_{3v}$ point group.

Anticipating the results of Problem 20.36, the translational motion is the sum of V_1, a one-dimensional representation given by

$$\hat{E} \longrightarrow [1] \quad \hat{C}_3 \longrightarrow [1] \quad \hat{C}_3^2 \longrightarrow [1] \quad \hat{\sigma}_{v1} \longrightarrow [1] \quad \hat{\sigma}_{v2} \longrightarrow [1] \quad \hat{\sigma}_{v3} \longrightarrow [1]$$

and V_3, a two-dimensional representation given by

$$\hat{E} \longrightarrow \begin{bmatrix} 1 & 0 \\ 0 & 1 \end{bmatrix} \quad \hat{C}_3 \longrightarrow \begin{bmatrix} -0.500 & -0.866 \\ 0.866 & -0.500 \end{bmatrix} \quad \hat{C}_3^2 \longrightarrow \begin{bmatrix} -0.500 & 0.866 \\ -0.866 & -0.500 \end{bmatrix}$$

$$\hat{\sigma}_{v1} \longrightarrow \begin{bmatrix} -1 & 0 \\ 0 & 1 \end{bmatrix} \quad \hat{\sigma}_{v2} \longrightarrow \begin{bmatrix} 0.500 & 0.866 \\ 0.866 & -0.500 \end{bmatrix} \quad \hat{\sigma}_{v3} \longrightarrow \begin{bmatrix} 0.500 & -0.866 \\ -0.866 & -0.500 \end{bmatrix}$$

Each of the symmetry operations leaves z invariant, which is the same as V_1. Thus z should be assigned to V_1 and (x, y) should be assigned to V_3.

20.18. Confirm the assignments of the R_i to the V_i shown in Table 20-7.

Applying the rules given in Sec. 20.21 to V_1, to which is assigned the z component of translation, gives

$$\chi(V_{R_z}, \hat{E}) = \chi(V_1, \hat{E}) = 1$$

$$\chi(V_{R_z}, 2\hat{C}_3(z)) = \chi(V_1, 2\hat{C}_3(z)) = 1$$

$$\chi(V_{R_z}, 3\hat{\sigma}_v) = -\chi(V_1, 3\hat{\sigma}_v) = -1$$

Table 20-7

$\mathscr{C}_{3v}$ Representation	$\hat{E}$	$2\hat{C}_3(z)$	$3\hat{\sigma}_v$	Translation and Rotation
$V_1 = A_1$	1	1	1	z
$V_2 = A_2$	1	1	-1	R_z
$V_3 = E$	2	-1	0	$(x, y), (R_x, R_y)$

which corresponds to V_2 in Table 20-7. Similarly, for V_3,

$$\chi(V_{R_{x,y}}, \hat{E}) = \chi(V_3, \hat{E}) = 2$$

$$\chi(V_{R_{x,y}}, 2\hat{C}_3(z)) = \chi(V_3, 2\hat{C}_3(z)) = -1$$

$$\chi(V_{R_{x,y}}, 3\hat{\sigma}_v) = -\chi(V_3, 3\hat{\sigma}_v) = 0$$

which corresponds to V_3 in Table 20-7.

20.19. CO_2 belongs to $\mathscr{D}_{\infty h}$, whose character table is Table 20-8. Discuss the normal modes of vibration for the molecule.

Table 20-8

$\mathscr{D}_{\infty h}$ Representation	$\hat{E}$	$2\hat{C}_\infty^\phi$	$\hat{\sigma}_v$	$\hat{i}$	$\hat{\sigma}_h$	$2\hat{S}_\infty^\phi$	$\infty\hat{C}_2$	Translation and Rotation
$V_1 = A_{1g} = \Sigma g^+$	1	1	1	1	1	1	1	
$V_2 = A_{1u} = \Sigma u^+$	1	1	1	-1	-1	-1	-1	
$V_3 = A_{2g} = \Sigma g^-$	1	1	-1	1	1	1	-1	R_z
$V_4 = A_{2u} = \Sigma u^-$	1	1	-1	-1	-1	-1	1	z
$V_5 = E_{1g} = \Pi g$	2	$2\cos\phi$	0	2	-2	$-2\cos\phi$	0	(R_x, R_y)
$V_6 = E_{1u} = \Pi u$	2	$2\cos\phi$	0	-2	2	$2\cos\phi$	0	(x, y)
$\vdots$	$\vdots$	$\vdots$	$\vdots$	$\vdots$	$\vdots$	$\vdots$	$\vdots$	$\vdots$

Using (20.7) gives

$$\chi(V_{3\Lambda}, \hat{E}) = 3(3) = 9 \qquad \chi(V_{3\Lambda}, \hat{C}_\infty^\phi) = 3[1 + 2\cos\phi] = 3 + 6\cos\phi$$

$$\chi(V_{3\Lambda}, \infty\hat{\sigma}_v) = 3 \qquad \chi(V_{3\Lambda}, \hat{i}) = -3(1) = -3 \qquad \chi(V_{3\Lambda}, \hat{\sigma}_h) = 1$$

$$\chi(V_{3\Lambda}, 2\hat{S}_\infty^\phi) = 1[-1 + 2\cos\phi] = -1 + 2\cos\phi$$

$$\chi(V_{3\Lambda}, \infty\hat{C}_2) = 1\left[1 + 2\cos\frac{2\pi(1)}{2}\right] = -1$$

Because $h = \infty$ for this group, (20.6) is not applicable and $V_{3\Lambda}$ must be resolved by trial and error, giving $V_{3\Lambda} = V_1 \oplus 2V_2 \oplus V_5 \oplus 2V_6$. Subtracting $V_2 \oplus V_5 \oplus V_6$, which represents the contributions to $V_{3\Lambda}$ for x, y, z, R_x, and R_y, gives

$$V_{\text{vib}} = V_1 \oplus V_2 \oplus V_6$$

The interpretation of this result is that there is one symmetrical stretch, one asymmetrical stretch, and one doubly degenerate bend (see Fig. 20-19).

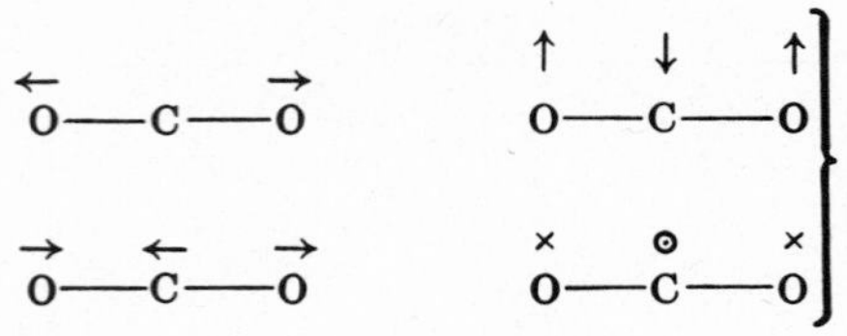

Fig. 20-19

Supplementary Problems

SYMMETRY ELEMENTS AND OPERATIONS

20.20. Construct perspective and orthographic drawings for C_1 (or 1). Is $\hat{C}_1$ distinct?

Ans.   See Fig. 20-2; no, $\hat{C}_1 = \hat{E}$.

20.21. Construct perspective sketches and orthographic projections for the equivalent points around the C_3 and 6 symmetry elements.

Ans.   Drawings are similar to Fig. 20-3e except that three and six points, respectively, appear.

20.22. List the distinct operations for $\hat{i}^k$.   *Ans.* $\hat{i}$

20.23. Equate $\bar{3}$ and $\bar{6}$ to $\tilde{3}$ and $\tilde{6}$.   *Ans.* $\bar{3} = \tilde{6}, \bar{6} = \tilde{3}$

20.24. Prepare a list of $\hat{S}_n^k$ that are distinct, where $n = 1, 2, 3, 4,$ and 6.

Ans. $\hat{S}_3, \hat{S}_3^5, \hat{S}_4, \hat{S}_4^3, \hat{S}_6, \hat{S}_6^5$

20.25. What type of array is formed after $\hat{t}'$ has operated on the result of $\hat{t}$ operating on a motif, assuming many such operations and that the axes along which $\hat{t}$ and $\hat{t}'$ are performed are not collinear?

Ans.   Two-dimensional array (net)

20.26. Construct perspective sketches for the 3_1 and 3_2 elements. What is the relation between these elements?

Ans.   See Fig. 20-20; 3_1 shows a clockwise rotation, and 3_2 simplifies to a counterclockwise rotation.

20.27. Identify the symmetry elements present in (*a*) a die; (*b*) the letter B, assuming both lobes to be the same size; (*c*) the sign $\times$; and (*d*) a softball, including the stitching.

Ans.   (*a*) E;   (*b*) $C_2, \sigma_v, \sigma_v', E$;   (*c*) $C_4, 2C_2, 2C_2', 2\sigma_v, 2\sigma_v', \sigma_h, i, E$;   (*d*) $S_4, \sigma_v, \sigma_v', C_2, i, E$

20.28. Give a geometrical proof that only 1-, 2-, 3-, 4-, and 6-fold axes are permitted in the solid state.

Ans.   Interior angle of regular n-gon $= \dfrac{(n-2)\pi}{n} = \dfrac{2\pi}{k}$

POINT GROUPS

20.29. Determine the symmetry elements present in a benzene molecule.

Ans. C_6; $3C_2$, which pass through the center of the molecule and through 2 C's and 2 H's; $3C_2'$, which pass through the center of the molecule and bisect the C—C bonds; σ_h; $3\sigma_v$ and $3\sigma_v'$, which bisect the angles formed by the C_2 and C_2' axes, respectively, and thus are $3\sigma_d$ and $3\sigma_d'$; i, E

20.30. Prepare multiplication tables for $\mathscr{C}_3$, $\mathscr{C}_{2h}$, and $\mathscr{D}_2$.

Ans.   See Tables 20-9, 20-10, 20-11.

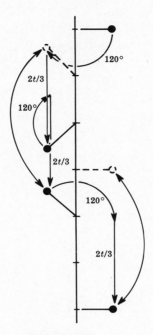
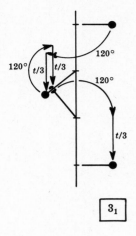

Fig. 20-20

Table 20-9

$\mathscr{C}_3$	$\hat{E}$	$\hat{C}_3$	$\hat{C}_3^2 = \hat{B}$
$\hat{A} = \hat{E}$	$\hat{E}$	$\hat{C}_3$	$\hat{C}_3^2$
$\hat{C}_3$	$\hat{C}_3$	$\hat{C}_3^2$	$\hat{E}$
$\hat{C}_3^2$	$\hat{C}_3^2$	$\hat{E}$	$\hat{C}_3$

Table 20-10

$\mathscr{C}_{2h}$	$\hat{E}$	$\hat{C}_2$	$\hat{\sigma}_h$	$\hat{i} = \hat{B}$
$\hat{A} = \hat{E}$	$\hat{E}$	$\hat{C}_2$	$\hat{\sigma}_h$	$\hat{i}$
$\hat{C}_2$	$\hat{C}_2$	$\hat{E}$	$\hat{i}$	$\hat{\sigma}_h$
$\hat{\sigma}_h$	$\hat{\sigma}_h$	$\hat{i}$	$\hat{E}$	$\hat{C}_2$
$\hat{i}$	$\hat{i}$	$\hat{\sigma}_h$	$\hat{C}_2$	$\hat{E}$

Table 20-11

$\mathscr{D}_2$	$\hat{E}$	$\hat{C}_2(x)$	$\hat{C}_2(y)$	$\hat{C}_2(z) = \hat{B}$
$\hat{A} = \hat{E}$	$\hat{E}$	$\hat{C}_2(x)$	$\hat{C}_2(y)$	$\hat{C}_2(x)$
$\hat{C}_2(x)$	$\hat{C}_2(x)$	$\hat{E}$	$\hat{C}_2(z)$	$\hat{C}_2(y)$
$\hat{C}_2(y)$	$\hat{C}_2(y)$	$\hat{C}_2(z)$	$\hat{E}$	$\hat{C}_2(x)$
$\hat{C}_2(z)$	$\hat{C}_2(z)$	$\hat{C}_2(y)$	$\hat{C}_2(x)$	$\hat{E}$

20.31. Classify the following molecules according to the point group:

(a) C_2H_4 (b) p-Dichlorobenzene   (c) SiF_6^{2-}
(d) H_2S (e) Naphthalene   (f) PCl_5
(g) Hexachlorobenzene   (h) HOCl   (i) I_3^-
(j) CH_2ClF (k) Coronene, $C_{24}H_{12}$ (l) CO_3^{2-}
(m) HCO_3^- (n) S_8, crown form   (o) H_2O_2
(p) C_2H_6, staggered   (q) C_2H_6, eclipsed

Ans.   (a) $\mathscr{D}_{2h}$ (b) $\mathscr{D}_{2h}$ (c) $\mathscr{O}_h$ (d) $\mathscr{C}_{2v}$
         (e) $\mathscr{D}_{2h}$ (f) $\mathscr{D}_{3h}$ (g) $\mathscr{D}_{6h}$ (h) $\mathscr{C}_s$
         (i) $\mathscr{D}_{\infty d}$ (j) $\mathscr{C}_s$ (k) $\mathscr{D}_{6h}$ (l) $\mathscr{D}_{3h}$
         (m) $\mathscr{C}_s$ (n) $\mathscr{D}_{4d}$ (o) $\mathscr{C}_2$ (p) $\mathscr{D}_{3d}$
         (q) $\mathscr{D}_{3h}$

REPRESENTATION OF GROUPS

20.32. Show that $\hat{E}$ and $\hat{\sigma}_v'$ can be expressed as

$$\begin{bmatrix} 1 & 0 & 0 \\ 0 & 1 & 0 \\ 0 & 0 & 1 \end{bmatrix} \quad \text{and} \quad \begin{bmatrix} -1 & 0 & 0 \\ 0 & 1 & 0 \\ 0 & 0 & 1 \end{bmatrix}$$

respectively, for $\mathscr{C}_{2v}$ by performing the respective operations on a point having coordinates x_1, y_1, and z_1 and converting to matrix notation.

20.33. Confirm the expressions for $\hat{E}$ and $\hat{\sigma}_v'$ given in Problem 20.32 by using Table 20-3.

20.34. Using the matrix expressions given in Example 20.13 for the operations in $\mathscr{C}_{2v}$, confirm the entries in Table 20-1 for $\hat{C}_2 \times \hat{\sigma}_v'$, $\hat{\sigma}_v \times \hat{C}_2$, $\hat{\sigma}_v \times \hat{\sigma}_v$, $\hat{\sigma}_v \times \hat{\sigma}_v'$, $\hat{\sigma}_v' \times \hat{C}_2$, and $\hat{\sigma}_v' \times \hat{\sigma}_v$.

20.35. Demonstrate that the elements of V_3 for the $\mathscr{C}_{2v}$ point group multiply as the symmetry operations of the group.

20.36. The six distinct operations in $\mathscr{C}_{3v}$ are $\hat{E}$, $\hat{C}_3$, $\hat{C}_3^2$, $\hat{\sigma}_{v1}$, $\hat{\sigma}_{v2}$, and $\hat{\sigma}_{v3}$, and they fall into the classes $\hat{E}$, $2\hat{C}_3$, and $3\sigma_v$. Show that the representation for this group provided by Table 20-3 is reducible. Determine the two obvious irreducible representations for the group and the dimension of the third.

Ans. $\hat{E} \longrightarrow \begin{bmatrix} 1 & 0 & 0 \\ 0 & 1 & 0 \\ 0 & 0 & 1 \end{bmatrix}$ $\hat{\sigma}_{v1} \longrightarrow \begin{bmatrix} -1 & 0 & 0 \\ 0 & 1 & 0 \\ 0 & 0 & 1 \end{bmatrix}$

$\hat{C}_3 \longrightarrow \begin{bmatrix} -0.500 & -0.866 & 0 \\ 0.866 & -0.500 & 0 \\ 0 & 0 & 1 \end{bmatrix}$ $\hat{\sigma}_{v2} \longrightarrow \begin{bmatrix} 0.500 & 0.866 & 0 \\ 0.866 & -0.500 & 0 \\ 0 & 0 & 1 \end{bmatrix}$

$\hat{C}_3^2 \longrightarrow \begin{bmatrix} -0.500 & 0.866 & 0 \\ -0.866 & -0.500 & 0 \\ 0 & 0 & 1 \end{bmatrix}$ $\hat{\sigma}_{v3} \longrightarrow \begin{bmatrix} 0.500 & -0.866 & 0 \\ -0.866 & -0.500 & 0 \\ 0 & 0 & 1 \end{bmatrix}$

$3^2 > 6$, therefore reducible; V_1 is the one-dimensional subset and V_3 is the two-dimensional subset of the matrices as blocked out above; $\ell = (6 - 1^2 - 2^2)^{1/2} = 1$

20.37. Find the characters of the irreducible representations blocked out in Problem 20.36. Recognizing that (20.3) is valid, find $\chi(V_2, \hat{R})$ for this group.

Ans. $\chi(V_2, \hat{E}) = 1$, $\chi(V_2, \hat{C}_3) = \chi(V_2, \hat{C}_3^2) = 1$, $\chi(V_2, \hat{\sigma}_{v1}) = \chi(V_2, \hat{\sigma}_{v2}) = \chi(V_2, \hat{\sigma}_{v3}) = -1$

20.38. Using V_3 as given in Problem 20.36, show that (20.5) is valid.

> *Ans.* $\chi(V_3, \hat{C}_3) = \chi(V_3, \hat{C}_3^2) = -1$, $\chi(V_3, \hat{\sigma}_{v1}) = \chi(V_3, \hat{\sigma}_{v2}) = \chi(V_3, \hat{\sigma}_{v3}) = 0$

20.39. For $\mathscr{C}_{3v}$, find the characters of the irreducible representations corresponding to the Mulliken symbols E, A_1, and A_2.   *Ans.*   See Table 20-7.

APPLICATIONS OF GROUP THEORY TO MOLECULAR PROPERTIES

20.40. The unit cell of cinnabar, HgS, belongs to the 3 2 point group. Is this mineral potentially optically active?   *Ans.*   Yes

20.41. Using symmetry arguments, discuss the dipole moment in NH_3.

> *Ans.*   It lies along C_3, which is the intersection of the σ_v's.

20.42. Determine which irreducible representations correspond to the translations along the z and x axes for a water molecule.   *Ans.*   See Table 20-4.

20.43. Verify the assignments of translation components given in Table 20-12 for $\mathscr{D}_{2h}$, e.g., C_2H_4.

Table 20-12

$\mathscr{D}_{2h}$ Representation	$\hat{E}$	$\hat{\sigma}(xy)$	$\hat{\sigma}(xz)$	$\hat{\sigma}(yz)$	$\hat{i}$	$\hat{C}_2(z)$	$\hat{C}_2(y)$	$\hat{C}_2(x)$	Translation and Rotation
$V_1 = A_g$	1	1	1	1	1	1	1	1	
$V_2 = A_u$	1	-1	-1	-1	-1	1	1	1	
$V_3 = B_{1g}$	1	1	-1	-1	1	1	-1	-1	R_z
$V_4 = B_{1u}$	1	-1	1	1	-1	1	-1	-1	z
$V_5 = B_{2g}$	1	-1	1	-1	1	-1	1	-1	R_y
$V_6 = B_{2u}$	1	1	-1	1	-1	-1	1	-1	y
$V_7 = B_{3g}$	1	-1	-1	1	1	-1	-1	1	R_x
$V_8 = B_{3u}$	1	1	1	-1	-1	-1	-1	1	x

20.44. Confirm the assignments of R_y and R_z for $\mathscr{C}_{2v}$.

20.45. Verify the assignments of rotation components given in Table 20-12 for $\mathscr{D}_{2h}$.

20.46. Discuss the symmetry of the vibrations of ONO, a nonlinear molecule.

> *Ans.* $\mathscr{C}_{2v}$ (see Fig. 20-11)

20.47. Describe the vibrational motion in the linear molecules HCN and NNO.

> *Ans.*   Both belong to $\mathscr{C}_{\infty v}$ (see Table 20-13) and can be shown to have a symmetrical stretch V_1, an asymmetrical stretch V_1, and a doubly degenerate bend V_3 (see Fig. 20-19).

20.48. Discuss the symmetry of the vibrations of (a) nonplanar AB_3, (b) C_2H_4, and (c) planar $WXYZ$ chain.

> *Ans.*   (a) $\mathscr{C}_{3v}$ (see Table 20-7) having a symmetrical stretch V_1, symmetrical deformation V_1, degenerate stretch V_3, and degenerate deformation V_3
>   (b) $\mathscr{D}_{2h}$ (see Table 20-12) having $3V_1 \oplus V_2 \oplus 2V_3 \oplus V_4 \oplus V_5 \oplus 2V_6 \oplus 2V_8$
>   (c) $\mathscr{C}_s$ (see Table 20-14) having $5V_1 \oplus V_2$

Table 20-13

$\mathscr{C}_{\infty v}$ Representation	$\hat{E}$	$2\hat{C}_{\infty}^{\phi}$	$\infty\hat{\sigma}_v$	Translation and Rotation
$V_1 = A_1 = \Sigma^+$	1	1	1	z
$V_2 = A_2 = \Sigma^-$	1	1	-1	R_z
$V_3 = E_1 = \Pi$	2	$2\cos\phi$	0	$(x, y), (R_x, R_y)$
$V_4 = E_2 = \Delta$	2	$2\cos 2\phi$	0	
$V_5 = E_3 = \Phi$	2	$2\cos 3\phi$	0	
$\vdots$	$\vdots$	$\vdots$	$\vdots$	

Table 20-14

$\mathscr{C}_s$ Representation	$\hat{E}$	$\hat{\sigma}(xy)$	Translation and Rotation
$V_1 = A'$	1	1	x, y, R_z
$V_2 = A''$	1	-1	z, R_x, R_y

Chapter 21

Intermolecular Bonding

Extended Covalent Bonding

21.1 COVALENT BONDING

Two types of substances have large networks of covalent bonds holding the atoms together in the condensed phases: (1) nonmetallic elements such as carbon for which the bonding arrangement satisfies electron deficiencies on the individual atoms and (2) compounds such as silicon dioxide that undergo energetically favorable rearrangements of the electrons in the molecules that are stable in the gas phase. These bonds are no different from those described in Chaps. 16 and 18, and these substances will have intermediate to relatively high values for the melting point, boiling point, heat of sublimation, and heat of vaporization, because these values depend on the thermal energy necessary to break the intermolecular bond.

If the covalent bonding is three-dimensional, and if all bonds are equivalent, the substance will be *isotropic*, i.e., its properties will be the same in all directions. If the bonding varies within the substance or is not three-dimensional, the substance will be *anisotropic*, i.e., certain properties will depend on the direction of observation. Lower melting and boiling points can be expected for anisotropic materials.

21.2 HYDROGEN BONDING

The electronegativity differences between some elements and hydrogen are large enough that the hydrogen will form a weak coordinate covalent bond with an unshared pair of electrons on another molecule, giving a chain or three-dimensional network of molecules. For compounds containing F, O, N, or Cl with H, these weak hydrogen bonds significantly increase the melting and boiling points of the compounds over the values predicted in the absence of these additional bonds.

Metallic Bonding

21.3 THE FREE-ELECTRON MODEL

In this theory of the bonding between the atoms in a metal, the kernels of the atoms are placed in the physical arrangement found in the solid state (e.g., hexagonal or cubic closest-packed crystal structure), and the electrons act as freely moving particles in a three-dimensional box of uniform potential energy. The metallic bonds formed between the positively charged kernels and negatively charged electrons are rather strong, giving intermediate to high values for the melting point, boiling point, heat of sublimation, and heat of vaporization.

At absolute zero, all the electrons will be in the lowest available quantum states with paired spins. The energy of the highest filled level predicted by (*14.45*) is

$$E_F = \frac{h^2}{8m_e}\left(\frac{3N}{\pi V_m}\right)^{2/3} \qquad (21.1)$$

where E_F is known as the *Fermi energy*, N is the number of conduction or valence electrons per mole of metal, and V_m is the molar volume of the metal. At temperatures above absolute zero, the electronic kinetic energy per mole is given by

$$E = \frac{3}{5} N E_F \left[1 + \frac{5\pi^2}{12} \left(\frac{kT}{E_F} \right)^2 + \cdots \right] \tag{21.2}$$

EXAMPLE 21.1. Find E_F for Na given $\rho = 0.97 \times 10^3$ kg m^{-3}. What is the electronic contribution to the thermal energy at 25 °C?

Substituting the molar volume,

$$V_m = \frac{M}{\rho} = \frac{(23.0 \text{ g mol}^{-1})[(10^{-3} \text{ kg})/(1 \text{ g})]}{0.97 \times 10^3 \text{ kg m}^{-3}} = 2.37 \times 10^{-5} \text{ m}^3 \text{ mol}^{-1}$$

into (21.1) gives

$$E_F = \frac{(6.626 \times 10^{-34} \text{ J s})^2}{8(9.11 \times 10^{-31} \text{ kg})} \left[\frac{3(6.022 \times 10^{23} \text{ mol}^{-1})}{\pi(2.37 \times 10^{-5} \text{ m}^3 \text{ mol}^{-1})} \right]^{2/3} = 5.04 \times 10^{-19} \text{ J}$$

The thermal energy (Sec. 3.3) becomes

$$E(\text{thermal}) = E - E_0 = \frac{\pi^2 N k^2 T^2}{4 E_F} \tag{21.3}$$

Substituting values for Na gives

$$E(\text{thermal}) = \frac{\pi^2 (6.022 \times 10^{23} \text{ mol}^{-1})(1.380\,6 \times 10^{-23} \text{ J K}^{-1})^2 (298 \text{ K})^2}{4(5.04 \times 10^{-19} \text{ J})} = 49.9 \text{ J mol}^{-1}$$

21.4 THE BAND THEORY

Instead of the uniform potential-energy function used in Sec. 21.3, the band theory assumes potential-energy wells centered at each kernel of the metal atom. For N atoms, bands of closely spaced energy levels are generated that are available for the valence electrons to occupy. Conduction occurs if a partially filled band passes over the walls of the barriers.

EXAMPLE 21.2. Prepare an energy diagram predicting the conductivity of Li.

Figure 21-1a shows the regularly spaced Li kernels and 1s electrons deep in the potential-energy well and the half-filled 2s band near the top of the well. The top of the band passes over the walls; hence Li should be a conductor.

EXAMPLE 21.3. Consider the n-type semiconductor diagramed in Fig. 21-2a. What does the vertical arrow represent? Describe the mechanism for conduction if the impurity has five electrons.

The arrow represents the promotion of the "loose" fifth electron of the impurity by thermal energy. If the band into which this electron is promoted lies above the walls of the wells, conduction occurs.

Ionic Bonding

21.5 BORN–HABER CYCLE

The lattice energy (see Sec. 4.10) between ions is strong enough that most ionic compounds have relatively high melting points, boiling points, heats of sublimation, and heats of vaporization.

If MX is the halide of the alkali metal family (e.g., Li, Na, K, and Rb), the *Born–Haber cycle*, shown in Fig. 21-3, gives the lattice energy as

$$\Delta_{\text{lat}} H(\text{MX}) = -\Delta_f H(\text{MX(s)}) + \Delta_{\text{sub}} H(\text{M}) + (\text{BE})/2 + I_M + E_A(\text{X}) \tag{21.4}$$

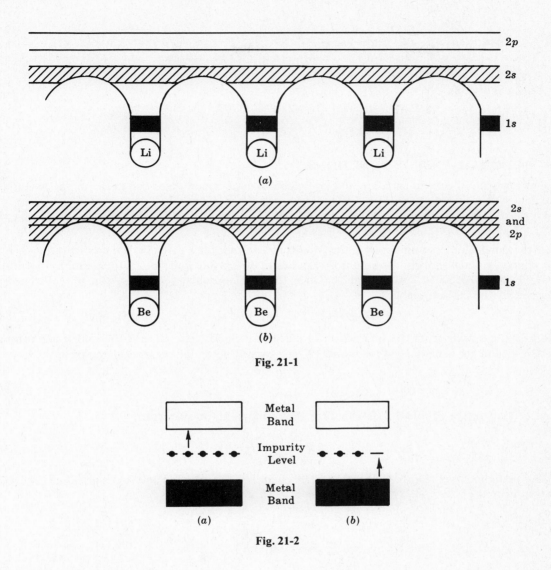

Fig. 21-1

Fig. 21-2

where BE is the bond energy of the halogen, I_M is the ionization energy of the metal, and $E_A(X)$ is the electron affinity of the halogen. If the dissociation energy is used instead of the bond energy for the halogen, the term $\frac{1}{2}RT$ must be added to the right side of (21.4). The equation is also valid for an oxide of the alkaline earth metals (e.g., Mg, Ca, Ba, and Sr).

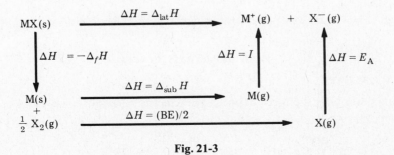

Fig. 21-3

EXAMPLE 21.4. Calculate the lattice energy for NaCl at 25 °C given $\Delta_f H°(NaCl(s)) = -411.153 \text{ kJ mol}^{-1}$, $\Delta_{sub} H(Na) = 107.32 \text{ kJ mol}^{-1}$, $BE = 243.358 \text{ kJ mol}^{-1}$ for Cl_2, $I_{Na} = 502.04 \text{ kJ mol}^{-1}$, and $E_A(Cl) = -354.81 \text{ kJ mol}^{-1}$ for Cl.

Using (21.4) gives

$$\Delta_{lat} H = -(-411.153) + 107.32 + (243.358)/2 + 502.04 + (-354.81)$$

$$= 787.38 \text{ kJ mol}^{-1}$$

21.6 POTENTIAL-ENERGY FUNCTIONS

The attraction between each ion and its neighbors in a crystal gives rise to the molar potential

$$U_{att} = -\frac{L \mathcal{M} q^2}{4\pi\varepsilon_0 r} \qquad (21.5)$$

where the dimensionless number $\mathcal{M}$ is the *Madelung constant* for a given crystal configuration, q is the ionic charge, r is the shortest cation-anion distance in the crystal, and $4\pi\varepsilon_0$ is the permittivity constant ($4\pi\varepsilon_0 = 1.112\,650\,056 \times 10^{-10} \text{ C}^2 \text{ N}^{-1} \text{ m}^{-2}$). Once the ion approaches another ion, a repulsive force described by the molar potential

$$U_{rep} = B\,e^{-r/\rho} \qquad \text{or} \qquad U_{rep} = B'r^{-n} \qquad (21.6)$$

takes effect, resulting from the interaction of electronic clouds, etc. Here B, B', and ρ are constants and n is an integer usually between 6 and 12. The overall potential energy is given by

$$U = -\frac{L \mathcal{M} q^2}{4\pi\varepsilon_0 r} + B\,e^{-r/\rho} \qquad (21.7)$$

It can be shown (see Problem 21.24) that for the sodium chloride structure

$$B = \frac{4L \mathcal{M} q^2 \rho\, e^{a/2\rho}}{4\pi\varepsilon_0 a^2} \qquad (21.8)$$

where a is the unit cell length. Combining (21.8) with (21.7) evaluated at $r = a/2$ gives for the sodium chloride structure

$$\Delta_{lat} H = \frac{2L \mathcal{M} q^2}{4\pi\varepsilon_0 a}\left(1 - \frac{2\rho}{a}\right) + 2RT \qquad (21.9)$$

where $2RT$ is added to convert the energy to an enthalpy.

EXAMPLE 21.5. Calculate $\Delta_{lat} H(NaCl)$ at 25 °C using (21.9), given $a = 0.564\,02 \text{ nm}$, $\rho = 0.034 \text{ nm}$, and $\mathcal{M} = 1.747\,56$. Compare this value to that determined in Example 21.4.

$$\Delta_{lat} H = \frac{2(6.022 \times 10^{23} \text{ mol}^{-1})(1.747\,56)(1.602\,2 \times 10^{-19} \text{ C})^2}{(1.112\,65 \times 10^{-10} \text{ C}^2 \text{ N}^{-1} \text{ m}^{-2})(5.640\,2 \times 10^{-10} \text{ m})}\left[1 - \frac{2(0.034)}{0.564\,02}\right]$$

$$+ 2(8.314 \text{ J K}^{-1} \text{ mol}^{-1})(298 \text{ K})$$

$$= (8.610 \times 10^5 \text{ J mol}^{-1})(1 - 0.120\,6) + 4\,955 \text{ J mol}^{-1} = 762.1 \text{ kJ mol}^{-1}$$

which agrees within 3.2% with the value calculated using the Born–Haber cycle.

Van der Waals Forces

21.7 DIPOLE MOMENTS

Coulomb's law describing the force between two electric charges separated by a distance r within a dielectric is given by

$$f = \frac{q_1 q_2}{4\pi\varepsilon r^2} = \frac{q_1 q_2}{(\varepsilon/\varepsilon_0) 4\pi\varepsilon_0 r^2} \qquad (21.10)$$

where $\varepsilon/\varepsilon_0$ is the *dielectric constant* or *relative permittivity*. The *molar polarization* of a substance is related to $\varepsilon/\varepsilon_0$ by

$$\mathcal{P} = \frac{\varepsilon/\varepsilon_0 - 1}{\varepsilon/\varepsilon_0 + 2}\left(\frac{M}{\rho}\right) \qquad (21.11)$$

The units of $\mathcal{P}$ are $m^3\,mol^{-1}$ in the SI system. Reported values of $\mathcal{P}$ in units of $C\,m^2\,V^{-1}\,mol^{-1}$ can be converted by dividing by $4\pi\varepsilon_0$. Considering the polarization of the molecule to result from permanent and induced dipole moments, it can be shown that

$$\mathcal{P} = \frac{L}{3\varepsilon_0}\left(\alpha + \frac{\mu^2}{3kT}\right) \qquad (21.12)$$

where α is the *polarizability* of the molecule in units of $C^2\,N^{-1}\,m$, μ is the permanent dipole moment of the molecule in units of $C\,m$, and ε_0 is the permittivity of a vacuum ($\varepsilon_0 = 8.854\,187\,18 \times 10^{-12}\,C^2\,N^{-1}\,m^{-2}$). Often the dipole moment is expressed in units of *debyes*, where $1\,D = 3.335\,64 \times 10^{-30}\,C\,m$. Reported values of α in units of m^3 can be converted by multiplication by ε_0. Equation (21.12) implies that a plot of $\mathcal{P}$ against $1/T$ will be linear and will have a slope proportional to μ and an intercept proportional to α.

Recalling the discussion given in Sec. 16.6 for diatomic molecules, a polyatomic molecule will have a dipole moment if it has the center of negative charge separated from the center of positive charge. Very frequently the determination of the molecular shape (see Secs. 18.9–18.11) is a prerequisite to predicting whether a substance is polar or not.

If the dipole moment for a particular bond in a molecule is known, it is often possible to predict semiquantitatively the dipole moment for a molecule containing two such bonds by vectorial analysis (see Problem 21.10).

EXAMPLE 21.6. The dielectric constant for liquid $CHCl_3$ at $20\,°C$ is 4.806. Given $-d\log(\varepsilon/\varepsilon_0)/dT = 0.160 \times 10^{-2}\,K^{-1}$, find α and μ. The density of $CHCl_3$ is $1.483\,2 \times 10^3\,kg\,m^{-3}$ at $20\,°C$.

At $20\,°C$ the polarization is given by (21.11) as

$$\mathcal{P} = \frac{4.806 - 1}{4.806 + 2}\frac{(119.38\,g\,mol^{-1})[(10^{-3}\,kg)/(1\,g)]}{1.483\,2 \times 10^3\,kg\,m^{-3}} = 4.501 \times 10^{-5}\,m^3\,mol^{-1}$$

The dielectric constant at $25\,°C = 298\,K$ is found by integrating the temperature-dependence function, giving

$$\log(\varepsilon/\varepsilon_0)_{298} = \log(\varepsilon/\varepsilon_0)_{293} - (0.160 \times 10^{-2})(298 - 293)$$

$$= \log(4.806) - (0.160 \times 10^{-2})(5) = 0.673\,8$$

or

$$(\varepsilon/\varepsilon_0)_{298} = 4.718$$

Assuming an insignificant change in density, the polarization at $25\,°C$, calculated as above, is $4.455 \times 10^{-5}\,m^3\,mol^{-1}$.

Although a plot of $\mathcal{P}$ against $1/T$ could be prepared, it is simpler to substitute the two values of $\mathcal{P}$ and T into (21.12) and solve for α and μ directly. The results are

$$\mu = 3.83 \times 10^{-30}\,C\,m = 1.15\,D \qquad \alpha = 7.77 \times 10^{-40}\,C^2\,N^{-1}\,m$$

21.8 POTENTIAL-ENERGY FUNCTIONS

The weak, attractive *London forces* between the atoms, molecules, or ions of any substance are described by the molar potential-energy function

$$U_L = \frac{(-1.8 \times 10^{-18}\,J)\alpha^2 L}{(4\pi\varepsilon_0)^2 r^6} \qquad (21.13)$$

These forces alone cause the condensation of a gas to a liquid or solid for substances such as the noble gases, which do not have additional potential energies resulting from dipole-dipole interactions, ionic bonding, metallic bonding, or covalent bonding. Such substances have relatively low melting and boiling points as well as small heats of sublimation and vaporization.

The potential energy for a polar substance is the sum of a term for the dipole-dipole interaction and a term for the induced dipole effect, giving

$$U_d = -\frac{2\mu^2[(\mu^2/3kT)+\alpha]L}{(4\pi\varepsilon_0)^2 r^6} \tag{21.14}$$

As in cases of ionic and covalent bonding, the potential-energy curve generated by the sum of (21.6), (21.13), and (21.14) passes through a minimum. Van der Waals radii have been assigned for several elements, which may be used to estimate the minimizing distance for different substances. Usually the sum of van der Waals radii is of the order of 0.35–0.55 nm.

EXAMPLE 21.7. Compare U_L and U_d for $CHCl_3$ at 25 °C using the data in Example 21.6.

Taking the ratio of (21.14) to (21.13) and substituting data gives

$$\frac{U_d}{U_L} = \frac{2\mu^2[(\mu^2/3kT)+\alpha]}{(1.8\times10^{-18})\alpha^2}$$

$$= \frac{2(3.83\times10^{-30})^2\{[(3.83\times10^{-30})^2/3(1.380\ 7\times10^{-23})(298)]+7.77\times10^{-40}\}}{(1.8\times10^{-18})(7.77\times10^{-40})^2}$$

$$= 0.056$$

Solved Problems

EXTENDED COVALENT BONDING

21.1. Several of the elements in the carbon family of the periodic table [e.g., C(dia), Si, Ge, and Sn(gray)] satisfy the four-electron deficiency on each atom by undergoing sp^3 hybridization and forming a three-dimensional network of covalent bonding in which each atom is bonded to four others. Given $\Delta_f H°/(kJ\ mol^{-1}) = 1.895$ for C(dia) and 716.682 for C(g) at 25 °C, calculate the average C—C bond energy in diamond. Compare this value to 331 kJ mol^{-1}, which is the average value in organic compounds.

The energy necessary to sublime 1 mol of diamond is given by (4.7) as

$$\Delta_{sub}H = [(1)(716.682)] - [(1)(1.895)] = 714.787 \text{ kJ}$$

On the average, two C—C bonds must be broken to sublime a given carbon atom, giving an average energy of $(714.787)/2 = 357.394$ kJ mol^{-1}, about the same as in organic compounds.

21.2. Several of the elements in the nitrogen family of the periodic table [e.g., P(red), P(black), As, Sb, and Bi] require three electrons to complete the "octet" of the outer-shell electrons. Describe the bonding in the condensed phases of these elements. Compare the melting points of these elements to those in the carbon family.

These elements form networks of puckered six-membered rings. Although the bonds between rings are very strong, the bonding between the networks is quite weak, and these substances will melt and boil at lower temperatures than the elements in the carbon family.

21.3. Assuming the absence of hydrogen bonding, predict the boiling points of HF and NH_3 by plotting boiling points against atomic number for the other compounds in each family having a similar formula and extrapolating to the proper atomic number. The actual boiling points are

19.54 °C for HF and −33.35 °C for NH_3. Which hydrogen bond is stronger? Is this in agreement with the electronegativities of the elements?

If a plot of −35.38 °C for HI, −67.0 °C for HBr, and −84.9 °C for HCl against atomic number were prepared and extrapolated to HF, the predicted boiling point would be about −100 °C. Likewise the value for NH_3 is about −100 °C as determined by plotting 22 °C for H_3Bi, −17 °C for SbH_3, −55 °C for H_3As, and −87.4 °C for PH_3. The increase for HF is about twice as much as that for NH_3, so the HF bond is stronger. The result agrees with the fact that F is more electronegative than N.

METALLIC BONDING

21.4. The band theory diagram for a *p*-type semiconductor is shown in Fig. 21-2*b*. Describe the mechanism for conduction if the impurity has three valence electrons.

The thermal energy promotes a metal electron to the impurity level, producing the required partially filled band, which can conduct if the energy level of the band is above the walls.

21.5. Find η in (*3.27*) for silver.

Using the molar mass and density for Ag, the molar volume is

$$V_m = \frac{(107.868 \text{ g mol}^{-1})[(10^{-3} \text{ kg})/(1 \text{ g})]}{10.5 \times 10^3 \text{ kg m}^{-3}} = 1.03 \times 10^{-5} \text{ m}^3 \text{ mol}^{-1}$$

which upon substitution into (*21.1*) gives

$$E_F = \frac{(6.626 \times 10^{-34} \text{ J s})^2}{8(9.11 \times 10^{-31} \text{ kg})} \left[\frac{3(6.022 \times 10^{23} \text{ mol}^{-1})}{\pi(1.03 \times 10^{-5} \text{ m}^3 \text{ mol}^{-1})} \right]^{2/3} = 8.80 \times 10^{-19} \text{ J}$$

Anticipating the results of Problem 21.17 gives

$$\eta = \frac{\pi^2 Rk}{2E_F} = \frac{\pi^2(8.314 \text{ J K}^{-1} \text{ mol}^{-1})(1.380\ 7 \times 10^{-23} \text{ J K}^{-1})}{2(8.80 \times 10^{-19} \text{ J})}$$

$$= 6.44 \times 10^{-4} \text{ J K}^{-2} \text{ mol}^{-1}$$

IONIC BONDING

21.6. Assuming the melting point to be proportional to $\Delta_{lat}H$, qualitatively compare the melting points for NaCl and BaO given $a = 0.564\ 02$ nm for NaCl and 0.550 nm for BaO. Both substances crystallize in the same configuration, giving identical values of $\mathcal{M}$ in (*21.19*). Assume $\rho = 0.034$ nm for both substances.

The difference in the values of $\Delta_{lat}H$ for these substances will arise primarily from the difference in q. Because q^2 is four times as large for BaO as for NaCl, BaO is predicted to have the higher melting point. The actual values are 801 °C for NaCl and 1 923 °C for BaO.

21.7. Find the expression for $\Delta_{lat}H$ using (*21.5*) and the second form for U_{rep} given in (*21.6*). Using $n = 6$ and 12, find $\Delta_{lat}H$ at 25 °C for NaCl, and compare these values to those found in Examples 21.4 and 21.5.

The overall potential energy is given by

$$U = -\frac{L\mathcal{M}q^2}{4\pi\varepsilon_0 r} + B'r^{-n}$$

For a minimum at $r = a/2$,

$$\left. \frac{\partial U}{\partial r} \right|_{r=a/2} = \frac{L\mathcal{M}q^2}{4\pi\varepsilon_0(a/2)^2} - nB'(a/2)^{-n-1} = 0$$

yielding

$$B' = \frac{L\mathcal{M}q^2(a/2)^{n-1}}{4\pi\varepsilon_0 n}$$

The expression for U becomes

$$U = -\frac{L\mathcal{M}q^2}{4\pi\varepsilon_0}\left[\frac{1}{r} - \frac{(a/2)^{n-1}}{nr^n}\right]$$

and

$$\Delta_{\text{lat}}H = \frac{L\mathcal{M}q^2}{4\pi\varepsilon_0}\left[\frac{1}{r} - \frac{(a/2)^{n-1}}{nr^n}\right] + 2RT$$

Upon substitution of $r = a/2$ and the data from Example 21.5, for $n = 6$,

$$\Delta_{\text{lat}}H = \frac{(6.022\times10^{23}\text{ mol}^{-1})(1.747\ 56)(1.602\ 2\times10^{-19}\text{ C})^2}{1.112\ 65\times10^{-10}\text{ C}^2\text{ N}^{-1}\text{ m}^{-2}}\left[\frac{5}{6}\left(\frac{1}{2.820\ 1\times10^{-10}\text{ m}}\right)\right]$$

$$+ 2(8.314\text{ J K}^{-1}\text{ mol}^{-1})(298\text{ K})$$

$$= 722.5\text{ kJ mol}^{-1}$$

and for $n = 12$, $\Delta_{\text{lat}}H = 794.2\text{ kJ mol}^{-1}$. Both values are close to those determined in Examples 21.4 and 21.5, with the higher value of n giving better agreement.

VAN DER WAALS FORCES

21.8. Predict which molecules in Fig. 18-16 will have significant dipole moments.

Using the three-dimensional sketches given in Fig. 18-17, the following can be predicted:
(*a*) The polar covalent bonds in $n\text{-}C_8H_{18}$ give rise to small dipole moments depending on the orientation of the carbon atoms, but the effect is not large.
(*b*) The planar molecule BF_3 has the bonds arranged such that the polar nature of the bonds cancels, giving a nonpolar molecule.
(*c*) The S_8 molecule has no polar nature.
(*d*) The planar molecule C_2H_4 is a nonpolar molecule.
(*e*) The linear molecule C_2H_2 is a nonpolar molecule.
(*f*) There is a displacement of positive and negative centers of charge in HCNO, giving a dipole moment.
(*g*) The linear molecule N_2O has a dipole moment.
(*h*) The octahedral symmetry of $CrCl_6^{3-}$ makes the center of the molecule the center of both the positive and negative charge, so that the molecule is nonpolar. However, the presence of an ionic charge makes this determination purely an academic exercise.
(*i*) The square planar configuration of XeF_4 is a nonpolar configuration.

21.9. In the gaseous state, $\mu/(D) = 1.08$ for HCl, 0.82 for HBr, and 0.44 for HI. The dielectric constants of the gases at 101 325 Pa are 1.004 6 for HCl at 0 °C and 1.003 13 for HBr at 20 °C. Find the values of α for these substances. Assuming α to be proportional to the number of electrons present, qualitatively predict α for HI. Check this prediction by calculating α from the dielectric constant of 1.002 34 at 0 °C for the gas.

Solving (*1.8*) for M/ρ and substituting into (*21.11*) gives for the gases

$$\mathscr{P} = \frac{\varepsilon/\varepsilon_0 - 1}{\varepsilon/\varepsilon_0 + 2}\left(\frac{RT}{P}\right) \tag{21.15}$$

Substituting the data for HBr gives

$$\mathscr{P} = \frac{1.003\ 13 - 1}{1.003\ 13 + 2}\left(\frac{(8.314\text{ m}^3\text{ Pa K}^{-1}\text{ mol}^{-1})(293\text{ K})}{101\ 325\text{ Pa}}\right)$$

$$= 2.51\times10^{-5}\text{ m}^3\text{ mol}^{-1}$$

Using (*21.12*) gives

$$2.51\times10^{-5} = \frac{6.022\times10^{23}}{3(8.854\times10^{-12})}\left(\alpha + \frac{(0.82)^2(3.335\ 64\times10^{-30})^2}{3(1.380\ 7\times10^{-23})(293)}\right)$$

which upon solving gives $\alpha = 4.9 \times 10^{-40} \, C^2 \, N^{-1} \, m$. Similarly for HCl, $\mathcal{P} = 3.43 \times 10^{-5}$, giving $\alpha = 3.7 \times 10^{-40}$. According to the number of electrons, the predicted value for HI should be greater than the values for HCl and HBr. Repeating the calculations confirms this prediction: $\mathcal{P} = 1.75 \times 10^{-5}$ and $\alpha = 5.81 \times 10^{-40}$.

21.10. The dipole moment of chlorobenzene is 1.69 D. Predict the dipole moments for *o-*, *m-*, and *p-*dichlorobenzene.

Consider the sketches of the molecules shown in Fig. 21-4. For *o-*dichlorobenzene, the vectorial summing process gives

$$\mu_x = 1.69 \cos 30° + 1.69 \cos 90° = 1.46$$

$$\mu_y = 1.69 \sin 30° + 1.69 \sin 90° = 2.54$$

$$\mu = (\mu_x{}^2 + \mu_y{}^2)^{1/2} = [(1.46)^2 + (2.54)^2]^{1/2} = 2.93 \text{ D}$$

which is a little higher than the actual 2.50 D. For *m-*dichlorobenzene,

$$\mu_x = 1.69 \cos(-30°) + 1.69 \cos 90° = 1.46$$

$$\mu_y = 1.69 \sin(-30°) + 169 \sin 90° = 0.85$$

$$\mu = [(1.46)^2 + (0.85)^2]^{1/2} = 1.69 \text{ D}$$

which is close to the actual value of 1.72 D. For *p-*dichlorobenzene, $\mu = \mu_x = \mu_y = 0$, which agrees with the experimental value.

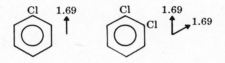

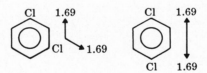

Fig. 21-4

21.11. The dielectric constant for He is 1.000 068 4 at 101 325 Pa and 140 °C, and for Ar it is 1.000 545 at 101 325 Pa and 23 °C. (*a*) Using (*21.12*) with $\mu = 0$, find α for these elements. (*b*) Assuming a van der Waals radius of 0.2 nm for each element, calculate U_L, and qualitatively compare the boiling points of these substances.

(*a*) Using (*21.15*) gives

$$\mathcal{P}_{He} = \frac{1.000\,068\,4 - 1}{1.000\,068\,4 + 2}\left(\frac{(8.314)(413)}{101\,325}\right) = 7.73 \times 10^{-7} \, m^3 \, mol^{-1}$$

$$\mathcal{P}_{Ar} = 4.41 \times 10^{-6} \, m^3 \, mol^{-1}$$

Then, from (*21.12*),

$$\alpha_{He} = \frac{3\varepsilon_0 \mathcal{P}}{L} = (4.411 \times 10^{-35})\mathcal{P} = (4.411 \times 10^{-35})(7.73 \times 10^{-7})$$

$$= 3.41 \times 10^{-41} \, C^2 \, N^{-1} \, m$$

a very low value, and $\alpha_{Ar} = 1.95 \times 10^{-40} \, C^2 \, N^{-1} \, m$.

(b)   With $r = 0.4$ nm, (21.13) gives

$$U_L(\text{He}) = \frac{(-1.8 \times 10^{-18} \text{ J})(3.41 \times 10^{-41} \text{ C}^2 \text{ N}^{-1} \text{ m})^2(6.022 \times 10^{23} \text{ mol}^{-1})}{(1.112\,65 \times 10^{-10} \text{ C}^2 \text{ N}^{-1} \text{ m}^{-2})^2(4 \times 10^{-10} \text{ m})^6}$$

$$= -24.9 \text{ J mol}^{-1}$$

and $U_L(\text{Ar}) = -813$ J mol^{-1}. The values of U_L indicate that much more energy is needed to boil Ar than He, neglecting repulsive forces, which agrees with the experimental values of $\Delta_{\text{vap}} H^\circ = 6\,519$ J mol^{-1} for Ar at 87.29 K and 84 J mol^{-1} for He at 4.22 K. Note that at 4.22 K,

$$24.9 \text{ J mol}^{-1} + 2RT = 95.1 \text{ J mol}^{-1}$$

for He, a value very close to the experimental value of 84 J mol^{-1}.

Supplementary Problems

EXTENDED COVALENT BONDING

21.12. Some of the elements in the oxygen family (e.g., S, Se, and Te) require two electrons to complete the outer-shell octet of electrons. Describe the bonding in the condensed phases of these elements. Compare the melting points of these elements to those in the carbon and nitrogen families.

Ans.   Networks of chains or single rings; lower, similar

21.13. Find the average Ge—Ge bond energy given that the heat of sublimation for Ge is 376.6 kJ mol^{-1}. Why would the energy of a Ge—Ge bond be predicted as less than that of a C—C bond?

Ans.   188.3 kJ mol^{-1}; bonding electrons farther from the nucleus in Ge

21.14. Would ΔH for the phase transition

$$\text{SiO}_2(\alpha\text{-qtz}) \longrightarrow \text{SiO}_2(\alpha\text{-tridymite})$$

be large? In both structures each Si is bonded to four O's in a large three-dimensional network.

Ans.   No

21.15. Discuss the bonding in diamond and graphite. Which substance will show the following properties: (*a*) very hard, (*b*) soft, (*c*) electrical insulator, (*d*) electrical conductor along two crystallographic axes, (*e*) absorption of gases, and (*f*) oxidation?

Ans.   Diamond has sp^3-hybridized atoms forming four equivalent bonds; graphite has sp^2 atoms forming covalent bonds in layers and weak forces between layers.   (*a*) Bonds in diamond are hard to break. (*b*) Bonds between graphite layers are easy to break.   (*c*) No available electrons in diamond.   (*d*) Electrons between graphite layers move easily.   (*e*) Layers of graphite are widely spaced, and electrons provide mechanism for absorption.   (*f*) Both oxidize under extreme conditions.

21.16. The boiling points for the hydrogen-containing compounds of the oxygen family are $-2\,^\circ$C for H$_2$Te, $-41.5\,^\circ$C for H$_2$Se, and $-68.7\,^\circ$C for H$_2$S. (*a*) Predict the boiling point for H$_2$O assuming the absence of hydrogen bonding. (*b*) Based on electronegativities, predict the relative strengths of hydrogen bonding in NH$_3$, H$_2$O, and HF. (*c*) Why does water seemingly violate this order?

Ans.   (*a*) About $-80\,^\circ$C.   (*b*) HF $>$ H$_2$O $>$ NH$_3$.   (*c*) H$_2$O can form three-dimensional networks of bonds, whereas HF and NH$_3$ can only form chains.

METALLIC BONDING

21.17. Find the electronic contribution to the heat capacity by differentiating (21.3) with respect to T. Evaluate η in (3.27) for Na.

Ans. $C_V(\text{electronic}) = \pi^2 RkT/2E_F$; using $E_F = 5.04 \times 10^{-19}$ J gives $\eta = 5.62 \times 10^{-4}$ J K^{-2} mol^{-1}

21.18. Find E_F and η for aluminum, and compare the latter value to that found in Problem 3.10. The density of Al is $2.702 \times 10^3 \text{ kg m}^{-3}$.

> *Ans.* $E_F = 8.98 \times 10^{-19} \text{ J}$, $\eta = 6.30 \times 10^{-4} \text{ J K}^{-2} \text{ mol}^{-1}$; about 54% low, but correct order of magnitude

21.19. Prepare an energy diagram demonstrating the conductivity of Be, given that the 2p band overlaps the 2s band.   *Ans.* See Fig. 21-1b.

21.20. A photoconductor has an unfilled band near a filled band. Describe the mechanism for conduction.

> *Ans.* Light quanta excite electrons, giving two partially filled bands; the upper band or both may conduct depending on the height of the barrier.

IONIC BONDING

21.21. Qualitatively predict the melting points for BaO and MgO given $a = 0.550$ nm and $0.421\ 3$ nm, respectively, and assuming equivalent crystal configurations.

> *Ans.* $\text{MP} \propto \Delta_{\text{lat}} H \propto q^2/a$, giving MgO > BaO based on a; actual values are $1\ 923\ °\text{C}$ and $2\ 800\ °\text{C}$

21.22. Using (21.4), calculate the lattice energy for periclase, MgO, at $25\ °\text{C}$ given $\Delta_f H°(\text{MgO(s)}) = -601.70 \text{ kJ mol}^{-1}$, $\Delta_{\text{sub}} H°(\text{Mg}) = 147.70 \text{ kJ mol}^{-1}$, $\text{BE} = 498.340 \text{ kJ mol}^{-1}$ for O_2, $I_{\text{Mg}} = 2\ 200.80 \text{ kJ mol}^{-1}$ for two electrons, and $E_A(\text{O}) = 735 \text{ kJ mol}^{-1}$ for two electrons. Given $a = 0.421\ 3$ nm, $\rho = 0.034$ nm, and $\mathcal{M} = 1.747\ 56$, calculate $\Delta_{\text{lat}} H$ using (21.9).   *Ans.* $3\ 934$ and $3\ 866 \text{ kJ mol}^{-1}$

21.23. Combining (21.7) and (21.8) gives

$$U = -\frac{2L\mathcal{M}q^2}{4\pi\varepsilon_0 a}\left\{\frac{a}{2r} - \frac{2\rho}{a} e^{[(a/2)-r]/\rho}\right\}$$

Prepare a plot of U against r for values of r between 0.1 nm and 1.0 nm for MgO, given $a = 0.421\ 3$ nm and $\rho = 0.034$ nm.   *Ans.* Typical potential-energy well with minimum of $-3\ 866 \text{ kJ mol}^{-1}$ at $0.421\ 3$ nm

21.24. Show that (21.8) is the correct expression for B for the halite structure by differentiating (21.7) with respect to r, setting the result equal to zero at $r = a/2$, and solving for B.

VAN DER WAALS FORCES

21.25. Predict which molecules in Fig. 18-22 will have significant dipole moments.

> *Ans.* H_2O_2, $SnCl_2$, $COCl_2$, $SeBr_4$

21.26. Calculate $\varepsilon/\varepsilon_0$ for water vapor at STP given $\alpha = 1.60 \times 10^{-40} \text{ C}^2 \text{ N}^{-1} \text{ m}$ and $\mu = 1.85$ D.

> *Ans.* $\mathcal{P} = 7.998 \times 10^{-5} \text{ m}^3 \text{ mol}^{-1}$; $\varepsilon/\varepsilon_0 = 1.010\ 7$

21.27. Given $\varepsilon/\varepsilon_0 = 2.238$ and $\rho = 1.595\ 4 \times 10^3 \text{ kg m}^{-3}$ for $CCl_4(\text{l})$ and 1.70 and $0.466 \times 10^3 \text{ kg m}^{-3}$ for $CH_4(\text{l})$. Find α for these substances. Arrange CCl_4, CH_4, and $CHCl_3$ (see Example 21.6) in order of increasing α. Predict where CH_3Cl and CH_2Cl_2 would fit into this sequence.

> *Ans.* $\mathcal{P}/(\text{m}^3 \text{ mol}^{-1}) = 2.816 \times 10^{-5}$ for CCl_4 and 6.5×10^{-6} for CH_4; $\alpha/(\text{C}^2 \text{ N}^{-1} \text{ m}) = 12.42 \times 10^{-40}$ for CCl_4 and 2.9×10^{-40} for CH_4; $CH_4 < CH_3Cl < CH_2Cl_2 < CHCl_3 < CCl_4$

21.28. Jona reported the following polarization-temperature data for $NH_3(\text{g})$:

$\mathcal{P}/(\text{mL mol}^{-1})$	57.57	55.01	51.22	44.99	42.51	39.59
$T/(\text{K})$	292.2	309.0	333.0	387.0	413.0	446.0

Determine α and μ for ammonia.

Ans.   Plot of $\mathcal{P}$ against $1/T$ is linear with slope $= 0.015\,236\,\mathrm{m^3\,mol^{-1}\,K}$ and intercept $= 5.543 \times 10^{-6}\,\mathrm{m^3\,mol^{-1}}$, giving $\alpha = 2.445 \times 10^{-40}\,\mathrm{C^2\,N^{-1}\,m}$ and $\mu = 5.277 \times 10^{-30}\,\mathrm{C\,m} = 1.582\,\mathrm{D}$

21.29.   The dipole moment of chlorobenzene is 1.69 D, and that of nitrobenzene is 4.22 D. Predict the dipole moments for *o*-, *m*-, and *p*-chloronitrobenzene using vectorial analysis.

Ans.   5.27, 3.67, and 2.53 (actual are 4.64, 3.73, and 2.83)

21.30.   The boiling point of n-C_5H_{12} is 36.07 °C, and that of neopentane is 9.5 °C. If neither molecule has any appreciable intermolecular bonding except London forces, why is there such a large difference in the boiling points?

Ans.   The "linear" n-C_5H_{12} molecules have large parallel overlap, allowing stronger London forces than the "spherical" neopentane molecules, which have very little overlap.

21.31.   The dielectric constant for $N_2(g)$ is 1.000 580 at 23 °C and 101 325 Pa, and that for $CO(g)$ is 1.000 70 under the same conditions. If the molecules in each gas are 0.3 nm apart, calculate α and the attractive potential energy for these substances. Which substance will have the higher boiling point and heat of vaporization? The dipole moment of CO is 0.112 D.

Ans. $\mathcal{P}/(\mathrm{m^3\,mol^{-1}}) = 4.69 \times 10^{-6}$ for N_2 and 5.7×10^{-6} for CO; $\alpha/(\mathrm{C^2\,N^{-1}\,m}) = 2.07 \times 10^{-40}$ for N_2 and 2.4×10^{-40} for CO; $U_L/(\mathrm{J\,mol^{-1}}) = -5\,150$ for N_2 and $-6\,900$ for CO, $U_d/(\mathrm{J\,mol^{-1}}) = -4.7$ for CO; $U/(\mathrm{J\,mol^{-1}}) = -5\,150$ for N_2 and $-6\,900$ for CO (the dipole moment contribution is insignificant); CO is predicted to have the higher boiling point and heat of vaporization (actual values are 6.042 kJ mol^{-1} at 81.66 K for CO and 5.577 kJ mol^{-1} at 77.34 K for N_2).

21.32.   Predict the major contributions to the intermolecular bonding in:

(*a*) Mg   (*b*) Br_2 (*c*) HF

(*d*) HBr   (*e*) C_2H_6 (*f*) AgCl

(*g*) Si   (*h*) PCl_5 (*i*) PCl_3

(*j*) H_2Te (*k*) SO_2 (*l*) NO_2

(*m*) *p*-Xylene   (*n*) Ne   (*o*) $SnBr_4$

(*p*) *trans*-$PtCl_2Br_2$ (*q*) *cis*-$PtCl_2Br_2$

Ans.   (*a*) Metallic   (*b*) London   (*c*) Hydrogen bonding,
                                                                           dipole, and London

(*d*) Dipole and London   (*e*) London   (*f*) Ionic

(*g*) Covalent   (*h*) London   (*i*) London

(*j*) Dipole and London   (*k*) Dipole and London   (*l*) Dipole (bent molecule)
                                                                           and London

(*m*) London   (*n*) London   (*o*) London

(*p*) London   (*q*) Dipole and London

Chapter 22

Crystals

Unit Cell

22.1 INTRODUCTION

The constituents of a crystalline solid, whether atoms, molecules, or ions, are arranged in an ordered, repetitive fashion in three dimensions. The array is called a (*space*) *lattice*. It is possible to choose a group of atoms to serve as a model of the crystal just as it is possible to select the repeated motif on wallpaper to serve as the representation of the entire roll. The representative atoms chosen for the model are collectively called the *unit cell* of the crystal. If the unit cell is properly chosen, it is possible to generate the entire lattice by repeating the structure of the unit cell.

Consider the three-dimensional arrangement of points in Fig. 22-1, which might correspond to the location of the constituents of a metallic crystal. An acceptable unit cell for this crystal is in bold outline. By translating this unit cell along the three axes shown in the diagram, the entire crystal pattern can be generated.

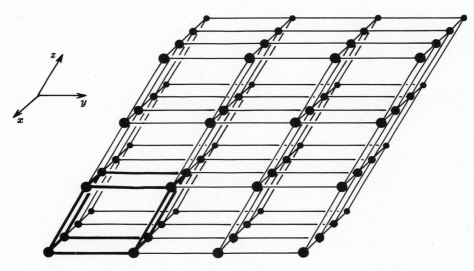

Fig. 22-1   (After C. R. Metz, "Models and the Crystalline State," STRC-343, *Modular Laboratory Program*, Chemical Education Resources, Inc., Palmyra, PA, 1988.)

Although any three noncoplanar rows will define a unit cell, by convention the preferred unit cell is one that represents the symmetry of the crystal, is as nearly orthogonalized as possible, and is minimal in content of atoms. The translational distances (Sec. 20.8) defining the lengths of the unit cell in the x, y, and z directions are designated a, b, and c, respectively, and the angles between these edges are α (between b and c), β (between a and c), and γ (between a and b). By convention, c usually lies parallel to the highest-order rotation axis in the unit cell, and a and b lie parallel or perpendicular to other symmetry elements, if present. Under these restrictions, usually $c \leq a \leq b$, and α and $\beta \geq 90°$.

The seven *crystal systems* (the rhombohedral system is sometimes listed as a subsystem of the hexagonal system, giving six) are given in Table 22-1. (Optional conventions are shown in parentheses.)

431

Table 22-1

Crystal System	Geometry of System	Bravais Lattices	Point Groups		Minimum Symmetry
Triclinic	$a \neq b \neq c$† $\alpha \neq \beta \neq \gamma$†	P	$\bar{1}$ 1	$\mathscr{C}_i$ $\mathscr{C}_1$	C_1
Monoclinic	$a \neq b \neq c$† $\alpha = \beta = 90°,\ \gamma > 90°$ $(\alpha = \gamma = 90°,\ \beta > 90°)$	$P;\ B$ or A $(C$ or $A)$	$2/m$ m 2	$\mathscr{C}_{2h}$ $\mathscr{C}_s$ $(\mathscr{C}_{1h}, \mathscr{C}_{1v}, \mathscr{C}_v)$ $\mathscr{C}_2$	C_2 or σ
Orthorhombic	$a \neq b \neq c$ $\alpha = \beta = \gamma = 90°$	$P;\ C,\ B$ or $A;$ $F;\ I$	$2/m\ 2/m\ 2/m\ (mmm)$ $2mm\ (mm2)$ 222	$\mathscr{D}_{2h}$ $\mathscr{C}_{2v}$ $\mathscr{D}_2$	$3 \perp C_2$ or C_2 and 2σ
Tetragonal	$a = b \neq c\ (a_1 = a_2 \neq c)$ $\alpha = \beta = \gamma = 90°$	$P;\ I$	$4/m\ 2/m\ 2/m\ (4/m\,mm)$ $422\ (42)$ $4mm$ $\bar{4}2m\ (\bar{4}m2)$ $4/m$ 4 $\bar{4}$	$\mathscr{D}_{4h}$ $\mathscr{D}_4$ $\mathscr{C}_{4v}$ $\mathscr{D}_{2d}$ $\mathscr{C}_{4h}$ $\mathscr{C}_4$ $\mathscr{S}_4$	C_4 or $\bar{4}$
Hexagonal‡	$a = b \neq c$ $(a_1 = a_2 = -a_3 \neq c)$ $\alpha = \beta = 90°,\ \gamma = 120°$	$P\ (H)$	$6/m\ 2/m\ 2/m\ (6/m\,mm)$ $622\ (62)$ $6mm$ $\bar{6}m2\ (\bar{6}\,2m)$ $6/m$ 6 $\bar{6}\ (3/m)$	$\mathscr{D}_{6h}$ $\mathscr{D}_6$ $\mathscr{C}_{6v}$ $\mathscr{D}_{3h}$ $\mathscr{C}_{6h}$ $\mathscr{C}_6$ $\mathscr{C}_{3h}\ (\mathscr{S}_3)$	C_6 or $\bar{6}$
Rhombohedral (trigonal)	$a = b = c\ (a_1 = a_2 = a_3)$ $120° > \alpha = \beta = \gamma \neq 90°$ $(a_1 = a_2 \neq c,\ \alpha = \beta = 90°,$ $\gamma = 120°)$	$R\ (P)$	$\bar{3}\ 2/m\ (\bar{3}m)$ 32 $3m$ $\bar{3}$ 3	$\mathscr{D}_{3d}$ $\mathscr{D}_3$ $\mathscr{C}_{3v}$ $\mathscr{C}_{3i}$ $\mathscr{C}_3$	C_3 or $\bar{3}$
Cubic (isometric)	$a = b = c\ (a_1 = a_2 = a_3)$ $\alpha = \beta = \gamma = 90°$	$P;\ F;\ I$	$4/m\ \bar{3}\ 2/m\ (m3m)$ $432\ (43)$ $\bar{4}3m$ $2/m\ \bar{3}\ (m3)$ 23	$\mathcal{O}_h$ $\mathcal{O}$ $\mathcal{T}_d$ $\mathcal{T}_h$ $\mathcal{T}$	$4C_3$ at $54°44'$

† Fortuitous equalities do not place the lattice in a system of higher symmetry.
‡ The hexagonal system is often overdetermined by including a_3, which is a linear combination of a_1 and a_2 (see Fig. 22-6).

The 14 *Bravais lattices* are also listed in Table 22-1. These are the only possible arrangements of identical points in space that retain the full symmetry of the system such that the entire crystal can be generated by translation. The symbol P means a primitive lattice in the shape defined by the geometric requirements of the system (R means a primitive rhombohedron, and H is sometimes used to describe a primitive hexagonal unit cell); C, B, or A means the centering of an atom only in the faces defined by the a and

b axes, the a and c axes, or the b and c axes, respectively; F means face-centering on all faces; and I means body-centering.

The point groups listed for each crystal system in Table 22-1 reflect the unique combinations of the various symmetry elements allowed in each system. Although each of these 32 point groups could be represented by a single symbol using the Schönflies system (see Chap. 20), crystallographers use the Hermann–Mauguin system, which usually consists of (1) a symbol describing the symmetry element along the c axis; (2) a symbol describing the symmetry element, if any, along one of the other axes or at an angle of 54°44′ to the c axis in the cubic system; and (3) a symbol describing the symmetry element, if any, along the third axis or at an angle of 30° or 45° to the second axis in the hexagonal and tetragonal systems, respectively. The equivalencies between the Schönflies and Hermann–Mauguin systems are given in Table 22-1, as well as the minimum symmetry requirements for a unit cell to belong to that system.

If the allowed symmetry elements include glide planes and screw axes, and if centering is permitted, 230 space groups are generated. The symbol for a space group consists of the centering followed by an abbreviated Hermann–Mauguin symbol; e.g., $P\,2_1 2_1 2_1$ is a primitive unit cell in the orthorhombic system in which the twofold axes are twofold screw axes, and $F m 3 m$ (complete symbol is $F\,4/m\,\bar{3}\,2/m$) is a face-centered unit cell in the cubic system. Because very careful X-ray analysis and other tests are required to distinguish the various space groups, they will not be considered further.

22.2 UNIT CELL CONTENT

The *unit cell content* (Z) is the number of points contained within the unit cell. A primitive unit cell has points only at the corners of the parallelepiped, whereas a multiple unit cell contains additional points, which are edge-centered, face-centered, or body-centered. A point on a corner is being shared by several unit cells and thus contributes only a fraction of its volume and mass to the unit cell under consideration; the total contribution from all corners is $Z = 1$. A point that is centered on a face of a unit cell is being shared by exactly two unit cells and thus contributes half of its volume and mass to the unit cell under consideration. A point that is centered within the unit cell is not being shared and thus contributes its entire volume and mass. A point that is centered along an edge of the unit cell is being shared by three or four unit cells; the total contribution of these points will be an integer.

EXAMPLE 22.1.   Determine the unit cell content for the primitive unit cell shown in Fig. 22-2.

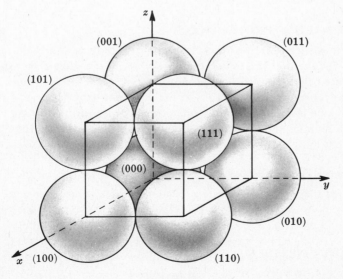

Fig. 22-2

Each corner is shared by eight unit cells, and thus a given corner atom contributes only one-eighth of its volume and mass to the unit cell under consideration (see Fig. 22-3). The corners contribute a total of $8(\frac{1}{8}) = 1 = Z$ for this unit cell.

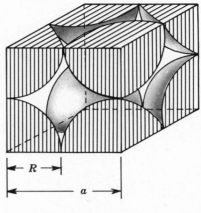

Fig. 22-3

22.3 UNIT CELL COORDINATES

Each atom of the unit cell may be located by assigning x, y, and z coordinates to the atom. Consider the primitive cubic unit cell shown in Fig. 22-2. If the atom in the rear lower left-hand corner of the unit cell is taken as the origin of a three-dimensional Cartesian coordinate system, its coordinates (given as multiples of a, b, and c) would be (000), those of the atom on the body diagonal would be (111), etc. The coordinates of all eight atoms are given in Fig. 22-2. The letters u, v, and w are often used to describe decimal fractions. A bar over a number or letter, e.g., $\bar{u}$, means the negative, which can be equally interpreted as $1 - u$.

By convention, the set of coordinates (000) stands for the locations of all eight corners, i.e., (100), (111), (101), (110), (001), (011), (010), and (000); the set of coordinates $(00\frac{1}{2})$ stands for $(00\frac{1}{2})$, $(10\frac{1}{2})$, $(01\frac{1}{2})$, and $(11\frac{1}{2})$; the set given by $(\frac{1}{2}\frac{1}{2}0)$ stands for $(\frac{1}{2}\frac{1}{2}0)$ and $(\frac{1}{2}\frac{1}{2}1)$; etc. The minimum number of coordinate sets necessary to express the location of all atoms in the unit cell will be equal to Z.

22.4 CRYSTALLOGRAPHIC PROJECTIONS

A crystallographic projection, or view of the unit cell looking along one of the crystallographic axes, shows the shape of the unit cell in two dimensions, with the three-dimensional information being given by the coordinate of each constituent along that axis. If the projection is made along the z axis, the combined symbol ① represents an atom in the xy plane (the plane of the paper) and an atom above the xy plane by one unit length c. Only one projection is required for a cubic crystal. For crystals in the hexagonal, monoclinic, and tetragonal systems, two projections are required as a minimum, and for crystals in the other systems, three projections are necessary.

EXAMPLE 22.2 Prepare a crystallographic projection for the primitive cubic unit cell (Fig. 22-2).

The two-dimensional shape will be a square with atoms at each corner. The three-dimensional information for the cell is contained in the symbol ①. See Fig. 22-4. Clearly, only one projection is required.

22.5 COORDINATION NUMBER

The *coordination number* (CN) of an atom in a crystal is the number of nearest-neighbor atoms. All atoms in the Bravais lattice have the same CN.

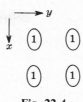

Fig. 22-4

EXAMPLE 22.3. Determine CN for an atom in the primitive cubic unit cell.

If the primitive cubic unit cell were translated to generate several unit cells as shown in Fig. 22-5, around any given atom there would be six equally spaced nearest-neighbor atoms at a distance a. Thus CN = 6. Any other atoms in the crystal lattice will be at a distance greater than a from the atom under consideration.

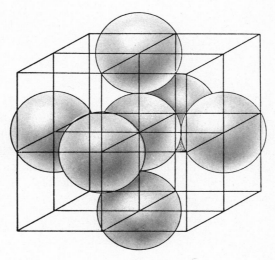

Fig. 22-5

22.6 THEORETICAL DENSITY

If the unit cell dimensions are known, the *theoretical density* for a substance can be calculated from

$$\rho = \frac{ZM}{LV} \tag{22.1}$$

where M is the molar mass, L is Avogadro's number, and

$$V = abc(1 - \cos^2 \alpha - \cos^2 \beta - \cos^2 \gamma + 2 \cos \alpha \cos \beta \cos \gamma)^{1/2} \tag{22.2a}$$

For unit cells having 90° angles between edges, (22.2a) simplifies to

$$V = abc \tag{22.2b}$$

EXAMPLE 22.4. Polonium is the only element known to crystallize in a primitive cubic unit cell under room conditions. Given $a = 0.336$ nm, find the theoretical density.

Using (22.2) gives $V = (3.36 \times 10^{-10} \text{ m})^3$ for the unit cell, which upon substitution into (22.1) gives

$$\rho = \frac{(1)(209 \text{ g mol}^{-1})[(10^{-3} \text{ kg})/(1 \text{ g})]}{(6.022 \times 10^{23} \text{ mol}^{-1})(3.36 \times 10^{-10} \text{ m})^3} = 9.15 \times 10^3 \text{ kg m}^{-3}$$

22.7   CRYSTAL RADII

If the spheres representing the atoms in a unit cell are assumed to touch along an edge, a face diagonal, a body diagonal, etc., the *crystal radius of the atom* (R) can be calculated from the unit cell dimensions.

EXAMPLE 22.5.   Find the relationship between a and R for a primitive cubic unit cell, and calculate R for Po, given $a = 0.336$ nm.

Figure 22-3 shows that $a = 2R$, giving $R = 0.168$ nm.

22.8   SEPARATION OF ATOMS

The distance (ℓ) between two atoms in a unit cell can be calculated from their coordinates $(x_1 y_1 z_1)$ and $(x_2 y_2 z_2)$ as

$$\ell = [a^2(x_2 - x_1)^2 + b^2(y_2 - y_1)^2 + c^2(z_2 - z_1)^2 - 2ab(x_2 - x_1)(y_2 - y_1) \cos \gamma$$

$$-2ac(x_2 - x_1)(z_2 - z_1) \cos \beta - 2bc(y_2 - y_1)(z_2 - z_1) \cos \alpha]^{1/2} \qquad (22.3a)$$

which simplifies to

$$\ell = [a^2(x_2 - x_1)^2 + b^2(y_2 - y_1) + c^2(z_2 - z_1)^2]^{1/2} \qquad (22.3b)$$

for an orthogonal unit cell and to

$$\ell = a[(x_2 - x_1)^2 + (y_2 - y_1)^2 + (z_2 - z_1)^2]^{1/2} \qquad (22.3c)$$

for a cubic unit cell.

EXAMPLE 22.6.   Find the distance between two Po atoms that lie along a body diagonal given $a = 0.336$ nm.

Substituting (000) and (111) into ($22.3c$) gives

$$\ell = a[(1 - 0)^2 + (1 - 0)^2 + (1 - 0)^2]^{1/2}$$

$$= a(3)^{1/2} = 0.336 \text{ nm}(3)^{1/2} = 0.582 \text{ nm}$$

Crystal Forms

22.9   METALLIC CRYSTALS

Most metals crystallize in either the *hexagonal closest-packed* or *cubic closest-packed* unit cells, where $CN = 12$. These unit cells are a hexagonal body-centered cell (see Example 22.7) and the cubic face-centered cell, respectively, and have the highest packing density for simple lattices, as well as a great stability because of the high CN. Several of the alkali metals crystallize in the body-centered cubic unit cell, where $CN = 8$. Very few metals use the remaining 12 Bravais lattices because of the inefficient packing.

EXAMPLE 22.7.   Calculate the efficiency of packing in the hexagonal closest-packed unit cell (see Fig. 22-6).

The diagrams in Fig. 22-6 show that $Z = 2$ for the unit cell. If a radius of R is assumed for the metal atom, the total volume occupied by the metal atoms is

$$V_{occ} = 2(\tfrac{4}{3}\pi R^3) = 8.38 R^3$$

The diagrams show that $a_1 = a_2 = 2R$. By careful inspection of the geometry of the system, it can be shown that $c = 2(\tfrac{2}{3})^{1/2}a = 1.633a$. Using ($22.2a$) for the volume of the unit cell gives

$$V_{cell} = a^2 c(1 - \cos^2 \gamma)^{1/2} = a^2 c \sin \gamma$$

$$= (2R)^2(1.633)(2R)(\sin 120°) = 11.31 R^3$$

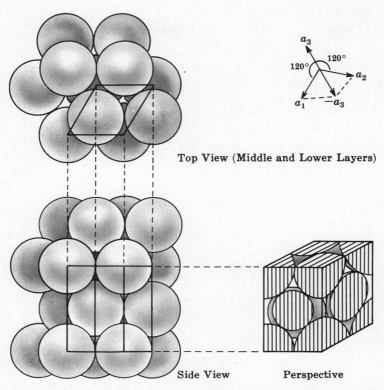

Top View (Middle and Lower Layers)

Side View   Perspective

Fig. 22-6   (After C. R. Metz, "Models and the Crystalline State," STRC-343, *Modular Laboratory Program*, Chemical Education Resources, Inc., Palmyra, PA, 1988.)

The packing efficiency is given by

$$\frac{V_{\text{occ}}}{V_{\text{cell}}} = \frac{8.38 R^3}{11.31 R^3} = 74.1\%$$

22.10 COVALENTLY BONDED CRYSTALS

As discussed in Sec. 21.1, the elements of the carbon family of the periodic table crystallize in networks of three-dimensional equivalent covalent bonds.

22.11 IONIC CRYSTALS

To this point, all unit cells have been assumed to have identical constituents. With some modifications, the previous treatment applies to ionic crystals, in which the constituents are charged ions. The content Z now refers to the number of ion groups present in the unit cell, as specified by the empirical formula for the substance; the ratio of Z_+ to Z_- will be the ratio of subscripts in the empirical formula. Sets of coordinates must be assigned to each type of ion present; crystallographic projections should distinguish between types of ions; the coordination numbers CN_+ and CN_- (each defined as the number of oppositely charged nearest neighbors) will be in the inverse ratio of the subscripts in the empirical formula; and the unit cell length will be related to two radii, R_+ and R_-.

For ionic substances with the empirical formula MX that crystallize in the cubic system, the unit cell can be predicted from the radius ratio R_+/R_- (see Table 22-2). Common noncubic unit cells for the empirical formula MX include the wurtzite and PbO structures in the hexagonal and tetragonal systems, respectively. Common unit cells for the empirical formula M_2X or MX_2 include the cuprite and fluorite structures in the cubic system and the rutile structure in the tetragonal system.

Table 22-2

Empirical Formula	Radius Ratio	CN	Structure if Cubic
MX	$0.225 < R_+/R_- < 0.414$	4	Sphalerite (wurtzite, if hexagonal)
	$0.414 < R_+/R_- < 0.732$	6	Halite
	$0.732 < R_+/R_- < 1.000$	8	CsCl
MX_2 or M_2X	$0.225 < R_+/R_- < 0.414$	4 and 2	SiO_2 and Cu_2O
	$0.414 < R_+/R_- < 0.732$	6 and 3	TiO_2, CdI_2, NiS_2, and FeS_2
	$0.732 < R_+/R_- < 1.000$	8 and 4	CaF_2

EXAMPLE 22.8. AgCl crystallizes in the NaCl structure (see Fig. 22-7). Describe the unit cell in terms of interpenetrating Bravais lattices. What are the coordinates of the ions? Given $R = 0.126$ nm for Ag^+ and 0.181 nm for Cl^-, show that this structure is correctly predicted by the radius ratio rule. What are the values of CN for each ion? Determine Z_+, Z_-, and Z for the unit cell, and calculate the theoretical density, given $a = 0.554\,91$ nm. Prepare a crystallographic projection for this unit cell.

AgCl can be described as two interpenetrating face-centered cubic structures, one of Ag^+ located at (000) and the second of Cl^- located at $(\frac{1}{2}00)$. The coordinates of the Ag^+ ions are identical to those of a face-centered cube: (000), $(\frac{1}{2}\frac{1}{2}0)$, $(\frac{1}{2}0\frac{1}{2})$, and $(0\frac{1}{2}\frac{1}{2})$. The coordinates of the Cl^- ions are $(\frac{1}{2}00)$, $(0\frac{1}{2}0)$, $(\frac{1}{2}\frac{1}{2}\frac{1}{2})$, and $(00\frac{1}{2})$.

The radius ratio is

$$\frac{R_+}{R_-} = \frac{0.126}{0.181} = 0.696$$

which falls into the NaCl structure range in Table 22-2. Around any given Ag^+, there are six Cl^- ions as nearest neighbors, giving $CN_+ = 6$. Likewise, around a given Cl^- there are six Ag^+ ions as nearest neighbors, giving $CN_- = 6$. The ratio CN_+/CN_- is $1:1$, which is inverse to the $1:1$ ratio in the empirical formula.

Because both sets of ions are face-centered cubic, $Z_+ = Z_- = 4$ and $Z = 4$. The ratio Z_+/Z_- is $1:1$ as in the empirical formula. The theoretical density is given by (22.1) and (22.2b) as

$$\rho = \frac{(4)(143.32)(10^{-3})}{(6.022 \times 10^{23})(5.549\,1 \times 10^{-10})^3} = 5.571 \times 10^3 \text{ kg m}^{-3}$$

Because there are two types of ions, the projection should distinguish between them. A common technique is to use the symbol * after the coordinate of one type of ion (Cl^- in Fig. 22-8).

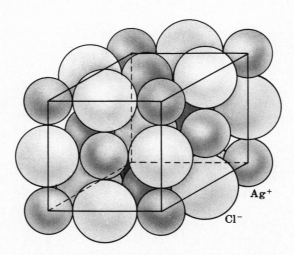

Ag^+

Cl^-

Fig. 22-7

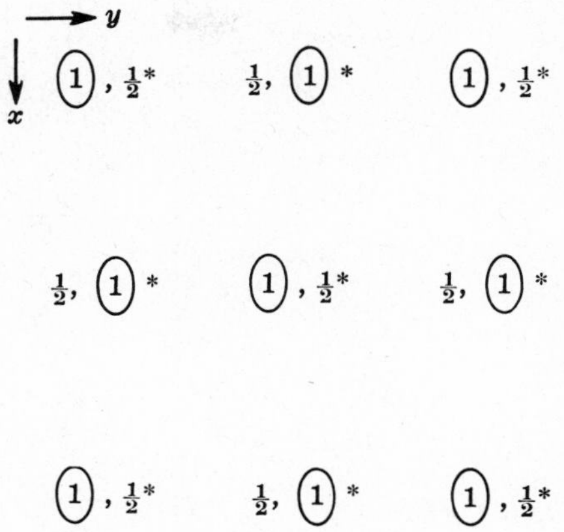

$$\xrightarrow{\quad} y$$

Fig. 22-8

22.12 MOLECULAR CRYSTALS

The molecules in these crystals are held together by van der Waals forces or hydrogen bonds (see Chap. 21).

Crystallography

22.13 MILLER INDICES

The *Miller indices* are a set of integers hkl [or $hk(\overline{h+k})l$ for hexagonal crystals] that are used to describe a given plane in a crystal. The procedure for determining the Miller indices for a plane is: (1) Prepare a three-column table with the unit cell axes at the tops of the columns; (2) enter in each column the intercept (expressed as a multiple of a, b, or c) of the plane with that axis; (3) invert all numbers; and (4) clear fractions to obtain h, k, and l.

From this it is easy to see that the Miller indices fix the direction of the plane. In particular, the normal to the plane has direction cosines proportional to h/a, k/b, and l/c.

EXAMPLE 22.9. Consider the plane shown in Fig. 22-9a that intersects the x, y, and z axes at a, b, and c, the unit cell dimensions, respectively. What are the Miller indices for this plane?

Preparing the table as described above:

a	b	c	
1	1	1	intercepts
1	1	1	reciprocals
1	1	1	clear fractions

gives the indices as 111.

22.14 d SPACINGS

When the intercepts of a plane are all doubled, tripled, etc., its Miller indices do not change. Therefore it is possible to construct a family of planes with identical values of hkl that are all parallel and separated by a constant distance d_{hkl}. This interplanar distance is related to the unit cell dimensions

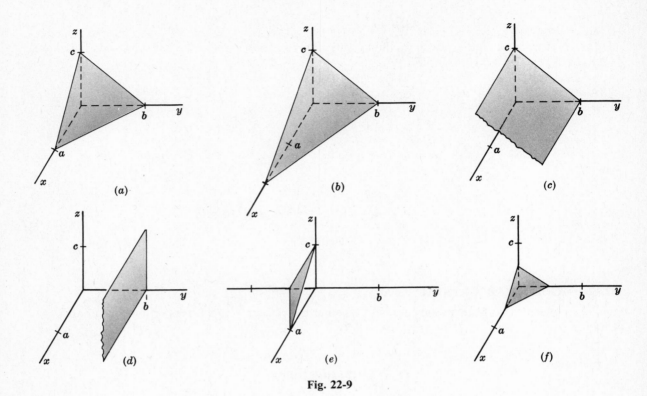

Fig. 22-9

and angles by

$$\frac{1}{d_{hkl}^2} = \frac{h^2 + k^2 + l^2}{a^2} \tag{22.4a}$$

$$\frac{1}{d_{hkl}^2} = \frac{h^2 + k^2}{a^2} + \frac{l^2}{c^2} \tag{22.4b}$$

$$\frac{1}{d_{hkl}^2} = \frac{h^2}{a^2} + \frac{k^2}{b^2} + \frac{l^2}{c^2} \tag{22.4c}$$

$$\vdots$$

$$\frac{1}{d_{hkl}^2} = \frac{\begin{array}{c}(h^2/a^2)\sin^2\alpha + (k^2/b^2)\sin^2\beta + (l^2/c^2)\sin^2\gamma + (2hk/ab)(\cos\alpha\cos\beta - \cos\gamma) \\ + (2kl/bc)(\cos\beta\cos\gamma - \cos\alpha) + (2lh/ca)(\cos\gamma\cos\alpha - \cos\beta)\end{array}}{1 - \cos^2\alpha - \cos^2\beta - \cos^2\gamma + 2\cos\alpha\cos\beta\cos\gamma} \tag{22.4g}$$

for cubic, tetragonal, orthohombic, ..., and triclinic crystals, respectively. Formula (22.4g) includes the others as special cases.

22.15 POINT GROUP SYMMETRY

Crystal faces on a macroscopic crystal that are related by symmetry operations are called *forms* (e.g., cubes, octahedra, and prisms). A crystal can be assigned to a point group by use of the flow chart in Fig. 20-14. The point group for the unit cell of the crystal will be the same if the crystal has developed enough forms to allow it to be classified in only one point group. For example, a cube of pyrite could be classified in any one of the cubic point groups, but the presence of an octahedron reduces the possible point groups to 3, and the presence of a pyritohedron reduces the possible point groups to 2, with only one in common: $2/m\overline{3}$. If a crystal has extensive twinning (composite crystals related by additional symmetry elements), it may be mistakenly classified in a point group of higher symmetry. To avoid these problems of pseudosymmetry, various crystals should be inspected, etchings made, X-ray data taken, etc.

EXAMPLE 22.10.   Determine the point groups for the etched cubes shown in Figs. 22-10a–d. All cubes but (d) have opposite faces the same.

For the cube shown in Fig. 22-10a, the following analysis can be made using Fig. 20-14: (1) Are there four C_3 axes at 54°44′? yes; (2) is there a C_4 axis? yes; (3) is there a σ_h perpendicular to C_4? yes, therefore $\mathcal{O}_h$ or $4/m\,\bar{3}\,2/m$.

For the cube shown in Fig. 22-10b: (1) Are there four C_3 axes at 54°44′? no; (2) is there at least one C_n where $n \geq 2$? yes, C_4; (3) is there an S_{2n} present? no; (4) are there n C_2 axes perpendicular to C_n? yes; (5) is there a σ_h perpendicular to C_n? yes, therefore $\mathcal{D}_{4h}$ or $4/m\,2/m\,2/m$.

For the cube shown in Fig. 22-10c: (1) Are there four C_3 axes at 54°44′? yes; (2) is there a C_4 axis? no; (3) is there an S_4 present? no; (4) is there a σ_h perpendicular to C_2? yes, therefore $\mathcal{T}_h$ or $2/m\,\bar{3}$.

For the cube in Fig. 22-10d: (1) Are there four C_3 axes at 54°44′? no; (2) is there a C_n axis where $n \geq 2$? no; (3) is there a σ present? no; (4) is i present? no, therefore $\mathcal{C}_1$ or 1.

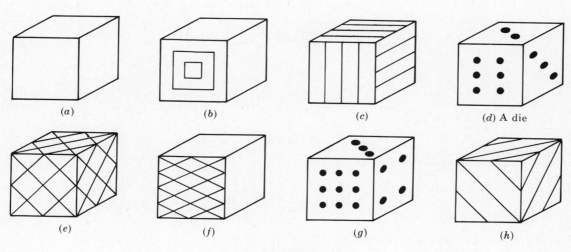

(a)	(b)	(c)	(d) A die
(e)	(f)	(g)	(h)

Fig. 22-10

X-Ray Spectra

22.16   BRAGG EQUATION

A crystal plane will "reflect" a beam of X-rays when

$$n\lambda = 2d_{hkl} \sin \theta \qquad (22.5)$$

where n is an integer known as the *order of reflection* and λ is the wavelength of the radiation. The angle θ is the angle of reflection from the hkl plane (although in most experiments the angle 2θ is measured). Usually n is reduced to unity by incorporating the order of reflection into the value of hkl. [According to (22.4), this can be done by multiplying h, k, and l by n.]

22.17   EXTINCTIONS

If a nonprimitive unit cell is used to describe a substance, certain reflections are not allowed. For example, if a body-centered unit cell is chosen, the values of hkl that are permitted are those that satisfy $h + k + l =$ even; if an end-centered cell designated as C is chosen, $h + k =$ even (with analogous criteria for A and B cells); and if a face-centered unit cell is chosen, all indices must be even or all must be odd. Additional extinctions are present for the various glide planes and screw axes.

EXAMPLE 22.11. Consider the C-centered orthorhombic unit cell, two of which are shown in Fig. 22-11. Prepare the matrix of transformation between the preferred orthorhombic unit cell and the primitive monoclinic unit cell indicated. Show that $h + k =$ even must be satisfied for a reflection to occur.

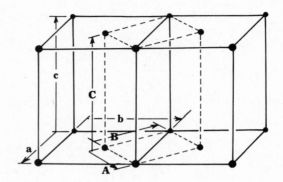

Fig. 22-11

Assuming the unit cell axes to be vectors,

$$\mathbf{a} = \mathbf{A} - \mathbf{B} \qquad \mathbf{b} = \mathbf{A} + \mathbf{B} \qquad \mathbf{c} = \mathbf{C}$$

which gives the following matrix of transformation:

$$\begin{bmatrix} 1 & -1 & 0 \\ 1 & 1 & 0 \\ 0 & 0 & 1 \end{bmatrix}$$

This matrix also describes the transformation of Miller indices giving

$$h = H - K \qquad k = H + K \qquad l = L$$

from which it can be seen that $h + k = 2H$, an even number.

22.18 METHOD OF ITO

The *method of Ito* is a technique for indexing the X-ray powder pattern of a substance and inferring the dimensions of the unit cell. If

$$Q_{hkl} = \frac{1}{d_{hkl}^2} \tag{22.6}$$

the Bragg equation becomes

$$Q_{hkl} = \frac{4 \sin^2 \theta}{\lambda^2} \tag{22.7}$$

For orthogonal crystal systems, (22.4) give

$$Q_{hkl} = h^2 a^{*2} + k^2 b^{*2} + l^2 c^{*2} \tag{22.8}$$

where $a^* = a^{-1}$, $b^* = b^{-1}$, and $c^* = c^{-1}$. Thus, by assigning values of hkl to the observed Q's (that is, to the observed θ's), a set of equations can be obtained for a^*, b^*, and c^*. See Problem 22.18.

22.19   INTENSITIES

The *structure factor* for a plane *hkl* is defined as

$$F(hkl) = \sum_j f_j \, e^{2\pi i (hx_j + ky_j + lz_j)} \tag{22.9}$$

where the summation is performed over all the atoms in the unit cell; f_j is the *scattering factor*, which is related to the number of electrons and $(\sin\theta)/\lambda$; and the x_j, y_j, and z_j are the unit cell coordinates of the atoms (Sec. 22.3). The intensity of the scattered X-ray beam is proportional to $F(hkl)^* F(hkl)$.

EXAMPLE 22.12.   Determine the structure factor for the 200 plane in NaCl assuming Na$^+$ ions at (000), $(\frac{1}{2}\frac{1}{2}0)$, $(\frac{1}{2}0\frac{1}{2})$, and $(0\frac{1}{2}\frac{1}{2})$ and Cl$^-$ ions at $(0\frac{1}{2}0)$, $(\frac{1}{2}00)$, $(00\frac{1}{2})$, and $(\frac{1}{2}\frac{1}{2}\frac{1}{2})$. For the experimental values of θ and λ, $f_+ = 8.8$ and $f_- = 13.7$.

Using (22.9) with $k = l = 0$ gives

$$F(200) = 8.8\{e^{2\pi i[(2)(0)]} + e^{2\pi i[(2)(1/2)]} + e^{2\pi i[(2)(1/2)]} + e^{2\pi i[(2)(0)]}\}$$

$$+ 13.7\{e^{2\pi i[(2)(1/2)]} + e^{2\pi i[(2)(0)]} + e^{2\pi i[(2)(1/2)]} + e^{2\pi i[(2)(0)]}\}$$

$$= 8.8(2 + 2e^{\pi i}) + 13.7(2 + 2e^{2\pi i}) = 45.0(1 + e^{2\pi i}) = 45.0(1 + 1) = 90.0$$

where $e^{2\pi i}$ was evaluated from

$$e^{ix} = \cos x + i \sin x \tag{22.10}$$

Solved Problems

UNIT CELLS

22.1.   Determine Z for the body-centered cubic unit cell shown in Fig. 22-12.

The contribution of the eight corners is $8(\frac{1}{8}) = 1$, and that of the body-centered atom is 1, giving $Z = 1 + 1 = 2$.

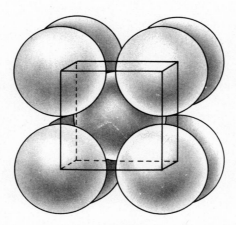

Fig. 22-12

22.2.   Determine the coordinates of the atoms shown in Fig. 22-12.

The eight corners will be represented by (000). The coordinates of the atom in the center of the unit cell are $(\frac{1}{2}\frac{1}{2}\frac{1}{2})$. The required two sets of coordinates agree with the unit cell content as determined in Problem 22.1.

22.3. Prepare a crystallographic projection for the body-centered cubic unit cell.

   The eight corner atoms will be represented by four ① symbols at the corners of a square, and the body-centered atom will be represented by a $\frac{1}{2}$ in the center of the square (see Fig. 22-13).

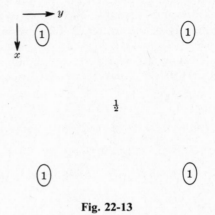

Fig. 22-13

22.4. Determine CN for an atom in the body-centered cubic unit cell.

   Considering the atom in the center of the unit cell shown in Fig. 22-12, it is surrounded by eight nearest-neighbor atoms, the corners of the cube. Thus $CN = 8$.

22.5. Sodium crystallizes in the body-centered cubic structure with $a = 0.424$ nm. Calculate the theoretical density of Na.

   From Problem 22.1, $Z = 2$. Using (22.2b) gives $V = (4.24 \times 10^{-10}\ \text{m})^3$, which upon substitution into (22.1) gives

$$\rho = \frac{(2)(23.0\ \text{g mol}^{-1})[(10^{-3}\ \text{kg})/(1\ \text{g})]}{(6.022 \times 10^{23}\ \text{mol}^{-1})(4.24 \times 10^{-10}\ \text{m})^3} = 1.00 \times 10^3\ \text{kg m}^{-3}$$

22.6. Find the relationship between a and R for a body-centered cubic unit cell, and calculate R for Na, given $a = 0.424$ nm.

   The spheres in this unit cell are touching along a body diagonal. The length of the body diagonal in terms of a is $a(3)^{1/2}$, and in terms of R it is $4R$, giving

$$R = \frac{a(3)^{1/2}}{4}$$

Substituting the data for Na gives

$$R = \frac{(0.424\ \text{nm})(3)^{1/2}}{4} = 0.184\ \text{nm}$$

22.7. Find the distance between the body-centered atom and one corner atom in Na, given $a = 0.424$ nm.

   Substitution of the coordinates (000) and $(\frac{1}{2}\frac{1}{2}\frac{1}{2})$ into (21.3c) gives

$$\ell = (0.424\ \text{nm})[(\tfrac{1}{2}-0)^2 + (\tfrac{1}{2}-0)^2 + (\tfrac{1}{2}-0)^2]^{1/2} = (0.424)(3/4)^{1/2} = 0.367\ \text{nm}$$

22.8. Metallic zinc crystallizes in a hexagonal closest-packed unit cell (see Fig. 22-6), with $a = 0.266\,5$ nm and $c = 0.494\,9$ nm. Give the coordinates of the atoms, and prepare crystallographic projections for the unit cell. What is CN? Find the theoretical density of the metal and the distances between atoms contained in the basal parallelogram.

By convention, the coordinates (000) account for all eight corners. The coordinates of the atom in the center can be found to be $(\frac{1}{3}\frac{1}{3}\frac{1}{2})$ by recognizing that the center of this atom lies at the centroid of the equilateral triangle formed by three of the atoms in the parallelogram. The projections are shown in Fig. 22-14. A given atom is touching six others in the same plane as well as three above and three below the plane, hence $CN = 12$ for this unit cell.

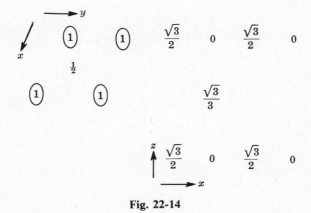

Fig. 22-14

Using (22.2a) gives

$$V = a^2 c(1 - \cos^2 \gamma)^{1/2} = a^2 c \sin \gamma$$

$$= (0.266\,5 \text{ nm})(0.494\,9 \text{ nm})(\sin 120°) = 30.44 \times 10^{-30} \text{ m}^3$$

which upon substitution into (22.1) gives

$$\rho = \frac{(2)(65.37 \times 10^{-3})}{(6.022 \times 10^{23})(30.44 \times 10^{-30})} = 7.132 \times 10^3 \text{ kg m}^{-3}$$

There are two interatomic distances in the basal parallelogram. The shorter is that between any three adjacent atoms and has the value $\ell = a = 0.266\,5$ nm. The longer is that between the far corners of the parallelogram and is given by (22.3a) (or the law of cosines) as

$$\ell = [a^2(1-0)^2 + a^2(1-0)^2 + c^2(0-0)^2 - 2a^2(1-0)(1-0)(\cos 120°) - 0 - 0]^{1/2}$$

$$= a(2 - 2\cos 120°)^{1/2} = a(3)^{1/2} = 0.461\,6 \text{ nm}$$

CRYSTAL FORMS

22.9. Diamond has C atoms located at (000), $(\frac{1}{2}\frac{1}{2}0)$, $(\frac{1}{2}0\frac{1}{2})$, $(0\frac{1}{2}\frac{1}{2})$, $(\frac{1}{4}\frac{1}{4}\frac{1}{4})$, $(\frac{1}{4}\frac{3}{4}\frac{3}{4})$, $(\frac{3}{4}\frac{1}{4}\frac{3}{4})$, and $(\frac{3}{4}\frac{3}{4}\frac{1}{4})$. Given $a = 0.356\,70$ nm, find the covalent radius of a carbon atom and the theoretical density of diamond.

The body diagonal of the unit cell is equal to $8R$, giving $a = 8R/(3)^{1/2}$, and hence

$$R = \frac{a(3)^{1/2}}{8} = \frac{(0.356\,70 \text{ nm})(3)^{1/2}}{8} = 0.077\,2 \text{ nm}$$

Recognizing that $Z = 8$, (22.1) and (22.2b) give

$$\rho = \frac{(8)(12.011 \times 10^{-3})}{(6.022 \times 10^{23})(3.567\,0 \times 10^{-10})^3} = 3.516 \times 10^3 \text{ kg m}^{-3}$$

22.10. The cuprite unit cell can be described as a face-centered cubic unit cell of Cu^+ ions with its origin at $(\frac{1}{4}\frac{1}{4}\frac{1}{4})$ interpenetrating a body-centered cubic unit cell of O^{2-} ions with its origin at (000). Only a portion of the Cu^+ cubic unit cell is contained in the cuprite unit cell and thus appears as a tetrahedron of Cu^+ ions. Prepare a crystallographic projection of this unit cell. Find the coordinates of all the ions. Given $a = 0.426\,96$ nm, determine the theoretical density of Cu_2O.

    The face-centered cubic structure implies that $Z_+ = 4$, and the body-centered structure implies that $Z_- = 2$, which is in agreement with the empirical formula. The coordinates of the Cu^+ are $(\frac{1}{4}\frac{1}{4}\frac{1}{4})$, $(\frac{1}{4}\frac{3}{4}\frac{3}{4})$, $(\frac{3}{4}\frac{3}{4}\frac{1}{4})$, $(\frac{3}{4}\frac{1}{4}\frac{3}{4})$, and those of O^{2-} are (000) and $(\frac{1}{2}\frac{1}{2}\frac{1}{2})$. $CN_+ = 2$ and $CN_- = 4$, which is in agreement with the empirical formula. See Fig. 22-15 for the projection (* indicates Cu^+).

    Using (22.1) and $(22.2b)$ gives

$$\rho = \frac{(2)(143.08 \times 10^{-3})}{(6.022 \times 10^{23})(4.269\,6 \times 10^{-10})^3} = 6.105 \times 10^3 \text{ kg m}^{-3}$$

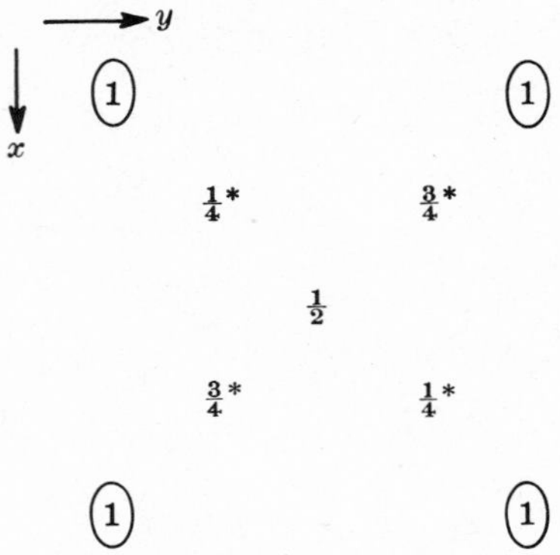

Fig. 22-15

22.11. Consider the portion of the unit cell for ice shown in Fig. 5-5. Each water molecule is held in place by hydrogen bonding between four additional molecules. Determine CN for the H and the O atoms, assuming all O—H distances to be the same.

    Around a given H there are two O's, and around a given O there are four H's, giving $CN(H) = 2$ and $CN(O) = 4$, which are in the inverse ratio of the formula subscripts.

CRYSTALLOGRAPHY

22.12. Determine the indices for the plane shown in Fig. 22-9c.

    The table,

a	b	c	
∞	1	1	intercepts
0	1	1	reciprocals
0	1	1	clear fractions

gives 011.

22.13. Indicate the intercepts of the plane having the indices $1\bar{2}1$.

    The value $h = 1$ means that the x axis is intersected at a, the value $k = \bar{2}$ means that the y axis is intersected at $-0.5b$, and the value $l = 1$ means that the z axis is intersected at c, giving the plane shown in Fig. 22.9e.

22.14. Find the spacing between the plane with indices 101 in NaCl given $a = 0.564\,02$ nm.

Using (*22.4a*) gives

$$\frac{1}{d_{101}^2} = \frac{1^2 + 0^2 + 1^2}{(0.564\,02 \text{ nm})^2} \quad \text{or} \quad d_{101} = \frac{0.564\,02}{(2)^{1/2}} = 0.398\,82 \text{ nm}$$

22.15. Using Fig. 20-14, determine the point groups for the crystals shown in Figs. 22-16*a–c*.

(*a*) Barite: (1) Are there four C_3 axes at 54°44'? no; (2) is there at least one C_n with $n \geq 2$? yes, C_2; (3) is there an S_{2n} present? no; (4) are there n C_2 axes perpendicular to C_n? yes; (5) is there a σ_h perpendicular to C_n? yes, therefore $\mathcal{D}_{2h}$ or $2/m\ 2/m\ 2/m$.

(*b*) Zircon: (1) Are there four C_3 axes at 54°44'? no; (2) is there at least one C_n with $n \geq 2$? yes, C_4; (3) is there an S_{2n} present? yes, but other elements too; (4) are there n C_2 axes perpendicular to C_n? yes; (5) is there a σ_h perpendicular to C_n? yes, therefore $\mathcal{D}_{4h}$ or $4/m\ 2/m\ 2/m$.

(*c*) Galena: (1) Are there four C_3 axes at 54°44'? yes; (2) is there a C_4 axis? yes; (3) is there a σ_h perpendicular to C_4? yes, therefore $\mathcal{O}_h$ or $4/m\ \bar{3}\ 2/m$

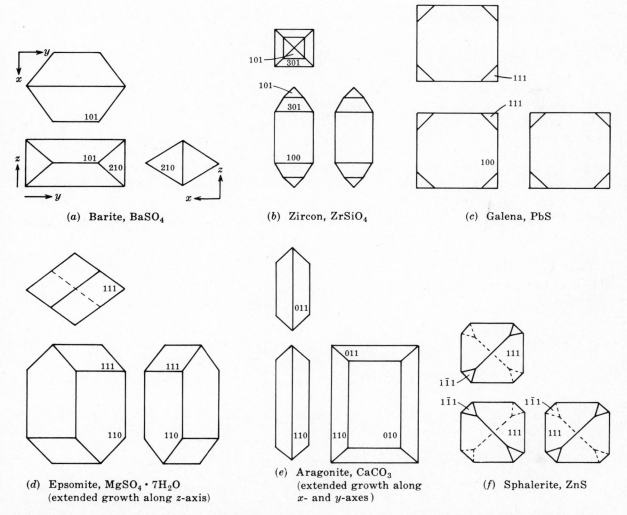

(*a*) Barite, $BaSO_4$

(*b*) Zircon, $ZrSiO_4$

(*c*) Galena, PbS

(*d*) Epsomite, $MgSO_4 \cdot 7H_2O$
(extended growth along z-axis)

(*e*) Aragonite, $CaCO_3$
(extended growth along
x- and y-axes)

(*f*) Sphalerite, ZnS

Fig. 22-16 (After L. G. Berry and B. Mason, *Mineralogy*, W. H. Freeman and Co., San Francisco, 1959.)

X-RAY SPECTRA

22.16. Given $\lambda = 0.154\,18$ nm for filtered Cu radiation, at what angle would the maximum reflection by the 200 plane of AgCl occur, assuming that $a = 0.554\,91$ nm?

Using (22.4) gives

$$\frac{1}{d_{200}^2} = \frac{2^2 + 0^2 + 0^2}{(0.554\,91\ \text{nm})^2} \qquad \text{or} \qquad d_{200} = 0.277\,46\ \text{nm}$$

Then (22.5) gives

$$\sin\theta = \frac{0.154\,18\ \text{nm}}{2(0.277\,46\ \text{nm})} = 0.277\,84 \qquad \text{or} \qquad \theta = 16.1°$$

22.17. Which of the following indices are allowed in the pattern of AgCl; 100, 010, 001, 200, 020, 002, 110, 101, 011, 120, 102, 012, 210, 201, 021, 220, 202, 022, 111, 222, 221, 212, 122, 211, 121, 112?

Recalling that AgCl has a face-centered cubic unit cell, the only allowed reflections are those in which all indices are even or all are odd, giving: 200, 020, 002, 220, 202, 022, 111, and 222. Because $a = b = c$, $200 = 020 = 002$ and $220 = 202 = 022$, so that really only four peaks will be observed: 200, 220, 111, and 222.

22.18. The powder pattern of halite using filtered Cu radiation ($\lambda = 0.154\,18$ nm) shows six peaks. The values of 2θ and the relative intensities for these spectral lines are 27.1°(10%), 31.5°(100%), 45.2°(45%), 53.6°(5%), 56.3°(10%), and 65.9°(5%). Given that halite is known to crystallize in the cubic system, determine a for the mineral, and index the lines.

Converting the various values of 2θ to Q using (22.7) gives

$$Q_1 = \frac{4\sin^2(27.1°/2)}{(0.154\,18)^2} = 9.24$$

$Q_2 = 12.40$, $Q_3 = 24.9$, $Q_4 = 34.2$, $Q_5 = 37.5$, and $Q_6 = 49.8$. Recalling the results of Sec. 22.17, none of these Q's corresponds to a $100 = 010 = 001$ d spacing. Assuming Q_2, the most intense peak, to be for the $200 = 020 = 002$ plane, (22.8) gives

$$Q_{100} = \frac{Q_{200}}{4} = \frac{12.40}{4} = 3.10$$

Using $Q_{100} = 3.10$, Q_{111} is given by

$$Q_{111} = (1^2 + 1^2 + 1^2)Q_{100} = 3(3.10) = 9.30 = Q_1$$

Continuing by trial and error gives $Q_{300} = 27.9$, $Q_{400} = 49.6 = Q_6$, $Q_{120} = 15.50$, $Q_{121} = 18.60$, $Q_{220} = 24.8 = Q_3$, $Q_{222} = 37.2 = Q_5$, and $Q_{113} = 34.1 = Q_4$. Summarizing, the peaks correspond to the 111, 200, 220, 113, 222, and 400 planes. Using (22.8) gives $a^{*2} = 3.10$, which converts to $a = 0.569$ nm. Based on the extinctions, the unit cell must be face-centered cubic.

22.19. Calculate the intensity of the 220 peak relative to that of the 200 peak for NaCl using the results of Example 22.12 and Problem 22.51.

Using $F(200) = 90.0$ and $F(220) = 69.6$, the relative intensity will be

$$\frac{I(220)}{I(200)} = \frac{F(220)^*F(220)}{F(200)^*F(200)} = \frac{(69.6)^2}{(90.0)^2} \times 100 = 59.8\%$$

The actual value is about 45%.

Supplementary Problems

UNIT CELLS

22.20. Determine Z for the face-centered cubic unit cell shown in Fig. 22-17.

    Ans. $8(\frac{1}{8}) + 6(\frac{1}{2}) = 4$

22.21. Determine the coordinates of the atoms shown in Fig. 22-17.

    Ans. (000), $(\frac{1}{2}0\frac{1}{2})$, $(\frac{1}{2}\frac{1}{2}0)$, $(0\frac{1}{2}\frac{1}{2})$

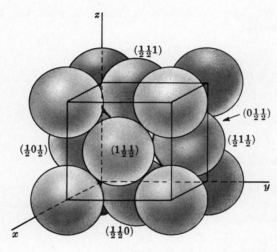

Fig. 22-17

22.22. Prepare a crystallographic projection for the face-centered cubic unit cell.

    Ans.   See Fig. 22-18.

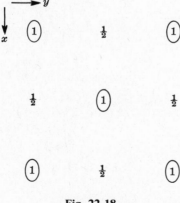

Fig. 22-18

22.23. Determine CN for the face-centered cubic unit cell.   *Ans.* 12

22.24. Platinum crystallizes in the face-centered cubic structure with $a = 0.392\ 3$ nm. Calculate the theoretical density of Pt.   *Ans.* 21.5×10^3 kg m^{-3}

22.25. Find the relation between a and R for a face-centered cubic unit cell. Calculate R for Pt, assuming $a = 0.392\,3$ nm.   *Ans.* $a(2)^{1/2}/4$; $0.138\,7$ nm

22.26. Find the distance between two of the face-centered atoms on adjacent sides of Pt given $a = 0.392\,3$ nm.

     Ans. $a/(2)^{1/2}$, $0.277\,4$ nm

22.27. Using the data in Problem 22.8, show that the lengths of the two types of body diagonals (000 to 111 and 100 to 011) in zinc are $0.676\,7$ nm and $0.562\,1$ nm. Show that in the ideal case $c = 2a(2/3)^{1/2}$.

CRYSTAL FORMS

22.28. Calculate the packing efficiency in the cubic closest-packed unit cell (face-centered cubic), and compare it to the answer for Example 22.7.

     Ans. $V_{cell} = 64R^3/2^{3/2}$, $V_{occ} = 16\pi R^3/3$; 74.1% (same)

22.29. Calculate the packing efficiencies in the primitive and body-centered cubic unit cells. Compare these answers to those found in Problem 22.28 for the other Bravais lattices.

     Ans. $\dfrac{\pi}{6} = 52.4\%$, $\dfrac{\pi(3)^{1/2}}{8} = 68.0\%$; FCC and HCP most efficient

22.30. Repeat Problem 22.9 for Si ($a = 0.543\,05$ nm).   *Ans.* $0.117\,7$ nm, 2.330×10^3 kg m^{-3}

22.31. The fluorite structure can be described as a face-centered cubic unit cell of Ca^{2+} ions interpenetrated by a complete primitive cube of F^-. What are the coordinates of the ions? Find CN_+ and CN_-. Prepare a crystallographic projection for this unit cell.

     Ans. (000), $(0\frac{1}{2}\frac{1}{2})$, $(\frac{1}{2}0\frac{1}{2})$, and $(\frac{1}{2}\frac{1}{2}0)$ for Ca^{2+} and $(\frac{1}{4}\frac{1}{4}\frac{1}{4})$, $(\frac{1}{4}\frac{3}{4}\frac{3}{4})$, $(\frac{3}{4}\frac{1}{4}\frac{3}{4})$, $(\frac{3}{4}\frac{3}{4}\frac{1}{4})$, $(\frac{3}{4}\frac{3}{4}\frac{3}{4})$, $(\frac{1}{4}\frac{1}{4}\frac{3}{4})$, $(\frac{1}{4}\frac{3}{4}\frac{1}{4})$, and $(\frac{3}{4}\frac{1}{4}\frac{1}{4})$
           for F^-; $CN_+ = 8$, $CN_- = 4$; see Fig. 22-19a.

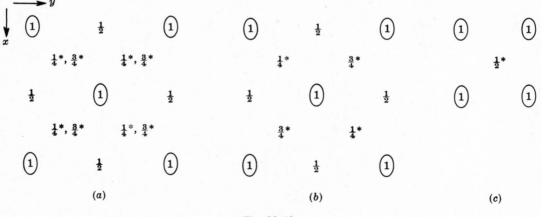

Fig. 22-19

22.32. The coordinates of the ions in the sphalerite unit cell are (000), $(\frac{1}{2}\frac{1}{2}0)$, $(0\frac{1}{2}\frac{1}{2})$, and $(\frac{1}{2}0\frac{1}{2})$ for Zn^{2+} and $(\frac{1}{4}\frac{1}{4}\frac{1}{4})$, $(\frac{1}{4}\frac{3}{4}\frac{3}{4})$, $(\frac{3}{4}\frac{1}{4}\frac{3}{4})$, and $(\frac{3}{4}\frac{3}{4}\frac{1}{4})$ for S^{2-}. Describe the unit cell in terms of interpenetrating Bravais lattices. What are Z and CN for the ions and the unit cell? Prepare a crystallographic projection for the unit cell.

     Ans.   Recalling that the tetrahedron of S^{2-} is an alternative way of describing a face-centered cube (see Problem 22.10), there are two face-centered cubes, one at (000) and the second at $(\frac{1}{4}\frac{1}{4}\frac{1}{4})$; $Z_+ = Z_- = Z = 4$; $CN_+ = CN_- = 4$; see Fig. 22-19b.

22.33. The coordinates of the ions in the CsCl unit cell are (000) for Cs^+ and $(\frac{1}{2}\frac{1}{2}\frac{1}{2})$ for Cl^-. Describe the unit cell in terms of interpenetrating Bravais lattices, and prepare a crystallographic projection.

    Ans.   Two primitive cubes, one of Cs^+ at (000) and the second of Cl^- at $(\frac{1}{2}\frac{1}{2}\frac{1}{2})$; see Fig. 22-19c.

22.34. Using the values of R_+ and R_- given below, predict the crystal structures for the alkali metal halides and the alkaline-earth chalcogenides, assuming the radius ratio rule to be valid and that the crystals are cubic.

Li^+	0.068 nm			O^{2-}	0.132 nm	F^-	0.133 nm
Na^+	0.097 nm	Mg^{2+}	0.066 nm	S^{2-}	0.184 nm	Cl^-	0.181 nm
K^+	0.133 nm	Ca^{2+}	0.099 nm	Se^{2-}	0.191 nm	Br^-	0.196 nm
		Sr^{2+}	0.112 nm			I^-	0.220 nm
Cs^+	0.167 nm	Ba^{2+}	0.134 nm				

    Ans.   Sphalerite structure: LiCl, LiBr, LiI, MgS, and MgSe. Halite structure: LiF, NaX, (KCl), KBr, KI, MgO, CaS, CaSe, SrS, SrSe, BaS, and BaSe. CsCl structure: KF, KCl, CsX, CaO, SrO, and BaO.

22.35. The red oxide of lead, litharge, crystallizes in the tetragonal crystal system with Pb^{2+} at $(0\frac{1}{2}u)$ and $(\frac{1}{2}0\bar{u})$, where $u = 0.24$, and O^{2-} at (000) and $(\frac{1}{2}\frac{1}{2}0)$. Describe the unit cell.

    Ans. O^{2-} in C-centered tetragonal with Pb^{2+} in tetragonal disphenoid (distorted tetrahedron) with the corners on the faces of the unit cell (see Fig. 22-20).

22.36. Ti_2O (rutile) crystallizes in the tetragonal system with Ti^{4+} at (000) and $(\frac{1}{2}\frac{1}{2}\frac{1}{2})$ and O^{2-} at $(\bar{u}u0)$, $(uu0)$, $(\frac{1}{2}+u\ \frac{1}{2}-u\ \frac{1}{2})$, and $(\frac{1}{2}-u\ \frac{1}{2}+u\ \frac{1}{2})$, where $u = 0.305$. Given $a = 0.459\ 37$ nm and $c = 0.296\ 18$ nm, calculate the theoretical density.   *Ans.* $Z = 2$; 4.246×10^3 kg m^{-3}

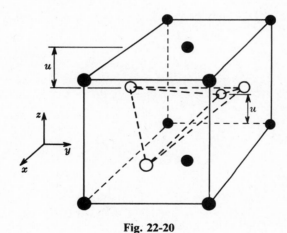

Fig. 22-20

22.37. ZnS is known to be polymorphic, crystallizing in both the sphalerite and wurtzite (zincite) structures. The wurtzite structure can be described as two interpenetrating hexagonal closest-packed unit cells, one of Zn^{2+} and the other of S^{2-}. The penetration is such that the coordinates of Zn^{2+} are (000) and $(\frac{1}{3}\frac{2}{3}\frac{1}{2})$ and those of S^{2-} are $(00u)$ and $(\frac{1}{3}\frac{2}{3}\frac{1}{2}+u)$, where $u = 3/8$. Calculate the theoretical densities of these minerals given $a = 0.540\ 93$ nm for sphalerite and given $a = 0.382\ 30$ nm and $c = 0.625\ 65$ nm for wurtzite.

    Ans. 4.09×10^3 kg m^{-3}, 4.09×10^3 kg m^{-3}

22.38. The coordinates of Ca^{2+} in the calcite unit cell are (000) and $(\frac{1}{2}\frac{1}{2}\frac{1}{2})$, those of C are $(\frac{1}{4}\frac{1}{4}\frac{1}{4})$ and $(\frac{3}{4}\frac{3}{4}\frac{3}{4})$, and those of O are $(\frac{1}{4}+u\ \frac{1}{4}\ \frac{1}{4}-u)$, $(\frac{1}{4}+u\ \frac{1}{4}-u\ \frac{1}{4})$, $(\frac{1}{4}\frac{1}{4}+u\ \frac{1}{4}-u)$, $(\frac{3}{4}-u\ \frac{3}{4}\ \frac{3}{4}+u)$, $(\frac{3}{4}\frac{3}{4}-u\ \frac{3}{4}+u)$, and $(\frac{3}{4}-u\ \frac{3}{4}+u\ \frac{3}{4})$, where

$u = 0.243$. Find Z_+ and Z_- from these coordinates. If the calcite structure is considered in terms of a hexagonal unit cell instead of the above rhombohedral cell, $Z = 6$ for the new cell with $a = 0.498\,99$ nm and $c = 1.706\,4$ nm. Calculate the density of calcite.

Ans. $Z_+ = 2$, $Z_- = 2$ for CO_3^{2-}; 2.710×10^3 kg m^{-3}

22.39. Calculate the theoretical density of (ortho)rhombic S given $a = 1.046\,46$ nm, $b = 1.286\,60$ nm, and $c = 2.448\,60$ nm. $Z = 128$ for this unit cell.   Ans. 2.067×10^3 kg m^{-3}

CRYSTALLOGRAPHY

22.40. Determine the Miller indices for the planes shown in Figs. 22-9b and d.   Ans.   122, 010

22.41. Sketch the plane having the indices 222.   Ans.   See Fig. 22-9f

22.42. Find the spacing between the family of planes having the indices 202 in NaCl given $a = 0.564\,02$ nm. Compare with the result of Problem 22.14.

Ans.   0.199 41 nm (half the 101 spacing)

22.43. Determine the point groups for the etched cubes shown in Figs. 22-10e–h.

Ans.   (e) $\mathscr{O}_h$ or $4/m\,\bar{3}\,2/m$; (f) $\mathscr{D}_{2h}$ or $2/m\,2/m\,2/m$; (g) $\mathscr{C}_{2h}$ or $2/m$; (h) $\mathscr{C}_{3v}$ or $3m$

22.44. Determine the point groups for the crystals shown in Figs. 22-16d–f.

Ans.   (d) $\mathscr{D}_2$ or 222; (e) $\mathscr{D}_{2h}$ or $2/m\,2/m\,2/m$; (f) $\mathscr{T}_d$ or $\bar{4}3m$

22.45. Identify the point groups to which the crystals shown in Fig. 22-21 belong. Note that other samples of the quartz shown in Fig. 22-21d do not show the σ_h and σ_d planes that this crystal does.

Ans.   (a) $\mathscr{T}_h$ or $2/m\,\bar{3}$; (b) $\mathscr{C}_{6v}$ or $6mm$; (c) $\mathscr{D}_{2h}$ or $2/m\,2/m\,2/m$; (d) Pseudosymmetry of $\mathscr{D}_{3h}$, correctly as $\mathscr{D}_3$ or 32

22.46. Four crystals were prepared as geometrical cubes, polished, and etched to develop the true symmetry. The results were as follows: (a) no etch pattern showing; (b) vertical striations on all four side faces and no pattern on the top and bottom; (c) vertical striations on the side faces, horizontal striations on the front and back, and no pattern on the top and bottom; and (d) square designs on a pair of parallel and opposite faces. Classify these crystals.

Ans.   (a) $\mathscr{O}_h$ or $4/m\,\bar{3}\,2/m$ (b) $\mathscr{D}_{4h}$ or $4/m\,2/m\,2/m$
        (c) $\mathscr{D}_{2h}$ or $2/m\,2/m\,2/m$ (d) $\mathscr{D}_{4h}$ or $4/m\,2/m\,2/m$

X-RAY SPECTRA

22.47. Repeat Problem 22.16 for the 111 plane.   Ans. $d_{111} = 0.320\,38$ nm; $\theta = 13.9°$

22.48. Which of the possible indices listed in Problem 22.17 are allowed for CsCl?

Ans.   two simple cubics; 100 = 010 = 001, 200 = 020 = 002, 110 = 101 = 011, 111, 210 = 120 = 102 = 201 = 012 = 021, 220 = 202 = 022, 222, and 211 = 121 = 112

22.49. The powder pattern for sylvite using filtered Cu radiation ($\lambda = 0.154\,18$ nm) shows six peaks. The values of 2θ and the relative intensities for these lines are 28.3°(100%), 40.5°(55%), 50.2°(20%), 58.6°(5%), 66.3°(15%), and 73.6°(10%). Given that sylvite is known to crystallize in the cubic system, determine a for the mineral, and index the lines.

Ans.   The values 100, 110, 111, 200, 210, and 211, which will work for the spectrum, must be discarded because they correspond to a primitive cubic cell, which is not possible for a compound. Hence 200, 220, 222, 400, 420, and 422 with $Q_{100} = 2.51$, giving $a = 0.631$ nm.

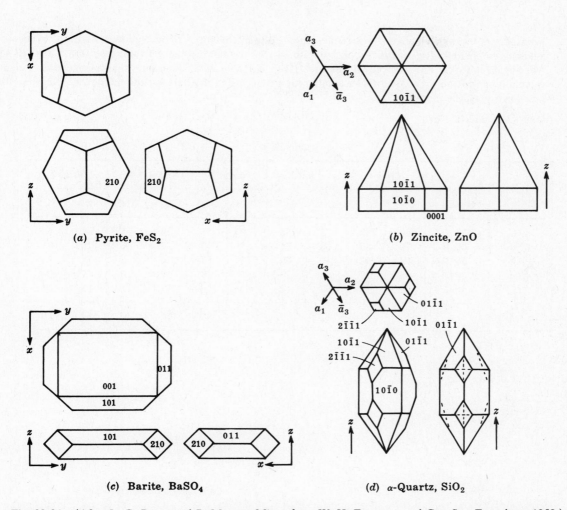

(a) Pyrite, FeS$_2$

(b) Zincite, ZnO

(c) Barite, BaSO$_4$

(d) α-Quartz, SiO$_2$

Fig. 22-21 (After L. G. Berry and B. Mason, *Mineralogy*, W. H. Freeman and Co., San Francisco, 1959.)

22.50. Sparteine sulfate pentahydrate, $C_{15}H_{26}N_2 \cdot H_2SO_4 \cdot 5H_2O$, crystallizes in the monoclinic crystal system with $a = 0.803$ nm, $b = 1.52$ nm, $c = 0.884$ nm, and $\beta = 91°30'$. Index the following intense lines of the powder pattern as reported by Metz: $d/(\text{nm}) = 0.894\,46$, $0.809\,32$, $0.767\,48$, $0.713\,20$, $0.579\,36$, $0.567\,55$, $0.547\,33$, $0.477\,14$, $0.467\,18$, $0.443\,13$, $0.432\,04$, $0.404\,58$, $0.391\,39$, $0.383\,71$, $0.370\,78$, $0.362\,01$, $0.350\,90$, $0.345\,28$, $0.335\,32$, and $0.326\,39$. Given $Z = 2$, calculate the theoretical density of this compound. The observed density is 1.28×10^3 kg m^{-3}.

Ans. 001; 100; 011 and 020; 110; 021 and 101; 11$\bar{1}$; 120 and 111; 12$\bar{1}$; 121; 002 and 031; 012 and 130; 200; 210, 13$\bar{1}$, and 10$\bar{2}$; 102, 11$\bar{2}$, 040, 131, and 022; 20$\bar{1}$ and 112; 201 and 21$\bar{1}$; 041, 211, and 220; 140 and 122; 032; 221; $d = 1.30 \times 10^3$ kg m^{-3}

22.51. Determine the structure factors for the 111 and 220 planes in NaCl (see Example 22.12). Assume that $f_+ = 7.4$ and $f_- = 10.0$ for the 220 plane and $f_+ = 9.2$ and $f_- = 14.4$ for the 111 plane. Find the ratio of the intensities of the 111 and 220 peaks.

Ans. $F(220) = 17.4(1 + 2e^{2\pi i} + e^{4\pi i}) = 69.6$,
$F(111) = 9.2(1 + 3e^{2\pi i}) + 14.4(3e^{\pi i} + e^{3\pi i}) = -20.8$; 8.9%

22.52. Quantitative X-ray analysis of mixtures can be performed if a calibration curve is determined. Using the $2\theta = 28.3°$ peak for sylvite and the $31.5°$ peak for halite, known samples were analyzed for peak heights five different times to obtain a statistical average. A plot of (height$_{28.3}$/height$_{31.5}$) or (height$_{31.5}$/height$_{28.3}$) against composition gives a calibration curve to which an unknown can be compared in the analysis by working backwards. For the known mixtures given in Table 22-3, determine the average peak height, calculate one of the ratios, and make a calibration plot. Using the data for the unknown, determine the composition of the mixture.

Ans.   49 mass% NaCl

Table 22-3

Mass % Halite	Mass % Sylvite	Trial 1		Trial 2		Trial 3		Trial 4		Trial 5	
		28.3	31.5	28.3	31.5	28.3	31.5	28.3	31.5	28.3	31.5
80.0	20.0	20	59	21	56	20	55	20	55	22	58
60.0	40.0	79	89	68	86	85	93	77	87	76	95
50.0	50.0	74	64	79	69	83	57	76	66	75	58
40.0	60.0	61	27	63	27	62	29	62	26	62	25
20.0	80.0	71	10	61	12	60	10	73	11	63	11
Unknown		59	41	52	37	52	39	52	40	57	40

Chapter 23

Phenomena at Interfaces

Surface Tension of Liquids

23.1 MEASUREMENT OF SURFACE TENSION

The surface tension of a liquid (γ) was defined in Table 3-3. The SI unit for surface tension is $N\,m^{-1}$ ($1\,N\,m^{-1} = 1\,J\,m^{-2}$), and the commonly used cgs unit is $dyn\,cm^{-1}$ ($1\,N\,m^{-1} = 10^3\,dyn\,cm^{-1}$).

In the *capillary tube method* of measuring γ, the liquid in the tube of radius r will rise or be depressed a distance h given by

$$h = \frac{2\gamma}{\rho r g} \tag{23.1}$$

where g is the gravitational constant. If a comparison method is used, then

$$\frac{\gamma_1}{\gamma_0} = \frac{\rho_1 h_1}{\rho_0 h_0} \tag{23.2}$$

In the *drop-weight method*, γ is determined from the mass of the drop that forms on the end of a capillary tube before falling off. In the *bubble-pressure method*, γ is determined by measuring the pressure required to produce a bubble of a gas in the liquid at the end of a capillary tube. In the *ring method*, γ is related to the force necessary to lift a ring from the surface of the liquid.

EXAMPLE 23.1. The surface tension of water at 20 °C is $72.75 \times 10^{-3}\,N\,m^{-1}$. A 33.24 vol % solution of ethanol has $\gamma = 33.24 \times 10^{-3}\,N\,m^{-1}$ at this same temperature. Given $\rho = 0.961\,4 \times 10^3\,kg\,m^{-3}$ for the solution and $0.998\,2 \times 10^3\,kg\,m^{-3}$ for water. How much less will the alcohol solution rise in the same capillary tube?

Using (23.2) gives

$$\frac{h_1}{h_0} = \frac{\gamma_1}{\gamma_0} \frac{\rho_0}{\rho_1} = \frac{33.24 \times 10^{-3}}{72.75 \times 10^{-3}} \left(\frac{0.998\,2 \times 10^3}{0.961\,4 \times 10^3} \right) = 0.474$$

Thus the solution will rise only 47.4% as far as pure water.

23.2 TEMPERATURE DEPENDENCE

The temperature dependence of the surface tension of a liquid is given by

$$\gamma = \gamma_0 \left(1 - \frac{T}{T_c} \right)^n \tag{23.3a}$$

where γ_0 is a constant for a given liquid, T/T_c is the reduced temperature, and n is a constant (≈ 1.2 for liquids that are not highly associated).

EXAMPLE 23.2. The critical temperature of water is 374.1 °C. Determine n and γ_0 in (23.3a) for water using the following data

$t/(°C)$	10	20	30	40	50
$\gamma/(10^{-3}\,N\,m^{-1})$	74.22	72.75	71.18	69.56	67.91

455

Equation (*23.3*) can be transformed into linear form by taking logarithms of both sides, giving

$$\log \gamma = \log \gamma_0 + n \log(1 - T/T_c) \tag{23.3b}$$

Thus a plot of $\log \gamma$ against $\log(1 - T/T_c)$ will be linear, with an intercept of $\log \gamma_0$ and a slope of n. From the plot (see Fig. 23-1), $n = 0.766$ and $\log[\gamma_0/(10^{-3}\ \text{N m}^{-1})] = 2.062\ 0$, giving $\gamma_0 = 115.35 \times 10^{-3}\ \text{N m}^{-1}$. Because water is a highly associated liquid, n would be expected to be considerably different from the value for nonassociated liquids.

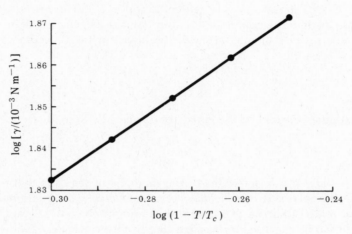

Fig. 23-1

23.3 VAPOR PRESSURE OF DROPLETS

For a very small droplet of radius r, the vapor pressure P_s is given by the *Kelvin equation* as

$$\ln \frac{P_s}{P} = \frac{2\gamma M}{r\rho RT} \tag{23.4}$$

where P is the vapor pressure of large (bulk) samples of the substance.

EXAMPLE 23.3. Determine P_s/P for a drop of water at 25 °C that has $r = 0.1$ cm. At this temperature, $\gamma = 71.97 \times 10^{-3}\ \text{N m}^{-1}$ and $\rho = 0.997\ 044 \times 10^3\ \text{kg m}^{-3}$.

Using (*23.4*) gives

$$\ln \frac{P_s}{P} = \frac{2(71.97 \times 10^{-3}\ \text{N m}^{-1})(18.01\ \text{g mol}^{-1})[(10^{-3}\ \text{kg})/(1\ \text{g})]}{(1 \times 10^{-3}\ \text{m})(0.997\ 044 \times 10^3\ \text{kg m}^{-3})(8.314\ \text{J K}^{-1}\ \text{mol}^{-1})(298\ \text{K})} = 1.05 \times 10^{-6}$$

Taking antilogarithms gives $P_s/P = 1.000\ 001\ 05$, an insignificant change.

23.4 PARACHOR

The *parachor* ($\{P\}$) is defined as

$$\{P\} = \frac{10^3 M (\gamma \times 10^3)^{1/4}}{\rho_{\text{liq}} - \rho_{\text{gas}}} \tag{23.5}$$

where M is the molar mass (g mol^{-1}), γ is the surface tension (N m^{-1}), and ρ is the density (kg m^{-3}). The parachor is an additive property depending on the elements present in the compound and the configuration of the atoms. By working from known values and from predicted values for $\{P\}$, it is often possible to determine certain structural properties of compounds.

EXAMPLE 23.4. Predict the parachor and γ for $CH_3C_6H_4CN$ given that the parachor equivalents are for C, 4.8; H, 17.1; N, 12.5; double bond, 23.2; triple bond, 46.6; and six-membered ring, 6.1. The average density of the three isomers is 0.985×10^3 kg m^{-3} for the liquids, and ρ_{gas} can be assumed to be negligible.

The predicted parachor is the sum of the equivalents, giving

$$\{P\} = 8(4.8) + 7(17.1) + 1(12.5) + 46.6 + 6.1 + 3(23.2) = 292.9$$

Using (23.5) gives

$$(\gamma \times 10^3)^{1/4} = \frac{\{P\}\rho}{M \times 10^3} = \frac{(292.9)(0.985 \times 10^3)}{(117.14)(10^3)} = 2.46$$

which gives $\gamma = 36.6 \times 10^{-3}$ N m^{-1}.

Surface Tension in Binary Systems

23.5 INTERFACIAL TENSION

When two immiscible or partially miscible liquids A and B are in contact, an *interfacial tension* (γ_{AB}) exists at the boundary between the two layers. The *spreading coefficient* of B on A (S_{BA}) is given by

$$S_{BA} = \gamma_A - \gamma_B - \gamma_{AB} \tag{23.6}$$

where values of $S_{BA} > 0$ imply that B will "wet" A by spreading out over the surface, and $S_{BA} < 0$ implies that B will not wet A and will form droplets on the surface.

For a drop of liquid B placed on a solid A, the *contact angle* (θ) is defined as the angle that the surface of the drop makes with the solid and is given by

$$\cos \theta = \frac{\gamma_A - \gamma_{AB}}{\gamma_B} \tag{23.7}$$

EXAMPLE 23.5. At 20 °C, $\gamma_{AB} = 8.5 \times 10^{-3}$ N m^{-1} between water and *n*-octyl alcohol. At this temperature, $\gamma/(10^{-3}$ N m$^{-1}) = 27.53$ for the alcohol and 72.75 for water. Predict whether or not wetting occurs for a drop of alcohol in contact with a pool of water.

The value of the spreading coefficient as given by (23.6) is

$$S_{\text{alc-wat}} = \gamma_{\text{wat}} - \gamma_{\text{alc}} - \gamma_{\text{alc-wat}}$$

$$= (72.75 \times 10^{-3} \text{ N m}^{-1}) - (27.53 \times 10^{-3} \text{ N m}^{-1}) - (8.5 \times 10^{-3} \text{ N m}^{-1})$$

$$= 36.7 \times 10^{-3} \text{ N m}^{-1}$$

Wetting will occur because $S_{\text{alc-wat}} > 0$.

23.6 SURFACE EXCESS CONCENTRATION

Many organic solutes that are either polar molecules or molecules containing both polar and nonpolar groupings have a tendency to concentrate at the surface of the solution, and thus values of γ decrease as the bulk concentration increases. Many electrolytic solutes tend to be depleted at the surface, and thus values of γ increase as the bulk concentration increases. These tendencies are discussed in terms of the *surface excess concentration* (Γ_2) defined by

$$\Gamma_2 = -\frac{C}{RT}\frac{d\gamma}{dC} \tag{23.8}$$

An approximate *effective thickness* (x) in which the actual concentration is significantly different from the bulk concentration is given by

$$x = |\Gamma_2 / C| \tag{23.9}$$

EXAMPLE 23.6. For a 1.00×10^{-4} M solution of n-butanoic acid in water at 25 °C, $d\gamma/dC = -0.080$ N m^2 mol^{-1}. Find Γ_2 for this solution, and determine the average surface area available to each molecule.

Using (23.8) gives

$$\Gamma_2 = -\frac{(1.00 \times 10^{-4} \text{ mol dm}^{-3})[(10 \text{ dm})/(1 \text{ m})]^3(-0.080 \text{ N m}^2 \text{ mol}^{-1})}{(8.314 \text{ J K}^{-1} \text{ mol}^{-1})(298 \text{ K})}$$

$$= 3.2 \times 10^{-6} \text{ mol m}^{-2}$$

The average surface area for each molecule is

$$\frac{1}{(3.2 \times 10^{-6} \text{ mol m}^{-2})(6.022 \times 10^{23} \text{ mol}^{-1})} = 5.2 \times 10^{-19} \text{ m}^2$$

Adsorption

23.7   ADSORPTION ISOTHERMS

The amount of substance adsorbed at a given temperature is described by various equations called *adsorption isotherms*. For example, at rather low pressures (P) the volume of adsorbed gas on a solid (V) is given by the *Freundlich isotherm* as

$$V = kP^n \qquad (23.10)$$

where k and n are constants with $n < 1$. More complete equations are the *Langmuir isotherm*,

$$V = V_{\text{mono}}\theta \qquad (23.11)$$

where V_{mono} represents the volume of gas to form a monolayer over the entire surface of the solid, where

$$\theta = \frac{bP}{1 + bP} \qquad (23.12)$$

and where b is the ratio of the rate constant for adsorption to that for desorption; and the *Brunauer–Emmett–Teller* (*BET*) *isotherm*,

$$V = \frac{V_{\text{mono}}c(P/P^\circ)}{1 - P/P^\circ}\left(\frac{1 - (n+1)(P/P^\circ)^n + n(P/P^\circ)^{n+1}}{1 + (c-1)(P/P^\circ) - c(P/P^\circ)^{n+1}}\right) \qquad (23.13)$$

Here n is the number of layers of adsorbed molecules, P° is the vapor pressure of the pure liquid, and

$$c = \exp\{[\Delta_{\text{ads}}H(\text{mono}) - \Delta_{\text{vap}}H]/RT\} \qquad (23.14)$$

where $\Delta_{\text{ads}}H(\text{mono})$ is the heat of vaporization for the first layer, assumed to be different from the heat of vaporization of the liquid, $\Delta_{\text{vap}}H$. For $n = \infty$, (23.13) becomes

$$V = \frac{V_{\text{mono}}cP}{(P^\circ - P)[1 + (c-1)(P/P^\circ)]} \qquad (23.15)$$

EXAMPLE 23.7. The Langmuir isotherm is useful in describing monolayer adsorption (*chemisorption*), but fails at higher pressures where a second layer begins to form (*physisorption*). Show that (23.11) qualitatively predicts the Freundlich isotherm at relatively low pressures and that V is pressure-independent at high pressures.

At low pressures, the denominator of (23.12) approaches unity, giving $\theta = bP$, and (23.11) becomes

$$V = V_{\text{mono}}bP$$

which agrees with (23.10) for $n = 1$ and $k = V_{\text{mono}}b$. At high pressures the denominator of (23.12) approaches bP, giving $\theta = 1$, and (23.11) becomes

$$V = \frac{V_{\text{mono}}bP}{bP} = V_{\text{mono}}$$

which is pressure-independent.

EXAMPLE 23.8. The mass (m) of adsorbed solute on a solid is given by a form of the Freundlich isotherm as

$$m = kC_2{}^n \qquad (23.16)$$

where C_2 is the concentration. If C_2 is expressed in molarity (mol dm^{-3}), the values of k and n in (23.16) that give m in units of (g acetic acid)(g blood charcoal)$^{-1}$ are $k = 0.160$ and $n = 0.431$. Find the amount of acetic acid that 1 kg of charcoal would adsorb from a 5.00 mass % (0.837 M) vinegar solution.

The mass of solute is calculated using (23.16) as

$$m = (0.160)(0.837)^{0.431} = 0.148 \text{ (g HAc)(g charcoal)}^{-1} = 148 \text{ (g HAc)(kg charcoal)}^{-1}$$

which upon converting to moles becomes

$$\frac{148 \text{ (g HAc)(kg charcoal)}^{-1}}{60.05 \text{ g mol}^{-1}} = 2.47 \text{ (mol HAc)(kg charcoal)}^{-1}$$

23.8 HETEROGENEOUS CATALYSIS

Heterogeneous catalysis occurs as the reactant collects or adsorbs on the surface of the catalyst. If more than one gas is adsorbed by the catalyst during a gaseous decomposition reaction, (23.12) becomes

$$\theta_i = \frac{b_i P_i}{1 + b_A P_A + b_B P_B + \cdots} \qquad (23.17)$$

for each gas, assuming simple adsorption. The adsorption of additional gases reduces the surface area available for the reaction under consideration and thus causes the inhibition (*poison* or retardation) of the reaction.

The general expression describing the rate of decomposition for the reaction

$$A \longrightarrow \text{ products}$$

is given by

$$\frac{d\xi}{dt} = k'\theta \qquad (23.18)$$

where k' is a proportionality constant and θ is the fraction of the surface covered.

EXAMPLE 23.9. Derive the integrated rate equation for the general case in which A is moderately adsorbed. Show that the following data of Stock and Bodenstein for the decomposition of SbH$_3$ on Sb at 25 °C satisfy this rate equation:

$t/(\text{s})$	0	300	600	900	1 200	1 500
$P(\text{SbH}_3)/(\text{bar})$	1.013	0.741	0.516	0.331	0.192	0.094

Substituting (23.12) into (23.18) gives

$$-\frac{d}{dt}[P(\text{SbH}_3)] = \frac{k'bP(\text{SbH}_3)}{1 + bP(\text{SbH}_3)} = \frac{kP(\text{SbH}_3)}{1 + bP(\text{SbH}_3)}$$

which upon rearrangement becomes

$$-k\,dt = \{[P(\text{SbH}_3)]^{-1} + b\}\,dP(\text{SbH}_3)$$

Integration and substitution of proper limits yield

$$\ln\left[\frac{P(\text{SbH}_3)_0}{P(\text{SbH}_3)}\right] + b[P(\text{SbH}_3)_0 - P(\text{SbH}_3)] = kt$$

Unless b is known, k cannot be determined graphically. However, both b and k can be determined numerically by substituting the data and solving the five equations to give an average of $b = 1.824 \, \text{bar}^{-1}$ and $k = 2.67 \times 10^{-3} \, \text{s}^{-1}$. Because of the difficulty in working with this equation, the Freundlich equation is sometimes used for these systems (see Problem 23.32).

EXAMPLE 23.10.   Derive the expression for θ describing the reaction

$$A_2 \longrightarrow 2A$$

where A is strongly adsorbed.

For simple adsorption, the rate of condensation is given by $k_1 P(1-\theta)$ and the rate of desorption by $k_2 \theta$. Setting these rates equal at equilibrium and solving for θ gives (23.12). When dissociation accompanies adsorption, the law of mass action gives the rate of condensation as $k_1 P(1-\theta)^2$ and the rate of desorption as $k_2 \theta^2$. Setting these equal for equilibrium and solving for θ gives

$$\theta = \frac{bP^{1/2}}{1 + bP^{1/2}}$$

where $b = (k_1/k_2)^{1/2}$.

Solved Problems

SURFACE TENSION OF LIQUIDS

23.1.   A capillary tube was calibrated at $20 \, ^\circ\text{C}$ using water, and the water rose 8.37 cm before it came to equilibrium. A sample of mercury was depressed 3.67 cm using the same capillary. Given $\rho = 0.998 \, 2 \times 10^3 \, \text{kg m}^{-3}$ for water and $13.593 \, 9 \times 10^3 \, \text{kg m}^{-3}$ for Hg, find γ for mercury given $\gamma = 72.75 \times 10^{-3} \, \text{N m}^{-1}$ for water. What is the nominal size of the capillary tubing used?

Using (23.2) gives

$$\gamma_1 = (72.75 \times 10^{-3} \, \text{N m}^{-1}) \frac{(13.593 \, 9 \times 10^3 \, \text{kg m}^{-3})(3.67 \, \text{cm})}{(0.998 \, 2 \times 10^3 \, \text{kg m}^{-3})(8.37 \, \text{cm})} = 0.434 \, \text{N m}^{-1}$$

Rearranging (23.1) gives

$$r = \frac{2\gamma}{\rho g h} = \frac{2(72.75 \times 10^{-3} \, \text{N m}^{-1})}{(0.998 \, 2 \times 10^3 \, \text{kg m}^{-3})(9.8 \, \text{m s}^{-2})(8.37 \times 10^{-2} \, \text{m})} = 1.78 \times 10^{-4} \, \text{m}$$

or a nominal size of 0.2 mm.

23.2.   Use (23.3a) to predict γ for water at $25 \, ^\circ\text{C}$.

Substituting the results of Example 23.2 into (23.3a) gives

$$\gamma = (115.35 \times 10^{-3} \, \text{N m}^{-1})\left(1 - \frac{298 \, \text{K}}{647.3 \, \text{K}}\right)^{0.766}$$

$$= 71.91 \times 10^{-3} \, \text{N m}^{-1}$$

The actual value is $71.97 \times 10^{-3} \, \text{N m}^{-1}$ at $25 \, ^\circ\text{C}$.

23.3.   The thermodynamic properties associated with a change in surface area are given by

$$\frac{G_\gamma}{A} = \gamma \qquad \frac{S_\gamma}{A} = -\frac{d\gamma}{dT} \qquad \frac{U_\gamma}{A} \approx \frac{H_\gamma}{A} = \gamma - T\frac{d\gamma}{dT}$$

Using $\gamma = 71.97 \times 10^{-3} \, \text{N m}^{-1}$ and $d\gamma/dT = -0.15 \times 10^{-3} \, \text{N m}^{-1} \, \text{K}^{-1}$ for water at $25 \, ^\circ\text{C}$, determine these thermodynamic properties.

Substituting the data into the equations gives

$$G_\gamma/A = (71.97 \times 10^{-3} \text{ N m}^{-1}) = 71.97 \times 10^{-3} \text{ J m}^{-2} = 71.97 \text{ mJ m}^{-2}$$

$$S_\gamma/A = -(-0.15 \times 10^{-3} \text{ N m}^{-1} \text{ K}^{-1}) = 0.15 \text{ mJ m}^{-2} \text{ K}^{-1}$$

$$U_\gamma/A \approx H_\gamma/A = (71.97 \text{ mJ m}^{-2}) - (298 \text{ K})(-0.15 \text{ mJ m}^{-2} \text{ K}^{-1})$$

$$= 117 \text{ mJ m}^{-2}$$

23.4. What is the minimum size of water droplets such that the vapor pressure does not differ by more than 1% from the bulk value?

Substituting $P_s = 1.01P$ and the necessary data for water into (23.4) gives

$$r = \frac{2(71.97 \times 10^{-3} \text{ N m}^{-1})(18.01 \text{ g mol}^{-1})[(10^{-3} \text{ kg})/(1 \text{ g})]}{(0.997\,044 \times 10^3 \text{ kg m}^{-3})(8.314 \text{ J K}^{-1} \text{ mol}^{-1})(298 \text{ K})(\ln 1.01)} = 1.05 \times 10^{-7} \text{ m}$$

23.5. Equation (23.4), with an appropriate sign change, can be used to describe bubble formation in a liquid. Calculate the pressure within a bubble of water vapor at 100 °C that contains about 50 molecules ($r \approx 1 \times 10^{-8}$ m). For water at 100 °C, $\gamma = 58.9 \times 10^{-3}$ N m^{-1} and $\rho = 0.958 \times 10^3$ kg m^{-3}.

The pressure within the bubble will be

$$\ln \frac{P_b}{1.013\,25 \text{ bar}} = -\frac{2(58.9 \times 10^{-3} \text{ N m}^{-1})(18.01 \times 10^{-3} \text{ kg mol}^{-1})}{(1 \times 10^{-8} \text{ m})(0.958 \times 10^3 \text{ kg m}^{-3})(8.314 \text{ J K}^{-1} \text{ mol}^{-1})(373 \text{ K})} = -0.07$$

$$P_b/(1.013\,25 \text{ bar}) = 0.85 \qquad P_b = 0.86 \text{ bar}$$

Note that this pressure is less than that exerted by the liquid, and thus the bubble will be "forced out of existence" by the liquid. Unless a larger bubble forms ("bumping") or an impurity (such as a boiling chip) is present, the liquid will superheat.

23.6. The density of paraldehyde is $0.994\,3 \times 10^3$ kg m^{-3}, $M = 132.16$ g mol^{-1}, and the surface tension is 25.9×10^{-3} N m^{-1}. Calculate $\{P\}$. Calculate the parachor for acetaldehyde using parachor equivalents of 4.8 for C, 17.1 for H, 20.0 for O, and 23.2 for a double bond. Given that paraldehyde is a trimer of acetaldehyde containing a six-membered ring (additional parachor equivalent of 6.1) of carbon and oxygen, describe the bonding ring.

Neglecting ρ_{gas}, the value of $\{P\}$ for paraldehyde is given by (23.5) as

$$\{P\} = \frac{10^3 (132.16)(25.9)^{1/4}}{0.994\,3 \times 10^3} = 300$$

Summing the contributions for acetaldehyde gives

$$\{P\} = 2(4.8) + 4(17.1) + 1(20.0) + 1(23.2) = 121.2$$

Assuming a trimer ring of acetaldehyde, $\{P\}$ would be $3(121.2) + 6.1 = 369.7$, which is too high. Assuming a trimer ring without double bonding, $\{P\}$ would be $3(121.2) + 6.1 - 3(23.2) = 300.1$, which agrees with the value determined using (23.5).

SURFACE TENSION IN BINARY SYSTEMS

23.7. Wetting for a liquid on a solid is arbitrarily defined as $\theta < 90°$. Determine the relationship between γ_A and γ_{AB} for wetting to occur.

For $\theta < 90°$, $\cos \theta > 0$, and so (23.7) gives

$$0 < \frac{\gamma_A - \gamma_{AB}}{\gamma_B}$$

or $\gamma_A > \gamma_{AB}$ (e.g., water on glass).

23.8. Use the following surface tension data for solutions of NaCl in water at 20 °C to determine the effective thickness of the surface excess concentration:

$C/(\text{mol dm}^{-3})$	0.496	0.963	1.922	2.829	4.523
$\gamma/(10^{-3}\ \text{N m}^{-1})$	73.75	74.39	76.05	77.65	80.95

A plot of γ against C is linear (see Fig. 23-2), giving $d\gamma/dC = 1.80 \times 10^{-6}$ N m² mol⁻¹. The effective thickness is given by (23.9) as

$$x = \left| \frac{-d\gamma/dC}{RT} \right| = \left| \frac{-(1.80 \times 10^{-6}\ \text{N m}^2\ \text{mol}^{-1})}{(8.314\ \text{J K}^{-1}\ \text{mol}^{-1})(298\ \text{K})} \right|$$

$$= 7.27 \times 10^{-10}\ \text{m}$$

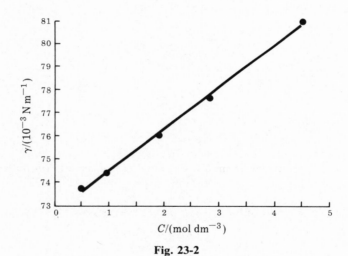

Fig. 23-2

ADSORPTION

23.9. The amount of a gas adsorbed on a solid has rather definite correlation with the critical point of the gas. Given that 1 kg of charcoal at 15 °C will adsorb 8.0×10^{-3} m³ (measured at STP) of N_2 and 380×10^{-3} m³ of SO_2, approximately what volumes of CO_2 and Cl_2 would be adsorbed under similar conditions? The critical temperatures for the gases are −146.9 °C for N_2, 31 °C for CO_2, 144 °C for Cl_2, and 157.4 °C for SO_2.

Based on the order of the critical temperatures, the volumes of CO_2 and Cl_2 should be greater than the volume of N_2 and less than the volume of SO_2. Less CO_2 should be adsorbed than Cl_2. The observed values are 48×10^{-3} m³ for CO_2 and 235×10^{-3} m³ for Cl_2, which agree with this prediction.

23.10. An equation similar to (4.20b) describes the temperature dependence of the amount of gas adsorbed on the surface of a solid. Find $\Delta_{ads}H$ for N_2 at 1.013 25 bar given that 155 cm^3 (measured at STP) is adsorbed by 1 g of charcoal at 88 K, and 15 cm^3 at 273 K.

Solving
$$\ln \frac{V_2}{V_1} = -\frac{\Delta_{ads}H}{R}\left(\frac{1}{T_2} - \frac{1}{T_1}\right) \qquad (23.19)$$

for $\Delta_{ads}H$ and substituting the data gives

$$\Delta_{ads}H = \frac{-R\ln(V_2/V_1)}{1/T_2 - 1/T_1}$$

$$= \frac{-(8.314\ \text{J K}^{-1}\ \text{mol}^{-1})\ln(155/15)}{1/88\ \text{K} - 1/273\ \text{K}} = -2.52\ \text{kJ mol}^{-1}$$

23.11. Show that for $n = 1$ the BET isotherm gives the Langmuir isotherm.

For $n = 1$, (23.13) becomes

$$V = \frac{V_{mono}c(P/P^\circ)}{1 - P/P^\circ}\left(\frac{1 - 2(P/P^\circ) + (P/P^\circ)^2}{1 + (c-1)(P/P^\circ) - c(P/P^\circ)^2}\right)$$

$$= \frac{V_{mono}c(P/P^\circ)}{1 - P/P^\circ}\left(\frac{(1 - P/P^\circ)^2}{[1 + c(P/P^\circ)](1 - P/P^\circ)}\right) = \frac{V_{mono}c(P/P^\circ)}{1 + c(P/P^\circ)} = \frac{V_{mono}bP}{1 + bP}$$

if $b = c/P^\circ$.

23.12. Langmuir reported the following results for the adsorption of N_2 on mica at 90 K:

$P/(\text{Pa})$	0.28	0.34	0.40	0.49	0.60	0.73	0.94	1.28	1.71	2.35	3.35
$V/(\text{mm}^3\ (20\ ^\circ\text{C},\ 1\ \text{atm})\ \text{g}^{-1})$	12.0	13.4	15.1	17.0	19.0	21.6	23.9	25.5	28.2	30.8	33.0

(a) Show that these data obey the Freundlich isotherm at low pressures. (b) Find the number of moles of N_2 equivalent to V_{mono} given $b = 1.56\ \text{Pa}^{-1}$. (c) What is the surface area of the mica phase?

(a)   Equation (23.10) can be transformed into a linear equation by taking logarithms of both sides, giving

$$\log V = \log k + n\log P$$

From a plot of $\log V$ against $\log P$ (see Fig. 23-3), the intercept is 1.423, giving

$$k = \text{antilog}(1.423) = 26.5$$

and the slope of the straight line through the low-pressure data is $0.617 = n$.

(b)   Eliminating θ in (23.11) and (23.12), solving the resulting equation for V_{mono}, and substituting the value of b and the data for 0.40 Pa gives

$$V_{mono} = \frac{V(1 + bP)}{bP}$$

$$= \frac{(15.1\ \text{mm}^3\ \text{g}^{-1})[1 + (1.56\ \text{Pa}^{-1})(0.40\ \text{Pa})]}{(1.56)(0.40)} = 39.3\ \text{mm}^3\ \text{g}^{-1}$$

which upon converting to moles becomes

$$n = \frac{(1.013\ 25\ \text{bar})(39.3 \times 10^{-9}\ \text{m}^3\ \text{g}^{-1})}{(8.314 \times 10^{-5}\ \text{bar m}^3\ \text{K}^{-1}\ \text{mol}^{-1})(293\ \text{K})} = 1.63 \times 10^{-6}\ (\text{mol } N_2)(\text{g mica})^{-1}$$

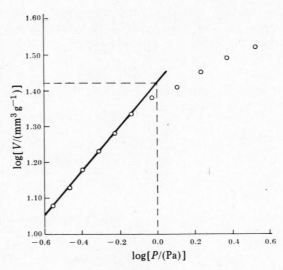

Fig. 23-3

(*c*) The number of molecules covering the surface of the mica is

$$[1.63 \times 10^{-6}\ (\text{mol N}_2)(\text{g mica})^{-1}](6.022 \times 10^{23}\ \text{mol}^{-1}) = 9.82 \times 10^{17}\ \text{g}^{-1}$$

Assuming a density of $0.808\ 1 \times 10^3\ \text{kg m}^{-3}$ for liquid N_2, the area covered by a single molecule is approximately

$$\left[\frac{28.0 \times 10^{-3}\ \text{kg mol}^{-1}}{(0.808\ 1 \times 10^3\ \text{kg m}^{-3})(6.022 \times 10^{23}\ \text{mol}^{-1})}\right]^{2/3} = 14.9 \times 10^{-20}\ \text{m}^2$$

The area of adsorption is the product of these results, i.e.,

$$(9.82 \times 10^{17}\ \text{g}^{-1})(14.9 \times 10^{-20}\ \text{m}^2) = 0.146\ \text{m}^2\ \text{g}^{-1}$$

23.13. Derive the integrated rate equation for the special case in which substance A is only slightly adsorbed. Show that the following data describing the decomposition of N_2O on Au at 900 °C satisfy this rate equation:

$t/(\text{s})$	0	1 800	4 800	7 200
$P(\text{N}_2\text{O})/(\text{Pa})$	26 700	18 100	9 300	5 900

Substituting (*23.12*) into (*23.18*) gives

$$-\frac{d}{dt}[P(\text{N}_2\text{O})] = \frac{k'bP(\text{N}_2\text{O})}{1 + bP(\text{N}_2\text{O})}$$

which upon making the approximation $1 \gg bP(\text{N}_2\text{O})$ becomes

$$-\frac{d}{dt}[P(\text{N}_2\text{O})] = k'bP(\text{N}_2\text{O}) = kP(\text{N}_2\text{O})$$

a first-order reaction. Integration and substitution of proper limits give

$$\ln[P(\text{N}_2\text{O})/P(\text{N}_2\text{O})_0] = -kt$$

Although the value of k could be determined from the slope of a plot of the logarithm against t, it is easier to calculate k for the three sets of data and find an average of these values. For $t = 1\,800$ s,

$$k = -\frac{\ln(18\,100/26\,700)}{1\,800\ \text{s}} = 2.14 \times 10^{-4}\ \text{s}^{-1}$$

and for the other data, $k = 2.19 \times 10^{-4}\ \text{s}^{-1}$ and $2.10 \times 10^{-4}\ \text{s}^{-1}$, giving an average of $2.14 \times 10^{-4}\ \text{s}^{-1}$.

23.14. Derive (23.17) for the case of two gases, A and B, being adsorbed.

Recognizing that the fraction of surface uncovered is $1 - \theta_A - \theta_B$, the rates of adsorption are

$$k_{1,A} P_A (1 - \theta_A - \theta_B) \qquad \text{and} \qquad k_{1,B} P_B (1 - \theta_A - \theta_B)$$

and the rates of desorption are $k_{2,A} \theta_A$ and $k_{2,B} \theta_B$. Equating these rates for equilibrium gives

$$\theta_A = \frac{b_A P_A}{1 + b_A P_A + b_B P_B} \qquad \theta_B = \frac{b_B P_B}{1 + b_A P_A + b_B P_B}$$

23.15. Hinshelwood and Burk reported the following data for the decomposition of NH_3 on Pt at $1\,138\ °C$ with $P(NH_3)_0 = 13\,300$ Pa:

$-[\Delta P(NH_3)/\Delta t]/(\text{Pa s}^{-1})$	36.7	17.7	11.1
$P(H_2)/(\text{Pa})$	6 700	13 300	20 000

Show that these data indicate that the H_2 is strongly adsorbed after it is formed.

The rate of decomposition of NH_3 is given by

$$-\frac{d}{dt}[P(NH_3)] = k'[1 - \theta(NH_3) - \theta(H_2) - \theta(N_2)]P(NH_3) \approx k'[1 - \theta(H_2)]P(NH_3)$$

assuming $\theta(H_2) \gg \theta(NH_3),\ \theta(N_2)$. Making the same assumption for the denominator of (23.17) gives

$$\theta(H_2) = \frac{b(H_2)P(H_2)}{1 + b(H_2)P(H_2)}$$

The expression for $1 - \theta(H_2)$ is

$$1 - \theta(H_2) = \frac{1}{1 + b(H_2)P(H_2)} = \frac{1}{b(H_2)P(H_2)}$$

where $b(H_2)P(H_2) \gg 1$. Substitution into the rate equation gives

$$-\frac{d}{dt}[P(NH_3)] = [k'/b(H_2)][P(NH_3)/P(H_2)] = kP(NH_3)/P(H_2)$$

The values of $k/(\text{Pa s}^{-1}) = [-\Delta P(NH_3)/\Delta t][P(H_2)/P(NH_3)]$ using the above data are 18.5, 17.7, and 16.7, which are reasonably constant and imply that the rate equation is valid.

Supplementary Problems

SURFACE TENSION OF LIQUIDS

23.16. Molten LiCl wets BN at $800\ °C$. At this temperature $\gamma = 12.5 \times 10^{-3}\ \text{N m}^{-1}$ and $\rho = 1.417 \times 10^3\ \text{kg m}^{-3}$. Describe what will happen if a BN tube having a radius of 1 mm and height of 1 cm is placed in a sample of LiCl(l).   *Ans.* $h = 1.80$ cm (fountain effect)

23.17. Equation (23.2) with $n = 1.2$ should be valid for benzene, which is not highly associated in the liquid state. Given $\gamma = 28.85 \times 10^{-3}$ N m^{-1} at 20 °C, determine γ_0 and predict γ at 30 °C. The critical temperature of C_6H_6 is 288.9 °C.   *Ans.* $\gamma_0 = 69.84 \times 10^{-3}$ N m^{-1}, $\gamma = 27.57 \times 10^{-3}$ N m^{-1} (actual is 27.56×10^{-3} N m^{-1})

23.18. Calculate G_γ/A, S_γ/A, and H_γ/A for benzene at 25 °C using $\gamma = 28.21 \times 10^{-3}$ N m^{-1} and $d\gamma/dT = -0.11 \times 10^{-3}$ N m^{-1} K^{-1}.   *Ans.* $G_\gamma/A = 28.21$ mJ m^{-2}, $S_\gamma/A = 0.11$ mJ m^{-2} K^{-1}, $H_\gamma/A = 61$ mJ m^{-2}

23.19. Using (23.3), derive the general expression for $d\gamma/dT$. Using the data of Problem 23.17, show that $d\gamma/dT = -0.11 \times 10^{-3}$ N m^{-1} K^{-1} for benzene at 25 °C.   *Ans.* $d\gamma/dT = (-\gamma_0/T_c)(1 - T/T_c)^{n-1}$

23.20. Determine P_s/P for a drop of water at 25 °C that has $r = 10^{-6}$ cm.   *Ans.*   1.111 (a significant change)

23.21. Pure water vapor at 25 °C must be compressed so that $P_s/P = 2.7$ before condensation will occur. What is the radius of the droplets that are formed, and approximately how many water molecules are in each droplet? See Example 23.3 for data.   *Ans.* $r = 1.1 \times 10^{-9}$ m, $N = \frac{4}{3}\pi r^3 \rho L/M = 190$ molecules

23.22. The solubility of small crystals of a solid in water (x) relative to the solubility of coarse crystals (x_0) is given by

$$\ln\frac{x}{x_0} = \frac{V_{m,s}}{RT}\left(\frac{6\gamma_s}{a}\right)$$

for a cubic crystal, where x is the mole fraction, $V_{m,s}$ is the molar volume of the pure solute, γ_s is the surface tension of the solid, and a is the length of the edge of the crystal. What is the relative solubility for $BaSO_4$ at 25 °C given $\gamma_s = 0.5$ N m^{-1} and $\rho = 4.5 \times 10^3$ kg m^{-3} for a crystal with $a = 1 \times 10^{-7}$ m? If there is no site for nucleation, what will happen if $x/x_0 = 1.5$ in a solution?

Ans. $x/x_0 = 1.9$; the solute will remain supersaturated

23.23. The parachor equivalent for C is 4.8; for H, 17.1; and for O, 20.0. (*a*) Predict the parachors for methanol and ethanol. (*b*) Calculate the parachors given $\gamma = 22 \times 10^{-3}$ N m^{-1} for each alcohol and $\rho/(\text{kg m}^{-3}) = 0.791\,4 \times 10^3$ for methanol and $0.789\,3 \times 10^3$ for ethanol. (*c*) The parachor defect in these substances is related to the amount of hydrogen bonding between molecules. In which substance is this effect more significant?

Ans.   (*a*) 93.2 for methanol and 132.2 for ethanol;   (*b*) 87.6 for methanol and 126.2 for ethanol;   (*c*) defect is 5.6 for methanol and 6.0 for ethanol, suggesting more hydrogen bonding in ethanol

SURFACE TENSION IN BINARY SYSTEMS

23.24. The interfacial tension between water and carbon tetrachloride is 45×10^{-3} N m^{-1} at 20 °C. At this temperature, $\gamma/(10^{-3}$ N m$^{-1}) = 26.95$ for CCl_4 and 72.75 for H_2O. Will wetting occur as a drop of water is placed on a pool of carbon tetrachloride? What about a drop of CCl_4 placed on a pool of water?

Ans. $S(H_2O\text{-}CCl_4) = -91 \times 10^{-3}$ N m^{-1}, no; $S(CCl_4\text{-}H_2O) = 1 \times 10^{-3}$ N m^{-1}, yes

23.25. The value of γ_{AB} for Hg on glass is large enough that $\gamma_A - \gamma_{AB}$ is negative. Will wetting occur?

Ans.   No: $\theta > 90°$

23.26. For a 1.61 M aqueous solution of ethanol, $d\gamma/dC = -6.27 \times 10^{-6}$ N m^2 mol^{-1} at 50 °C. Determine the surface excess concentration and the effective thickness.   *Ans.* $\Gamma_2 = 3.76 \times 10^{-6}$ mol m^{-2}, $x = 2.34 \times 10^{-9}$ m

ADSORPTION

23.27. Using (23.19) and the data in Problem 23.10, predict the amount of N_2 that could be adsorbed under similar conditions at 15 °C.   *Ans.* $V_2 = 14$ cm^3 (at STP) g^{-1} or 14×10^{-3} m^3 kg^{-1}

23.28. If 1 kg of charcoal at 25 °C will adsorb 0.62 mol of acetic acid from a 0.031 M aqueous solution and 2.48 mol from a 0.882 M solution, verify the constants given in Example 23.8.

   Ans. $m = 3.72 \times 10^{-2}$ (g HAc) (g charcoal)$^{-1}$ at 0.031 M and 1.49×10^{-1} at 0.882 M; (*23.16*) then gives $n = 0.412$ and $k = 0.156$.

23.29. Show that (*23.11*) can be put into the linear form

$$\frac{P}{V} = \frac{1}{V_{mono}b} + \frac{P}{V_{mono}}$$

Prepare a plot of P/V against P for the data in Problem 23.12, and confirm the given values of b and V_{mono}.

   Ans.   Slope $= 2.59 \times 10^{-2}$ mm^{-3} g, from which $V_{mono} = 38.6$ mm^3 g$^{-1} = 1.60 \times 10^{-6}$ (mol N$_2$) (g charcoal)$^{-1}$; intercept $= 0.016\ 3$ Pa mm^{-3} g, giving $b = 1.59$ Pa^{-1}

23.30. The following data were obtained for the adsorption of N$_2$ on Al$_2$O$_3$ at 77.3 K:

$P/(\text{Pa})$	4 230	8 600	12 900	17 170	22 570
$n/[10^{-4}$ (mol N$_2$)(g Al$_2$O$_3$)$^{-1}]$	8.31	9.03	9.85	10.45	11.18

(*a*) Prepare a plot of $P/[V(P° - P)]$ against $P/P°$, assuming $P° = 101\ 190$ Pa. The intercept and slope give $1/V_{mono}c$ and $(c-1)/V_{mono}c$, respectively. Find V_{mono}, expressed in (mol N$_2$) g^{-1}, and c. (*b*) Using $\Delta_{vap}H = 5.6$ kJ mol^{-1}, find $\Delta_{ads}H$(mono). (*c*) Assuming the area covered by one nitrogen molecule is 16.2×10^{-20} m^2, find the surface area of the alumina.

   Ans.   (*a*) Intercept $= 7$ g mol^{-1}, slope $= 1\ 120$ g mol^{-1}; $c = 161$; $V_{mono} = 8.87 \times 10^{-4}$ mol g^{-1}; (*b*) 8.9 kJ mol^{-1}; (*c*) 86.5 m^2(g alumina)$^{-1}$

23.31. The constant b in (*23.12*) can be estimated from statistical mechanics by

$$b/(\text{Pa}) = (7.23 \times 10^{22})q_{AS} \frac{e^{-\Delta_{ads}U/RT}}{Tq_A q_S}$$

where $q_S = 1$ because the solid adsorbent is rigid; q_A is given by

$$q_A = (2\pi mkT/h^2)^{3/2} q_{A,int}$$

where $q_{A,int}$ would contain the internal contributions given by (*8.19*), (*8.21*), (*8.24*), and (*8.26*); q_{AS} is given by

$$q_{AS} = (1 - e^{-h\nu_z/kT})^{-1}(1 - e^{-h\nu_x/kT})^{-1}(1 - e^{-h\nu_y/kT})^{-1}$$

where the adsorbed molecule might have three different vibrational motions; and $\Delta_{ads}U$ is the internal energy change for the adsorption process. If $\Delta_{ads}U = 3.0$ kJ mol^{-1} for Ar at 200 K and $\nu_x = \nu_y = \nu_z = 5 \times 10^{12}$ s^{-1}, find b and estimate θ at 1 atm.   *Ans.* $b = 1.30 \times 10^{-12}$ Pa^{-1}, $\theta = 1.32 \times 10^{-7}$

23.32. The rate equation for the decomposition described in Example 23.9 is provided by the Freundlich isotherm

$$-\frac{d}{dt}[P(\text{SbH}_3)] = k[P(\text{SbH}_3)]^n$$

Evaluate k and n from two points on a plot of $P(\text{SbH}_3)$ against t.

   Ans. 8.02×10^{-4} bar s^{-1} at 0.709 bar and 4.17×10^{-4} bar s^{-1} at 0.203 bar; $n = 0.523$; $k = 9.60 \times 10^{-4}$

23.33. Substitute (*23.12*) into (*23.18*), and derive the integrated rate equation for the special case in which A is strongly adsorbed. Show that the following data of Hinshelwood and Burk for the decomposition of NH$_3$ on W at 856 °C satisfy this rate equation:

$t/(s)$	0	100	200	300
$P(NH_3)/(Pa)$	26 700	24 800	23 100	21 600

Note that at longer times the hydrogen that is formed reduces the available area on the surface of the catalyst and this rate equation fails (see Problem 23.15).

Ans. $b(NH_3)P(NH_3) \gg 1$, $P(NH_3) = [P(NH_3)]_0 - k't$; slope of plot of $P(NH_3)$ against t gives $k' = 17.0 \text{ Pa s}^{-1}$.

23.34. Find the expressions for θ_A and θ_B for the case of two gases being adsorbed on a surface such that one of them is diatomic and decomposes upon adsorption.

Ans. $k_{1,A}P_A(1 - \theta_A - \theta_B) = k_{2,A}\theta_A$, $k_{1,B}P_B(1 - \theta_A - \theta_B)^2 = k_{2,B}\theta_B^2$; $\theta_A = b_A P_A/[1 + b_A P_A + (b_B P_B)^{1/2}]$; $\theta_B = b_B P_B/[1 + b_A P_A + (b_B P_B)^{1/2}]$

23.35. Hinshelwood and Prichard reported the following data at 741 °C for the decomposition of N_2O on Pt:

$t/(s)$	315	750	1 400	2 250	3 450	5 150
$P(O_2)/(Pa)$	1 330	2 670	4 000	5 330	6 670	8 000

where $P(N_2O)_0 = 12 670$ Pa and $P(N_2O) = 12 670 - P(O_2)$. Assuming retardation by O_2, which is moderately adsorbed on Pt, show that

$$\{1 + b(O_2)[P(N_2O)]_0\}\ln \frac{[P(N_2O)]_0}{[P(N_2O)]_0 - P(O_2)} = k't + b(O_2)P(O_2)$$

and find b and k'.

Ans. Rate $= k'[1 - \theta(O_2)]P(N_2O)$, $\theta(O_2) = b(O_2)P(O_2)/[1 + b(O_2)P(O_2)]$; plot of $P(O_2)/t$ against $(1/t)\ln\{[P(N_2O)]_0/\{[P(N_2O)]_0 - P(O_2)\}\}$ is linear with intercept -1.70 Pa s^{-1} and slope 16 760 Pa; $b(O_2) = 2.45 \times 10^{-4} \text{ Pa}^{-1}$; $k' = 4.17 \times 10^{-4} \text{ s}^{-1}$

Chapter 24

Macromolecules

Molar Mass

24.1 AVERAGE MOLAR MASS

Most polymers are mixtures with a wide distribution of molar masses (M_i). The fraction of polymer chains [$f(M_i)$] having molar masses between M_1 and M_2 ($M_1 \leq M_i \leq M_2$) is given by

$$f(M_i) = \int_{M_1}^{M_2} N(M_i) \, dM_i \qquad (24.1)$$

where $N(M_i)$ is the distribution function describing the mixture.

Properties that are dependent on the number of particles but independent of the size of particles (e.g., colligative properties of solutions) are best described using the *number-average molar mass* ($\bar{M}_n$):

$$\bar{M}_n = \frac{\sum n_i M_i}{\sum n_i} = \sum x_i M_i \qquad (24.2)$$

where n_i is the number of moles and x_i is the mole fraction of component i in the mixture. Properties that are dependent on the contribution of each particle relative to its mass (e.g., light scattering and ultracentrifugation) are best described using the *mass-average molar mass* ($\bar{M}_m$):

$$\bar{M}_m = \frac{\sum n_i M_i^2}{\sum n_i M_i} \qquad (24.3)$$

Other average molar masses can be defined in a similar way. For example, the *z-average molar mass* ($\bar{M}_z$), which is determined from equilibrium ultracentrifugation, is defined as

$$\bar{M}_z = \frac{\sum n_i M_i^3}{\sum n_i M_i^2} \qquad (24.4)$$

and the *viscosity molar mass* ($\bar{M}_v$) is given by

$$\bar{M}_v = \left[\frac{\sum n_i M_i^{a+1}}{\sum n_i M_i} \right]^{1/a} \qquad (24.5)$$

The *polydispersity* of a mixture of polymers is often numerically determined by the ratio of $\bar{M}_m$ to $\bar{M}_n$:

$$\text{Polydispersity index} = \frac{\bar{M}_m}{\bar{M}_n} \qquad (24.6)$$

EXAMPLE 24.1. A protein sample consists of an equimolar mixture of ribonuclease ($M = 13.7 \text{ kg mol}^{-1}$), hemoglobin ($M = 15.5 \text{ kg mol}^{-1}$), and myoglobin ($M = 17.2 \text{ kg mol}^{-1}$). Calculate the number-average and mass-average molar masses.

Applying (24.2) to a system of three components gives the average molar mass as

$$\bar{M}_n = \frac{(0.333\,3)(13.7 \text{ kg mol}^{-1}) + (0.333\,3)(15.5 \text{ kg mol}^{-1}) + (0.333\,3)(17.2 \text{ kg mol}^{-1})}{0.333\,3 + 0.333\,3 + 0.333\,3}$$

$$= 15.5 \text{ kg mol}^{-1}$$

The mass-average molar mass is given by (24.3) as

$$\bar{M}_m = \frac{(0.333\,3)(13.7)^2 + (0.333\,3)(15.5)^2 + (0.333\,3)(17.2)^2}{(0.333\,3)(13.7) + (0.333\,3)(15.5) + (0.333\,3)(17.2)}$$

$$= 15.6 \text{ kg mol}^{-1}$$

24.2 DISTRIBUTION FUNCTIONS

The molar mass distribution function $N(M_i)$ defined in (24.1) is usually expressed either in terms of the *mole fraction* or *number distribution* $[X(x)]$ or in terms of the *mass distribution* $[W(x)]$, where x is the number of monomers that have formed the polymer. If p represents the extent of reaction (the fraction of monomers that have reacted) for a polymer formed by condensation, then

$$X(x) = p^{x-1}(1-p) \qquad\qquad (24.7)$$

$$W(x) = xp^{x-1}(1-p)^2 \qquad\qquad (24.8)$$

EXAMPLE 24.2.   The number-average and mass-average molar masses are related to p by

$$\bar{M}_n = \frac{M_0}{1-p} \qquad\qquad (24.9)$$

$$\bar{M}_m = M_0 \frac{1+p}{1-p} \qquad\qquad (24.10)$$

where M_0 is the molar mass of the monomer. What is the maximum value of the polydispersity index for polymers formed by condensation?

Substituting (24.9) and (24.10) into (24.6) gives

$$\frac{\bar{M}_m}{\bar{M}_n} = \frac{M_0[(1+p)/(1-p)]}{M_0/(1-p)} = 1+p \qquad\qquad (24.11)$$

Substituting the maximum value of $p = 1$ gives the maximum value of $\bar{M}_n/\bar{M}_m$ as 2.

EXAMPLE 24.3.   Use (24.7) to determine the probability of finding a polymer containing 25 monomers after the reaction is 99.5% complete. How does this value change after the reaction is 99.9% complete?

Substituting $x = 25$ and $p = 0.995$ into (24.7 gives

$$X(x) = (0.995)^{25-1}(1-0.995) = 4.43 \times 10^{-3}$$

The probability for the second set of conditions is

$$X(x) = (0.999)^{25-1}(1-0.999) = 9.76 \times 10^{-4}$$

which is only 22.0% of the first value.

Solutions of Macromolecules

24.3 THERMODYNAMIC AND COLLIGATIVE PROPERTIES

The properties of solutions of macromolecules are described using equations similar to those for "regular" solutions (see Secs. 11.10–11.16). Because the molar masses of macromolecular solutes are high compared to ordinary molecular and ionic solutes, it is often difficult to prepare a solution containing a high concentration expressed in terms of the amount of solute (e.g., mole fraction, molarity, and molality) because of the limited solubility of the macromolecule.

EXAMPLE 24.4. Because of the large difference between the sizes of the solvent molecules and the macromolecular solute molecules, (5.13) is often written

$$\Delta_{\text{soln}}S = -R \sum_i^{\text{components}} n_i \ln \phi_i \qquad (24.12)$$

where ϕ_i is the volume fraction of component i in the solution given by

$$\phi_i = \frac{n_i V_{m,i}^\circ}{\sum n_i V_{m,i}^\circ} \qquad (24.13)$$

where $V_{m,i}^\circ$ is the molar volume of the pure component. Derive an equation relating $\Delta_{\text{soln}}S$ for 1 mol of a binary mixture as a function of the ratio of molar volumes of pure solute to pure solvent ($\rho = V_{m,2}^\circ / V_{m,1}^\circ$) and the volume fraction of solute.

For the binary solution, (24.12) gives

$$\Delta_{\text{soln}}S = -R(n_1 \ln \phi_1 + n_2 \ln \phi_2)$$
$$= -R(x_1 \ln \phi_1 + x_2 \ln \phi_2)$$

where x_i is the mole fraction. The mole fraction of solute is related to ϕ_2 using (24.13):

$$\phi_2 = \frac{n_2 V_{m,2}^\circ}{n_1 V_{m,1}^\circ + n_2 V_{m,2}^\circ} = \frac{n_2\rho}{n_1 + n_2\rho} = \frac{x_2\rho}{1 + (\rho - 1)x_2}$$

giving $x_2 = \phi_2/[\rho + \phi_2(1 - \rho)]$ and

$$x_1 = 1 - x_2 = 1 - \frac{\phi_2}{\rho + \phi_2(1 - \rho)} = \frac{\rho(1 - \phi_2)}{\rho + \phi_2(1 - \rho)}$$

The volume fraction of solvent can be expressed in terms of ρ and ϕ_2 by using (24.13) and the expression for x_1:

$$\phi_1 = \frac{n_1 V_{m,1}^\circ}{n_1 V_{m,1}^\circ + n_2 V_{m,2}^\circ} = \frac{n_1}{n_1 + n_2\rho} = \frac{x_1}{x_1 + x_2\rho} = 1 - \phi_2$$

The desired equation for $\Delta_{\text{soln}}S$ is

$$\Delta_{\text{soln}}S = -R\left[\frac{\rho(1 - \phi_2)}{\rho + \phi_2(1 - \rho)} \ln(1 - \phi_2) + \frac{\phi_2}{\rho + \phi_2(1 - \rho)} \ln \phi_2\right] \qquad (24.14)$$

24.4 OSMOTIC PRESSURE

The osmotic pressure (Π) of solutions of macromolecules is given by the *osmotic virial equation*

$$\Pi = RT\left(\frac{C'}{\bar{M}_n} + B_2 C'^2 + B_3 C'^3 + \cdots\right) \qquad (24.15)$$

where the various virial coefficients are related to excluded volumes, shapes of solute molecules, and interactions between solute molecules and C' is the concentration expressed in (kg solute)(m³ soln)$^{-1}$ = (g solute)(dm³ soln)$^{-1}$.

EXAMPLE 24.5. Use the data of Boedtker and Doty for solutions of collagen in citrate buffer at 25 °C to determine the molar mass and the second virial coefficient.

$C'/(\text{kg m}^{-3})$	2.4	4.1	5.0	5.5	6.4
$(\Pi/C'RT)/(10^{-3}\,\text{mol kg}^{-1})$	3.7	4.2	4.2	4.2	4.9

Equation (*24.15*) can be put into the form

$$\frac{\Pi}{C'RT} = \frac{1}{\bar{M}_n} + B_2 C' + B_3 C'^2 + \cdots \tag{24.16}$$

which implies that a plot of $\Pi/C'RT$ against C' will be linear with $1/\bar{M}_n =$ intercept and $B_2 =$ slope. From Fig. 24-1,

$$\bar{M}_n = \frac{1}{3.1 \times 10^{-3} \text{ mol kg}^{-1}} = 320 \text{ kg mol}^{-1}$$

$$B_2 = 2.5 \times 10^{-4} \text{ mol m}^3 \text{ kg}^{-2}.$$

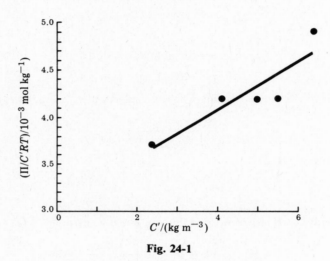

Fig. 24-1

24.5 VISCOSITY

The determination of viscosity was presented in Sec. 2.6. Solutions of macromolecules are often described by the *relative viscosity* (η_r), the *specific viscosity* (η_{sp}), and the *relative specific viscosity* (η_{sp}/C'') defined as

$$\eta_r = \frac{\eta}{\eta_1} \tag{24.17}$$

$$\eta_{sp} = \frac{\eta - \eta_1}{\eta_1} = \frac{\eta}{\eta_1} - 1 = \eta_r - 1 \tag{24.18}$$

where η is the viscosity of the solution, η_1 is the viscosity of the pure solvent, and C'' is the concentration of solute in units of kg m^{-3}. The *intrinsic viscosity* ($[\eta]$) of a polymer is

$$[\eta] = \lim_{C'' \to 0} \left(\frac{\eta_{sp}}{C''} \right) \tag{24.19}$$

and is related to the viscosity molar mass ($\bar{M}_v$) by the *Mark–Houwink equation*

$$[\eta] = K\bar{M}_v^a \tag{24.20}$$

where K and a are constants for a given polymer-solvent system.

EXAMPLE 24.6. Everett and Foster reported the following molar mass-viscosity data for solutions of amylose in 0.33 M KCl:

$\bar{M}_v/(\text{kg mol}^{-1})$	270	552	847	1 050	1 350	2 220
$[\eta]/(10^{-3}\,\text{m}^3\,\text{kg}^{-1})$	60.0	90.0	115	126	152	173

Determine the constants K and a in (24.20) for this system.

Equation (24.20) can be rewritten in linear form by taking logarithms as

$$\log[\eta] = \log K + a \log \bar{M}_v$$

A plot of $\log\{[\eta]/(10^{-3}\,\text{m}^3\,\text{kg}^{-1})\}$ against $\log \bar{M}_v$ will be linear, giving $\log K = $ intercept and $a = $ slope. From Fig. 24-2,

$$a = 0.52 \qquad K = \log^{-1} 0.528 = 3.37$$

giving $$[\eta] = (3.37 \times 10^{-3}\,\text{m}^3\,\text{kg}^{-1})[\bar{M}_v/(\text{kg mol}^{-1})]^{0.52}$$

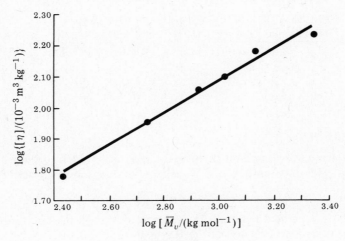

Fig. 24-2

24.6   ULTRACENTRIFUGATION

Sedimentation of suspended macromolecules under the force of gravity is a very slow process. However, the use of an *ultracentrifuge*, a very high speed centrifuge capable of producing forces in the order of 10^6 the force of gravity, will permit suspended molecules to be sedimented in relatively short periods of time.

The molar mass of the suspended particles can be determined by ultracentrifugation using two different approaches. In the *sedimentation velocity method*, the rate dr/dt at which the particles move from the center of rotation at a distance (r) from the center of rotation divided by the angular velocity (ω) is defined as the *sedimentation coefficient* (S)

$$S = \frac{dr/dt}{\omega^2 r} \qquad\qquad (24.21)$$

and is related to the mass-average molar mass by

$$\bar{M}_m = \frac{RTS}{D(1 - v\rho)} \qquad\qquad (24.22)$$

where D is the diffusion coefficient, ρ is the density of the solutions, and v is the specific volume of the polymer.

The alternative method, the *sedimentation equilibrium method*, is based on an equilibrium being established between the rate at which the solute is driven outward and the rate at which it diffuses inward as a result of the concentration gradient. Under these conditions,

$$\bar{M}_z = \frac{2RT \ln(C_2/C_1)}{(1 - v\rho)\omega^2(r_2{}^2 - r_1{}^2)} \qquad (24.23)$$

where C_i is the concentration of the polymer at a distance r_i.

EXAMPLE 24.7. What will be the ratio of the equilibrium concentrations of a polymer with $\bar{M}_z = 50.0$ kg mol^{-1} at a distance of 5.00 cm compared to a distance of 4.00 cm in an ultracentrifuge operating at 1.0×10^4 min^{-1}? Assume typical values of $v = 8.0 \times 10^{-4}$ m^3 kg^{-1} and $\rho = 1.02 \times 10^3$ kg m^{-3}.

Solving (24.23) for $\ln(C_2/C_1)$ and substituting the data gives

$$\ln\left(\frac{C_2}{C_1}\right) = \frac{\bar{M}_z(1 - v\rho)\omega^2(r_2{}^2 - r_1{}^2)}{2RT}$$

$$= \frac{\left\{ \begin{array}{c} (50.0 \text{ kg mol}^{-1})[1 - (8.0 \times 10^{-4} \text{ m}^3 \text{ kg}^{-1})(1.02 \times 10^3 \text{ kg m}^{-3})] \\ [(1.0 \times 10^4 \text{ min}^{-1})[(1 \text{ min})/(60 \text{ s})](2\pi)]^2[(5.00 \times 10^{-2} \text{ m})^2 - (4.00 \times 10^{-2} \text{ m})^2] \end{array} \right\}}{2(8.314 \text{ kg m}^2 \text{ s}^{-2} \text{ K}^{-1} \text{ mol}^{-1})(298 \text{ K})}$$

$$= 1.83$$

$$C_2/C_1 = 62$$

24.7 LIGHT SCATTERING

Light is scattered in all directions from an incident beam as it travels through a solution of suspended macromolecules. This phenomenon is called *Rayleigh scattering* and is described by

$$I = I_0 \, e^{-\tau l} \qquad (24.24)$$

where the intensity of the incident beam (I_0) is reduced to I upon passing through the distance l and τ is the *turbidity*. The turbidity is related to the Rayleigh ratio observed at 90° to the beam (R_{90}) by

$$\tau = \frac{16\pi}{3} R_{90} \qquad (24.25)$$

Debye extended the theory of light scattering and related τ to the index of refraction (n), concentration, and the mass-average molar mass by

$$\tau = H\bar{M}_m C'' \qquad (24.26)$$

where

$$H = \frac{32\pi^3 n_0{}^2 (dn/dC'')^2}{3L\lambda^4} \qquad (24.27)$$

Equation (24.26) is often written in the form

$$\frac{HC''}{\tau} = \frac{1}{\bar{M}_m} + BC'' + \cdots \qquad (24.28)$$

to correct for nonideal behavior.

EXAMPLE 24.8. The *scattering factor* (K) defined as

$$K = \frac{2\pi^2 n_0{}^2 (dn/dC'')^2}{L\lambda^4} \qquad (24.29)$$

is used in discussing Rayleigh scattering. What is the relation of K to H?

Dividing (24.27) by (24.29) gives $H/K = 16\pi/3$, the same factor that appears in (24.25).

Thermodynamic Properties

24.8 GENERAL PROPERTIES

The work of stretching an *elastomer* is

$$w = \int F \, dl \tag{24.30}$$

where F is the restoring force and l is the displacement. Elastomers typically do not follow Hooke's law except for very small displacements. A general equation of state for elastomers is given by

$$F = \left(\frac{\partial H}{\partial l}\right)_{T,P} + T\left(\frac{\partial F}{\partial T}\right)_{P,l} \tag{24.31}$$

where $(\partial H / \partial l)_{T,P} = 0$ for an *ideal rubber*.

For reversible processes having expansion and stretching work, (6.5) becomes

$$dG = V \, dP - S \, dT + F \, dl \tag{24.32}$$

EXAMPLE 24.9.   Show that $-(\partial S/\partial l)_{T,P} = (\partial F/\partial T)_{P,l}$.

Because G is a state function,

$$\left[\frac{\partial}{\partial T}\left(\frac{\partial G}{\partial l}\right)_{T,P}\right]_{l,P} = \left[\frac{\partial}{\partial l}\left(\frac{\partial G}{\partial T}\right)_{l,P}\right]_{T,P}$$

Substituting $F = (\partial G/\partial l)_{T,P}$ and $-S = (\partial G/\partial T)_{l,P}$, which can be derived from (24.32), gives the desired Maxwell relation

$$\left(\frac{\partial F}{\partial T}\right)_{l,P} = -\left(\frac{\partial S}{\partial l}\right)_{T,P} \tag{24.33}$$

24.9 FUSION OF POLYMERS

Both amorphous and crystalline polymers are glasses at low temperatures. Upon heating, glassy polymers undergo a change in properties at the *glass transition temperature* (T_g). An amorphous polymer becomes rubbery above T_g and, upon continued heating, will form a gum and finally a viscous liquid. A crystalline polymer becomes thermoplastic in nature (e.g., more elastic and more flexible) above T_g, and upon continued heating will melt sharply at the *crystalline melting temperature* (T_m) to form a viscous liquid.

EXAMPLE 24.10.   Calculate the entropy change for polystyrene, which melts at 239 °C given $\Delta_{fus}H = 80.3$ J/g.

For the fusion process, (5.9) gives

$$\Delta_{fus}S = \frac{\Delta_{fus}H}{T_m} = \frac{80.3 \text{ J/g}}{512 \text{ K}} = 0.167 \text{ J K}^{-1} \text{ g}^{-1}$$

Solved Problems

MOLAR MASS

24.1.   Signer and Gross reported the following mass fraction-molar mass range data for a sample of polystyrene:

Mass fraction	0.002	0.017	0.036	0.084	0.200	0.238
$M/(\text{kg mol}^{-1})$	25–35	35–45	45–55	55–65	65–75	75–85

0.202	0.104	0.060	0.033	0.017	0.005	0.002
85–95	95–105	105–115	115–125	125–135	135–145	145–155

Prepare a plot of the mass fraction against the molar mass. Calculate $\bar{M}_n$ and $\bar{M}_m$, and indicate these values on the graph.

The plot of mass fraction against M is shown in Fig. 24-3. To keep the graph simple, the midrange value of M was used in each case. The number-average and mass-average molar masses are given by (24.2) and (24.3), respectively, as

$$\bar{M}_n = \frac{(0.002)(30.\ \text{kg mol}^{-1}) + (0.017)(40.\ \text{kg mol}^{-1}) + \cdots}{0.002 + 0.017 + \cdots} = 83\ \text{kg mol}^{-1}$$

$$\bar{M}_m = \frac{(0.002)(30.\ \text{kg mol}^{-1})^2 + (0.017)(40.\ \text{kg mol}^{-1})^2 + \cdots}{(0.002)(30.\ \text{kg mol}^{-1}) + (0.017)(40.\ \text{kg mol}^{-1}) + \cdots} = 87\ \text{kg mol}^{-1}$$

The values of $\bar{M}_n$ and $\bar{M}_m$ are shown in the figure.

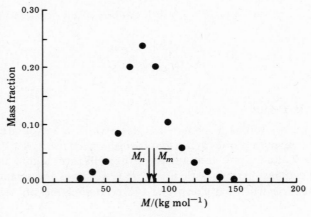

Fig. 24-3

24.2. The polydispersity of a mixture of polymers can be estimated using the *root mean square spread* (ΔM) given by

$$\Delta M = [\overline{(M_i - M_n)^2}]^{1/2} = [(\bar{M}_m - \bar{M}_n)\bar{M}_n]^{1/2} \qquad (24.34)$$

Calculate ΔM for the polystyrene mixture in Problem 24.1.

Substituting the values of $\bar{M}_m$ and $\bar{M}_n$ into the right-hand form of (24.34) gives

$$\Delta M = \{[(87\ \text{kg mol}^{-1}) - (83\ \text{kg mol}^{-1})](83\ \text{kg mol}^{-1})\}^{1/2}$$

$$= 18\ \text{kg mol}^{-1}$$

24.3. A polymer was prepared by condensing the monomer $HO-(CH_2)_5-COOH$. The mass-average molar mass of the polymer was $44.1\ \text{kg mol}^{-1}$, and the number-average molar mass was

22.1 kg mol^{-1}. Determine the extent of reaction, and prepare a plot of the mass distribution for this polymer.

Solving (24.11) for p and substituting the values of $\bar{M}_m$ and $\bar{M}_n$ gives the extent of reaction as

$$p = \frac{\bar{M}_m}{\bar{M}_n} - 1 = \frac{44.1 \text{ kg mol}^{-1}}{22.1 \text{ kg mol}^{-1}} - 1 = 0.995$$

Substituting $p = 0.995$ and values of $x \le 500$ into (24.8) gives the plot shown in Fig. 24-4.

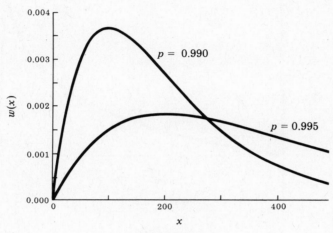

Fig. 24-4

24.4. Show that $X(x)$ is normalized.

The sum of all the probabilities must be equal to unity if the distribution function is normalized.

$$\sum_{x=1}^{\infty} X(x) = \sum_{x=1}^{\infty} p^{(x-1)}(1-p) = (1-p) \sum_{x=1}^{\infty} p^{(x-1)}$$

$$= (1-p)(1+p+p^2+\cdots) = \frac{1-p}{1-p} = 1$$

where $1+p+p^2+\cdots$ is recognized as the series expansion for $(1-p)^{-1}$.

SOLUTIONS OF MACROMOLECULES

24.5. Use (24.14) to calculate $\Delta_{\text{soln}} S$ for a series of polystyrene solutions in toluene for $0 \le \phi_2 \le 1$. Assume $M = 50.0$ kg mol^{-1} and $\rho = 0.903 \times 10^3$ kg m^{-3} for polystyrene and $\rho = 0.866\,9 \times 10^3$ kg m^{-3} for toluene.

For this system,

$$V_{\text{m,1}}^{\circ} = \frac{(92.15 \text{ g mol}^{-1})[(10^{-3} \text{ kg})/(1 \text{ g})]}{0.866\,9 \times 10^3 \text{ kg m}^{-3}} = 1.063 \times 10^{-4} \text{ m}^3 \text{ mol}^{-1}$$

$$V_{\text{m,2}}^{\circ} = \frac{50.0 \text{ kg mol}^{-1}}{0.903 \times 10^3 \text{ kg m}^{-3}} = 5.54 \times 10^{-2} \text{ m}^3 \text{ mol}^{-1}$$

giving $\rho = 521$. As a sample calculation, assume $\phi_2 = 0.500$

$$\Delta_{\text{soln}}S = -(8.314 \text{ J K}^{-1}\text{ mol}^{-1})\left[\frac{(521)(1-0.500)}{521+(0.500)(1-521)}\ln(1-0.500)+\frac{0.500}{521+(0.500)(1-521)}\ln(0.500)\right]$$

$$= 5.76 \text{ J K}^{-1}\text{ mol}^{-1}$$

Figure 24-5 shows $\Delta_{\text{soln}}S$ as a fraction of ϕ_2. At $\phi_2 = 0$ and 1, $\Delta_{\text{soln}}S = 0$.

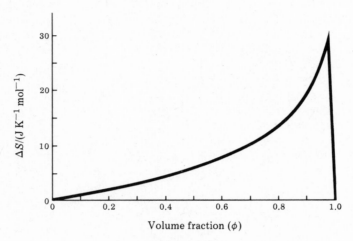

Fig. 24-5

24.6. Calculate the vapor pressure lowering at 25 °C, boiling point elevation, and the osmotic pressure for a 1.00 mass % solution of polystyrene ($\bar{M}_n = 50.0 \text{ kg mol}^{-1}$) in toluene. For pure toluene, the vapor pressure is 3 760 Pa at 25 °C and $K_{\text{bp}} = 3.33 \text{ K m}^{-1}$.

The given concentration can easily be changed into mole fraction, molarity, and molality by assuming 100.00 g of solution.

$$x_2 = \frac{(1.00 \text{ g})[(1 \text{ mol})/(50.0 \times 10^3 \text{ g})]}{(1.00 \text{ g})[(1 \text{ mol})/(50.0 \times 10^3 \text{ g mol}^{-1})]+(99.00 \text{ g})[(1 \text{ mol})/(92.15 \text{ g mol}^{-1})]}$$

$$= 1.86 \times 10^{-5}$$

$$C \approx m = \frac{(1.00 \text{ g})[(1 \text{ mol})/(50.0 \times 10^3 \text{ g})]}{0.099 \; 00 \text{ kg solvent}} = 2.02 \times 10^{-4} \text{ m}$$

Using (*11.12*), (*11.15*), and (*11.21*) gives for the solution

$$\Delta P = x_2 P_1^\circ = (1.86 \times 10^{-5})(3\;760 \text{ Pa}) = 6.99 \times 10^{-2} \text{ Pa}$$

$$\Delta T_{\text{bp}} = K_{\text{bp}}m = (3.33 \text{ K m}^{-1})(2.02 \times 10^{-4} \text{ m}) = 6.73 \times 10^{-4} \text{ K}$$

$$\Pi = CRT = (2.02 \times 10^{-4} \text{ mol dm}^{-3})(0.083 \; 14 \text{ dm}^3 \text{ bar K}^{-1}\text{ mol}^{-1})(298 \text{ K})[(10^6 \text{ Pa})/(1 \text{ bar})]$$

$$= 5.00 \times 10^3 \text{ Pa}$$

Note that changes in vapor pressure, boiling point, and freezing point are quite small for solutions of polymers and would be difficult to determine experimentally. However, values of osmotic pressure are sufficiently large to allow experimental determination.

24.7. Equation (*24.16*) implies that a plot of $\Pi/C'RT$ against C' will be linear unless the higher terms in the series are important, which would result in a nonlinear plot. A linear plot over a larger range of C' is often realized by plotting $(\Pi/C'RT)^{1/2}$ against C' as predicted by the equation

$$\left(\frac{\Pi}{C'RT}\right)^{1/2} = \left(\frac{1}{\bar{M}_n}\right)^{1/2} + \frac{1}{2} B_2 \bar{M}_n^{1/2} C' + \cdots \qquad (24.35)$$

which can be derived from (24.16) by using the series expansion $(1+x)^{1/2} = 1 + x/2 + \cdots$. Prepare plots of (24.16) and (24.35) for the following data given by Masson and Melville for poly(vinyl acetate) solutions in benzene at 20 °C,

$C'/(\text{kg m}^{-3})$	3.6	6.0	7.0	9.3	13.8
$(\Pi/C')/(\text{Pa m}^3 \text{ kg}^{-1})$	24.5	27.4	28.8	32.0	40.9

and calculate $\bar{M}_n$ from the intercepts.

The plots are shown in Fig. 24-6. Note that the (Π/C') plot shows considerably more curvature than the $(\Pi/C')^{1/2}$ plot. The intercepts are $\Pi/C' = 17.9$ Pa m^3 kg^{-1} and $(\Pi/C')^{1/2} = 4.39$ Pa$^{1/2}$ m$^{3/2}$ kg$^{-1/2}$, giving

$$\bar{M}_n = \frac{RT}{\Pi/C'} = \frac{(8.314 \text{ Pa m}^3 \text{ K}^{-1} \text{ mol}^{-1})(293 \text{ K})}{17.9 \text{ Pa m}^3 \text{ kg}^{-1}} = 136 \text{ kg mol}^{-1}$$

and

$$\bar{M}_n^{1/2} = \frac{(RT)^{1/2}}{(\Pi/C')^{1/2}} = \frac{[(8.314)(293)^{1/2}]}{4.39} = 11.2 \text{ kg}^{1/2} \text{ mol}^{-1/2}$$

or $\bar{M}_n = 126$ kg mol^{-1}.

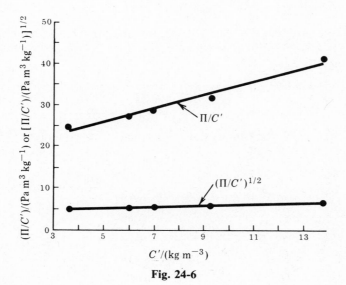

Fig. 24-6

24.8. Use the following relative viscosity data for solutions of polystyrene in toluene at 25 °C to determine $[\eta]$:

$C''/(10^3 \text{ kg m}^{-3})$	0.002 0	0.004 0	0.006 0	0.008 0	0.010 0
η_r	1.102	1.208	1.317	1.430	1.548

Calculate $\bar{M}_v/(\text{kg mol}^{-1})$ for this polymer using $a = 0.69$ and $K = 1.7 \times 10^{-3}$ m^3 kg^{-1} in (24.20).

The intrinsic viscosity is the intercept of a plot of $(\eta_r - 1)/C''$ against C'' (see Fig. 24-7), giving $[\eta] = 50.1 \times 10^{-3} \text{ m}^3 \text{ kg}^{-1}$. Solving (24.20) for $\bar{M}_v$ gives

$$\bar{M}_v/(\text{kg mol}^{-1}) = \left(\frac{\eta}{K}\right)^{1/a} = \left(\frac{50.1 \times 10^{-3} \text{ m}^3 \text{ kg}^{-1}}{1.7 \times 10^{-3} \text{ m}^3 \text{ kg}^{-1}}\right)^{1/0.69}$$

$$= 130$$

$$\bar{M}_v = 130 \text{ kg mol}^{-1}$$

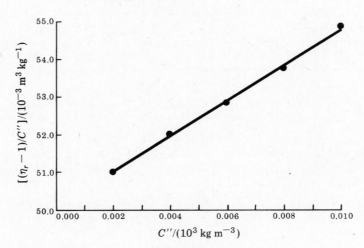

Fig. 24-7

24.9. The relative displacement (r/r_0) of bovine serum albumin was observed as a function of time:

$t/(\text{s})$	700	3 580	4 540	5 020
r/r_0	1.012 9	1.067 9	1.087 1	1.096 5

Given $\omega = 6\,260 \text{ s}^{-1}$, find the sedimentation coefficient. Assuming $v = 7.34 \times 10^{-3} \text{ m}^3 \text{ kg}^{-1}$, $\rho = 9.93 \times 10^2 \text{ kg m}^{-3}$, $D = 6.97 \times 10^{-11} \text{ m}^2 \text{ s}^{-1}$ at 25 °C, determine the molar mass of the sample.

In terms of relative displacement, (24.21) can be written as

$$S = \frac{d(r/r_0)/dT}{\omega^2(r/r_0)}$$

which upon rearrangement and integration yields

$$\frac{d(r/r_0)}{r/r_0} = \omega^2 S \, dt$$

$$\ln(r/r_0) = \omega^2 St + \text{constant}$$

Thus a plot of $\ln(r/r_0)$ against t will be linear, giving $S = (\text{intercept})/\omega^2$. From Fig. 24-8,

$$S = \frac{1.84 \times 10^{-5} \text{ s}^{-1}}{(6\,260 \text{ s}^{-1})^2} = 4.70 \times 10^{-13} \text{ s}$$

and from (24.22),

$$\bar{M}_m = \frac{(8.314 \text{ kg m}^2 \text{ s}^{-2} \text{ K}^{-1} \text{ mol}^{-1})(298 \text{ K})(4.70 \times 10^{-13} \text{ s})}{(6.97 \times 10^{-11} \text{ m}^2 \text{ s}^{-1})[1 - (7.34 \times 10^{-3} \text{ m}^3 \text{ kg}^{-1})(9.93 \times 10^2 \text{ kg m}^{-3})]}$$

$$= 61.6 \text{ kg mol}^{-1}$$

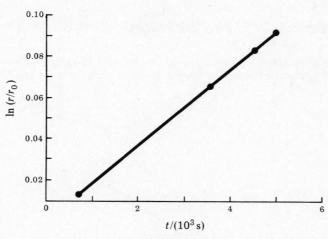

Fig. 24-8

24.10. Dandliker reported the limiting value of $C''/R_{90} = 219$ kg m^{-2} at 435.8 nm at $C'' = 0$ for bovine plasma albumin in water at 25 °C. Using $dn/dC'' = 1.97 \times 10^{-4}$ m^3 kg^{-1} and $n_0 = 1.333$, calculate the molar mass.

At $C'' = 0$, (*24.28*) gives, upon substitution of (*24.25*) and (*24.29*),

$$\bar{M}_m = \frac{\tau}{HC''} = \frac{(16\pi/3)R_{90}}{HC''} = \frac{R_{90}}{KC''} = \frac{1}{K(C''/R_{90})}$$

The value of K for this system is given by (*24.29*) as

$$K = \frac{2\pi^2(1.333)^2(1.97 \times 10^{-4} \text{ m}^3 \text{ kg}^{-1})^2}{(6.022 \times 10^{23} \text{ mol}^{-1})(435.8 \times 10^{-9} \text{ m})^4}$$

$$= 6.27 \times 10^{-5} \text{ m}^2 \text{ mol kg}^{-2}$$

and the value of the molar mass is

$$\bar{M}_m = \frac{1}{(6.27 \times 10^{-5} \text{ m}^2 \text{ mol kg}^{-2})(219 \text{ kg m}^{-2})} = 72.8 \text{ kg mol}^{-1}$$

THERMODYNAMIC PROPERTIES

24.11. The work required to adiabatically stretch a small segment of a rubber band is 0.05 J cm^{-1}. Given that the heat capacity of the segment is 0.05 J K^{-1}, calculate the temperature change of the rubber band upon a 2-cm expansion.

Equating the work to ΔU gives

$$\Delta T = \frac{w}{C} = \frac{(0.05 \text{ J cm}^{-1})(2 \text{ cm})}{0.05 \text{ J K}^{-1}} = 2 \text{ K}$$

24.12. Show that a rubber band will shorten when heated.

Using (*6.49*) for $(\partial F/\partial T)_{l,P}$ gives

$$\left(\frac{\partial F}{\partial T}\right)_{l,P} = -\left(\frac{\partial F}{\partial l}\right)_{T,P}\left(\frac{\partial l}{\partial T}\right)_{F,P} \tag{24.36}$$

Because $(\partial F/\partial T)_{l,P}$ and $(\partial F/\partial l)_{T,P}$ are both positive, $(\partial l/\partial T)_{F,P}$ must be negative. In other words, l and T are inversely related.

24.13. Prepare a sketch of molar volume against temperature for polystyrene. For this plastic, $T_g = 100\,°C$ and $T_m = 240\,°C$.

At the glass transition temperature, the slope of the volume-temperature curve increases, and at the crystalline melting temperature there is a large isothermal increase in volume (see Fig. 24-9).

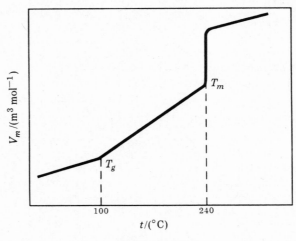

Fig. 24-9

Supplementary Problems

MOLAR MASS

24.14. If $N(M_i)$ is a normalized distribution function, what will (24.1) give for $f(M_i)$ when the integration is done over the limits of 0 to ∞?   *Ans.* $f(M_i) = 1$

24.15. Show that $\bar{M}_n = \bar{M}_m = \bar{M}_z = M_i$ if the distribution of M_i is *monodisperse* $(x_i = 1)$.

24.16. Because the numerator of (24.2) is the summation of $n_i M_i$, the value of $\bar{M}_n$ is said to be "sensitive" to low values of M_i. Is the value of $\bar{M}_m$ "sensitive" to the low or to the high values of M_i? Explain your answer.

Ans.   High values; the numerator of $n_i M_i^2$ is affected more by the square of the high values than by the square of the low values

24.17. Calculate the polydispersity index for the polystyrene mixture given in Problem 24.1.   *Ans.* 1.05

24.18. What is the relation between $\bar{M}_v$ and $\bar{M}_m$? Which is greater if typical values of a vary from 0.5 to 0.9?

Ans. $\bar{M}_v = \bar{M}_m$ for $a = 1$; $\bar{M}_v < \bar{M}_m$ for values of $a < 1$

24.19. A common working definition of a polymer produced by condensation is $x \geq 100$. Show that (24.9) predicts that $p = 0.99$ for $\bar{M}_n = 100 M_0$. Prepare a plot of the mass distribution for $p = 0.99$.   *Ans.*   See Fig. 24-4

SOLUTIONS OF MACROMOLECULES

24.20. Figure 24-5 shows that $\Delta_{soln} S$ undergoes a maximum for $0.99 < \phi_2 < 1.00$. Find the maximum $\Delta_{soln} S$ for the polystyrene-toluene mixture described in Problem 24.5.

Ans.   32.45 J K^{-1} mol^{-1} at $\phi_2 = 0.992\,5$

24.21. Use (*6.12*) and (*24.12*) to derive an expression for $\Delta_{soln}G$ for solution of polymers. Calculate $\Delta_{soln}G$ for the $\phi_2 = 0.500$ solution of polystyrene described in Problem 24.5.

   Ans. $\Delta_{soln}G = RT(n_1 \ln \phi_1 + n_2 \ln \phi_2)$, $-1\,716\,\text{J mol}^{-1}$

24.22. The limiting values of $(\Pi/C')/(\text{Pa m}^3\,\text{kg}^{-1})$ at 25 °C for two fractions of polystyrene in toluene were reported by Krigbaum and Flory as 62.5 and 21.8. What are the molar masses of these fractions?

   Ans. $39.6\,\text{kg mol}^{-1}$, $114\,\text{kg mol}^{-1}$

24.23. For solutions of polystyrene in benzene, $K = 0.95 \times 10^{-3}\,\text{m}^3\,\text{kg}^{-1}$ and $a = 0.74$ at 25 °C. Predict the intrinsic viscosity of a solution containing a sample of polystyrene with $\bar{M}_v = 50.0\,\text{kg mol}^{-1}$.

   Ans. $0.017\,2\,\text{m}^3\,\text{kg}^{-1}$

24.24. Tanford, Kawahara, and Lapanje reported the following viscosity data for a series of polypeptide chains:

Polypeptide	Ribonuclease	Hemoglobin	Myoglobin	β-Lactoglobulin
$\bar{M}_v/(\text{kg mol}^{-1})$	13.68	15.5	17.2	18.4
$[\eta]/(10^{-3}\,\text{m}^3\,\text{kg}^{-1})$	16.0	18.9	20.9	22.8

Determine K and a in (*24.20*) for this series of macromolecules.

   Ans. $a = 1.17$, $K = 7.48 \times 10^{-4}\,\text{m}^3\,\text{kg}^{-1}$

24.25. At 20 °C, $D = 6.9 \times 10^{-11}\,\text{m}^2\,\text{s}^{-1}$, $v = 7.49 \times 10^{-4}\,\text{m}^3\,\text{kg}^{-1}$, and $\rho = 9.97 \times 10^2\,\text{kg m}^{-3}$ for hemoglobin. Using a sedimentation coefficient of $4.48 \times 10^{-13}\,\text{s}$, determine the molar mass of hemoglobin.   *Ans.* $62\,\text{kg mol}^{-1}$

24.26. Equation (*24.23*) can be written in the form

$$\ln C = \frac{1-v\rho}{2RT}\,\omega^2 \bar{M}_z r^2 + \text{constant}$$

Using the following data for β-lactoglobulin reported by Pederson, determine $\bar{M}_z$ at 25 °C:

$r/(\text{mm})$	49.0	49.5	50.0	50.5	51.0	51.5
$C/(\text{kg m}^{-3})$	1.30	1.46	1.64	1.84	2.06	2.31

Assume $v = 7.514 \times 10^{-4}\,\text{m}^3\,\text{kg}^{-1}$, $\rho = 1.034 \times 10^3\,\text{kg m}^{-3}$, and $\omega = 1\,150\,\text{s}^{-1}$.

   Ans.   Slope $= 2.29 \times 10^3\,\text{m}^{-2}$, $38.6\,\text{kg mol}^{-1}$

24.27. The intensity of light passing through a 1.00-cm sample of a solution of polystyrene in methyl ethyl ketone was observed to decrease by 0.9%. Calculate the turbidity of the solution. The concentration of the solution was $10.6\,\text{kg m}^{-3}$. Given $\lambda = 546.1\,\text{nm}$, $n_0 = 1.377$, and $dn/dC'' = 2.20 \times 10^{-4}\,\text{m}^3\,\text{kg}^{-1}$, calculate H for the system. Calculate the molar mass of the polystyrene.

   Ans. $\tau = 0.904\,\text{m}^{-1}$, $H = 5.67 \times 10^{-4}\,\text{m}^2\,\text{mol kg}^{-2}$, $\bar{M}_m = 150.\,\text{kg mol}^{-1}$

THERMODYNAMIC PROPERTIES

24.28. Show that the work needed to stretch a rubber band can be determined by a graphical integration of stress ($\sigma = F/A$) plotted against strain [$\varepsilon = (l - l_0)/l_0$].

   Ans. $w = \int F\,dl = l_0 A_0 \int \sigma\,d\varepsilon$

24.29. Show that the entropy of a rubber band decreases as it is stretched. *Hint*: Equate (*24.33*) and (*24.36*).

    Ans. $(\partial S/\partial l)_{T,P}$ is negative

24.30. Show that the Clapeyron equation for the stretching of an elastomer is $\partial(F/T)/\partial(1/T) = \Delta H/\Delta l$.

24.31. The entropy change for the fusion of polypropylene is $0.577\ \text{J K}^{-1}\ \text{g}^{-1}$, and the enthalpy change is $259\ \text{J g}^{-1}$. What is T_m for this plastic?   *Ans.*   176 °C

Appendix

The following tables list the SI base and derived units for many of the physical quantities encountered in physical chemistry. Also given are the units (with conversion factors) in the modified cgs system that have been used for many years in most of the older published research work and tables of data.

Base Units

Physical Quantity and Unit		Conversion Factor[a,b]
Length (l)		
meter	m	
angstrom	Å	10^{-10} *
Mass (m)		
kilogram	kg	
atomic mass unit	u, amu	$1.660\ 565\ 5(86) \times 10)^{-27}$
Time (t)		
second	s	
Electric current (I)		
ampere	A	
Temperature (T)		
kelvin	K	
degree centigrade, degree Celsius	°C	$T/(\text{K}) = t/(°\text{C}) + 273.15$*
Amount of substance (n)		
mole	mol	
Luminous intensity (I_v)		
candela	cd	

[a] Multiply by conversion factor to change to SI. Asterisk (*) denotes exact conversion factor.

[b] The numbers in parentheses represent the standard deviation in the last significant figure(s) cited. Values based on *J. Phys. Chem. Ref. Data*, **2**, 663–734 (1973).

Derived Units

Physical Quantity and Unit		Conversion Factor[a,b] or Equivalent Units
Density (ρ)	kg m^{-3}	
	g cm^{-3}	10^3 *
Dipole moment (μ)	C m	
debye ($= 10^{-18}$ esu cm)	D	$3.335\ 641(14) \times 10^{-30}$
Electronic charge (Q)		
coulomb	C	$=$ A s
Electric conductance (κ)		
siemens	S	$= \Omega^{-1} =$ A V^{-1} $=$ A^2 s^3 m^{-2} kg^{-1}
Electric potential difference (E)		
volt	V	$=$ J A^{-1} s^{-1} $=$ kg m^2 A^{-1} s^{-3}
Electric resistance (R)		
ohm	Ω	$=$ V A^{-1} $=$ kg m^2 A^{-2} s^{-3}
Energy (U, E)		
joule	J	$=$ N m $=$ Pa m^3 $=$ kg m^2 s^{-2}
erg ($=$ g cm^2 s^{-2})	erg	10^{-7} *
calorie	cal	4.184*
electronvolt	eV	$1.602\ 189\ 2(46) \times 10^{-19}$
liter atmosphere	L atm	$1.013\ 25 \times 10^2$ *
wave number	cm^{-1}	$1.986\ 477(10) \times 10^{-23}$
atomic mass unit	u, amu	$1.492\ 442(6) \times 10^{-10}$
Enthalpy (H)—*see* Energy		
Entropy (S)	J K^{-1}	
entropy unit, gibb	cal K^{-1}	4.184*
Force (F)		
newton	N	$=$ Pa m^{-2} $=$ kg m s^{-2}
dyne ($=$ g cm s^{-2})	dyn	10^{-5} *
Frequency (ν)		
hertz	Hz	$=$ s^{-1}
cycle per second	cps	1*
Heat (q)—*see* Energy		
Inductance (L)		
henry	H	$=$ V A^{-1} s $=$ kg m^2 A^{-2} s^{-2}
Magnetic field strength (H)	A m^{-1}	
oersted	Oe	$7.957\ 747\ 2 \times 10^1$
Magnetic flux (Φ)		
weber	Wb	$=$ V s $=$ kg m^2 A^{-1} s^{-2}
Magnetic flux density (B)		
tesla	T	$=$ V s m^{-2} $=$ kg A^{-1} s^{-2}
gauss	G	10^{-4} *
Power (P)		
watt	W	$=$ J s^{-1} $=$ kg m^2 s^{-3}
horsepower	hp	7.46×10^2 *
Pressure (P)		
pascal	Pa	$=$ N m^{-2} $=$ kg m^{-1} s^{-2}
atmosphere	atm	$1.013\ 25 \times 10^5$ *
bar	bar	10^5 *
pound per square inch	psi	$6.894\ 757\ 293\ 167 \times 10^3$
torr, mmHg	torr	$1.333\ 223\ 684\ 21 \times 10^2$
Viscosity (η)	Pa s	$=$ N s m^{-2} $=$ kg m^{-1} s^{-1}
poise ($=$ g cm^{-1} s^{-1})	P	10^{-1} *
Work (w)—*see* Energy		

[a] Multiply by conversion factor to change to SI. Asterisk (*) denotes exact conversion factor.
[b] The numbers in parentheses represent the standard deviation in the last significant figure(s) cited. Values based on *J. Phys. Chem. Ref. Data*, **2**, 663–734 (1973).

Fundamental Constants

Symbol	Quantity	Value[a]
a_0	Bohr radius	$5.291\,770\,6(44) \times 10^{-11}$ m
c	Velocity of light	$2.997\,924\,58(1) \times 10^8$ m s^{-1}
e	Elementary charge	$1.602\,189\,2(46) \times 10^{-19}$ C
F	Faraday constant	$9.648\,456(27) \times 10^4$ C mol^{-1}
g	Gravitational acceleration	$9.806\,65$ m s^{-2}
G	Gravitational constant	$6.672\,0(27) \times 10^{-11}$ m^3 kg^{-1} s^{-2}
h	Planck constant	$6.626\,176(36) \times 10^{-34}$ J s
$\hbar$	$h/2\pi$	$1.054\,588\,7(57) \times 10^{-34}$ J s
k	Boltzmann constant	$1.380\,662(44) \times 10^{-23}$ J K^{-1}
L	Avogadro constant	$6.022\,045(31) \times 10^{23}$ mol^{-1}
m_e	Electron rest mass	$9.109\,534(47) \times 10^{-31}$ kg
m_n	Neutron rest mass	$1.674\,954\,3(86) \times 10^{-27}$ kg
m_p	Proton rest mass	$1.672\,648\,5(86) \times 10^{-27}$ kg
P_0	Standard state pressure	$1.013\,25 \times 10^5$ Pa
R	Gas constant	$8.314\,41(26)$ J K^{-1} mol^{-1}
$\mathscr{R}$	Rydberg constant	$1.097\,373\,177(83) \times 10^7$ m^{-1}
V_m	Ideal gas molar volume	$2.241\,383(70) \times 10^{-2}$ m^3 mol^{-1}
ε_0	Permittivity of vacuum	$8.854\,187\,82(5) \times 10^{-12}$ C^2 N^{-1} m^{-2}
$4\pi\varepsilon_0$	Permittivity constant	$1.112\,650\,056(6) \times 10^{-10}$ C^2 N^{-1} m^{-2}
μ_B	Bohr magneton	$9.274\,078(36) \times 10^{-24}$ J T^{-1}
μ_e	Electron magnetic moment	$9.284\,832(36) \times 10^{-24}$ J T^{-1}
μ_N	Nuclear magneton	$5.050\,824(20) \times 10^{-27}$ J T^{-1}
μ_0	Permeability of vacuum	$4\pi \times 10^{-7}$ kg m A^{-2} s^{-2}

[a] The numbers in parentheses represent the standard deviation in the last significant figure(s) cited. Values based on *J. Phys. Chem. Ref. Data*, **2**, 663–734 (1973) and *Pure Appl. Chem.*, **51**, 1–41 (1979).

PERIODIC TABLE OF THE ELEMENTS

(Atomic masses are based on ^{12}C. Atomic masses in parentheses are for the most stable isotope.)

Legend:
- 6 — Atomic number
- C — Symbol
- 12.011 — Atomic mass

Groups / Periods

Period	1	2												3	4	5	6	7	8
1	1 H 1.0079																		2 He 4.00260
2	3 Li 6.941	4 Be 9.01218												5 B 10.81	6 C 12.011	7 N 14.0067	8 O 15.9994	9 F 18.998403	10 Ne 20.179
3	11 Na 22.98977	12 Mg 24.305												13 Al 26.98154	14 Si 28.0855	15 P 30.97376	16 S 32.06	17 Cl 35.453	18 Ar 39.948
4	19 K 39.0983	20 Ca 40.08	21 Sc 44.9559	22 Ti 47.90	23 V 50.9415	24 Cr 51.996	25 Mn 54.9380	26 Fe 55.847	27 Co 58.9332	28 Ni 58.70	29 Cu 63.546	30 Zn 65.38		31 Ga 69.72	32 Ge 72.59	33 As 74.9216	34 Se 78.96	35 Br 79.904	36 Kr 83.80
5	37 Rb 85.4678	38 Sr 87.62	39 Y 88.9059	40 Zr 91.22	41 Nb 92.9064	42 Mo 95.94	43 Tc (98)	44 Ru 101.07	45 Rh 102.9055	46 Pd 106.4	47 Ag 107.868	48 Cd 112.41		49 In 114.82	50 Sn 118.69	51 Sb 121.75	52 Te 127.60	53 I 126.9045	54 Xe 131.30
6	55 Cs 132.9054	56 Ba 137.33	57 La* 138.9055	72 Hf 178.49	73 Ta 180.9479	74 W 183.85	75 Re 186.207	76 Os 190.2	77 Ir 192.22	78 Pt 195.09	79 Au 196.9665	80 Hg 200.59		81 Tl 204.37	82 Pb 207.2	83 Bi 208.9804	84 Po (209)	85 At (210)	86 Rn (222)
7	87 Fr (223)	88 Ra 226.0254	89 Ac† 227.0278	104 Unq (261)	105 Unp (262)	106 Unh (263)													

*Lanthanide series

58 Ce 140.12	59 Pr 140.9077	60 Nd 144.24	61 Pm (145)	62 Sm 150.4	63 Eu 151.96	64 Gd 157.25	65 Tb 158.9254	66 Dy 162.50	67 Ho 164.9304	68 Er 167.26	69 Tm 168.9342	70 Yb 173.04	71 Lu 174.967

†Actinide series

90 Th 232.0381	91 Pa 231.0359	92 U 238.029	93 Np 237.0482	94 Pu (244)	95 Am (243)	96 Cm (247)	97 Bk (247)	98 Cf (251)	99 Es (252)	100 Fm (257)	101 Md (258)	102 No (259)	103 Lr (260)

Adapted from Wolfe: *General, Organic, and Biological Chemistry*, McGraw-Hill Book Company, 1986.

INDEX

ASK FOR THE *SCHAUM'S* SOLVED PROBLEMS SERIES AT YOUR LOCAL BOOKSTORE OR CHECK THE APPROPRIATE BOX(ES) ON THE PRECEDING PAGE AND MAIL WITH THIS COUPON TO:

MCGRAW-HILL, INC.
ORDER PROCESSING S-1
PRINCETON ROAD
HIGHTSTOWN, NJ 08520

OR CALL
1-800-338-3987

_____

NAME (PLEASE PRINT LEGIBLY OR TYPE)

_____

ADDRESS (NO P.O. BOXES)

_____

CITY   **STATE**   **ZIP**

ENCLOSED IS   ❑ A CHECK   ❑ MASTERCARD   ❑ VISA   ❑ AMEX   (✓ ONE)

ACCOUNT # _____   **EXP. DATE** _____

SIGNATURE _____

MAKE CHECKS PAYABLE TO MCGRAW-HILL, INC. <u>PLEASE INCLUDE LOCAL SALES TAX AND $1.25 SHIPPING/HANDLING</u>
PRICES SUBJECT TO CHANGE WITHOUT NOTICE AND MAY VARY OUTSIDE THE U.S. FOR THIS INFORMATION,
WRITE TO THE ADDRESS ABOVE OR CALL THE 800 NUMBER.